# of Tumour Viruses

Cold Spring Harbor Laboratory
1973

COLD SPRING HARBOR MONOGRAPH SERIES

The Lactose Operon
The Bacteriophage Lambda
The Molecular Biology of Tumour Viruses

THE MOLECULAR BIOLOGY OF TUMOUR VIRUSES

International Standard Book Number 0-87969-108-5
Library of Congress Catalog Card Number 74-154474

Printed in the United States of America

Orders should be addressed to Cold Spring Harbor Laboratory
P.O. Box 100, Cold Spring Harbor, N.Y. 11724

Contributors

D. Baltimore *Massachusetts Institute of Technology*
T. L. Benjamin *Public Health Research Institute of the City of New York*
M. M. Burger *Biocenter, University of Basel*
L. V. Crawford *Imperial Cancer Research Fund*
W. Eckhart *Salk Institute*
R. L. Erikson *University of Colorado Medical Center*
A. M. Fried *Imperial Cancer Research Fund*
T. Friedman *Salk Institute*
B. Hirt *Swiss Cancer Research Institute*
C. Mulder *Cold Spring Harbor Laboratory*
B. Ozanne *Cold Spring Harbor Laboratory*
U. Pettersson *University of Uppsala*
R. Pollack *Cold Spring Harbor Laboratory*
B. Roizman *University of Chicago*
J. Sambrook *Cold Spring Harbor Laboratory*
H. M. Temin *University of Wisconsin*
G. J. Todaro *National Cancer Institute*
J. Tooze *Imperial Cancer Research Fund*
J. D. Watson *Harvard University* and *Cold Spring Harbor Laboratory*
R. A. Weiss *Imperial Cancer Research Fund*
H. Westphal *National Institute of Child Health and Development*
J. A. Wyke *Imperial Cancer Research Fund*

Contents

Foreword

While today research on tumour viruses occupies the center stage of cancer research, this was not always so. In fact, until the past decade, the concept that human cancer might frequently have a viral origin was not popular, and many of the discoveries now viewed as epochal were made by tenacious men whose necessary persistence was often considered a form of paranoia, only to be tolerated because, then as now, there almost always has been more "cancer money" available than first-rate ideas in need of exploitation. As seen from a distance and even close up, much past tumour virus research has appeared hopelessly disconnected, with its literature frequently impenetrable, burdened by oddly named viruses and host cells, each with their own misunderstood immunology.

Hence, our thought that today's tumour virus research would move faster if there existed a concise book to help introduce outsiders to the world of animal cell culture and tumour virus experimentation. Initially we imagined the writing task to be a simple affair, flowing smoothly out of lecture notes transcribed during the 1969 and 1970 Cold Spring Harbor tumour virus workshops. But too soon we began to appreciate the multitude of facts which had to be noted, if not explained, if we were to be of real value to the many, many interested biologists still lacking any practical feeling for the world of animal cell culture. So we steadily widened the scope of the book, hoping that the added material would simplify, not complicate, the thought processes of an audience already overwhelmed by the information explosion of the past two decades of the new biology.

Naturally complicating our task has been the frantic pace at which new data about tumour viruses now appear. Too clearly our final product will be somewhat out-of-date and in need of revision before the printer's ink has dried. But we comfort ourselves with the thought that a slightly dated book on tumour virus molecular biology is better than no book at all. So despite the ever receding publication date, we knew we would see the job through to the end. That we could continuously be so optimistic owes much to the fact that we had John Tooze, not only to coordinate and edit, but also to write long sections of the final manuscript. I am now most pleased with the final outcome and suspect that with its appearance, John's most needed contribution to tumour virus research will be universally seen as a job very well done.

J. D. WATSON

April 1973

Preface

Although this book has never before been published, you are looking at the second edition. The first, which lacked the first two chapters of this edition, was completed in November 1970, was set in type but, for reasons which are of only parochial interest, reached no more than the galley proof stage. The galleys were, however, copied and the copies enjoyed a limited circulation. In the summer of 1972 enough enthusiasm was rekindled in various of the contributors to make an attempt at a revised edition seem worthwhile. This is the result of their work superimposed on the original 1970 text.

From the outset we decided to try to produce a text book rather than a collection of review articles, which explains why individual chapters are not attributed to a particular author or authors and why some of those who contributed text may no longer immediately recognize those sections for which they were initially responsible.

Most of the data discussed have been reported since 1960 and we have made a deliberate attempt to select that work which seemed to us likely to have a lasting interest. As a result we have not cited every publication relating to tumour virology that has appeared since 1960, and obviously we have not attempted to produce an exhaustive history of tumour virology, not least because Ludwik Gross has chronicled the subject in the second edition of *Oncogenic Viruses*.

We have also taken full advantage of the current practise of disseminating important information in advance of its formal publication.

Where we knew such information we have cited it. We realize that this may seem unfair, especially to those who may have pertinent data that we have failed to mention because we had no knowledge of them. We apologize to such persons but our only alternative was to ignore important results, of which we had knowledge, simply because of the inevitable delays which publication entails. On balance we believe it was in the best interests of any readers for us to use such information and risk a charge of bias.

Finally, it is my privilege on behalf of all the contributors to thank everyone who has helped us produce this book. With one exception I do this collectively, trusting that all those who in diverse ways have helped will not think our thanks are any the less sincere because they are not individually acknowledged. To do that would mean printing a list of names covering more than one page. I must, however, make one exception. Without the patience and diligence of Miss Judy Gordon of the Cold Spring Harbor Laboratory, who prepared the manuscript for the printers, this book might well never have been produced.

J. Tooze

March 1973

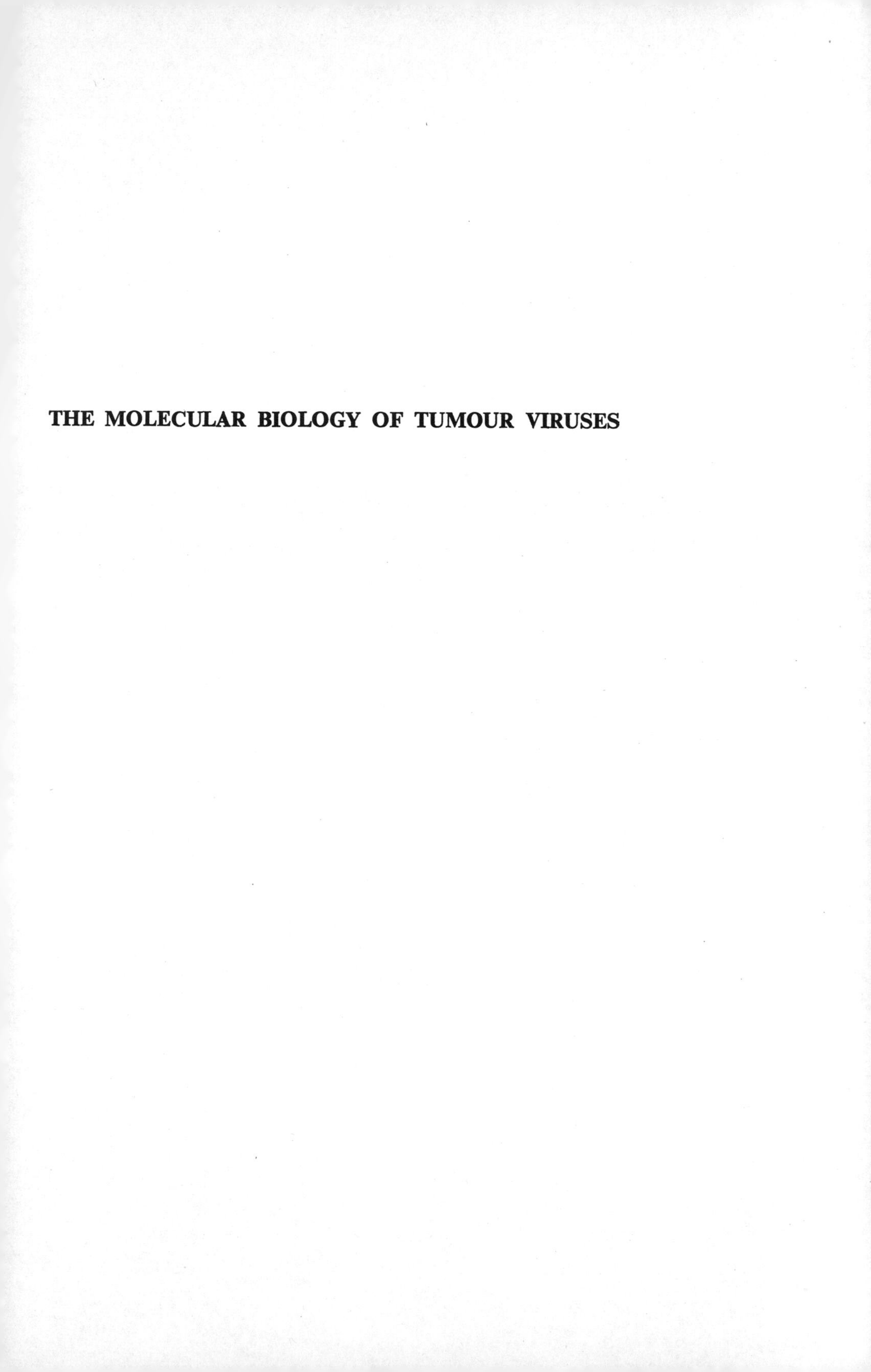

THE MOLECULAR BIOLOGY OF TUMOUR VIRUSES

I

Origins of Contemporary Tumour Virus Research

THE CANCER CELL DEFINED

The evolution of unicellular into multicellular organisms has demanded the emergence of very precise rules to govern both the spacial siting and the multiplication rates of their respective differentiated cells. A given specialized cell does not grow at any site in its multicellular host but is restricted to have only certain cell types for its neighbours and to proliferate only upon the receipt of a message signifying that more of its kind are needed for the orderly growth or maintenance of the complete organism. Given the existence of so many types of multicellular organisms, each with its unique patterns of differentiation, there are likely to exist many, many molecular pathways along which these signals are transmitted. Any of them in turn may fail and lead to one of the many abnormal forms of growth that collectively are called cancer.

Cancer, thus, is not a unitary category but a wide, to some, hopelessly wide collection of different abnormal conditions, each displaying its own specific properties. In thinking how to study cancer, most people quickly see the need to distinguish the property of excessive cell division from the property of abnormal cellular affinities. The quality of uncontrolled growth is what leads to the localized masses of single cell types often referred to as tumours. When these tumours grow only at the sites of origin, they do not necessarily upset the functioning of their host and frequently are called benign tumours.

When, however, a cell that has lost the ability to position itself correctly begins to grow in an inappropriate region, the resulting tumour often quickly leads to disease symptoms and culminates in the eventual death of the afflicted organism. Such growths are known as malignant tumours and quite frequently are the only tumours given the name cancer. Now there is general agreement that the loss of ability to establish normal cellular affinities reflects fundamental differences between the outer surfaces of a cancer cell and its normal counterpart. But as we shall see later, what these changes are at the molecular level remains almost a total mystery. So for the time being, all our operational definitions of a cancer cell are biological, generally being some measure of its ability to grow into a tumour when injected into a suitable recipient organism.

As far as we can tell, cancer is a disease at the cellular level, that starts when a hereditary change somehow transforms a single normal cell into a cancer cell (Lindner and Gartler, 1965; Fialkow et al., 1967). When the newly created cancer cell divides, both of its descendants carry the cancer property that they, in turn, pass on to their progeny. How these changes are induced is still a very murky subject, with many highly diverse external agents (e.g., certain viruses, ionizing radiations, ultraviolet light, various chemical compounds), collectively called carcinogens, all increasing the frequency of the cancerous transformation. Given the hereditary property of cancer, it has seemed simplest to believe that most carcinogens act directly by altering the genetic makeup of one or more chromosomes.

Still very debatable is the number of changes necessary to transform a normal cell into a typical cancer cell. Most likely most of the cells isolated from visible tumours are the products of more than one genetic event (Armitage and Doll, 1957), and there is general agreement that the increased virulence of many advanced tumours is the result of a progressive series of changes, each of which leads to a greater degree of uncontrolled growth (Burnet, 1957).

As almost all types of differentiated cells give rise to cancer cells, the variety of different cancers, each with their own peculiar biochemical features, is much larger than we would wish to face. Hopefully, however, common features relate many types of cancer so that in focusing upon the origin of one specific cancer, we will obtain information directly applicable to many others. That this may be so is strongly suggested by the fact that a given tumour virus is often able to transform more than one type of differentiated cell into its cancerous counterpart. For example, the polyoma virus causes tumours of both the salivary and prostrate glands of mice. This may mean that these salivary and prostate cancers, despite their very different morphological appearances, have lost control over cell division and cellular interactions through the misfunctionings of the same molecular processes.

TUMOUR IMMUNOLOGY

Cancer research and the study of the immune response have been interconnected closely for about as long as they have been seriously carried out. As soon as the idea arose that the immunological response protected vertebrates from disease-producing microorganisms, the question "what is the precise role played by the immunological response during the development and spread of cancer?" could be asked. Since cancer cells were thought possibly to possess surface antigenic determinants not present in their normal counterparts, the possibility existed that cancer cells, like microorganisms, might routinely be recognized as foreign and might stimulate production of antibodies that react with them. If so, the outcome of the appearance of a new cancer cell might be determined by the relative rates at which the cancer cell divided and an effective immunological attack was generated.

Answering this question, however, has not been a simple matter; and even today the role of the immunological response as a defense mechanism against cancer is far from understood. Certainly when the problem was first posed, immunology as a science was not advanced enough to give meaningful answers; in fact, as we shall see, one of the prime stimulants for development of modern immunology has been a desire to understand the antigenic properties of tumour cells.

Transplantation of Tumours

By the last part of the 19th century it was possible to transplant tumours from one animal to another. Small portions of preexisting spontaneous tumours were implanted under the skin of a similar animal and observed to see if a new tumour grew at the site of inoculation. The first successful such experiment was done in 1876 by the Russian Novinsky, who transplanted the cancer of a four-year-old dog. Over the next two decades experiments in a variety of European laboratories revealed that tumours of many other mammals, including the mouse and rat, were occasionally transplantable (see review in Gross, 1970). On the whole, however, only the exceptional transplantation experiments succeeded; most tries were total failures. Sometimes the transplanted tumour grew briefly and then gradually regressed, a response we now realize to be immunological.

Once the primary transplant had taken, secondary transplants could also be achieved, occasionally yielding tumour cells highly adapted for serial transfer. The most successful early practitioner of this art was Paul Ehrlich (1907), who in his Frankfort laboratory established a number of transplantable mouse mammary carcinomas, several of which eventually became adapted to grow well in almost any mouse strain irrespective of its genetic constitution.

Subsequent study of the variables controlling the outcome of transplantation revealed that younger hosts were more receptive than older animals and that newborn animals were the most likely to accept a tumour. Also, it emerged that the closer the relationship between donor and recipient animals, the greater the chance that the transplanted tumour would survive. Transplants between species almost always failed, although the use of newborn animals allowed an occasional success.

The route of administration of the tumour cell was also found to affect the frequency of takes. Direct inoculation into the peritoneum usually worked better than implantation under the skin, and cell suspensions were more easily transplanted than solid tumour masses. Most important, the use of cell suspensions permitted more quantitative experiments—the number of cells necessary to form a tumour being related to the cancerous quality (oncogenicity) of that tumour.

Histocompatibility Loci

The rejection of tumour cells transplanted from one adult animal to another unrelated animal of the same species is not a peculiarity of cancer cells, for normal cells transplanted between unrelated adult animals are also regularly rejected. As we now know, chiefly from extensive experiments in which skin was grafted from one mouse to another, this rejection is an immunological phenomenon. Grafts between individuals of the same species are called homografts (Medawar, 1958), whereas those which cross species lines are called heterografts. Homografts between pairs of unrelated animals usually fail because unrelated animals very rarely have exactly the same constellation of antigens on their cell surfaces. The surface of any cell is made up of many different kinds of molecules, many of which are highly antigenic; one set of these molecules, called the histocompatibility antigens (Snell, 1948), gives all the cells in any organism an antigenic identity. Each histocompatibility antigen is specified by a corresponding histocompatibility gene, and defining the antigenic differences between individuals of an outbred population usually requires very extensive genetic analysis.

Such genetic experiments are much easier to interpret when there exist highly inbred strains obtained by many generations of brother-sister matings (Little, 1941; Strong, 1942). In these strains virtually all genes are present in the homozygous state, and so each histocompatibility gene is present in two identical copies. The product of a cross between two inbred strains with different histocompatibility antigens is thus an F_1 hybrid generation in which each individual contains one copy of each of the histocompatibility genes present in both parents. As a result, F_1 hybrids accept grafts from either of the parental strains. By contrast, each of the parents lacks half the histocompatibility antigens present in the F_1 hybrids and so rejects their grafts.

Many years of doing crosses between various inbred mouse strains have revealed the existence of a number of different genes located on several chromosomes, each of which affects the antigenic properties of cell surfaces (Snell, 1957; Gorer, 1956, 1961). These histocompatibility (h) genes do not affect the rejection response equally. Some surface components are much more antigenic than others and correspondingly have received more attention from immunogeneticists. In particular, attention has been focused on the H2 genetic locus (Gorer et al., 1948) that exists in a large number of different allelic forms and which provides the strongest barrier to transplantation (Snell, 1957, 1963). Multiple alleles also exist at many other histocompatibility loci, and so by chance two unrelated mice almost never possess identical sets of histocompatibility genes.

The speed with which grafts are rejected usually relates to the number and type of histocompatibility genes by which the two individuals differ. If they differ in only one or two "weak" genes, the rejection may be very slow indeed. But if they vary in one "strong" or many "weak" genes, rejection will be rapid.

As far as we can tell, extensive histocompatibility diversity also exists in all other vertebrate forms (Mann and Fahey, 1971). Indeed, the histocompatibility genetics of humans is strikingly similar to that of mice, so that more than a half-century of basic mouse genetics has direct clinical applications.

Until recently, many people thought that cancer cells lack the normal histocompatibility antigens, but now it is certain that they retain many of the surface antigens present in their normal progenitors. This means that the immunological properties of tumours must be studied with highly inbred strains. When tumours are transplanted between animals with different histocompatibility properties, it is impossible to distinguish rejection because of the presence of a possible tumour-specific antigen from rejection because of differences in normal histocompatibility antigens. Until this variable was removed, experiments on tumour immunology were impossible to interpret.

Immunological Tolerance

How immunological systems can distinguish foreign objects from normal cellular constituents still is very much a mystery. We do know, however, that the designation as normal or foreign is first made at the time in an individual's life when the immunological system begins to work (Burnet and Fenner, 1949). A competent immunological system does not exist early in embryological development, but only appears after an individual already possesses all the types of differentiated cells that it will need for its later existence. Then, somehow, the immunologically competent cells come into contact with all the cell types that they will see later in life so that they cannot later respond to them.

This form of immunological memory must be a very reliable process, for on its faithful functioning rests the normal inability of an organism to produce antibodies against its own cellular constituents. Conversely any cell type not present early in life is quickly marked out as foreign and destroyed. These characteristics of immune responses, first clearly elucidated through the work of Burnet (1961) and Medawar (1961), explain why tumour transplants, or grafts of normal tissue, are rejected much less efficiently by very young animals. Young and especially newborn animals are not only very inefficient in mounting immune responses against foreign cells, but also sometimes accept foreign cells as "self" and at a later date tolerate further grafts or injections of the same foreign cells without any immunological complications; hence the term "immunological tolerance" (Billingham et al., 1953).

Cell-mediated Immunological Response

For a long time the rejection of homografts was believed to be mediated by the same type of freely circulating antibodies that participate in the immunological response to foreign microorganisms. Somehow the specific binding of free antibodies to surface antigens was thought to lead to cellular death. But in 1953, N. A. Mitchison, first working at Oxford and then at the Jackson Laboratory in Bar Harbor, Maine, discovered that the active agent was not free antibodies but cell-bound receptors, presumably antibodies, present on the surfaces of lymphocytes. His experiments (Mitchison, 1954, 1955; Mitchison and Dube, 1955) exploited the earlier observation (Medawar, 1946) that when an animal rejects a homograft, it acquires specific immunity to subsequent grafts of the same genetic origin, which are rejected much faster than the first graft. Mitchison's primary finding was that transfer of lymphocytes from an immune animal to a non-immune recipient also transferred specific homograft immunity. In contrast, injection of freely circulating antibodies did not transfer any immunity. Mere combination of specific antibodies with their respective surface antigens does not lead to cell death. Killing, by a process still to be understood, only occurs when a lymphocyte makes contact with an appropriate target cell (Winn, 1959; Rosenau and Moon, 1961). The chief role of the lymphocyte-bound antibodies thus may be to lock the respective lymphocytes on the surface of the correct "foreign cell." Whether after specific contact has been made the antibody-antigen complexes play any further role in the killing process is not known (Perlmann and Holm, 1969).

The action of lymphocytes in killing specific target cells can now be easily observed in vitro by colony inhibition tests (Hellström and Sjögren, 1965; Hellström, 1965). Addition of lymphocytes from specifically immune animals will prevent an appropriate cell from forming a colony when placed on a petri dish containing suitable culture medium. Under identical conditions addition of free antibodies from the same donor causes no inhibition of target cell multiplication.

Specificity of the "T Lymphocytes" for Immunological Surveillance

Over the past several years, there has arisen increasing evidence that lymphocytes fall into two functional groups. Lymphocytes of the "B" group have their origin in the bone marrow (bursa in fowl), whereas the thymus is the source of the "T" lymphocytes (Roitt et al., 1969). The primary function of the "B" lymphocytes is to serve as the progenitor of the plasma cells that make almost all the circulating antibodies (Glick et al., 1956; Warner et al., 1962). In contrast, "T" lymphocytes never give rise to cells that have extensive endoplasmic reticulum and make large amounts of antibodies. Instead their main role is that of surveillance—to attach to foreign target cells and somehow kill them. Evidence for this comes from a variety of experiments which follow the consequences of surgical removal of the thymus very early in life (Miller, 1962; Cooper et al., 1965), or which observe specific mutants of the mouse (Flanagan, 1966; Pantelouris, 1968; Raff, 1969) and man (DiGeorge, 1968; Huber, 1968) that never develop a thymus. Animals deprived of a thymus continue to make freely circulating antibodies but lack the homograft rejection reaction.

Tumour-specific Antigens

More and more evidence is accumulating that most, if not all, cancer cells are antigenically distinct from their normal progenitors (Gross, 1943; Foley, 1953; Prehn and Main, 1957; Klein et al., 1960; Old and Boyse, 1964). The surface changes that give cancer cells altered cellular affinities so modify their antigenic properties that they become recognized as "foreign" and generate a cell-mediated immunological response. Most newly arising cancer cells may multiply only a few times before they become engulfed and destroyed by lymphocytes bearing the appropriate surface antibodies (Klein et al., 1960). Indeed the main function of the immunological response may be to serve as the vertebrate surveillance system against spontaneously arising or carcinogen-induced cancers. Certainly animals whose immunological response has been weakened by an immunosuppressive agent or by removal of an immune organ (Allison et al., 1967; Dent et al., 1968) not only much less efficiently reject tumour grafts, but also have a greatly increased incidence of spontaneous tumours. The possibility thus exists that the cell-mediated immunological response evolved primarily as a defense against cancer (Burnet, 1964, 1970) and the appearance of the freely circulating antibodies as a means to fight microbial infections may have been a later evolutionary event.

Examination of a large number of skin tumours induced by the carcinogen methylcholanthrene gave the surprising answer that each was antigenically distinct (Prehn and Main, 1957). This finding says that there is no unique surface configuration universal to all cancer cells, and it further suggests that any of a large number of different surface modifications can lead to that loss of normal cellular affinities which is characteristic of cancer cells. But in complete contrast to tumours induced by chemical carcinogens, tumours induced by any particular tumour virus all seem to carry the same tumour-specific antigens (Sjögren, 1965; Klein, 1966). Probably a chemical carcinogen can modify cellular genetic material at a number of different chromosomal sites, changes at any of which lead to the cancer cell phenotype. Tumour viruses, on the other hand, may cause changes at only a few sites, perhaps even at a unique site in the cell genome, and so induce only one characteristic antigenic modification.

Not all tumour-specific antigens are located on the cell surface. For example, in addition to tumour-specific antigens, cells transformed by both polyoma virus and SV40 contain in their nuclei components called T (tumour) antigens. How these internal antigens are recognized by an immunological system is not known. Perhaps trace amounts are in fact located in external surface positions, or alternatively T antigens may induce an immunological response only after the transformed cells are killed by a cell-mediated response against their surface antigens.

Blocking Antibodies

The fact that all tumour cells possess unique antigens that should mark them out as "foreign cells," to be swiftly destroyed by an immunological response, raises the question of how the occasional cancer cell none the less manages to grow into a clinically detectable tumour. The obvious explanation, that the host organism fails to produce the specific antibodies needed to combine with the tumour specific-antigens, may not be correct. Largely through the work of K. E. and I. G. Hellström (1969, 1970), Swedish immunologists working at the University of Washington in Seattle, we know that many tumours grow in the presence of large numbers of lymphocytes bearing antibodies against the respective tumour antigens. The Hellströms detected such lymphocytes by their in vitro colony inhibition tests. When tumour cells and pure lymphocytes from the same animal are mixed in vitro, the tumour cells do not form colonies but are killed.

This finding raises the question of why such lymphocytes do not act this way in vivo. Surprisingly the answer proved to lie in the existence of circulating antibodies (or possibly antigen-antibody complexes) with the same specificity as the lymphocytes. This emerged from experiments in which free antibodies as well as lymphocytes were simultaneously added to the tumour cells (Moller, 1965; Hellström et al., 1969). Under these circumstances the tumour cells were not killed. The blocking factor prevents lymphocytes from interacting with the tumour cells either by preferentially covering the tumour cell surfaces so that the lymphocytes cannot recognize the tumour cells, or by blocking the killer lymphocytes themselves. Antibodies acting in this fashion are called *blocking antibodies*.

The existence of blocking antibodies helps explain the phenomenon of immunological enhancement (Kaliss and Molomut, 1952; Kaliss, 1958; Snell, 1954). When free antibodies against the appropriate H2 histocompatibility antigens were injected into mice on which homografts were growing, they strikingly promoted homograft growth instead of hastening their rejection, as had been anticipated. Effective immunological defense against cancer may, therefore, depend upon the primary response being the production of lymphocyte-bound antibody with little, if any, synthesis of free antibodies against the tumour surface antigens. How this preferential synthesis is normally achieved is not known, neither do we yet know anything about the factors that might upset this balance and thereby lead to clinical cancer.

RNA TUMOUR VIRUSES

Rous Sarcoma Virus

The first tumour virus to be studied seriously was found in 1911 by Peyton Rous working in New York at the Rockefeller Institute. There for several years he had been studying a spontaneous chicken sarcoma (tumour of connective tissue) that he had been passaging through closely related Plymouth Rock chickens (Rous, 1910). With each passage the tumour acquired heightened transplantability and showed greater tendency to spread from the original graft site. Rous then tested to see whether cell-free filtrates could also induce a tumour to grow at the inoculation site. Although previous experiments with extracts of transplantable tumours of rats, mice and dogs had failed, Rous immediately succeeded, using material that passed through ordinary filter paper. Then more careful experiments were done using filters known to hold back bacteria, again with positive results, thereby establishing a virus as the etiological cause of the tumour (Rous, 1911). Within the next two years Rous isolated new viruses from several other spontaneous chicken tumours and so viral etiology of a large fraction of spontaneously arising chicken tumours became a real possibility. Now the name Rous sarcoma virus (RSV) is given not only to the original virus isolated by Rous but to a large number of independently isolated chicken viruses that induce similar sarcomas; the general name avian sarcoma virus (ASV) is also used.

Regressions Due to Immunological Responses

In Rous's original experiments very large numbers of virus particles were inoculated for every visible tumour produced. Tumours arose most frequently when young chickens were inoculated, and especially large numbers of virus particles had to be inoculated in order to induce tumours in adult chickens. Moreover, when adult chickens were used as recipients, the tumours frequently regressed; as we now realize the high frequency of takes in young chickens probably reflected their lesser ability to mount a cell-mediated immunological response. Most likely for every visible tumour produced, even in very young chickens, thousands of cells are transformed into cancer cells and only a small fraction somehow overcome immunological attack. Already in 1913 Rous could distinguish antibodies against the infectious virus particles from those against the tumour cells, but the immature state of immunology at that time prevented a real understanding of what was happening (Rous, 1913; Rous and Murphy, 1914).

The first isolated strains of RSV had restricted host ranges; they would induce tumours in only a few strains of chicken. Continued passage of such RSV, however, often has led to a much wider host range. At first, RSV infection seemed limited to chickens, but in 1928 Fujinami in Japan (Fujinami and Suzue, 1928) showed that a strain of RSV (Fujinami ASV) could cause tumours in ducks, and that virus recovered from duck tumours could again induce tumours in chickens. Soon afterwards, tumours were induced with RSV in young turkeys, guinea fowl, and pigeons, though regression was common in pigeons. All this time, it was thought that RSV would not grow in mammals, but in 1957 Zilber and Svet-Moldavsky in Moscow showed that infection by the Carr-Zilber RSV strain induced a fatal hemorrhagic disease in adult rats (Zilber and Kriukova, 1957; Svet-Moldavsky, 1957). The following year Svet-Moldavsky (1958) observed the formation of sarcomas after inoculating newborn rats with the same RSV strain. The potential host range of some strains of RSV may even extend to humans since the viruses transform normal human cells in culture.

Growth of RSV in Embryonated Eggs

Already in 1911 Rous and Murphy realized that RSV would grow in embryonated chicken eggs. Both the embryo itself and its surrounding membranes contain cells that can be transformed by the virus. Most important, at this stage no immunological response exists and tumour regression does not complicate studies of cell transformation. But it was not until 1938 that Keogh in England employed growth of RSV on the chorioallantoic membrane as a quantitative assay for RSV (Keogh, 1938). Following direct inoculation on the membrane, small tumours developed and their number was directly proportional to the concentration of the virus suspension. This result demonstrated the very important conclusion that infection by a single RSV particle can transform a normal cell into a cancer cell.

Morphology of RSV Particles

Although Claude et al. (1947) first observed RSV in the electron microscope, the morphology of the virus particle was not learned until methods for examining thin sections of cells were developed (Gaylord, 1955; Bernhard et al., 1956). Sections of cells infected with RSV often reveal the presence on the cell surface membranes of large numbers of spherical particles, some 70 to 80 mμ in diameter. Each such particle is surrounded by an external lipid-containing membrane within which is a centrally located body called the nucleoid. Subsequent biochemical characterization showed that the genetic material located in the nucleoid is RNA (Crawford and Crawford, 1961); each RSV particle contains perhaps as much as 10^7 daltons of RNA.

The RSV particles seen on the outer surfaces of infected cells are progeny virus particles just ready to be released. Prior to this stage of the viral life cycle, the nascent RNA-containing nucleoids move into contact with the cell membrane to begin an enveloping process, at the conclusion of which mature RSV particles bud off from the cell membrane. Budding somehow occurs such that the integrity of the outer membrane is maintained even though many new particles are released every hour.

Transformation of Cultured Cells

Over the past decade most work with RSV has involved the use of cultured cells. In secondary cultures of chicken fibroblasts, RSV grows to high titres and morphologically altered transformed cells become visible several days after infection (Manaker and Groupé, 1956). When chick fibroblasts growing as monolayers on petri dishes are infected with RSV, the transformed sarcoma cells stand out as easily countable foci, the number of foci being directly proportional to the concentration of virus added (Temin and Rubin, 1958).

Occasionally fibroblasts prepared from a chicken embryo proved totally resistant to RSV. Such resistant cells frequently contain closely related leukemia viruses that multiply in chicken fibroblasts without causing obvious morphological changes (Rubin, 1960, 1961). A number of different strains of such avian leukemia viruses exist, each conferring a distinctive pattern of resistance to infection by RSV (Vogt and Ishizaki, 1966). The resistance or sensitivity of chick cells to infection by particular strains of RSV is also controlled by chromosomal genes which are inherited in a simple Mendelian fashion. Sensitivity is dominant and usually reflects the presence at the cell surface of specific sites that allow particular strains of RSV to bind to and penetrate the cells.

Defectiveness and Helper Viruses

In the 1950's and early 1960's many investigators concentrated their work on an RSV strain, obtainable in very high titres, that was isolated by W. R. Bryan of the National Institutes of Health. High yields of virus were obtained when each cell was infected with large numbers of virus particles. But when the cells were infected instead with only single virus particles, they did not seem to yield any progeny particles even though they became morphologically transformed and could initiate tumours (Prince, 1959). Several years later this phenomenon was shown by Temin (1962) at Madison, Wisconsin, and Vogt and Rubin at Berkeley to involve the simultaneous presence of a related leukemia virus. Only cells simultaneously infected with both the Bryan RSV strain and the leukemia virus were thought to produce progeny particles containing an RSV genome. The Bryan strain of RSV was thus considered a defective virus, genetically unable to manufacture its outer membrane, and so dependent upon a "helper" leukemia virus to provide membrane components to surround its RNA genomes. As we shall explain later in this book, after much controversy recent experiments have proved that the Bryan strain is indeed defective; the progeny virus particles specified by the genome of the Bryan strain lack a surface component necessary for infection. This component as we now know can be supplied either by a coinfecting avian leukemia virus or by an endogenous avian leukemia virus genome inherited by chicks through the egg and sperm (see below). It is important to remember, however, that the Bryan RSV strain can transform cells once it has entered them; its defectiveness has nothing to do with the ability to transform cells per se, and other strains of RSV are nondefective for replication as well as transformation.

Evidence for a DNA Provirus

As soon as cultured cells were available, it became possible to study the effect of a variety of metabolic inhibitors upon the various stages of RSV multiplication. These experiments did not give the answer expected for a virus with an RNA genome. Infection by RNA viruses should not be directly influenced by compounds which inhibit DNA synthesis. Yet, if the DNA inhibitor bromodeoxyuridine is present during the first twelve hours of infection, it blocks both transformation by, and multiplication of, RSV (Bader, 1964, 1965). Likewise, addition of actinomycin D, a drug which specifically inhibits DNA-dependent RNA polymerase, immediately blocks formation of new progeny particles by transformed cells (Temin, 1963). To explain these results, Temin (1964) proposed that the infecting RNA genomes were used as templates to make DNA genomes which in turn were then integrated as proviruses into one or more chicken chromosomes. Almost no one liked this idea until in the spring of 1970 Mizutani and Temin and, independently, Baltimore of MIT located RNA-dependent DNA polymerase within mature particles of RSV and RSV-like tumour viruses (Mizutani and Temin, 1970; Baltimore, 1970). As we shall tell in great detail below, study of the role of proviruses dominates much current work on RNA tumour viruses.

Avian Leukemia Viruses

Leukemia and other related diseases of the blood-forming tissues in chickens are caused by viruses. This was first realized by Ellermann and Bang (1908, 1909) who, working in Denmark just after the turn of the century, transmitted chicken leukemia by inoculating an extract of leukemia cells that had been filtered to remove bacteria. Because leukemia was not then clearly designated as a form of cancer, their discovery did not create the major impact which followed Rous's (1911) proof that a virus could cause a solid tumour. Collectively, these viruses are sometimes referred to as fowl leukosis viruses. Technically the term leukemia should refer only to diseases characterized by pathologically excessive numbers of circulating blood cells. The name avian leukemia virus (ALV) is, however, often used in a taxonomic sense to describe viruses that are obviously related to authentic avian leukemia virus even though they may cause other diseases of the blood.

The main clinical manifestations of these viruses are (1) lymphoblastosis, a disease in which cancerous lymphocytes massively infiltrate organs like the spleen, liver, and lung; (2) erythroblastosis, a condition leading to excessive numbers of immature erythrocytes in the blood; and (3) myeloblastosis, a disease characterized by large numbers of myeloblasts circulating in the blood. While sometimes these diseases are seen in relatively pure form, more frequently a given fowl shows manifestations of more than one response. For example, the term erythro-myeloblastosis refers to the condition where large numbers of erythroblasts and myeloblasts simultaneously circulate in the blood.

Most likely a given avian leukemia virus has the potential to transform more than one type of normal blood cell precursor. The precise form of a bird's disease may depend not only on the cell type initially transformed to the cancerous state, but also upon which other types of cells subsequently become transformed as the disease progresses. This point, however, remains basically a conjecture because tissue culture systems to study cell transformation by ALV are not yet available. A further complication is that many isolates of avian leukemia viruses may be mixtures of different viruses, each perhaps capable of producing only one type of leukemic response.

These uncertainties notwithstanding, there do exist strains of ALV which characteristically produce predominantly one type of disease. One such strain, AMV (avian myeloblastosis virus), has been intensively studied since 1952 by Beard and his colleagues at Duke University (Beard et al., 1952). For over 150 passages through uninfected chickens, it has maintained the property of inducing the formation of very large numbers of myeloblasts. Cell counts as high as 2×10^6/ml are found in diseased blood, and attendant viral titres frequently are in the order of 10^{11} particles/ml (Sharp and Beard, 1952). AMV-diseased chickens now provide the most convenient source of material for biochemical study of avian leukemia viruses, yielding much greater amounts of virus than are obtained from RSV-induced tumours.

In the work with AMV the advantages of inoculating very young (three-day-old) chicks was quickly appreciated. Most likely their lack of immunological responsiveness at this age makes them much more susceptible to infection. Even, however, with such young chicks very large numbers of virus particles must be added to transmit the disease. Usually between 10^6 to 10^9 particles are added to insure a large fraction of takes. Whether this means that most physical particles have lost their infectivity or whether primary cell transformation events are very rare is still unknown.

AMV Production by Cultured Myeloblasts

AMV-infected myeloblasts can be taken from diseased chickens and maintained in tissue culture under conditions where they can divide and produce progeny AMV. Such cells provide ideal material for electron microscope study of the viral life cycle (Bernhard et al., 1958; Bernhard, 1958). AMV, like all other known avian leukemia viruses with the exception of Marek's disease virus (MDV), was found to be morphologically identical to RSV, containing RNA as its genetic material (Allison and Burke, 1962). Maturation of the virus occurs on the cell surface, and so far as we can tell, RSV and all forms of ALV are very closely related (Bonar et al., 1959).

In Vitro Cell Transformations

Detailed analysis of how leukemia viruses act as carcinogens has been greatly hindered by the failure so far to culture chicken blood-cell progenitors. These are presumably the cells that ALV transforms in vivo, and until they can be cultivated, this transformation cannot be studied at the cellular and molecular levels. It has been reported that after infection in vitro with AMV, chicken blood-cell progenitors begin to divide (Baluda and Goetz, 1961; Baluda et al., 1964); but these experiments are not easy to duplicate and they have not yet led to the development of a reproducible in vitro assay system for transformation by AMV or ALV.

Inapparent Infections

ALV particles are not only found in diseased animals; in fact, many embryos and chickens carry these viruses as apparently harmless passengers without any evidence of disease (Rubin, 1960; Vogt and Rubin, 1962). The viruses apparently multiply in a variety of normal cells without regularly transforming them into their cancerous counterparts; conceivably the rare normal cell is transformed but then quickly destroyed by an immunological response. Within a few months of hatching, however, the birds develop leukemia and eventually die of the disease.

When cells from carrier embryos are used for experiments, the viruses they harbour may interfere with infections by closely related RSV. Indeed, one strain of ALV was first noticed by Rubin because of its ability to inhibit multiplication of RSV, and the virus was therefore named RIF (Rous interfering factor).

Vertical Transmission

Much past work on ALV has been generated by a desire to find a way to decrease the incidence of the various chicken leukemias. Lymphoblastosis in particular is of great economic importance. The densely crowded conditions in contemporary poultry farms clearly favour spread of the virus from one chicken to another and so the disease often reaches epidemic proportions. Infection (horizontal transmission) is not, however, the only way in which chickens become infected by ALV. Some birds inherit the viral genome through eggs and sperms of their parents, a phenomenon known as vertical transmission. Embryos that inherit ALV vertically often contain multiplying virus particles (Cottral et al., 1954). When the immune system comes into existence, the ALV particles are recognized as "normal" cell compounds and so antibodies against ALV are never made throughout the life of the bird (Rubin et al., 1961, 1962). Such tolerant birds, however, retain their capacity to mount mediated immune responses against any transformed cancerous cells which may arise after hatching.

Murine Leukemia Viruses

The possibility that leukemia in mice as well as in birds had a viral origin was seriously considered long before 1951 when Ludwik Gross of the Bronx Veterans Hospital did the first definitive experiments. During the early 1930's, several inbred mouse strains were noticed to have high frequencies of spontaneous leukemia. Particularly useful in subsequent work have been the C58 strain developed by MacDowell at Cold Spring Harbor and the various Ak strains bred by Jacob Furth, then of the Cornell University Medical College (Richter and MacDowell, 1929; Furth et al., 1933). Leukemia in these strains tends to develop between 6 and 18 months of age, with as many as 85 percent of the mice developing detectable disease before the time of natural death. The leukemias that develop are usually of the non-disseminating variety characterized by massive infiltrations of the spleen and thymus with cancerous lymphocytes. Much less often the afflicted mice have large numbers of cancerous myeloblasts circulating in their blood. During the same period a number of mouse strains, that were developed for other reasons, were found to have relatively low (1–2%) leukemia incidences during their normal life spans. One of these strains, C3H, was initially bred at the Jackson Laboratory for its high incidence (90%) of mammary cancer (Strong, 1935). An increased tendency to develop one type of cancer thus need not be correlated with an increased tendency to develop a different cancer.

The existence of these strains led to experiments in which leukemic cells were inoculated into individuals of both the same strain and unrelated strains (Richter and MacDowell, 1929; Korteweg, 1929; Furth and Strumia, 1931). In general, leukemia developed in the recipient animals only when they were of the same strain as the donor, though occasional takes between unrelated strains did occur. If, however, inoculation is made into newborn mice, then strain barriers always break down with, for example, newborn C3H mice readily accepting Ak leukemic cells (Gross, 1950). The Ak cells that grow in newborn C3H mice retain their own specificity however, because they remain unable to grow after subsequent transfer to adult C3H mice while retaining their transplantability to Ak mice.

When leukemia cells are serially passaged through mice, they frequently become progressively more virulent so that leukemia develops more quickly in recipients. These "transplantable leukemias" also tend to lose some of their immunological specificity, occasionally becoming able to multiply in unrelated adult mice. In most experiments with transplantable leukemia, the number of leukemic cells that must be inoculated into a single animal in order to insure a high probability of take is very large. Probably very large cell numbers are usually needed to overcome the cell-mediated immunological response that must be generated following appearance of the "foreign" leukemia cells. This argument, however, does not explain the experiments by Furth and Kahn (1937) in which they occasionally were able to transmit leukemia following the inoculation of only single cells. How such cells escape immunological surveillance is not at all clear. Perhaps they reflect the tendency of clinically advanced leukemia to acquire, by a series of mutational events, cell surfaces that somehow fail to generate strong cell-mediated immunological responses.

Throughout the decade that transplantable leukemia strains were being developed, there were frequent attempts to pass the disease by means of extracts of leukemia cells and, thereby, implicate viruses as the ultimate cause of murine leukemia. In such experiments, extracts made from leukemic cells arising in a high leukemia strain were inoculated into adult mice from strains that normally have very low incidence. All the well-done experiments, however, consistently gave negative results (Furth et al., 1933; MacDowell et al., 1939; Engelbreth-Holm, 1948). As we now realize, this was inevitable for two reasons. First and foremost, the use of adult recipients guaranteed that any newly transformed leukemia cells would stimulate cellular immune responses. Second, the murine leukemia viruses have different host ranges; some mice resist infection by some strains of virus but are susceptible to infection by other strains. These phenomena had not, of course, been even dreamed of at the time.

Necessity of Using Very Young Mice

Gross's search for leukemia viruses also was done without foreknowledge of the discovery of immunological tolerance. Fortunately he started his experiments in 1945 with young mice, guessing that previous failures might have resulted from the use of adult mice as recipients. Because Andervont and Bryan (1944) and Bittner (1944) had caused mammary tumours to develop by inoculating the Bittner strain of mouse mammary tumour virus (see below) into 7 to 21-day-old mice, Gross began using mice of this age. He failed, however, to transmit leukemia to 7 to 21-day-old mice with cell-free extracts of leukemia cells. Success did not come until 1950 when he switched to even younger mice and inoculated day-old suckling C3H mice with cell-free material from Ak leukemia cells (Gross, 1951). Then over 50 percent of the inoculated mice developed leukemia in comparison to a control frequency of less than 1 percent. As proof that his extracts were actually cell free, he showed that the newly developed leukemia cells had the immunological specificity of the recipient C3H mice, not that of the donor Ak cells used to make the cell-free extracts.

Not all of Gross's subsequent experiments, however, gave the same high incidence of induced leukemia. Apparently the leukemia virus he first studied was easily inactivated as he prepared the filtered extracts. Continued passage, however, of this virus through many generations of suckling mice gradually produced a much more stable virus that regularly induced leukemia $2\frac{1}{2}$ to $3\frac{1}{2}$ months after inoculation (Gross, 1957). In contrast, the virus preparations used in his first successful experiments did not produce leukemia until some 6 to 12 months after inoculation.

Isolation of Other Murine Leukemia Viruses

The isolation of the Gross leukemia virus quickly led to attempts to find viruses responsible for other form of cancer in mice. But when extracts made from a variety of different transplantable sarcomas and carcinomas (cancers of epithelial tissue) were inoculated into newborn mice, the cancers that occasionally arose were usually leukemias, not the form of cancer used to prepare the cell-free extracts. In contrast to the Gross virus, several of these more recently isolated mouse leukemia viruses (MLV) usually cause disseminating myeloid leukemias but they may also lead to the lymphatic form. Probably, like their avian counterparts, each isolate of MLV has the capacity to transform more than one type of normal cell and may indeed be a mixture of different viruses. Frequently, however, stocks of MLV preferentially transform one particular cell type.

In the absence of any rational taxonomic nomenclature, the various isolates of MLV are named after the persons who first isolated them. The Graffi virus (Graffi et al., 1955), the Moloney virus (Moloney, 1960), the Friend virus (Friend, 1957), and the Rauscher virus (Rauscher, 1962) have been studied in some detail. At first there was a temptation to believe that each of these viruses was the causative agent of the sarcoma or carcinoma from which it was isolated. But very similar viruses have now been isolated from normal cells as well as from transplantable tumour cells, and so it may well be that these various strains of MLV were passengers and perhaps not the causative agents of the tumours from which they were obtained.

C-Type Particles

The murine leukemia and sarcoma viruses and their avian and feline counterparts have very similar morphologies and gross chemical compositions. Each virus comprises a nucleoid, containing the RNA genome associated with protein, enclosed in a lipid-containing membrane called the envelope. All these RNA viruses replicate without killing the host cell, which continues to multiply after infection, and in thin sections of infected cells in the electron microscope progeny virus can be seen budding from the cell surface.

Frequently mature RNA sarcoma and leukemia viruses are collectively described as "C-type particles"; this term and two others, "A-type particles" and "B-type particles," were coined by Bernhard in 1960 to distinguish the different sorts of particles he had observed in thin sections of infected cells in the electron microscope. C-type particles have a centrally placed and spherical nucleoid, whereas A-type particles have an annular centrally placed nucleoid.

In sections of cells infected by RNA leukemia or sarcoma viruses, the A-type particles are most numerous at or very close to the cell surface, whereas C-type particles occur in intercellular spaces in the culture medium. It seems therefore that A-type particles are precursors of C-type particles.

In B-type particles the spherical nucleoid is excentric and very often it is surrounded by an inner membrane. Of the known RNA tumour viruses, only mouse mammary tumour virus particles (MMTV) have this morphology.

MLV and Spontaneous Leukemias

Spontaneous leukemias of low incidence mouse strains also were found to contain large numbers of MLV particles. They can be seen in EM thin sections budding off from leukemic cells as well as from a variety of normal cells. Thus, all murine leukemias, no matter whether they develop in a high incidence or low incidence strain, most likely have a viral origin. Parallel examination of healthy cells from leukemia-free animals revealed very few MLV-like particles and many cells seemed to lack any traces of them (Feldman and Gross, 1966). This fact led Gross to postulate that the MLV genome can also exist in a latent form undetectable by electron microscope examination, in which case the essential difference between high and low incidence strains would be the frequency at which the latent virus can be converted to an active form capable of causing leukemia.

Gross attempted to show that MLV isolated from high leukemic strains causes leukemia more readily than virus from low incidence strains, and by 1966 he had obtained what seemed to be positive results. We now know, however, that the ease with which MLV can be induced in murine cells is determined by at least two loci in the mouse genome and that the resistance/susceptibility of mice to infection by particular strains of MLV is also under genetic control.

MLV and Induced Leukemias

When mice of a low leukemia strain are exposed to large sublethal X-ray doses, they frequently develop leukemia (Krebs et al., 1930;

Kaplan, 1947; Kaplan and Brown, 1952). For example, 400 rad of whole body radiation to several-week-old C3H mice leads to leukemia in up to 60 percent of the exposed mice (Gross et al., 1959). From many of these X-ray-induced leukemias, infectious MLV particles have been isolated (Gross, 1959; Gross et al., 1959). In contrast, extracts of tissue from non-irradiated animals fail to yield any infectious material. Thus, somehow, the radiation treatment seems to bring about the appearance of MLV particles.

Most important, viruses are seen in the electron microscope in thin sections within several days of the radiation treatment and several months before the appearance of appreciable numbers of leukemic cells (Gross and Feldman, 1968). This hints that the first step in X-ray carcinogenesis is the conversion of a latent leukemia virus to a form that multiplies as C-type particles within normal cells. Then, sometime later, some of these newly produced MLV particles presumably transform the appropriate blood precursors into leukemic cells.

Support for a two-stage process comes from experiments in which removal of the thymus markedly lowers the frequency of X-ray-induced leukemia (Kaplan, 1950). Surprisingly, however, an infected thymus need not be present at the time of irradiation. Grafting a new thymus as late as eight days after irradiation of thymectomized mice restores a high leukemia frequency (Carnes et al., 1956). This tells us that the cells immediately affected by the X ray are not in the thymus. Conceivably, a large number of different cell types harbor latent MLV that is activated by the X-ray treatment to begin multiplying. Some of these new MLV particles then move to the thymus where the target cells for transformation are located.

This general scheme provides an explanation for the failure of X rays to influence the development of leukemia in high leukemia strains (Gross et al., 1959). The same X-ray dose which causes a leukemia frequency of greater than 50 percent in a low leukemia strain does not influence either the final incidence or the time taken for leukemia to develop in high leukemia strains. This is what we would expect if the activation of the latent MLV of high leukemic strains occurs spontaneously sometime early in adult life. Irradiation of young adults would thus not be needed for the activation process.

Injection of several chemical carcinogens also increases the incidence of leukemia in low leukemic strains. Extracts made from such leukemia cells generally yield infectious MLV particles, leading to the suggestion that chemical carcinogens also can activate latent leukemia viruses. Among the carcinogens that produce infectious extracts are methylcholanthrene (Irino et al., 1963), dimethylbenzanthracene (Zilber and Postnikova, 1966) and urethan (Ribacchi and Giraldo, 1966). All these agents also induce other forms of cancer; it is not known whether or not they act indirectly by activating RNA tumour viruses.

Inheritance through Germ Cells

MLV normally is acquired vertically through the germ cells, not horizontally by infection from other mice (Gross and Dreyfuss, 1967). Crosses of high leukemic with low leukemic mice strains yield F_1 progeny that behave largely like the high leukemic parent and develop leukemia independently of whether the male or female parent was of the high leukemic line (MacDowell and Richter, 1935; Cole and Furth, 1941). Thus, MLV is transmitted through both the egg and the sperm or possibly the seminal fluid. The frequency of leukemia in the progeny, however, was always slightly higher when the female parent came from the high leukemic strain. In one such experiment when the mother came from the high leukemic strain C58, 62 percent of the F_1 progeny developed leukemia compared to 42 percent when the male parent was of strain C58. The reason for these differences is far from clear. They may reflect the greater likelihood that MLV will infect an egg than it will successfully penetrate a sperm. Or the fact that MLV particles are visible in the milk of leukemic mice may mean that some progeny which do not acquire MLV in the germ cells acquire it later through the milk (Gross, 1962; Feldman et al., 1963; Dmochowski et al., 1963). But unlike the situation with the mouse mammary virus, milk does not seem to be an important source of infection. Experiments in which the suckling progeny from high and low incidence parents were fed by the genetically opposing foster mothers gave only marginally different results from those obtained using the natural parents (Furth et al., 1942; Fekete and Otis, 1954).

The above data suggested that MLV might be passed from parent to offspring in a latent proviral form but did not exclude the possibility that C-type particles were transmitted in the sex cells. In the past two years, however, genetic data and biochemical data have been obtained which indicate that a DNA provirus of MLV is present in mouse cell chromosomes.

MLV in Embryonic Tissue

MLV particles multiply in embryonic and neonatal tissue without seeming to affect later development. Their presence in such tissues was first shown by extraction from embryos of infectious material that transmitted leukemia to day-old mice. Later large numbers of C-type particles were seen in thin sections of apparently healthy embryonic cells from both low and high incidence strains (Dmochowski et al., 1963; Feldman et al., 1963). Thus, like their avian counterparts, MLV particles multiply in embryos and young mice without causing cancer and only act as a disease-causing agent in middle-aged mice.

The fact that C-type particles are so ubiquitous in embryonic life, yet are virtually undetectable in most healthy adult cells (Feldman and Gross, 1966), must have some significance. One possible explanation is that they play some essential role in cell differentiation, possibly transferring genetic information from one cell to another. If so, the induction of cancer by RNA tumour viruses late in life could reflect a failure of some control process that only infrequently goes awry in populations of wild mice.

Sarcoma Viruses from Leukemic Cells

The first murine sarcoma virus (MSV) to be found was isolated in London by Jennifer Harvey (1964). After passaging the Moloney strain of MLV in rats, she obtained a virus preparation that produced pleomorphic sarcomas as well as leukemias. Subsequently, Moloney (1966) reported the induction of multiple rhabdomyosarcomas by injecting high doses of the Moloney MLV into newborn Balb/c mice, and Kirsten and Mayer (1967) isolated a murine sarcoma virus in stocks of the Kirsten MLV strain that had been passed in rats.

Attempts to separate the sarcoma-inducing capacity from the ability to induce leukemia consistently failed, leading to the recognition by Hartley and Rowe (1966) that the mouse sarcoma virus is a defective virus that multiplies only in the presence of a related leukemic virus. Pure MSV preparations, therefore, cannot be obtained and all MSV preparations also contain morphologically identical MLV particles. In other words, strains of MSV resemble the Bryan strain of RSV (see above); they can transform cells in the absence of a leukemia helper virus, but they cannot multiply to produce infectious progeny in the absence of a helper virus.

Murine Osteosarcoma and an MLV-like Virus

It has long been known that sarcomas of the bone (osteosarcomas) can be induced by the localization of radioactive atoms within bone-forming cells. To test the possibility that a virus might be involved, Finkel et al. (1966), of the Argonne National Laboratory, attempted to recover infectious virus from four spontaneous and seven radiation-induced osteosarcomas. Only one tumour, that a spontaneous one, yielded extracts that produced new osteosarcomas after inoculation into newborn CFI mice. Extracts of these new tumours, when pooled and injected into newborn mice, again induced osteosarcomas, as did virus isolated during twenty-two consecutive passages. No other tumour form was ever induced and so this virus most likely does not need the simultaneous presence of a leukemic virus in order to multiply. In the electron microscope, C-type particles are seen within the tumour cells and so this virus is probably closely related to MLV and MSV. Further work is still needed to show whether radiation-induced osteosarcomas also contain a similar virus.

Adaptation of MLV to Cross Species Barriers

With time, the number of species in whose cells MLV and MSV can grow has steadily widened. Graffi and Gimmy (1957) first showed that MLV induced leukemia in newborn rats, and subsequent work has shown after only several rat-to-rat passages it grows in rats as well as, if not better than, in mice. After such passages it can induce leukemia in greater than 95 percent of the rats tested within several months after inoculation (Gross, 1963). Rat-adapted MLV does not lose its ability to grow in mice; its capacity to cause leukemia remains unaltered after many rat passages.

Recently MSV and MLV have been found to grow, albeit poorly, in human fibroblasts. Continued passage of the resulting viruses in human cells, however, leads to strains which grow best in human cells, losing at the same time the capacity to grow in mice (Aaronson, 1971). These adapted strains are antigenically distinct from the original mouse viruses. One way to interpret these findings is to postulate that genetic recombination may have occurred between a latent human RNA virus and the superinfecting mouse viruses. But the data do not rule out a more prosaic interpretation, namely, that passage through human cells selects out a preexisting variant virus which is present in the original stocks at very low concentration but which can grow to high titres in human cells.

Feline Leukemia and Sarcoma Viruses

Leukemia is one of the most frequent cancers of cats. It usually takes the form of a generalized lymphosarcoma, though late in the disease cancerous circulating blood cells are sometimes seen. From these leukemias it has been proven relatively easy to isolate infectious virus that transmit leukemia to other cats. Jarret and his colleagues isolated in 1964 the first feline leukemia virus. In the following years several new isolations of FeLV were reported and it seems likely that all cat leukemia may be of viral origin (Kawakami et al., 1967; Rickard et al., 1969). EM examination has revealed that FeLV is a typical C-type particle.

More recently an infectious agent has also been isolated from several feline fibrosarcomas that will cause new fibrosarcomas when inoculated into newborn cats (Snyder and Theilen, 1969; Gardner et al., 1970). Stocks of feline sarcoma virus (FeSV) contain an excess of FeLV and in this respect resemble murine sarcoma viruses (Sarma et al., 1971). Interestingly, stocks of FeSV not only infect newborn cats, but also induce solid tumours when injected into adult cats. Such tumours, however, generally regress, most likely because of cell-mediated immunological responses.

Ability to Cross Species Barriers

FeLV and FeSV induce leukemia and sarcomas when injected into neonatal dogs (Rickard et al., 1969; Theilen et al., 1970). FeSV also induces solid tumours in young rabbits, marmosets, and monkeys. In the rabbit and monkey the tumours regress (Theilen et al., 1970), but in the marmoset they can grow until they kill their host (Deinhardt, 1970).

These experiments with various species of animals clearly indicate that FeSV and FeLV are no respectors of species barriers; both viruses will grow in human cells. Sarma et al. (1970), for example, reported that FeSV morphologically transforms human embryo cells in culture, and in the previous year Jarrett et al. (1969) had reported that FeLV grows to high titres in human cells without prior adaptation. This finding raised the possibility that some cases of human leukemia might originate by infection with FeLV. Epidemiological studies (Dorn et al., 1970) suggest there is no connection between the occurrence of leukemia and the presence of cats in the affected households, but the possibility deserves further investigation. And in the laboratory, FeLV and FeSV should be treated with caution.

Bovine Lymphosarcoma

Lymphosarcoma in cattle presents a general pathological condition similar to that of the mouse and feline leukemias (Jarrett, 1970). In some regions its occurrence is frequent enough to be economically important, especially since it tends to occur in clusters, and large fractions of given herds have been sacrificed in attempts to confine the disease (Marshak et al., 1962; Bendixen, 1965). Calves of leukemic parents often tend to acquire it, and very convincing evidence exists that transmission frequently occurs through the bull. This suggests that if bovine lymphosarcoma is a virus disease, the virus may well be transmitted vertically.

C-type particles are seen in bovine leukemia cells (Dutcher et al., 1964) but proof that these particles cause the disease will depend on successful infectivity experiments. Only a very few attempts to transmit the disease with filtered cell-free extracts have been made (Hoflund et al., 1963) and these have given ambiguous results. One complication, of course, is that cattle live much longer than mice or cats and the incubation period of the putative bovine lymphosarcoma virus may be commensurately longer that the incubation periods of the murine and feline cancer viruses.

Mouse Mammary Tumour Virus

The discovery that a virus is the cause of the mammary tumours that "spontaneously" develop in female mice had its origin in studies of the formal genetics of the mouse that began in the early 1900's at W. B. Castle's laboratory in the Bussey Institute of Harvard University. There C. C. Little, a student of Castle, began developing various inbred strains of mice by brother-sister matings. This endeavour he continued when he moved to Cold Spring Harbor to work (1919–1925) at the Department of Genetics of the Carnegie Institute of Washington, and later (1925–1929) when he was at the University of Michigan as its President. During these years several strains were selectively inbred for high incidence of mammary cancers (see review by Little, 1947). Particularly important were the C3H and A strains selected by Strong (1935, 1936, 1942), a mouse geneticist who first joined Little at Ann Arbor, going with him to Bar Harbor, Maine, where in 1929 Little created the Jackson Laboratory for the study of fundamental cancer biology.

Transmission through a Milk Factor

By 1933 sufficient crosses had been done between the high (C3H) and the low (C57) incidence strains to show that the pattern of inheritance of susceptibility to mammary tumours depended upon whether the male or female parent carried the high incidence trait. When the females were from the C3H strain, greater than 90 percent of the female progeny developed mammary cancers, whereas if the C3H parent was the male, less than 10 percent of the female progeny developed mammary tumours (staff of the R. B. Jackson lab, 1933). These results led the Bar Harbor group to conclude that a non-chromosomal factor was transmitted from parent to offspring through the female parent. Soon afterwards, Korteweg (1934, 1936) in Holland, using mice from Bar Harbor, arrived at the same conclusion.

One conceivable explanation for such maternal inheritance was transmission of a factor through the cytoplasm of the egg, while another possibility was intra-uterine infection. A third possibility, the one quickly shown to be correct, was transmission through the milk of the mother. In experiments done at Bar Harbor between 1934 and 1936 by John Bittner, newborn mice were removed from their high incidence A mothers and subsequently nursed by mothers of the low incidence CBA strain. The resulting incidence of cancer was much lower than expected if their natural mothers had been the nurse (Bittner, 1936). Correspondingly, when newborn mice of a low incidence strain were nursed by a high incidence strain, most of the female offspring developed mammary tumours. Subsequent experiments showed that a few drops of milk were sufficient to transmit the high incidence trait. Newborn mice, therefore, must be transferred immediately to foster parents if they are to remain free of the milk factor.

Hormonal-dependent Tumour Growth

Mammary cancer normally develops only in female mice of susceptible strains. Injection of female hormones (estrogens) into male mice of the same strain causes the disease (Lacassagne, 1932). In contrast, similar treatment to males of low incidence strains never results in mammary tumours (Bittner, 1942). Thus, estrogens are carcinogenic only when the milk factor is present. How they act still remains a puzzle, although work with tissue cultures shows that mammary cells often multiply only in the presence of suitable estrogens (Furth, 1953).

B-Type Particles in Mice with Milk Factor

In 1942 Bittner found that the milk factor was a virus when he passed infectious material through Seitz filter pads that trap bacteria (Bittner, 1942; Andervont and Bryan, 1944). Today Bittner's virus and other closely related viruses discovered subsequently are called mouse mammary tumour viruses (MMTV). The first convincing electron micrographs of these MMTV were taken by Dmochowski (1954) and Bernard and Bauer (1955). They saw large numbers of 75-mμ diameter bodies in the cytoplasm of mammary tumour cells, while budding off from the cell surfaces were very distinctive 105-mμ diameter bodies (B-type particles) with excentrically placed nucleoids. The intracytoplasmic particles seem to be precursors that eventually move to surface sites where they acquire lipid-containing membranes and become the mature B-type particles. Whether the excentric location of nucleoids in B-type particles reflects a fundamental difference between MMTV and the more symmetrical C-type viruses is not known. But other lines of evidence suggest that MMTV are not closely related to MSV and MLV. For example, although all these viruses have an RNA genome, there is no detectable homology between the base sequences of MMTV and MLV RNA, and the MMTV do not have the interspecies specific antigen which is present in all known mammalian C-type leukemia and sarcoma viruses.

In high incidence strains, MMTV particles are present in a variety of different tissues, with very large numbers in normal mammary tissue as well as in the milk. Purified virus is not infectious when injected into very young mice, but concentrated preparations occasionally cause tumours in adults (Andervont, 1945a; Bittner, 1952a).

Lack of Cultured Cells That Can Be Infected by MMTV

Mouse mammary tumour cells can be cultured and in culture they continue to release B-type MMTV particles. The replication of MMTV in such cells can, of course, be studied by electron microscopy. The virus particles that are released into the culture medium have been concentrated and some of their biochemical and biophysical properties have been determined. But no one has yet succeeded in isolating untransformed mammary epithelial cells which can be productively infected or transformed in vitro by MMTV from any source. The biological activity of stocks of MMTV can only be assayed in mice. Progress in understanding the biology of these viruses is certain, therefore, to be comparatively slow.

Several Forms of MMTV

The simple idea that MMTV is transmitted only through the milk quickly led to the observation that when males of the high incidence C3H strain are mated to females of certain low incidence strains (e.g., Balb/c), the female progeny frequently develop mammary tumours (Andervont, 1945b; Mühlbock, 1950, 1952; Bittner, 1952b). At first the possibility was considered that the female parents became infected by

MMTV during mating. This hypothesis, however, could not account for the fact that supposedly MMTV-free C3H males nursed on MMTV-free milk also pass the disease to their progeny (Andervont and Dunn, 1948). Thus, for many years the significance of the Bittner agent remained controversial.

Recently, the MMTV story was clarified when it was realized that more than one strain of MMTV exists (Mühlbock and Bentvelzen, 1969), and that the C3H strain of mice originally studied at Bar Harbor contains two types of MMTV. One, the original Bittner milk factor, is now called MMTV-S. When it is present, mammary carcinomas develop early in life. MMTV-S never passes though the egg or sperm; milk is the sole vehicle of transmission. But C3H mice also carry the much less virulent MMTV-L. When C3H progeny are foster-nursed on an MMTV-free strain, they lose MMTV-S but retain MMTV-L (Pitelka et al., 1964). As far as is known, MMTV-L passes equally well through both the sperm and the egg but never through the milk. This virus went unnoticed for a long time because of its low oncogenic potential. The tumour it induces usually takes over a year to become visible and only seldom becomes malignant. Electron microscopic examination of MMTV-L-induced tumours revealed B-type particles morphologically identical to those seen in MMTV-S-induced tumours (Pitelka et al., 1964; Calafat and Hageman, 1968). Attempts to purify infectious stocks of MMTV-L so far have failed (Hageman et al., 1968); and it may be that MMTV-L is always transmitted from one cell to another in a proviral form on a chromosome (Moore, 1963; Daams et al., 1968). Whether the latent provirus has to be activated to the B-type particle form in order to induce a tumour is not at all clear.

A third form of MMTV called MMTV-P, which was first isolated from the European mouse strain GR (Mühlbock, 1965), is almost as virulent as MMTV-S. It induces pregnancy-dependent neoplastic lesions which start out as plaque-like growths (hence the name MMTV-P) that become noticeable within the first 100 days of life. Morphologically identical to other forms of MMTV, the virus is easily transmitted both through the milk and through the eggs and sperm of the GR strain (Bentvelzen, 1968; Mühlbock and Bentvelzen, 1969). Crosses of GR mice with mice of other strains reveal that GR mice contain a specific genetic allele MS^e that is necessary both for the maintenance and sex cell transmission of MMTV-P. In the absence of the MS^e allele MMTV-P particles cannot infect cells, perhaps because they are unable to establish a proviral form on one of their host cell chromosomes.

Induction of Mammary Tumours in MMTV-free Females

Certain strains of mice like C57BL and 020 normally show no evidence of harbouring any MMTV genomes. But Timmermans et al. (1969) have found that application of X ray or urethane to 020 mice activates a highly virulent MMTV. One way to interpret this result is to postulate that these mice strains contain a latent MMTV provirus held in a inactive form by the presence of a repressor-like molecule that prevents transcription of the proviral DNA into an RNA product. The existence of such a provirus-repressor system might explain why the C57BL and 020 strains do not permit multiplication of MMTV-L.

In any case, the much earlier results of Dmochowski and Orr (1949), who induced mammary tumours by applying the carcinogen methylcholanthrene to foster-nursed C3H mice, most likely also involved the activation of an MMTV-L provirus. Likewise, when Heston et al. (1950) bred C3H mice free of Bittner agent and observed tumour incidences of up to 40 percent, they may have been observing provirus activation. Clearly many more experiments have to be done before these speculations harden into facts.

Other C-Type RNA Viruses

All the sarcoma and leukemia viruses of chicks, laboratory mice and domestic cats have the same morphology—they are all C-type particles—and similar biochemical and biophysical properties. But although it is true to say that all known sarcoma and leukemia viruses are C-type particles, it does not follow that the converse is true. Typical C-type RNA viruses have now been isolated from species belonging to three orders of vertebrates: reptiles, birds and mammals, including both old- and new-world primates (Todaro and Huebner, 1972). Whether or not all these viruses act as carcinogens in their natural host species or in other species maintained in the laboratory has yet to be determined.

A typical example of one of these C-type viruses, which may or may not be oncogenic, is that discovered in a spontaneous mammary tumour of an eight-year-old Rhesus monkey by Chopra and Mason (1970). In the electron microscope they saw in thin sections of this tumour typical C-type particles. Furthermore, the virus liberated by the tumour cells was shown to be able to multiply in a variety of monkey cells in culture as well as in one line of human leukocytes (Jensen et al., 1970; Chopra et al., 1971; Ahmed et al., 1971). Cultivated cells infected by this virus do not, however, become morphologically transformed. Thus there is no indication as yet that this virus is the cancer virus that was responsible for the development of the mammary tumour in the Rhesus monkey. Proof for such a role can only come from experiments which show that injecting this virus into susceptible animals leads to mammary cancers. Such experiments are under way. But given the relatively long life span of monkeys, tumours, if they are to develop, may require an incubation period of several years.

Every C-type RNA virus that is discovered must, of course, be considered as a potential tumour virus and should be tested for ability to transform cells in vitro, and, more importantly, ability to induce tumours in vivo.

DNA TUMOUR VIRUSES

Infectious Agents from Papillomas (Warts)

Warts (papillomas) are commonly found in a many species of mammals. They arise by excessive proliferation of epithelial tissue; they are usually benign growths that only rarely give rise to invasive cancers. They are often infectious, and the knowledge that human warts are transmitted by viruses dates to the early part of this century. The first papilloma virus to be well characterized was isolated by Shope (1933) of the Rockefeller Institute from wild cottontail rabbits. Cell-free extracts made from their warts, after passage through filters, caused new warts to appear on the skin of rabbits that had previously been wart free. The Shope papilloma virus, as it is now usually called, is very specific, infecting only scarified rabbit skin and never causing tumours when inoculated internally.

Purification of the infectious material from the cottontail warts revealed that the virus is relatively stable, spherical in shape with a diameter of 50 mμ and present in very large amounts within each wart. The Shope papilloma virus became the first tumour virus to be studied in detail at the molecular level (Beard et al., 1939), and by the late 1940's the molecular weight of the virus and of the viral DNA, a single molecule of 5×10^5 daltons, had been determined (Beard, 1948).

Absence of Viral Particles from Proliferating Tissue

Early in his work with this virus, Shope made the then very puzzling observation that no infectious virus particles could be recovered from the warts that appeared on inoculated domestic rabbits. Only after infection of wild cottontail rabbits did the resulting warts produce virus particles in significant numbers (Shope, 1933; Shope and Hurst, 1933). Much subsequent effort went into attempting to detect the virus in warts of domestic rabbits using immunological methods. While traces of viral-specific antigens were found in most warts, sufficient negative results cropped up for people to start suggesting that the Shope virus might completely disappear after transforming a cell or that it might exist in a latent non-infectious proviral state (Shope, 1937; Kidd and Rous, 1940).

Partial clarification came first from electron microscopic studies that revealed the complete absence of mature virus particles or their precursors in the proliferating cells of the basal layer (Noyes and Mellors, 1957). Virus multiplication seemed to occur only in older cells that had become keratinized and were never destined to multiply again. Progeny virus particles assembled in the cell nuclei which often became filled with a million or so mature particles that stuck to each other in crystalline arrays. Multiplication of the Shope papilloma virus thus quickly leads to cellular death.

Most people now suspect that the proliferating cells of a wart contain the Shope virus in a proviral form and that it is normally only converted to the mature form when its host cell starts to become keratinized. At this stage a metabolic signal must convey the message that the host cell is about to die and, by a process conceivably analogous to the induction of lysogenic phage, the provirus leaves the chromosomal site and begins a lytic reproductive cycle. Given the general correctness of this scheme it might further be supposed that the Shope provirus, normally being highly adapted to the cells of a cottontail rabbit, does not receive a meaningful inducing signal when it is present in the cells of the genetically distinct domestic rabbit.

Lack of Culture Systems to Study Cell Transformation

The papilloma viruses rarely have any cytopathic effect on cells in tissue culture. Human papilloma virus occasionally induces a cytopathic effect in monkey kidney cells (Mendelson and Kligman, 1961) and in human and murine fetal skin cells (Oroszlan and Rich, 1964), but there is no reliable assay based on cytopathogenicity for any papilloma virus. Bovine papilloma virus transforms bovine and murine cells, as does bovine papilloma DNA; but with both intact virus and viral DNA the efficiency of transformation is extremely low (Black et al., 1963; Thomas et al., 1963, 1964). Human papilloma virus also transforms human cells but again at an extremely low efficiency (Noyes, 1965). Presumably each papilloma virus has a restricted host range which does not include any of the cultivated cells tested so far, added to which, the temperature of incubation (37–39°C) which has been used may be too high to allow these viruses to multiply (Fenner, 1968).

Immunologically Induced Regression

After an initial period of rapid growth the warts of many rabbits simultaneously cease multiplying and begin to regress due to a cell-mediated immunological response (Beard and Rous, 1934). In other rabbits, the warts do not regress and frequently become transformed into active carcinomas that eventually kill the animals (Rous and Beard, 1934; Syverton and Berry, 1935). These differences now are thought to reflect the existence of circulating "blocking antibodies" that circumvent killing by sensitized lymphocytes. Why some rabbits respond by making excessive amounts of blocking antibodies is totally obscure.

Human Wart Virus

The virus responsible for the familiar human contagious warts is morphologically identical to the Shope papilloma virus (Strauss et al.,

1949; Williams et al., 1961; Crawford, 1965). It is also completely absent from dividing cells while present in very large numbers in the outer keratinized tissue. In contrast to rabbit warts, human warts almost always regress, never giving rise to carcinomas.

Formation of a single human wart most likely starts by the transformation of a single basal cell into a cell which has lost its normal control over cell division. A wart is probably a clone of descendants from a single transformed cell. This conclusion comes from experiments (Murray et al., 1971) in which the glucose-6-phosphate dehydrogenase of single warts of women heterozygous for two electrophoretic variants (isozymes) of this enzyme was characterized. The gene for this enzyme is on the X chromosome and the cells of heterozygous women have the genetic potential to specify both isozymes; but, in fact, only one isozyme is made in any one cell because in each cell one X chromosome is at random prevented from functioning. Individual warts were found always to contain one, but not both, isozyme; this strongly suggests that each wart is a clone from one transformed cell rather than a collection of cells independently transformed by progeny virus released from the first cell to be infected.

Polyoma Virus

The first person to know about polyoma virus was Ludwik Gross. In some of his first successful experiments with MLV he observed that a few mice inoculated with extracts from leukemic Ak mice developed salivary gland (parotid) adenocarcinomas while remaining free of any traces of leukemia. The sharp difference between these two forms of cancer suggested that his extracts contained two different tumour viruses, one a leukemia virus, the other specifically inducing adenocarcinomas, which he called the parotid agent (Gross, 1953a, b). His ability to separate these two activities on the basis of their filtration behaviour favoured the two-virus hypothesis. Filters with larger porosity passed both activities, whereas those of smaller porosity preferentially held back MLV activity, suggesting that the virus causing the parotid tumours was of smaller size. This was soon proved by centrifugation experiments that showed the leukemia virus sedimented faster than the parotid agent. Even clearer separation of the two viruses was achieved by heating cell-free extracts from Ak mice to 65°C for 30 minutes. This treatment completely inactivated MLV while the ability to induce parotid tumours was unaffected.

Origin of the Name Polyoma Virus

While parotid tumours are the most frequent consequence of inoculation by the parotid agent, a variety of other tumours occasionally develop. These include medullary adrenal tumours, epithelial thymic tumours, mammary gland carcinomas, renal carcinomas, liver hemangiomas, and subcutaneous fibrosarcomas (Stewart et al., 1958). Because the parotid agent was able to transform so many different types of cells, Stewart and Eddy proposed the name "polyoma virus" and this is now universally accepted.

Polyoma Virus in Mice

Polyoma virus can be found in a large fraction of both wild and laboratory-bred mice. Its original isolation from a leukemia mouse was fortuitous as Gross discovered when he made extracts of the low leukemia C3H mice and observed that they also induced parotid tumours. Polyoma virus does not apparently produce a detectable disease when it multiplies in adult mice; it is generally regarded as a harmless passenger. The virus produces tumours only when it is inoculated into newborn laboratory mice and so in nature it probably acts as a carcinogen only on those rare occasions when a mouse is infected at the moment of birth.

While polyoma virus was easily isolated from extracts of whole mice, extracts of parotid tumours induced by the virus had virtually no infectivity, suggesting that the tumour cells themselves do not contain mature virus particles (Gross, 1955; Stewart et al., 1957). Subsequent work confirmed that this is the case and also established that the morphology of polyoma virus is very similar to that of the papilloma viruses. These findings led to the speculation that polyoma virus and the papilloma viruses are present in tumour cells as proviruses.

Polyoma Virus in Cultured Cells

A major breakthrough in understanding the replication of polyoma virus came through the efforts of Stewart and Eddy, at the National Institutes of Health, who were the first to develop cultures of mouse embryo cells in which polyoma virus replicates as a lytic virus (Stewart et al., 1957, 1958). Possession of cultures of these permissive cells made it possible to grow amounts of polyoma virus sufficient for chemical analysis and so opened the way for investigation of the molecular structure of the virus particle. Equally important, the various stages in the lytic cycle could be analyzed in detail; previously, studies had had to be confined to infected animals and the lytic cycle had remained a complete black box.

Soon afterwards several groups devised culture systems in which polyoma virus transformed cells. Cultures of both mouse and hamster cells were found after infection to give rise to stably transformed variants that proved to have many of the properties of the cells in tumours induced by polyoma virus (Sachs and Winocour, 1959; Vogt and Dulbecco, 1960). Attention, however, soon focused on the hamster cells because they are non-permissive and do not allow polyoma virus to replicate; but they are transformed by the virus at a higher frequency than are mouse cells, most of which allow the virus to replicate in a lytic cycle. The hamster cells, therefore, offered two great advantages to experimentalists intent on elucidating the mechanism of transformation: they could be transformed at a high frequency and the primary events of transformation could be studied without the complication of a preponderance of lytic events going on simultaneously in the culture.

Antigens of Polyoma-transformed Cells

By early 1961, work in Stockholm (Sjögren et al., 1961) and at the NIH (Habel, 1961) clearly defined the existence of a new antigen on the the surface of polyoma-transformed cells that prevents their transplantation into genetically equivalent adult mice. This transplantation antigen (TSTA) is specific for polyoma virus; antibodies against it show no cross reaction with tumours induced by SV40 (see below). Most likely TSTA is the antigen recognized by the immunological surveillance system that normally prevents cells transformed by polyoma virus from multiplying in adult mice. A second antigen, called the "T" (tumour) antigen, was recognized soon after in polyoma-transformed cells (Habel, 1965). It is preferentially found in the nucleus, thought to be completely distinct from TSTA, and not involved in tumour regression.

Small Number of Polyoma Genes

The excitement created by the discovery of good culture systems for studying polyoma virus intensified when biochemical analysis revealed that this virus has a very small double-helical DNA chromosome. Early measurements suggested the molecular weight of polyoma virus DNA to be about 3×10^6 daltons (4800 nucleotide pairs), a value since confirmed by a variety of physical techniques (Crawford, 1963). It was immediately realized that the genome of polyoma virus is as small as that of the coliphage ϕX174, which had already attracted the attention of molecular biologists because of the small number of genes it possesses. Tumour virologists were, therefore, greatly excited by the prospect of determining not only the exact number of polyoma virus genes but also their function; presumably one or more of these genes is responsible for transforming a cell to the cancerous state.

Polyoma-specific RNA in Transformed Cells

The possibility of pinpointing the gene(s) in polyoma virus that transforms cells obviously caused great excitement. But what if this gene product acted like a chemical carcinogen in a hit and run manner, randomly causing some genetic change that transformed a cell and then disappearing from the cell? If that were the case, the chances of elucidating the molecular biology of transformation would be close to zero. On the other hand, if the viral gene product(s) were required not only to initiate transformation but also to keep the cell transformed continuously thereafter, elucidating the molecular biology of transformation could be entirely feasible. Fortunately Benjamin (1966), working at the Salk Institute, was able to show by DNA-RNA hybridization measurements that each transformed cell produces a small, but clearly detectable, amount of messenger RNA transcribed off polyoma virus DNA. The primary transformation event is thus characterized by the insertion of some, if not all, of the polyoma genome into the host cell in a form capable of regular self-reproduction. Subsequent experiments showed that the polyoma virus DNA was in fact integrated into chromosomal DNA, thereby putting the provirus hypothesis on a very firm basis.

SV40 Virus

The discovery of SV40 virus was essentially a by-product of work done with poliomyelitis virus. Vaccines against polio virus usually were made using monkey kidney cells; at that time use of human cells was prohibited because, it was argued, they might harbour undetected human cancer viruses. Systematic studies of the indigenous viruses in monkey cells were therefore initiated to ensure that they were not present in any vaccine used to protect human populations. Serial numbers were assigned to these monkey viruses, SV1 (Simian Virus 1), SV2, etc. SV40 was not discovered until 1960 because it does not produce easily detectable cytopathic effects when it multiplies in the cells of the Rhesus monkey, the source of most cells used to produce vaccines. Only when African green monkeys of the genus *Cercopithecus* were used as a source of test cells did the existence of SV40 come to light (Sweet and Hilleman, 1960). Then it was quickly realized that many batches of polio vaccine, both of the killed and live attenuated varieties, contained infectious SV40 particles, often in considerable numbers. This was a most worrisome finding since immediately after its discovery SV40 was found to induce tumours when injected into newborn hamsters (Eddy et al., 1962; Girardi et al., 1962). Moreover, experiments soon revealed that SV40 can transform certain cells. Subsequent studies of the cancer incidence in those persons inoculated with batches of contaminated vaccine have revealed no increase in cancer frequency over that found in a control population. But considering the millions of people injected with such material, the ease with which SV40 initially was overlooked has many disquieting overtones.

Similarity between SV40 and Polyoma

Almost from the moment of its discovery SV40 has been intensively studied. Cell culture systems for studying either the lytic replication of the virus or transformation were quickly devised (Todaro and Green, 1964; Black, 1966) as were procedures for purifying the virus particles (Black et al., 1964), and it soon became obvious that SV40 is very similar to polyoma virus. The two viruses have identical structures, their chemical compositions are very similar and so is the molecular biology of their interactions with permissive and nonpermissive host cells.

To date, SV40 has not been associated with any disease in adult monkeys and its relationship with its natural hosts closely parallels the relationship between polyoma virus and the mouse. It would, of course, be rash to say categorically that SV40 never causes cancers in monkeys, but it seems most unlikely that the virus is an important carcinogen of its natural host.

Permissive vs. Nonpermissive Cells

Cells that support the complete replication of polyoma virus or SV40, and as a result suffer lysis, are called permissive cells, whereas cells that do not support the replication of these viruses but may be stably transformed by them are called nonpermissive cells. Some cells for example, mouse cells including those of stable lines such as 3T3, are nonpermissive for SV40, but permissive for polyoma virus; other cells, for example, hamster cells, are nonpermissive for both of these viruses. However, cells permissive for both SV40 and polyoma virus have never been described.

We do not know why some cells are permissive while others are nonpermissive. Hybrid cells formed by fusing permissive mouse and nonpermissive hamster cells are productively infected by polyoma virus; the permissiveness of such hybrids to infection by polyoma virus behaves as a dominant character contributed by the murine chromosomes (Basilico, 1971). This result suggests that permissiveness may be the consequence of the production by mouse cells of a factor or factors necessary for the replication and/or release of polyoma virus. Hamster cells are presumably nonpermissive, not because they synthesize repressors which specifically inhibit the multiplication of polyoma virus and SV40, but because they fail to make the necessary permissive factor(s). The nature of these putative factors and how they act is unknown.

Because permissive cells are killed by SV40 and polyoma virus, they can be used in plaque-assay tests to titrate the number of infectious particles in stocks of these viruses. Confluent monolayers of cells are first infected with various dilutions of virus and then they are overlayed with medium containing about 0.9% agar. After incubation at 37°C for several days, the monolayers are stained with neutral red, a dye which is taken up only by living cells. Virus plaques appear as unstained areas in a background of stained cells. The total number of virus particles in a virus preparation can be counted by electron microscopy; it exceeds the number of infectious units by a factor or 100 or more.

With both polyoma virus and SV40 it is possible to isolate the very rare transformed cell from populations of productively infected permissive cells. For example, mouse 3T3 cells transformed by polyoma virus (Py3T3) and African green monkey kidney cells transformed by SV40 (SV40-AGMK) have been isolated. These rare transformants are thought to arise either from permissive cells which have been infected by a defective virus particle that cannot complete its replication but is capable of transforming, or from rare variant cells in the population which cannot support a productive infection.

Rescue of SV40 from Transformed Cells

For several years, much uncertainty surrounded the observation that cultures of SV40-transformed cells often yielded a few infectious SV40 particles. Since polyoma virus never could be recovered from the cells it transforms, airborne contamination was considered a distinct possibility. Another explanation seriously proposed was that populations of nonpermissive cells might regularly give rise to a very small fraction of permissive variants. By postulating that each time a nonpermissive cell divided there was a very low but finite probability that one of the daughter cells would be a permissive variant able to support the replication of SV40, it was possible to account for the persistence of a low and fairly constant titre of SV40 particles in each culture of transformed cells. Yet another way to explain this phenomenon was to postulate that the SV40 particles arise after the activation of a SV40 provirus which then begins to multiply.

It was realized that this last explanation was indeed correct when Gerber (1966) showed that the infectious SV40 particles were only detected in plaque assay tests which involved plating large numbers of SV40-transformed cells on top of a lawn of permissive monkey cells. By contrast, infectious SV40 particles were never detected when cell-free extracts from transformed cells were added to lawns of permissive cells. This indicated that some interaction between a permissive and a nonpermissive cell was necessary for the virus to be produced; subsequent experiments not only showed that direct contact between cells was necessary for the virus to be produced, but also suggested that virus was not released until a permissive cell fused with a nonpermissive cell.

As soon as it was discovered that inactivated Sendai virus causes cells to fuse, this procedure was used to fuse SV40-transformed nonpermissive cells with uninfected permissive cells; infectious virus was shown to be released from a large fraction of these heterokaryons (see Chapter 2). Among other things this result proved that many, if not all, cells transformed by SV40 contain at least one copy of the complete SV40 genome in a proviral form (Koprowski et al., 1967; Watkins and Dulbecco, 1967).

Human Papova Viruses

The papilloma viruses, polyoma virus and SV40 are classified together as the papova viruses (Melnick, 1962). Recently viruses with either a morphology similar to the morphologies of SV40 and polyoma virus or antigens which cross react with antisera against SV40 or both have been detected in human tissues (Padgett et al., 1971; Weiner et al., 1972a,b). The agent Padgett et al. isolated from brain tissue of a patient with progressive multifocal leukoencephalopathy (PML), by inoculating cultures of human fetal brain cells with extracts of the brain tissue, has the morphology of a papova virus. As judged by immunofluorescent staining it is not, however, antigenically related to SV40, polyoma virus or human papilloma virus.

The virus isolated by Weiner et al. (1972a,b) from brain tissue of patients with the same disease not only has the same morphology as

SV40 but also, as neutralization and immunofluorescent studies show, is antigenically related to SV40. Moreover, this virus will grow in primary cultures of African green monkey cells as well as in human fetal brain cells. Whether or not these newly discovered human papova viruses cause PML, which is a subacute demyelinating disease, or any other disease in man remains to be seen.

Cell Transformation by Human Adenoviruses

The discovery that certain human adenoviruses induced tumours following injection into newborn hamsters was made soon after the discovery of the oncogenic potential of SV40. The announcement by Trentin's group (Trentin et al., 1962) at the Baylor University Medical School created much excitement, since adenovirus infections are very common in human populations and most humans contain antibodies directed against one or more of these viruses (Rowe et al., 1953; Huebner et al., 1954). Initial suspicion that the stocks of adenoviruses with oncogenic potential might have been contaminated with either SV40 or polyoma virus were short-lived, and it was soon established that several types of adenoviruses act as carcinogens when inoculated into newborn rodents (Huebner et al., 1962).

The adenoviruses are DNA viruses; their chromosome is a single, linear, double-stranded DNA molecule of about 23×10^6 daltons, which is some 7–8 times larger than the chromosomes of polyoma virus or SV40 (Green et al., 1967). The lytic replication cycle of adenoviruses can easily be studied in a variety of human and animal cells in culture, but studies of transformation have been seriously hampered by the lack of cultivated cells which are nonpermissive and can be transformed in vitro. Most studies of transformed cells have therefore involved use of cell lines that were derived from tumours induced in newborn hamsters. No trace of mature adenovirus particles was found in any transformed cell, but the presence of some adenovirus DNA integrated into cellular DNA was shown by DNA-RNA hybridization experiments. After its insertion, some of this viral DNA serves as a template for RNA transcription and as much as 1 percent of the total messenger of transformed cells can be adeno-specific (Fuginaga and Green, 1966).

The presence of so much adenovirus mRNA within transformed hamster cells raised the possibility that adenovirus mRNA might be found in certain human tumours if these were caused by human adenoviruses. So far, however, tests on over 200 different human cancers by Green and his collaborators at St. Louis have given completely negative answers. But as these tests would not have detected small amounts of adeno-specific mRNA, the possibility that adenoviruses are a significant factor in human cancer remains.

Adeno–SV40 Hybrids

The frequent occurrence of adenovirus infections among many American military personnel led to the preparation of attenuated live vaccines to protect against these viruses. Because of the then current prohibition against using human cells to grow any vaccine intended for human use, the attenuated adenovirus strains were routinely grown in monkey kidney cells. Later, when the existence of SV40 became known, tests were made to check the possibility that SV40 was present in some batches of these vaccines. To everyone's dismay, it was found that SV40 had been present as a major contaminant in all of the adenovirus vaccines that had been prepared. Unfortunately, by then the vaccines had been injected into over a million soldiers. Quickly it was determined that the presence of SV40 was far from a matter of chance. When antibodies against SV40 were added to eliminate specifically the SV40 contaminants of the adenovirus stocks, the multiplication of the adenoviruses was blocked. It was then discovered that monkey cells are, in fact, nonpermissive for human adenoviruses. They will, however, support human adenovirus multiplication if they are infected simultaneously with SV40 (Rabson et al., 1964). The product of one or more SV40 genes is thus necessary for some specific step in the replication of human adenoviruses in monkey cells. SV40 is frequently referred to as a helper virus when, with human adenoviruses, it coinfects monkey cells.

Subsequent study of these mixed adenovirus-SV40 populations revealed that genetic recombination occasionally occurred between the human adenoviruses and SV40, producing hybrid adenovirus genomes into which had been inserted all or a portion of an SV40 genome (Rowe and Baum, 1964; Rapp et al., 1964; Huebner et al., 1964). Usually the formation of these hybrid DNA molecules involved the deletion of certain adenovirus genes, thereby making the particles carrying these genomes defective (Rowe and Baum, 1965); that is, they only multiply in cells simultaneously infected with a normal adenovirus. Stocks of such adeno-SV40 hybrids thus obligatorily also contain normal human adenoviruses. Both viruses are necessary for their joint multiplication in monkey cells, the hybrid genome supplying the crucial SV40 gene(s), the complete adenovirus genome coding for those adeno-specific proteins not coded for by the hybrid genome.

Subsequently other adeno-SV40 hybrids were discovered that retain all the adenovirus genes necessary for replication in human cells as well as some SV40 genetic material (Lewis et al., 1969; Crumpacker et al., 1971). These viruses are not defective and multiply in both human and monkey cells in the absence of SV40 or adenovirus helpers. These nondefective hybrids may prove very useful for elucidating the functions of SV40 genes, hopefully including those responsible for transformation. Already, for example, a series of nondefective hybrids with different amounts of SV40 DNA have been isolated, and by observing which SV40 functions are expressed in cells transformed by these viruses, it is possible to map parts of the SV40 genome.

Proliferations Induced by Poxviruses

Infection by any virus belonging to the pox group initially results in cell proliferation. The pox produced, for example, during smallpox infection can be regarded as small tumours that quickly regress because of cell death caused either directly by reproduction of the virus or by inflammatory responses against the proliferating cells. With other poxviruses the proliferative lesions persist much longer, often for periods of several months. In general, those viruses that quickly kill cells are the most virulent and the proliferative responses they elicit are not usually the cause of disease symptoms. In nature it appears likely that most poxviruses are so adapted to their natural hosts that they seldom produce a fatal disease, and the only manifestations of their presence are superficial masses of proliferating cells called fibromas. When, however, poxviruses infect an unnatural host, their multiplication often can result in a fulminating necrotic disease that culminates in death of their host.

One of the best-understood poxviruses is that which causes the fatal rabbit disease myxomatosis. Originally this virus was confined to a group of South American rabbits where its presence induced the formation of benign fibromas at the site of infection. Its spread to an unnatural host was first noticed when it infected domestic rabbits descended from the common European rabbit. These rabbits came down with a fatal, highly contagious disease that was named myxomatosis by the Italian microbiologist Sanarelli, who then (1896) directed the Hygiene Institute in Montevideo. He suspected that it was a viral disease, proof for which came later from experiments with filtered cell extracts. Subsequent work revealed this virus would cause myxomatosis in any rabbit of European origin; this led to the idea that its deliberate introduction into wild populations of European rabbits might control their numbers where they were present as major pests. The first major attempt at eradication began in 1949 in Australia where the rabbits are all descended from European rabbits, introduced early during colonization, which quickly multiplied to become the major economic pest of that subcontinent. Almost immediately the eradication program was a great success and within several years the rabbit was virtually exterminated from large parts of Australia. Subsequently, rabbit populations with greater genetic resistance have appeared, and myxomatosis may no longer be a major factor in keeping the Australian rabbit population within a tolerable size (Fenner and Ratcliff, 1965).

The Shope Fibroma Virus

In 1932 Richard Shope isolated a virus from a fibroma on a wild cottontail rabbit that he had shot near his home in Princeton, New Jersey (Shope, 1932). Now known as the Shope fibroma virus, it is the cause of benign fibromas found in wild cottontail rabbits over much of the United States. It will also grow in domestic rabbits where the lesions it produces are also benign, regressing, in fact, somewhat faster than in the cottontail. In nature it is spread by insect bites, particularly those of the mosquito which, also, is thought to be the major vector in the spread of myxomatosis virus.

The Shope fibroma virus is morphologically identical to the myxomatosis virus, both of which are oval in shape and about 230 μ in diameter (Bernhard et al., 1954; Lloyd and Kahler, 1955). Their genetic material is DNA, of which 1.6×10^8 daltons is present in each particle. Unlike polyoma virus, SV40 and the adenoviruses, the poxviruses are not assembled in the nucleus, and all their reproductive stages occur in the cytoplasm (Fenner, 1968). In so restricting themselves to cytoplasmic locations, they have evolved specific poxvirus polymerases to make their DNA and RNA and their multiplication does not automatically interfere with normal synthesis of cellular nucleic acid.

Viral Production in Proliferating Cells

Progeny poxvirus may appear as soon as 12 hours after infection with the number of mature particles steadily increasing until as many as 1000 new particles are present in the cytoplasmic regions of infected cells. In cell cultures the multiplication of most poxviruses usually is a cytocidal process with most infected cells eventually disintegrating after their cytoplasm becomes filled with progeny virus. Some actively proliferating cells, however, do contain small numbers of mature virus particles and it is possible that a large fraction of the proliferating cells in the nascent poxes (fibromas) of infected animals contain multiplying virus. This certainly seems to be the case with the Shope fibroma virus for, even in tissue culture, infected rabbit kidney cells show a proliferative response, continuing to divide after the appearance of progeny virus (Kilham, 1956).

There is no good evidence for transformations induced by poxviruses that give rise to genetically stable cancer cells. Neither is there evidence that fibromas ever give rise to invasive cancer cells. They tend to remain at the site of infection, never spreading to distant locations within their host. As yet, immunological studies of cells induced to proliferate by poxviruses are very incomplete, and the only evidence for new surface antigens is the regularity with which the fibromas regress.

Why the presence of multiplying poxviruses so universally induces the proliferation of their host cells still remains to be worked out. Conceivably the virus-specific enzymes involved in promoting the synthesis of the poxvirus nucleic acids produce regulatory molecules that normally are absent from cells and when such molecules appear they might trigger a new round of mitosis.

Regression of Yaba Virus Tumours

Examination of some superficial tumours on the hands and limbs of Rhesus monkeys living near Yaba, Nigeria, revealed that these growths were caused by a highly infectious poxvirus, now called the Yaba monkey virus (Niven et al., 1961). Within a week after inoculation of infectious virus, visible tumours were seen that continued to grow for 4 to 6 weeks, by which time some were as large as 5 cm in diameter. Then the tumours began to regress, a process sometimes taking a month to complete. While other monkey species are generally resistant to this virus, it has produced lesions in several human volunteers. The lesions reached several centimeters in diameter before they began to regress. Several lines of primate cells are known to support the multiplication of this virus (Yohn et al., 1966), and using these cultivated cells, it may be possible to determine how Yaba virus induces cell proliferation.

Molluscum Contagiosum as a Proliferative Disease

A poxvirus was also found to be the cause of the human skin disease molluscum contagiosum. Infection by this virus leads some weeks later to the appearance of localized, small, white nodules of 2-mm diameter that result from proliferation of epithelial cells. Many of the cells within the lesions are of giant size, being filled with the so-called molluscum bodies, some 20–30 μ in diameter, that in turn contain large numbers of typical poxvirus particles. After some months the nodules usually regress, probably because of immunological attack. Considering that this virus causes a human disease, it is somewhat surprising that it is almost totally ignored.

Herpesviruses in Renal Tumours of Frogs

Kidney tumours are very frequently found in the leopard frogs (*Rana pipiens*) of northern New England. In the 1930's, Benjamin

Lucké's experiments suggested these tumours might have a viral origin, for when he injected glycerinated tumour extracts into apparently healthy young frogs, the frequency of renal tumours significantly increased over that found in uninoculated control animals (Lucké, 1938). More convincing evidence of a viral etiology came from the experiments of Duryee (1956, 1965), who passed tumour extracts through filters that trapped bacteria and then observed that tumours developed in over a third of the frogs injected with the extracts. Histological examination of these tumours frequently revealed intranuclear inclusion bodies similar to those found in cells infected with DNA-containing herpesviruses. But it was not until 1956 that, with the electron microscope, Fawcett unambiguously revealed the presence of typical DNA-containing herpesviruses within these inclusion bodies and their complete absence from normal kidney cells. Confusing these first electron microscopic studies, however, was the finding that some tumours were completely free of any sign of herpesviruses, hinting that perhaps in such tumours the virus was present in proviral form.

Virus Production at Low Temperatures

In his early work, Lucké noticed larger numbers of intranuclear inclusion bodies in the tumours of frogs collected in the winter and spring than in the tumours of frogs examined in the summer and fall. This observation was extended by Rafferty (1964) and Mizell et al. (1968), who found that when frogs bearing apparently virus-free tumours are incubated at 8°C, multiplying herpesviruses soon appear in many of the tumour cells.

The route by which this virus is naturally transmitted is still obscure. Since it appears to persist in a latent or proviral state, the viral genome might conceivably be passed through the germ cells. On the other hand, horizontal infection probably occurs in the very abundant winter populations of frogs in many lakes in the northern United States. A natural population of frogs may well contain infected animals in which the virus is overtly replicating, infected animals in which the virus is present in a proviral form hard to detect, and uninfected animals. This would certainly account in a simple way for the observation that injection of the virus into "tumour-free animals" leads regularly to an increase in the frequency of tumours over the spontaneous frequency. In any case, the phenomenon of low temperature induction strongly suggests that the Lucké virus is present in a proviral form in the multiplying tumour cells.

Unrelated Viruses from Renal Tumour Extracts

The complete failure to find a tissue culture system in which the Lucké virus will multiply seriously hampers further research. While a number of viruses that will grow in cultures of frog cells have been isolated from frog tumours (Granoff et al., 1966), they all proved to be different from the Lucké agent. Most important, none of these viruses, given the names FV (frog virus) 1, 2, 3, 4, induce kidney tumours when inoculated into young frogs. Moreover, electron microscopic examinations (Darlington et al., 1966) have revealed that some of these viruses (e.g., FV1) are morphologically different from typical herpesviruses and replicate exclusively in the cytoplasm, whereas all known herpesviruses multiply in the nucleus. Although FV4 has the morphology typical of herpesviruses, it has significantly different GC content (54%) from the Lucké virus (45%), and DNA-DNA hybridization experiments indicate that the base sequences of the genomes of these two viruses are unrelated (Gravel, 1971). Clearly much work must be done before it is proved that the renal tumours of *Rana pipiens* are caused by the Lucké virus and not by a still undetected virus. Although recent studies (Tweedell, 1967; Mizell et al., 1969) with centrifuged extracts show a parallel between the frequency of herpes-like particles and their tumour-inducing ability, more highly purified virus preparations must be carefully characterized at the molecular level if the considerable doubts about the viral etiology of these frog kidney tumours are to be erased. See reviews by Granoff (1972), Mizell (1972) and Rafferty (1972).

Herpesvirus and Marek's Disease

For a very long time, the causative agent of avian neurolymphomatosis, or as it is commonly called Marek's disease, was thought to be an RNA virus. This belief arose because this highly contagious and economically important fowl disease, marked by paralytic lymphocytic infiltrations of nerve trunks, is also characterized by the massive lymphocytic invasions of internal organs that mark most avian leukemias. However, the histology of Marek's disease led to the realization that it is a distinct disease, especially in its acute form which can kill chicks within 6–8 weeks of hatching. During such infections, cell death caused by multiplying virus particles is an equal, if not a more important, cause of disease symptoms than virus-induced cell proliferation. This suggests that in some cells the virus multiplies in a lytic way, while in others it exists in a proviral form.

Only in 1967 did it become possible to commence serious work on the causative virus. Then Churchill and Biggs (1967) succeeding in growing the Marek virus (MDV) in cultures of chicken kidney cells. From such cells virus particles were isolated (Epstein et al., 1968; Nazerian et al., 1968) that had the morphology of typical herpesviruses, and which after inoculation into susceptible young chicks led to typical cases of Marek's disease (Cook and Sears, 1970). Furthermore, attenuated strains of MDV, as well as a herpesvirus indigenous to turkeys, can be used to vaccinate chickens against the disease. To date, however, it has not been possible to transform cells in culture with MDV and this remains a chief research objective.

Difficulties in Showing That Herpes-like Viruses Induce Human Cancers

The assertion that a human cancer is caused by a specific virus is likely to rest on circumstantial evidence, barring some unforeseen circumstance in which a virus preparation, previously not thought to be oncogenic, is inoculated into people and widespread cancer cases then develop. Most future assertions are likely to be based upon one or more of the following criteria:

(a) Infection with the virus precedes the development of the cancer.

(b) Virus-specific products, nucleic acids, etc., specified by the virus continue to be made in the tumour cells.

(c) The virus or viruses of the same group are able to induce tumours in animals.

(d) The virus can transform cells in tissue culture.

(e) Vaccination against the virus of a population in high risk areas lowers the incidence of cancer.

(f) The epidemiology of the cancer is that expected of an infectious disease.

These are difficult criteria to establish especially where the herpesviruses are concerned because of two highly characteristic properties of these viruses (Roizman, 1969). First, herpesviruses are ubiquitous; at least one and sometimes as many as four distinct herpesviruses have been found in virtually every species of animal which has been examined, and natural populations are widely infected. Antibody to some herpesvirus is present in most adult members of most animal populations. Second, the herpesviruses tend to persist, very often, for the life of the infected individual; periodically they may manifest themselves in man and in animals which have circulating anti-herpesvirus antibody.

Squamous Cell Carcinomas and Herpes Simplex Infections

Squamous cell carcinoma is uncommon and little is known about the neoplastic cells. The association between this cancer and herpes simplex virus stems from the observation that squamous cell carcinomas occasionally develop at the site of recurrent herpetic infections (Wyburn-Mason, 1957; Kvasnicka, 1963, 1964, 1965). However, since the population with recurrent herpetic infections uniformly contains antibody to the virus, and since adults without antibody to this virus are very rare, sero-epidemiologic investigation can yield at best only suggestive evidence.

Cervical Carcinoma and Herpesvirus

Cervical carcinoma occurs most frequently, but not exclusively, in promiscuous women practicing poor personal hygiene or in women with promicuous partners (Beilby et al., 1968); for many years it has been alleged that this cancer is triggered by some venereally transmitted factor. Two recent observations suggest that this factor may be a variant of herpes simplex virus designated subtype 2. First, women with cervical carcinoma more often have antiserum to subtype 2 virus than matched controls (Aurelian et al., 1970; Nahmias et al., 1969, 1970; Naib et al., 1969; Rawls et al., 1969; Royston et al., 1970). Second, cervical smears from patients with carcinoma contain cells which react in immunofluorescence tests with antisera to subtype 2 virus, whereas smears from matched controls do not (Royston and Aurelian, 1970).

This serological evidence correlating cervical carcinoma with infections by herpes simplex virus subtype 2 is not entirely convincing, however, for two reasons. First, herpes simplex subtype 2 infections are doubtless not the only thing women with cervical carcinoma may have in common. Hormonal factors or infections with bacteria or other parasites may be more significant. Second, the serological data currently available could be interpreted in other ways. For example, cervical carcinoma might stimulate the production of antibody to herpes simplex virus or predispose a person to infection by this virus.

EB Virus and Burkitt's Lymphoma

Dennis Burkitt (1962a, b), an English surgeon then at the Medical School at Kampala, Uganda, first suggested that the restricted geographical distribution of the lymphoma which commonly afflicts children in regions of East Africa (Burkitt, 1958) is related to climatic factors. This suggestion raised the possibility that the lymphoma is caused by a virus transmitted by an arthropod vector. Subsequent epidemiological evidence (Haddow, 1964), however, seemed to rule against the possibility that this cancer is spread from children with the disease to normal children by a vector, and even today epidemiologists disagree amongst themselves about whether or not Burkitt's lymphoma is an infectious disease. Pike et al. (1967) have claimed that the disease occurs in time-space clusters and shows epidemic drift, which is characteristic in infectious diseases. Moreover, the pattern of occurrence of the Burkitt tumour in New Guinea suggests its appearance there also depends on climatic factors (Booth et al., 1967). On the other hand, cases of the disease have been reported from outside the putative areas of endemicity (O'Connor et al., 1965; Wright, 1966); in fact the disease has been found, albeit at extremely low frequencies, in virtually every area in which a search has been made.

Before Burkitt's hypothesis had been challenged on epidemiological grounds, a herpesvirus was found in a line of Burkitt lymphoma cells in culture (Epstein et al., 1964). Since 1964 this virus, commonly called the EB (Epstein-Barr) virus, or viruses which are closely related antigenically have been isolated from cultures of lymphoblasts obtained from human lymphomas and post-nasal carcinomas (see review by Epstein, 1970). Though antigenically they are quite distinct from the herpes simplex and herpes zoster groups, morphologically they are identical, containing a very regular internal capsid composed of 162 structurally equivalent protein subunits.

Virus Production after Metabolic Deprivation

EB virus is almost invariably present in cultures of cells from Burkitt lymphomas; cultures obtained from single tumour cells, cloned in the presence of EB virus antibody, invariably contain infected cells. The virus becomes apparent only after the lymphoma cells are cultivated, and usually populations of rapidly growing cells contain few cells that produce virus. The number of virus-producing cells can sometimes be increased by a period of arginine deprivation (Henle and Henle, 1968) or by maintaining the cells in spent medium at 33 to 37°C for prolonged periods. Even after such manipulations, no more than about 20 percent of the cells, and usually less, produce viral products, and virus production coincides with or results in irreversible cytopathic changes.

The absence of detectable herpes multiplication within most dividing Burkitt cells strongly hints that in these cells the viral genome is present in a proviral form. Direct evidence favouring this conjecture comes from recent DNA hybridization experiments (Zur Hausen and Schulte-Holthausen, 1970; Zur Hausen et al., 1970) that show the presence of 1 to 26 genome equivalents of EB DNA in cells from ten Burkitt lymphomas and ten nasopharyngeal carcinomas.

Leukocyte Transformations Induced by EB Virus

There is preliminary evidence for cell transformation by EB virus in vitro. When normal leukocytes from the peripheral blood are exposed to EB virus, the cells, previously unable to grow in culture, change into lymphoblast-like cells capable of continuous culture; and when they are starved of arginine, they begin to release EB virus (Henle et al., 1967; Pope et al., 1968; Gerber et al., 1969). Moreover, these "transformed" cells acquire a chromosomal abnormality, a constriction of the long arm of human chromosome 10, which also occurs in many virus-producing cells in cultures of Burkitt's lymphoma.

But, suggestive though this evidence is, one overriding question remains: If EB virus is a causative factor in Burkitt's lymphoma and post-nasal carcinomas, why are these tumours common in some parts of the world and extremely rare in others when EB virus is ubiquitous? Either EB virus has no causative role or the virus can induce cancers only when other accessory factors are also present. The nature of the putative accessory factors is unknown although host gene distribution, chronic malaria in Africa, and exposure to burnt incense in the Orient are often mentioned.

EB-like Virus and Mononucleosis

Greatly confusing early work with EB virus was the observation that a large number, if not a majority, of adults contain antibodies against EB, yet they had not suffered a virus disease which could be ascribed to it. Then suddenly in the Henles' Laboratory in Philadelphia, a technician whose sera previously had been EB negative became EB positive. Soon they learned that between the tests of her sera she had developed a clinical case of mononucleosis, a disease with lymphoma-like qualities, characterized by a transient excessive proliferation of lymphatic tissue. This suggested that mononucleosis, for a long time thought to be a disease caused by an undiscovered virus, was due to either EB or a very closely related virus. Subsequent work has confirmed this conjecture, and now it is guessed (Niederman et al., 1968) that most humans with anti-EB antibodies have suffered mononucleosis infections of varying severity, the great majority being so mild that they went unnoticed.

Lymphoblast cells from mononucleosis patients, like the Burkitt cells, are easy to adapt to growth in culture and release EB-like virus when starved of arginine. Whether the virus released from the cells induced to proliferate by mononucleosis is identical to that obtained from Burkitt cells is not clear, though certainly they are closely related antigenically. Today the only way to obtain EB virus is to induce its release from transformed cells in culture, and only comparatively small amounts of virus can be collected in this way. As a result, characterizing the virus at the molecular level is proving a painfully slow and difficult task and will remain so until cultivated cells which support extensive replication of the virus are selected.

Virulent Lymphoma and Herpesvirus Saimiri

In 1968 Melendez and his coworkers at the New England Primate Center isolated a herpesvirus from a primary kidney culture obtained from a healthy squirrel monkey (*Saimiri sciureus*). As long as it multiplies in this specific monkey host, herpesvirus Saimiri causes no virulent disease and most likely coexists as a harmless passenger. But in the owl monkey, or the marmoset, it rapidly induces a very virulent lymphoma which can lead to death within a month after inoculation (Melendez et al., 1969, 1970). Most important, unlike most tumour viruses, herpesvirus Saimiri induces tumours in adult, as well as in newborn, animals. During the course of the disease, many necrotic cells and proliferating lymphoblasts are seen (Hunt et al., 1970), a fact that suggests that herpesvirus Saimiri multiplies lytically in some cells, but in others induces cell proliferation and may exist as a provirus.

More recent experiments show that this virus also multiplies very well in cultures of human lung cells, and given its striking capacity for causing lymphoma in several monkey species, the possibility that it might do likewise in humans must be given very serious consideration. A conjecture to be emphasized is that viruses that normally do not multiply in man may be much more likely to act as human tumour viruses than viruses which have a long history of coevolution with man, for selective forces naturally lead to the emergence of viruses that seldom, if ever, cause the death of their host.

DO HUMAN TUMOUR VIRUSES EXIST?

RNA tumour viruses are known to cause sarcomas and leukemias in chickens, mice, and cats and to cause a carcinoma, breast cancer, in mice. The existence of these animal viruses does not, of course, prove that human RNA tumour viruses exist. However, it is difficult not to believe that such human viruses await discovery and that they will prove to show many of the biochemical, biophysical, and biological characteristics of the animal C-type and B-type viruses.

Whether the small DNA tumour viruses (with the exception of the papilloma viruses) and the adenoviruses ever cause tumours outside the laboratory is open to doubt, but at least one herpesvirus, Marek's disease virus, indisputably is a natural carcinogen. That fact alone forces us to seriously consider certain human herpesviruses as potential carcinogens. In the remainder of this book we summarize what is now known about the animal tumour viruses in the hope that we shall stimulate further research which may ultimately answer the question, "Do human tumour viruses exist?"

Literature Cited

AARONSON, S. A. 1971. Common genetic alterations of RNA tumour viruses grown in human cells. Nature *230:* 445.

AHMED, M., S. A. MAYYASI, H. C. CHOPRA, I. ZELLJADT and E. M. JENSEN. 1971. Mason-Pfizer monkey virus isolated from spontaneous mammary carcinoma of a female monkey. I. Detection of virus antigen by immunodiffusion, immunofluorescent, and virus agglutination techniques. J. Nat. Cancer Inst. *46:* 1325.

ALLISON, A. C. and D. C. BURKE. 1962. The nucleic acid contents of viruses. J. Gen. Microbiol. *27:* 181.

ALLISON, A. C., L. D. BERMAN and R. H. LEVEY. 1967. Increased tumour induction by adenovirus type 12 in thymectomized mice and mice treated with anti-lymphocyte serum. Nature *215:* 185.

ANDERVONT, H. B. 1945a. Susceptibility of young and of adult mice to the mammary tumor agent. J. Nat. Cancer Inst. *5:* 397.

———. 1945b. Fate of the C3H milk influence in mice of strains C and C57 Black. J. Nat. Cancer Inst. *5:* 383.

ANDERVONT, H. B. and W. R. BRYAN. 1944. Properties of the mouse mammary-tumor agent. J. Nat. Cancer Inst. *5:* 143.

ANDERVONT, H. B. and T. B. DUNN. 1948. Mammary tumors in mice presumably free of the mammary tumor agent. J. Nat. Cancer Inst. *8:* 227.

ARMITAGE, P. and R. A. DOLL. 1957. A two-stage theory of carcinogenesis in relation to the age distribution of human cancer. Brit. J. Cancer *11:* 161.

AURELIAN, L., I. ROYSTON and H. J. DAVIS. 1970. Antibody to genital herpes simplex virus: Association with cervical atypia and carcinoma *in situ*. J. Nat. Cancer Inst. *45:* 455.

BADER, J. P. 1964. The role of deoxyribonucleic acid in synthesis of Rous sarcoma virus. Virology *22:* 462.

———. 1965. The requirement for DNA synthesis in the growth of Rous sarcoma and Rous-associated virions. Virology *26:* 253.

BALTIMORE, D. 1970. Viral RNA-dependent DNA polymerase. Nature *226:* 1209.

BALUDA, M. A. and I. E. GOETZ. 1961. Morphological conversion of cell cultures by avian myeloblastosis virus. Virology *15:* 185.

BALUDA, M. A., C. MOSCOVIVI and I. E. GOETZ. 1964. Specificity of the *in vitro* inductive effect of avian myeloblastosis virus. Nat. Cancer Inst. Monogr. *17:* 449.

BASILICO, C. 1971. The multiplication of polyoma virus in mouse-hamster somatic hybrids. Lepetit Colloq. Biol. Med. *2:* 12. North-Holland, Amsterdam.

BEARD, J. W. 1948. Review: purified animal viruses. J. Immunol. *58:* 49.

BEARD, J. W. and P. ROUS. 1934. A virus-induced mammalian growth with the characters of a tumor (the Shope rabbit papilloma). II. Experimental alterations of the growth on the skin. Morphological considerations: the phenomena of retrogression. J. Exp. Med. *60:* 723.

BEARD, J. W., W. R. BRYAN and R. W. G. WYCKOFF. 1939. Isolation of the rabbit papilloma virus protein. J. Infec. Dis. *65:* 43.

BEARD, J. W., D. G. SHARP, E. A. ECKERT, D. BEARD and E. B. MOMMAERTS. 1952. Properties of the virus of the fowl erythromyeloblastic disease. Proc. Second Nat. Cancer Conf. *2:* 1396.

BEILBY, J. O. W., C. H. CAMERON, R. D. CATTERALL and D. DAVIDSON. 1968. Herpes-virus hominis infection of the cervix associated with gonorrhoea. Lancet *1:* 1065.

BENDIXEN, H. J. 1965. Bovine enzootic leukosis. Advances in Veterinary Science *10:* 129. Academic Press, New York.

BENJAMIN, T. L. 1966. Virus-specific RNA in cells productively infected or transformed by polyoma virus. J. Mol. Biol. *16:* 259.

BENTVELZEN, P. 1968. Genetical control of the vertical transmission of the Mühlbock mammary tumour virus in the GR mouse strain. Hollandia Publishing House, Amsterdam.

BERNHARD, W. 1958. Electron microscopy of tumor cells and tumor viruses. A review. Cancer Res. *18:* 491.

———. 1960. The detection and study of tumor viruses with the electron microscope. Cancer Res. *20:* 712.

BERNHARD, W. and A. BAUER. 1955. Mise en évidence de corpuscules d'aspect virusal dans des tumeurs mammaires de la souris. Étude au microscope électronique. Compt. Rend. Acad. Sci. *240:* 1380.

BERNHARD, W., A. BAUER, J. HAREL and C. OBERLING. 1954. Les formes intracytoplasmiques du virus fibromateux de Shope. Étude de coupes ultrafines au microscope életronique. Bull. de Cancer *41:* 423.

BERNHARD, W., C. OBERLING and P. VIGIER. 1956. Ultrastructure de virus dans le sarcome de Rous, leur rapport avec le cytoplasme des cellules tumorales. Bull. de Cancer *43:* 407.

BERNHARD, W., R. A. BONAR, D. BEARD and J. W. BEARD. 1958. Ultrastructure of viruses of myeloblastosis and erythroblastosis isolated from plasma of leukemic chickens. Proc. Soc. Exp. Biol. Med. *97:* 48.

BILLINGHAM, R. E., L. BRENT and P. B. MEDAWAR. 1953. 'Actively acquired tolerance' of foreign cells. Nature *172:* 603.

BITTNER, J. J. 1936. Some possible effects of nursing on the mammary gland tumour incidence in mice. Science *84:* 162.

———. 1942. Milk-influence of breast tumours in mice. Science *95:* 462.

———. 1944. Inciting influences in the etiology of mammary cancer in mice. Res. Conf. on Cancer, Amer. Ass. Adv. Sci., p. 63.

———. 1952a. Tumor-inducing properties of the mammary tumor agent in young and adult mice. Cancer Res. *12:* 510.

BITTNER, J. J. 1952b. Transfer of the agent for mammary cancer in mice by the male. Cancer Res. *12:* 387.
BLACK, P. H. 1966. Transformation of mouse cell line 3T3 by SV40: Dose response relationship and correlation with SV40 tumor antigen production. Virology *28:* 760.
BLACK, P. H., J. W. HARTLEY, W. P. ROWE and R. J. HUEBNER. 1963. Transformation of bovine tissue culture cells by bovine papilloma virus. Nature *199:* 1016.
BLACK, P. H., E. M. CRAWFORD and L. V. CRAWFORD. 1964. The purification of simian virus 40. Virology *24:* 381.
BONAR, R. A., D. F. PARSONS, G. S. BEAUDREAU, C. BECKER and J. W. BEARD. 1959. Ultrastructure of avian myeloblasts in tissue culture. J. Nat. Cancer Inst. *23:* 199.
BOOTH, K., D. P. BURKITT, D. J. BASSETT, R. A. COOKE and J. BIDDULPH. 1967. Burkitt lymphoma in Papua, New Guinea. Brit. J. Cancer *21:* 657.
BURKITT, D. P. 1958. A sarcoma involving the jaws in African children. Brit. J. Surg. *46:* 218.
———. 1962a. A children's cancer dependent on climatic factors. Nature *194:* 232.
———. 1962b. Determining the climatic limitations of a children's cancer common in Africa. Brit. Med. J. *27:* 1019.
BURNET, F. M. 1957. Cancer: a biological approach. Brit. Med. J. *I*, p. 779, Processes of control; p. 782, Significance of the somatic mutation; p. 841, Viruses associated with neoplastic conditions; p. 844, Practical applications.
———. 1961. Immunological recognition of self. Science *133:* 307.
———. 1964. *Cellular Immunology*. Cambridge University Press.
———. 1970. *Immunological Surveillance*. Pergamon Press.
BURNET, F. M. and F. FENNER. 1949. *The Production of Antibodies*, 2nd ed. Macmillan, Melbourne.
CALAFAT, J. and P. HAGEMAN. 1968. Some remarks on the morphology of virus particles of the type B and their isolation from mammary tumors. Virology *36:* 308.
CARNES, W. H., H. S. KAPLAN, M. B. BROWN and B. B. HIRSCH. 1956. Indirect induction of lymphomas in irradiated mice. III. Role of the thymic graft. Cancer Res. *16:* 429.
CHOPRA, H. C. and M. M. MASON. 1970. A new virus in a spontaneous mammary tumor of a Rhesus monkey. Cancer Res. *30:* 2081.
CHOPRA, H. C., I. ZELLJADT, E. M. JENSEN, M. M. MASON and N. J. WOODSIDE. 1971. Infectivity of cell cultures by a virus isolated from a mammary carcinoma of a Rhesus monkey. J. Nat. Cancer Inst. *46:* 127.
CHURCHILL, A. E. and P. M. BIGGS. 1967. Agent of Marek's disease in tissue culture. Nature *215:* 528.
CLAUDE, A., K. R. PORTER and E. G. PICKELS. 1947. Electron microscope study of chicken tumor cells. Cancer Res. *7:* 421.
COLE, R. K. and J. FURTH. 1941. Experimental studies on the genetics of spontaneous leukemia in mice. Cancer Res. *1:* 957.
COOK, M. K. and J. F. SEARS. 1970. Preparation of infectious cell-free herpes type virus associated with Marek's disease. J. Virol. *5:* 258.
COOPER, M. D., R. D. A. PETERSON and R. A. GOOD. 1965. Delineation of the thymic and bursal lymphoid systems in the chicken. Nature *205:* 143.
COTTRAL, G. E., B. R. BURMESTER and N. F. WATERS. 1954. Egg transmission of avian lymphomatosis. Poultry Sci. *33:* 1174.
CRAWFORD, L. V. 1963. The physical characteristics of polyoma virus. II. The nucleic acid. Virology *19:* 279.
———. 1965. A study of human papilloma virus DNA. J. Mol. Biol. *13:* 362.
CRAWFORD, L. V. and E. M. CRAWFORD. 1961. The properties of Rous sarcoma virus purified by density gradient centrifugation. Virology *13:* 227.

CRUMPACKER, C. S., P. H. HENRY, T. KAKEFUDA, W. P. ROWE, M. J. LEVIN and A. M. LEWIS, JR. 1971. Studies of nondefective Ad. 2-SV40 hybrid viruses. III. Base composition, molecular weight and conformation of the $Ad2^{+}ND_1$ genome. J. Virol. *7:* 352.

DAAMS, J. H., A. TIMMERMANS, A. VAN DER GUGTEN and P. BENTVELZEN. 1968. Genetical resistance of mouse strain C57BL to mammary tumour viruses. II. Resistance by means of a repressed, related provirus. Genetica *38:* 400.

DARLINGTON, R. W., P. E. GRANOFF and D. C. BREEZE. 1966. Viruses and renal cancer of *Rana pipiens:* Ultrastructural studies. Virology *29:* 149.

DEINHARDT, F. 1970. Induction of tumors in marmoset monkeys with ST-feline fibrosarcoma virus. Comp. Leukemia Res. 1969, p. 401.

DENT, P. B., R. D. A. PETERSON and R. A. GOOD. 1968. The relationship between immunologic function and oncogenesis. *Immunologic Deficiency Diseases of Man* (D. Bergsma and R. A. Good, ed.,), p. 443. National Foundation, New York.

DIGEORGE, A. M. 1968. Congenital absence of the thymus and its immunologic consequences: Concurrence with congenital hypoparathyroidism. *Immunologic Deficiency Diseases of Man* (D. Bergsma and R. A. Good, ed.), p. 116. National Foundation, New York.

DMOCHOWSKI, L. 1954. Discussion in: Proceedings, Symposium on 25 years of progress in mammalian genetics and cancer. J. Nat. Cancer Inst. *15:* 785.

DMOCHOWSKI, L. and J. W. ORR. 1949. Chemically induced breast tumours and the mammary tumour agent. Brit. J. Cancer *3:* 520.

DMOCHOWSKI, L., C. E. GREY, F. PADGETT and J. A. SYKES. 1963. Studies on the structure of the mammary tumor-inducing virus (Bittner) and of leukemia virus (Gross). *Viruses, Nucleic Acids, and Cancer*, p. 85. 17th Ann. Symp. Fund. Cancer Res. Williams & Wilkins, Baltimore, Md.

DORN, C. R., D. O. N. TAYLOR and R. SCHNEIDER. 1970. Epidemiology of canine leukemia and lymphoma. Comp. Leukemia Res. 1969, p. 403.

DURYEE, W. R. 1956. Precancer cells in amphibian adenocarcinoma. Ann. N. Y. Acad. Sci. *63:* 1280.

———. 1965. Factors influencing development of tumors in frogs. Ann. N. Y. Acad. Sci. *126:* 59.

DUTCHER, R. M., E. P. LARKIN and R. R. MARSHAK. 1964. Virus-like particles in cow's milk from a herd with a high incidence of lymphosarcoma. J. Nat. Cancer Inst. *33:* 1055.

EDDY, B. E., G. S. BORMAN, G. E. GRUBBS and R. D. YOUNG. 1962. Identification of the oncogenic substance in Rhesus monkey kidney cell cultures as Simian Virus 40. Virology *17:* 65.

EHRLICH, P. 1907. Experimentelle Studien an Mäusentumoren. Z. Krebsforsch. *5:* 59.

ELLERMANN, V. and O. BANG. 1908. Experimentelle Leukämie bei Hühnern. Zentralbl. Bakteriol., Abt. I (Orig.) *46:* 595.

———. 1909. Experimentelle Leukämie bei Hühnern. Z. Hyg. Infekt. *63:* 231.

ENGELBRETH-HOLM, J. 1948. Is it possible to transmit or accelerate the development of mouse leukemia by tissue extracts? Blood *3:* 862.

EPSTEIN, M. A. 1970. Aspects of the EB virus. Advance. Cancer Res. *13:* 383.

EPSTEIN, M. A., B. G. ACHONG and Y. M. BARR. 1964. Virus particles in cultured lymphoblasts from Burkitt's lymphoma. Lancet *2:* 702.

EPSTEIN, M. A., B. G. ACHONG, A. E. CHURCHILL and P. M. BIGGS. 1968. Structure and development of the herpes-type virus of Marek's disease. J. Nat. Cancer Inst. *41(3):* 805.

FAWCETT, D. W. 1956. Electron microscope observations on intracellular virus-like particles associated with the cells of the Lucké renal adenocarcinoma. J. Biophys. Biochem. Cytol. *2:* 725.

FEKETE, E. and H. K. OTIS. 1954. Observations on leukemia in AKR mice born from transferred ova and nursed by low leukemic mothers. Cancer Res. *14:* 445.

FELDMAN, D. G. and L. GROSS. 1966. Electron microscopic study of the distribution of the mouse leukemia virus (Gross) in organs of mice and rats with virus-induced leukemia. Cancer Res. *26:* 412.

FELDMAN, D. G., L. GROSS and Y. DREYFUSS. 1963. Electron microscope study of the passage A mouse leukemia virus in mammary glands of pregnant, virus-injected C3H(f) mice. Cancer Res. *23:* 1604.

FENNER, F. 1968. *The Biology of Animal Viruses.* Vol. I. Molecular and Cellular Biology. Academic Press, New York.

FENNER, F. and F. N. RATCLIFF. 1965. *Myxomatosis*, p. 379. Cambridge Univ. Press, London.

FIALKOW, P. J., S. M. GARTLER and A. YOSHIDA. 1967. Clonal origin of chronic myelocytic leukemia in man. Proc. Nat. Acad. Sci. *58:* 1468.

FINKEL, M. P., B. O. BISKIS and P. B. JINKINS. 1966. Virus induction of osteosarcomas in mice. Science *151:* 698.

FLANAGAN, S. P. 1966. "Nude," a new hairless gene with pleiotropic effects in the mouse. Genet. Res. *8:* 295.

FOLEY, E. J. 1953. Antigenic properties of methylcholanthrene-induced tumors in mice of the strain of origin. Cancer Res. *13:* 835.

FRIEND, C. 1957. Cell-free transmission in adult Swiss mice of a disease having the character of a leukemia. J. Exp. Med. *105:* 307.

FUGINAGA, K. and M. GREEN. 1966. The mechanism of viral carcinogenesis by DNA mammalian viruses: Viral-specific RNA in polyribosomes of adeno-virus tumor and transformed cells. Proc. Nat. Acad. Sci. *55:* 1567.

FUJINAMI, A. and L. SUZUE. 1928. Contribution to the pathology of tumor growth. Experiments on the growth of chicken sarcoma in the case of heterotransplantation. Trans. Japan. Pathol. Soc. *18:* 616.

FURTH, JACOB. 1953. Conditioned and autonomous neoplasms: A review. Cancer Res. *13:* 477.

FURTH, J. and M. C. KAHN. 1937. The transmission of leukemia of mice with a single cell. Amer. J. Cancer *31:* 276.

FURTH, J. and M. STRUMIA. 1931. Studies on transmissible lymphoid leukemia of mice. J. Exp. Med. *53:* 715.

FURTH, J., H. R. SEIBOLD and R. R. RATHBONE. 1933. Experimental studies on lymphomatosis of mice. Amer. J. Cancer *19:* 521.

FURTH, J., R. K. COLE and M. C. BOON. 1942. The effect of maternal influence upon spontaneous leukemia of mice. Cancer Res. *2:* 280.

GARDNER, M. B., R. W. RONGEY, P. ARNSTEIN, J. D. ESTES, P. SARMA, R. J. HUEBNER and C. G. RICKARD. 1970. Experimental Transmission of feline fibrosarcoma to cats and dogs. Nature *226:* 807.

GAYLORD, W. H. 1955. Virus-like particles associated with the Rous sarcoma as seen in sections of the tumor. Cancer Res. *15:* 80.

GERBER, P. 1966. Studies on the transfer of subviral infectivity from SV40-induced hamster tumor cells to indicator cells. Virology *28:* 501.

GERBER, P., J. WHANG-PENG and J. H. MONROE. 1969. Transformation and chromosome changes induced by Epstein-Barr virus in normal human leukocyte cultures. Proc. Nat. Acad. Sci. *63:* 740.

GIRARDI, A. J., B. H. SWEET, V. B. SLOTNICK and M. R. HILLEMAN. 1962. Development of tumors in hamsters inoculated in the neo-natal period with vacuolating virus, SV40. Proc. Soc. Exp. Biol. Med. *109:* 649.

Glick, B., T. S. Chang and R. C. Jaap. 1956. The bursa of Fabricius and antibody formation. Poult. Sci. *35:* 224.

Gorer, P. A. 1956. Some recent work on tumor immunity. Advance. Cancer Res. *4:* 149.

———. 1961. The antigenic structure of tumors. Advance. Immunol. *1:* 345.

Gorer, P. A., S. Lyman and G. D. Snell. 1948. Studies on the genetic and antigenic basis of tumour transplantation. Linkage between a histocompatibility gene and "fused" in mice. Proc. Roy. Soc. (London), Ser. B *135:* 499.

Graffi, A. and J. Gimmy. 1957. Erzeugung von Leukosen bei der Ratte durch ein leukämogenes Agens der Maus. Naturwissenschaften *44:* 518.

Graffi, A., H. Bielka, F. Fey, F. Scharsach and R. Weiss. 1955. Gehaüftes Auftreten von Leukämien nach Injektion von Sarkom-Filtraten. Wiener klin. Wochenschr. *105:* 61.

Granoff, A. 1972. Lucké tumour-associated viruses—A review. In *Oncogenesis* and *Herpesviruses*, p. 171. Int. Agency for Res. on Cancer, Lyon.

Granoff, A., P. E. Came and D. C. Breeze. 1966. Viruses and renal carcinoma of *Rana papiens.* I. The isolation and properties of viruses from normal and tumor tissue. Virology *29:* 133.

Gravel, M. 1971. Comparison of Herpes-type viruses associated with Lucké tumor bearing frogs. Virology *43:* 730.

Green, M., M. Pina, R. Kimes, P. Wensink, L. MacHattie and C. A. Thomas, Jr. 1967. Adenovirus DNA: I. Molecular weight and conformation. Proc. Nat. Acad. Sci. *57:* 1302.

Gross, I. 1943. Intradermal immunization of C3H mice against a sarcoma that originated in an animal of the same line. Cancer Res. *3:* 326.

Gross, L. 1950. Susceptibility of suckling-infant and resistance of adult mice of the C3H and the C57 lines to inoculation with Ak leukemia. Cancer *3:* 1073.

———. 1951. "Spontaneous" leukemia developing in C3H mice following inoculation, in infancy, with Ak-leukemic extracts, or Ak-embryos. Proc. Soc. Exp. Biol. Med. *76:* 27.

———. 1953a. A filterable agent, recovered from Ak leukemic extracts, causing salivary gland carcinomas in C3H mice. Proc. Soc. Exp. Biol. Med. *83:* 414.

———. 1953b. Neck tumors, or leukemia, developing in adult C3H mice following inoculation, in early infancy, with filtered (Berkefeld N), or centrifuged (144,000 × *g*), Ak-leukemic extracts. Cancer *6:* 948.

———. 1955. Induction of parotid carcinomas and/or subcutaneous sarcomas in C3H mice with normal C3H organ extracts. Proc. Soc. Exp. Biol. Med. *88:* 362.

———. 1957. Development and serial cell-free passage of a highly potent strain of mouse leukemia virus. Proc. Soc. Exp. Biol. Med. *94:* 767.

———. 1959. Serial cell-free passage of a radiation-activated mouse leukemia agent. Proc. Soc. Exp. Biol. Med. *100:* 102.

———. 1962. Transmission of mouse leukemia virus through milk of virus-injected C3H female mice. Proc. Soc. Exp. Biol. Med. *109:* 830.

———. 1963. Serial cell-free passage in rats of the mouse leukemia virus. Effect of thymectomy. Proc. Soc. Exp. Biol. Med. *112:* 939.

———. 1966. Are the common forms of spontaneous and induced leukemia and lymphomas in mice caused by a single virus? Nat. Cancer Inst. Monogr. *22:* 407.

———. 1970. *Oncogenic Viruses.* Pergamon Press, New York and London.

Gross, L. and Y. Dreyfuss. 1967. How is the mouse leukemia virus transmitted from host to host under natural life conditions? *Carcinogenesis: A Broad Critique*, p. 9. 20th Ann. Symp. Fund. Cancer Res. Williams & Wilkins Co., Baltimore, Md.

GROSS, L. and D. G. FELDMAN. 1968. Electron microscopic studies of radiation-induced leukemia in mice. Virus release following total-body X-ray irradiation. Cancer Res. *28:* 1677.

GROSS, L., B. ROSWIT, E. R. MADA, Y. DREYFUSS and L. A. MOORE. 1959. Studies on radiation-induced leukemia in mice. Cancer Res. *19:* 316.

HABEL, K. 1961. Resistance of polyoma virus immune animals to transplanted polyoma tumors. Proc. Soc. Exp. Biol. Med. *106:* 722.

———. 1965. Specific complement-fixing antigens in polyoma tumors and transformed cells. Virology *25:* 55.

HADDOW, A. J. 1964. Age incidence in Burkitt's lymphoma syndrome. East African Med. J. *41:* 1.

HAGEMAN, P. C., J. LINKS and P. BENTVELZEN. 1968. Biological properties of B particles from C3H and C3Hf mouse milk. J. Nat. Cancer Inst. *40:* 1319.

HARTLEY, J. W. and W. P. ROWE. 1966. Production of altered cell foci in tissue culture by defective Moloney sarcoma virus particles. Proc. Nat. Acad. Sci. *55:* 780.

HARVEY, J. J. 1964. An unidentified virus which causes the rapid production of tumours in mice. Nature *204:* 1104.

HELLSTRÖM, I. 1965. Distinction between the effects of antiviral and anticellular polyoma antibodies on polyoma tumour cells. Nature *208:* 652.

HELLSTRÖM, I. and H. O. SJÖGREN. 1965. Demonstration of H_2 isoantigens and polyoma specific tumour antigens by measuring colony formation *in vitro*. Exp. Cell Res. *40:* 212.

HELLSTRÖM, K. E. and I. HELLSTRÖM. 1969. Cellular immunity against tumor antigens. Advance. Cancer Res. *XII*: 167. Academic Press, New York.

HELLSTRÖM, I. and K. E. HELLSTRÖM. 1970. Immunological enhancement. Ann. Rev. Microbiol. *24:* 373.

HELLSTRÖM, I., K. E. HELLSTRÖM, C. A. EVANS, G. H. HEPPNER, G. E. PIERCE and J. P. S. YANG. 1969. Serum-mediated protection of neoplastic cells from inhibition by lymphocytes immune to their tumor specific antigens. Proc. Nat. Acad. Sci. *62:* 362.

HENLE, W. and G. HENLE. 1968. Effect of arginine-deficient media on the herpes-type virus associated with cultured Burkitt tumor cells. J. Virol. *2:* 182.

HENLE, W., V. DIEHL, G. KOHN, H. ZUR HAUSEN and G. HENLE. 1967. Herpes-type virus and chromosome marker in normal leukocytes after growth with irradiated Burkitt cells. Science *157:* 1064.

HESTON, W. E., M. K. DERINGER, T. B. DUNN and W. D. LEVILLAIN. 1950. Factors in the development of spontaneous mammary gland tumors in agent-free strain C3Hb mice. J. Nat. Cancer Inst. *10:* 1139.

HOFLUND, S., B. THORELL and G. WINQVIST. 1963. Experimental transmission of bovine leukosis. (Abstract) Proc. Int. Symp. Comp. Leukemia Res., Doc. III, 2. Hannover, Germany.

HUBER, J. 1968. Experience with various immunologic deficiencies in Holland. *Immunologic Deficiency Diseases in Man* (D. Bergsma and R. A. Good, ed.), p. 53. National Foundation, New York.

HUEBNER, R. J., W. P. ROWE, T. G. WARD, R. H. PARROTT and J. A. BELL. 1954. Adenoidal-pharyngeal-conjunctival agents. A newly recognized group of common viruses of the respiratory system. New Eng. J. Med. *251:* 1077.

HUEBNER, R. J., W. P. ROWE and W. T. LANE. 1962. Oncogenic effects in hamsters of human adenovirus types 12 and 18. Proc. Nat. Acad. Sci. *48:* 2051.

HUEBNER, R. J., R. M. CHANOCK, B. A. RUBIN and M. J. CASEY. 1964. Induction by adenovirus type 7 of tumors in hamsters having the antigenic characteristics of SV40 virus. Proc. Nat. Acad. Sci. *52:* 1333.

HUNT, R. D., L. V. MELENDEZ, N. W. KING, C. E. GILMORE, M. D. DANIEL, M. E. WILLIAMSON and T. C. JONES. 1970. Morphology of a disease with features of malignant lymphoma in marmosets and owl monkeys inoculated with *Herpesvirus saimiri*. J. Nat. Cancer Inst. *44:* 447.

IRINO, S., Z. OTA, T. SEZAKI, M. SUZAKI and K. HIRAKI. 1963. Cell-free transmission of 20-methylcholanthrene-induced RF mouse leukemia and electron microscopic demonstration of virus particles in its leukemia tissue. Gann *54:* 225.

JARRETT, O. 1970. Evidence for the viral etiology of leukemia in the domestic mammals. Advance. Cancer Res. *13:* 39. Academic Press, New York.

JARRETT, W. F. H., W. B. MARTIN, G. W. CRIGHTON, R. G. DALTON and M. F. STEWART. 1964. Leukaemia in the cat: Transmission experiments with leukaemia (lymphosarcoma). Nature *202:* 566.

JARRETT, O., H. M. LAIRD and D. HAY. 1969. Growth of feline leukemia virus in human cells. Nature *224:* 1208.

JENSEN, E. M., I. ZELLJADT, H. C. CHOPRA and M. M. MASON. 1970. Isolation and propagation of a virus from a spontaneous mammary carcinoma of a Rhesus monkey. Cancer Res. *30:* 2388.

KALISS, N. 1958. Immunological enhancement of tumor homografts in mice. A review. Cancer Res. *18:* 992.

KALISS, N. and N. MOLOMUT. 1952. The effect of prior injections of tissue antiserum on the survival of cancer homografts in mice. Cancer Res. *12:* 110.

KAPLAN, H. S. 1947. Observations on radiation-induced lymphoid tumors of mice. Cancer Res. *7:* 141.

———. 1950. Influence of thymectomy, splenectomy and gonadectomy on incidence of radiation-induced lymphoid tumors in strain C57 black mice. J. Nat. Cancer Inst. *11:* 83.

KAPLAN, H. S. and M. B. BROWN. 1952. Protection against radiation-induced lymphoma development by shielding and partial-body irradiation of mice. Cancer Res. *12:* 441.

KAWAKAMI, T. G., G. H. THEILEN, D. L. DUNGWORTH, S. G. BEALL and R. J. MUNN. 1967. "C"-type viral particles in plasma of feline leukemia. Science *158:* 1049.

KEOGH, E. V. 1938. Ectodermal lesions produced by the virus of Rous sarcoma. Brit. J. Exp. Pathol. *19:* 1.

KIDD, J. G. and P. ROUS. 1940. A transplantable rabbit carcinoma originating in a virus-induced papilloma and containing the virus in masked or altered form. J. Exp. Med. *71:* 813.

KILHAM, L. 1956. Propagation of fibroma virus in tissue cultures of cottontail testes. Proc. Soc. Exp. Biol. Med. *92:* 739.

KIRSTEN, W. H. and L. A. MAYER. 1967. Morphologic responses to a murine erythroblastosis virus. J. Nat. Cancer Inst. *39:* 311.

KLEIN, G. 1966. Tumor antigens. Ann. Rev. Microbiol. *20:* 223.

KLEIN, G., H. O. SJÖGREN, E. KLEIN and K. E. HELLSTRÖM. 1960. Demonstration of resistance against methylcholanthrene-induced sarcomas in the primary autochthonouse host. Cancer Res. *20:* 1561.

KOPROWSKI, H., F. C. JENSEN and Z. S. STEPLEWSKI. 1967. Activation of production of infectious tumor virus SV40 in heterokaryon cultures. Proc. Nat. Acad. Sci. *58:* 127.

KORTEWEG, R. 1929. Eine überimpfbare Leukosarkomatose bei der Maus. Z. Krebsforsch. *29:* 455.

———. 1934. Proefondervindelijke onderzoekingen aangaande erfelijheid van kanker. Nederland. Tijdschr. Geneesk. *78:* 240.

———. 1936. On the manner in which the disposition to carcinoma of the mammary gland is inherited in mice. Genetics *18:* 350.

KREBS, C., H. C. RASK-NIELSEN and A. WAGNER. 1930. The origin of lymphosarcomatosis and its relation to other forms of leucosis in white mice. Acta Radiol., Suppl. 10, p. 1–53.

KVASNICKA, A. 1963. Relationship between herpes simplex and lip carcinoma: II. Antiherpetic antibodies in patients with lip cancer. Neoplasma Bratislava *10:* 82.

———. 1964. Relationship between herpes simplex and lip carcinoma. III. Neoplasma Bratislava *10:* 199.

———. 1965. Relationship between herpes simplex and lip carcinoma. IV. Neoplasma Bratislava *12:* 61.

LACASSAGNE, A. 1932. Apparition de cancers de la mammelle chez la souris mâle, soumise à des injections de folliculine. Compt. Rend. Acad. Sci. *195:* 630.

LEWIS, A. M., JR., M. J. LEVIN, W. H. WIESE, C. S. CRUMPACKER and P. H. HENRY. 1969. A non-defective (competent) adenovirus-SV40 hybrid isolated from the AD2-SV40 hybrid population. Proc. Nat. Acad. Sci. *63:* 1128.

LINDNER, D. and S. M. GARTLER. 1965. Glucose-6-phosphate dehydrogenase mosaicism: Utilization as a cell marker in the study of leiomyomas. Science *150:* 67.

LITTLE, C. C. 1941. The genetics of tumor transplantation. *Biology of the Laboratory Mouse* (G. D. Snell, ed.), p. 279. McGraw-Hill Book Co., Inc., New York.

———. 1947. The genetics of cancer in mice. Biol. Rev. *22:* 315.

LLOYD, B. J. and H. KAHLER. 1955. Electron microscopy of the virus of rabbit fibroma. J. Nat. Cancer Inst. *15:* 991.

LUCKÉ, B. 1938. Carcinoma of the kidney in the leopard frog: The occurrence and significance of metastasis. Amer. J. Cancer *34:* 15.

MACDOWELL, E. C. and M. N. RICHTER. 1935. Mouse leukemia. IX. The role of heredity in spontaneous cases. Arch. Pathol. *20:* 709.

MACDOWELL, E. C., J. S. POTTER, M. BOVARNICK, M. N. RICHTER, M. J. TAYLOR, E. N. WARD, T. LAANES and M. P. WINTERSTEINER. 1939. *Experimental Leukemia*. Yearbook No. 38, Carnegie Inst. of Wash., p. 191.

MCALLISTER, R. M., J. E. FILBERT, M. O. NICOLSON, R. W. RONGEY, M. B. GARDNER, R. V. GILDEN and R. J. HUEBNER. 1971. Transformation and productive infection of human osteosarcoma cells by a feline sarcoma virus. Nature *230:* 279.

MANAKER, R. A. and V. GROUPÉ. 1956. Discrete foci of altered chicken embryo cells associated with Rous sarcoma virus in tissue culture. Virology *2:* 838.

MANN, D. L. and J. L. FAHEY. 1971. Histocompatibility antigens. Ann. Rev. Microbiol. *25:* 679.

MARSHAK, R. R., L. L. CORIELL, W. C. LAWRENCE, J. E. CROSHAW, JR., H. F. SCHRYVER, K. P. ALTERA and W. W. NICHOLS. 1962. Studies on bovine lymphosarcoma. I. Clinical aspects, pathological alterations, and herd studies. Cancer Res. *22:* 202.

MEDAWAR, P. B. 1946. Immunity to homologous grafted skin. I. The suppression of cell division in grafts transplanted to immunized animals. Brit. J. Exp. Pathol. *27:* 9.

———. 1958. The homograft reaction. Proc. Roy. Soc. (London) Ser. B *148:* 145.

———. 1961. Immunologic tolerance. Science *133:* 303.

MELENDEZ, L. V., R. D. HUNT, M. D. DANIEL, F. G. GARCIA and C. E. O. FRASER. 1969. *Herpes saimiri.* II. Experimentally induced malignant lymphoma in primates. Lab. Animal Care *19:* 378.

MELENDEZ, L. V., M. D. DANIEL, R. D. HUNT, C. E. O. FRASER, F. G. GARCIA, N. W. KING and M. E. WILLIAMSON. 1970. *Herpesvirus saimiri.* V. Further evidence to consider this virus as the etiological agent of a lethal disease in primates which resembles a malignant lymphoma. J. Nat. Cancer Inst. *44:* 1175.

MELNICK, J. L. 1962. Papova virus group. Science *135*: 1128.

MENDELSON, C. G. and A. M. KLIGMAN. 1961. Isolation of wart virus in tissue culture. Arch. Dermatol. *83:* 559.

MILLER, J. F. A. P. 1962. Effect of neo-natal thymectomy on the immunological responsiveness of the mouse. Proc. Roy. Soc. (London) Ser. B *156:* 415.

MITCHISON, N. A. 1954. Passive transfer of transplantation immunity. Proc. Roy. Soc. (London) Ser. B *142:* 72.

———. 1955. The role of lymph node cells in conferring immunity by adoptive transfer. J. Exp. Med. *102:* 157.

MITCHISON, N. A. and O. L. DUBE. 1955. Studies on the immunological response to foreign tumor transplants in the mouse. J. Exp. Med. *102:* 179.

MIZELL, M. 1972. The Lucké tumour herpesvirus—Its presence and expression in tumour cells. In *Oncogenesis* and *Herpesviruses,* p. 206. Int. Agency for Res. on Cancer, Lyon.

MIZELL, M., C. W. STACKPOLE and S. HELPEREN. 1968. Herpes-type virus recovery from "virus-free" frog kidney tumors. Proc. Soc. Exp. Biol. Med. *127:* 808.

MIZELL, M., I. TOPLIN and J. J. ISAACS. 1969. Tumor induction in developing frog kidneys by a zonal centrifuge purified fraction of the frog herpes-type virus. Science *165:* 1134.

MIZUTANI, S. and H. M. TEMIN. 1970. RNA-dependent DNA polymerase in virions of Rous sarcoma virus. Nature *226:* 1211.

MOLLER, E. 1965. Antagonistic effects of humoral isoantibodies on the *in vitro* cytotoxicity of immune lymphoid cells. J. Exp. Med. *122:* 11.

MOLONEY, J. B. 1960. Biological studies on a lymphoid leukemia virus extracted from sarcoma S.37. I. Origin and introductory investigations. J. Nat. Cancer Inst. *24:* 933.

———. 1966. A virus-induced rhabdomyosarcoma of mice. Nat. Cancer Inst. Monogr. *22:* 139.

MOORE, D. H. 1963. Mouse mammary tumour agent and mouse mammary tumours. Nature *198:* 429.

MÜHLBOCK, O. 1950. Mammary tumor agent in the sperm of high-cancer-strain male mice. J. Nat. Cancer Inst. *10:* 861.

———. 1952. Studies on the transmission of the mouse mammary tumor agent by the male parent. J. Nat. Cancer Inst. *12:* 819.

———. 1965. Note on a new inbred mouse strain GR/A. Europe. J. Cancer *1:* 123.

MÜHLBOCK, O. and P. BENTVELZEN. 1969. The transmission of the mammary tumor viruses. Perspect. Virol. *6:* 75.

MURRAY, R. F., J. HUBBS and B. PAYNE. 1971. Possible clonal origin of common warts (*Verruca vulgaris*). Nature *232:* 50.

NAHMIAS, A. J., Z. M. NAIB and W. E. JOSEY. 1969. Herpesvirus hominis type 2 infection: Association with cervical cancer and perinatal disease. Perspectives in Virology, vol. 8.

NAHMIAS, A. J., W. E. JOSEY, Z. M. NAIB, C. LUCE and B. GUEST. 1970. Antibodies to herpesvirus hominis types 1 and 2 in humans: II. Women with cervical cancer. Amer. J. Epidemiol. *91:* 547.

NAIB, Z. M., A. J. NAHMIAS, W. E. JOSEY and J. H. KRAMER. 1969. Genital herpetic infection: Association with cervical dysplasia and carcinoma. Cancer *23:* 940.

NAZERIAN, K., J. J. SOLOMON, R. WITTER and B. R. BURMESTER. 1968. Studies on the etiology of Marek's disease. II. Finding of a herpesvirus in cell culture. Proc. Soc. Exp. Biol. Med. *127(1):* 177.

NIEDERMAN, J. C., R. W. MCCOLLUM, G. HENLE and W. HENLE. 1968. Infectious mononucleosis: Clinical manifestations in relation to EB virus antibodies. J. Amer. Med. Assoc. *203:* 205.

NIVEN, J. S. F., J. A. ARMSTRONG, C. H. ANDREWES, H. G. PEREIRA and R. C. VALENTINE. 1961. Subcutaneous "growths" in monkeys produced by a poxvirus. J. Pathol. Bacteriol. *81:* 1.

NOYES, W. F. 1965. Studies on the human wart virus. II. Changes in primary human cell cultures. Virology *25:* 358.

NOYES, W. F. and R. C. MELLORS. 1957. Fluorescent antibody detection of the antigens of the Shope papilloma virus in papillomas of the wild and domestic rabbit. J. Exp. Med. *106:* 555.

O'CONNOR, G. T., H. RAPPAPORT and E. B. SMITH. 1965. Childhood lymphoma resembling "Burkitt tumor" in the United States. Cancer *18:* 411.

OLD, L. J., and E. A. BOYSE. 1964. Immunology of experimental tumors. Ann. Rev. Med. *15:* 167.

OROSZLAN, S. and M. A. RICH. 1964. Human wart virus: *in vitro* cultivation. Science *146:* 531.

PADGETT, B. L., D. L. WALKER, G. M. ZURHEIN, R. J. ECKROADE and B. H. DESSEL. 1971. Cultivation of papova-like virus from human brain with progressive multifocal leukoencephalopathy. Lancet *1:* 1257.

PANTELOURIS, E. M. 1968. Absence of thymus in a mouse mutant. Nature *217:* 370.

PERLMANN, P. and G. HOLM. 1969. Cytotoxic effects of lymphoid cells *in vitro*. Advance. Immunol. *11:* 117.

PIKE, M. C., E. H. WILLIAMS and B. WRIGHT. 1967. Burkitt's tumour in the West Nile district of Uganda. Brit. Med. J. *2:* 395.

PITELKA, D. R., H. A. BERN, S. NANDI and K. B. DEOME. 1964. On the significance of virus-like particles in mammary tissues of C3Hf mice. J. Nat. Cancer Inst. *33:* 867.

POPE, J. H., M. K. HORNE and W. SCOTT. 1968. Transformation of foetal human leukocytes *in vitro* by filtrates of a human leukaemic cell line containing herpes-like virus. Int. J. Cancer *3:* 857.

PREHN, R. T. and J. M. MAIN. 1957. Immunity to methylcholanthrene-induced sarcomas. J. Nat. Cancer Inst. *18:* 769.

PRINCE, A. M. 1959. Quantitative studies on Rous sarcoma virus. IV. An investigation of the nature of "noninfective" tumors induced by low doses of virus. J. Nat. Cancer Inst. *23:* 1361.

RABSON, A. S., G. T. O'CONNOR, I. K. BEREZESKY and F. J. PAUL. 1964. Enhancement of adenovirus growth in African green monkey cell cultures by SV40. Proc. Soc. Exp. Biol. *116:* 187.

RAFF, M. C. 1969. Theta isoantigen as a marker of thymus-derived lymphocytes in mice. Nature *224:* 378.

RAFFERTY, K. A. 1964. Kidney tumors of the leopard frog: A review. Cancer Res. *24*: 169.

———. 1972. Pathology of amphibian renal carcinoma—A review. In *Oncogenesis* and *Herpesviruses*, p. 159. Int. Agency for Res. on Cancer, Lyon.

RAPP, F., J. L. MELNICK, J. S. BUTEL and T. KITAHARA. 1964. The incorporation of SV40 genetic material into adenovirus 7 as measured by intranuclear synthesis of SV40 tumor antigen. Proc. Nat. Acad. Sci. *52:* 1348.

RAUSCHER, F. J. 1962. A virus-induced disease of mice characterized by erythrocytopoiesis and lymphoid leukemia. J. Nat. Cancer Inst. *29:* 515.

RAWLS, W. E., W. TOMKINS and J. L. MELNICK. 1969. The association of herpesvirus type 2 and carcinoma of the cervix. Amer. J. Epidemiol. *89:* 547.

RIBACCHI, R. and G. GIRALDO. 1966. Leukemia virus release in chemically or physically induced lymphomas in BALB/c mice. Nat. Cancer Inst. Monogr. *22:* 701.

RICHTER, M. N. and E. C. MACDOWELL. 1929. The experimental transmission of leukemia in mice. Proc. Soc. Exp. Biol. Med. *26:* 362.

RICKARD, C. G., J. E. POST, F. NORONHA and L. M. BARR. 1969. A transmissible virus-induced lymphatic leukemia of the cat. J. Nat. Cancer Inst. *42:* 987.

ROITT, I. M., M. F. GREAVES, J. BROSTOFF and J. H. L. PLAYFAIR. 1969. The cellular basis of immunological responses. Lancet *II:* 367.

ROIZMAN, B. 1969. The herpesviruses: A biochemical definition of the group. Current Topics Microbiol. Immunol. *49:* 1.

ROSENAU, W. and H. D. MOON. 1961. Lysis of homologous cells by sensitized lymphocytes in tissue culture. J. Nat. Cancer Inst. *27:* 471.

ROUS, P. 1910. A transmissible avian neoplasm: Sarcoma of the common fowl. J. Exp. Med. *12:* 696.

———. 1911. A sarcoma of the fowl transmissible by an agent separable from the tumor cells. J. Exp. Med. *13:* 397.

———. 1913. Resistance to a tumor-producing agent as distinct from resistance to the implanted tumor cells. Observations with sarcoma of the fowl. J. Exp. Med. *18:* 416.

ROUS, P. and J. W. BEARD. 1934. Carcinomatous changes in virus-induced papillomas of the skin of the rabbit. Proc. Soc. Exp. Biol. Med. *32:* 578.

ROUS, P. and J. B. MURPHY. 1914. On immunity to transplantable chicken tumors. J. Exp. Med. *20:* 419.

ROWE, W. P. and S. G. BAUM. 1964. Evidence for a possible genetic hybrid between adenovirus type 7 and SV40 viruses. Proc. Nat. Acad. Sci. *52:* 1340.

———. 1965. Studies of adenovirus-SV40 hybrid viruses. II. Defectiveness of the hybrid particles. J. Exp. Med. *122:* 955.

ROWE, W. P., R. J. HUEBNER, L. K. GILLMORE, R. H. PARROTT and T. G. WARD. 1953. Isolation of a cytogenic agent from human adenoids undergoing spontaneous degeneration in tissue culture. Proc. Soc. Exp. Biol. Med. *84:* 570.

ROYSTON, I. and L. AURELIAN. 1970. Immunofluorescent detection of herpesvirus antigens in exfoliated cells from human cervical carcinoma. Proc. Nat. Acad. Sci. *67:* 204.

ROYSTON, I., L. AURELIAN and H. J. DAVIS. 1970. Genital herpesvirus findings in relation to cervical neoplasia. J. Reprod. Med. *4:* 109.

RUBIN, H. 1960. A virus in chick embryos which induces resistance *in vitro* to infection with Rous sarcoma virus. Proc. Nat. Acad. Sci. *46:* 1105.

———. 1961. The nature of a virus-induced cellular resistance to Rous sarcoma virus. Virology *13:* 200.

RUBIN, H., A. CORNELIUS and L. FANSHIER. 1961. The pattern of congenital transmission of an avian leukosis virus. Proc. Nat. Acad. Sci. *47:* 1058.

RUBIN, H., L. FANSHIER, A. CORNELIUS and W. F. HUGHES. 1962. Tolerance and immunity in chickens after congenital and contact infection with an avian leukosis virus. Virology *17:* 143.

SACHS, L. and E. WINOCOUR. 1959. Formation of different cell-virus relationships in tumour cells induced by polyoma. Nature *184:* 1702.

SARMA, P. S., R. J. HUEBNER, J. F. BASKER, L. VERNON and R. V. GILDER. 1970. Feline leukemia and sarcoma viruses: Susceptibility of human cells to infection. Science *168:* 1098.

SARMA, P. S., R. V. GILDEN and R. J. HUEBNER. 1971. Complement fixation test for feline leukemia and sarcoma viruses (the COCAL test). Virology *44:* 137.

SHARP, D. G. and J. W. BEARD. 1952. Counts of virus particles by sedimentation on agar and electron micrography. Proc. Soc. Exp. Biol. Med. *81:* 75.

SHOPE, R. E. 1932. A filterable virus causing tumor-like condition in rabbits and its relationship to virus myxomatosum. J. Exp. Med. *56:* 803.

———. 1933. Infectious papillomatosis of rabbits. J. Exp. Med. *58:* 607.

———. 1937. Immunization of rabbits to infectious papillomatosis. J. Exp. Med. *65:* 219.

SHOPE, R. E. and E. W. HURST. 1933. Infectious papillomatosis of rabbits. J. Exp. Med. *58:* 607.

SJÖGREN, H. O. 1965. Transplantation methods as a tool for detection of tumor-specific antigens. Prog. Exp. Tumor Res. *6:* 289.

SJÖGREN, H. O., I. HELLSTRÖM and G. KLEIN. 1961. Transplantation of polyoma virus-induced tumors in mice. Cancer Res. *21:* 329.

SNELL, G. D. 1948. Methods for the study of histocompatibility genes. J. Genet. *49:* 87.

———. 1954. The enhancing effect (or actively acquired tolerance) and the histocompatibility-2 locus in the mouse. J. Nat. Cancer Inst. *15:* 665.

———. 1957. The homograft reaction. Ann. Rev. Microbiol. *11:* 439.

———. 1963. The immunology of tissue transplantation. Conceptual Advance. Immunol. Oncol., p. 323. Hoeber-Harper, Inc., New York.

SNYDER, S. P. and G. H. THEILEN. 1969. Transmissible feline fibrosarcoma. Nature *221:* 1074.

Staff of Roscoe B. Jackson Memorial Laboratory, per C. C. Little, Director. 1933. The existence of non-chromosomal influence in the incidence of mammary tumors in mice. Science *78:* 465.

STEWART, S. E., B. E. EDDY, A. M. GOCHENOUR, N. G. BORGESE and G. E. GRUBBS. 1957. The induction of neoplasms with a substance released from mouse tumors by tissue culture. Virology *3:* 380.

STEWART, S. E., B. E. EDDY and N. BORGESE. 1958. Neoplasms in mice inoculated with a tumor agent carried in tissue culture. J. Nat. Cancer Inst. *20:* 1223.

STRAUSS, M. J., E. W. SHAW, H. BUNTING and J. L. MELNICK. 1949. "Crystalline" virus-like particles from skin papillomas characterized by intranuclear inclusion bodies. Proc. Soc. Exp. Biol. Med. *72:* 46.

STRONG, L. C. 1935. The establishment of the C3H inbred strain of mice for the study of spontaneous carcinoma of the mammary gland. Genetics *20:* 586.

———. 1936. The establishment of the "A" strain of inbred mice. J. Hered. *27:* 21.

———. 1942. The origin of some inbred mice. Cancer Res. *2:* 531.

SVET-MOLDAVSKY, G. J. 1957. Development of multiple cysts and of haemorrhagic affections of internal organs in albino rats treated during the embryonic or new-born period with Rous sarcoma virus. Nature *180:* 1299.

———. 1958. Sarcoma in albino rats treated during the embryonic stage with Rous virus. Nature *182:* 1452.

SWEET, B. H. and M. R. HILLEMAN. 1960. The vacuolating virus, SV40. Proc. Soc. Exp. Biol. Med. *105:* 420.

SYVERTON, J. T. and G. P. BERRY. 1935. Carcinoma in the cottontail rabbit following spontaneous virus papilloma (Shope). Proc. Soc. Exp. Biol. Med. *33:* 399.

TEMIN, H. 1962. Separation of morphological conversion and virus production in Rous sarcoma virus infection. Cold Spring Harbor Symp. Quant. Biol. *27:* 407.

———. 1963. The effects of actinomycin D on growth of Rous sarcoma virus *in vitro*. Virology *20:* 577.

———. 1964. Nature of the provirus of Rous sarcoma. Nat. Cancer Inst. Monogr. *17:* 557.

TEMIN, H. M. and H. RUBIN. 1958. Characteristics of an assay for Rous sarcoma virus and Rous sarcoma cells in tissue culture. Virology *6:* 669.

THEILEN, G. H., S. P. SNYDER, L. G. WOLFE and J. C. LANDON. 1970. Biological studies with viral induced fibrosarcomas in cats, dogs, rabbits and non-human primates. Comp. Leukemia Res. 1969, p. 393.

THOMAS, M., J. P. LEVY, J. TANZER, M. BOIRON and J. BERNARD. 1963. Transformation *in vitro* de cellules de peau de veau embryonnaire sous l'action d'extraits accelulaires de papillomes bovins. Compt. Rend. Acad. Sci. *257:* 2155.

THOMAS, M., M. BOIRON, J. TANZER, J. P. LEVY and J. BERNARD. 1964. *In vitro* transformation of mice cells by bovine papilloma virus. Nature *202:* 709.

TIMMERMANS, A., P. BENTVELZEN, P. C. HAGEMAN and J. CALAFAT. 1969. Activation of a mammary tumour virus in 020 strain mice by X-irradiation and urethan. J. Gen. Virol. *4:* 619.

TODARO, G. J. and H. GREEN. 1964. An assay for cellular transformation by SV40. Virology *23:* 117.

TODARO, G. J. and R. J. HUEBNER. 1972. The viral oncogene hypothesis: New evidence. Proc. Nat. Acad. Sci. *69:* 1009.

TRENTIN, J. J., Y. YABE and G. TAYLOR. 1962. The quest for human cancer viruses. Science *137:* 835.

TWEEDELL, K. S. 1967. Induced oncogenesis in developing frog kidney cells. Cancer Res. *27:* 2042.

VOGT, M. and R. DULBECCO. 1960. Virus-cell interaction with a tumor-producing virus. Proc. Nat. Acad. Sci. *46:* 365.

VOGT, P. K. 1967. A virus released by "non-producing" Rous sarcoma cells. Proc. Nat. Acad. Sci. *58:* 801.

VOGT, P. K. and R. ISHIZAKI. 1966. Patterns of viral interference in the avian leukosis and sarcoma complex. Virology *30:* 368.

VOGT, P. K. and H. RUBIN. 1962. The cytology of Rous sarcoma virus infection. Cold Spring Harbor Symp. Quant. Biol. *27:* 395.

WARNER, N. L., A. SZENBERG and F. M. BURNET. 1962. The immunological role of different lymphoid organs in the chicken. I. Dissociation of immunological responsiveness. Aust. J. Exp. Biol. Med. Sci. *40:* 373.

WATKINS, J. F. and R. DULBECCO. 1967. Production of SV40 virus in heterokaryons of transformed and susceptible cells. Proc. Nat. Acad. Sci. *58:* 1396.

WEINER, L. P., R. M. HERNDON, O. NARAYAN, R. T. JOHNSON, K. SHAH, L. J. RUBINSTEIN, T. J. PREZIOSI and F. K. CONLEY. 1972a. Virus related to SV40 in patients with progressive multifocal leukoencephalopathy. New Eng. J. Med. *286:* 385.

WEINER, L. P., R. M. HERNDON, O. NARAYAN and R. T. JOHNSON. 1972b. Further studies of a SV40-like virus isolated from human brain. J. Virol. *10:* 147.

WILLIAMS, M. G., A. F. HOWATSON and J. D. ALMEIDA. 1961. Morphological characterization of the virus of the human common wart. Nature *189:* 895.

WINN, H. J. 1959. The immune response and the homograft reaction. Nat. Cancer Inst. Monogr. *2:* 113.

WRIGHT, D. H. 1966. Burkitt's tumour in England: A comparison with childhood lymphosarcoma. Int. J. Cancer *1:* 503.

WYBURN-MASON, R. 1957. Malignant change following herpes simplex. Brit. Med. J. *2:* 615.

YOHN, D. S., V. A. HAENDIGES and J. T. GRACE. 1966. Yaba tumor pox virus synthesis *in vitro*. J. Bacteriol. *91:* 1977.

ZILBER, L. A. and I. N. KRIUKOVA. 1957. Haemorrhagic disease of rats caused by Rous sarcoma virus. Voprosy Virusologii *2:* 239.

ZILBER, L. A. and Z. A. POSTNIKOVA. 1966. Induction of a leukemogenic agent by a chemical carcinogen in inbred mice. Nat. Cancer Inst. Monogr. *22:* 397.

ZUR HAUSEN, H. and H. SCHULTE-HOLTHAUSEN. 1970. Presence of EB virus nucleic acid homology in a "virus-free" line of Burkitt tumour cells. Nature *227:* 245.

ZUR HAUSEN, H., H. SCHULTE-HOLTHAUSEN, G. KLEIN, W. HENLE, G. HENLE, P. CLIFFORD and L. SANTESSON. 1970. EBV DNA in biopsies of Burkitt tumours and anaplastic carcinomas of the nasopharynx. Nature *228:* 1056.

2

The Culture of Mammalian Cells

CELL CULTURE

Since the earliest days of experimental cancer research, attempts have constantly been made to simplify the study of the disease by cultivating normal and cancerous cells and tissues outside the organism. Comparison of normal and cancerous cells in culture, it has been hoped, would lead to the elucidation of their essential differences.

But cancer is a disease of whole organisms. It is not the tumour cells that are diseased, but their bearer whose otherwise normal life is disrupted by the presence and multiplication of the tumour cells. In concentrating on model systems—cultivated cells and the tumour viruses which infect them—cancer researchers have been obliged to make the assumption, usually unspoken, that neither cancer cells nor normal cells change when they are cultivated. Otherwise, their discoveries might well be applicable only to cultivated cells, and not to the diseased organism.

In discussing the strategy of science, Bernal points out an essential difference between discovery and problem-solving. "The essential feature of a strategy of discovery lies in determining the sequence of choice of problems to solve. Now it is in fact very much more difficult to see a problem than to find a solution for it. The former requires imagination, the latter only ingenuity." (Bernal, 1971).

Two model systems have come to dominate cancer research: cell culture and transformation by tumour viruses. The assumption of relevance is still being tested, but we believe that, in Bernal's terms, these model systems have already served us well: they have permitted us to pose questions about cancer cells that can be answered by experimentation.

This chapter describes the first of these model systems, cell culture.

Early Cell Culture

The first media used to support the growth of cells in vitro were the body fluids blood and lymph. Tissue culture began in 1907, when Harrison noticed that nervous tissue explanted from frog embryos into dishes of frog lymph developed axonal processes. In 1912, at about the time that Rous found a filterable avian sarcoma virus, Alexis Carrel grew bits of chick heart in vitro by putting them into drops of horse plasma. When the plasma clotted, the piece of heart was fixed in place, and when cells at the edge of the explant grew out into the clot, they could be counted with the aid of a microscope (Carrel, 1912). But bits of tissue could be kept alive outside the animal only if extreme precautions were taken to prevent bacterial infections. Single cells could not be kept alive at all and cell culture remained tissue culture. Few experiments comparing tumour cells with normal cells were possible, not least because plasma clot cultures died in a matter of days.

Long Term Culture

Carrel was the first to show that explants of chick heart could be kept alive and growing over extended periods if they were fed and subdivided regularly. Carrel reasoned that the explants died because the clot could not be feeding them sufficiently; by contrast a chick embryo, as a rapidly dividing cell population, must be well fed by its own fluid. When he added a few drops of sterile aqueous extract of whole chick embryos to the explants, he found that the cultured tissue would remain alive indefinitely, so long as he replaced the embryo extract every few days and periodically subdivided the growing clump of fibroblastic chick cells (Carrel, 1913). As a surgical *tour de force*, Carrel kept one such chick embryo explant alive from 1912 to 1935, without the help of antibiotics (Cameron, 1935).

During this time all in vitro work had to be done with bits of tissue, since all attempts to grow cultures from single cells in serum clots failed (Willmer, 1933; Paul, 1970). Even with this limitation, however, in vitro models for carcinogenesis were constructed using long term cultures of mouse tissue. The best of these systems was devised by Earle (1943), who in a definitive paper reported experiments using a series of explants from one mouse of the C3H strain. When the explants were incubated from 6 to 406 days with a suspension of the chemical carcinogen methylcholanthrene and replanted in C3H mice, there was a clear, positive correlation between the total in vitro dose of carcinogen and the number and malignancy of the resulting tumours.

Early Cloning

In 1948 Earle and his collaborators announced that, after five years of work with one of these carcinogen-treated explants, they were able to grow single cells completely isolated from other cells and to obtain from them cultures (now called clones) comparable in size to those derived from explants (Sanford et al., 1948). Cell lines (L-cells) descended from these first clones are still in use.

The first in vitro technique for obtaining clones was exceedingly difficult. Earle drew a single living mouse cell from the edge of a growing explant into a microcapillary tube, broke off the tip of the tube into a fresh serum clot, and waited. As fresh serum slowly diffused into the tube, the single cell was gradually introduced to a very small volume of new medium, and often (not always) cultures grew out from the microcapillary tubes into the serum clot (Sanford et al., 1948; *cf.* Lwoff et al., 1950).

With this technique Earle had effectively disposed of the hypothesis, believed at that time for all cells and accepted even today for some specialized cells, that the tissue cell has only limited autonomy as a physiological unit and that "initiation of cell division in a tissue depends on a system of internal exchange between cells through direct protoplasmic bridges; for proliferation of tissue cells these cellular connections must be intact" (Fischer, 1946). While in a tissue this may be true, clearly Earle's mouse cell clones began from single cells that could divide without cellular connections.

Trypsin to Dissociate Tissues

The first, and still the best, method of dissociating tissues into viable single cells was to treat them with proteolytic enzymes. In 1916 Rous had achieved the release of cells from the outgrowth of plasma clot cultures by digesting the clot and culture with trypsin (Rous and Jones, 1916). Later the enzymic dissociation of tissues was studied in more detail by Earle and by Moscona and Rinaldini (Shannon et al., 1952; Moscona, 1952; Rinaldini, 1959). Trypsin breaks peptide bonds next to arginine and lysine. Since trypsin inactivated by diisopropylfluorophosphate cannot dissociate cells, it is likely that trypsin dissociates cells by its enzymic activity (Easty and Mutolo, 1960); the fact that other proteolytic enzymes, for example papain and ficin, can also be used to dissociate cells supports this claim.

Primary and Secondary Cultures

The cells that grew out from an explant into a serum clot were called primary cells. In 1952 Earle's laboratory developed a trypsinization technique that permitted primary cultures to be prepared from whole chick embryos (Shannon et al., 1952). These primary cultures contained macrophage-like cells, contractile muscle fibres, and melanin pigmented cells.

With trypsin the cells in primary cultures could be dissociated from each other and from the clot and replaced in fresh dishes, resulting in a secondary culture. Secondary cultures contained fewer cell types than primary cultures because they were derived from only those cells that had multiplied in the primary culture. In particular, many differentiated cell types did not replate, and the secondary culture was made almost entirely of spindle-shaped cells. Since these spindle-shaped cells were similar in morphology and growth rate to the fibroblasts obtained by culturing connective tissue (e.g., cartilage), they were called fibroblasts (Dulbecco, 1952).

The HeLa Line

A human tumour was first successfully grown in culture in 1952 by George Gey (Gey et al., 1952). Gey, a physician at Johns Hopkins in Baltimore, minced cervical carcinomas removed at surgery and grew the cells in serum-clot culture. The cells of one such surgical specimen, which were called HeLa after the first letters of the patient's name, have been maintained in culture ever since.

Because HeLa and L-cells divide rapidly, they are ideal for biochemical investigations which require very large numbers of cells. In addition, HeLa and L-cells have served particularly well as hosts in which to propagate human and mouse viruses. (See Cold Spring Harbor Symposium, Volume 27 [1962].)

Cells That Grow in Suspension

Observing that monolayer cultures of HeLa shed free cells into the medium, Gey attempted to culture these separately from the adherent mass population by rotating the tubes in which the cells were growing. In Gey's roller tubes the cells that became detached from the monolayer continued to divide in suspension (Gey, 1954; Owens et al., 1954; Earle et al., 1956). Gey, Earle, and others eventually isolated pure suspension cultures of HeLa and L-cells (McLimmans et al., 1957).

While only rare cells from a culture of fibroblastic cells will grow in suspension, other cell types such as lymphocytes grow readily without attaching to a solid surface. Why some cells grow in culture only when they are anchored to a solid substate while others grow in suspension is totally obscure.

Current Cell Culture

After World War II the introduction of antibiotics simplified long term cell culture: sterile technique was no longer an absolute prerequisite, although overgrowth of cultures by yeasts and molds still followed careless work. However, culture media remained arbitrary mixtures of horse or bovine serum with embryo extract, sometimes diluted with sterile salt solution and glucose.

The current era of cell culture began in the 1950's with the development of semi-defined media and simple cloning techniques. The laboratories of Puck in Denver and Eagle in Bethesda succeeded in developing quantitative methods for the culture of HeLa and L-cells.

Easy Cloning Procedures

Puck's group succeeded in growing clones from single HeLa cells with 100 percent efficiency (Puck and Marcus, 1955, 1956). Taking advantage of the ability of large inocula of HeLa cells to grow on glass, they covered dishes with HeLa cells and then "sterilized" the layer of cells with large doses of X-irradiation. No HeLa cell in the layer could divide after irradiation, but the layer was still metabolically active, for the cells continued to enlarge. When single HeLa cells were than added to the irradiated "feeder" layer, every added cell grew into a colony.

Soon after devising it, Puck showed that the feeder layer was dispensable; so long as the HeLa cells that were to be cloned were dispersed gently by digestion with very dilute trypsin, they would grow into colonies on glass surfaces (Puck et al., 1956, 1958). Once the colonies had grown to about a millimetre across, they were ringed with a steel cylinder, trypsin was put inside the cylinder, and the suspended cells were pipetted into a new dish. This early work has not been superseded: trypsin is still used to remove cells from monolayers, and feeder layers are still used to persuade single, fastidious normal cells to grow into colonies after being plated at high dilution. Today, however, plastic culture dishes with many small wells (microwell dishes) are being used to clone cells.

A single fastidious cell will often grow to a colony that fills a small well of about 50 μlitres; the colony can then be transferred to a larger dish. Microwell dishes have a distinct advantage over the more common cloning cylinders. By first making appropriate dilutions of the cells in suspension, one can inoculate each well with an average of less than one cell. Then, every colony that grows is likely to be a clone, since the probability is low that any well will have received two cells (Poisson distribution). Large numbers of clones can be obtained easily in this way and microwell dishes ought also to serve well for replica plating. So far this technique is not in common use, although success at replica plating has been reported (Goldsby and Zipser, 1969; Suzuki et al., 1971).

Instability of Clones

Given a simple cloning procedure, a sensible next step was to isolate colonies formed by single HeLa or L-cells, subculture and subclone these, and compare the properties of different subclonal lines. When this was done, a surprising variability was found among different subclones from the same clonal lines.

Earle's group, for example, grew up in microcapillary tubes subclones from two sister cells of an L-cell culture. Cells of the clonal cultures were then injected into mice. Cells from one L subclone produced sarcomas in 97 percent of the mice injected; cells of the other L subclone produced sarcomas in only 1 percent of the mice injected (Sanford et al., 1954). Furthermore, Puck's group, among others, found that different HeLa subclones had different requirements for serum and different colony morphologies (Puck and Fisher, 1956; Chu and Giles, 1958; Puck, 1958; Vogt, 1958; Ely and Gray, 1961).

Both sets of experiments led to the same conclusion: the clonal descendents of single cultured cells were not identical; therefore, either the original L and HeLa cell populations were genetically mixed, or growth in culture induced genetic instability, or both (Hollaender, 1958).

Partially Defined Media

The need for a synthetic medium was recognized by early tissue culture workers (Fischer, 1948; Morgan et al., 1950; Waymouth, 1955), including, of course, Earle (Evans et al., 1956).

Many of the defined media that are used today stem from the discerning work performed by Eagle at the National Institutes of Health. In the 1950's as penicillin and streptomycin were shown to be harmless to cultured cells, Eagle decided to systematically determine the metabolic requirements of L and HeLa cells in long term culture (Eagle, 1955a,b,c,d). First he showed that cells of these two lines would form colonies when fed an arbitrary (*sic*) mixture of salts, amino acids vitamins, cofactors and carbohydrates plus a small amount (1%) of dialyzed horse serum, and that the cells grew as well in this almost entirely defined medium as in the complex mixtures of salts, sera, and embryonic tissue extracts used by Earle and Gey. Then the essential constituents for each cell type were quickly found by elimination: omission of a single essential component resulted in early death of the cultures.

Table 2.1. BASAL MEDIA FOR CULTIVATION OF THE HELA CELL AND MOUSE FIBROBLAST

L-Amino Acid (mM)		Vitamins (mM)		Salts (mM)	
Arginine	0.1	Biotin	10^{-3}	NaCl	100
Cystine	0.05	Choline	10^{-3}	KCl	5
Glutamine	2.0	Folic acid	10^{-3}	$NaH_2PO_4 \cdot H_2O$	1
Histidine	0.05	Nicotinamide	10^{-3}	$NaHCO_3$	20
Isoleucine	0.2	Pantothenic acid	10^{-3}	$CaCl_2$	1
Leucine	0.2	Pyridoxal	10^{-3}	$MgCl_2$	0.5
Lysine	0.2	Thiamine	10^{-3}		
Methionine	0.05	Riboflavin	10^{-4}		
Phenylalanine	0.1				
Threonine	0.2				
Tryptophan	0.02				
Tyrosine	0.1				
Valine	0.2				

Miscellaneous	
Glucose	5mM
Penicillin	0.005%
Streptomycin	0.005%
Phenol red	0.0005%

For studies of cell nutrition
- Dialyzed horse serum, 1%
- Dialyzed human serum, 5%

For stock cultures
- Whole horse serum, 5%
- Whole human serum, 10%

By 1955 Eagle had developed the first optimal cell culture medium; no more than 0.5–2 percent of dialyzed horse serum was needed to permit long term growth of both L-cells and HeLa cells in an otherwise completely defined medium, which is still in use as Basal Medium-Eagle (BME) (Eagle, 1955d). The contents of BME are listed in Table 2.1. Note that 13 amino acids are essential for the growth of L-cells and HeLa cells. Thus, while humans need only eight amino acids from food (Lys, Trp, Phe, Thr, Val, Met, Leu, and Ile), medium for HeLa cells must provide these and five more (Glu or Gln, Tyr, His, Arg and Cys) (Rose et al., 1955). Many media are now available that are optimal for the growth of other cell lines. Most formulas are more complex and richer than BME; apparently mammalian cells in vitro cannot survive on a less complex diet than BME (Eagle, 1956, 1959; Eagle et al., 1958; Morton, 1970).

Metabolism of Cells in Culture

Earle's culture medium, a mixture of horse serum and chick embryo extract diluted with saline, supported luxuriant growth of explants and mass cultures of mouse cells, but was far from optimal for the growth of a single cell or of a few cells. In the late 1940's Earle opened the subject of the metabolic requirements of cloned cells when he found that by preincubating medium with explants, the medium was "physically or chemically conditioned or altered by the cells," such that it would serve to support the growth of single cells (Sanford et al., 1948).

In his studies of the serine requirement of cultured cells, Eagle found that at low density, HeLa and L-cells leaked so much of their newly made serine into the medium that they could not support their own metabolism; but at high initial cell densities, the leakage amounted to cross-feeding and so did not prevent the cells from growing. Thus, cultures of these cells can have a variable requirement for metabolites which they are synthesizing, and this requirement is revealed at low population densities (Eagle et al., 1961). Subsequently Eagle showed that serine, asparagine, glutamine, inositol, and pyruvate are all lost to the medium at high cell dilution in amounts exceeding the biosynthetic capacity of one or another cell line (Eagle and Piez, 1962; Eagle et al., 1966). Cloning probably exerts a selection pressure for the survival of the rare cell with a diminished leakiness, so that cloned cells are likely to have membranes that are less leaky than cells in vivo.

Leakage of essential nutrients explained the conditioning of medium observed by Earle and Puck, and in fact, the addition of serine permitted colonies of HeLa cells to grow without feeders from single cells. Ham, in Puck's laboratory in Denver, extended Eagle's work on nutritional requirements and developed a completely defined medium in which Chinese hamster ovary cells and L-cells grew into colonies from single cells (Ham, 1962, 1963, 1965).

Morphology of Cells in Culture

The establishment of a permanent cell line from a primary culture containing many different types of cells involves the selection of those rare cells which are able to grow indefinitely in the conditions of cell culture. Cells which grow in culture assume a wide range of shapes that depend, among other things, on how well they anchor to glass or plastic surfaces, how rapidly they migrate, and on the composition of the culture medium. In other words, the shape a cell assumes when it is growing in vitro may not be the same as the shape its ancestors had when they were growing in an animal. Nevertheless, cells growing in culture are often described as being epithelial or fibroblastic or lymphoid if their shape resembles that of the cells of one or the other of these tissues.

Epithelial cells in primary cultures grow to a single cell layer and viewed from above an epithelial cell culture has a cobblestone appearance; in colonies epithelial cells adhere tightly to one another. HeLa cells are of epithelial origin; the line was derived from a human tumour, a carcinoma of the cervical epithelium (Figure 2.1), and in culture HeLa cells grow into layers many cells thick because they have lost the ability to control their growth.

Fibroblastic cells in primary cultures grow to form a layer usually 3–4 cells rather than one cell thick. Fibroblastic cells tend to move around the culture dish by the extension of processes with ruffled edges, and as a result, in sparse culture small colonies are dispersed.

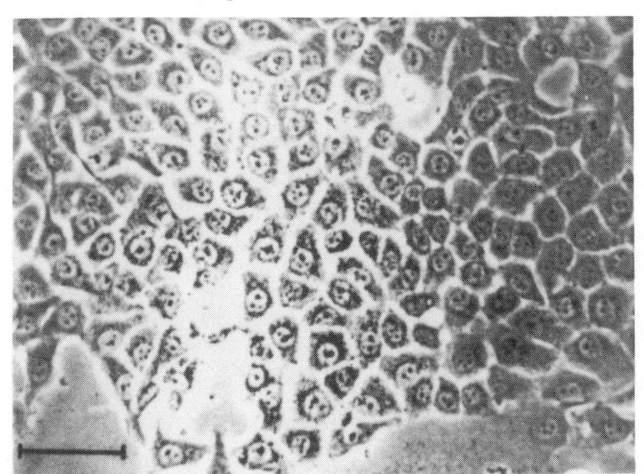

Figure 2.1. Morphology of epithelial cells in culture. Phase-contrast photomicrograph of cells at the edge of a HeLa colony which developed over a feeder layer. Colonies of such cells generally are fairly close packed, although more diffuse colonies can also be found. Bar represents 100 μ. Reprinted with permission from Puck et al., *J. Exp. Med.* 103, 273, 1956.

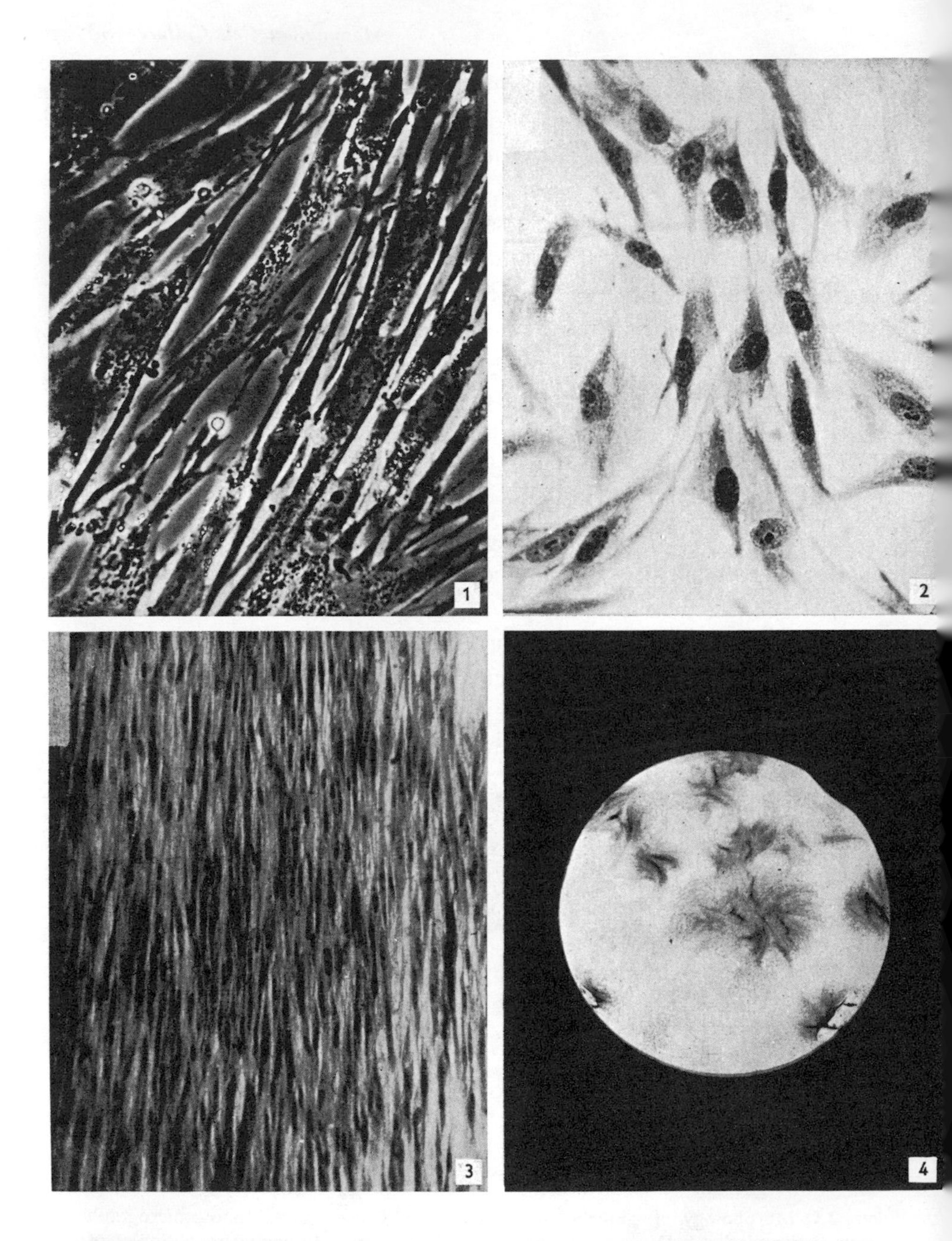

Figure 2.2. Morphology of fibroblast cells in culture, strain WI-1. (1) Diploid human fetal lung cells after 35 subcultures and 9 months in vitro. Phase contrast (× 360). (2) Diploid human fetal lung cells after 35 subcultivations and 9 months in vitro. Stained with May-Grünwald Giemsa (× 300). (3) Directional orientation of cells after 3 days in culture. Stained with May-Grünwald Giemsa (× 100). (4) Colony development from single cells planted in a 50-mm petri dish after 35 days in culture. Colonies are rough and hairy in appearance. May-Grünwald Giemsa. Reprinted with permission from Hayflick and Moorhead, *Exp. Cell Res.* 25, 592, 1961.

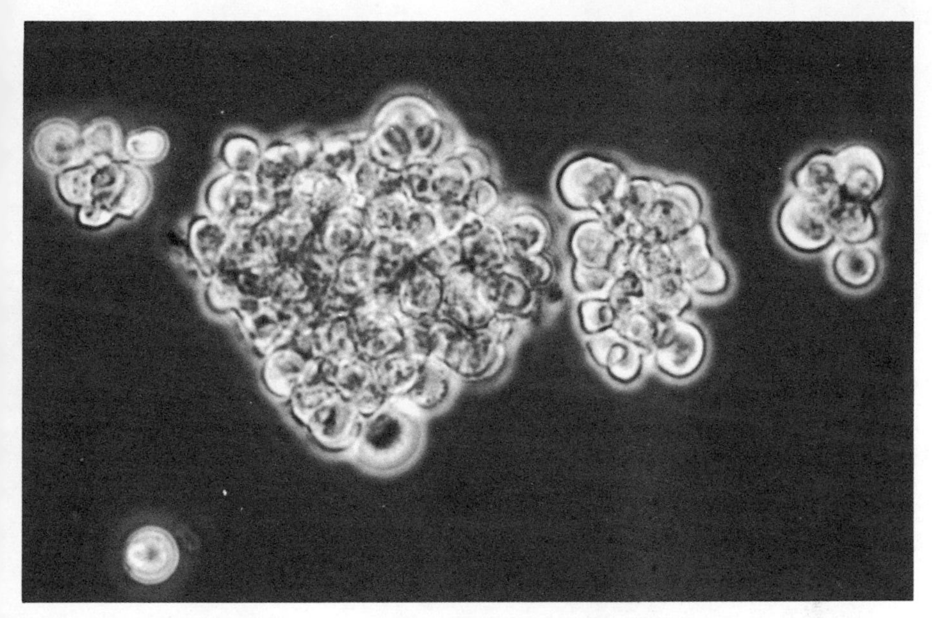

Figure 2.3. Morphology of lymphoid cells in culture. SK-L7 line derived from human lymphoid cells. Cells grow in clumps but do not stick to the plate. Photo courtesy of Dr. John Hlinka, Sloan-Kettering Institute (× 250).

These ruffled edges apparently have a sensory role, for when a fibroblastic cell migrates along the plate, its leading edge is ruffled until contact is made with another cell. Contact is followed by immediate cessation of both ruffling and cell migration. As a consequence, fibroblastic cells line up next to each other so that older colonies contain oriented streams of cells (Abercrombie and Heaysman, 1954; Abercrombie and Ambrose, 1958) (Figure 2.2).

Unlike epithelial and fibroblastic cells, typical lymphoid cells adhere poorly or not at all to solid surfaces. However, lymphoid cells grow well in suspension, either as a collection of separate cells, or more frequently in clumps (Armstrong, 1966; Moore et al., 1966, 1968) (Figure 2.3).

It is important to remember, however, that the morphology of a cell in culture depends on the culture conditions. For example, in high concentrations of human serum HeLa cells assume the stretched, elongate shape typical of fibroblasts, whereas in the absence of serum they resemble epithelial cells and pack closely together (Puck et al.,

1956; see *cAMP* and *Transformation*, this chapter). Moreover, polio-resistant subclones of HeLa cells have been isolated which have a fibroblastic morphology even in high concentrations of serum (Vogt, 1958).

Embryonic Origin of Cell Types

The earliest stages of development of all vertebrate embryos are strikingly similar. The fertilized egg cell divides many times to form a hollow ball of cells one-cell thick called a blastula. At a point on the surface of the blastula the sheet of cells tucks into itself, producing a two-cell thick dome called a gastrula. At the lip of this dome cells migrate and form tubules between the inner and outer layers. These three primitive cell layers, the outside of the ball, the inside and the tubules, called the ectoderm, endoderm and mesoderm, respectively, give rise to the various tissues and organs of the animal (see Table 2.2).

In later development of the embryo, interaction of cells of these three germ layers is essential for normal development of all tissues. The need for this interaction, called induction, was demonstrated experimentally by embryologists who in many cases could prevent embryonic rudiments from differentiating by transplanting them far from the embryonic tissues to which they are usually adjacent (Sussman, 1969). But despite this requirement for constant cell and tissue interaction during embryogenesis, the descent of cell types within a tissue from one of the three ancestral germ layers is usually obvious. When, however, cells are explanted and cultivated in vitro, their morphology becomes only a rough indicator of their origin in vivo, as Table 2.2 shows. For example, a clone of cells which in culture assumes an epithelial morphology may be of mesodermal or ectodermal origin.

Somatic Variation

By 1960 both normal fibroblasts and tumour cells could be cloned from single cells, and the metabolic requirements of clonal lines were becoming clear. Yet the instability of the growth properties of fibroblastic cells in culture continued to plague the field and make experiments on any hereditary alteration in cell-cell interaction impossible to interpret; however, this fact did not deter attempts to elucidate the mechanism of growth control in fibroblastic cells.

Table 2.2. Embryogenesis of Some Differentiated Cells

Ancestor Cell Type	Differentiated Cell	Morphology of Normal Cultured Cell
Ectoderm	Epidermis	Not grown
	Hair	Not grown
	Nails	Not grown
	Brain	Not grown
	Nervous system	Not grown
Endoderm	Inner lining of digestive tract	Not grown
	Liver	Epithelial
	Glands, e.g. pancreas	Not grown
Mesoderm	Notochord (sheath of spinal cord)	Not grown
	Connective tissue	Fibroblastic
	Blood vessels	Not grown
	Muscle	Fibroblastic
	Outer covering of digestive and respiratory tracts and other internal organs	Not grown
	Excretory system—kidney	Fibroblastic and epithelial
	Dermis	Epithelial
	Blood cells	Not grown
	Leukocytes	Lymphoid
	Lymphatic cells	Lymphoid

In the first months after explantation, normal fibroblastic cells grow rapidly in culture, but only to low cell density; the cells cease multiplying when they have formed a layer a few cells thick. Cells of these "monolayers" resume growth upon trypsinization, dilution and transfer, so normal fibroblasts are evidently under a form of reversible growth control, which is maintained even in the richest known medium. By contrast, tumour cells and cells of old established lines such as HeLa and L grow into layers many cells thick and then detach and sometimes even continue to grow in suspension. Such observations lead to the belief that cells that exhibit controlled growth in vitro and are not highly malignant when inoculated into appropriate animals are comparable to nonmalignant cells in vivo.

But if growth control is a sign of normality, what can be made of the fact that established cultures constantly adapt to the conditions of culture and acquire the ability to grow to increasing density? In the late 1950's it became clear that many "normal" cultivated cells had spontaneously acquired the ability to form tumours in vivo after growth for a few months in culture. These cells, called permanent cell lines, were like tumour cells in another way: they had ceased to contain the normal number of chromosomes for their species.

Establishment of "Normal" Fibroblast Cell Lines

When viruses such as polyoma virus, which can induce tumours in vivo, were shown to rapidly transform early-passage "normal" cells into malignant cell lines, it was impossible to tell whether the virus was speeding up an inevitable spontaneous "adaptation" or introducing new virus-coded alterations in growth behaviour. The study of adaptation to conditions of culture had become the study of "spontaneous" transformation to malignancy, and the question was, were any cultured cells "normal"?

To answer this fundamental question it was first necessary to make the establishment of cell lines a reproducible phenomenon, and then to dissociate establishment, the acquired ability to grow in culture, from transformation, the acquired ability to grow into a tumour when injected into a susceptible animal. The simplest way to show that nonmalignant cells can grow indefinitely in culture would be to establish permanent lines of fibroblastic cells that would not induce tumours when injected into susceptible animals.

Several attempts to select permanent fibroblastic cell lines whose cells are not malignant have met with success, but others have failed. For reasons that are totally obscure, it seems that fibroblasts of some species, for example, man and the domestic chicken, are intolerant of the conditions imposed by cell culture and do not survive to give rise to permanent lines. Fibroblasts of the Syrian hamster, the Chinese hamster and the mouse, on the other hand, appear to be much less fastidious and survive in culture to give rise to permanent cell lines. Whether or not the failure to select permanent cell lines of chick and human fibroblasts reflects the use of inappropriate media rather than some intrinsic property of the biochemistry of these cells is unknown.

Hamster Cell Lines: CHO and BHK

By 1958 Puck's group had succeeded in cultivating normal fibroblasts from tissues of the Chinese hamster using culture conditions that were slight variations of those optimal for the growth of HeLa cells (Puck et al., 1958). It is interesting in light of the recent work on the effects of fluctuation in pH (Eagle, 1971; Ceccarini and Eagle, 1971a, b) that one crucial step taken by Puck was to prevent the pH from deviating by more than 0.2 units during the first days of culture.

The Chinese hamster cells grew well, could be cloned, and the clones were found to grow indefinitely without any deviation in the cell chromosome number or change in the morphology of the colonies. One Chinese hamster fibroblast culture, derived from an explanted ovary, yielded an especially vigorous clone called CHO. A subclone of CHO was subsequently found to require proline for growth and to be capable of spontaneous reversion to proline independence (Ham, 1963). This subclone, called CHO Pro^{-}, has since been the parent line used in a series of elegant genetic studies by Puck and his colleagues.

By 1961 it had been shown that the majority of mouse embryo cells infected in vitro by polyoma virus died as a result of the infection even though injecting the virus into newborn mice frequently led to tumours (Vogt and Dulbecco, 1960; Sachs and Medina, 1961). When Syrian hamster fibroblasts in culture were found to show no obvious cytopathic effect after infection by polyoma virus, but were transformed, the Syrian hamster became the species of choice for studying transformation by polyoma virus, and Stoker and Macpherson at the Institute of Virology in Glasgow established the Syrian hamster line BHK21 (Stoker and Abel, 1962; Vogt and Dulbecco, 1962; Stoker and Macpherson, 1961).

Kidney cells from the twenty-first litter of a series of 1-day-old baby hamsters (hence, BHK21) had established themselves spontaneously as a rapidly growing cell line after about two months in culture. Subclones of the line, especially clone 13, have been used extensively since 1961 for studies of polyoma virus transformation (Stoker and Macpherson, 1964).

By 1964 variant sublines had been derived from BHK; cells of these sublines were malignant in vivo and grew both in suspension and in agar. All these changes were brought about reproducibly by infecting BHK21 cells with polyoma virus, but it is important to remember that they also arose spontaneously in the line, albeit at a low frequency (Stoker and Macpherson, 1964).

Human Cell Lines

In the conditions that Ham devised for culturing normal hamster cells, normal human fibroblasts survive only a few months; Hayflick and Moorhead (1961) found that human fibroblastic cells die off within a year of being placed in culture. Merz and Ross (1969) later measured the fraction of living cells in cultures of human skin fibroblasts and found first, that the percentage of cells capable of division decreased throughout the entire life span of the culture, and second, that the decrease in the viable fraction of cells was exponential with time.

Precisely why human fibroblastic cells are so intolerant of conditions in which fibroblasts of other mammalian species grow indefinitely remains obscure, but fluctuation in the pH of the culture may be an important factor, for Ceccarini and Eagle (1971a) have reported recently that it is possible to establish lines of normal human fibroblasts if the pH of the culture is closely regulated.

Chick Cells in Culture

No permanent cloned lines of chick cells exist; neither virally nor chemically transformed cells in primary and secondary cultures, nor "spontaneous" tumour cells can live for more than a few dozen generations in culture. For this reason most in vitro studies of the chick tumour viruses have been confined to the use of primary or secondary cultures of chick fibroblasts. To obtain these, a chick embryo is minced and dispersed into a single cell suspension and plated at high cell density: all the cells that stick, spread and survive make up the primary culture; those that survive resuspension and replating form the secondary culture.

Primary and secondary cultures of chick embryo fibroblasts behave differently in many ways from mammalian cell cultures. The chick cells will not grow in sparse culture, so they cannot be cloned, and in dense culture, chick fibroblasts continue to divide until they have depleted their medium (Rubin, 1971).

Mouse Cell Lines

Mouse cells respond even more variably to culture than do hamster cells: the cells readily adapt to permanent culture, but in the process they always suffer changes in their social behaviour or in their chromosome composition or in some other important property. In 1963 Green and Todaro provided both a rational procedure for establishing permanent lines of mouse cells that retain growth control and the beginning of an understanding of growth control in vitro, when they established the 3T3 line, the first mouse fibroblast line which was not spontaneously transformed (Todaro and Green, 1963).

Previous attempts to establish fibroblastic cell lines from mouse tissues (such as the early work of Earle) had led to the conclusion that a crisis inevitably struck the cultured cells within months of explantation. Cells grew rapidly for a few months only, then they entered a critical period in which, despite all attempts to stimulate division, most cells failed to divide and eventually died. The established lines that could sometimes be recovered from crisis usually began as dense foci of dividing cells that appeared on a background of moribund cells. As Earle had first shown, upon cloning, these cells usually grew to high densities and usually proved to be malignant in vivo.

Green and Todaro proposed that these foci were dense because they arose from a subset of the cells that were able to survive crisis. Cells of this subset were able to divide despite the presence all around of normal, nondividing cells. If their hypothesis were true, establishment of lines of mouse cells, such as L-cells, had involved unwitting selection for this type of cell.

To test whether or not the ability to grow in the presence of cell-cell contact was necessary for cells to survive crisis, Green and Todaro chose to carry a mouse embryo explant through crisis without ever letting the cells remain in close contact for long periods of time. In this way they hoped to avoid providing a selective advantage to the post-crisis cell that could overgrow nondividing cells, and to give equal opportunity for the establishment of a line from any post-crisis cell that had retained growth control. Such a cell would not divide when in contact with other cells, and so would never be recovered from a culture that was carried through crisis at high cell density.

A million mouse cells just about cover the surface of a 30-cm^2 dish; in other words, they form a monolayer. A mouse embryo was minced and the cells were dissociated with trypsin; 3×10^5 cells were plated and the cultures were repeatedly trypsinized and diluted to 3×10^5, 6×10^5, or 12×10^5 cells/plate every three days, through early rapid growth, through crisis and after for more than a year in all. At first the cells grew into a monolayer again in three days from all the initial densities, but as crisis approached, at least 6×10^5 cells were needed to reach confluence before the next transfer; therefore the transfer at 3×10^5 cells/plate prevented cell-cell contact during crisis, while the transfers at 6 and 12×10^5 cells/plate permitted it.

The results were clear cut. Cells in all three regimens survived crisis and gave rise to established lines which were named 3T3, 3T6, and 3T12, respectively. While 3T6 and 3T12 cells grew to the high cell density characteristic of all other established mouse lines, 3T3 cells ceased dividing at a low density as they formed a monolayer. By greatly reducing cell-cell contact during establishment, Todaro and Green had selected a permanent line exhibiting growth control. They had demonstrated in one experiment that cell-cell contact plays a critical role in growth control and that the ability to grow despite cell-cell contact is distinct and separable from the ability to grow per se. They had isolated a line of cells which, despite its capacity to grow in culture indefinitely, was sensitive to contact inhibition of division, which is, we must assume, a characteristic of most cells in vivo.

Growth Control of Mouse Fibroblasts and Malignancy

The standard assay of the tumourigenicity of a cell line is to inject different numbers of cells into a series of animals that will not immunologically reject the cells. The fewer the cells and the shorter the time needed to give a tumour, the more tumourigenic the line. Unfortunately, the mice from which the 3T3 and 3T6 cell lines were derived by Todaro and Green were not inbred, and so they could not be tested for tumourigenicity. Seven years later Aaronson, working with Todaro, made 3T3-like, 3T6-like, and 3T12-like cell lines from Balb/c inbred mouse embryos and found an excellent correlation between the maximum saturation densities of the lines in vitro and their tumourigenicity in Balb/c mice (Figure 2.4) (Aaronson and Todaro, 1968). 3T12 cells, which grew to high saturation densities in vitro, were highly tumourigenic, whereas 3T3 cells, which grew to low saturation densities, had a very low tumourigenicity. In a subject open to much ambiguity of interpretation, this correlation between saturation density in vitro and initiation of tumour formation in vivo stands out for its clarity.

The "Normality" of 3T3 Fibroblasts

The 3T3 cells transformed by DNA tumour viruses have saturation densities characteristic of 3T12 cells and are almost as malignant as 3T12 cells; as a result the 3T3 cell is often spoken of as a "normal" cell

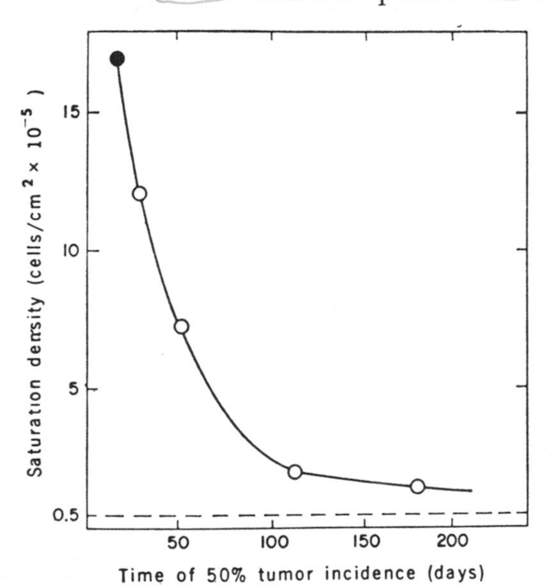

Figure 2.4. Relationship between saturation density and tumour-forming ability. 10^7 cells were injected. (○) Balb/3T12 cells. (●) SV-Balb/3T3 cells. Dashed line represents saturation density of Balb/3T3 cells. Reprinted with permission from Aaronson and Todaro, *Science* 162, 1024, 1968.

by those studying transformation (Todaro and Green, 1964; Todaro et al., 1964). It is important to realize, however, that 3T3 cells are different from the fibroblasts in any mouse in several respects. 3T3 cells divide essentially forever (certainly longer than the lifetime of any mouse) and have lost the perfect diploid assortment of chromosomes. Futhermore, the maximum density to which 3T3 cells will grow is lower than that of primary mouse embryo fibroblasts (Stoker, 1967), and 3T3 cells are somewhat less readily agglutinated by the lectins concanavalin A and wheat germ agglutinin than are the cells of most mouse tissues.

Serum and Growth Control

The isolation of the 3T3 lines proved that one form of growth control operates in vitro through cell-cell contact. Presumably cell-cell contact also plays a role in regulating cell multiplication in vivo, while in addition in vivo humoral signals (e.g., hormones in the blood) regulate cell multiplication. It is not surprising, therefore, that serum contains compounds that can initiate cell division in nondividing cultures. In fact, most mammalian cells need some serum to grow in culture, and all cells respond to added serum by growing to higher densities (Rubin, 1971; Holley and Kiernan, 1968). Spontaneous or viral transformation lowers the serum requirement for cell division; the high saturation density of transformed cells in culture is a reflection both of a loss of contact inhibition of division and of an increase in the efficiency with which serum is used by the cells.

Topoinhibition

The opposing contributions of cell-cell contact and the several growth-stimulating factors in serum have recently been investigated by Dulbecco (1970a). In 1967 H. Green's group showed directly that loss of cell-cell contact was sufficient to bring about cell division in a culture of stationary 3T3 cells. A narrow scratch in the monolayer caused all the cells at the edge of the wound to divide. Most cells in the rest of the monolayer did not divide, even though all cells were bathed in the same medium at all times (Todaro et al., 1967).

To quantitate the wound experiment, Dulbecco used autoradiography to compare the fraction of cells in DNA synthesis along the edge of the wound with the fraction of cells in the untouched monolayer that were making DNA. The wounding experiment was repeated at several different serum concentrations with cells of several different lines, including 3T3 and its transformed derivatives.

Dulbecco's results may be summarized as follows. With cultures of untransformed BHK21 cells, Balb-3T3 cells and BSC1 cells, and primary mouse embryo fibroblasts the fraction of radioactive nuclei in the wound was higher than the fraction of radioactive nuclei in the unwounded cell layer at all serum concentrations from 0 to 12 percent. With increasing concentrations of serum the proportion of cells with labeled nuclei in the wound and in the monolayer rose at different rates. Furthermore, each of the four sorts of untransformed cells responded in a quantitatively different way to the increasing concentration of serum; the concentration of serum at which the ratio of labeled wound cells to labeled monolayer cells reached a maximum was characteristic for each of the four sorts of cells. By contrast, cultures of various transformed cells showed little or no difference between the number of cells in the wound and in the monolayer synthesizing DNA; moreover, the fraction of cells making DNA was not markedly dependent on the concentration of serum.

In brief, cells of different lines differed in their serum requirement for DNA synthesis. This requirement (WSR, for wound serum requirement) measured the true serum requirement of a cell line, since all contact-dependent inhibition was absent from cells in the wound. WSR was higher for the untransformed cells than for the transformed cells, demonstrating that viral transformation increases the ability of each cell to respond to serum.

Dulbecco coined the term topoinhibition for that property of a cell line which, at low concentrations of serum, caused the cells in the monolayer to have a lower proportion of labeled nuclei than the cells in the wound. Topoinhibition was absent in the transformed lines, since their cells showed as high a fraction of labeled nuclei in the monolayer as in the wound.

Dulbecco concluded his analysis by listing four requirements for serum exhibited by growing fibroblastic cells: serum is needed to keep the cells viable, to enable them to initiate DNA synthesis (WSR), to enable them to complete mitosis, and to counteract topoinhibition (CTI). It remains to be seen whether or not these multiple roles correlate in a simple way with the multiple peaks of growth-stimulatory activity recently detected in fractions of serum obtained by Sephadex chromatography (Holley and Kiernan, 1968; Paul et al., 1971; Lipton et al., 1971).

Serum Substitutes—Fully Defined Media

Currently, minimal but only partly defined media are available for cultivating cells of several species and tissues (Table 2.1). With some small variation, the defined fraction (90–95%) of these media comprises vitamins, amino acids, salts, pH buffer, glucose as a carbon source, and trace metals; the undefined 5–10 percent is still serum (Morton, 1970). Experiments would be much simpler if cells could be grown routinely in defined media made entirely of precise amounts of pure chemicals (Eagle, 1960; Fisher et al., 1958; Lockhart and Eagle, 1959; Ham, 1965; Birch and Pirt, 1971). Some success has been reported in producing an autoclavable, serum-free, chemically defined medium which supports the growth of L-cells in suspension (Nagle and Brown, 1971). Methylcellulose is used as an inert macromolecule to provide the osmotic pressure usually provided by serum proteins. So far the chief difficulty seems to be the requirement of cells for serum for clonal growth and for growth in sparse culture, where the cells are under greatest stress.

pH Regulation

Since the vertebrate physiological buffer is CO_2-bicarbonate, it is not surprising that Eagle found this buffer to be optimal for the growth of cells in culture when he developed BME. Unfortunately, CO_2-bicarbonate is in gas-liquid equilibrium, so cells must be grown in incubators that are flushed with a mixture of air and CO_2. Tris and phosphate buffers, which have the advantage of not involving a gas phase, do not support indefinite cell growth. A class of nongaseous buffers (Ceccarini and Eagle, 1971a) has been developed in which mass cultures will grow, but colonies will grow from single cells with high efficiency only if the CO_2-bicarbonate is also present.

During experiments cells which are in petri dishes are repeatedly taken from the incubator. The pH of the culture may rise by a unit or more as CO_2 is lost to the air and fall again as CO_2 is regained in the incubator. Furthermore, as cells grow they synthesize lactic acid, causing the pH of the medium to drop even while the dishes remain in the incubator. By adding both organic buffers and CO_2-bicarbonate to the medium, variations in the pH of a culture can be kept to within 0.2 of a unit/day, and striking improvements in cell growth have been reported (Eagle, 1971). Since the pH in vivo is highly regulated, it is safe to assume that these improvements reflect a closer approximation to in vivo conditions than is usually achieved.

Many cancer cells continue to grow as the pH of their medium varies over a wide range; they are less susceptible to the growth-inhibitory effect of pH fluctuations that are untransformed cells. But cultivated cancer cells and cells transformed by tumour viruses grow to higher cell densities than untransformed cells even when the latter are maintained at their optimal pH (Ceccarini and Eagle, 1971b). Thus the observation that transformation reduces a cell's requirement for serum and reduces its susceptibility to contact inhibition of division cannot be explained simply by saying that transformation renders a cell more tolerant of fluctuations of the pH of its medium.

Lymphoid Cells in Culture

While the impetus to develop fibroblastic cell lines that exhibit growth control came from microbiologists who wished to quantitate the growth properties of animal cells, interest in lymphoid cell lines had a clinical origin. It was anticipated that it would be difficult to establish cultures of human leukocytes because leukocytes in the peripheral circulation of healthy persons never divide, whereas peripheral leukocytes of patients with leukemia divide as they circulate in the blood. Thus the failure of a leukocyte to divide in vivo was taken to be a sign of normality by clinicians.

Phytohemagglutinin

In 1960, however, Nowell showed that normal human leukocytes could indeed divide in vitro if phytohemagglutin (PHA), extracted from the pokeweed plant, was added to the cells. After PHA stimulation, it was easy to maintain leukocytes from many normal and leukemic persons in long term cultures. Moore pioneered in the large scale culture of leukocytes; with PHA and his medium (RPMI 1940) cultures can be obtained routinely from normal and leukemic leukocytes (Moore et al., 1966, 1967, 1968). And because leukocytes grow in suspension, Moore was able to scale up culture volumes so that grams of cells could be obtained for biochemical studies. Lymphoid cell lines can also be established from antibody-forming and blood-forming cells (lymphocytes) of the spleen and marrow by PHA stimulation.

The mitogenic action of PHA on leukocytes was completely unexpected. Nowell's discovery opened a new mode of diagnosis, the role of chromosomal abnormality in disease. For with PHA the clinician could examine the chromosomes of the leukocytes recovered from a few millilitres of a patient's blood (Moore et al., 1967; Broder et al., 1970).

Lymphocytes from persons with mononucleosis and Burkitt's lymphoma are especially easy to grow in culture (Klein et al., 1968) and cell lines have been established without the aid of PHA. Histologically these permanent lymphoid lines resemble the immature "blast-like" transformed cells seen following the initial in vitro stimulation of peripheral small lymphocytes by antigens or by PHA. Immunoglobulins, interferon, components of complement, enzymes and soluble histocompatibility antigens have been found in the media of lymphoid cultures (Fahey et al., 1966; Glade and Broder, 1971).

Human fibroblastic and lymphoid cells show the two extremes of growth in culture: the fibroblastic cells grow well at first but inevitably die after many months without ever giving rise to established lines, whereas lymphoid cells must at first be coaxed with mitogens, but then can be subcultured indefinitely (Moore and McLimmans, 1968).

Recently, cloned lines of human lymphoid cells have been established from single leukocytes stimulated by PHA. Choi and Bloom (1970) isolated single lymphocytes in microwells, while McCredie et al. (1971) grew lymphoid colonies in soft agar. Together these techniques should provide a series of cloned human cell lines whose many secretion products, rapid growth, and lack of crisis offer new opportunities to the biochemist wishing to study differentiated human cells.

TRANSFORMATION

Injecting a tumour virus into the appropriate host will, by definition, cause a tumour to grow in the injected animal. Excised and placed in culture, the cells of the tumour will differ in many ways from the cells of neighbouring healthy regions of the same tissue. Many tumour viruses will in vitro rapidly and reproducibly alter fibroblasts of the appropriate species so that they come to resemble in many ways the cells of a tumour (Eagle et al., 1970). After being altered in vitro in this way, a cell is said to be transformed. Of course one need not use a virus to transform a cell, just as one need not inject a virus to get a tumour, as smokers should know. Chemical and physical carcinogens, such as dimethylbenzanthracene (DMBA) (Chen and Heidelberger, 1969), smog (Freeman et al., 1971), and X rays (Sachs, 1967; DiPaola et al., 1971) can transform cultivated cells as well as induce tumours in animals. Since almost all tumour viruses and carcinogens have been tested on fibroblasts, transformation is defined by a set of differences in the growth properties of fibroblastic cells in culture.

In 1958 Rubin and Temin developed the first reproducible assay for a tumour virus, the focus-forming unit (FFU) assay for Rous sarcoma virus in primary culture of chick embryo fibroblast (Temin and Rubin, 1958). Manaker and Groupé (1956) had noted that after primary cultures of chick cells were infected with high concentrations of RSV, the cells in infected cultures became swollen and eventually masses of rounded cells detached from the dish. At high dilution of infecting virus, the infected cells were recognizable as discrete foci of rounded cells. Encouraged by this report that foci develop in cultures of infected cells, Temin and Rubin spent three years developing a quantitative assay for RSV. By using better culture media, by plating chick cells at a lower density than that used by Manaker and Groupé so that the cells were able to grow for a few days after infection, and by adding agar to the medium to keep foci in one place, they finally arrived at a focus-forming unit (FFU) assay that showed that a single RSV particle can cause a focus to develop in vitro. They also confirmed Rubin's previous finding (Rubin, 1955) that infected cells continuously release virus as they multiply; and finally, they concluded from the kinetics of virus release that the assembly of virus is completed at the cell surface and then the virus is released to the medium (Temin and Rubin, 1958).

Assays for In Vitro Transformation

Transformation of chick fibroblasts by RSV is still assayed by the FFU method of Temin and Rubin, as is transformation of mouse fibroblasts by the murine sarcoma viruses. Other assays have since been developed to detect differences between other transformed and untransformed cells. Some of these assays are outlined below.

Overgrowth in Excess Serum

In H. Green's laboratory Temin and Rubin's focus-formation assay was modified to assay the transformation of mouse 3T3 cells. Todaro and Green found that, even when plated in excess serum, untransformed 3T3 cells stopped dividing at a low cell density. Upon infection with a tumour virus, such as SV40, variant 3T3 colonies appeared, the cells of which continued to grow after the culture had reached a monolayer. The frequency of appearance of these colonies of transformed cells was proportional to the dose of SV40 (Todaro and Green, 1964, 1966; Todaro et al., 1964; Black, 1966).

Cloned descendants of these colonies contained the appropriate virus-specific tumour antigen and grew to high saturation densities (Figure 2.5a,b). Furthermore, polyoma-transformed and SV40-transformed 3T3 cells could be distinguished from one another because of differences in the colony morphologies (Figure 2.5c,d,e).

Figure 2.5. **(a)** Electron micrograph of vertical section through full thickness of saturation density culture of 3T3, showing two flattened cells and a cytoplasmic process from a third. Such cells may measure over 75 μ in length. The nuclei (N) are unusually close together for this line. The cell layer measures 3.1 μ in thickness. The black line indicated by arrow marks the plane of cleavage from the plastic petri dish ($\times$ 6500).
(b) Vertical section through full thickness of a polyoma-transformed colony. Extensive nuclear and cytoplasmic overlapping is present. The cell layer measures 27 μ in thickness ($\times$ 2800).
(c) Saturation density culture of 3T3, 17 days following inoculation of 1000 cells. Note evenness of the monolayer, separation of cell nuclei, pale poorly defined cytoplasm. Hematoxylin stain ($\times$ 150).

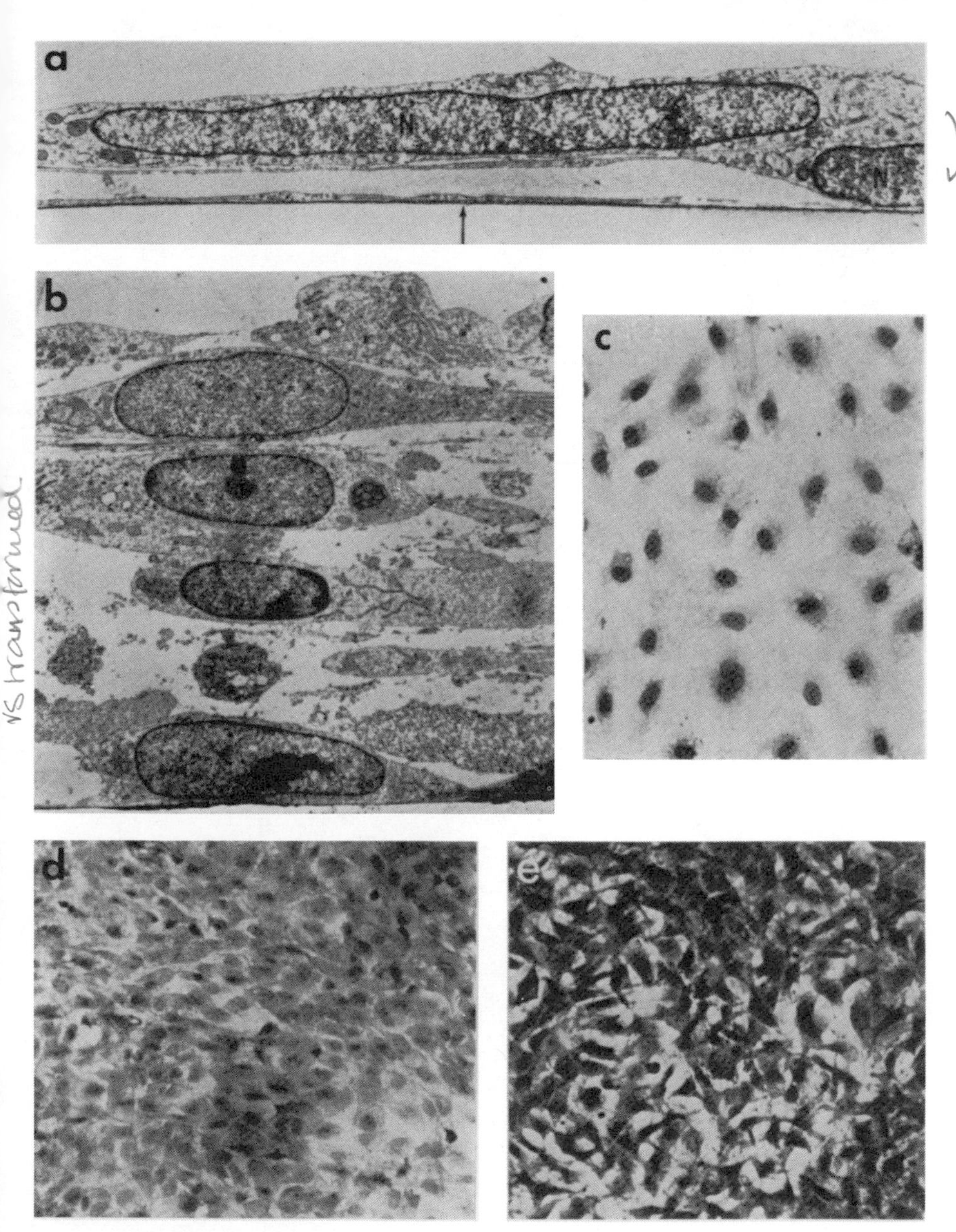

(d) Central portion of SV40-transformed colony 14 days after infection of 3T3 cells and before saturation density was attained. Note the close packing of cells, their tendency to mosaic arrangement, and their epithelioid appearance. Hematoxylin stain (× 150).
(e) Central portion of polyoma virus-transformed colony, arising from 3T3 cells infected 18 days previously. Note spindle shape of cells and pattern of cell layering and interlacing. Compare with Figure (d). Hematoxylin stain (× 150). Figures 2.5 a–e reprinted with permission from Todaro et al., *Proc. Nat. Acad. Sci.* 51, 66, 1964.

Growth in Agar Suspension

Fibroblasts of primary cultures and cells of most fibroblastic lines must attach to a solid surface before they can divide. They fail to grow when they are suspended in fluid or in a gel of agar or methylcellulose, a property which Stoker has termed anchorage dependence of multiplication (Stoker et al., 1968).

Macpherson and Montagnier (1964) used growth in agar as a selective assay for the transformation of BHK fibroblasts by polyoma virus. Only the stably transformed cells grew into large colonies that were easy to pick out of the agar with a pipette and replate. Cells that grew in agar were found to carry viral antigens, to grow to high saturation densities, to have a reduced serum requirement, and to form tumours upon injection into appropriate animals (Stoker, 1964; Clarke et al., 1970) (Figure 2.6). The number of colonies of transformed cells growing in agar was proportional to the dose of virus. This assay selects cells which are able to grow when they are suspended in agar, whereas Temin and Rubin used agar to immobilize foci and restrict the diffusion of progeny RSV; the foci of chick fibroblasts transformed by

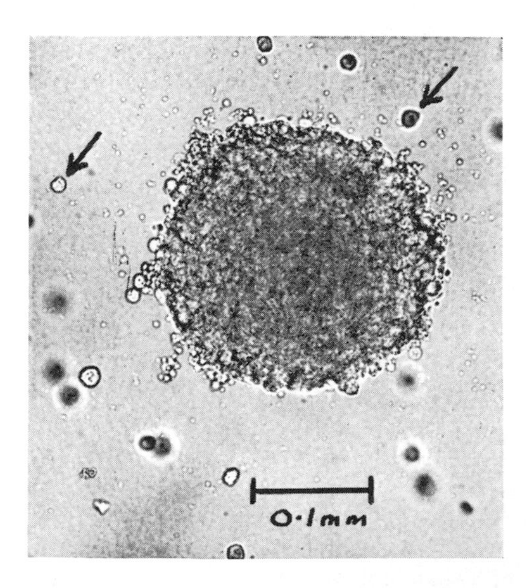

Figure 2.6. A colony of transformed cells in agar medium 7 days after seeding the culture with BHK21/13 cells infected with polyoma virus. Single nondividing cells are arrowed. Reprinted with permission from Macpherson and Montagnier, *Virology* 23, 291, 1964.

RSV were in contact with the surface of their dishes and were not selected for the ability to grow in the absence of anchorage.

Reduced Serum Requirement

Most mammalian cells in culture need some serum to divide and untransformed cells require higher concentrations of serum than transformed cells (Jainchill and Todaro, 1970). For example, 3T3 cells will not grow optimally unless the serum concentration is greater than 5 percent. Because 3T3 cells will not grow in 0.5% serum, it was possible to select for SV40-transformed 3T3 cells at this low serum concentration (Scher and Todaro, 1971). After SV40 infection, a small fraction of 3T3 cells in 0.5% serum divided and formed colonies which were cloned (Smith et al., 1971). Most clones carried SV40-specific T antigen, and most, but not all, had a high saturation density.

Each of these assays for transformation selects for cells with different virus-induced alterations. A cell which possesses any one of these alterations does not inevitably possess all the others. For example, SV40-transformed 3T3 cells selected by virtue of their ability to grow in 0.5% serum do not necessarily grow in agar. In other words, most transformed cells exhibit only a subset of the set of detectable alterations which can be induced by any one tumour virus.

Mesodermal Origin of Transformable Cells

All early cell culture was concerned with fibroblasts, cells of mesodermal origin. In 1960 Nowell, by culturing normal human leukocytes, introduced a second major mesodermal cell type into in vitro studies. Tumours derived from a liver cell, an adrenal-cortex cell, or a neuronal cell have occasionally been placed in culture, but the fact remains that almost all cultivated "normal" cells are of mesodermal origin.

Since mesodermal cells grow in culture so much better than other cells, it is not surprising that only those tumour viruses which induce tumours of mesodermal cells, the sarcomas and leukemias, have been studied in vitro. However, the great majority of malignant cancers in man occurs in cells that descend from the ectoderm and endoderm: carcinomas of man are far more common than sarcomas and leukemias. It remains an open question whether or not the common carcinomas of the skin, lung, breast, cervix, prostrate, colon, and liver are caused by viruses, not least because the putative host cells cannot be grown in vitro. Attempts to culture endodermal and ectodermal cell types free of contaminating mesodermal fibroblasts are being made, and it would be surprising if successful culture were not followed by isolation of new classes of tumour viruses.

Hormones and Differentiated Cell Tumours

The effect of hormones on tissue growth has interested physiologists for many decades (Cannon, 1929), and the repeated failures to cultivate cell types other than leukocytes and fibroblasts may well reflect our inability to formulate media which meet the hormonal requirements of fastidious cells. Those cells of the body that respond to hormones secreted by the pituitary gland are of particular interest, because since the 1950's the laboratories of Jacob Furth and Gordon Sato have succeeded not only in creating tumours of cells sensitive to pituitary hormone and of cells which secrete hormone, but also in growing these tumour cells in culture, without loss of hormone response or hormone secretion. Furthermore, Furth showed that a defect in a tissue's ability to respond to or secrete certain hormones is sufficient to lead to the eventual malignant transformation of cells in that tissue.

Homeostasis in the Ovary

In the normal female mammal the ovaries and the pituitary are linked by a feedback loop: the pituitary secretes the hormone gonadotropin into the blood; gonadotropin stimulates cells in the ovary (but not the ova themselves) to synthesize estrogens, the estrogens enter the blood and thereby reach the pituitary, where they slow down gonadotropin secretion. Thus, the concentrations of both ovarian estrogens and pituitary gonadotropin in the blood will normally be in equilibrium, the concentration of each hormone being determined by the concentration of the other.

In 1944 M. and G. Biskind showed that the pituitary-ovary feedback loop could be interrupted, or opened, by transplanting the ovary to the spleen of castrated females. Once in the spleen (Figure 2.7) the ovarian cells still receive gonadotropin, but because all blood from the

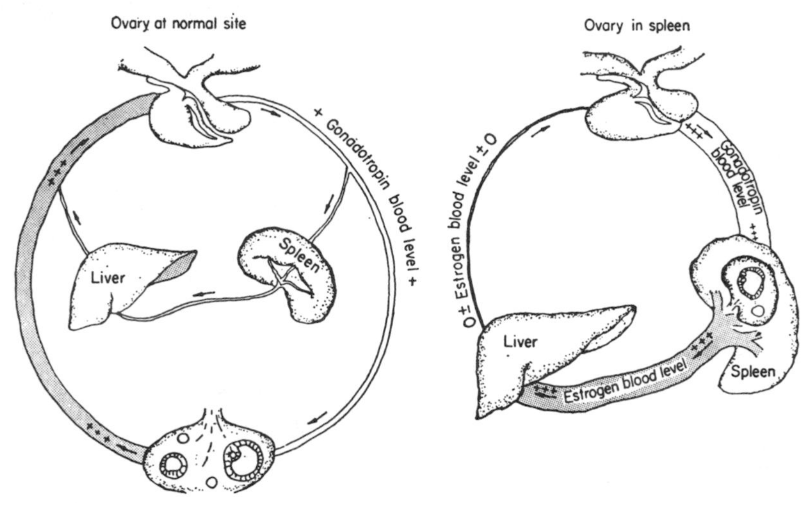

Figure 2.7. Mechanism of induction of ovarian tumours by autotransplantation of ovaries in the spleen of castrates. Reprinted with permission from Furth, *Harvey Lectures* 63, 47, 1967.

spleen is shunted to the liver, the estrogens made by ovary cells are inactivated in the liver and never reach the pituitary. The Biskinds found that when the ovary was transplanted, its cells, under unremitting gonadotropin stimulation, multiplied rapidly and the ovary increased beyond its normal size. The enlarged transplanted ovaries were composed of histopathologically normal cells.

In 1947 Furth derived a metastasizing tumour from such enlarged but nonmalignant ovaries by transplanting them serially, first into the spleens of ovarectomized mice and then subcutaneously into normal mice. Despite their invasive malignant character, the cells of the resulting tumour continued to secrete estrogens in response to gonadotropin. Furth had apparently found that when normal cells (in this case from the ovary) are caused to divide continually because of constant abnormal hormonal stimulation, they eventually became neoplastic: a new type of cell appeared that was able to kill the host (Furth and Sobel, 1947; Furth, 1953, 1967).

Many other sets of hormones which stimulate division of specialized cells are locked into feedback loops. For example, the pituitary also secretes thyrotropic hormone, which causes the thyroid to secrete thyroid hormone. If synthesis of thyroid hormone is blocked by propylthiouracil, synthesis of thyrotropic hormone ceases to be regulated, and the pituitary tissue and thyroid tissue both increase in size and eventually become malignant (Furth and Clifton, 1957; Furth, 1967).

When a loop is opened as, for example, when a tissue shows a decreased sensitivity to a hormone, an excess of that hormone will be secreted. Abnormally high concentrations of hormone lead to proliferation either of the cells synthesizing the hormone, or of the cells responding to it, or of both. Eventually these hyperplastic cells become able to continue to multiply even after they are transplanted into animals with a normal hormonal balance, at which point the cells have become neoplastic. This type of progression to malignancy occurs in the absence of any exogenous carcinogens.

A gradation of increasing abnormality resembles, at least superficially, the series of changes that, for example, avian leukemia viruses cause in avian leukocytes, rendering them able to multiply and eventually able to kill their hosts. So, until proven otherwise, when tumour viruses are found in tissues that have a hyperplastic response to hormone imbalance, both the virus and the hormones must be assumed to be cocarcinogens (Milo et al., 1972). For example, Table 2.3, from an excellent review by Furth (1967), demonstrates the necessity for both a carcinogen and an excess of the pituitary hormone prolactin during the induction of breast cancer in rodents.

Table 2.3. BREAST CANCER INDUCTION WITH SUBCARCINOGENIC DOSES OF CARCINOGENS

	Radiation	Chemical*	Virus
Dose	50 rad	10 mg	0.1 ml milk
Species	rat	rat	mouse
Carcinogen alone	0	0	0
Prolactin alone	0	0	0
Carcinogen and prolactin	58%	85%	40%

* 3-methylcholanthrene

Differentiated Tumours

Many types of tumours have been created by Furth's technique of disrupting hormonal feedback loops. All share an important property: the tumours continue to synthesize and secrete the hormone(s) that are secreted by the tissue from which they arise. Thus, cells of tumours of the adrenal cortex secrete steroid hormones (estrogens) in response to pituitary ACTH (Cohen et al., 1957), and testicular Leydig tumour cells secrete other steroids (androgens) in response to pituitary luteinizing hormone (Shin, 1967; Shin et al., 1968).

Hormone synthesis is not the only differentiated function that tumours can retain. The cells of certain brain tumours (neuroblastomas) have the excitable membranes characteristic of neurones (Augusti-Tocco and Sato, 1969; Minna et al., 1971), while others (glial cell tumours) synthesize a protein made only by the glial cells of the brain (Benda et al., 1968).

At first sight, tumours of differentiated cells would seem to be ideal sources of differentiated cell lines since such tumour cells, because of their unbridled growth in vivo, might be expected to grow in culture more rapidly than differentiated cells from normal tissues. However, initial attempts to cultivate differentiated tumour cells failed: the cultures were nearly always overgrown by fibroblasts. Fibroblasts are, of course, always explanted with the tumour because every tumour is interlaced with connective tissue and blood vessels, and apparently fibroblasts grow faster in culture than differentiated tumour cells (Sato et al., 1960; Zaroff et al., 1961).

Cell Lines from Differentiated Tumours

Sato overcame the problem of fibroblast overgrowth and apparent "dedifferentiation" of explanted tumour cells with a straightforward but ingenious technique, animal–culture alternate passage, that has allowed him to adapt many differentiated tumour cells to growth in culture. When he excised the tumours and plated the cells in culture, they attached but grew poorly, and within days the fibroblasts became the predominant cell type. At that point, Sato put the entire culture into suspension and injected it into a second animal. Only the tumour cells that had remained alive in culture, and not the first animal's fibroblasts, could form a tumour in the second animal. The differentiated cells of this second tumour adapted more readily to growth in culture than differentiated cells from the first tumour, presumably because they descended from the minority of tumour cells that survived the first culture (Buonassisi et al., 1962).

By repeating cycling through culture (to enrich for tumour cells able to grow in vitro) and through animals (to remove the fibroblasts) Sato's group and others succeeded in getting a wide variety of differentiated cell types to grow in culture (Buonassisi et al., 1962; Sato and Buonassisi, 1963, 1964; Stollar et al., 1964; Yasumura et al., 1966; Laskov and Scharff, 1970; Posner et al., 1971; Benda et al., 1968; Tashjian et al., 1968; Augusti-Tocco and Sato, 1969; Shin et al., 1968; Shin, 1967; Richardson et al., 1969).

Differentiated Hormone-dependent Normal Cells

Furth had shown that during the hormonal induction of a tumour, the stimulated organ first passed through a stage of hyperplasia; the cells were induced to divide but the tissue retained its normal structure: the stimulated organ was simply larger than usual (Furth and Sobel, 1947). Moreover, transplantation experiments showed that hyperplastic cell division depended on the continued presence of the stimulating hormone. For example, continued growth of hyperplastic ovaries in vivo depended on the continued presence of the two gonadotropins, luteinizing hormone (LH) and follicle-stimulating hormone (FSH).

Previously, Sato had shown that stable lines of cells from the tumours which eventually arose from hyperplastic ovaries could be selected by animal–culture alternate passage. Cells of these lines still secreted the hormones characteristic of ovarian tissue but they had lost their dependence on LH or FSH. Is it possible to culture differentiated cells that have retained not only the ability to secrete a hormone, but also the ability to respond to a hormone? Apparently so, for when the hyperplastic ovarian cells are placed in culture, they will grow only if LH and FSH are added to the medium (Clark et al., 1972).

The successful culture of hormone-dependent lines is remarkable for two reasons. First, it should now be possible to isolate stable lines

Table 2.4. Clonal, Functionally Differentiated Cell Lines from Tumours

Species	Tumour	Tissue	Function	Reference
Mouse	neuroblastoma	brain	transmitter synthesis; excitable membranes	Augusti-Tocco and Sato (1969)
Rat	glioma	brain	glucocorticoid response	Benda et al. (1968)
Rat	hepatoma	liver	albumin synthesis and secretion	Richardson et al. (1969)
Mouse	Leydig cell	testicle	cyclic AMP response; steroid secretion	Shin (1967)
Rat	Leydig cell	testicle	steroid secretion	Shin et al. (1968)
Rat	pituitary		growth hormone secretion	Yasumura et al. (1966)
Rat	pituitary		ACTH secretion	Sato and Yasumura (1966)
Rat	ovary		division dependent on gonadotropin, LH or FSH	Clark et al. (1972)
Mouse	myeloma cells	marrow	secretion of gamma globulin	Laskov and Scharff (1970)
Hamster	melanoma	skin	pigment production	Moore (1964)

Adapted from Posner et al., 1971.

from any embryonic differentiated cells that are dependent upon growth hormone (GH). Second, LH-dependent cell lines can be obtained from explants of normal tissues without passing through a crisis period.

Table 2.4 lists some cloned lines of differentiated cells isolated by Sato's method.

Cyclic AMP

Hormonal stimulation causes many cells to differentiate or to secrete specialized products. In many cases, the mechanism by which a hormone effects this response involves adenosine cyclic 3′,5′-monophosphate, or as it is commonly called, cyclic AMP (cAMP). Cyclic AMP was first isolated and purified by Sutherland and his colleagues (Rall et al., 1957), who showed that the release of glycogen from liver cells, stimulated by adrenaline, was further enhanced by the addition of cAMP. Later, they showed that adrenaline itself stimulates the accumulation of cAMP in liver cells by accelerating the reaction catalyzed by adenyl cyclase: ATP → cAMP + PPi (Rall and Sutherland, 1958).

Cyclic AMP seems to be a general intracellular second messenger acting, upon hormone stimulation, as an intermediate which elicits the cell's characteristic response. For example, when ACTH from the pituitary reaches the surfaces of adrenal cells, cAMP is synthesized in the cells, and this then causes the cell to synthesize its characteristic corticosteroids. When ACTH reaches fat cells, these cells also make cAMP, but in them cAMP stimulates not steroid synthesis, but lipolysis (Pastan and Perlman, 1971).

Working in Sato's laboratory, Schimmer demonstrated that cAMP stimulates lines of rat and mouse tumour cells that respond to gonadotropin to secrete steroids; this in vitro system mimics the effect of gonadotropin on testicular Leydig cell tumours (Schimmer et al., 1968). Furthermore, Schimmer isolated mutants of a mouse tumour line that no longer responded to gonadotropin, but still responded to cAMP by making steroids (Schimmer, 1969). These mutant cell lines are the first of what should become an interesting class of mutant phenotypes, blocks at points along the chain of events from hormonal stimulation to cellular response.

cAMP and Transformation

Adenyl cyclase is localized at the plasma membranes of animal cells (DeRobertis et al., 1967). Another enzyme, a phosphodiesterase found in the "soluble" cytoplasm of the cell, hydrolyzes cAMP to 5′-adenosine monophosphate. This enzyme is inhibited by methylxanthines, such as theophylline, which are therefore agents that raise the intracellular concentration of cAMP (Sutherland and Rall, 1958). Cyclic AMP itself is not sufficiently hydrophobic to enter some cells, and so for in vitro studies the derivative compound $N^6,O^{2'}$-dibutyryl adenosine 3′:5′-cyclic monophosphoric acid (Bu_2-cAMP) has been used instead. Bu_2-cAMP is active in many cells where cAMP is not, and it is resistant to degradation by diesterase (Pasternak et al., 1962).

Recent work with more familiar cell lines suggests that Bu_2-cAMP, and therefore presumably cAMP, may play an important role in transformation by tumour viruses. In 1971 Hsie and Puck reported that the morphology of cells of the permanent Chinese hamster ovary line CHO was altered by Bu_2-cAMP. In the absence of the agent, CHO cells were compact, well separated, and poorly oriented. When Bu_2-cAMP was added, the cells acquired the morphology of differentiated fibroblasts and lined up in parallel arrays (Hsie and Puck, 1971). This alteration was completely reversed by removal of Bu_2-cAMP and was prevented by low concentrations of colcemid and vinblastine, two inhibitors of the assembly of cellular microtubules. Presumably, Bu_2-cAMP stimulates the reorganization or assembly of microtubular proteins within the cell, since these proteins, along with actin, give a cell its characteristic shape (Fine and Bray, 1971; Wessells et al., 1971).

Also in 1971 Pastan and his colleagues reported that both Earle's L-cells and XC-cells, a line obtained from a rat sarcoma induced by RSV, were morphologically altered by Bu_2-cAMP. In the presence of the agent, L-cells aligned in parallel and regained the spindly appearance of normal mouse fibroblasts (Johnson et al., 1971). Their most striking observation, however, was that both XC-cells and L-cells showed a decreased tendency to grow over one another in the presence of Bu_2-cAMP. This observation is consonant with Bürk's (1968) report that theophylline decreases the growth of transformed BHK21 cells.

Sheppard (1971) extended these observations by showing that the apparent saturation densities of polyoma-transformed 3T3 cells and "spontaneously" transformed 3T6 cells were reduced by 0.1 mM of Bu_2-cAMP almost to the saturation density of 3T3 cells. If Sheppard's finding proves to be generally valid, then growth control, as measured by saturation density, may be a differentiated response of normal fibroblasts which is mediated through cAMP, just as synthesis of steroid hormone is a cAMP-mediated response of adrenal cortex cells (Perry et al., 1971). It is not clear, however, whether or not Bu_2-cAMP is merely slowing down the rate at which cells divide rather than lowering their saturation densities per se (Hsie et al., 1971; Johnson and Pastan, 1972).

THE CELL CYCLE

Generation Time

The time it takes for an entire culture to double in number when all cells are dividing is called the generation time. Generation times of common cell lines range from 10 to 50 hours. When all cells are dividing, the cell number increases exponentially with time, and the average doubling time of a single cell can be derived from a plot of the log of cell number vs. time, which will be a straight line.

Cell Cycle Time

Measuring the doubling time of single fibroblastic cells is more difficult. Under the microscope the cells of any growing culture oscillate between two forms: cells in mitosis are rounded up (Figure 2.8a); cells

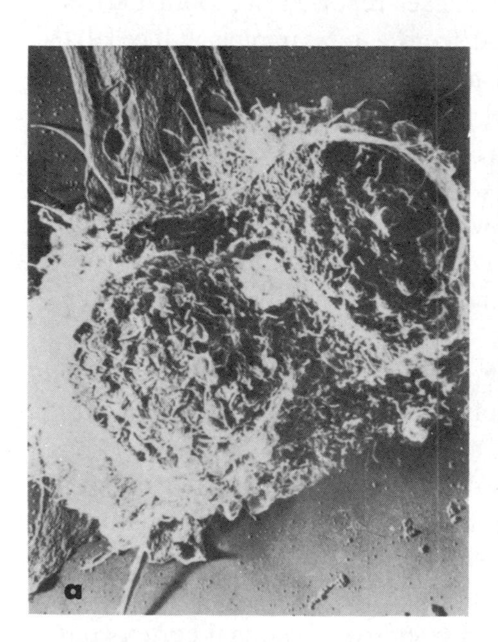

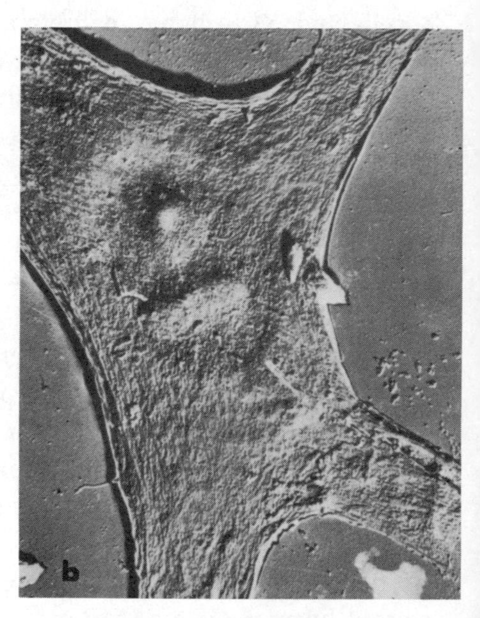

Figure 2.8. **(a)** Surface replica of dividing cell. Many microvilli can be distinguished on the daughter cells as well as bubble-like structures (×2600). Reprinted with permission from Follett and Goldman, *Exp. Cell Res.* 59, 124, 1970.
(b) Shadowed, carbon replica of surface of a fully spread BHK21/C13 fibroblast. No microvilli can be distinguished on the surface of the cell (×3900). Reprinted with permission from Follett and Goldman, *Exp. Cell Res.* 59, 124, 1970.

in the remainder of the cycle, commonly called the interphase cells, are flattened out (Figure 2.8b). These changes in shape accompanying cell multiplication occur cyclically, and the time it takes a single cell to go from mitosis to mitosis is called the cell cycle time. The cell cycle times of individual HeLa cells from one clone were first directly measured from cinéfilms (Hsu, 1955, 1960). These difficult studies showed that the cell cycle time varied greatly from cell to cell, although the average cell cycle time was the same as the population's doubling time.

In a population of dividing cells, the fraction of cells within each stage at any time is about equal to the fraction of the average cycle time taken up by that stage. For example, about four percent of the cells in an asynchronous log-phase cell culture are in mitosis at any one time, so mitosis must take about four percent of the average cycle time, or about 60 minutes for a population that doubles once a day.

DNA Synthesis

The cell replicates its genome during interphase. Proof of this was first obtained by Swift (1950, 1953). He chemically coupled Feulgen dye to the DNA of interphase cells and measured the amount of DNA by measuring the amount of green light absorbed per nucleus. He found that the amount of DNA in interphase cells had a bimodal distribution, with peaks at the diploid and tetraploid DNA values. But microspectroscopy cannot reveal the fraction of interphase time occupied by DNA synthesis or the rate of DNA synthesis.

When cultures of dividing cells are fed a short pulse of tritiated thymidine, radioactivity is localized in the nuclei of only those cells that were making DNA during the pulse. After the cells are fixed, the radioactive DNA can be detected by coating the culture with X-ray film. After allowing time for radioactive decay to expose overlaying regions of the film, the film is developed. The fraction of nuclei overlaid by grains can be counted easily. If the thymidine is given for a very small fraction of the cycle time, the fraction of labeled cell nuclei will be the fraction of the cycle occupied by DNA synthesis.

One of the first studies by tritiated thymidine autoradiography of cells in culture showed that the cells that were in mitosis at the time of the pulse were not labeled, a simple proof that DNA synthesis occurred only during interphase and not during mitosis (Firket and Verly, 1958). By chasing out the pulse of labeled thymidine with an excess of cold thymidine, it was found that neither cells in which mitosis began in the first few hours after the pulse, nor interphase cells finishing mitosis just before the pulse, were labeled (Hsu, 1965; Stanners and Till, 1960; Sisken and Kinosita, 1961). Thus DNA synthesis in interphase is bracketed by gaps before and after mitosis.

Howard and Pelc (1953) named the four stages of the cell cycle mitosis (M) → first gap (G_1) → DNA synthesis (S) → second gap (G_2) → mitosis again (M). S is relatively constant for mammalian cells of different cycle times (about 7 hours) and M is about 3–4 percent of the cell cycle. In an elegant experiment Sisken and Kinosita (1961) measured the cycle time of individual labeled cells by microcinematography, and then by autoradiography measured DNA synthesis in the same cells. They found that differences between the cycle times of individual cells occur in G_1.

Synchronization

The questions of how, when, and where various constituents of dividing cells are synthesized can be approached in two different ways: either single cells can be studied by UV absorption, cytochemistry or autoradiography; or masses of synchronously dividing cells can be studied by conventional biochemical assays. Once techniques to synchronize cells became available, the latter approach predominated.

The first attempts to synchronize mammalian cell cultures involved the use of cold shocks (Newton and Wildy, 1959). After one hour at 4°C, the HeLa cells in an exponentially growing culture failed to divide for 16–20 hours, and then about 80 percent of the cells divided within 4 hours. Apparently, the cold treatment set every cell back to the stage just after mitosis (G_1), since the doubling time of an untreated culture was about 20 hours.

Self-Synchronization

After lines of cells exhibiting growth control became available, H. Green's group found that monolayers of such cells were self-synchronized: adding serum to contact-inhibited cells induced the population to divide synchronously about 30 hours later. Since a burst of DNA synthesis occurred about 20 hours after the addition of the serum, the contact-inhibited cells must have been blocked in early G_1, probably just after mitosis (Nilhausen and Green, 1965; Todaro et al., 1965).

Previously a technique had been developed to synchronize populations of cells that lack growth control; this technique took advantage of the fact that cells in mitosis, especially those in metaphase, are only loosely attached to their culture dishes. Simple mechanical manipulations, such as pipetting fresh medium over the growing culture or shaking the culture flasks, were enough to detach the subpopulation of cells in mitosis. These detached cells could be recovered by gentle centrifugation of the medium, and they then went through another full division cycle synchronously (Terasima and Tolmach, 1963).

Drug Synchronization

One limitation of these self-synchronization techniques is that only a small number of synchronized cells can be obtained at any one time. By contrast, drugs can be used to synchronize an entire culture. There are many nontoxic agents that prevent cells from passing a particular stage of the cell cycle. A dose of one of these drugs permits unsynchronized cells to move through the cycle until they reach the blocked stage. When the drug is washed out, the population is partially synchronized (Petersen et al., 1968).

For example, in 1962 Xeros showed that an excess (2 mM) of thymidine stops cells at any point in S, presumably because the thymidine acts as a feedback inhibitor of enzymes responsible for the synthesis of precursors of DNA. When a growing cell culture is given large doses of thymidine for, say, 24 hours, cells entering or in S are blocked and, upon removal of the thymidine, a partially synchronized population is obtained. The synchrony is only partial because S phase takes 6 to 7 hours and therefore some cells will be up to 7 hours further through the cycle than others.

In 1964 Puck and Bootsma independently reported that a more complete synchrony could be obtained by using two pulses of thymidine (Puck, 1964a; Petersen and Anderson, 1964; Bootsma et al., 1964).

Removal of the first thymidine pulse permitted the partially synchronized population to continue through the cycle. About 8 hours after the removal of the thymidine, all the partially synchronized cells must have completed S. A second pulse of thymidine at this time stopped all these cells at one point in the cycle, namely, the subsequent G_1/S interphase. Removing the thymidine allowed the cycle to begin again in full synchrony (see Figure 2.9). In this way grams of cells were obtained which were synchronized to within 5 percent of the cell cycle time (Puck, 1964a).

Just as excess thymidine prevents cells from moving through the S period, certain drugs that dissociate the mitotic spindle prevent cells from completing mitosis. This class of drugs includes colchicine, colcemid, vincristine, and vinblastine, all of which have been used to synchronize cell populations (Biesele, 1958; Puck, 1959, 1964b; Puck and Steffen, 1963).

In 1967 Pfeiffer and Tolmach combined two methods of synchronization, blocking with vinblastine and shaking loose and recovering mitotic cells. The mitotic cells that accumulated in vinblastine-treated cultures were discarded. Two vinblastine pulses left only a few cells remaining on the plate. All of these were in a small segment, or "window," of the cycle that could be adjusted by varying the time between pulses. This technique had the advantage of providing a synchronized population which had not been affected by the drug.

Synchronization by Physical Separation

Synchronous populations of cells can be obtained using methods that take advantage of differences in the size and the density of cells at different stages of the cell cycle (Sinclair and Bishop, 1965). The volume of a cell doubles as it moves from early G_1 to M. Since the density of the cell remains constant, the sedimentation velocity increases $2^{2/3} =$ 1.59 times during the same period, and cells in different parts of the cycle can be separated by their sedimentation velocity through a density gradient (Miller and Phillips, 1969; Anderson et al., 1970; Schindler et al., 1970; Sitz et al., 1970).

Shall and McClelland (1971) reported what must be the easiest and least traumatic way yet devised to obtain a synchronized population of

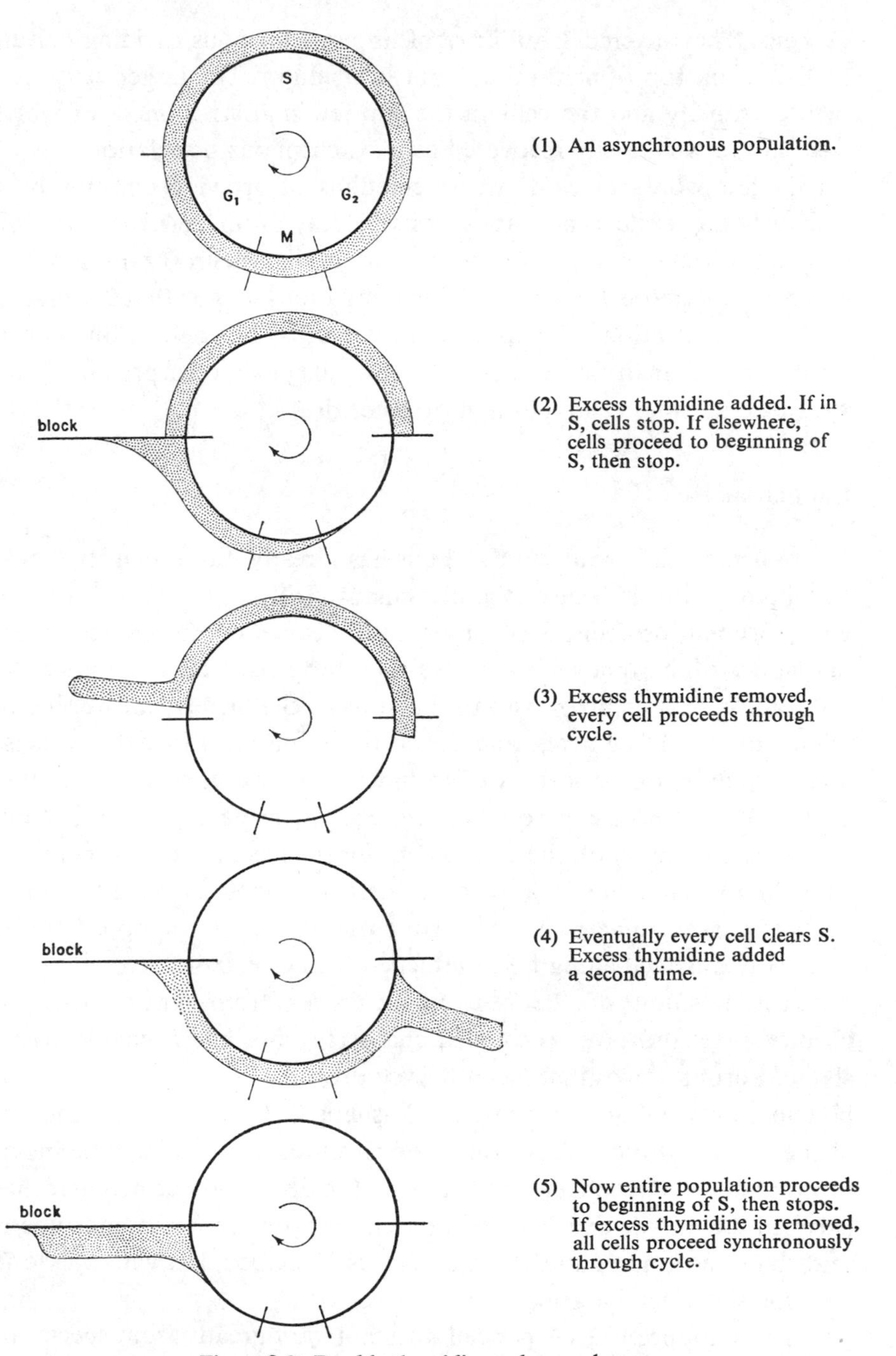

Figure 2.9. Double thymidine pulse synchrony.

G_1 cells. They layered 1 millilitre of an asynchronous dividing culture of L-cells on top of a 12-ml column of medium. The larger cells sedimented rapidly and the cells in the top few millilitres, most of which were in G_1, were easily recovered as a synchronous population.

A few words of caution: as cinéfilms of growing cultures have revealed, the cycle time varies considerably from cell to cell. This inherent variation imposes a limitation to all synchronization procedures. Furthermore, Potter's (1971) investigations suggest that thymidine, at the concentration used to synchronize cells, has deleterious effects on primary human fibroblasts, and it would not be surprising if this were true of other cell types and of other drugs.

Chromosomes

When a cell is ready to divide, it has already duplicated its DNA. Just before mitosis begins, the chromatin (DNA, histones, and other chromosomal proteins) condenses into visible chromosomes. Condensation of chromosomes marks the prophase, or first stage, of mitosis. As mitosis progresses to metaphase, the nuclear membrane of the dividing cell fragments, and the chromosomes continue to condense so that each becomes clearly visible under the microscope. In metaphase each chromosome is composed of two equivalent parts, called chromotids, joined in one spot, the centromere. In the past, the centromere was called the kinetochore because, as mitosis continues, each chromosome parts at the centromere and the chromatids are pulled to opposite poles of the dividing cell along fibres attached to the centromeres.

The position of the centromere varies from chromosome to chromosome and gives each metaphase chromosome a characteristic shape. Chromosomes that have their centromere at one end are said to be telocentric and at metaphase are V-shaped. When the centromere is in the middle of the chromosome, the chromosome is said to be metacentric and at metaphase is X-shaped. Finally, if the centromere lies between the middle and one end of a chromosome, the chromosome is said to be acrocentric and at metaphase is X-shaped, but with one long and one short pair of arms.

The amount of DNA per cell does not vary greatly from species to species among the mammals, ranging only from 5 to 7 picograms/cell for *Trichosurus* (marsupial), mouse, rat, Guinea pig, rabbit, pig, cat, dog, sheep, horse, ox, and man (Sober, 1968). Yet despite the relatively constant amount of DNA in mammalian cells, the number of chromosomes and their shape differs greatly from species to species.

Ploidy

The somatic cells of eukaryotes contain sets of pairs of chromosomes and therefore are said to be diploid. One chromosome of each pair is inherited from the male parent while its partner is inherited from the female parent. The two chromosomes in each pair have an identical size and shape with one exception, the pair of sex chromosomes, which, depending on the sex of the organism, may differ in size and shape. Female mammals, for example, have two identical sex chromosomes (XX) but male mammals have an X chromosome and an nonidentical Y chromosome. Sperm and ova (the gametes) by contrast have only one copy of each chromosome and are said to be haploid.

During S phase the DNA and chromosomal proteins are duplicated; at the beginning of mitosis, therefore, each visible chromosome (a pair of identical chromatids) has twice as much DNA as the corresponding daughter chromosome (a single chromatid) at the end of mitosis. So from the end of S until the daughter cells separate, a diploid cell actually has four copies of every gene and is transiently tetraploid.

If a diploid cell replicates its DNA but fails to divide, it becomes tetraploid and has two copies of every maternal and every paternal chromosome, or four copies of each chromosome in all. When such a tetraploid cell replicates its DNA, it becomes transiently octoploid and the two daughter cells arising after mitosis are each tetraploid. (Also see Defendi and Stoker, 1973.)

If maternal and paternal chromosomes, including both sex chromosomes, are all identical in number and shape to a standard set called the karyotype of the species, then the cells are said to be euploid. Abnormal cells are aneuploid if they have the wrong number of chromosomes, and heteroploid if the chromosome number differs from cell to cell. Thus a person who has an extra G_{22} chromosome (a mongoloid) is aneuploid but not heteroploid.

Chromosome Preparations and Karyotyping

Because the cells of most mammalian and avian species have large numbers of chromosomes that crowd together in mitosis, counting and observing individual chromosomes in mitotic cells is very difficult. This difficulty was partly overcome by Hughes and Hsu, who used hypotonic salt solutions to swell mitotic cells; some of the cells in such preparations had chromosomes that were separate and distinct (Hughes, 1952; Hsu, 1952).

Hughes had been systematically altering the tonicity of Tyrode saline, studying the effect of the solutions on cultures of chick tissue, and found that cells in mitosis were especially sensitive to hypotonicity. He noticed that in hypotonic solutions mitotic cells were blocked in metaphase, when the chromosomes were most easily stained. He did not develop this finding into a technique for studying chromosomes because he was more concerned with the course of normal mitosis and the mode of action of mitotic inhibitors. Hsu, on the other hand, was seeking clear chromosome preparations and took advantage of an accidental switch of saline solutions in his laboratory: when cultures of embryonic human skin were washed in hypotonic saline instead of an isotonic saline before fixation, the chromosomes in the cells that had been in mitosis were much more distinct and easy to count than they were in the usual preparations.

The osmotic swelling method enabled Hsu to draw up a human somatic cell karyotype (Hsu, 1952). Ninety-one (73%) of the mitotic skin cells he examined had 48 chromosomes. This sample was statistically significant, but unfortunately Hsu's chromosome counts proved to be wrong. Apparently, hypotonic shock partially separated the chromosomes, but chromosomes of cells caught early in mitosis were not easily comparable to those caught late in mitosis, and chromosomes sometimes overlapped, making the counting very difficult (Hsu, 1952; Hsu and Pomerat, 1953).

Three years after Hsu published his data on the human karyotype, Tjio and Levan found that colchicine could be used to block cells at metaphase (Levan, 1956; Tjio and Levan, 1956). This permitted counts to be made of condensed, separated chromosomes. Levan, Tjio and Puck counted 261 cells from four cultures of human embryonic tissue treated with colchicine and "were surprised to find that the chromosome number 46 predominated in the tissue cultures from all four embryos, only single cases deviating from this number." Forty-six is now the accepted human diploid number (Tjio and Puck, 1958; Human Chromosome Study Group, 1960).

Chromosomes of metaphase mouse and Chinese hamster cells can be isolated and separated by sedimentation velocity into size classes (Figure 2.10) (Somers et al., 1963; Salzman and Mendelsohn, 1968; Maio and Schildkraut, 1966, 1967, 1969; Huberman and Attardi, 1966). Once preparations of isolated metaphase chromosomes are available in highly purified form, it should be possible to apply a variety of analytical techniques to answer questions concerning their structure and composition (Prescott, 1970).

Currently a slight modification of Levan's colchicine technique is used to prepare a karyotype. Preparations of chromosomes are stained, usually with Giemsa dye, and photographed at 1,000-fold magnification. The print is enlarged, and the photographs of the chromosomes are cut out and arranged in order of size and shape. Figure 2.11 illustrates the euploid karyotypes of somatic cells from the mouse, Chinese or golden hamster, and man (Hsu and Benirschke, 1967).

The Human Karyotype

Giemsa-stained preparations of human cells blocked at metaphase have, of course, been intensively studied. The euploid human somatic cell has 2 sex chromosomes (X and Y chromosomes) and 44 autosomes, of which 34 are metacentric and 10 are acrocentric. A male has one X and one Y chromosome whereas a female has two X chromosomes; the X chromosome is metacentric and the Y chromosome is acrocentric. There are no telocentric chromosomes in the human karyotype, but telocentric chromosomes occur in some stable lines of cultivated human cells.

Until 1970 when methods for identifying individual chromosomes became available, the 23 pairs of human chromosomes could be sorted into seven classes on the basis of the position of the centromere and the length of the chromosome. With the exception of the Y-sex chromosome, however, it was difficult or impossible to distinguish individual chromosomes in Giemsa-stained cells. It was, of course, at least as difficult to

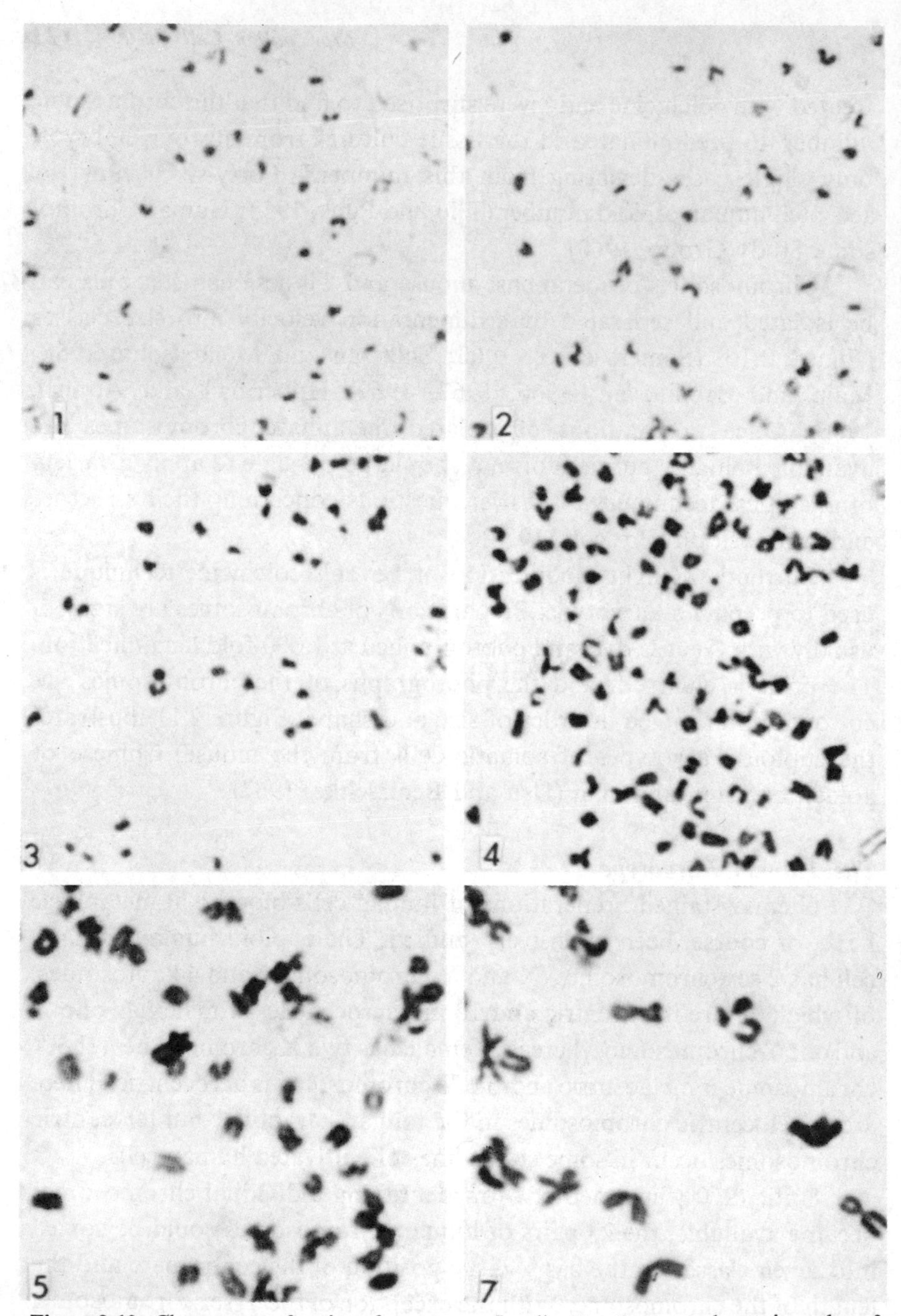

Figure 2.10. Chromosome fractions from mouse L-cells. Fractions are shown in order of increasing chromosome length starting at the upper left photograph. Unna methylene blue stain (×1250). Reprinted with permission from Maio and Schildkraut, *J. Mol. Biol.* 40, 203, 1969.

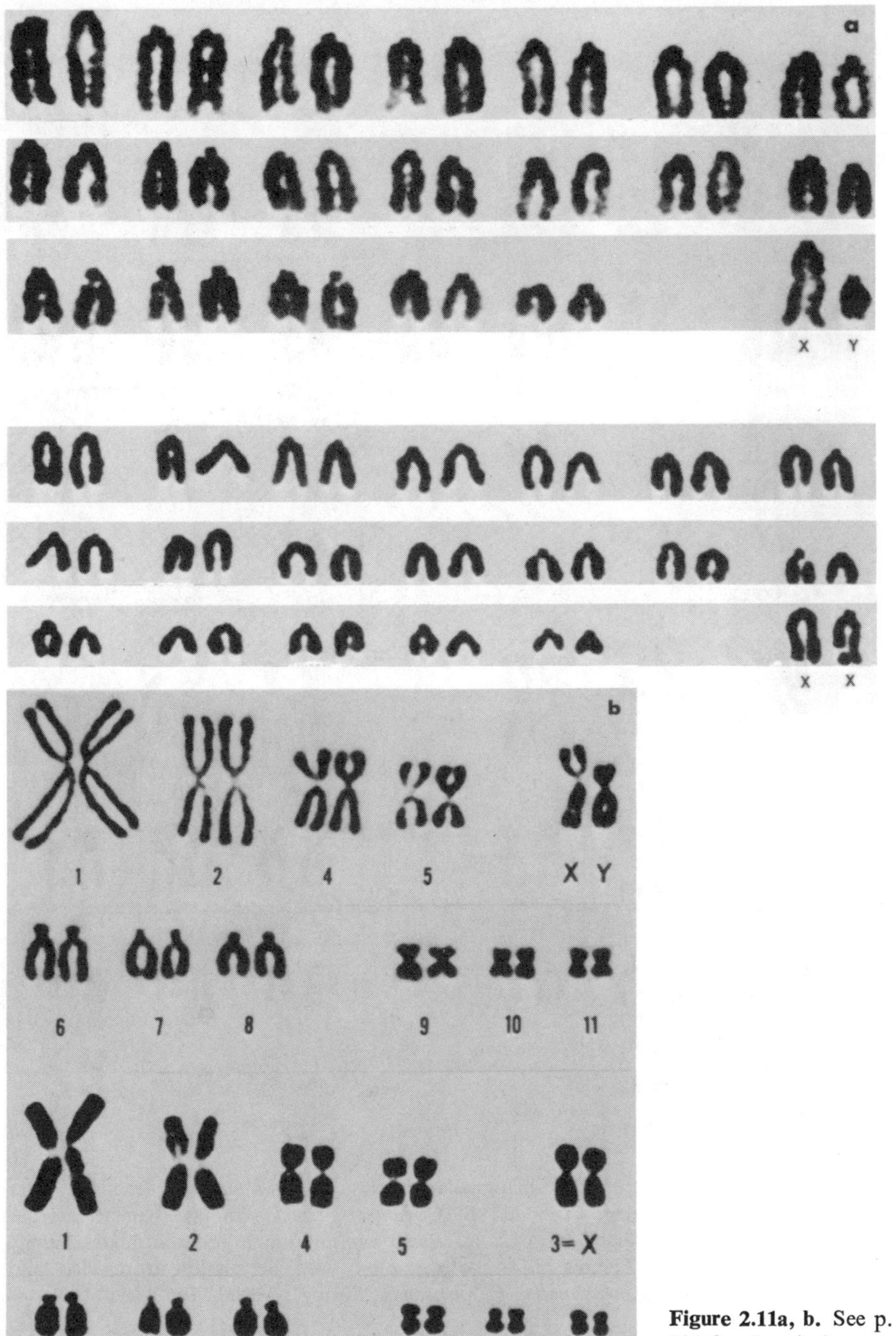

Figure 2.11a, b. See p. 124 for description.

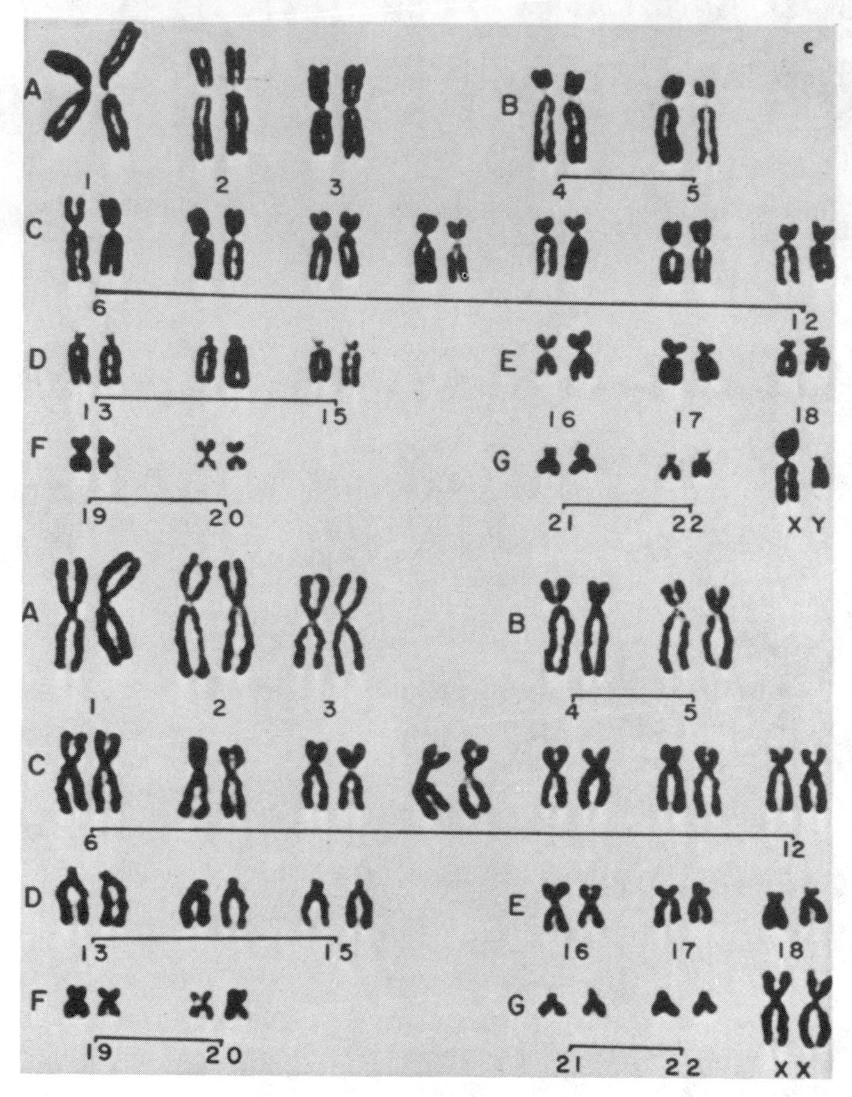

Figure 2.11. Karyotypes of some mammalian species. **(a)** *Mus musculus* (mouse), order Rodentia, family Muridae; 2n = 40. **(b)** *Cricetulus griseus* (Chinese hamster), order Rodentia, family Cricetidae; 2n = 22. **(c)** *Homo sapiens* (man), order Primates, family Hominidae; 2n = 46. Figures 2.11 a–c reprinted with permission from Hsu and Benirschke, *An Atlas of Mammalian Chromosomes*, Springer-Verlag, 1967.

distinguish individual chromosomes of many other mammalian species. The mouse, for example, has a karyotype composed of 40 telocentric and acrocentric chromosomes which differ only in overall length.

Identifying Individual Chromosomes

Caspersson and Zech of the Karolinska Institute, using stains suggested by Modest of the Children's Hospital of Harvard Medical School, were the first people to distinguish each unique pair of human chromosomes. Quinacrine and its derivatives quinacrine mustard and quinacrine dihydrochloride (atabrine) bind to specific regions of each chromosome (Caspersson et al., 1970b). When preparations of metaphase chromosomes stained by these dyes are examined under indirect UV illumination, the fluorescense of the quinacrine dye causes bands to appear across the arms of the chromosomes.

Caspersson and his group have shown that all 46 human chromosomes can be paired because each pair of chromosomes has a distinct banding pattern, and the Y chromosome is so intensely fluorescent that it can be seen even in interphase somatic cells (Pearson et al., 1970; George, 1970). The resolution of the quinacrine staining method is indisputable because, for example, in spontaneous translocations of a piece of chromosome 13 to either chromosome 14 or chromosome 21, the number, location and width of the fluorescent bands characteristic of the chromosome 13 fragment are preserved on the new chromosome (Caspersson et al., 1971a).

Recently it has become evident that the chromosomal bands which distinguish one chromosome from another can also be revealed with Giemsa stain, provided that the chromosomal DNA is denatured and renatured before staining (Pardue and Gall, 1970; Arrighi and Hsu, 1971; Gagne et al., 1971; Sumner et al., 1971; Schnedl, 1971a, b; Patil et al., 1971; Lomholt and Mohr, 1971; Evans et al., 1971). If the DNA of fixed chromosomes is denatured with heat or alkali and is then allowed to renature before the preparations are stained with Giemsa, transverse densely staining bands are revealed on each chromosome (Pardue and Gall, 1970; Schnedl, 1971a; Lomholt and Mohr, 1971); the pattern of these bands is specific for each chromosome. The bands include the regions to either side of almost every centromere and also most of the bands revealed by the quinacrine fluorescence method. Since the regions adjacent to centromeres, which are stained by both these procedures, are known to contain repetitious DNA, it has been suggested that most if not all of the bands revealed in chromosomes by quinacrine and Giemsa staining contain repetitious (satellite) DNA (Gall and Pardue, 1969; Pardue and Gall, 1970; Patil et al., 1971; Jones and Corneo, 1971).

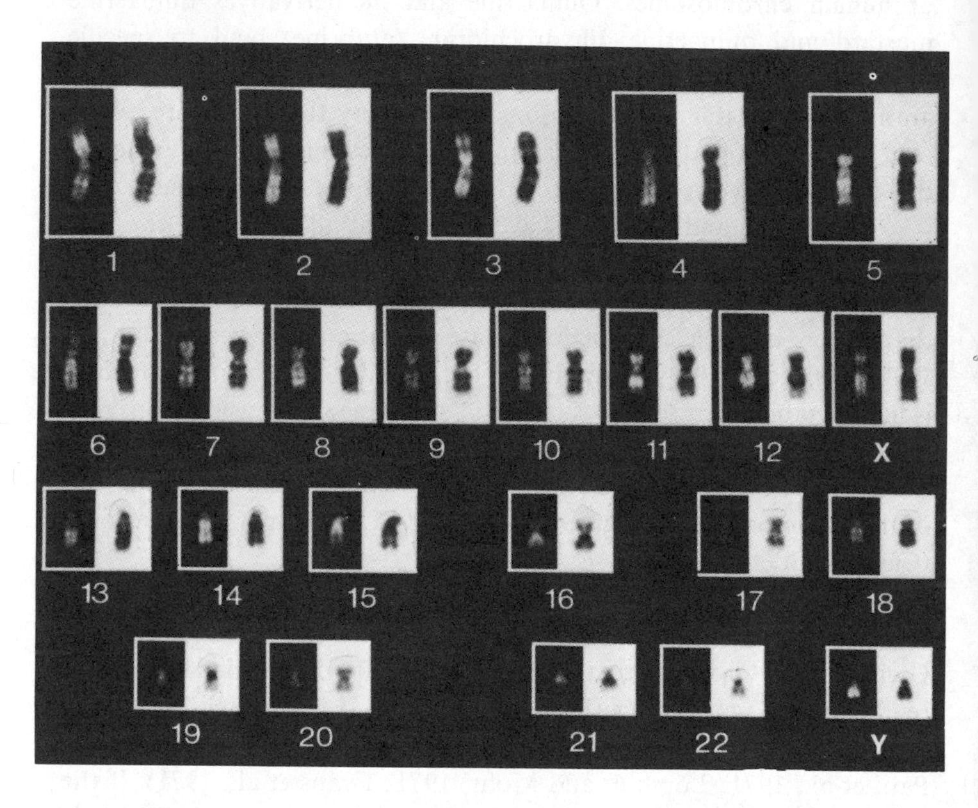

Figure 2.12. Composite karyotype with left-hand member of each chromosome pair from a cell stained with quinacrine dihydrochloride and fluoresced, and the right-hand member from a different cell treated with the acetic-saline-Giemsa. Note the correspondence between many of the fluorescent Q bands and the Giemsa-staining G bands. Reprinted with permission from Evans et al., *Chromosoma* 35, 310, 1971.

Various modifications of this Giemsa staining method have been developed and two of them yield excellent banding patterns: these two methods involve either cooling mitotic cells before fixing and staining them or exposing fixed cells to dilute solutions of trypsin before staining with Giemsa (Shiraishi, 1970; Wang and Federoff, 1972).

Figure 2.12 illustrates the normal human karyotype. The banding patterns revealed by quinacrine and Giemsa are generally similar, but the acetic-saline-Giemsa staining technique (Evans et al., 1971) clearly resolves many more bands than does quinacrine staining.

As an example of the power of these techniques to resolve long standing errors of identification, quinacrine staining has shown that the chromosome which is present in three copies in myeloid leukemia cells (the so-called Philadelphia chromosome) is not the same as the chromosome that is present in three copies in Down's syndrome (mongolism). These chromosomes had long been equated since mongoloids are much more susceptible to myeloid leukemia than are normal persons (O'Riordan et al., 1971).

Constancy of Chromosome Number—Heteroploidy

Immediately after being placed in culture, most cells are euploid and they remain so until they reach crisis. During crisis most cells die, and the clonal lines that emerge are almost always heteroploid; that is, the chromosome number and karyotype differ from cell to cell. Heteroploid cell lines grow indefinitely, so euploidy appears to, if anything, restrict rather than stimulate growth of cells in culture.

Most cultivated heteroploid cells have a median chromosome number three to four times the haploid number of their species. For instance, normal mouse cells have 40 chromosomes but 3T3 cells have a median chromosome number of about 70; normal human cells have 46 chromosomes but HeLa cells have a median chromosome number of about 66.

The heteroploidy that accompanies the establishment of cells in culture strikingly resembles the heteroploidy of cells in tumours. Histograms of the chromosome number of primary cultures of normal cells are extremely narrow, with a mode at the diploid chromosome number; histograms of the chromosome numbers of tumour cell populations and of established cell lines are usually very broad with an aneuploid (subtetraploid) mode (Hauschka et al., 1957; Hsu and Klatt, 1958; Hsu et al., 1961; Hsu, 1961; Harris, 1964; Kirkland et al., 1967). BHK21 is an established hamster cell line that is exceptional: it is aneuploid but not heteroploid, with a mean chromosome number of 43 instead of the diploid number, 44 (Stoker and Macpherson, 1964).

Aberrant nuclear morphology, tripolar and tetrapolar mitosis and giant cells are the pathologist's hallmarks of malignancy; they are also histological indicators of genetic instability and it is not surprising, therefore, that the cells of tumours and tumour cell lines are heteroploid. Apparently selection in vivo for malignancy and the ability to metastasize often leads to the emergence of heteroploid tumour cells. It may, therefore, be no coincidence that selection for the ability to divide in vitro almost invariably also leads to the emergence of heteroploid cells.

DNA Constancy

According to Kraemer et al. (1971) the DNA content of many heteroploid lines is constant from cell to cell despite the broad variation in the number of chromosomes per cell. Kraemer and his colleagues measured the DNA content of fixed interphase cells by binding fluorescent Feulgen dye to cell's DNA, then passing single cells quickly through a very narrow beam of blue laser light and recording the pulse of fluorescence from each cell; control experiments showed that the pulse height was proportional to DNA content. The DNA histograms of heteroploid cells obtained in this way were as narrow as the DNA histograms of euploid cells and much narrower than their chromosome histograms (Figure 2.13).

In the heteroploid cells studied by Kraemer, the DNA content of G_1 cells was 1.5 times that of euploid cells of the same species. If this difference in amount of DNA between heteroploid and euploid cells proves to be universal, then DuPraw's (1970) speculative hypothesis offers an explanation of the constancy of DNA in variously heteroploid

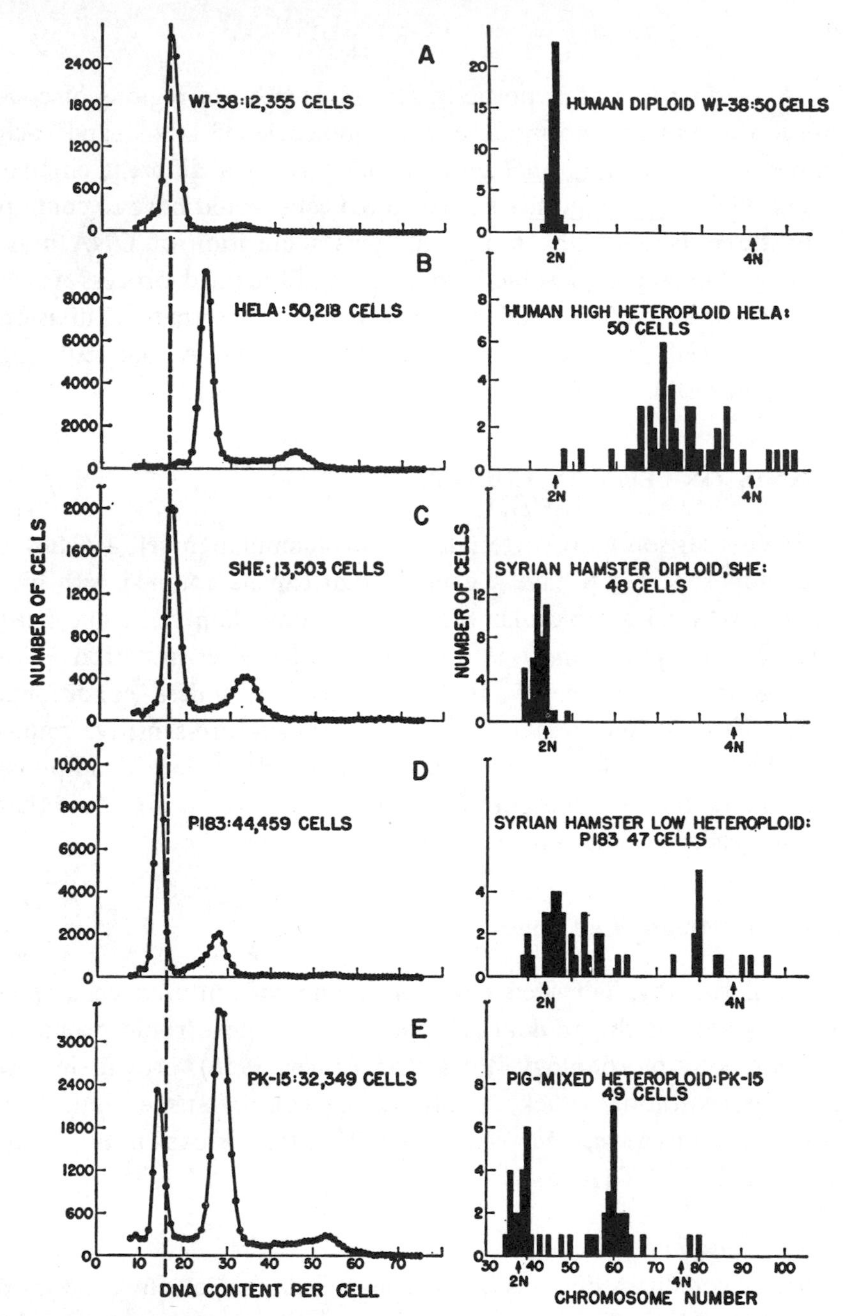

Figure 2.13. The distribution of DNA content (in arbitrary units) from flow microfluorometry and chromosome number histograms of growing cell lines: A, WI-38; B, HeLa; C, Syrian hamster embryo; D, P183 (polyoma-transformed derivative of BHK21/C13); and E, porcine PK-15. The dashed line at the left represents the average 2C DNA content of diploid mammalian cells. The abscissal scales for DNA content and chromosome number are not comparable. Reprinted with permission from Kraemer et al., *Science* 174, 714, 1971.

cells. According to this hypothesis, chromosomes are regions of condensation within an enormous circular molecule of DNA, and each member of a pair of normal chromosomes is on a different circular molecule. To fit this hypothesis heteroploid cells would have to contain exactly three haploid DNAs per cell. This acquisition of DNA must also be accompanied by some change in the ill-defined process which causes chromosomes to condense. The net result is that on division heteroploid cells receive the same amount of DNA but differing numbers of visible chromosomes.

MUTANTS OF CELLS IN CULTURE

By comparison with bacterial genetics, mammalian cell genetics is all but nonexistent. To have a genetics that can be coupled with biochemistry to yield a molecular biology of mammalian cells, one must have a mating system and one must be able to select mutations with reversion frequencies below 10^{-7}. Furthermore, unless conditional lethal mutations (suppressor-sensitive or temperature-sensitive mutations which are phenotypically wildtype under permissive conditions) are available, the only mutations which can be selected are those that occur in genes inessential for survival in culture.

Origins of Mutant Cell Lines

Until recently, inherited changes in lines of cultured cells were invariably seen as the gradual gain or loss of a pleiotropic property, such as colony morphology (Puck and Fisher, 1956), growth in low serum concentration (Puck, 1958), or growth in suspension. One possible exception was M. Vogt's (1958, 1959) isolation of polio-resistant sublines of HeLa.

Natural Mutations

Many hereditary diseases result from the absence of single enzymes (Table 2.5). These diseases are inherited as recessive mutations: the heterozygous person, who has inherited the mutant allele for one of these genes on one chromosome of a pair and the wild-type allele on

Table 2.5. SOME GENETIC ABNORMALITIES THAT AFFECT SPECIFIC MOLECULES PRESENT IN THE CULTURED HUMAN MONONUCLEAR CELL

Disease or Variant	Affected Molecule	Genetics
"A" electrophoretic variant of glucose-6-phosphate dehydrogenase (G-6-PD)	Glucose-6-phosphate dehydrogenase	Sex-linked co-dominant
Acatalasia I (Japanese variant)	Catalase	Autosomal recessive
Acatalasia II (Swiss variant)	Catalase	Autosomal recessive
Arginosuccinic aciduria	Arginosuccinase	Autosomal recessive
Branched-chain ketonuria	Uncertain	Autosomal recessive
Citrullinemia	Arginosuccinate synthetase	Autosomal recessive
Cystinosis	Uncertain	Autosomal recessive
Electrophoretic variant of phosphoglucomutase	Phosphoglucomutase	Autosomal codominant
Electrophoretic variant of lactic dehydrogenase	Lactic acid dehydrogenase	Autosomal codominant
Electrophoretic variants of 6-phosphoglucuronic acid dehydrogenase	6-Phosphoglucuronic acid	Autosomal codominant
Familial nonspherocytic hemolytic anemia	Glucose-6-phosphate dehydrogenase	Sex-linked recessive
GM_1 Gangliosidosis Type I (Generalized Gangliosidosis)	Absence of β-galactosidase A, B, C	Autosomal recessive
GM_1 Gangliosidosis Type II (Juvenile GM_1 Gangliosidosis)	Absence of β-galactosidase B and C	Autosomal recessive
GM_2 Gangliosidosis Type I (Tay-Sachs Disease)	Absence of hexosaminidase A	Autosomal recessive
GM_2 Gangliosidosis Type II (Sandhoff's Disease)	Absence of hexosaminidase A and B	Autosomal recessive
GM_2 Gangliosidosis Type III (Juvenile GM_2 Gangliosidosis)	Partial deficiency of hexosaminidase A	Autosomal recessive
G-6-PD deficiency (Mediterranean type)	Glucose-6-phosphate dehydrogenase	Sex-linked recessive
G-6-PD deficiency (Negro type)	Glucose-6-phosphate dehydrogenase	Sex-linked recessive
Galactokinase deficiency	Galactokinase	Autosomal recessive
Galactosemia	Uridine diphosphogalactose transferase	Autosomal recessive
Gaucher's disease	Glucoceribosidase	Autosomal recessive
Glycogen storage disease (Cori type II)	Lysosomal α-1,4-glucosidase	Autosomal recessive
Glycogen storage disease (Type III)	Amylo-1-6-glucosidase	Probably autosomal recessive

From Krooth and Sell (1970).

Table 2.5.—*Continued*

Disease or Variant	Affected Molecule	Genetics
Homocystinuria	Cystathionine synthetase	Autosomal recessive
Hunter's syndrome	Uncertain (Macromolecule which influences cellular content of mucopolysaccharides)	Sex-linked recessive
Hurler's syndrome	Uncertain (Macromolecule which influences cellular content of mucopolysaccharides)	Autosomal recessive
"I-cell disease"	Uncertain (beta-glucuronidase activity decreased; mucopolysaccharide content increased)	Probably autosomal recessive
Infantile metachromatic leukodystrophy	Arylsulfatase A	Autosomal recessive
Lesch-Nyhan syndrome	Hypoxanthine-guanine xanthine phosphoribosyltransferase	Sex-linked recessive
Madison electrophoretic variant of glucose-6-phosphate dehydrogenase	Glucose-6-phosphate dehydrogenase	Sex-linked codominant
Methylmalonic aciduria	Uncertain (Defective vitamin B_{12} synthesis)	Uncertain
Niemann-Pick disease	Uncertain (Increased cellular level of sphingomyelin)	Autosomal recessive
Orotic aciduria	Orotidine-5′-monophosphate pyrophosphorylase and orotidine-5′-monophosphate decarboxylase	Autosomal recessive
Refsum's disease	Phytanic acid α-oxidase	Autosomal recessive
Sanfilippo syndrome	Uncertain (Macromolecule which influences cellular content of mucopolysaccharides)	Autosomal recessive
Scheie syndrome	Uncertain (Macromolecule which influences cellular content of mucopolysaccharides)	Autosomal recessive
Xeroderma pigmentosa	Uncertain (Defective excision repair of DNA) perhaps an endonuclease	Autosomal recessive

the other chromosome, will show little or no sign of the disease, presumably because the single wild-type gene in the heterozygote's genome can satisfy the person's or organism's requirement for that particular gene product. The homozygous person, who has inherited the mutant allele on both chromosomes of a pair, will show the symptoms of the disease.

As techniques for the cultivation of primary cells were developed, it became apparent that some of the mutant characters first recognized through hereditary diseases in the whole organism were also expressed by cultivated cells obtained from affected individuals (Boyle and Raivio, 1970; O'Brien et al., 1971). Other genetic abnormalities have no clinical symptoms but give rise to functional molecules that can be distinguished from the normal molecules by biochemical techniques. Table 2.5 lists some genetic abnormalities in which a deficient activity of a specific enzyme has been demonstrated.

Inactivation of X Chromosomes

If a mutation is located on the X chromosome, then males, who have only one X chromosome and therefore carry single alleles for all X-linked genes, will necessarily have the mutant phenotype. Remarkably, however, females are mosaic in their expression of any X-linked gene for which they are heterozygous (Russell, 1963); for example, tortoise-shell cats are always female and heterozygous; and their patches of black and brown fur are each descended from genetically different skin cells. In a normal female mammal the two X chromosomes in each cell become differentiated at an early stage in embryonic development: one remains genetically active and identical to the single male X chromosome, whereas the other is condensed, genetically inert, and among the last of the chromosomes of the karyotype to be replicated (Lyon, 1971).

As expected, primary cultures of biopsies from women heterozygous for X-linked genes, such as those for glucose-6-phosphate dehydrogenase (G6PD) (Davidson et al., 1963) and hypoxanthine-guanine phosphoribosyl transferase (HGPRT) (Migeon et al., 1968), yield two clonal phenotypes, wildtype and mutant.

In Vitro Selection of Mutant Cells

The chief obstacle to the selection of cultivated cells with point mutations is their ploidy. In bacteria, which have a single copy of each gene, the rate of spontaneous mutation in any single gene is of the order of one mutation/10^8 cell generations (Luria and Delbrück, 1943). Most vertebrate cell lines are at least diploid: normal somatic cells have two copies of each gene, and the established heteroploid lines ought to have at least three copies of most genes although this has not been shown directly (Kraemer et al., 1971). One would expect the rate at which a recessive mutation becomes detectable in a cell culture to be the square or cube of the mutation rate in haploid cells, in other words, about one in 10^{16} cells if we extrapolate from observations of microorganisms.

Since 10^{16} cells would weigh about 10 tons, spontaneous recessive mutations should never be detected in cell lines. Paradoxically, however, somatic mammalian cell lines lacking various enzymic activities have been isolated at frequencies approaching one in 10^6 cells. For example, HGPRT-negative cells from Lesch-Nyhan patients are resistant to the drug 8-azaguanine, whereas wild-type cells use HGPRT to incorporate the drug into nucleic acids and are killed by it. Recently Harris (1971) used fluctuation analysis (Luria and Delbrück, 1943) to examine the frequency of spontaneous appearance of cells resistant to 8-azaguanine in diploid, tetraploid and octoploid populations of Chinese hamster cells. He found that the rate at which mutant cells were detected was essentially the same for all three populations, even though they should have had two, four and eight copies of the HGPRT gene, respectively!

At present it is far from obvious why a presumably recessive mutation of a gene present in as many as eight copies per cell can be detected. For this reason many cytogeneticists often describe cells with apparently mutant phenotypes as variants rather than mutants, since a mutant must have an altered genotype as well as an altered phenotype.

Negative Selection

Penicillin lyses only growing bacteria and has no effect on resting populations. When penicillin is added to a mixture of wild-type and mutant bacteria in a minimal medium in which the mutant cannot grow, the penicillin kills only the growing cells and thereby enriches the culture for the mutants. Removal of the toxic drug permits testing of the presumptive mutant cell and its descendants. This is called negative selection since it selects for mutants indirectly by killing normal cells (Davis, 1948; Lederberg and Zinder, 1948). Negative selection works well for mammalian cells, if the appropriate combination of toxic drug and selective conditions are applied, and permits the isolation of mutant lines descended from one cell in a million.

Auxotrophs

Cells that grow in minimal medium are called prototrophs; auxotrophs are variant cell lines which are dependent on specific additional nutrients for growth.

By giving a culture minimal medium and adding a drug toxic to dividing cells, one can kill all prototrophs and then, by removing the drug and changing to supplemented medium, recover auxotrophs. The serum requirement of most cells complicates this experiment, since even when dialyzed serum from which low molecular weight metabolites have been removed is used, degradation of macromolecules in the dialyzed serum results in the release of metabolites into the medium.

Growing cells can incorporate the thymidine analogue bromodeoxyuridine (BrdU) into their DNA. Mammalian cells that have incorporated BrdU are much more sensitive to the lethal effect of UV and blue-light irradiation than cells whose DNA contains thymidine (Djordjevic and Szybalski, 1960). Puck has isolated proline-requiring and glycine-requiring auxotrophs from the Chinese hamster line CHO by negative selection, using BrdU plus blue light as a killing agent (Puck and Kao, 1967).

CHO cells, which can grow without added glycine, were given BrdU in medium that lacked glycine. The BrdU was incorporated into the DNA of all the cells, since all could grow. However, when a rare mutant arose that needed glycine to grow, that cell could not multiply and so would not incorporate BrdU. After a time all cells were exposed to blue light. The growing cells were killed, but the rare, non-dividing Gly^- mutants survived. Then the culture was changed to medium lacking BrdU and containing glycine, and the Gly^- cells were recovered (Kao and Puck, 1968).

The CHO subline Puck's group used is monosomic for one chromosome (Kao and Puck, 1967). Possibly the mutations they recovered are in that chromosome since they have the expected haploid reversion frequency of 10^{-6} to 10^{-8} (Kao and Puck, 1968); also certain chemical and physical mutagens (nitrosoguanidine, X rays, ethylmethylsulfonate) increased the rate of forward mutation in this system, and the mutation frequency per rad of X rays was comparable to that of haploid *E. coli* (Kao and Puck, 1969). This work suggests that ethylmethylsulfonate and nitrosoguanidine will be effective mutagens for vertebrate haploid cells when these are cultured successfully (Freed and Metzger-Freed, 1970).

Drug Selection

In mammalian cells three of the four nucleotides that are precursors to DNA have two synthetic pathways. A loss of either synthetic pathway for any one nucleotide is not lethal to the cell, so a class of mutants can be isolated which lacks one of the pathways.

Purine pathways to DNA. The major pathway of synthesis of dATP and dGTP constructs the purine ring on a phosphoribosyl backbone, giving nucleoside monophosphates. In synthesizing the purine base, enzymes in this pathway transfer two methyl groups from dihydrofolic acid into the purine ring, so the pathway is blocked by antagonists of folic acid, such as aminopterin (Figure 2.14).

The second pathway salvages the purine bases that are liberated by catabolism. It makes dAMP and dGMP from the completed purine ring by the addition of phosphoribose to the bases adenine, hypoxanthine and guanine (Figure 2.15). This pathway uses the "salvage" enzymes hypoxanthine-guanine phosphoribosyl transferase (HGPRT) and adenine phosphoribosyl transferase (APRT). Thus cells from patients with the Lesch-Nyhan syndrome, which lack HGPRT (Seegmiller et al., 1967), must make all their purine precursors of DNA directly via the folic acid pathway. Recently H. Green's group has selected mouse cells lacking APRT by negative selection with the drug alanosine (Kusano et al., 1971).

Since the second pathway bypasses the folic acid-dependent methylation step, it can operate in the presence of aminopterin provided that one of the salvage bases (hypoxanthine, adenine or guanine) is added to the medium. The base analogues thioguanine and azaguanine are incorporated by and are toxic to cells which can salvage purines (Brockman and Stutts, 1960), and so mutational loss of HGPRT renders the cell resistant to azaguanine and thioguanine (Szybalski

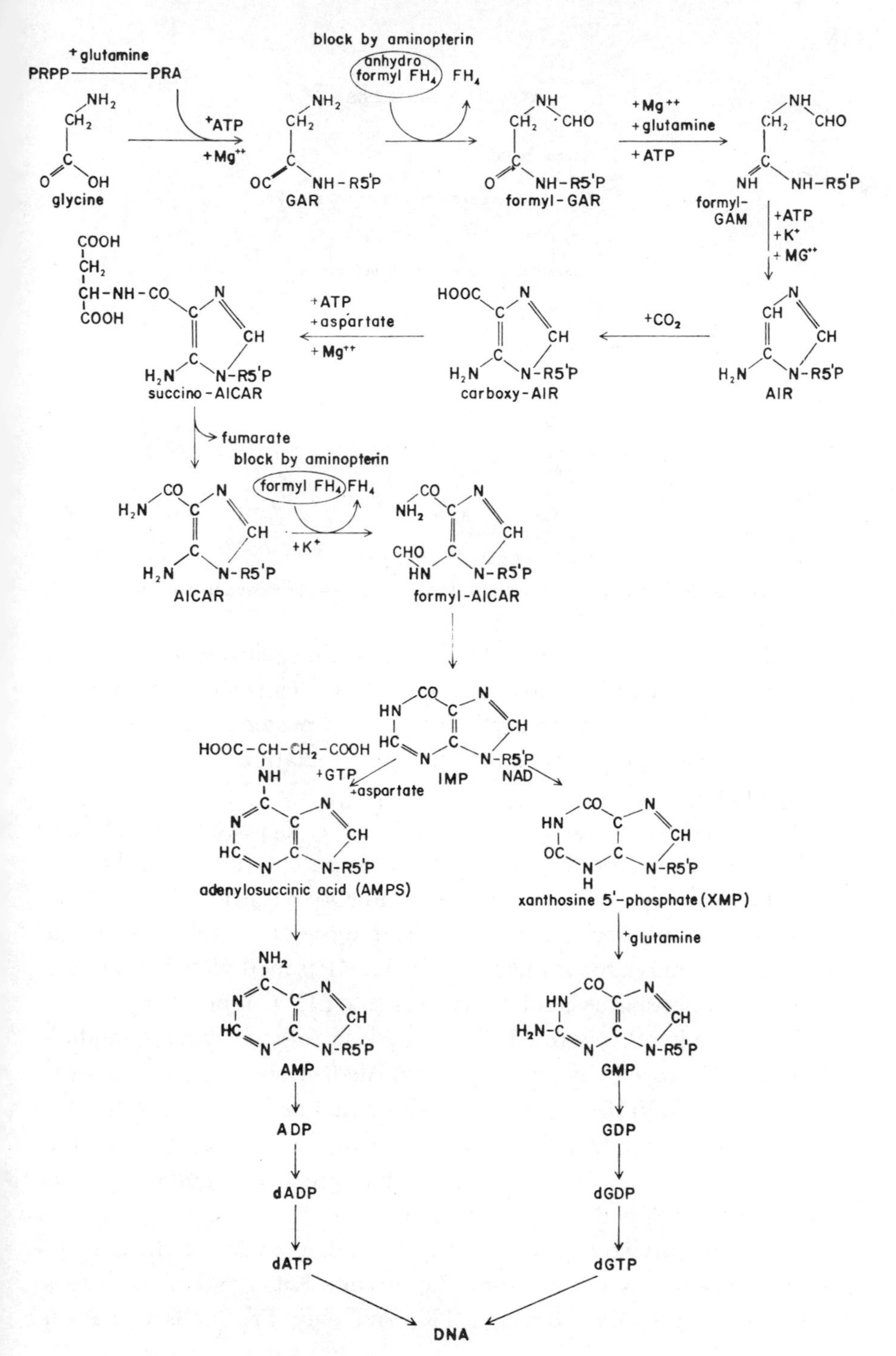

Figure 2.14. Pathways of purine synthesis for DNA, I: endogenous route.

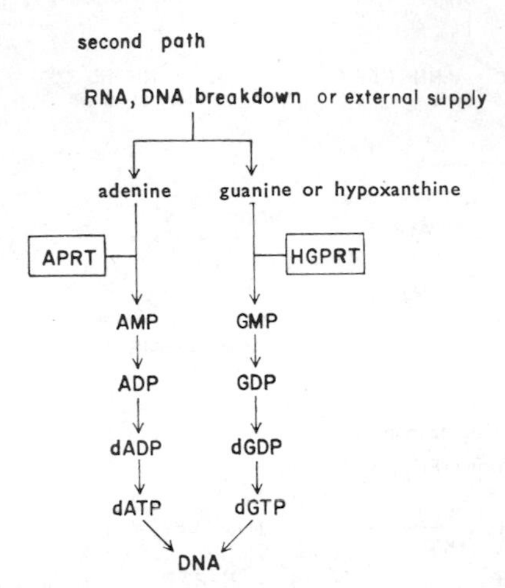

Figure 2.15. Pathways of purine synthesis for DNA, II: scavenger route.

et al., 1962; Littlefield, 1963). At the same time, however, the same mutation renders a cell sensitive to aminopterin. Thus with appropriate drugs survival can be made dependent on the presence or absence of HGPRT (Table 2.6) and forward and back mutation frequencies can be measured.

Pyrimidine pathways to DNA. Likewise it is possible to mutate cells for the presence or absence of one of the pathways that synthesizes thymidine triphosphate. The major synthetic pathway to dTTP is through the uridine derivative dUMP. In the presence of folic acid and the enzyme thymidylate synthetase (TS), dUMP is methylated to dTMP, which is then phosphorylated by kinases to dTTP (Figure 2.16).

The second pathway of dTTP synthesis uses the enzyme thymidine kinase (TK) to recycle thymidine (thymine deoxyriboside), a breakdown product of DNA. In the presence of TK, thymidine is converted directly to dTMP which reenters DNA (Figure 2.16). A cell deprived of the main dTTP pathway by aminopterin but given thymidine survives because of the TK pathway (Table 2.6).

TK$^-$ cells survive exposure to the thymidine analogue BrdU (Hsu and Somers, 1962; Kit et al., 1963; Dubbs and Kit, 1964). This drug is incorporated into DNA through TK, and only TK$^-$ cells can form

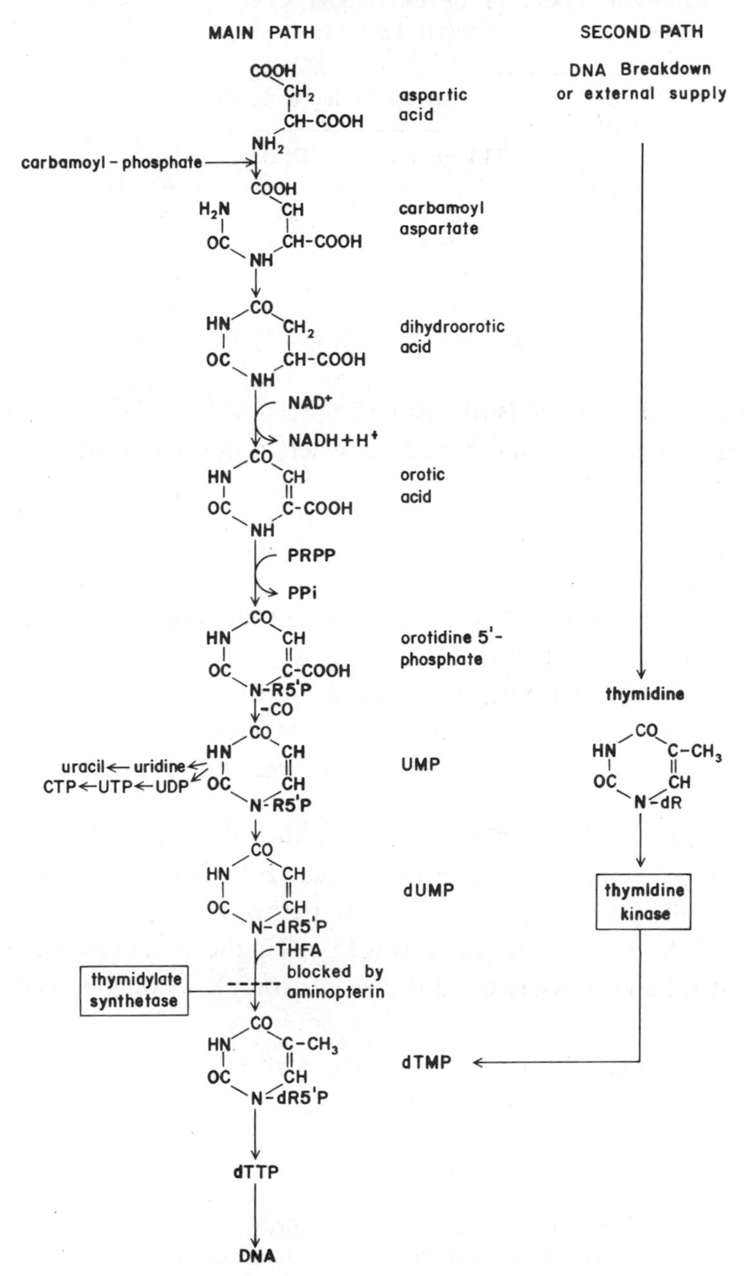

Figure 2.16. Two pathways of thymidine synthesis for DNA.

Table 2.6. VIABILITY OF DIFFERENT CELLS IN MEDIUM WITH DRUGS

Cell	Drugs Added to Medium		
	Thioguanine	BrdU	HAT*
Wild type	dies	dies	lives
TK^-	dies	lives	dies
$HGPRT^-$	lives	dies	dies

* Aminopterin with thymidine and hypoxanthine

colonies in the presence of BrdU. But a TK^- cell will be killed in medium containing aminopterin and thymidine, whereas normal fibroblasts will survive (Table 2.6).

HAT Selection

TK-deficient and HGPRT-deficient cell lines were selected by Szybalski and Littlefield (Szybalska and Szybalski, 1962; Littlefield, 1963) who, taking advantage of enzyme redundancy, developed a combination of drugs which permits selection for or against the presence of these enzymes. This is called HAT selection (Szybalski et al., 1962).

Both TK^- and $HGPRT^-$ cells will die in medium containing hypoxanthine, aminopterin, and thymidine (HAT medium), but wild-type cells will survive on their scavenger pathways. Survivors in HAT medium will be the rare TK^+ or $HGPRT^+$ cell in the mutant population. Thus forward and back selection is possible for TK as well as HGPRT

Table 2.7. DRUG SELECTION OF DIFFERENT CELLS

Selection	Drug Used
Wild type to TK^-	BrdU
Wild type to $HGPRT^-$	thioguanine
TK^- to wild type	HAT
$HGPRT^-$ to wild type	HAT

(Szybalski et al., 1962; Littlefield, 1963, 1965). Table 2.7 summarizes the selections that are possible with these drugs.

Negative Selection for Reversion of Transformation

The clinical course of a malignant tumour is almost always in the direction of increased malignancy with the passage of time. Cultures of tumour cells and cells transformed by tumour viruses, however, throw off variants which are less malignant than the parental cells and in some cases exhibit the phenotype of a normal cell (Macpherson, 1965). This strongly suggests that the progression to increased malignancy in vivo is not the result of some irreversible genetic change.

In 1968 Pollack et al. isolated revertants of virus-transformed mouse and hamster cells and of hamster tumour cells by negative selection with fluorodeoxyuridine (FdU), which like BrdU kills only the cells that are making DNA (Ruechert and Mueller, 1960; Reyes and Heidelberger, 1965). A pulse of FdU killed all the transformed cells that continued to divide in dense cultures. Some of the surviving cells grew into colonies that had a low saturation density and looked flat. These fiat revertants of hamster cells transformed by polyoma virus and of 3T3 cells transformed by polyoma virus and SV40 retained viral T antigen and the ability to grow in low concentrations of serum (Jainchill and Todaro, 1970). Flat colonies isolated from hamster tumour cells were found to be less tumourigenic than the parental cells (Pollack and Teebor, 1969). Flat cells were also less readily agglutinated by the lectin wheat germ agglutinin (WGA) than their transformed parents (Pollack and Burger, 1969).

Over long periods lectins are toxic to transformed cells. Ozanne and Sambrook (1971) cultured SV40-transformed 3T3 cells in media containing the lectin concanavalin A and isolated the cells that survived. These cells proved to be resistant to the toxic effect of the lectin, and most grew to low saturation densities.

Flat cells containing SV40 T antigen and SV40-specific DNA sequences can be isolated directly from an SV40-infected culture of 3T3 cells and need not be derived from a transformed cell line (Smith and Scher, 1971; Scher and Nelson-Rees, 1971). Cells that grow into colonies in media containing only 0.5% serum have SV40 T antigen as well as a reduced serum requirement. Some of the colonies look flat and when cloned have a low saturation density. Wyke (1971a) has used Puck's negative selection by BrdU and light to select for polyoma-transformed BHK cells that fail to divide in agar. These cells, like Pollack's revertants, retain viral T antigen.

The saturation density of 3T3 cells transformed by polyoma virus is lowered by Bu_2-cAMP or inhibitors of phosphodiesterase or both (Sheppard, 1971). This reversion is lost when the cAMP is removed. Stable revertants isolated by selection with FdU have more cAMP than their transformed parents (Sheppard, 1972) or their untransformed grandparents. This leads to the interesting hypothesis that negative selection with FdU might yield a cell with increased growth control resulting from a mutation to constitutive synthesis of cAMP.

Conditional Lethal Mutants

To date no one has obtained a suppressor-sensitive mutation in a mammalian cell, or for that matter, in a mammalian virus; however, a few temperature-sensitive (*ts*) mutants have been isolated.

Mammalian cells cease to divide when cultivated at temperatures below about 31°C or above about 40°C. Thompson et al. (1970) coupled negative selection with temperature shifts to select temperature-sensitive mutants of mouse L-cells growing in Spinner culture. The cells were grown at 33°C and mutagenized with nitrosoguanidine. They were then raised to 39°C for 3 days and returned to 33°C for 20 days. While at 39°C, the cells were exposed to a concentration of tritiated thymidine that kills dividing cells. This cycling from 33 to 39°C and back, with toxic doses of tritiated thymidine at the high temperature, was repeated four times. Then, to kill cells that might have become unable to take up thymidine, rather than those temperature sensitive for DNA synthesis, the survivors were cycled between 39°C and 33°C twice with cytosine arabinoside present at the high temperature. When the final survivors were cloned, Thompson's group found that the cells could grow into colonies at 33°C with about 80 percent efficiency, but at 39°C they formed colonies at much lower efficiencies, ranging from 0.001 to 0.1 percent. Of course, the parental L-cell line could form colonies equally well at both temperatures. One clone, *ts*-Al, was temperature sensitive for DNA synthesis but not for RNA or protein

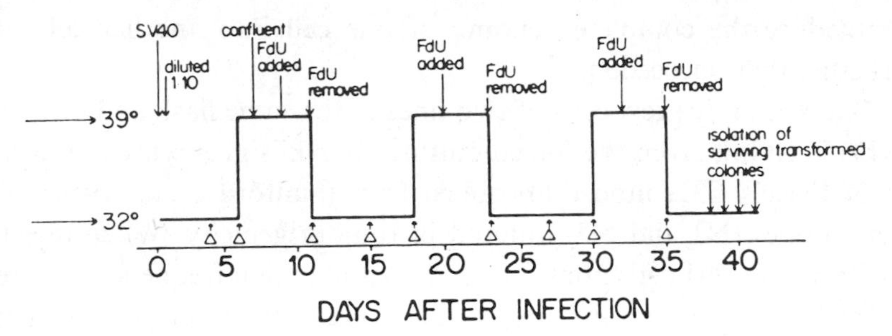

Figure 2.17. Schematic outline of the selection procedure used to isolate *ts*-SV3T3 cells. Medium containing 10% calf serum was changed at the times indicated (△). Reprinted with permission from Renger and Basilico, *Proc. Nat. Acad. Sci.* 69, 109, 1972.

synthesis. A similar selection has been carried out on BHK cells using FdU at the high temperature. In these experiments, too, cells were recovered that were temperature sensitive for cell division at any cell density (Miess and Basilico, 1972).

Recently a protocol of temperature cycling and density-dependent negative selection with FdU on SV40-transformed mouse cells (SV3T3) (Figure 2.17) resulted in a set of cell lines that seem to be temperature sensitive for growth control rather than for growth (Renger and Basilico, 1972). These *ts*-SV3T3 cells grew equally well at 32°C and 39°C when they were kept sparse. As the cultures reached confluence, the one kept at 39°C stopped dividing, whereas the one kept at 32°C continued to increase in cell number until it reached ten times the cell density at confluence. The phenotype of these cells is similar to that of a BHK cell transformed by the mutant polyoma virus *ts*-3. Surprisingly, SV40 recovered from *ts*-SV3T3 cells is wild type, so the mutation is likely to be cellular despite the fact that the SV40 stock used to make SV3T3 for these experiments was first mutagenized with nitrosoguanidine.

HYBRIDIZATION OF MAMMALIAN SOMATIC CELLS

Films of living fibroblasts had shown that two cells could exchange cytoplasm or even fuse together, that the nuclei of fused cells divided synchronously, and that when this happened, at anaphase of the following mitotis all the chromosomes would assemble into one nucleus (Hsu, 1960). Barski et al. (1960, 1961) therefore tried deliberately to introduce the complete genome of one cell into another cell. Remarkedly, they succeeded.

Barski simply grew cells of two lines in the same flask and obtained a hybrid cell line from the mixed culture. Barski's group used clones of two of Earle's C3H mouse fibroblast lines (Sanford et al., 1954). The parental lines (N1 and N2) differed in tumourigenicity. When injected into C3H mice, N1 cells grew into tumours at a high frequency, whereas N2 cells made fewer tumours, which, furthermore, could be distinguished from N1 tumours. The two parental lines also had different karyotypes: N1 cells had 55 telocentric chromosomes and N2 cells had 60 chromosomes, 13 of which were metacentrics.

Mixtures of N1 and N2 cells were grown together in many flasks, and every few weeks a culture was assayed for tumourigenicity and chromosome counts were made of a few hundred cells. For the first two months cells with the N2 karyotype predominated, but then a new cell type, called M, appeared and soon overgrew the culture. M cells had 115 chromosomes, including between 9 and 15 metacentrics; in other words the chromosomes of M cells equaled the sum of the chromosomes of one N1 cell and one N2 cell. After six months the mixed cultures contained essentially only M and N2 cells; apparently the highly malignant N1 cells had been overgrown by a mixture of N2 cells and the putative hybrid M cells. This was fortunate, for it meant that any increase in the tumourigenicity of the cells in these cultures over the tumourigenicity of N2 cells could be ascribed to the N1-cell genes retained in the hybrid M cells.

In fact, after six months the mixed populations were more tumourigenic than N2 cells, and the tumours formed by M cells were round-cell sarcomas that were clearly different from both the spindle-cell sarcomas formed by N1 cells and the anaplastic small-cell tumours formed by N2 cells (Barski et al., 1961).

Barski's first hybrids were between two cell lines from the same strain of mouse. Sorieul and Ephrussi (1961) hybridized unrelated mouse cells, thereby showing that gross genotypic differences were no obstacle to the formation of hybrids. As in Barski's case, the hybrid cells had a selective advantage in vitro, and eventually they multiplied to become an appreciable fraction of the mixed culture.

HAT Selection for Hybrids

In 1964 Littlefield used Szybalski's HAT system as a selective technique to detect hybrids that would not otherwise have a growth advantage over the parents. He mixed $HGPRT^-$ L-cells (A9) and TK^- L-cells (CL1D), added HAT medium, and waited. Both parental cells were killed but some survivors grew to colonies; these were colonies of hybrid cells. The survival of these hybrids in HAT medium provided evidence of genetic complementation in hybrid cells, since at least one gene of each parent (the TK gene of A9 and the HGPRT gene of CL1D) must have been functioning to keep the hybrids alive (Kao et al., 1968).

The HAT system was also used by Ephrussi and his students to make hybrids from a mixture of two cell lines of different species (Weiss and Ephrussi, 1966; Scaletta et al., 1967; Davidson and Ephrussi, 1965, 1970). Weiss and Ephrussi used Littlefield's HAT-sensitive CL1D and A9 cells to make interspecific hybrids: they crossed each mouse line separately with wild-type, diploid rat cells. A great excess of the mouse cells was mixed with rat cells and placed in HAT medium. Most of the mouse cells were killed, but some hybridized with the rat cells and the hybrid cells overgrew the unhybridized, diploid rat cells (Weiss and Ephrussi, 1966) (Figure 2.18).

The first human/non-human hybrid cells were also made by Weiss (Weiss and Green, 1967), who again used cells of the CL1D mouse line (TK^-) and HAT medium. The diploid fibroblast line W138 was the human parental cell. All CL1D cells died in the HAT medium, but dense colonies of hybrid cells grew out from the W138 monolayer; the hybrids were sustained in HAT medium by the presence of a functional human TK gene. Since most human chromosomes can be distinguished from mouse chromosomes, the karyotypes of these hybrids showed unambiguously that chromosomes of both parents were present.

Hybrids and Genetic Mapping

The human/mouse hybrid cells that Weiss and Green isolated had a remarkable peculiarity: whereas the chromosome number of most other interspecific hybrid cells was close to the sum of the chromosome numbers of the two parental cells, the human/mouse hybrid cells had

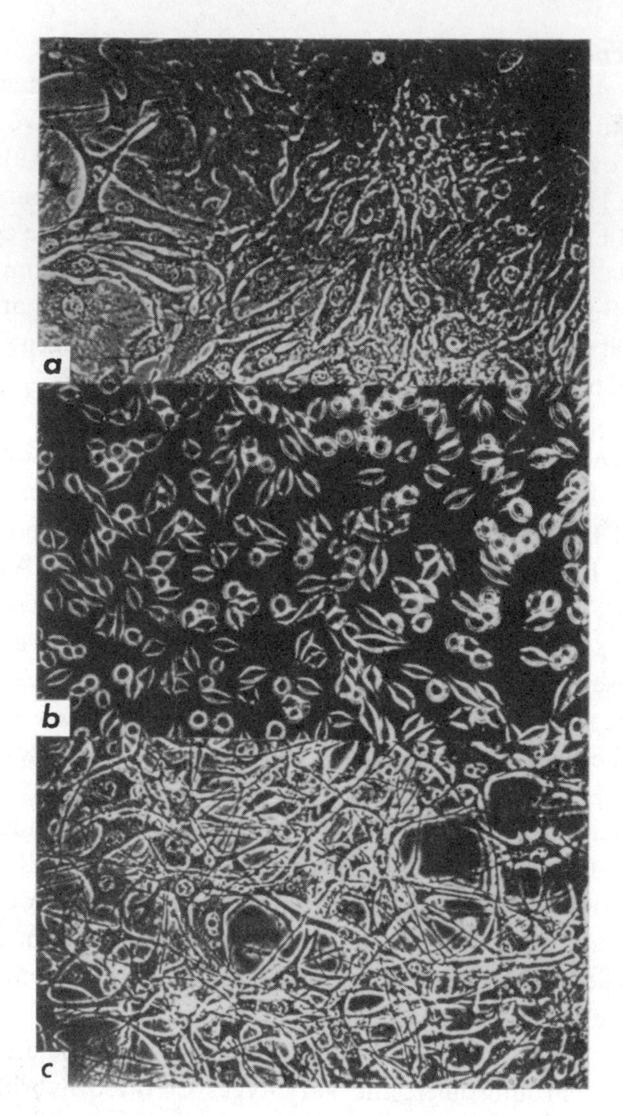

Figure 2.18. **(a)** Photomicrograph of living rat cells R_1. Notice the flat appearance of the cells. **(b)** Living cells of the mouse cell line, 1 D. Notice the highly refractile appearance of the cells due to non-spreading on the plastic surface. **(c)** Photomicrograph of living cells of MAT. The morphology of these hybrid cells is clearly intermediate between the parents, R_1 and 1 D, shown above. All three photographs were taken with phase contrast. Reprinted with permission from Weiss and Ephrussi, *Genetics* 54, 1095, 1966.

twice the expected number of mouse chromosomes, and no hybrid cell had more than about 20 human chromosomes. Weiss and Green grasped the significance of this and cultured the hybrids for a few months, carefully monitoring their chromosome constitution. Eventually all but one human chromosome disappeared from hybrid cells cultured in HAT medium, and the remaining human chromosome, which was always a small metacentric, was rapidly lost when the hybrids were put in medium containing BrdU. Their conclusion, later verified by enzyme assays, was that the human gene for TK was localized on this small metacentric chromosome, later identified as chromosome E_{17} (Migeon and Miller, 1968; Miller et al., 1971; Green et al., 1971). Thus, not only complementation, but also genetic mapping was shown to be possible without recourse to the usual genetic crosses (Green, 1969; Kao et al., 1969, Littlefield and Goldstein, 1970; Kusano et al., 1971).

The new techniques that reveal specific patterns of bands on individual chromosomes should permit rapid identification of human chromosomes in hybrid lines. First attempts have been successful (Caspersson et al., 1971a) but the full force of the technique has not yet been felt.

Littlefield and Marin used the HAT selection system to isolate revertants of hamster cells transformed by polyoma virus, apparently by causing the transformed cells to lose all the polyoma genes. They made hamster/hamster hybrid cells by fusing cells of TK^- and $HGPRT^-$ BHK sublines in HAT medium. These hybrids were transformed with polyoma virus and gained the ability to grow in agar. In time the tetraploid transformants shed their excess hamster chromosomes, and concurrently some colonies of revertants appeared that could no longer grow in agar. Such hybrid revertants could be retransformed by polyoma virus, they had lost viral T antigen, and they were less tumourigenic than transformed hybrids that could still grow in agar (Marin and Littlefield, 1968). Clearly using this experimental approach it may be possible to identify the hamster cell chromosome(s) into which polyoma virus DNA integrates during transformation.

Since 1967 very many hybrid cells have been isolated and studied (see Table 2.8 and reviews by Defendi, 1969, 1971).

Table 2.8. Expression of Parental Genes in Hybrid Cells

Function	Specific Gene(s)	Expression in Hybrid	Reference
Differentiated products	Collagen	+	Green et al. (1966)
	Hyaluronate	+	Green et al. (1966)
	Melanin, dopaoxidase	−	Davidson and Yamamoto (1968)
		−	Silagi (1967)
	ES 2 kidney esterase	−	Klebe et al. (1970)
	Serum albumin	+	Schneider and Weiss (1971)
	Growth hormone	−	Sonnenschein et al. (1968)
	S-100 protein	±	Benda and Davidson (1971)
	Macrophage properties (cell membrane ATPase, surface receptors, phagocytosis in heterokaryons)	−	Gordon and Cohn (1970)
	Immunoglobulin	−	Coffino et al. (1971)
Uninduced level of inducible enzymes	Tyrosine amino transferase	+	Schneider and Weiss (1971)
	Glycerol-3-phosphate dehydrogenase	+	Davidson and Benda (1970)
Induced level of inducible enzymes	Tyrosine amino transferase	−	Schneider and Weiss (1971)
	Glycerol-3-phosphate dehydrogenase	−	Davidson and Benda (1970)
Virus-induced products	Polyoma antigens (T antigen + transplant)	+	Defendi et al. (1964) Defendi et al. (1967)
	SV40 T antigen	+	Weiss et al. (1968a)
	Interferon synthesis	+	Carver et al. (1968)
Growth control (failure to produce tumours)		+	Harris et al. (1969)
Growth control		+	Klein et al. (1971)
Growth control		+	Bregula et al. (1971)
Growth control		+	Wiener et al. (1971)

Adapted from a table by H. Green (unpublished).

Table 2.8.—*Continued*

Function	Expression	Reference
Growth control	−	Barski et al. (1961)
Growth control	−	Scaletta and Ephrussi (1965)
Growth control	−	Silagi (1967)
Growth control (contact inhibition to culture)	+	Weiss et al. (1968b)
Capacity to respond to interferon	+	Carver et al. (1968)
Electrical excitability	+	Minna et al. (1971)
Permissiveness for viral functions	+	Wang et al. (1970)
Permissiveness for viral functions	+	Basilico et al. (1970)
Non-permissiveness for viral functions	−	Pollack et al. (1971)
Non-permissiveness for viral functions	−	Basilico et al. (1970)
Capacity to differentiate (teratoma)	−	Finch and Ephrussi (1967)

Heterokaryons

The frequency of appearance of hybrid colonies in a HAT selective system is about one to ten colonies per million parental cells. This low frequency of emergence of viable hybrids is probably the product of some number of infrequent intermediate steps in hybrid formation, and the first of these must be fusion of the parental cells. This first step can be enhanced so that cells with two or more nuclei can be made at will from mixtures of different cells. Cells with more than one nucleus are called polykaryons, and if the nuclei come from cells of different species, such cells are called heterokaryons.

Cell Fusion Mediated by Sendai Virus

In 1962 Okada reported that the myxovirus HVJ (Sendai) would fuse mouse tumour cells. Three years later Harris and Watkins used UV irradiation to kill the Sendai virus and found that the inactivated virus would still induce the fusion of cells of the same or different species. After labeling the nuclei of one population of parental cells with tritiated thymidine, Harris and Watkins (1965) showed by autoradiography that heterokaryons containing labeled and unlabeled nuclei, as well as polykaryons, were formed. In 1966 Yerganian and Nell showed that Harris and Watkins's Sendai fusion technique also increased the frequency of formation of viable hybrids in a mixture of cells from two species of hamsters. A few years later Coon and Weiss (1969a, b) compared hybrid formation with and without Sendai, and found that adding inactivated Sendai virus increased the frequency of viable hybrids about 100-fold, to about 10^{-3} of the mixed cells.

Sendai-mediated fusion is now a routine technique, but it is worrisome to use the virus as adjuvant, not least because Sendai virus is usually grown in chick embryos and may well become contaminated with avian leukemia viruses (Enders and Neff, 1968). Fortunately, Koprowski et al. have recently shown that lysolecithin can fuse cells, thus obviating the need for adding an extraneous, even if dead, virus (Croce et al., 1971).

Heterokaryons do not divide as such, presumably since division requires both nuclei to be in phase at mitosis, and when they are, the likelihood is that all chromosomes will be found within one nuclear membrane at anaphase. Such a cell then becomes a hybrid rather than a heterokaryon.

Genetic Dominance

Hybrids and heterokaryons are useful for studies of dominance of gene expression in differentiation; one can take a differentiated and a non-differentiated cell, fuse them, and ask whether or not the resulting heterokaryons, or after some weeks hybrid cells, are differentiated. As expected, no absolute rule about expression of parental genes has emerged; usually, however, if one parental cell secretes a specialized compound, the secretion product is made by at least some hybrids. Heterokaryons have also been used to show that DNA synthesis is initiated by a trans-dominant factor, because the nuclei of synchronized G_1 cells begin to make DNA as soon as the cells are fused with cells in S phase (Johnson and Rao, 1971).

The production of relatively pure populations of enucleated cells with the drug cytochalasin B1 (Carter, 1967; Prescott et al., 1972) (see Chapter 3) offers the possibility of creating a degenerate hybrid cell with nucleus and cytoplasm of two different origins (Poste and Reeve, 1972; Ladda and Estensen, 1970). Apparently enucleated monolayers retain a great deal of biochemical integrity, for infectious poliovirus can be recovered after infection of a pure (>99%) enucleated population of BSC1 cells (Pollack and Goldman, 1973).

Virus Expression

One can infect cells of a hybrid line with viruses that are lytic or transforming for one or the other parental cell and ask which parental phenotype dominates. Basilico and Green have shown that the infection of hamster/mouse hybrids with polyoma virus is fully productive, and they conclude that the failure of hamster cells to support replication of polyoma virus probably does not result from the presence of a soluble repressor (Basilico et al., 1970). Provided that they retain most of the parental chromosomes, hybrid cells seem to have host ranges and susceptibilities equal to the sum of those of their parents. An interesting exception to this is the finding that the nonpermissive mouse genes permit synthesis of SV40 T antigen but block the replication of SV40 in primate/mouse hybrids (Swetly et al., 1969; Pollack et al., 1971; Knowles et al., 1971). And Basilico has described a single hamster/mouse hybrid clone that encapsidates cellular DNA at an abnormally high rate (Basilico and Burstin, 1971).

Recovery of SV40 from Transformed 3T3 Cells

As early as 1962 Gerber was able to demonstrate the presence of SV40 in SV40-induced hamster brain tumours (Gerber and Kirschstein, 1962) by cocultivating tumour cells with permissive monkey kidney cells. If, however, the tumour cells were killed before cocultivation, no virus was recovered. In 1967 Koprowski et al. showed that the efficiency of recovery of SV40 from transformed cells was increased significantly by using Sendai virus to mediate the fusion of the transformed cells with permissive monkey kidney cells; and Watkins and Dulbecco (1967) found that with this procedure it was possible to recover SV40 from transformed 3T3 cells from which the virus could not be recovered by any other means.

This remarkable discovery, besides showing that the entire SV40 genome was present in the transformed cell, suggests a novel selection for mutant SV40. Possibly there are SV40 genes that can be expressed only in transformed cells and not in the lytic cycle; mutations in such genes could be detected by mutagenesis of the transformed cell followed by recovery of the virus from the transformants. The mutants could then be assayed in untransformed nonpermissive cells.

The Malignancy of Hybrids

Are hybrids between malignant and normal cells malignant or normal? Since there are many different spontaneously and virus-transformed lines and many primary and established nonmalignant lines, a clear answer might be expected from cell hybridization experiments; but in fact controversy surrounds this problem even after five years of work. Initially malignancy appeared to be dominant; the first hybrids isolated by Barski, Sorieul and Cornifert arose spontaneously from a mixture of a highly malignant, variant L-cell subline and a weakly malignant, variant L-cell subline. The hybrids were almost as malignant as the highly malignant parents but gave tumours that differed histologically from those of the malignant parent.

Ephrussi and his colleagues showed that malignancy persists in hybrids derived from crosses between malignant L-cells and normal mouse cells and between polyoma-transformed and nontransformed mouse cells (Scaletta and Ephrussi, 1965; Defendi et al., 1967). The interpretation Ephrussi's group placed on these data at the time was that the abnormal ability of a cell to grow into a lethal tumour was not suppressible by any contribution from a nonmalignant cell.

However, heteroploid cell lines and hybrid cells are genetically unstable, and most hybrids lose some chromosomes with time in culture (Marin and Littlefield, 1968; Green, 1969). This instability prevents any simple interpretation of experiments showing dominance of malignancy in hybrids; if the hybrid cells were to lose some critical chromosomes of the normal parent during the course of their isolation, cloning, amplification and injection, then their malignancy would not prove that normal cells lack genes capable of suppressing the malignant phenotype.

In a massive series of studies using inactivated Sendai virus to generate many dozens of hybrids between the highly malignant Ehrlich ascites cell and a series of L-cell derivatives or normal diploid cells,

Harris and Klein have recently shown that tumours arising from hybrid cell lines are always composed of hybrid cells with fewer chromosomes than the sum of the parental chromosomes (Klein et al., 1971; Bregula et al., 1971; Wiener et al., 1971). This group concluded that a hybrid cell has little capacity for progressive growth in vivo if it retains the complete chromosome set of both parents, since they did not see a single tumour whose cells contained all of the parental chromosomes.

Most phenotypic revertants of virus-transformed cells show an increase in chromosome number, whereas their malignant back-revertants have lost chromosomes and returned to the subtetraploid mode (Rabinowitz and Sachs, 1968, 1970; Pollack et al., 1970; Hitotsumachi et al., 1971; Wyke, 1971a, b; Culp et al., 1971). These revertants resemble nonmalignant hybrids and premalignant cervical carcinoma cells (Kirkland and Stanley, 1971) in that loss of chromosomes accompanies loss of growth control in all these cases.

CONCLUSION

Table 2.9 lists the properties of some cell lines that have been or should be especially useful for studying cell-virus interactions.

In a series of lectures on Novel Genetic Systems given in 1956, four years before Barski's discovery of hybrid mammalian cells, the geneticist Pontecorvo spoke of the usefulness of filamentous fungi. These fungi regularly hybridize and shed chromosomes, thus bypassing the sexual cycle in what Pontecorvo termed a parasexual cycle. In concluding he said, "The colony (of fungi) is far from limited to mutation in a haploid clone as its only means of genetic variation. It can receive and absorb immigrant nuclei and cytoplasm from neighbouring colonies, it can try out all sorts of combinations and ratios of different nuclei for vegetative adaptation, it can shelter recessives both in heterokaryotic condition and in heterozygotic conditions, and it can try out all sorts of combinations of genes in diploid and haploid condition" (Pontecorvo, 1958).

Pontecorvo, of course, had no reason to mention mammalian cells in his lecture, but with hybrids and heterokaryons, mammalian cells too can safely be said to have at least the foundation of a parasexual

Table 2.9. Cell Lines of Interest

Part I. History of Cells

	Origin				
	Species	Strain	Tissue	Derivation	Original Reference
3T3	mouse	Swiss	embryo	post-crisis	Todaro and Green (1963)
A9	mouse	C3H	connective	line L 929	Littlefield (1963)
B10	cow	?	lens	post-crisis	Macintyre and Pontén (1967)
Balb 3T3	mouse	Balb/c	embryo	post-crisis	Aaronson and Todaro (1968)
BHK21	hamster	Golden	kidney	long-term culture	Stoker and Macpherson (1964)
BSC1	monkey	African green	kidney	long-term culture	Hopps et al. (1963)
C1300	mouse	?	brain	neuroblastoma, alternate passage	Augusti-Tocco and Sato (1969)
CHO	hamster	Syrian	ovary	post-crisis	Kao and Puck (1968)
Clone 1 D	mouse	C3H	connective	line L 929	Hsu and Somers (1962)
CV-1	monkey	?	kidney	?	Jensen et al. (1964)
Daudi	human	?	lymphatic	Burkitt Lymphoma	Klein et al. (1968)
HeLa	human	?	cervix	carcinoma, long term	Gey et al. (1952)
L 929	mouse	C3H	connective	culture in serum clot	Sanford et al. (1948)
L60I-A1	mouse	C3H	connective	line L 929	Thompson et al. (1970)
L.N.F.	human	?	skin	Strain: Lesch-Nyhan male	Seegmiller et al. (1967)
MPC-11	mouse	?	marrow	myeloma	Laskov and Scharff (1970)
RAG	mouse	Balb/c	kidney	adenocarcinoma, alternate passage	Klebe et al. (1970)
RECL-1	rat	?	skin	strain	Petursson et al. (1964)
RPC-5	mouse	?	marrow	myeloma	Mohit and Fan (1971)
RPH-2A	frog	?	egg	long-term culture	Freed and Metzger-Freed (1970)
WI38	human	?	skin	strain	Hayflick and Moorhead (1961)
Y1	mouse	?	adrenal cortex	adrenal tumour, alternate passage	Yasumura et al. (1966)

These cell lines are listed in alphabetical order. Some cells have given rise to many derivative lines. These are not listed if their common names carry the name of the parental line. Thus, an SV40-transformed 3T3 line, SV3T3, is not listed.

Table 2.9. Cell Lines of Interest

Part II. Growth Properties

	Characteristics			Retains Growth Control assayed by				Karyotype	
	grows without serum	plating efficiency	grows in suspension	serum	density	attach	tumour	ploidy	mode
3T3	no	50%	no	+	+	+	?	hetero	65
A9	yes	90%	yes	−	−	−	−	hetero	55
B10	no	low	no	+	+	+	+	eu	60
Balb3T3	no	50%	no	+	+	+	+	hetero	?
BHK21	no	20%	yes	+	−	+	+	di	44
BSC1	no	low	no	+	+	+	?	di	42?
C1300	yes	80%	yes	−	−	−	−	hetero	?
CHO	no	95%	yes	+	−	+	?	pseudo-di	21, 22
Clone 1 D	yes	90%	yes	−	−	−	−	hetero	53
CV-1	no	low	no	+	+	+	?	?	?
Daudi	no	none	yes	+	−	−	−	di	46
HeLa	yes	95%	yes	−	−	−	−	hetero	65
L	yes	90%	yes	−	−	−	−	hetero	?
L60I-A1	no	yes	yes	?	−	−	−	hetero	?
L.N.F.	no	low	no	+	+	+	+	eu	46
MPC-11	no	none	yes	?	−	−	−	hetero	?
RAG	no	yes	?	?	−	−	−	pseudo-di	45
RECL-1	no	yes	no	+	+	+	+	pseudo-di	42
RPC-5	no	none	yes	−	−	−	−	hetero	?
RPH-2A	no	10%	no (?)	+	+	+	+	haploid!	13
WI38	no	low	no	+	+	+	+	di	46
Y1	no	yes	yes	+	−	−	−	pseudo-di	39

Table 2.9. CELL LINES OF INTEREST

Part III. Phenotype

	Response to Tumour Virus*		Phenotype		
	Transformed by	Killed by	Morphology	Nutritional	Mutant Sublines
3T3	a, b	b	fibroblast	proto	$APRT^-$, TK^-, $HGPRT^-$
A9	?	?	fibro	$HGPRT^-$	none
B10	f	?	epithelial	proto	none
Balb3T3	a, b, e	b	fibro	proto	none
BHK21	b, f	?	fibro	proto	$HGPRT^-$, TK^-
BSC1	a	a	epith	proto	none
C1300	?	?	makes neurones	proto	none
CHO	?	?	epith	proto	Pro^-, Gly^-, $Folate^-$
Clone 1 D	?	?	fibro	TK^-	none
CV-1	?	b	epith	proto	none
Daudi	d	?	lympho	proto	none
HeLa	?	a	epith	proto	$HGPRT^-$, TK^-
L	?	?	fibro	proto	A9, Clone 1 D, L60I-A1
L60I-A1	?	?	TS-growth	proto	none
L.N.F.	a	?	fibro	$HGPRT^-$	none
MPC-11	?	?	makes IgG	proto	none
RAG	?	?	epith	$HGPRT^-$	none
RECL-1	b	?	fibro	proto	none
RPC-5	?	?	makes IgG	proto	none
RPH-2A	g (?)	?	epith	proto	none
WI38	a, h, i	?	fibro	proto	none
Y1	?	?	ACTH → steroids	proto	cAMP → steroids

* The viruses are: (a) SV40; (b) polyoma; (c) adenovirus; (d) herpes II (EB); (e) murine LV, SV; (f) avian LV, SV; (g) Lucké frog virus; (h) herpesvirus (squirrel monkey); (i) rat sarcoma virus.

genetics, and it is clear that one major use of this parasexual genetic system will be to analyze by mutation and hybridization the genetics of the transformed state (Pontecorvo, 1971).

Literature Cited

AARONSON, S. A. and G. J. TODARO. 1968. Basis for the acquisition of malignant potential by mouse cells cultivated *in vitro*. Science *162:* 1024.

ABERCROMBIE, M. and E. AMBROSE. 1958. Interference microscope studies of cell contacts in tissue culture. Exp. Cell Res. *15:* 332.

ABERCROMBIE, M. and J. HEAYSMAN. 1954. Observations on the social behavior of cells in tissue culture. Exp. Cell Res. *6:* 293.

ANDERSON, E., D. PETERSEN and R. TOBEY. 1970. Density invariance of cultured Chinese hamster cells with stage of the mitotic cycle. Biophys. J. *10:* 630.

ARMSTRONG, D. 1966. Serial cultivation of human leukemic cells. Proc. Soc. Exp. Biol. Med. *122:* 475.

ARRIGHI, F. and T. HSU. 1971. Localization of heterochromatin in human chromosomes. Cytogenics *10:* 81.

AUGUSTI-TOCCO, G. and G. SATO. 1969. Establishment of functional clonal lines of neurons from mouse neuroblastoma. Proc. Nat. Acad. Sci. *64:* 311.

BARSKI, G., S. SORIEUL and F. CORNEFERT. 1960. Production dans des cultures *in vitro* de deux souches cellulaires en association, de cellules de caractère "hybride." Comptes Rendus *251:* 1825.

———. 1961. "Hybrid" type cells in combined cultures of two different mammalian cell strains. J. Nat. Cancer Inst. *26:* 1269.

BASILICO, C. and S. J. BURSTIN. 1971. Multiplication of polyoma virus in mouse-hamster somatic hybrids: A hybrid cell line which produces viral particles containing predominantly host DNA. J. Virol. *7:* 802.

BASILICO, C., Y. MATSUYA and H. GREEN. 1970. The interaction of polyoma virus with mouse-hamster somatic hybrid cells. Virology *41:* 295.

BENDA, P. and R. DAVIDSON. 1971. Regulation of specific functions of glial cells in somatic hybrids. I. Control of S 100 proteins. J. Cell. Physiol. *78:* 209.

BENDA, P., J. LIGHTBODY, G. SATO, L. LEVINE and W. SWEET. 1968. Differentiated rat glial cell strain in tissue culture. Science *161:* 370.

BERNAL, J. D. 1971. Science in History. Vol. I: The Emergence of Science, p. 39. MIT Press, Cambridge, Mass.

BIESELE, J. 1958. Mitotic Poisons and the Cancer Problem, p. 121–126. Elsevier Publishing Co., New York.

BIRCH, J. and S. PIRT. 1971. The quantitative glucose and mineral nutrient requirement of mouse LS (suspension) cells in chemically defined medium. J. Cell Sci. *8:* 693.

BISKIND, M. S. and G. S. BISKIND. 1944. Development of tumors in the rat ovary after transplantation into the spleen. Proc. Soc. Exp. Biol. Med. *58:* 176.

BLACK, P. 1966. Transformation of mouse cell line 3T3 by SV40: Dose response relationship and correlation with SV40 tumor antigen production. Virology *28:* 760.

BOOTSMA, D., L. BUDKE and O. VOS. 1964. Studies on synchronous division of tissue culture cells initiated by excess thymidine. Exp. Cell Res. *33:* 301.

BOYLE, J. and K. RAIVIO. 1970. Lesch-Nyhan syndrome: Preventive control by prenatal diagnosis. Science *169:* 688.

BREGULA, U., G. KLEIN and H. HARRIS. 1971. The analysis of malignancy by cell fusion. II. Hybrids between Ehrlich cells and normal mouse cells. J. Cell Sci. *8:* 673.

BROCKMAN, R. W. and P. STUTTS. 1960. A mechanism of resistance to 6-thioguanine. Fed. Proc. *19:* 313.

BRODER, S. W., P. R. GLADE and K. HIRSCHHORN. 1970. Establishment of long-term lines from small aliquots of normal lymphocytes. Blood *35:* 539.

BUONASSISI, V., G. SATO and A. I. COHEN. 1962. Hormone-producing cultures of adrenal and pituitary tumor origin. Proc. Nat. Acad. Sci. *48:* 1184.

BÜRK, R. 1968. Reduced adenyl cyclase activity in a polyoma virus-transformed cell line Nature *219:* 1272.

CAMERON, G. 1935. Essentials of Tissue Culture Technique. Farrar & Rinehart, New York.

CANNON, W. 1929. Organization for physiological homeostasis. Physiol. Rev. *9:* 399.

CARREL, A. 1912. The permanent life of tissue outside of the organism. J. Exp. Med. *15:* 516.

———. 1913. Artificial activation of the growth *in vitro* of connective tissue. J. Exp. Med. *17:* 14.

CARTER, S. B. 1967. Effects of cytochalasins on mammalian cells. Nature *213:* 261.

CARVER, I., D. SETO and B. MIGEON. 1968. Interferon production and action in mouse, hamster and somatic hybrid mouse-hamster cells. Science *160:* 558.

CASPERSSON, T., L. ZECH and C. JOHANSSON. 1970a. Analysis of human metaphase chromosome set by aid of DNA-binding fluorescent agents. Exp. Cell Res. *62:* 490.

CASPERSSON, T., L. ZECH, C. JOHANSSON and E. J. MODEST. 1970b. Identification of human chromosomes by DNA-binding fluorescent agents. Chromosoma *30:* 215.

CASPERSSON, T., M. HULTEN, J. LINDSTEN, A. THERKELSEN and L. ZECH. 1971a. Identification of different Robertsonian translocations in man by quinacrine mustard fluorescence analysis. Hereditas *67:* 213.

CASPERSSON, T., L. ZECH, H. HARRIS, F. WIENER and G. KLEIN. 1971b. Identification of human chromosomes in a mouse-human hybrid by fluorescence techniques. Exp. Cell Res. *65:* 475.

CECCARINI, C. and H. EAGLE. 1971a. pH as a determinant of cellular growth and contact inhibition. Proc. Nat. Acad. Sci. *68:* 229.

———. 1971b. Induction and reversal of contact inhibition of growth by pH modification. Nature New Biol. *233:* 271.

CHEN, T. and C. HEIDELBERGER. 1969. Quantitative studies on malignant transformation of mouse prostate cells by carcinogenic hydrocarbons *in vitro*. Int. J. Cancer *4:* 166.

CHOI, K. W. and A. D. BLOOM. 1970. Cloning human lymphocytes *in vitro*. Nature *227:* 171.

CHU, E. H. Y. and N. H. GILES. 1958. Comparative chromosomal studies on mammalian cells in culture. I. The HeLa strain and its mutant clonal derivatives. J. Nat. Cancer Inst. *20:* 383.

CLARK, J. L., K. L. JONES, D. GOSPODAROWICZ and G. H. SATO. 1972. Growth response to hormones by a new rat ovary cell line. Nature New Biol. *236:* 180.

CLARKE, G., M. STOKER, A. LUDLOW and M. THORNTON. 1970. Requirements of serum for DNA synthesis in BHK21 cells: Effects of density, suspension and virus transformation. Nature *227:* 798.

COFFINO, P., B. KNOWLES, S. NATHENSON and M. SCHARFF. 1971. Suppression of immunoglobulin synthesis by cellular hybridization. Nature New Biol. *231:* 87.

COHEN, A. I., J. FURTH and R. F. BUFFETT. 1957. Histologic and physiologic characteristics of hormone-secreting transplantable adrenal tumors in mice and rats. Amer. J. Pathol. *33:* 631.

COON, H. G. and M. C. WEISS. 1969a. A quantitative comparison of formation of spontaneous and virus-produced viable hybrids. Proc. Nat. Acad. Sci. *62:* 852.

COON, H. and M. WEISS. 1969b. Sendai-produced somatic cell hybrids between L-cell strains and between liver and L cells. Wistar Inst. Symp. Monogr. *9:* 83.

CROCE, C. M., W. SAWICKI, D. KRITCHEVSKY and H. KOPROWSKI. 1971. Induction of homokaryocyte, heterokaryocyte and hybrid formation by lysolecithin. Exp. Cell Res. *67:* 427.

CULP, L., W. GRIMES and P. BLACK. 1971. Contact-inhibited revertant cell lines isolated from SV40-transformed cells. I. Biologic, virologic and chemical properties. J. Cell Biol. *50:* 682.

DAVIDSON, R. and P. BENDA. 1970. Regulation of specific functions of glial cells in somatic hybrids. II. Control of inducibility of glycerol-3-phosphate dehydrogenase. Proc. Nat. Acad. Sci. *67:* 1870.

DAVIDSON, R. and B. EPHRUSSI. 1965. A selective system for the isolation of hybrids between L cells and normal cells. Nature *205:* 1169.

———. 1970. Factors influencing the "effective mating rate" of mammalian cells. Exp. Cell Res. *61:* 222.

DAVIDSON, R. and K. YAMAMOTO. 1968. Regulation of melanin synthesis in mammalian cells as studied by somatic hybridization. Proc. Nat. Acad. Sci. *60:* 894.

DAVIDSON, R., H. NITOWSKY and B. CHILDS. 1963. Demonstration of two populations of cells in the human female heterozygous for glucose-6-phosphate dehydrogenase variants. Proc. Nat. Acad. Sci. *50:* 481.

DAVIS, B. 1948. Isolation of biochemically deficient mutants of bacteria by penicillin. J. Amer. Chem. Soc. *70:* 4267.

DEFENDI, V., ed. 1969. Heterospecific Genome Interaction. Wistar Inst. Symp. Monogr. *9.*

———. 1971. Pathology society symposium on mammalian cell hybridization. Fed. Proc. *30:* 192.

DEFENDI, V. and M. STOKER. 1973. General polyploidisation produced by cytochalasin B. Nature New Biol. In press.

DEFENDI, V., B. EPHRUSSI and H. KOPROWSKI. 1964. Expression of polyoma-induced cellular antigen(s) in hybrid cells. Nature *203:* 495.

DEFENDI, V., B. EPHRUSSI, H. KOPROWSKI and M. YOSHIDA. 1967. Properties of hybrids between polyoma-transformed and normal mouse cells. Proc. Nat. Acad. Sci. *57:* 299.

DEROBERTIS, E., G. ARNAZ, M. ALBERICI, R. BUTCHER and E. SUTHERLAND. 1967. Subcellular distribution of adenyl cyclase and phosphodiesterase in rat brain cortex. J. Biol. Chem. *242:* 3487.

DIPAOLO, J., P. DONOVAN and R. NELSON. 1971. *In vitro* transformation of hamster cells by polycyclic hydrocarbons; factors influencing the number of cells transformed. Nature New Biol. *230:* 240.

DJORDJEVIC, B. and W. SZYBALSKI. 1960. Genetics of human cell lines. III. Incorporation of 5-bromo- and 5-iododeoxyuridine into the DNA of human cells and its effect on radiation sensitivity. J. Exp. Med. *112:* 509.

DUBBS, D. and S. KIT. 1964. Effect of halogenated pyrimidines and thymidine on growth of L cells and a subline lacking thymidine kinase. Exp. Cell Res. *33:* 19.

DULBECCO, R. 1952. Production of plaques in monolayer tissue cultures by single particles of an animal virus. Proc. Nat. Acad. Sci. *38:* 747.

DULBECCO, R. 1970a. Topoinhibition and serum requirement of transformed and untransformed cells. Nature *227:* 802.

———. 1970b. Behavior of tissue culture cells infected with polyoma virus. Proc. Nat. Acad. Sci. *67:* 1214.

DULBECCO, R. and M. STOKER. 1970. Conditions determining initiation of DNA synthesis in 3T3 cells. Proc. Nat. Acad. Sci. *66:* 204.

DUPRAW, E. 1970. DNA and Chromosomes. Holt & Rinehart, New York.

EAGLE, H. 1955a. The specific amino acid requirements of a human carcinoma cell strain HeLa in tissue culture. J. Exp. Med. *102:* 37.

———. 1955b. The specific amino acid requirements of a mammalian cell (strain L) in tissue culture. J. Biol. Chem. *214:* 839.

———. 1955c. The minimum vitamin requirement of the L and HeLa cells in tissue culture. The production of specific vitamin deficiencies and their cure. J. Exp. Med. *102:* 595.

———. 1955d. Nutrition needs of mammalian cells in tissue culture. Science *122:* 501.

———. 1956. The salt requirements of mammalian cells in tissue culture. Arch. Biochem. Biophys. *61:* 356.

———. 1959. Amino acid metabolism in mammalian cell cultures. Science *130:* 432.

———. 1960. The sustained growth of human and animal cells in a protein-free environment. Proc. Nat. Acad. Sci. *46:* 427.

———. 1971. Buffer combinations for mammalian cell culture. Science *174:* 500.

EAGLE, H. and K. PIEZ. 1962. Population-dependent requirement by cultured mammalian cells for metabolites which they can synthesize. J. Exp. Med. *116:* 29.

EAGLE, H., K. PIEZ and V. OYAMA. 1961. The biosynthesis of cystine in human cell cultures. J. Biol. Chem. *236:* 1425.

EAGLE, H., S. BARBAN, M. LEVY and H. O. SCHULZE. 1958. The utilization of carbohydrates by human cell cultures. J. Biol. Chem. *233:* 551.

EAGLE, H., C. WASHINGTON, M. LEVY and L. COHEN. 1966. The population-dependent requirement by cultured mammalian cells for metabolites which they can synthesize. J. Biol. Chem. *241:* 4994.

EAGLE, H., G. E. FOLEY, H. KOPROWSKI, H. LAZARUS, E. M. LEVINE and R. A. ADAMS. 1970. Growth characteristics of virus-transformed cells. Maximum population density, growth in soft agar, and xenogeneic transplantability. J. Exp. Med. *131:* 863.

EARLE, W. 1943. Production of malignancy *in vitro*. IV. The mouse fibroblast cultures and changes seen in the living cells. J. Nat. Cancer Inst. *4:* 165.

EARLE, W. R., J. C. BRYANT, E. L. SCHILLING and V. J. EVANS. 1956. Growth of cell suspensions in tissue culture. Ann. N. Y. Acad. Sci. *63:* 666.

EASTY, G. and V. MUTOLO. 1960. The nature of the intercellular material of adult mammalian tissues. Exp. Cell Res. *21:* 374.

ELY, J. O. and J. H. GRAY. 1961. Chromosome number of *in vivo* and *in vitro* cultured Krebs-2 carcinoma of mice: The selective property of the *in vitro* culture medium. Cancer Res. *21:* 1002.

ENDERS, J. and J. NEFF. 1968. Cytopathogenicity in monolayer hamster cell cultures fused with β-propiolactone-inactivated Sendai virus. Proc. Soc. Exp. Biol. Med. *127:* 260.

EVANS, H. J., K. E. BUCKTON and A. T. SUMNER. 1971. Cytological mapping of human chromosomes: Results obtained with quinacrine fluorescence and the acetic-saline-Giemsa techniques. Chromosoma *35:* 310.

EVANS, V., J. BRYANT, M. FIORAMONTI, W. MCQUILKAN, K. SANFORD and W. EARLE. 1956. Studies of nutrient media for tissue cells *in vitro*. Cancer Res. *16:* 77; 87.

FAHEY, J. L., I. FINEGOLD, A. S. RABSON and R. A. MANAKER. 1966. Immunoglobulin synthesis *in vitro* by established human cell lines. Science *152:* 1259.

FINCH, B. and B. EPHRUSSI. 1967. Retention of multiple development of potentialities by cells of a mouse testicular terato-carcinoma during prolonged culture *in vitro* and their extinction upon hybridization with cells of permanent lines. Proc. Nat. Acad. Sci. *57:* 615.

FINE, R. and D. BRAY. 1971. Actin in growing nerve cells. Nature New Biol. *234:* 115.
FIRKET, H. and W. VERLY. 1958. Autoradiographic visualization of synthesis of DNA in tissue culture with tritium-labeled thymidine. Nature *181:* 274.
FISCHER, A. 1946. The Biology of Tissue Cells. G. E. Stechert Co., New York.
———. 1948. Amino acid metabolism of tissue cells *in vitro*. Biochem. J. *43:* 491.
FISHER, H. W., T. T. PUCK and G. SATO. 1958. Molecular growth requirements of single mammalian cells. Proc. Nat. Acad. Sci. *44:* 4.
FOLLETT, E. and R. GOLDMAN. 1970. The occurrence of microvilli during spreading and growth of BHK21/C13 fibroblasts. Exp. Cell Res. *59:* 124.
FREED, J. and L. METZGER-FREED. 1970. Stable haploid cultured cells from frog embryos. Proc. Nat. Acad. Sci. *65:* 337.
FREEMAN, A., P. PRICE, R. BRYAN, R. GORDON, R. GILDEN, G. KELLOFF and R. HUEBNER. 1971. Transformation of rat and hamster embryo cells by extracts of city smog. Proc. Nat. Acad. Sci. *68:* 445.
FURTH, J. 1953. Conditioned and autonomous neoplasms: A review. Cancer Res. *13:* 477.
———. 1967. Pituitary cybernetics and neoplasia. Harvey Lectures *63:* 47. Academic Press, New York.
FURTH, J. and K. CLIFTON. 1957. Experimental pituitary tumors and the role of pituitary hormones in tumourigenesis of the breast and thyroid. Cancer *10:* 842.
FURTH, J. and H. SOBEL. 1947. Neoplastic transformation of granulosa cells in grafts of normal ovaries into spleens of gonadectomized mice. J. Nat. Cancer Res. *8:* 7.
GAGNÉ, R., R. TANGUAY and C. LABERGE. 1971. Differential staining patterns of heterochromatin in man. Nature New Biol. *232:* 29.
GALL, J. G. and M. L. PARDUE. 1969. Formation and detection of RNA-DNA hybrid molecules in cytological preparations. Proc. Nat. Acad. Sci. *63:* 378.
GEORGE, K. 1970. Cytochemical differentiation among human chromosomes. Nature *226:* 80.
GERBER, P. and R. KIRSCHSTEIN. 1962. SV40-induced ependymomas in newborn hamsters. Virology *18:* 582.
GEY, G. 1954. Some aspects of the constitution and behavior of normal and malignant cells maintained in continuous culture. Harvey Lectures *50:* 154. Academic Press, New York.
GEY, G., W. COFFMAN and M. KUBICECK. 1952. Tissue culture studies of the proliferative capacity of cervical carcinoma and normal epithelium. Cancer Res. *12:* 264.
GLADE, P. R. and S. W. BRODER. 1971. Preparation and care of established human lymphoid cell lines. *In* In Vitro Methods in Cell-Mediated Immunity, p. 561. Academic Press, New York.
GOLDSBY, R. and E. ZIPSER. 1969. The isolation and replica plating of mammalian cell clones. Exp. Cell Res. *54:* 271.
GORDON, S. and Z. COHN. 1970. Macrophage-melanocyte heterokaryons. J. Exp. Med. *131:* 981.
GREEN, H. 1969. Prospects for the chromosomal localization of human genes in human-mouse somatic cell hybrids. Wistar Inst. Symp. Monogr. *9:* 51.
GREEN, H., B. EPHRUSSI, M. YOSHIDA and D. HAMERMAN. 1966. Synthesis of collagen and hyaluronic acid by fibroblast hybrids. Proc. Nat. Acad. Sci. *55:* 41.
GREEN, H., R. WANG, O. KEHINDE and M. MEUTH. 1971. Multiple human TK chromosomes in human-mouse somatic hybrids. Nature New Biol. *234:* 138.
HAKOMORI, S. I. 1970. Cell density-dependent changes of glycolipid concentrations in fibroblasts, and loss of its response in virus-transformed cells. Proc. Nat. Acad. Sci. *67:* 1741.
HAKOMORI, S. I. and W. T. MURAKAMI. 1968. Glycolipids of hamster fibroblasts and derived malignant-transformed cell lines. Proc. Nat. Acad. Sci. *59:* 254.

HAM, R. G. 1962. Clonal growth of diploid Chinese hamster cells in a synthetic medium supplemented with purified protein fractions. Exp. Cell Res. *28:* 489.

———. 1963. Albumin replacement by fatty acids in clonal growth of mammalian cells. Science *140:* 802.

———. 1965. Clonal growth of mammalian cells in a chemically defined synthetic medium. Proc. Nat. Acad. Sci. *53:* 288.

HARRIS, H. and J. F. WATKINS. 1965. Hybrid cells derived from mouse and man: Artificial heterokaryons of mammalian cells from different species. Nature *205:* 640.

HARRIS, H., O. MILLER, G. KLEIN, P. WORST and T. TACHIBANA. 1969. Suppression of malignancy by cell fusion. Nature *223:* 363.

HARRIS, M. 1964. Variation in chromosome patterns. *In* Cell Culture and Somatic Variation, Chap. 4. Holt, Rinehart & Winston, New York.

———. 1971. Mutation rates in cells at different ploidy levels. J. Cell. Physiol. *78:* 177.

HARRISON, R. G. 1907. Observations on the living developing nerve fiber. Proc. Soc. Exp. Biol. Med. *4:* 140.

HAUSCHKA, T. S., S. T. GRINNELL, L. REVESZ and G. KLEIN. 1957. Quantitative studies on the multiplication of neoplastic cells *in vivo*. IV. Influence of doubled chromosome number on growth rate and final population size. J. Nat. Cancer Inst. *19:* 13.

HAYFLICK, L. and P. MOORHEAD. 1961. The serial cultivation of human diploid cell strains. Exp. Cell Res. *25:* 585.

HITOTSUMACHI, S., Z. RABINOWITZ and L. SACHS. 1971. Chromosomal control of reversion in transformed cells. Nature *231:* 511.

HOLLAENDER, A., ed. 1958. Symposium on genetic approaches to somatic cell variation. J. Cell. Comp. Physiol. *52:* suppl. 1.

HOLLEY, R. and J. KIERNAN. 1968. "Contact-inhibition" of cell division in 3T3 cells. Proc. Nat. Acad. Sci. *60:* 300.

HOPPS, H., B. BERNHEIM, A. NISALAK, J. TJIO and J. SMADEL. 1963. Biological characteristics of a continuous kidney cell line derived from the African green monkey. J. Immunol. *91:* 416.

HOWARD, A. and S. PELC. 1953. Synthesis of DNA in normal and irradiated cells and its relation to chromosome breakage. Heredity (suppl.) *6:* 261.

HSIE, A. W. and T. T. PUCK. 1971. Morphological transformation of Chinese hamster cells by dibutyryl adenosine cyclic 3′,5′-monophosphate and testosterone. Proc. Nat. Acad. Sci. *68:* 358.

HSIE, A. W., C. JONES and T. T. PUCK. 1971. Further changes in differentiation state accompanying the conversion of Chinese hamster cells to fibroblastic form by dibutyryl adenosine cyclic 3′,5′-monophosphate and hormones. Proc. Nat. Acad. Sci. *68:* 1648.

HSU, T. C. 1952. Mammalian chromosomes *in vitro*. I. The karyotype of man. J. Hered. *43:* 167.

———. 1955. Mammalian chromosomes *in vitro*. VI. Observation on mitosis with phase cinematography. J. Nat. Cancer Inst. *16:* 691.

———. 1960. Generation time of HeLa cells determined from line records. Texas Rep. Biol. Med. *18:* 31.

———. 1961. Chromosomal evolution in cell populations. Int. Rev. Cytol. *12:* 69.

———. 1965. Genetic cytology. *In* Cells and Tissues in Culture (E. Willmer, ed.), Vol. 1, p. 396. Academic Press, New York.

HSU, T. C. and K. BENIRSCHKE. 1967. An Atlas of Mammalian Chromosomes. Springer-Verlag, New York.

HSU, T. C. and O. KLATT. 1958. Mammalian chromosomes *in vitro*. IX. On genetic polymorphism in cell populations. J. Nat. Cancer Inst. *21:* 437.

HSU, T. and C. POMERAT. 1953. Mammalian chromosomes *in vitro*. II. A method for spreading the chromosomes of cells in tissue culture. J. Hered. *44:* 23.

HSU, T. C. and C. E. SOMERS. 1962. Properties of L-cells resistant to 5-bromodeoxyuridine. Exp. Cell Res. *26:* 404.

HSU, T. C., D. BILLEN and A. LEVAN. 1961. Mammalian chromosomes *in vitro*. XV. Patterns of transformation. J. Nat. Cancer Inst. *27:* 515.

HUBERMAN, J. and G. ATTARDI. 1966. Isolation of metaphase chromosomes from HeLa cells. J. Cell Biol. *31:* 95.

HUGHES, A. 1952. Some effects of abnormal tonicity on dividing cells in chick tissue cultures. Quart. J. Microscop. Sci. *93:* 207.

Human Chromosome Study Group. 1960. A proposed standard system of nomenclature of human mitotic chromosomes. J. Hered. *11*(5): 214.

JAINCHILL, J. L. and G. J. TODARO. 1970. Stimulation of cell growth *in vitro* by serum with and without growth factor; relation to contact inhibition and viral transformation. Exp. Cell Res. *59:* 137.

JENSEN, F., A. GIRARDI, R. GILDEN and H. KOPROWSKI. 1964. Infection of human and Simian tissue cultures with Rous sarcoma virus. Proc. Nat. Acad. Sci. *52:* 53.

JOHNSON, G. S. and I. PASTAN. 1972. The role of 3′,5′-adenosine monophosphate in the regulation of morphology and growth of transformed and normal fibroblasts. J. Nat. Cancer Inst. *48:* 1377.

JOHNSON, G., R. FRIEDMAN and I. PASTAN. 1971. Restoration of several morphological characteristics of normal fibroblasts in sarcoma cells treated with adenosine-3′,5′ cyclic monophosphate and its derivatives. Proc. Nat. Acad. Sci. *68:* 425.

JOHNSON, R. and P. RAO. 1971. Nucleo-cytoplasmic interactions in the achievement of nuclear synchrony in DNA synthesis and mitosis in multinucleate cells. Biol. Rev. *46:* 97.

JONES, K. W. and G. CORNEO. 1971. Location of satellite and homogeneous DNA sequences on human chromosomes. Nature New Biol. *233:* 268.

KAO, F. and T. T. PUCK. 1967. Genetics of somatic mammalian cells. IV. Properties of Chinese hamster cell mutants with respect to the requirement for proline. Genetics *55:* 513.

———. 1968. Genetics of somatic mammalian cells. VII. Induction and isolation of nutritional mutants in Chinese hamster cells. Proc. Nat. Acad. Sci. *60:* 1275.

———. 1969. Genetics of somatic mammalian cells. IX. Quantitation of mutagenesis by physical and chemical agents. J. Cell. Physiol. *74:* 245.

KAO, F., L. CHASIN and T. T. PUCK. 1969. Genetics of somatic mammalian cells. X. Complementation analysis of glycine-requiring mutants. Proc. Nat. Acad. Sci. *64:* 1284.

KAO, F., R. JOHNSON and T. PUCK. 1968. Complementation analysis on virus-fused Chinese hamster cells with nutritional markers. Science *164:* 312.

KIRKLAND, J. A. and M. A. STANLEY. 1971. Chromosomes of cancer cells. Nature *232:* 632.

KIRKLAND, J., M. STANLEY and K. CELLIER. 1967. Comparative study of histologic and chromosomal abnormalities in cervical epithelium. Cancer *20:* 1934.

KIT, S., D. R. DUBBS, L. J. PIETARSKI and T. C. HSU. 1963. Deletion of thymidine kinase activity from L-cells resistant to bromodeoxyuridine. Exp. Cell Res. *31:* 297.

KLEBE, R., C. CHEN and F. RUDDLE. 1970. Mapping of a human genetic regulator element by somatic cell genetic analysis. Proc. Nat. Acad. Sci. *66:* 1220.

KLEIN, E., G. KLEIN, J. NADKARNI, H. WIGZELL and P. CLIFFORD. 1968. Surface IgM Kappa specificity on a Burkett lymphoma cell *in vivo* and in derived culture lines. Cancer Res. *28:* 1300.

KLEIN, G., U. BREGULA, F. WIENER and H. HARRIS. 1971. The analysis of malignancy by cell fusion. I. Hybrids between tumor cells and L-cell derivatives. J. Cell Sci. *8:* 659.

KNOWLES, B., G. BARBANTI-BRENTANO and H. KOPROWSKI. 1971. Susceptibility of primate-mouse hybrid cells to SV40. J. Cell. Physiol. *78:* 1.

KOPROWSKI, H., F. JENSEN and Z. Steplewski. 1967. Activation of production of infectious tumor SV40 in heterokaryon cultures. Proc. Nat. Acad. Sci. *58:* 127.

KRAEMER, P., D. PETERSEN and M. VANDILLA. 1971. DNA constancy in heteroploidy and the stem line theory of tumors. Science *174:* 714.

KROOTH, R. and E. SELL. 1970. The action of Mendelian genes in human diploid cell strains. J. Cell. Physiol. *76:* 311.

KUSANO, T., C. LONG and H. GREEN. 1971. A new reduced human-mouse somatic cell hybrid containing the human gene for adenine phosphoribosyl transferase. Proc. Nat. Acad. Sci. *68:* 82.

LADDA, R. L. and R. D. ESTENSEN. 1970. Introduction of a heterologous nucleus into enucleated cytoplasms of cultured mouse L-cells. Proc. Nat. Acad. Sci. *67:* 1528.

LASKOV, R. and M. SCHARFF. 1970. Synthesis, assembly and secretion of gamma globulin by mouse myeloma cells. J. Exp. Med. *131:* 515.

LEDERBERG, J. and N. ZINDER. 1948. Concentration of biochemical mutants of bacteria with penicillin. J. Amer. Chem. Soc. *70:* 4267.

LEVAN, A. 1956. Chromosome studies on some human tumors and tissues of normal origin, grown *in vivo* and *in vitro* at the Sloan-Kettering Institute. Cancer *9:* 648.

LIPTON, A., I. KLINGER, D. PAUL and R. HOLLEY. 1971. Migration of mouse 3T3 fibroblasts in response to a serum factor. Proc. Nat. Acad. Sci. *68:* 2799.

LITTLEFIELD, J. W. 1963. The inosinic acid pyrophosphorylase activity of mouse fibroblasts partially resistant to β-azaguanine. Proc. Nat. Acad. Sci. *50:* 568.

———. 1964. Selection of hybrids from matings of fibroblasts *in vitro* and their presumed recombinants. Science *145:* 709.

———. 1965. Studies on thymidine kinase in cultured mouse fibroblasts. Biochim. Biophys. Acta *95:* 14.

LITTLEFIELD, J. and S. GOLDSTEIN. 1970. Some aspects of somatic cell hybridization. In Vitro *6:* 21.

LOCKHART, R. Z., JR., and H. EAGLE. 1959. Requirements for growth of single human cells. Science *129:* 252.

LOMHOLT, B. and J. MOHR. 1971. Human karyotyping by heat-Giemsa staining and comparison with fluorochrome techniques. Nature New Biol. *234:* 109.

LURIA, S. and M. DELBRÜCK. 1943. Mutations of bacteria from virus sensitivity to virus resistance. Genetics *28:* 491.

LWOFF, A., L. SIMINOVITCH and N. KJELDGAARD. 1950. Induction de la production de bacteriophages chez une bacteria lysogene. Ann. Inst. Pasteur *79:* 815.

LYON, M. 1971. Possible mechanism of X chromosome inactivation. Nature NB *232:* 229.

MACINTYRE, E. and J. PONTÉN. 1967. Interaction between normal and transformed fibroblasts in culture. I. Cells transformed by Rous sarcoma virus. J. Cell Sci. *2:* 309.

MACPHERSON, I. 1965. Reversion in hamster cells transformed by Rous sarcoma virus. Science *148:* 1731.

MACPHERSON, I. and L. MONTAGNIER. 1964. Agar suspension culture for the selective assay of cells transformed by polyoma virus. Virology *23:* 291.

MAIO, J. J. and C. L. SCHILDKRAUT. 1966. A method for the isolation of mammalian metaphase chromosomes. Methods in Cell Physiol. *2:* 113. Academic Press, New York.

———. 1967. Isolated mammalian metaphase chromosomes. I. General characteristics of nuclei acids and proteins. J. Mol. Biol. *24:* 29.

MAIO, J. and C. SCHILDKRAUT. 1969. Isolated mammalian metaphase chromosomes. II. Fractionated chromosomes of mouse and Chinese hamster cells. J. Mol. Biol. *40:* 203.

MANAKER, R. and V. GROUPÉ. 1956. Discrete foci of altered chicken embryo cells associated with Rous sarcoma virus in tissue culture. Virology *2:* 838.

MARIN, G. and J. LITTLEFIELD. 1968. Selection of morphologically normal cell lines from Py BHK hybrids. J. Virol. *2:* 69.

MCCREDIE, K. B., E. M. HERSH and E. J. FREIREICH. 1971. Cells capable of colony formation in the peripheral blood of man. Science *171:* 293.

MCLIMMANS, W. F., E. V. DAVIS, F. L. GLOVER and G. W. RAKE. 1957. The submerged culture of mammalian cells: The spinner culture. J. Immunol. *79:* 428.

MERZ, G. and J. ROSS. 1969. Viability of human diploid cells as a function of *in vitro* age. J. Cell. Physiol. *74:* 219.

MIESS, H. and C. BASILICO. 1972. Isolation of temperature-sensitive mutants of BHK21 cells. Nature New Biol. *239:* 66.

MIGEON, B. and C. MILLER. 1968. Human-mouse somatic cell hybrids with single human chromosomes (group E): Link with thymidine kinase activity. Science *162:* 1005.

MIGEON, B., V. DERKALONSTIAN, W. HYHAN, W. YOUNG and B. CHILDS. 1968. X-linked hypoxanthine-guanine phosphoribosyl transferase deficiency: Heterozygote has two clonal populations. Science *160:* 425.

MILLER, O., P. ALLDERDICE, D. MILLER, W. BREY and B. MIGEON. 1971. Human thymidine kinase gene locus: Assignment to chromosome 17 in a hybrid of man and mouse cells. Science *173:* 244.

MILLER, R. and R. PHILLIPS. 1969. Separation of cells by velocity sedimentation. J. Cell. Physiol. *73:* 191.

MILO, G. E., J. P. SCHALLER and D. S. YOHN. 1972. Hormonal modification of adenovirus transformation of hamster cells *in vitro*. Cancer Res. *32:* 2338.

MINNA, J., P. NELSON, J. PEACOCK, D. GLAZER and M. NIRENBERG. 1971. Genes for neuronal properties expressed in neuroblastoma × L-cell hybrids. Proc. Nat. Acad. Sci. *68:* 234.

MOHIT, B. and K. FAN. 1971. Hybrid cell line from a cloned immunoglobulin producing mouse myeloma and a nonproducing mouse myeloma. Science *171:* 75.

MOORE, G. 1964. *In vitro* cultures of a pigmented hamster melanoma cell line. Exp. Cell Res. *36:* 422.

MOORE, G. E. and W. F. MCLIMMANS. 1968. The life span of the cultured normal cell: Concepts derived from studies of human lymphoblasts. J. Theoret. Biol. *20:* 217.

MOORE, G. E., R. E. GERNER and H. A. FRANKLIN. 1967. Culture of normal human leukocytes. J. Amer. Med. Ass. *199:* 519.

MOORE, G. E., H. KITAMURA and S. TOSHIMA. 1968. Morphology of cultured hematopoietic cells. Cancer *22:* 245.

MOORE, G. E., E. ITO, K. ULRICH and A. A. SANDBERG. 1966. Culture of human leukemia cells. Cancer *19:* 713.

MORA, P., F. CUMAR and R. BRADY. 1971. A common biochemical change in SV40 and polyoma virus-transformed mouse cells coupled to control of cell growth in culture. Virology *46:* 60.

MORA, P., R. BRADY, R. BRADLEY and V. MCFARLAND. 1969. Gangliosides in DNA virus-transformed and spontaneously transformed tumorigenic mouse cell lines. Proc. Nat. Acad. Sci. *63:* 1290.

MORGAN, J., H. MORTON and R. PARKER. 1950. Nutrition of animal cells in tissue culture. Proc. Soc. Exp. Biol. Med. *73:* 1.

MORTON, H. 1970. A survey of commercially available tissue culture media. In Vitro *6:* 89.

MOSCONA, A. 1952. Cell suspensions from organ rudiments of chick embryos. Exp. Cell Res. *3:* 535.

NAGLE, S. and B. BROWN. 1971. Improved heat-stable glutamine-free chemically defined medium for growth of mammalian cells. J. Cell. Physiol. *77:* 259.

NEWTON, A. and P. Wildy. 1959. Parasynchronous division of HeLa cells. Exp. Cell Res. *16:* 624.

NILHAUSEN, K. and H. GREEN. 1965. Reversible arrest of growth in G1 of an established fibroblast line (3T3). Exp. Cell Res. *40:* 166.

NOWELL, P. C. 1960. Phytohemagglutinin: An initiator of mitosis in cultures of normal human leukocytes. Cancer Res. *20:* 462.

O'BRIEN, J., S. OKADA, D. FILLERUP, M. VEATH, B. ADORNATO, P. BRENNAN and J. LEROY. 1971. Tay-Sachs disease: Prenatal diagnosis. Science *172:* 61.

OKADA, Y. 1962. Analysis of giant polynuclear cell formation caused by HVJ virus from Ehrlich's ascites tumor cells. Exp. Cell Res. *26:* 98.

O'RIORDAN, M., J. ROBINSON, K. BUCKTON and H. EVANS. 1971. Distinguishing between the chromosomes involved in Down s syndrome (Trisomy 21) and chronic myeloid leukemia (Ph^1) by fluorescence. Nature *230:* 167.

OWENS, O., M. K. GEY and G. O. GEY. 1954. Growth of cells in agitated fluid medium. Ann. N.Y. Acad. Sci. *58:* 1039.

OZANNE, B. and J. SAMBROOK. 1971. Isolation of lines of cells resistant to agglutination by concanavalin A from 3T3 cells transformed by SV40. *In* The Biology of Oncogenic Viruses (L. Silvestri, ed.), p. 248. North-Holland, Amsterdam.

PARDUE, M. and J. GALL. 1970. Chromosomal localization of mouse satellite DNA. Science *170:* 1356.

PASTAN, I. and R. L. PERLMAN. 1971. Cyclic AMP in metabolism. Nature New Biol. *229:* 5.

PASTERNAK, T., E. SUTHERLAND and W. HENION. 1962. Derivatives of cyclic 3′,5′-adenosine monophosphate. Biochim. Biophys. Acta *65:* 558.

PATIL, S., S. MERRICK and H. LUBS. 1971. Identification of each human chromosome with a modified Giemsa stain. Science *173:* 821.

PAUL, D., A. LIPTON and I. KLINGER. 1971. Serum factor requirements of normal and SV40-transformed 3T3 mouse fibroblasts. Proc. Nat. Acad. Sci. *68:* 645.

PAUL, J. 1970. Cell and Tissue Culture. E. Livingstone, London.

PEARSON, P., M. BOBROW and C. VOSA. 1970. Technique for identifying Y chromosomes in human interphase nuclei. Nature *226:* 78.

PERRY, C. V., G. S. JOHNSON and I. PASTAN. 1971. Adenyl cyclase in normal and transformed fibroblasts in tissue culture. Biochemistry *246:* 5785.

PETERSEN, D. F. and E. C. ANDERSON. 1964. Quantity production of synchronized mammalian cells in suspension culture. Nature *203:* 642.

PETERSEN, D., E. ANDERSON and R. TOBEY. 1968. Mitotic cells as a source of synchronized cultures. Meth. Cell Physiol. *3:* 347.

PETURSSON, G., J. COUGHLIN and C. MEYLAN. 1964. Long-term cultivation of diploid rat cells. Exp. Cell Res. *33:* 60.

PFEIFFER, S. E. and L. J. TOLMACH. 1967. Selecting synchronous populations of cells: Review of many methods. Nature *213:* 139.

POLLACK, R. and R. GOLDMAN 1973. Synthesis of infective poliovirus in BSCl monkey cells enucleated with cytochalasin B. Science. In press.

POLLACK, R. E. and M. M. BURGER. 1969. Surface-specific characteristics of a contact-inhibited cell line containing the SV40 genome. Proc. Nat. Acad. Sci. *62:* 1074.

POLLACK, R. E. and G. W. TEEBOR. 1969. Relationship of contact inhibition to tumor transplantability, morphology and growth rate. Cancer Res. *29:* 1770.

POLLACK, R., H. GREEN and G. TODARO. 1968. Growth control in cultured cells: Selection of sublines with increased sensitivity to contact inhibition and decreased tumor-producing ability. Proc. Nat. Acad. Sci. *60:* 126.

POLLACK, R., S. WOLMAN and A. Vogel. 1970. Reversion of virus-transformed cell lines: Hyperploidy accompanies retention of viral genes. Nature *228:* 938.

POLLACK, R., J. SALAS, R. WANG, T. KUSANO and H. GREEN. 1971. Human-mouse hybrid cell lines and susceptibility to species specific viruses. J. Cell. Physiol. *77:* 117.

PONTECORVO, G. 1958. Trends in Genetic Analysis. Columbia Univ. Press, New York.

———. 1971. Induction of directional chromosome elimination in somatic cell hybrids. Nature *230:* 367.

POSNER, M., J. NOVE and G. SATO. 1971. Selection procedure for mammalian cells in monolayer cultute. In Vitro *6:* 253.

POSTE, G. and P. REEVE. 1972. Enucleation of mammalian cells by cytochalasin B. II. Formation of hybrid cells and heterokaryons by fusion of anucleate and nucleated cells. Exp. Cell Res. *73:* 287.

POTTER, C. 1971. Induction of polyploidy by concentrated thymidine. Exp. Cell Res. *68:* 442.

PRESCOTT, D. 1970. The structure and replication of eukaryotic chromosomes. Adv. Cell Biol. *1:* 57.

PRESCOTT, D., D. MYERSON and J. WALLACE. 1972. Enucleation of mammalian cells with cytochalasin B. Exp. Cell Res. *71:* 480.

PUCK, T. T. 1958. Growth and genetics of somatic mammalian cells *in vitro*. J. Cell. Comp. Physiol. *52:* 287.

———. 1959. Quantitative studies on mammalian cells *in vitro*. Rev. Mod. Physiol. *31:* 433.

———. 1964a. Phasing, mitotic delay and chromosomal aberrations in mammalian cells. Science *144:* 565.

———. 1964b. Studies of the life cycle of mammalian cells. Cold Spring Harbor Symp. Quant. Biol. *29:* 167.

PUCK, T. T. and H. W. FISHER. 1956. Genetics of somatic mammalian cells. I. Demonstration of the existence of mutants with different growth requirements in a human cancer cell strain (HeLa). J. Exp. Med. *104:* 427.

PUCK, T. and F. KAO. 1967. Genetics of mammalian cells. V. Treatment with 5-bromodeoxyuridine and visible light for isolation of nutritionally deficient mutants. Proc. Nat. Acad. Sci. *58:* 1227.

PUCK, T. T. and P. I. MARCUS. 1955. A rapid method for viable cell titration and clone production with HeLa cells in tissue culture. Proc. Nat. Acad. Sci. *41:* 432.

———. 1956. Action of X rays on mammalian cells. J. Exp. Med. *103:* 653.

PUCK, T. T. and J. STEFFEN. 1963. Life cycle analysis of mammalian cells. Biophys. J. *3:* 379.

PUCK, T. T., S. J. CIECIURA and A. ROBINSON. 1958. Genetics of somatic mammalian cells. III. Long-term cultivation of euploid cells from human and animal subjects. J. Exp. Med. *108:* 945.

PUCK, T., P. MARCUS and S. CIECIURA. 1956. Clonal growth of mammalian cells *in vitro*. J. Exp. Med. *103:* 273.

RABINOWITZ, Z. and L. SACHS. 1968. Reversion of properties in cells transformed by polyoma virus. Nature *220:* 1203.

———. 1970. Control of the reversion of properties in transformed cells. Nature *225:* 136.

RALL, T. and E. SUTHERLAND. 1958. Formation of a cyclic adenine ribonucleotide by tissue particles. J. Biol. Chem. *232:* 1065.

RALL, T., E. SUTHERLAND and J. BERTHET. 1957. The relationship of epinephrine and glucagon to liver phosphorylase. J. Biol. Chem. *224:* 463.

RENGER, H. and C. BASILICO. 1972. Mutation causing temperature-sensitive expression of cell transformation by a tumor virus. Proc. Nat. Acad. Sci. *69:* 109.

REYES, P. and C. HEIDELBERGER. 1965. Fluorinated pyrimidine. XXVI. Mammalian thymidylate synthetase: Its mechanism of action and inhibition by fluorinated nucleotides. Mol. Pharmacol. *1:* 14.

RICHARDSON, U., A. TASHJIAN and L. LEVINE. 1969. Establishment of a clonal strain of hepatoma cells which secrete albumin. J. Cell Biol. *40:* 236.

RINALDINI, L. 1959. An improved method for the isolation and quantitative cultivation of embryonic cells. Exp. Cell Res. *16:* 477.

ROSE, W., R. WIXOM, H. LOCKHART and G. LAMBERT. 1955. The amino acid requirements of man. XV. The valine requirement: Summary and final observations. J. Biol. Chem. *217:* 987.

ROUS, P. and F. JONES. 1916. A method for obtaining suspensions of living cells of the fixed tissues and for the plating out of individual cells. J. Exp. Med. *23:* 546.

RUBIN, H. 1955. Quantitative relations between causative virus and cell in the Rous No. 1 chicken sarcoma. Virology *1:* 445.

———. 1971. Growth regulation in cultures of chick embryo fibroblasts. Ciba Found. Symp. Growth Control in Cell Cultures, p. 127 (G. E. W. Wolstenholme and J. Knight, ed.). Churchill Livingstone, London.

RUECKERT, R. and G. MUELLER. 1960. Studies on unbalanced growth in tissue culture. I. Induction and consequences of thymidine deficiency. Cancer Res. *20:* 1584.

RUSSELL, L. B. 1963. Mammalian X-chromosome action: Inactivation limited to spread and in region of origin. Science *140:* 976.

SACHS, L. 1967. An analysis of the mechanism of neoplastic cell transformation by polyoma virus, hydrocarbons, and X-irradiation. *In* Current Topics in Dev. Biol. (A. A. Moscona and A. Monroy, ed.), Vol. 2, p. 129. Academic Press, New York.

SACHS, L. and D. MEDINA. 1961. *In vitro* transformation of normal cells by polyoma virus. Nature *189:* 457.

SALZMAN, N. P. and J. MENDELSOHN. 1968. Isolation and fractionation of metaphase chromosomes. Meth. Cell Physiol. *3:* 277. Academic Press, New York.

SANFORD, K. K., W. R. EARLE and G. D. LIKELY. 1948. The growth *in vitro* of single isolated tissue cells. J. Nat. Cancer Inst. *9:* 229.

SANFORD, K., G. LIKELY and W. EARLE. 1954. The development of variations in transplantability and morphology within a clone of mouse fibroblasts transformed to sarcoma-producing cells *in vitro*. J. Nat. Cancer Inst. *15:* 215.

SATO, G. H. and V. BUONASSISI. 1963. Hormone-secreting cultures of endocrine tumor origin. Nat. Cancer Inst. Monogr. *13:* 81.

———. 1964. Hormone synthesis in dispersed cell cultures. Wistar Inst. Symp. Monogr. *1:* 27.

SATO, G. H. and Y. YASUMURA. 1966. Retention of differentiated function in dispersed cell culture. Trans. N.Y. Acad. Sci. *28:* 2063.

SATO, G., L. ZAROFF and S. MILLS. 1960. Tissue culture populations and their relation to the tissue of origin. Proc. Nat. Acad. Sci. *46:* 963.

SCALETTA, L. and B. EPHRUSSI. 1965. Hybridization of normal and neoplastic cells *in vitro*. Nature *205:* 1169.

SCALETTA, L., N. RUSHFORTH and B. EPHRUSSI. 1967. Isolation and properties of hybrids between somatic mouse and Chinese hamster cells. Genetics *57:* 107.

SCHER, C. D. and W. A. NELSON-REES. 1971. Direct isolation and characterization of "flat" SV40-transformed cells. Nature New Biol. *233:* 263.

SCHER, C. and G. TODARO. 1971. Selective growth of human neoplastic cells in medium lacking serum growth factor. Exp. Cell Res. *68:* 479.

SCHIMMER, B. 1969. Phenotypically variant adrenal tumor cell cultures with biochemical lesions in the ACTH-stimulated steroidogenic pathway. J. Cell. Physiol. *74:* 115.

SCHIMMER, B., K. VEDA and G. SATO. 1968. Site of action of ACTH in adrenal cell cultures. Biochem. Biophys. Res. Commun. *32:* 806.

SCHINDLER, R., L. RAMSEIER, J. SCHAERF and A. GRIEDER. 1970. Studies on the division cycle of mammalian cells. III. Preparation of synchronously dividing cell populations by isotonic sucrose gradient centrifugation. Exp. Cell Res. *59:* 90.

SCHNEDL, W. 1971a. Banding pattern of human chromosomes. Nature New Biol. *233:* 93.

———. 1971b. Analysis of the human karyotype using a reassociation technique. Chromosoma *34:* 448.

SCHNEIDER, J. A. and M. C. WEISS. 1971. Expression of differentiated functions in hepatoma cell hybrids. I. Tyrosine aminotransferase in hepatoma-fibroblast hybrids. Proc. Nat. Acad. Sci. *68:* 127.

SEEGMILLER, J. E., F. M. ROSENBLOOM and W. N. KELLEY. 1967. Enzyme defect associated with a sex-linked human neurological disorder and excessive purine synthesis. Science *155:* 1682.

SHALL, S. and A. MCCLELLAND. 1971. Synchronization of mouse fibroblast LS cells grown in suspension culture. Nature New Biol. *229:* 59.

SHANNON, J., W. EARLE and H. WALTS. 1952. Massive tissue cultures prepared from whole chick embryos planted as a cell suspension on glass substrate. J. Nat. Cancer Inst. *13:* 349.

SHEPPARD, J. R. 1971. Restoration of contact-inhibited growth to transformed cells by dibutyryl adenosine 3′,5′-cyclic monophosphate. Proc. Nat. Acad. Sci. *68:* 1316.

———. 1972. Difference in the cyclic adenosine 3′,5′-monophosphate levels in normal and transformed cells. Nature New Biol. *236:* 14.

SHIN, S. 1967. Studies of interstitial cells in tissue culture. Steroid biosynthesis in monolayers of mouse testicular interstitial cells. Endocrinology *81:* 440.

SHIN, S., Y. YASUMURA, L. LEVINE and G. SATO. 1968. Studies on interstitial cells in tissue culture. Steroid biosynthesis by a clonal line of rat testicular interstitial cell. Endocrinology *82:* 614.

SHIRAISHI, Y. 1970. The differential reactivity of human peripheral leukocyte chromosomes induced by low temperature. Japan J. Genet. *45:* 429.

SILAGI, S. 1967. Hybridization of a malignant melanoma cell line with L cells *in vitro*. Cancer Res. *27:* 1953.

SINCLAIR, D. and D. BISHOP. 1965. Synchronous culture of strain-L mouse cells. Nature *205:* 1272.

SISKEN, J. E. and R. KINOSITA. 1961. Timing of DNA synthesis in the mitotic cycle *in vitro*. J. Biophys. Biochem. Cytol. *9:* 509.

SITZ, T. O., A. B. KENT, H. A. HOPKINS and R. R. SCHMIDT. 1970. Equilibrium density-gradient procedure for selection of synchronous cells from asynchronous cultures. Science *168:* 1231.

SMITH, H. and C. SCHER. 1971. Cell division in medium lacking serum growth factor: Comparison of lines transformed by different agents. Nature *232:* 558.

SMITH, H., C. SCHER and G. TODARO. 1971. Induction of cell division in medium lacking serum growth factor by SV40. Virology *44:* 359.

SOBER, H., ed. 1968. Handbook of Biochemistry, p. H-58. Chemical Rubber Co., Cleveland, Ohio.

SOMERS, C., A. COLE and T. HSU. 1963. Isolation of chromosomes. Exp. Cell Res. *9* (suppl.): 220.

SONNENSCHEIN, C., A. TASHJIAN and U. RICHARDSON. 1968. Somatic cell hybridization: Mouse-rat hybrid cell line involving a growth-hormone producing parent. Genetics *60:* 227.

SORIEUL, S. and B. EPHRUSSI. 1961. Karyological demonstration of hybridization of mammalian cells *in vitro*. Nature *190:* 653.

STANNERS, C. P. and J. E. TILL. 1960. DNA synthesis in individual L-strain mouse cells. (^{3}H-TdR incorporation demonstration of discontinuous DNA synthesis). Biochim. Biophys. Acta *37:* 406.

STOKER, M. 1964. Regulation of growth and orientation in hamster cells transformed by polyoma virus. Virology *24:* 165.

———. 1967. Contact and short-range interactions affecting growth of animal cells in culture. *In* Current Topics in Dev. Biol. (A. A. Moscona and A. Monroy, ed.), Vol. 2, p. 108. Academic Press, New York.

STOKER, M. and P. ABEL. 1962. Conditions affecting transformation by polyoma virus. Cold Spring Harbor Symp. Quant. Biol. *27:* 375.

STOKER, M. and I. MACPHERSON. 1961. Studies on transformation of hamster cells by polyoma virus *in vitro*. Virology *14:* 359.

———. 1964. Syrian hamster fibroblast cell line BHK21 and its derivatives. Nature *203:* 1355.

STOKER, M., C. O'NEILL, S. BERRYMAN and V. WAXMAN. 1968. Anchorage and growth regulation in normal and virus-transformed cells. Int. J. Cancer *3:* 683.

STOLLAR, V., V. BUONASSISI and G. SATO. 1964. Studies on hormone-secreting adrenocorticoid tumor in tissue culture. Exp. Cell Res. *35:* 608.

SUMNER, A., H. EVANS and R. BUCKLAND. 1971. New technique for distinguishing between human chromosomes. Nature New Biol. *232:* 31.

SUSSMAN, M. 1969. Growth and Development, 2nd Ed. Prentice-Hall, Englewood Cliffs, New Jersey.

SUTHERLAND E. and T. RALL. 1958. Fractionation and characterization of a cyclic adenine ribonucleotide formed by tissue particles. J. Biol. Chem. *232:* 1077.

SUZUKI, F., M. KASHIMOTO and M. HORIKAWA. 1971. A replica plating method of cultured mammalian cells for somatic cell genetics. Exp. Cell Res. *68:* 476

SWETLY, P., G. BRODANO, B. KNOWLES and H. KOPROWSKI. 1969. Response of SV40-transformed cell lines and cell hybrids to superinfection with SV40 and its DNA. J. Virol. *4:* 348.

SWIFT, H. 1950. The DNA content of animal nuclei. Physiol. Zool. *23:* 169.

———. 1953. Quantitative aspects of nuclear nucleoproteins. Int. Rev. Cytol. *2:* 1.

SZYBALSKA, E. and W. SZYBALSKI. 1962. Genetics of human cell lines. IV. DNA-mediated hereditable transformation of a biochemical trait. Proc. Nat. Acad. Sci. *48:* 2026.

SZYBALSKI, W., E. H. SZYBALSKA and G. RAGNI. 1962. Genetic studies with human cell lines. Nat. Cancer Inst. Monogr. *7:* 75.

TASHJIAN, A. H., JR., Y. YASUMURA, L. LEVINE, G. H. SATO and M. L. PARKS. 1968. Establishment of clonal strains of rat pituitary tumor cells that secrete growth hormone. Endocrinology *82:* 342.

TEMIN, H. M. and H. RUBIN. 1958. Characteristics of an assay for Rous sarcoma virus and Rous sarcoma cells in tissue culture. Virology *6:* 669.

TERASIMA, T. and L. J. TOLMACH. 1963. Growth and nucleic acid synthesis in synchronously dividing populations of HeLa cells. Exp. Cell Res. *30:* 344.

THOMPSON, L. H., R. MANKOVITZ, R. M. BAKER, J. E. TILL, L. SIMINOVITCH and G. F. WHITMORE. 1970. Isolation of temperature-sensitive mutants of L-cells. Proc. Nat. Acad. Sci. *66:* 377.

TJIO, J. and A. LEVAN. 1956. The chromosome number of man. Hereditas *42:* 1.

TJIO, J. H. and T. T. PUCK. 1958. The somatic chromosomes of man. Proc. Nat. Acad. Sci. *44:* 1229.

TODARO, G. and H. GREEN. 1963. Quantitative studies on the growth of mouse embryo cells in culture and their development into established lines. J. Cell Biol. *17:* 299.

———. 1964. An essay for transformation by SV40. Virology *23:* 117.

———. 1966. High frequency of SV40 transformation of mouse cell line 3T3. Virology *28:* 756.

TODARO, G., H. GREEN and B. GOLDBERG. 1964. Transformation of properties of an established cell line by SV40 and polyoma virus. Proc. Nat. Acad. Sci. *51:* 66.

TODARO, G. J., G. K. LAZAR and H. GREEN. 1965. The initiation of cell division in a contact-inhibited mammalian cell line. J. Cell. Comp. Physiol. *66:* 325.

TODARO, G., Y. MATSUYA, S. BLOOM, A. ROBBINS and H. GREEN. 1967. Stimulation of RNA synthesis and cell division in resting cells by a factor present in serum. Wistar Inst. Symp. Monogr. *7:* 87.

VOGT, M. 1958. A genetic change in a tissue culture line of neoplastic cells. J. Cell. Comp. Physiol. *52:* (suppl. 1) 271.

———. 1959. A study of the relationship between karyotype and phenotype in cloned lines of strain HeLa. Genetics *44:* 1257.

VOGT, M. and R. DULBECCO. 1960. Virus-cell interaction with a tumor-producing virus. Proc. Nat. Acad. Sci. *46:* 365.

———. 1962. Properties of cells transformed by polyoma virus. Cold Spring Harbor Symp. Quant. Biol. *27:* 367.

WANG, H. C. and S. FEDEROFF. 1972. Banding in human chromosomes treated with trypsin. Nature New Biol. *235:* 52.

WANG, R., R. POLLACK, T. KUSANO and H. GREEN. 1970. Human-mouse hybrid cell lines and susceptibility to polio virus. I. Conversion from polio sensitivity to polio resistance accompanying loss of human gene-dependent polio receptors. J. Virol. *5:* 677.

WATKINS, J. and R. DULBECCO. 1967. Production of SV40 virus in heterokaryons of transformed and susceptible cells. Proc. Nat. Acad. Sci. *58:* 1369.

WAYMOUTH, C. 1955. Simple nutrient solutions for animal cells. Texas Rep. Biol. Med. *13:* 522.

WEISS, M. and B. EPHRUSSI. 1966. Studies of interspecific (rat and mouse) somatic hybrids. I. Isolation, growth and evolution of the karyotype. Genetics *54:* 1095.

WEISS, M. and H. GREEN. 1967. Human-mouse hybrid cell lines containing partial complements of human chromosomes and functioning human genes. Proc. Nat. Acad. Sci. *58:* 1104.

WEISS, M., B. EPHRUSSI and L. SCALETTA. 1968a. Loss of T-antigen from somatic hybrids between mouse cells and SV40-transformed human cells. Proc. Nat. Acad. Sci. *59:* 1132.

WEISS, M. C., G. J. TODARO and H. GREEN. 1968b. Properties of a hybrid between lines sensitive and insensitive to contact inhibition of cell division. J. Cell. Physiol. *71:* 105.

WESSELLS, N., B. SPOONER, J. ASH, M. BRADLEY, M. LUDUENA, E. TAYLOR, J. WRENN and K. YAMADA. 1971. Microfilaments in cellular and developmental processes. Science *171:* 135.

WIENER, F., G. KLEIN and H. HARRIS. 1971. The analysis of malignancy by cell fusion. I. Hybrids between diploid fibroblasts and other tumor cells. J. Cell Sci. *8:* 681.

WILLMER, E. 1933. Studies on the growth of tissues *in vitro*. III. An analysis of the growth of chick heart fibroblasts in flask cultures in a plasma coagulum. J. Exp. Biol. *10:* 340.

WYKE, J. 1971a. A method of isolating cells incapable of multiplication in suspension culture. Exp. Cell Res. *66:* 203.

———. 1971b. Phenotypic variation and its control in polyoma-transformed BHK21 cells. Exp. Cell Res. *66:* 209.

XEROS, N. 1962. Deoxyriboside control and synchronization of mitosis. Nature *194:* 682.

YASUMURA, Y., V. BUONASSISI and G. SATO. 1966. Clonal analysis of differentiated function in animal cell cultures. I. Possible correlated maintenance of differentiated function and the diploid karyotype. Cancer Res. *26:* 529.

YERGANIAN, G. and M. B. NELL. 1966. Hybridization of dwarf hamster cells by UV-inactivated Sendai virus. Proc. Nat. Acad. Sci. *55:* 1066.

ZAROFF, L., G. SATO and S. MILLS. 1961. Single cell platings from freshly isolated mammalian tissues. Exp. Cell Res. *23:* 565.

3

The External Surfaces of Cells in Culture

For an all too long period, it has been obvious that we should try to discover the molecular composition and architecture of the outer surfaces of cells. Forearmed with this knowledge we might begin to understand why a given cell can have only certain cells as neighbours and what goes wrong when a cell becomes cancerous and loses its normal affinities. But until recently most articles on cell surfaces went little beyond the seminal formulation of Gorter and Grendel (1925), namely, that the essence of a membrane was a bimolecular layer of lipids. Over the past several years however, membranes have become much more accessible to meaningful experimentation, and there seems to be no good reason why the essential molecular features of the surfaces of several cell types should not be elucidated within this decade.

CHEMICAL COMPOSITION OF MEMBRANES

An essential function of cell surface membranes is to maintain the large differences between the concentrations of molecules inside and outside cells. To perform this function membranes must be semi-permeable, if not effectively impermeable, to many compounds and this explains why membranes contain so much lipid. For lipids, which are not soluble in water, provide a barrier against the movement of water-soluble compounds. But at the same time cell surface membranes must allow the rapid entry into the cell of particular water-soluble molecules, for example, foods such as amino acids and simple sugars. To meet this demand enzymes have evolved which, situated in surface membranes, pump water-soluble molecules into the cell against concentration gradients by an energy-consuming process called "active transport." Other proteins in surface membranes confer on cells their antigenic identities and presumably some of the proteins of surface membranes confer on cells their tissue specificity and dictate where in the body they can reside.

The Unit Membrane

By the end of the 19th century, Overton (1895), after observing that fat-soluble molecules passed readily in and out of cells, postulated that the surface membranes of cells must be formed chiefly by water-insoluble lipids. A quarter of a century later, Fricke (1925), who measured the electrical capacitance of erythrocytes, calculated that the lipid layer at the cell surface could only be about 30 Å thick. In the same year Gorter and Grendel (1925) measured the absolute lipid content of erythrocyte membranes and concluded that there was enough lipid for each erythrocyte to be bounded by a lipid film only two molecules thick (i.e., 30–40 Å). Because many of the lipid chains in cell membranes carry phosphate groups at one end, Gorter and Grendel postulated that the molecules were arranged in two layers, with the hydrophilic phosphate groups on the outside and the hydrophobic lipid chains pointing inside (Figure 3.1). This arrangement was more precisely formulated by Danielli and Davson (1935, 1943), who added a refinement. They postulated that pleated sheets of membrane protein

Figure 3.1. A phospholipid bilayer: schematic cross-sectional view. The filled circles represent the ionic and polar head groups of the phospholipid molecules, which make contact with water; the wavy lines represent the fatty acid chains. Reprinted with permission from Singer and Nicolson, *Science* 175, 720, 1972.

Figure 3.2. Portion of a human red blood cell, fixed with permanganate and sectioned, showing the unit membrane structure bounding the cell. 210,000×. Reprinted with permission from J. D. Robertson, *Cellular Membranes in Development*, Academic Press, 1964.

were associated with the charged phosphate groups of the phospholipids. Later, to explain observed permeability constants, Stein and Danielli (1956) suggested that some proteins might form channels through the lipid bilayer to allow the passage of selected small molecules.

By the mid 1950's techniques for thin-sectioning cells for electron microscopy had been developed and new evidence in support of the lipid bilayer hypothesis of membrane structure was obtained (see reviews by Robertson, 1959, 1960, 1964). After appropriate staining, cross sections of virtually all cell membranes revealed two thin, electron-opaque lines, each about 20 Å thick, separated by about 35 Å (Figure 3.2). Such 75 Å thick images were called by Robertson "unit membranes" and were interpreted, "Danielli and Davson" fashion, as showing a lipid bilayer 35 Å thick, coated on its inner and outer surfaces by layers of membrane protein about 20 Å thick.

Sources of Membranes for Chemical Analysis

All membranes are made up of several species of lipids and proteins, and the reported relative proportions of protein and of lipid in

Table 3.1. Estimates of Percent Compositions of Some Membrane Preparations

Preparation[a]	Bovine myelin	Retinal ROS	Human RBC	Liver plasma membrane	Guinea pig brain synaptic vesicles	Rat liver micro-somes	Rabbit muscle micro-somes	Bovine heart mito-chondria	Bovine kidney mito-chondria	Guinea pig kidney mito-chondria
P	22	59	60	60	66	62	54	76	76	86
TL	78	41	40	40	34	32	22	24	24	—
PL	33	27	24	26	28	25	11	22.5	22	14
NL	26	2	9.2	13	5.6	7.5	—	1.5	1.9	—
PC	7.5	13	6.7	8	11.5	12	8.3	9.3	8.8	—
PE	11.7	6.5	3.4	—	4.2	4.8	1.4	8.4	8.4	3.9
PS	7.1	2.5	2.4	—	3.3	2.1	Trace	Trace	Trace	Trace
PI	0.6	0.4	Trace	—	1.3	2.5	Trace	0.75	0.75	1.0
CL	—	0.4	—	—	Trace	—	—	4.3	4.2	3.1
SM	6.4	0.5	3.6	4.0	3.0	1.5	0.8	—	—	—
AP	10.3	Trace	2.4	—	—	—	0.5	—	—	—
PA	—	0.8	—	—	Trace	—	—	—	—	—
TG	—	—	Trace	Trace	Trace	3.5	—	—	—	—
GL	22.0	9.5	Trace	—	Trace	—	—	Trace	Trace	Trace
CS	17.2	—	—	—	Trace	—	—	—	—	—
S	3.5	—	—	—	Trace	—	—	—	—	—
C	17.0	2.0	9.2	13.0	5.6	4.0	—	0.24	1.2	—

Adapted from Dewey and Barr, 1970.

The calculations used in the preparation of this table have often required the making of assumptions that could not be verified from the papers from which the data were drawn. Because of this, and because of the variations in experimental results, the reader is advised that the tabulated values should be considered only as estimates presented to emphasize the similarities and differences among the different preparations. Values are based on relationship of weight of component to dry weight of preparation.

[a] P, protein; TL, total lipid; PL, phospholipid; NL, neutral lipid; PC, phosphatidylcholine; PE, phosphatidylethanolamine; PS, phosphatidylserine; PI, phosphatidylinositol; CL, cardiolipid; SM, sphingomyelin; AP, acidic phosphatides; PA, phosphatidic acid; TG, triglyceride; GL, glycolipid; CS, cerebroside; S, sulfatide; C, cholesterol.

membranes from different sources differ markedly. Table 3.1 lists the putative composition of several membranes that have been studied in some detail. Unfortunately, it frequently is impossible to obtain a clean separation of the various types of membranes within a given cell, and so many of the reported differences in composition of membranes are impossible to interpret. Although it would be surprising if surface and internal membranes from different sources did not differ in their chemical compositions, often we have no way of telling which differences we read about are real and which are artifactual.

Fortunately this need not be the case when we consider work on mammalian red blood cell (erythrocyte) membranes. These cells, which lack both nuclei and internal membranes, can readily be hemolysed to yield ghosts which are essentially pure cell surface membranes. For this reason they have attracted the attention of many biochemists, and much of the precise knowledge that exists about membrane composition is a result of analysis of erythrocyte ghosts. Myelin sheaths around axons, which are formed from the surface membranes of Schwann cells (Geren, 1954), are another source of comparatively pure cell surface membranes. Yet another excellent source of pure membranes is the retina, which contains only one major protein constituent, the visual protein rhodopsin (Blasie et al., 1969; Blaurock and Wilkins, 1969; Heitzmann, 1972). An almost equally simplified membrane is the sarcoplasmic reticulum, the name given to the smooth endoplasmic reticulum of muscle (see review by Martonosi, 1971). This membranous system surrounds the myofibrils and is especially adapted for the binding and release of Ca^{++}. It contains only two proteins tightly bound in large amounts, a Ca^{++}-dependent ATPase and a proteolipid (MacLennan et al., 1973).

Unfortunately, though considerable work and ingenuity is now going toward developing methods for the purification of membranes from other types of cells (see review by Warren and Glick, 1971), most reported analyses of allegedly "pure surface membranes" from cells grown in culture cannot yet be taken at their face value. They usually can only tell us what sorts of molecules are present in large amounts in these membranes, and caution must be taken not to place undue importance on minor quantitative differences.

Membrane Lipids

Cell surface membranes, or plasma membranes as they were called by 19th century cytologists, contain significant amounts of four phospholipids: phosphatidylcholine, phosphatidylenthanolamine, phosphatidylserine and sphingomyelin (Figure 3.3). Most cell membranes also generally contain significant amounts of the cerebrosides, a group of glycolipids that have one to several "head" sugar residues (usually glucose and galactose) attached to a hydrophobic "tail" constructed from the long-chain amino alcohol sphingosine and a fatty acid with 24 carbon atoms. The most abundant of the cerebroside fatty acids are nervonic acid, cerebronic acid and lignoceric acid.

Many cell types, including most lines of cultured cells used in the studies of transformation, also contain considerable amounts of gangliosides. These highly complex glycolipids are formed from the cerebrosides by the addition of one to several molecules of *N*-acetylneuraminic acid (sialic acid) (Figure 3.4). Because of the variety of ways the carbohydrate substitutions can be made, a multitude of different gangliosides exists.

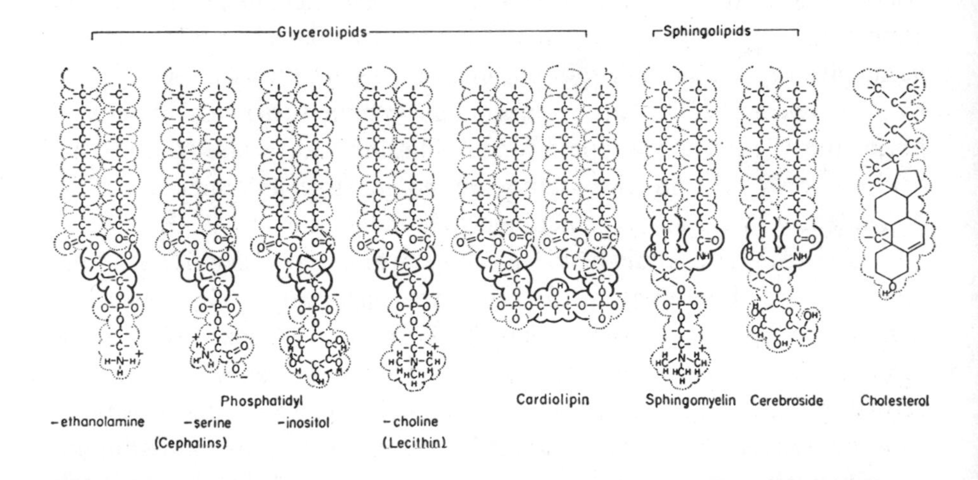

Figure 3.3. Structural formulas and approximate molecular outlines of some structurally important lipid molecules. Reprinted with permission from J. B. Finean, *Chemical Ultrastructure in Living Tissues*, Thomas Publishing Co., 1961.

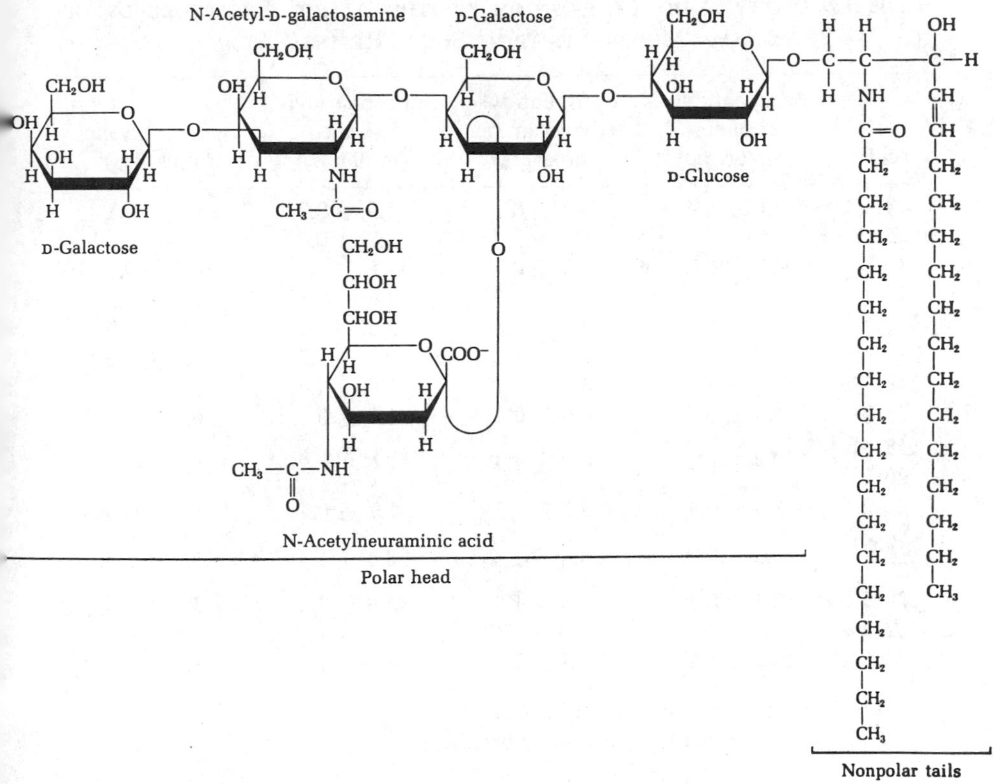

Figure 3.4. A ganglioside (GM_1). The hydrophilic head of gangliosides is relatively large and complex. Reproduced from Lehninger, 1970.

The steroid cholesterol is another major constituent of all animal plasma membranes. Because of its semirigid ring structure, cholesterol has a more defined shape than the other membrane lipids. Where it is present in membranes in large amounts, cholesterol can cause adjacent hydrocardon residues to assume extended configurations. This in turn causes the hydrocarbon residues to form hexagonally close-packed, crystalline domains. Membranes rich in cholesterol are therefore likely not only to be more stable than those which contain comparatively small amounts of this constituent, but also to be less permeable.

The stability of membranes is also influenced by the exact types of hydrocarbon chains found in the lipid bilayer. Table 3.2 lists the fatty acid moieties of the phospholipids of the squid retinal axon membrane.

Table 3.2. FATTY ACID COMPOSITION OF THE MAJOR PHOSPHOLIPIDS OF PLASMA MEMBRANES FROM SQUID RETINAL AXON

Fatty acid[a]	Phosphatidyl-choline (molar %)	Phosphatidyl-ethanolamine (molar %)	Phosphatidyl-serine (molar %)	Sphingomyelin (molar %)
14:0	1.0 ± 0.4	0.2 ± 0.1	1.9 ± 0.7	1.1 ± 0.1
15:0	—	—	0.8 ± 0.4	—
16:0	44.7 ± 0.2	3.0 ± 0.9	11.8 ± 1.9	16.3 ± 0.7
16:1	—	—	2.6 ± 0.1	1.7 ± 0.2
17:0	1.8 ± 0.7	2.3 ± 0.8	2.8 ± 0.5	4.5 ± 2.0
18:0	9.4 ± 0.2	29.5 ± 3.6	39.8 ± 7.7	42.4 ± 6.0
18:1	9.9 ± 1.1	8.6 ± 1.1	11.6 ± 4.2	7.2 ± 1.5
18:2	5.5 ± 0.8	0.6 ± 0.2	1.5 ± 0.8	0.6 ± 0.1
18:3, 20:1	3.4 ± 0.1	4.8 ± 1.3	2.0 ± 0.7	6.5 ± 1.1
20:2	1.6 ± 0.4	5.2 ± 1.3	2.7 ± 0.7	1.0 ± 0.6
20:4, 22:1	9.3 ± 2.0	18.6 ± 1.4	8.3 ± 1.4	5.5 ± 2.0
20:5	2.3 ± 0.1	6.6 ± 1.3	5.9 ± 2.4	3.3 ± 1.3
22:4	—	—	—	0.5 ± 0.2
22:6	10.2 ± 1.1	20.5 ± 2.9	8.2 ± 1.4	9.3 ± 1.1

Modified from Zambrano et al., 1971.
[a] Number carbons: number double bonds.

The large variations in chain lengths and degree of saturation are typical of most, if not all, membranes. Why such large variations occur is obscure, but there is good evidence that the greater the degree of unsaturation of the fatty acids, the less the molecular order in the interior of the lipid bilayer. Thus, those membranes that possess greater than average amounts of unsaturated fatty acids are easily modified to permit the flow of ions and small water-soluble molecules.

Asymmetric Location of Lipids

Evidence is now accumulating that the various phospholipids in surface membranes are not present in equal amounts on both sides of the bilayer (Bretscher, 1972). Phosphatidylcholine and sphingomyelin probably reside largely in the outer lipid layer, while the two chief aminophospholipids, phosphatidylserine and phosphatidylenthanolamine, probably face the cytoplasm. Such a distribution would explain the fact that the reagent [^{35}S]formylmethionyl (sulfone) methylphosphate, which is capable of reacting well with free lipid amino groups, reacts only weakly with the amino groups of phosphatidylserine or phosphatidylethanolamine in intact erythrocyte membranes; by contrast, when the red cells are hemolysed to form ghosts, these phospholipids become much more susceptible to chemical modification by this reagent. Furthermore, only phosphatidylcholine is split when limited amounts of a specific phospholipase that hydrolyzes phosphatidylcholine, phosphatidylserine and phosphatidylethanolamine are added to intact erythrocytes.

Cholesterol, by contrast, is probably located in both sides of the lipid bilayers in at least some surface membranes. In myelin sheaths, for example, cholesterol accounts for 40 percent of the total lipid; so much cholesterol is unlikely to be located in just one of the two lipid layers, not least because experiments with various artificial lipid bilayers have shown that at saturation the concentration of cholesterol never exceeds 50 moles percent of the total lipid.

Direct Proof of Bilayer Structure

The idea that membranes contain lipid bilayers in which the individual lipid molecules are arranged so that their hydrophilic groups face outwards was first proposed on theoretical grounds. And because of the lack of direct experimental evidence for the existence of such lipid bilayers in cell membranes, Korn in 1966 suggested that they may not in fact exist. However, X-ray crystallographers have now proved that the lipids in membranes are arranged in a bilayer structure; their work is an important milestone in our understanding of cell membranes.

While X-ray diffraction analysis had been done in natural membranes for over forty years (Schmitt et al., 1935, 1941), only in the past several years has the analysis reached the point where the "phases" of the X-ray reflections could be determined to yield Fourier diagrams (electron density profiles) (Finean and Burge, 1963; Moody, 1963; Blaurock, 1971; Blaurock and Wilkins, 1969; Wilkins et al., 1971; Caspar and Kirschner, 1971). The 10-Å resolution Fourier diagram of the myelin membrane shown in Figure 3.5 represents the highest resolution so far accomplished with any cellular membrane. The diagram

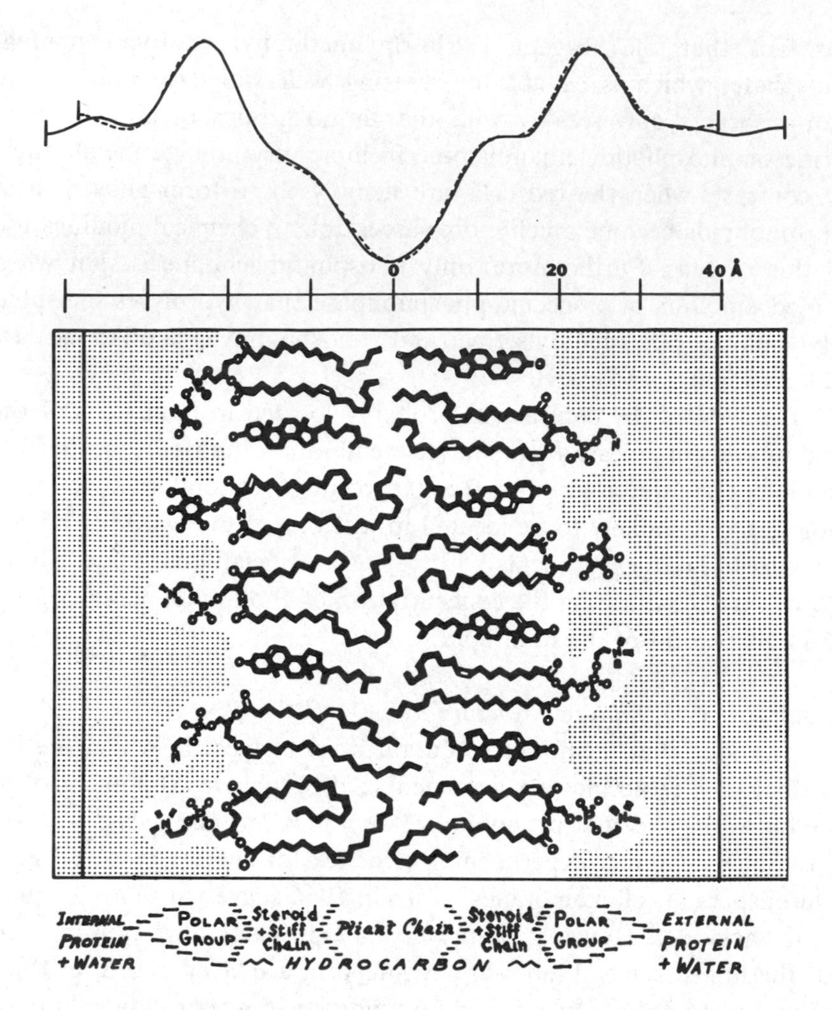

Figure 3.5. Schematic illustration of myelin membrane structure. The density profiles of rabbit optic (dotted) and sciatic myelin are shown above. The boundaries of the membrane unit of sciatic and optic myelin correspond to the wider and narrower margin, respectively. The distinctive portions of the structure which we identify with features of the density profile are disignated below the figure. For clarity the lipid molecules are represented as all lying in the same plane. This schematized arrangement is intended to indicate the radial orientation and liquid-like packing of the hydrocarbon portions of the lipid molecules. The composition illustrated [6 cholesterols; 5 glycerolipids (2 lecithins, 2 ethanolamine plasmalogens, 1 serine glycerol phosphatide); 4 sphingolipids (2 sphingomyelins, 2 cerebrosides)] corresponds approximately to the molar ratios measured for the principal lipid components of mammalian peripheral nerve myelins. The lipid composition of central nerve myelin is similar except that the sphingolipids consist predominantly of cerebrosides. Reprinted with permission from Caspar and Kirschner, *Nature New Biol.* 231, 46, 1971.

reveals strong peaks of electron density some 25 Å from the center of the bilayer. This result, together with those obtained with other membranes, unambiguously demonstrates that the electron-dense phosphate groups of all membranes are placed on the outside of the bilayer. Correspondingly, the trough of electron density at the center of the bilayer can only arise from the disordered configuration of the opposing hydrocarbon tails of lipid molecules.

Because myelin sheaths contain relatively little protein (25%), current X-ray diffraction techniques cannot reveal where it is located; therefore Figure 3.5 shows only the lipid components of the bilayer.

Membrane Proteins

With the exception of some specialized membranes (e.g., myelin), the amount of protein in cell surface membranes approximately equals the amount of lipid. How this protein is arranged is only now becoming clear. The early idea (Robertson, 1959) that most membrane protein existed as an extended β sheet covering both sides of the lipid bilayer is now known to be wrong. Many membrane proteins have globular shapes and contain significant α-helical sections (Ke, 1965; Lenard and Singer, 1966; Wallach and Zahler, 1966). For a long time, study of these proteins seemed insuperably difficult because when they are removed from their normal lipid environment, they usually prove to be difficult, if not impossible, to solubilize. However, with the development of new techniques for solubilizing such proteins, in particular the use of sodium dodecyl sulfate (SDS) (Shapiro et al., 1967; Weber and Osborn, 1969), it should soon be possible to estimate accurately the complexity of the protein compositions of many plasma membranes (see review by Guidotti, 1972).

Erythrocyte Proteins

To date most of our precise data about membrane proteins come from studies of the red blood cell membrane (Rosenberg and Guidotti, 1968, 1969; Winzler, 1969; Lenard, 1970; Trayer et al., 1971; Fairbanks et al., 1971; Steck et al., 1971). Figure 3.6 shows a separation of the proteins of the human erythrocyte membrane by electrophoresis in

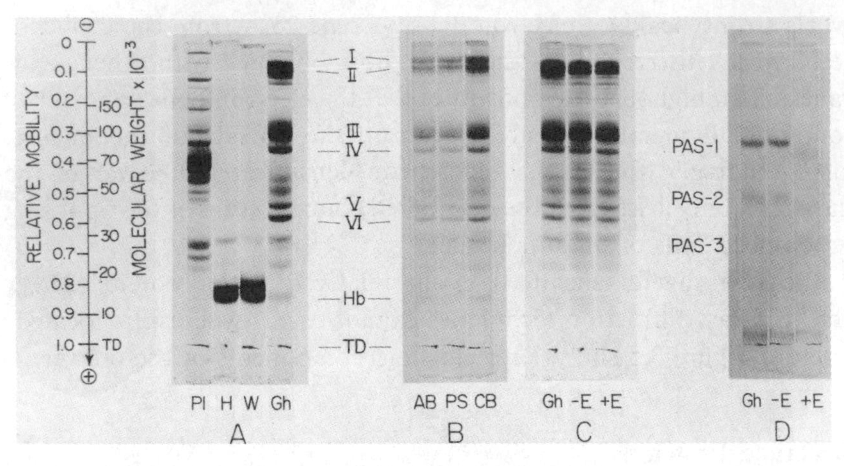

Figure 3.6. Electrophorograms of ghost membranes and other blood fractions. The relationship between relative mobility and polypeptide molecular weight is shown at the left. India ink marks inserted before fixation and staining show the final position of the pyronin Y tracking dye (*TD*) in each gel. (**A**) Electrophoretic analysis of proteins from plasma, red cell cytoplasm, and washed ghosts. *Pl* = plasma, 20 μl of a 4/50 dilution of the first supernatant. *H* = hemolysate supernatant, 25 μl of a 4/50 dilution. *W* = wash, 100 μl of a 4/10 dilution of the supernatant from the first ghost wash. *Gh* = ghosts, 25 μl of a 4/10 dilution of packed, washed ghosts (about 40 μg of protein). (**B**) Electrophorograms of ghost membrane proteins: comparison of Amido Black (*AB*), Ponceau S (*PS*), and Coomassie blue (*CB*) staining. The amount of protein applied to each gel was 28 μg. (**C**) The effect of sialidase on ghost membrane proteins: electrophorograms stained with Coomassie blue. One sample of ghosts (*Gh*) was prepared for electrophoresis by the standard procedure; a second portion ($-E$) was an incubation control for the sialidase treatment; the third ($+E$) was treated with the *Vibrio cholerae* sialidase. The amount of protein applied to gel *Gh* was 34 μg; gels $-E$ and $+E$ each contained 23 μg. (**D**) Effect of sialidase on ghost membrane proteins: replicates of gels in (C) stained for carbohydrate by the PAS procedure and photographed through a blue-green filter. Reproduced with permission from Fairbanks et al., *Biochemistry* 10, 2606, 1971.

an SDS-polyacrylamide gel system. Approximately 30 percent of the proteins have molecular weights greater than 150,000 daltons. Most of these large molecules can be selectively removed from the membrane by low salt extraction (Mazia and Ruby, 1968; Marchesi and Steers, 1968; Hoogeveen et al., 1970; Marchesi et al., 1970; Furthmayr and Timpl, 1970; Rosenthal et al., 1970).

Two species of polypeptides with molecular weights of ~220,000 and ~200,000 daltons account for most of this material. These two polypeptides, collectively called tektins by Mazia and Ruby (1968) and spectrin by Marchesi and his collaborators (1970), are present in equimolar amounts, and because they are easily cross-linked to each other, they may be components of a single functional molecule (Clarke, 1971; Hulla and Gratzer, 1972). In the electron microscope they appear as long, thin structures with diameters of between 40–50 Å and lengths of ~2100 Å. Because these molecules have a Mg^{++}-inhibited, Ca^{++}-activated ATPase activity (Rosenthal et al., 1970), it has been suggested (Clarke and Griffith, 1972) that they may in some way be related to myosin of striated muscle.

The fact that these very large molecules can be extracted from erythrocyte ghosts but never from intact cells suggests that they are situated exclusively on the inner surface of the membrane. And the further observation that they are never seen in the supernatants remaining after centrifugation of lysed red cells argues that they must be true membrane molecules, not cytoplasmic components that attach to the membrane during hemolysis.

Only two of the major membrane proteins are exposed on the outer surface of erythrocytes. One of these constitutes some 25 percent of the total membrane protein and has a molecular weight of ~100,000 daltons. The other, somewhat smaller, protein (molecular weight about 30–40,000 daltons) contains substantial amounts of carbohydrate (Winzler, 1969). Both proteins can be labeled by reagents which do not pass through intact erythrocyte membranes and which, therefore, selectively label those proteins that are exposed on the outer surface. The first such reagent to be so used was the diazonium salt of [^{35}S]-sulfanilic acid (Berg, 1969). Another such reagent, [^{35}S]formylmethionyl (sulfone) methylphosphate, has been developed recently by Bretscher (1971a). It is a powerful acylating agent that adds formylmethionyl sulfone groups onto amino (and possibly hydroxyl) groups. Another useful reagent is the enzyme lactoperoxidase, which in the presence of H_2O_2 places ^{125}I atoms on exposed tyrosine and histidine residues (Phillips and Morrison, 1970). When any of these reagents is added to an erythrocyte ghost, all the membrane proteins become labeled; but if they are added to intact red blood cells, only the two proteins with molecular weights of 100,000 daltons and 30–40,000 daltons are labeled (Bretscher, 1971a; Phillips and Morrison, 1971).

Proteins that Penetrate the Bilayer

Both these two proteins that are exposed on the outer surface of the lipid bilayer also extend through the bilayer to its inner surface. This was shown by comparing the fingerprints of these two proteins after they had been labeled by exposing intact erythrocytes or erythrocyte ghosts to [^{35}S]formylmethionyl (sulfone) methylphosphate. After labeling from the outside only, few tryptic peptides were labeled; after labeling simultaneously from the outside and the inside, many more tryptic peptides contained label and so molecules of these two proteins probably extend through the lipid bilayer (Bretscher, 1971b,c).

Most importantly, only the smaller of these two proteins contains much carbohydrate residue. Recently Marchesi and Andrews (1971) have found that this glycoprotein can be quantitatively extracted in a water-soluble form with lithium diiodosalicylate. More than half (~60%) of the molecule is carbohydrate, and there are indications that the polypeptide chain may contain as few as 200 amino acids (Segrest et

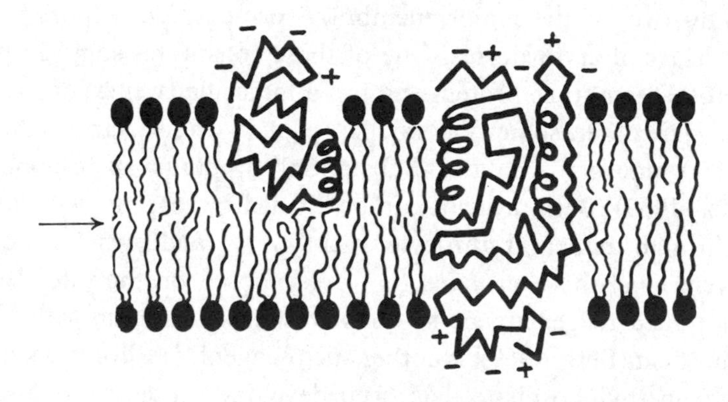

Figure 3.7. The lipid-globular protein mosaic model of membrane structure: schematic cross-sectional view. The phospholipids are depicted as in Figure 3.1 and are arranged as a discontinuous bilayer with their ionic and polar heads in contact with water. Some lipid may be structurally differentiated from the bulk of the lipid, but this is not explicitly shown in the figure. The integral proteins, with the heavy lines representing the folded polypeptide chains, are shown as globular molecules partially embedded in, and partially protruding from, the membrane. The protruding parts have on their surfaces the ionic residues (— and +) of the protein, while the nonpolar residues are largely in the embedded parts; accordingly, the protein molecules are amphipathic. The degree to which the integral proteins are embedded and, in particular, whether they span the entire membrane thickness depend on the size and structure of the molecules. The arrow marks the plane of cleavage to be expected in freeze-etching experiments. Reproduced with permission from Singer and Nicolson, *Science* 175, 722, 1972.

al., 1971). This means that each polypeptide chain may have bound to it as many as 100–200 different sugar residues. This glycoprotein may possess, therefore, a bewildering collection of different antigenic specificities. This finding explains how this single glycoprotein can carry MN blood group determinants, as well as the receptors for influenza virus, phytohemagglutinin and wheat germ agglutinin.

Both of the major surface proteins are subjected to enzymic cleavage when intact red blood cells are exposed to proteolytic enzymes. Addition of pronase cleaves off fragments from the 100,000 dalton component as well as from the glycoprotein. Approximately one-third of the large protein is split off, leaving a ~70,000 dalton fragment bound into the bilayer in a way that leaves it still accessible to attack by [^{35}S]sulfanilic acid diazonium salt (Bender et al., 1971; Bretscher, 1971b). Such experiments show that the integrity of the bilayer does not depend upon the intactness of the surface protein components and suggest that much membrane protein extends free from the bilayer (see Figure 3.7).

Membrane Enzymes

It is not known whether either the major erythrocyte glycoprotein or the ~100,000 dalton component have any enzymic role. Each red cell contains some 5×10^5 copies of the latter component, and the suggestion has been made (Bretscher, 1973) that they may serve as the channel for the active transport of bicarbonate ions. Various other enzymes are known to be firmly imbedded in the red cell membrane, but only comparatively small numbers of such molecules are likely to be present. At least one such enzyme, acetylcholinesterase, is inactivated by treatment of intact cells with pronase, so parts of it must extend through to the outside surface (Bender et al., 1971).

A preliminary estimate exists of the number (~200) of the Na^+-(K^+) stimulated "ATPase" molecules involved in the active transport of K^+ and Na^+ across the erythrocyte membrane. This specific "ATPase," unlike that bound to the long thin 200,000 (220,000) dalton fibers, is globular in shape and is stimulated, not inhibited, by Mg^{++}. It maintains the high concentration of K^+ within erythrocytes; experiments have shown that the rate of breakdown of ATP is inversely related to the external concentration of K^+ (see reviews by Skou, 1965; Whittam, 1967; Lin, 1971). During the hydrolysis of ATP, K^+ ions move into the red cell and Na^+ ions out, with approximately three ions moving inward (outward) for every ATP hydrolyzed to ADP and free phosphate. The fact that K^+ activates from the outside, while Na^+ stimulates ATP breakdown only when present inside the cell, argues that this protein extends through the bilayer. Hopefully, detailed structural analysis may soon yield information at the molecular level on how this "ATPase" facilitates selective ion transport.

Detailed understanding of how ions are actively transported across membranes should also come from further study of the proteins of the sarcoplasmic reticulum. These internal membranes of muscle have very active Ca^{++} transport systems that regulate the flow of Ca^{++} to and from myofibrils at the time of muscle contraction (Ebashi et al., 1969; Martonosi, 1971; Weber, 1966). The seven proteins which are involved in Ca^{++} transport comprise almost all the protein of these membranes (Figure 3.8) (Martonosi and Halpin, 1971; MacLennan and Wong, 1971; MacLennan et al., 1973). Two of these seven proteins are soluble only in the presence of deoxycholate and most probably they are "intrinsic" membrane components tightly bound within the lipid bilayer (MacLennan, 1970). One, a Ca^{++}-stimulated "ATPase" with a molecular weight of 102,000 daltons, is believed to extend through the lipid bilayer; the other is a proteolipid with a molecular weight of 12,000

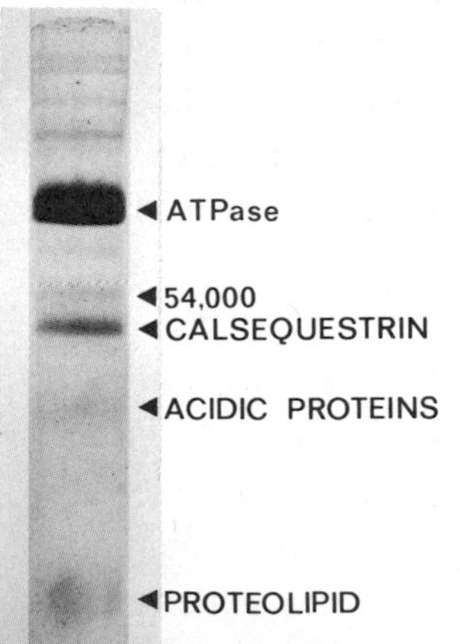

Figure 3.8. Disc gel electrophoretic profile of rabbit skeletal muscle sarcoplasmic reticulum after dissolution in sodium dodecyl sulfate (SDS). The polycrylamide gel was prepared in SDS according to the method of Weber and Osborn (1969) and stained with Coomassie blue. Reproduced with permission from MacLennan et al., *Cold Spring Harbor Symp. Quant. Biol.*, Vol. 37, 1973.

daltons. The other five proteins are referred to as the "extrinsic" proteins since they appear to be much less integrated into the lipid bilayer and become water soluble after the sarcoplasmic reticulum is disrupted. The two most abundant of these are calsequestrin, a Ca^{++}-binding protein with a molecular weight of 44,000 daltons, and an unnamed acidic protein with a molecular weight of 54,000 daltons. This acidic protein also binds Ca^{++}, though less efficiently than calsequestrin. The remaining three "extrinsic" proteins (22,000 to 32,000 daltons) also bind Ca^{++} and therefore are also likely to be involved in some aspect of Ca^{++} movement.

The sarcoplasmic reticulum normally sequesters large amounts of Ca^{++} and the release of only a small amount to the neighbouring myofibrils triggers contraction. It is believed that both calsequestrin and the 54,000 dalton acidic protein are localized on the interior surface of the reticulum, functioning to sequester Ca^{++} until it is needed to trigger contraction. By contrast, the other smaller Ca^{++}-binding proteins are thought to lie on the exterior surface of the reticulum next to the myofibrils. Conceivably they act after contraction to concentrate free Ca^{++} ions before they are pumped back into the interior of the sarcoplasmic reticulum by Ca^{++}-specific "ATPase" molecules.

Unfortunately, virtually nothing is yet known about the molecular biology of the various membrane-bound adenyl cyclases. These key enzymes, which make cyclic AMP, appear to be located exclusively on membranes. After their isolation from membranes they lose their biological activity, and so the exact ways these enzymes function to control the concentration of cyclic AMP remain to be worked out. For example, many of the effects of insulin are thought to involve its direct inhibition of adenyl cyclase. Until this enzyme can be purified, however, this conjecture will remain speculative.

Fluid Mosaic Model of Membrane Structure

The essence of the original Danielli-Davson model of the membrane stood unchanged for almost 25 years after its formulation in the late 1930's. It envisaged membranes as essentially static structures penetrated by channels through which small molecules in aqueous solution enter cells. During the last decade, however, a series of increasingly important objections began to be raised against the idea of a static lipid bilayer covered by proteins arranged in pleated sheets. To begin with, the objection was raised that this structure maximizes neither the hydrophobic interactions between the proteins and lipids, nor the hydrophilic interactions between the polar groups of the phospholipids and the aqueous environment. It does not, therefore, provide the lowest possible free energy for a lipid–protein membrane. Furthermore, evidence began to accumulate that many proteins in surface membranes have a globular conformation, not the pleated sheets initially postulated. And finally, experiments were carried out which showed that the lipids and proteins of membranes are not in static arrays but are in constant motion. For these several reasons the Danielli-Davson model has been superseded by the so-called Fluid Mosaic Model of membrane structure. In this model (see review by Singer and Nicolson, 1972) the lipids of the bilayer are fluid, not crystalline, and the proteins are globular, not extended pleated sheets, and are envisaged as floating in and on the surface of the fluid lipid bilayer.

Visualization of Globular Proteins

The existence of globular proteins embedded in the lipid bilayer is directly revealed by electron microscopy of replicas of membranes that have been subjected to freeze fracture. This procedure cleaves membranes between the two layers of phospholipid and exposes material embedded in the bilayer (Branton, 1966). Virtually all membranes examined so far possess globular particles (Branton, 1969), the diameters of which are a characteristic of that particular membrane (Figure 3.9). Splitting of the lipid bilayer of erythrocyte membranes reveals some 4.5×10^5 particles, each 85 Å in diameter (Pinto da Silva and Branton, 1970). The estimated number of these particles, which are believed to be protein, is so close to Bretscher's (1971a) estimate of the number of molecules of the larger protein on the external surface of erythrocytes that it is tempting to assume that they are the same molecules.

Electron micrographs of replicas of the fracture faces suggest that the globular particles lie within the lipid bilayer, but this appearance could be artifactual. In intact cells such proteins may well float within the bilayer, as shown in Figure 3.10, with their polar and nonpolar

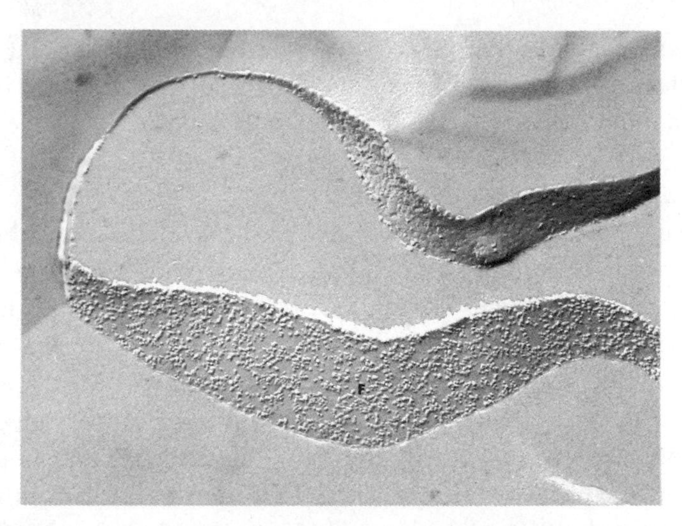

Figure 3.9. Electron micrograph of a red blood cell membrane prepared by the freeze etching procedure. The surface labeled *F* is an internal hydrophobic face of the membrane- The particles seen on this face average 85 Å in diameter. 35,000×. Reproduced with permission from Pinto da Silva and Branton, *J. Cell. Biol.*, 45, 598, 1970.

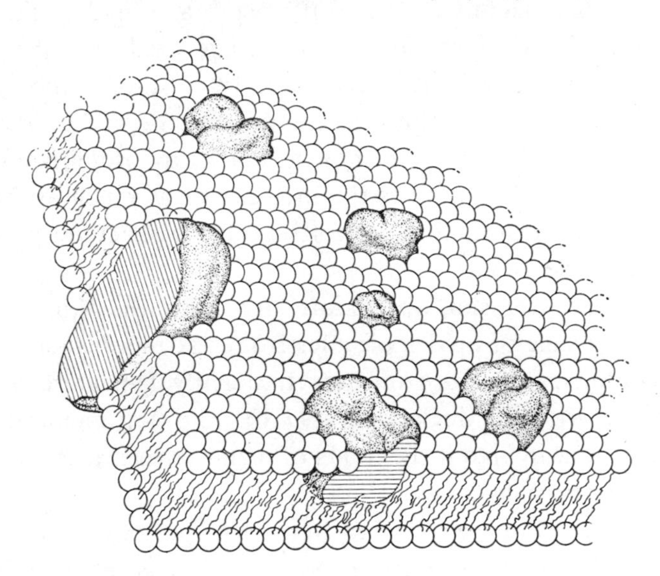

Figure 3.10. The lipid-globular protein mosaic model with a lipid matrix (the fluid mosaic model); schematic three-dimensional and cross-sectional views. The solid bodies with stippled surfaces represent the globular integral proteins, which at long range are randomly distributed in the plane of the membrane. At short range, some may form specific aggregates, as shown. Reproduced with permission from Singer and Nicolson, *Science* 175, 723, 1972.

amino acid side groups so distributed that only their nonpolar groups make effective contact with the hydrophobic hydrocarbon interior. This is thermodynamically the most stable structure.

Fluidity of the Plasma Membrane

The presence of large amounts of globular protein in membranes is not consistent with the Danielli and Davson model. Neither is the discovery that membranes are in effect two-dimensional fluids in which the spacial pattern at any given moment depends on the random motions of the constituents within a semiviscous hydrocarbon environment. The first clear indication of such fluid motion came from electron spin resonance studies and differential calorimetry of the lipid components of membranes (Hubbell and McConnell, 1968; Keith et al., 1968; Steim et al., 1969; Melchior et al., 1970; Kornberg and McConnell, 1971a). These various measurements revealed effective membrane viscosities some 1000 times greater than the viscosity of water. Given viscosities of this magnitude, a plasma membrane lipid could move in "random walks" from one side of an average-sized animal cell to the opposite side in approximately an hour. Interestingly, the viscosities recorded for natural membranes are not much greater than those observed for artificial lipid bilayers which lack any protein components (Tourtellotte et al., 1970). Thus the presence of as much as 50 percent protein within a membrane does not seriously restrict the mobility of the adjacent lipid residues.

All this lipid diffusion occurs laterally within the plane of a single monolayer. Passage of lipid molecules from one monolayer to another (flip-flop) occurs at least 10^{-8} times less frequently than lateral diffusional events (Kornberg and McConnell, 1971b). Flip-flop movements must obviously be rare in vivo if the lipids of membranes are asymmetrically distributed in the bilayer.

Convincing evidence of the movement of proteins within the lipid bilayer has come from several sorts of experiments. Brown (1972) and Cone (1972) studied the orientation of rhodopsin molecules within frog retinal membranes by measuring the dichroism induced by a brief flash of polarized light. Such flashes when directed perpendicular to the parallel rod membranes bleach those molecules aligned with the plane of polarization, thereby creating a measurable dichroism. However, such dichroism lasts for only a very brief period because the rhodopsin molecules undergo rotational diffusion. Accurate measurements of the rate at which the induced dichroism decays (Cone, 1972) indicate that the effective viscosity of the retinal membrane is about two poise, which is comparable to the viscosity of a light oil such as olive oil.

Similar values for the viscosity of the plasma membranes of animal cells growing in culture were obtained by Frye and Edidin (1970), who fused mouse cells with human cells to produce heterokaryons bearing both mouse and human transplantation antigens. The location of these antigens at various times after cell fusion was observed by fluorescent antibody techniques. Within 40 minutes of fusion, the two sets of antigens had become totally intermixed. Various metabolic inhibitors were then used to show that this redistribution was not the result of the rapid turnover of surface antigens, but that the cell membrane behaved as a fluid allowing relatively rapid diffusion of surface antigens.

The observations of Taylor et al. (1971), Unanue et al. (1972) and Davis (1972) are equally relevant. They independently showed that the addition of specific antibodies to cells bearing the corresponding surface antigens induced a redistribution of the antigen-antibody complexes. Davis traced the distribution of H2 isoantigens using ferritin-conjugated antibody. When he added ferritin-labeled anti-H2 antibody to appropriate mouse cells, H2 antigen-antibody complexes were observed to be randomly distributed on the cell surface. But when mouse cells were treated first with mouse H2 antibodies and then with ferritin-labeled rabbit antibody specific for mouse antibody, all the ferritin label was observed in discrete patches that ranged in size from several square microns to half the surface area of the plasma membrane. Davis concluded that H2 antigens have no fixed positions in cell membranes, but that in the presence of a cross-linking reagent (e.g., divalent antibodies) they become locally concentrated.

The same conclusion comes from experiments (Taylor et al., 1971; Unanue et al., 1972) in which the locations of antibodies on the surfaces of mouse lymphocytes were revealed by the appropriate rabbit anti-mouse globulin. Addition of such antibodies to lymphocytes leads

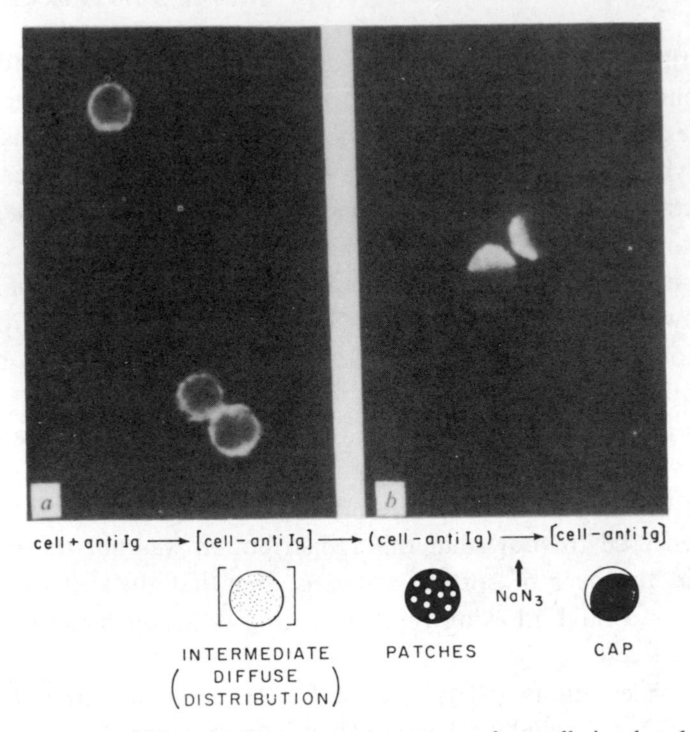

Figure 3.11. Pattern of immunofluorescence in mouse spleen cells incubated in rabbit anti-mouse Ig-Fl for 30 minutes (**a**) at 0°C ("ring" pattern), (**b**) at room temperature ("cap" pattern). The cells were washed and examined at room temperature in the presence of 3×10^{-3} M sodium azide which prevents any subsequent aggregation during microscopic examination. Reproduced with permission from Taylor et al., *Nature New Biol.* 233, 225, 1971.

Diagram is a schematic model for patch and cap formation. Reproduced with permission from Yahara and Edelman, *Proc. Nat. Acad. Sci.* 69, 611, 1972.

within several minutes of redistribution of the antigen-antibody complexes into discrete patches (Figure 3.11) that eventually coalesce into a single "cap." The "cap" is then taken into the interior of the cell by pinocytosis. Patch formation does not occur, however, if univalent Fab antibody fragments are used. Again two conclusions must be drawn: first, the fluidity of the plasma membrane permits rapid diffusion of its protein constituents, and second, the aggregation of antigens depends upon cross-linking by divalent antibodies.

The rate at which "diffusion" occurs is strongly influenced by metabolic considerations. "Cap" formation is inhibited when the generation of ATP is lowered by the presence of dinitrophenol (or NaCN + NaF), conceivably because cells so treated become unable to form the ruffled membranes (see next section) involved in directed cell movement (Edidin and Weiss, 1972). Most likely the rapid flow of membranous material that accompanies cell movement effectively speeds up the "diffusion" of membrane proteins.

The Continuously Changing Outline of Individual Cells

Rearrangement of membrane components also results from the active processes that cause cells to move and that lead to the engulfment of food substances (pinocytosis). Many types of cells in culture regularly send out extended pseudopodia-like processes (ruffled edges) that bring about movement in a given direction (Abercrombie et al., 1970a, b, c, 1971; Ingram, 1969). Most pseudopodia (ruffled edges) have a very fleeting existence, disappearing within minutes, if not seconds, of their formation (Figure 3.12). Thus the way they form and disappear is best seen in time-lapse cinéfilms of cells.

The idea that such ameboid-like movement is mediated by actin and myosin-like molecules is not new (see review by Hoffman-Berling, 1964), but only recently have such contractile proteins been isolated

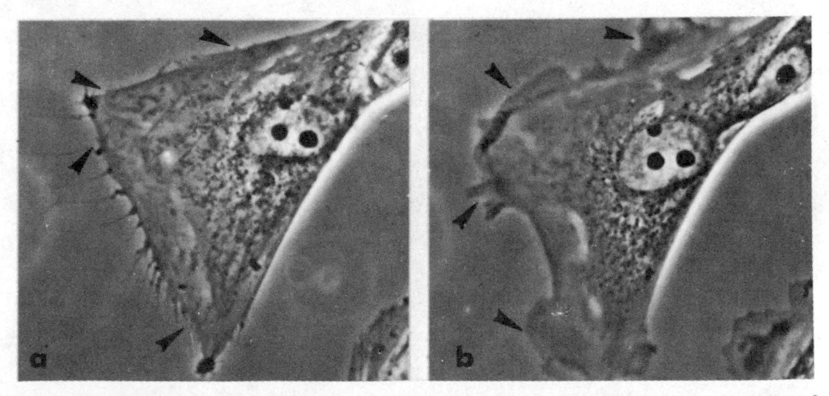

Figure 3.12. **(a)** Light micrograph of a BHK cell which has spread out on a solid surface in the presence of cytochalasin B. No ruffled edge is present. **(b)** The same cell 60 seconds following removal of cytochalasin. Note the immediate formation of expanding and ruffling membranes (compare arrows in both micrographs). Reproduced with permission from Goldman and Knipe, *Cold Spring Harbor Symp. Quant. Biol.*, Vol. 37, 1973.

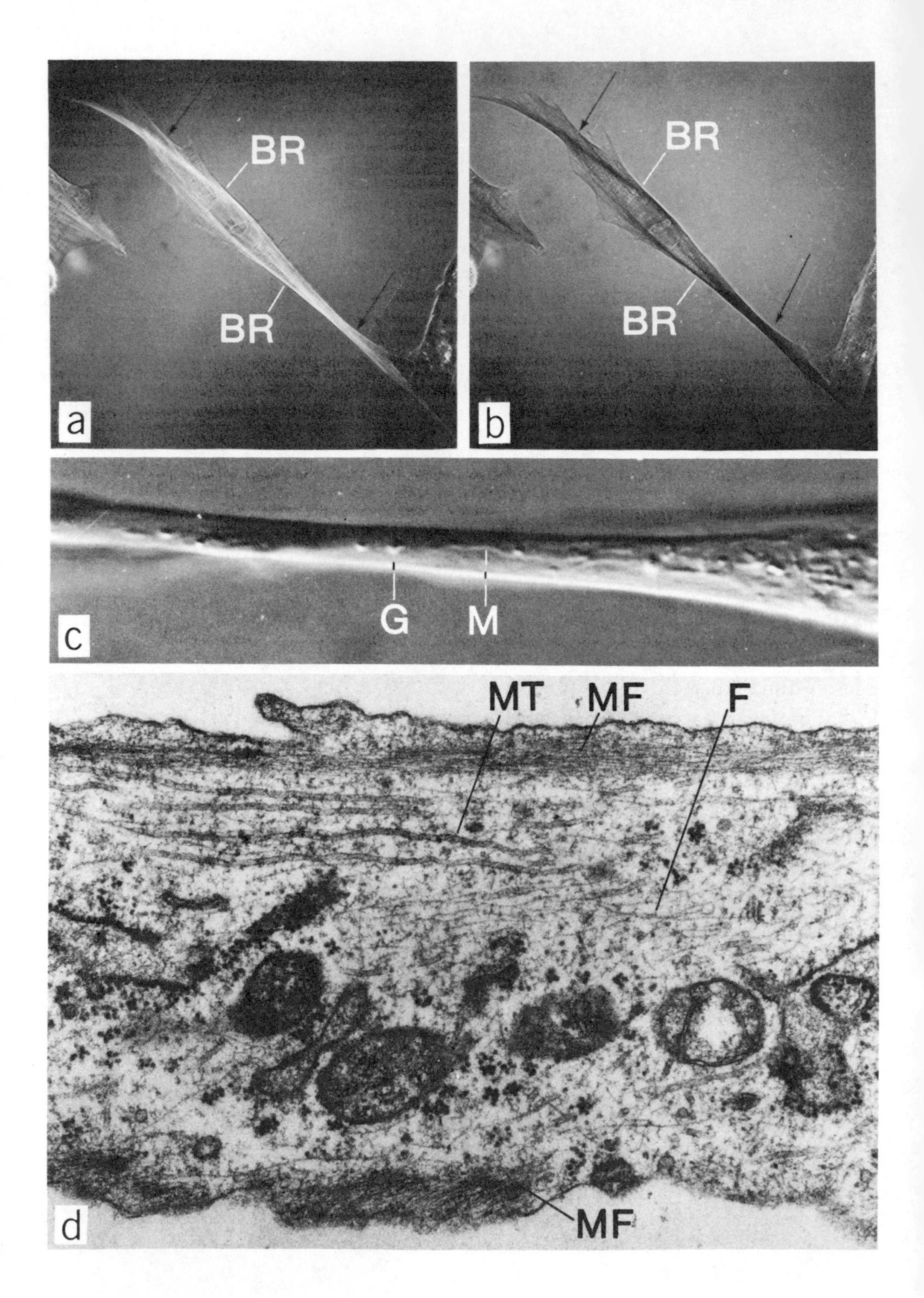
BR
BR
a
BR
BR
b
G
M
c
MT
MF
F
MF
d

from cells grown in culture. Actin, in particular, has been well characterized; it is one of the most abundant proteins of many cultivated cells (Fine and Bray, 1971; Anderson and Gesteland, 1972). For example, in chicken fibroblasts growing as a monolayer, actin comprises almost 10 percent of the total cell protein, with most of it present as polymerized microfilaments some 60 Å in diameter. Most microfilaments lie just beneath the plasma membrane (Figure 3.13), a fact which suggests that they may directly interact with specific membrane proteins (Goldman, 1971, 1972; Tilney and Mooseker, 1971). In addition to actin, myosin has recently been detected in blood platelets (Bettex-Galland and Lüscher, 1965; Booyse et al., 1971) as well as in cultured cells of several species (Adelstein et al., 1971; Adelstein and Conti, 1973). A sliding filament mechanism may, therefore, underlie the rapid protrusion and retraction of pseudopodia and ruffled edges.

Cytochalasin has proved very useful in probing the role of microfilaments in cells (Carter, 1967; Schroeder, 1970). Addition of cytochalasin almost immediately stops cell movement and causes the partial depolymerization of the actin microfilaments (Spooner et al., 1971; Wessels et al., 1971; Goldman, 1972; Allison et al., 1971; Goldman and Knipe, 1973). Microfilaments by themselves, however, are not sufficient for cell-directed movement. The much thicker microtubules (250 Å in diameter) also are necessary, most probably to provide a cell with a skeletal framework (Tilney and Porter, 1965; Tilney, 1968). For example, it is the preferential orientation of microtubules which gives fibroblasts their characteristic spindle shapes. And when colchicine, a specific inhibitor of the formation of microtubules, is added to fibroblasts, the cells round up (Goldman, 1971; Hsie and Puck, 1971; Puck et al., 1972).

Figure 3.13. **(a)** and **(b)**. A fully spread BHK21 fibroblast-like cell observed at opposite compensator settings with polarized light optics. Note the presence of two major cell processes (arrows) and birefringent fibres or streaks (*BR*). 40×.
(c) A Nomarski differntial interference micrograph of a major cell process viewed with an oil immersion lens (100×). Filamentous mitochondria (*M*) and granules (*G*) move bidirectionally in oriented paths along the length of these cell extensions.
(d) An electron micrograph of a section cut longitudinally through a major cell process similar to the ones designated by arrows in **(a)** and **(b)**. Most of the 250 Å microtubules (*MT*), 40–60 Å microfilaments (*MF*), and 100–120 Å filaments (*F*) are oriented along the long axis of the major cell process. 38,700×. Reproduced with permission from Goldman, *J. Cell Biol.* 51, 752, 1971.

BIOSYNTHESIS OF MEMBRANE CONSTITUENTS

We do not know how any cell membrane is assembled. A priori, we might wish to distinguish between a self-assembly process and growth by piecemeal addition to a preexisting membrane, but most probably we will learn that most membranes grow by both mechanisms. It seems unlikely that surface (plasma) membranes usually enlarge by the insertion of single phospholipid, glycolipid or protein (glycoprotein) molecules. It seems more likely that they grow by the fusion of preexisting membranous sacs (organelles) formed within cytoplasm (Figure 3.14). Numerous examples of such fusion processes have been described by electron microscopists over the past 20 years. For example, the

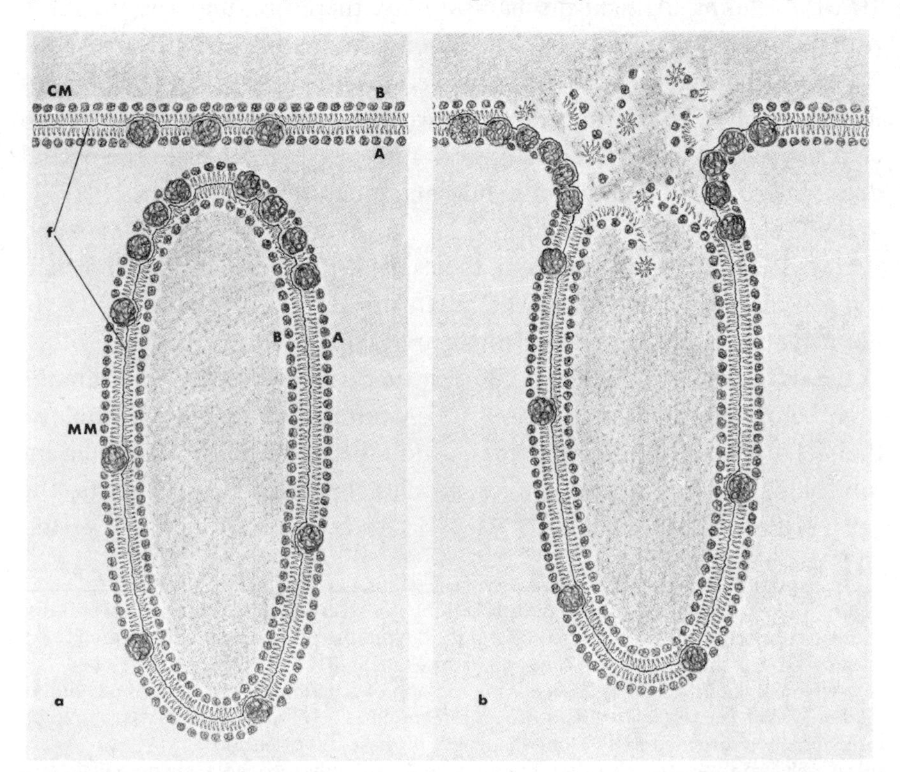

Figure 3.14. Diagram of membrane fusion model. **(a)** Undischarged mucocyst, just prior to fusion of cell (*CM*) and mucocyst (*MM*) membranes. Dark line (*f*) indicates fracture plane separating *A* and *B* faces. **(b)** Discharging mucocyst showing reorganization of the membranes. Reproduced with permission from Satir et al., *Nature* 235, 53, 1972.

secretion of the hydrolytic enzymes found initially within a zymogen granule in pancreatic cells involves fusion of the outer membrane of the granule with an adjacent region of the cell's plasma membrane.

Such fusion events (frequently called reverse pinocytosis) lead not only to the release (secretion) of the contents of the respective organelles, but also to the transfer of organelle membranes to the external plasma membrane. In this process the polarity of the membrane bilayer becomes reversed. What was on the outside of the membrane of the organelle forms part of the inner face of the plasma membrane, and vice-versa (see Figure 3.14). For example, any enzyme, which initially was preferentially bound to the inner surface of a Golgi body, after reverse pinocytosis will face outward and so be able to catalyze reactions between extracellular molecules.

Equally important may be the consequences of ordinary pinocytosis (Figure 3.15) which brings membrane-bound food vacuoles into the interior of cells. Within the cell the vacuoles fuse with lysosomes, thereby confining digestive events to within the membrane sacs. If pinocytosis occurs very frequently, it will tend to equalize the concentration of specific membrane constituents between the plasma membrane and membranes of intracellular organelles.

Synthesis of Membrane Lipids

The synthesis of the phospholipid (glycolipid) components of animal cell membranes may arbitrarily be divided into several steps (see review by Vagelos, 1971; Rossiter, 1968; Stoffel, 1971; Roseman, 1968; Ginsburg and Kobata, 1971). The first step consists of the construction of the fatty acid chains, a process which begins on multienzyme fatty acid synthetase complexes that are water soluble. The nascent fatty acid chains never exist free, but always remain bound to the multienzyme synthetase complexes to which they are attached by their acyl-carrier protein component. The end product of the fatty acid synthetase complex is usually the 16-carbon-atom molecule palmitic acid. The subsequent elongation–modification (oxidation) steps necessary to transform palmitic acid into the large variety of known saturated and unsaturated higher fatty acids are catalyzed by enzymes found in the mitochondria or the endoplasmic reticulum (ER).

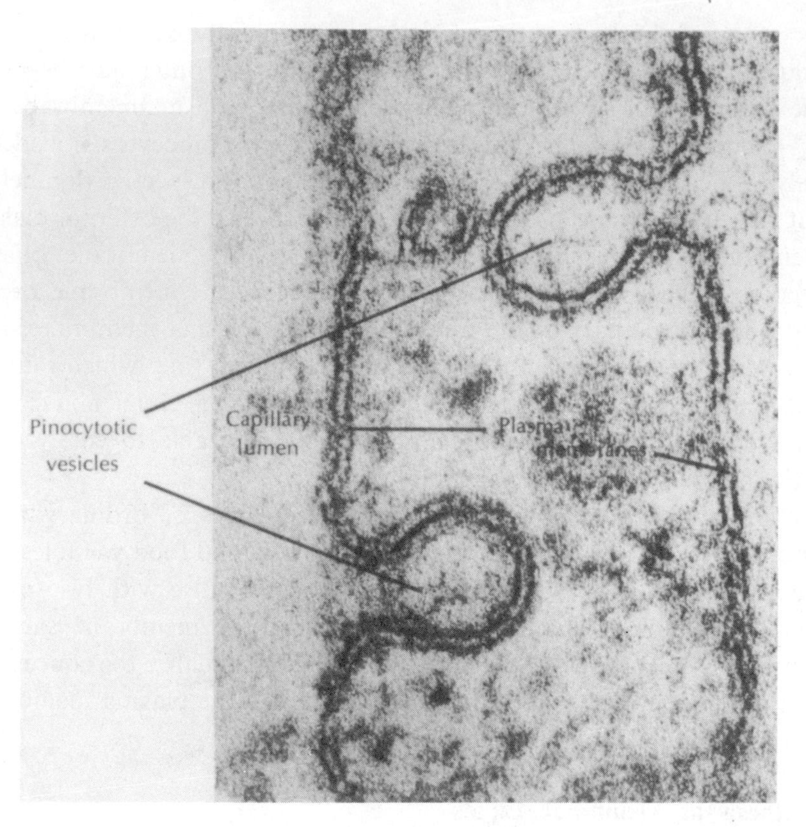

Figure 3.15. Pinocytotic vesicles. Blood capillary cell, showing the entire width of the cell (315,000×). Reproduced with permission from Loewy and Siekevitz, *Cell Structure and Function*, Holt, Rinehart, and Winston, 1969.

Synthesis of the phosphoglycerides themselves commences with union of glycerol-3-phosphate and two moles of fatty acid CoA to form the precursor phosphatidic acid. This can be converted by reacting with CDP-choline to phosphatidylcholine. Alternatively it can combine with CDP-enthanolamine to yield phosphatidylethanolamine. This in turn can react with serine to give phosphatidylserine (Figure 3.16). All these steps are catalyzed by enzymes which reside exclusively in membranes, in particular the membranes of the ER.

The synthesis of sphingomyelin and the various glycolipids starts with the combination of palmitoyl CoA and serine to form dihydrosphingosine. This intermediate almost instantly is oxidatively transformed

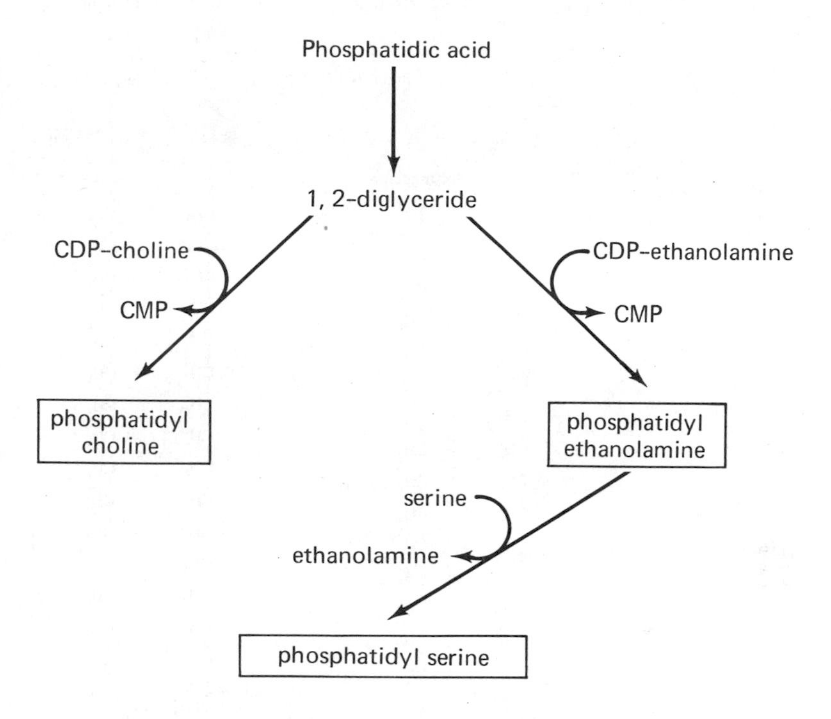

Figure 3.16. Pathways of biosynthesis of major phosphoglycerides within animal cells.

to sphingosine. Sphingosine in turn combines with a fatty acid CoA molecule to form a ceramide (Figure 3.17). These molecules are the usual raw materials for the synthesis of sphingomyelin as well as the cerebrosides, the hexose-containing neutral glycolipids. Cerebrosides may also be formed by an alternative pathway in which sphingosine first is substituted with galactose to form psychosine, which in turn is acylated to form the monocerebroside.

Further substitution of the simple monocerebrosides with neutral hexoses leads to the formation of di-, tri-, tetra- and pentahexose cerebrosides. The carbohydrate donor is thought usually to be a UDP (CDP, GDP) derivative with a specific enzyme needed for each specific substitution. Unique enzymes are likewise thought to be necessary for the synthesis of the various gangliosides, the cerebroside derivatives

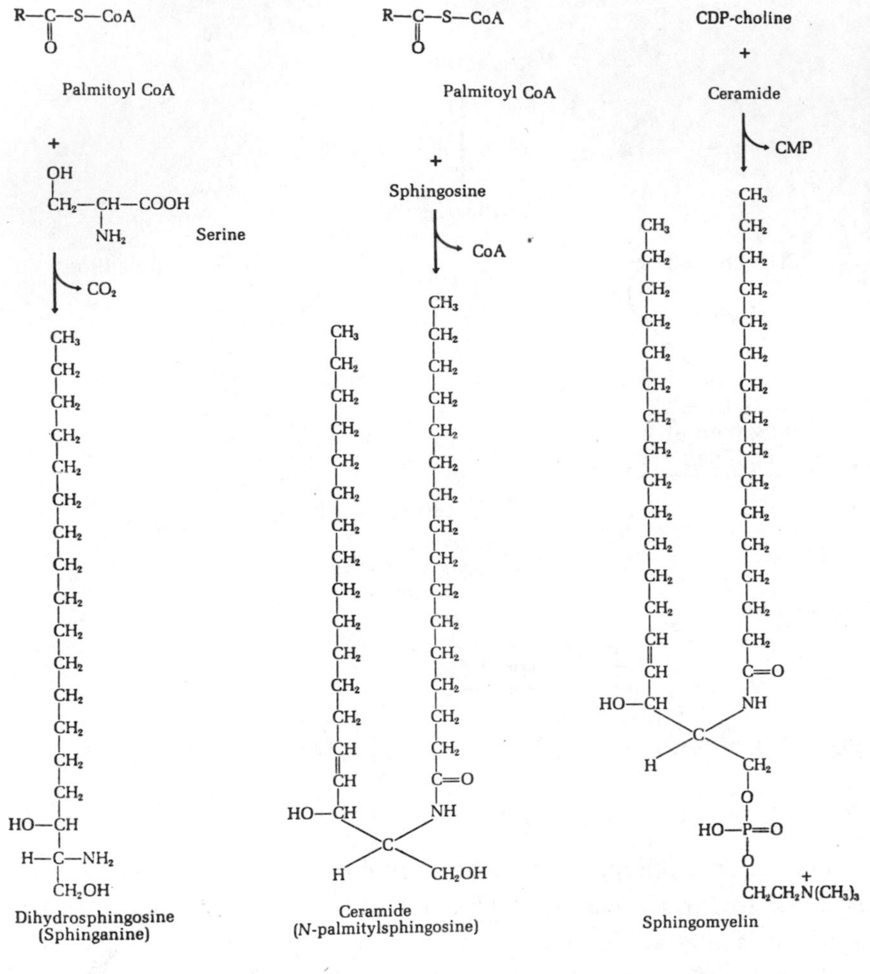

R—C(=O)—S—CoA

Palmitoyl CoA

+

Serine ($HOCH_2$—CH(NH_2)—COOH)

→ CO_2

Dihydrosphingosine
(Sphinganine)

R—C(=O)—S—CoA

Palmitoyl CoA

+

Sphingosine

→ CoA

Ceramide
(N-palmitylsphingosine)

CDP-choline

+

Ceramide

→ CMP

Sphingomyelin

that are substituted with one or more sialic acid residues (Figure 3.18). As far as we know, none of these biosynthetic enzymes exist free; they are all bound to some cellular membrane.

It appears, therefore, that the various membrane lipids are always synthesized in close proximity to the site at which they are inserted into a functional membrane. How this insertion happens, however, is still very mysterious, and at present it is impossible to decide whether or not any internal cellular membrane starts its existence de novo rather than growing by insertion into preexisting membranes.

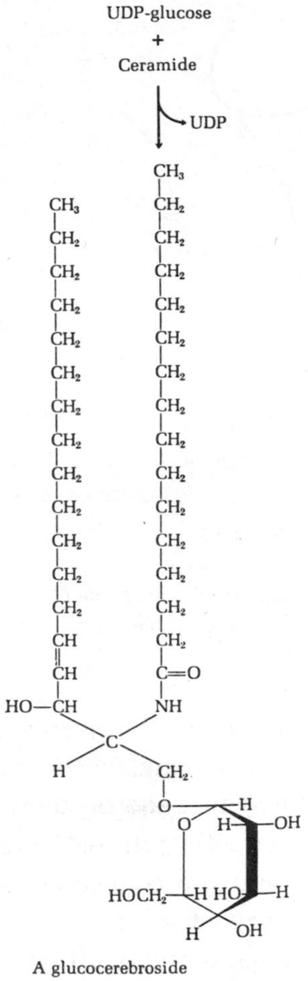

UDP-glucose
+
Ceramide
UDP
A glucocerebroside

Figure 3.17. Biosynthetic steps in the formation of sphingomyelin and glucosylceramide. Reproduced from Lehninger, 1970.

Synthesis of Membrane Proteins

Ribosomes engaged in protein synthesis can exist as free polysomes or as polysomes attached to cellular membranes. In bacteria most (80–90%) polysomes exist free, with a much smaller fraction (10–20%) bound to the plasma membrane. By contrast, most polysomes in many types of animal and plant cells are bound to the outer surfaces of the system of membrane sacs that constitute the granular ER. Because the granular ER is most highly developed in cells which are actively

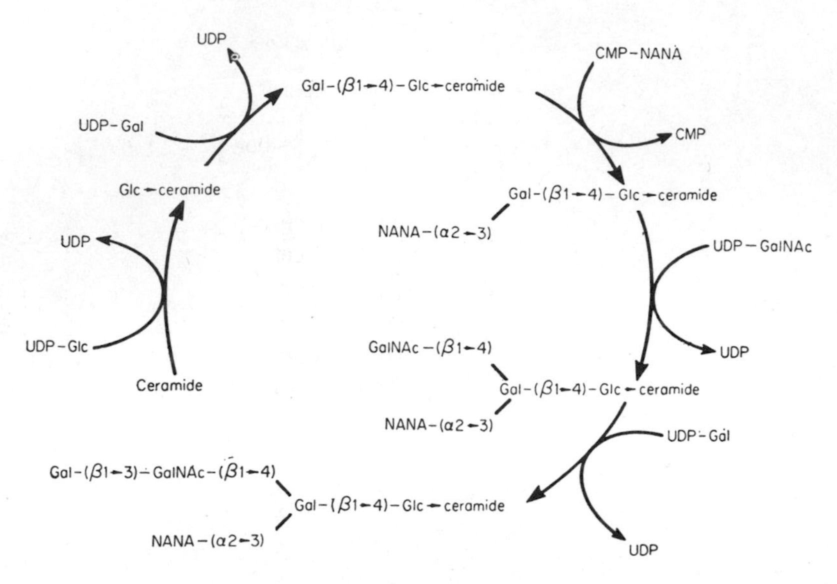

Figure 3.18. Biosynthesis of ganglioside GM_1. Reproduced with permission from S. Roseman, *Biochemistry of Glycoproteins and Related Substances*, S. Karger, Basel, 1968.

secreting proteins (e.g., the liver and the pancreas), it has generally been assumed (Siekevitz and Palade, 1968; Dallner et al., 1966; Jamieson and Palade, 1968) that the proteins synthesized by membrane-bound ribosomes are directly released into the internal cavity of the ER, later to move into the connecting sacs and vesicles of the "smooth ER," which is devoid of ribosomes, where they can be further processed for eventual secretion (Figure 3.19). The way ribosomes bind to the ER favours this idea (Sabatini et al., 1966). Attachment always occurs through the 60 S ribosomal subunit, the subunit to which the growing polypeptide chain is attached. More direct evidence comes from studies of protein synthesis in vitro, where only the membrane-bound ribosomes make serum glycoproteins (Ganoza and Williams, 1969; Redman, 1969). By contrast, unattached polysomes synthesize proteins like ferritin and hemoglobin that function in the surrounding cytoplasm and are not secreted from the cell.

It has been proposed that as membrane-bound ribosomes elongate polypeptide chains, the NH_2-terminal ends of the polypeptides begin to move into the adjacent lipid bilayer. So by the time its synthesis is completed, a considerable fraction of a newly synthesized polypeptide chain may have passed into the internal cavity of the ER. This conception certainly holds for the synthesis of antibodies. The ER of mature plasma cells that are actively making antibodies is bloated by the accumulation of vast numbers of antibody molecules.

There is, however, no reason to presume that all the protein made on the ER is immediately released into the inner cavity of the sacs. Many polypeptide chains might remain attached in the lipid bilayer with their NH_2 ends protruding inward while their COOH ends face outward. Whether a given protein chain is secreted or remains trapped might,

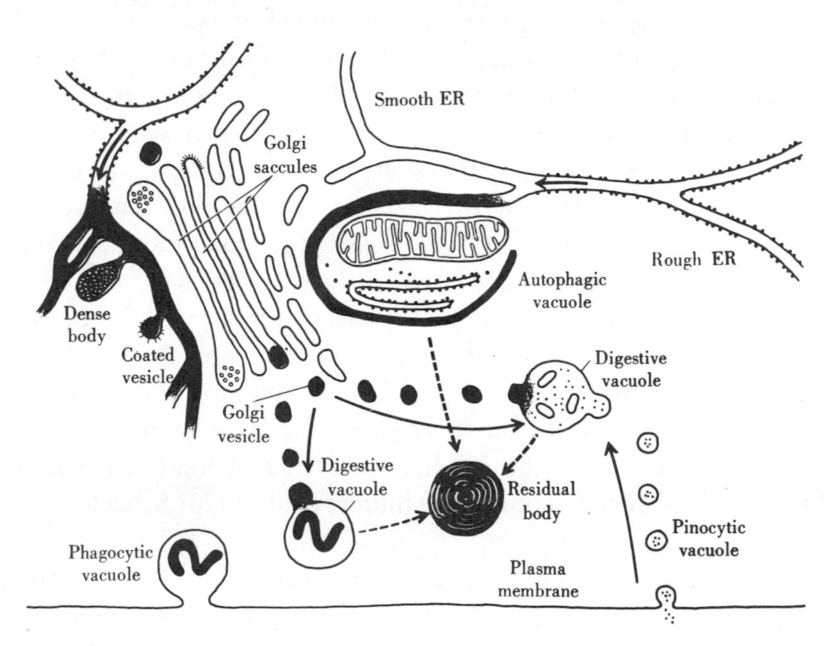

Figure 3.19. A schematic view of the endoplasmic reticulum (ER) showing probable relationships between the rough ER, the smooth ER, the Golgi apparatus, lysosomes, and the several secretory granules. Reproduced with permission from Novikoff and Holtzman, *Cells and Organelles*, Holt, Rinehart, and Winston, 1970.

one can speculate, depend upon its amino acid sequence. Polypeptide chains with a large number of contiguous hydrophobic residues might be trapped in the hydrophobic center of the lipid bilayer. Until recently this possibility was never seriously considered because it was argued that such "trapped" proteins would never reach their final destination. But now that we know that most, if not all, membranes are two-dimensional fluids, it is easy to imagine a "trapped" protein diffusing to the site at which it will function.

Membrane-associated Transferases

The enzymes which add sugar groups to newly synthesized proteins are found either within or bound to the smooth (free of ribosomes) portion of the ER or to the extensions of it that are called the Golgi bodies. While a few sugar groups may be added to nascent polypeptide chains, most substitution occurs after the polypeptide chains have been synthesized and moved into the smooth ER (Rambourg et al., 1969; Bennett and Leblond, 1970). The sugar donor in each case is a sugar (UDP) (CMP) nucleotide with each transfer requiring a glycosyl transferase, specific both for the sugar nucleotide and the accepting molecule (Roseman, 1968, 1970). Given the complexity of many membrane glycoproteins, some of which contain as many as 100–200 different hexose groups, the number of enzymes involved must be large, and at present we have only a very preliminary idea of the enzymology of glycoprotein synthesis. It is thought likely that most oligosaccharide chains are constructed by multienzyme complexes (a multiglycosyl-transferase system) with the product of each reaction serving as the substrate for the next reaction (Bouchilloux et al., 1970; Schacter et al., 1970).

Many of the specific oligosaccharide sequences found in glycoproteins are also found in glycolipids, and evidence is now accumulating that given glycosyltransferases are often used to put together the oligosaccharide components of both glycolipids and glycoproteins. For example, the sequence GalNAc-(α1 $\rightarrow$ 3)-Gal . . . occurs in erythrocyte glycolipids, in mucous glycoproteins and in free oligosaccharides of urine and milk of blood group A individuals. It is not found, however, in material from blood groups B or 0 individuals (Ginsburg and

Kobata, 1971). Given that the inheritance of blood type A is controlled by one gene, the *N*-acetylgalactosamine transferase coded for by this gene is likely to be involved in the synthesis of many different molecules.

Whether the various glycosyltransferases exist free within the cavities of the ER sacs or whether they are normal membrane constituents with their active sites facing inward is not known. Hints that they are fairly tightly bound to the membrane of the ER come from recent work demonstrating that such enzymes can also be found attached to the external plasma membranes (Bosmann, 1971; Roth et al., 1971a, b). Here they catalyze the addition of new hexose groups to preexisting plasma membrane components. This location is not surprising since an enzyme that faces toward the inner side of an ER sac must face outward from the cell if reverse pinocytosis transfers it to the cell surface.

The possibility has been raised (Roth et al., 1971b; Roth and White, 1972) that certain surface glucosyltransferases may only catalyze sugar transfer to acceptor molecules on adjacent cells ("trans" glycosylation), with such transfers playing a vital role in cell adhesion. Under this hypothesis these enzymes would act as "locks" in a lock and key model for intercellular adhesive recognition. The corresponding keys would be the glycosyl acceptors. Such a mechanism might be the explanation for the fact (see next section) that the surfaces of cells growing confluently have more complex glycolipid patterns than surfaces of sparsely growing cells. Conceivably the terminal hexoses are normally added by enzymes with "trans" specificities.

CHEMICAL CHANGES IN TRANSFORMED CELL SURFACES

Although there have been many studies of the differences between the chemical compositions of the surface membranes of transformed and untransformed cells, there is still no clear picture of the changes in composition and architecture of the cell surface that must occur during transformation. The reasons for our ignorance are obvious enough: First, the problem is intrinsically difficult and answering it depends on defining the composition, structure and functions of virtually every component of cell surface membranes; second, many of the experiments done so far were poorly planned by the investigators. For example, many workers have compared the surface chemistry of a particular transformed cell with that of a nontransformed but also nonisologous cell. It is therefore impossible to decide whether any differences they observed are a consequence of transformation rather than the result of genetic differences. Other investigators have compared the surface chemistry of cells which were not grown in identical conditions and any differences they found may have been caused by manipulations such as harvesting the cells. At present we have no way of knowing whether trypsinization, for example, has the same effect on the surfaces of transformed and untransformed cells. Finally investigators, especially those who have failed to detect differences between the surface chemistries of transformed and untransformed cells, have often failed to present data (saturation densities to which the cells were grown, the tumourigenicity of the cells, their susceptibility to agglutination by lectins) sufficient to convince their audience that the untransformed cells really were untransformed and that their transformed cells really were transformed.

But despite these several provisos the following observations seem generally to be accepted:

1. Histochemical stains specific for carbohydrate-containing components of membranes stain the surfaces of tumour cells and transformed cells more intensely than the surfaces of untransformed cells (Defendi and Gasic, 1963; Rambourg and Leblond, 1967; Montagnier, 1971). This finding was first interpreted as showing that there is more carbohydrate at the surface of a tumour cell than at the surface of a normal cell. But, in light of contradictory data obtained by more direct measurements (see below), we must conclude that these differences in the intensity of staining reflect the increased accessibility of carbohydrates to the stain, and not an increase in the absolute amount of carbohydrate at the surface. Later we shall see that the same conclusion has been reached from experiments which show that most glycolipids in normal cells are unable to combine with their respective antibodies directed against their haptenic groups.

2. Direct chemical measurements have shown that cells transformed by polyoma virus and SV40 contain less *N*-acetylneuraminic acid (NANA) (Ohta et al., 1968; Wu et al., 1969; Hakomori and Murakami,

1968; Kornfeld, 1969; Mora et al., 1969; Grimes, 1970) and galactosamine (Wu et al., 1969) than their untransformed counterparts. Neuraminidase releases about 40 percent less NANA from the surfaces of cells of the six or seven lines of virus-transformants tested so far than from the surfaces of the corresponding untransformed cells. Nearly all the NANA and galactosamine in animal cells occur in glycolipids and glycoproteins, and reports exist that both of these groups of compounds isolated from transformed cells contain less of the two sugars than comparable fractions obtained from untransformed cells (Hakomori and Murakami, 1968; Wu et al., 1969; Mora et al., 1969). Concomitant with the decrease in the amount of NANA and galactosamine, there is a reciprocal increase in the relative amount of glucosamine in cells transformed by SV40 (Wu et al., 1969). These results are unlikely to arise from artifactual differences created when the cells are detached from the petri dishes on which they are growing, because when Burger (1971a) compared a BHK line that grows in suspension with its polyoma-transformed derivative, he likewise observed 30 percent less sialic acid in the transformed cells.

Equally important are measurements of flat cell revertants that arise in populations of transformed cells but have regained the growth properties of untransformed cells. These contain amounts of sialic acid comparable to the amounts in untransformed cells (Culp et al., 1971).

3. Chicken fibroblasts transformed by RSV also contain some 25 percent less sialic acid than their untransformed counterparts. By contrast, when a non-transforming leukemia virus (RAV) multiplies in chicken fibroblasts, the amount of sialic acid in the cells does not decrease (Perdue et al., 1971, 1972).

Unfortunately, interpretation of the above data in terms of the molecular architecture of the cell surface is impossible because sialic acid is a major constituent of both glycolipids and glycoproteins. Measurements of changes in total cellular concentrations of sialic acid can never reveal whether or not the amounts of only one or both of these components have changed. Direct measurements, therefore, must be made on separated glycolipid and glycoprotein fractions.

Variations in the Glycolipids of Cultured Cells

Extensive variation can occur in the glycolipid composition of cultured cells (Hakomori, 1970). In normal fibroblasts from mice and chickens, as well as from animals of many other species, the principal glycolipids are sets of several types of gangliosides (Mora et al., 1969). Hamster fibroblasts, by contrast, have only a single major glycolipid that contains sialic acid. It is called hematoside or GM_3 (for terminology used here see Table 3.3). The other glycolipids in hamster fibroblasts are a series of neutral molecules whose carbohydrate chains range in length from the simple monohexosylceramide (GL-1) to the pento-hexoside GL-5 that frequently goes by the name "Forssman antigen" (Hakomori, 1971; Hakomori et al., 1971a; Siddiqui and Hakomori, 1971; Sakiyama et al., 1972). Even within one cloned cell line, extensive variations can occur in the amounts and types of glycolipids present (Renkonen et al., 1972; Sakiyama et al., 1972). Table 3.4, for example, shows the glycolipids present in several derivatives of the hamster NIL cell. No correlation has yet been found between the presence or absence of any of these glycolipids and any physiological parameter such as saturation density, cell size or cell shape.

Cell Density-dependent Extension Response

The longer neutral glycolipids GL-3, GL-4 and GL-5 are usually present in larger amounts (Table 3.4) in hamster cells grown to confluence than in the cells of sparse cultures (Hakomori, 1970; Robbins and Macpherson, 1971; Kijimoto and Hakomori, 1972; Sakiyama et al., 1972). Conceivably cell-cell contact is necessary for the synthesis of these molecules. A similar situation holds for cultured human diploid fibroblasts; in these cells concentrations of monosialohematoside (GM_3), disialohematoside and monosialoganglioside (GM_1) all increase as cultures become dense and reach confluence (Figure 3.20).

Correspondingly the activity of the biosynthetic enzyme responsible for the addition of the terminal hexose of ceramide trihexoside (UDP-galactose: lactosylceramide alpha-galactosyltransferase) increases 2- to 3-fold as the cultures reach confluence (Kijimoto and Hakomori,

1971). By contrast, the activity of the enzyme responsible for the synthesis of GL_2 does not appear to be dependent on the cell density of the culture.

Lack of Extension Response in Transformed Cells

Similar experiments with cells of several different transformed cell lines indicate that transformed cells lack this glycosyl extension response. In SV40-transformed 3T3 cells (SV3T3) the amounts of disialo- and monosialogangliosides are much less than the amounts in untransformed cells; also the amounts of these gangliosides are not dependent on the density to which the cells are grown (Mora et al., 1969; Sheinin et al., 1971). Correspondingly the amount of UDP-*N*-acetylgalactosamine: hematoside *N*-acetylgalactosaminyltransferase activity is lower in SV40-transformed mouse cells than in untransformed mouse cells (Cumar et al., 1970), but in flat SV3T3 revertants it is as high as in untransformed cells (Mora et al., 1971). Similarly in the hamster system, BHK and NIL cells transformed by polyoma virus contain less of the larger GM_2, GL-5, GL-4 and GL-3 glycolipids than untransformed cells, but transformation barely changes the amounts of GL-1 and GL-2. Moreover, no significant differences in glycolipid concentrations are found between transformed hamster cells growing in sparse and in confluent cultures (Hakomori and Murakami, 1968; Hakomori et al., 1968; Hakomori, 1970, 1971; Sakiyama et al., 1972).

In BHK cells transformed by polyoma virus (PyBHK) the activity of the enzyme sialyltransferase, which is responsible for the last step in hematoside synthesis, is only 15 percent of the activity in untransformed BHK cells (Den et al., 1971). This enzyme, which transfers sialic acid from CMP-NANA to lactosylceramide, is also present in reduced amounts (between 30 and 50 percent of the normal cell level) in SV40-transformed 3T3 cells (Grimes, 1970).

When RSV is added to chicken fibroblasts, over 90 percent of the cells can be transformed within two days. In this system glycolipid changes can be studied without the complications of prolonged periods of cell culture that might produce further genetic heterogeneity unrelated to the primary transformation event. Transformation is accompanied by significant changes in the cell's glycolipid content which

Table 3.3. SOME MAJOR GLYCOLIPIDS OF ANIMAL CELLS

Series	Structure	Name
1. Simple basic structure	$\text{Gal} \xrightarrow{\beta} \text{Cer}$	Cerebroside
	$\text{Gal} \longrightarrow \text{Gal} \longrightarrow \text{Cer}$	
	$\text{Glu} \longrightarrow \text{Cer}$	CMH (GL-1)
	$\text{Gal 1} \longrightarrow \text{4 Glu} \longrightarrow \text{Cer}$	CDH (GL-2)
2. Sulfatides	$HSO_3 \longrightarrow \text{3 Gal} \longrightarrow \text{Cer}$	Sulfatide
	$HSO_3 \longrightarrow \text{3 Gal 1} \longrightarrow \text{4 Glu} \longrightarrow \text{Cer}$	
3. Globoside series	$\text{Gal 1} \xrightarrow{\alpha} \text{4 Gal 1} \xrightarrow{\beta} \text{4 Glu} \longrightarrow \text{Cer}$	CTH (GL-3)
	$\text{GalNAc 1} \xrightarrow{\beta} \text{3 Gal 1} \xrightarrow{\alpha} \text{4 Gal 1} \xrightarrow{\beta} \text{4 Glu} \longrightarrow \text{Cer}$	Globoside (GL-4)
	$\text{GalNAc 1} \xrightarrow{\alpha} \text{3 GalNAc 1} \xrightarrow{\beta} \text{3 Gal 1} \xrightarrow{\alpha} \text{4 Gal 1} \xrightarrow{\beta} \text{4 Glu-Cer}$	FORSSMAN (GL-5)
4. Hematoside series	$\text{NANA 2} \xrightarrow{\alpha} \text{3 Gal 1} \xrightarrow{\beta} \text{4 Glu} \xrightarrow{\beta} \text{Cer}$	Monosialohematoside
	$\text{NGNA 2} \xrightarrow{\alpha} \text{3 Gal 1} \xrightarrow{\beta} \text{4 Glu} \xrightarrow{\beta} \text{Cer}$	(GM_3)
	$\text{NANA 2} \xrightarrow{\alpha} \text{8 NANA 2} \xrightarrow{\alpha} \text{3 Gal 1} \xrightarrow{\beta} \text{4 Glu} \xrightarrow{\beta} \text{Cer}$	Disialohematoside
	$\text{NGNA 2} \xrightarrow{\alpha} \text{8 NGNA 2} \xrightarrow{\alpha} \text{3 Gal 1} \xrightarrow{\beta} \text{4 Glu} \xrightarrow{\beta} \text{Cer}$	
	$\text{NGNA 2} \xrightarrow{\alpha} \text{3 Gal 1} \xrightarrow{\beta} \text{4 Glu} \xrightarrow{\beta} \text{Cer}$ ↑ AcO	
	$\text{NANA 2} \longrightarrow \text{3 Gal 1} \xrightarrow{\beta} \text{4 Glu} \xrightarrow{\beta} \text{Cer}$ ↑ AcO	

5. Ganglioside series

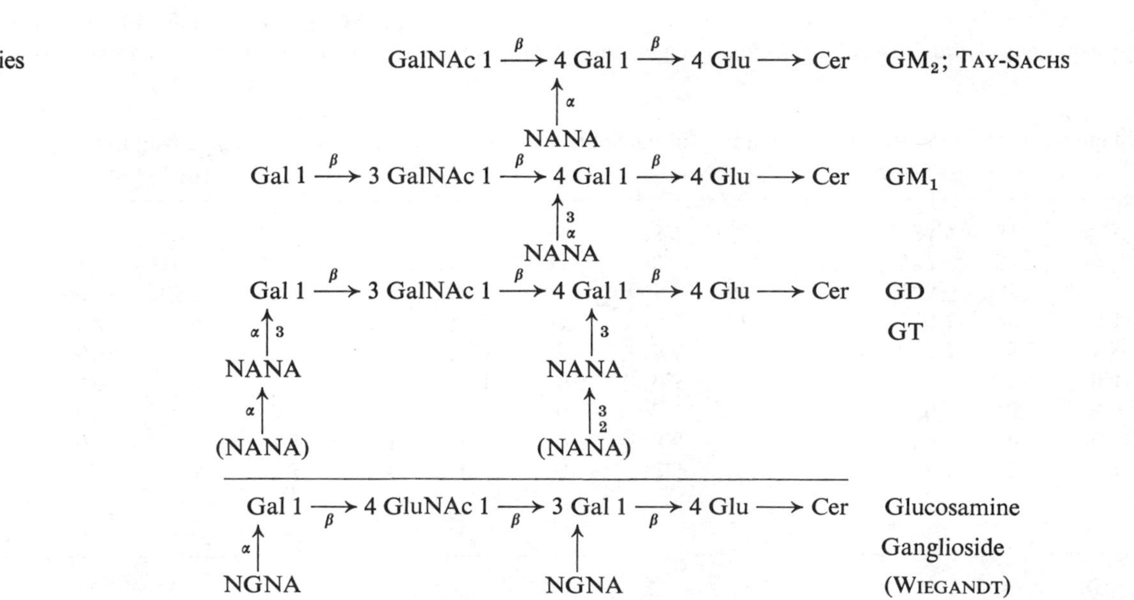

From Hakomori, 1971.

Table 3.4. SYNTHESIS OF LIPIDS BY SEVERAL NIL CELLS AT HIGH AND LOW DENSITY

NIL cell line	B1		2e		1c1		2c1		2c2		2c3	
Saturation density[a]	4.8×10^6		4.2×10^6		2.5×10^6		7.0×10^6		3.5×10^6		3.5×10^6	
Growth phase	G	C	G	C	G	C	G	C	G	C	G	C
Incorporation of [^{14}C]palmitate[b]												
GL-5	5.1	10.6	4.8	6.4	0	0	12.3	33	2.0	8.5	0.8	2.7
GL-4	0	0	3.8	10.0	0	0	3.6	16.5	0	0	0	0
GL-3	0	0	0	0	0	0	0	2.6	0	0	0	0
GL-2	3.5	1.1	0	0.5	11.4	10.5	0.5	1.3	0.6	0.5	0.4	0.4
GL-1	6.7	5.8	3.2	1.6	5.9	8.1	5.6	5.1	1.2	3.2	3.2	1.8
GM_3	17.4	31.4	7.9	4.1	27.5	53.2	19.3	18	9.9	10.4	16.6	6.5
PC + SM	790	765	800	725	706	694	695	660	734	706	755	761
PE	131	116	132	113	117	127	158	154	146	147	137	90.5
PI	11.6	30.1	14	38.5	31.6	36	38.7	44.8	35.9	54.5	31	48.3

Modified from Sakiyama et al., 1972.

G, growing culture; C, confluent culture; SM, sphingomyelin; PC, phosphatidylcholine; PE, phosphatidylethanolamine; PI, phosphatidylinositol.

[a] Number of cells per 5-cm plastic plate.

[b] Incorporation into individual lipids relative to incorporation into all lipids. The quantities represent cpm × 1000 divided by total cpm incorporated into phospholipids and glycolipids.

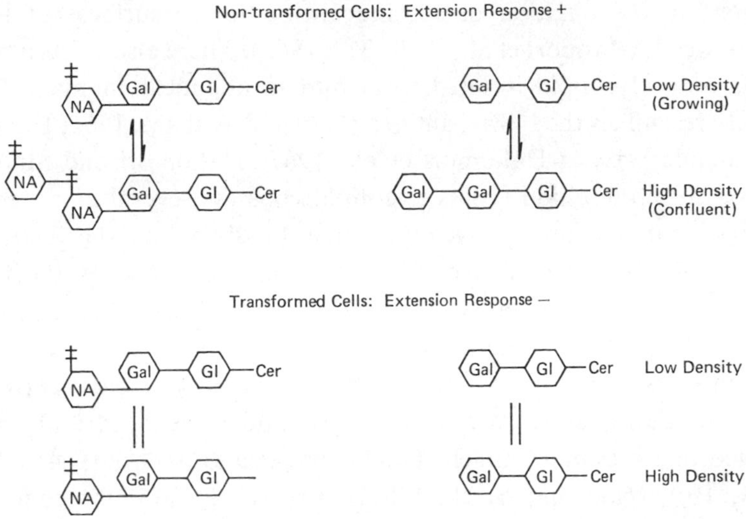

Figure 3.20. Schematic expression of the presence of "glycosyl extension response" in normal cells and absence of it in transformed cells. This scheme shows how, in normal fibroblastic cells, hematoside or CDH could be converted on cell-to-cell contact to disialohematoside and CTH at high cell population density, while this ability is lacking in the transformed cells and the quantity of hematoside or CDH is constant, irrespective of the cell population density. Reproduced with permission from S. Hakomori, *The Dynamic Structure of Cell Membranes*, Springer-Verlag, 1971.

become detectable within 60 hours of virus infection. Throughout the infection the amounts of detectable hematoside, disialohematoside and a monosialoganglioside progressively decrease (Hakomori et al., 1971b). By contrast, the amounts of GL-1 and a ceramide increase. None of these changes are observed when the cells are infected with an avian leukemia virus which grows well in chicken fibroblasts, but never transforms them. This finding confirms that there is a correlation between the lack of glycosyl extension response and transformation.

Glycolipids of Human Tumour Cells

The surfaces of a variety of human tumour cells possess exposed glycolipid groups that are not exposed at the surfaces of normal cells and that lead to the production of specific antibodies directed against the tumour cells. Free lactosylceramide acts as a competitive hapten, preventing combination of specific antibodies to surfaces of human tumour cells (Rapport et al., 1959; Tal, 1965). There also exists a fucose-containing glycolipid isolated from human adenocarcinoma cells that closely resembles the Lewis[a] antigen (Kay and Wallace, 1961; Hakomori and Jeanloz, 1964; Hakomori et al., 1967; Hakomori and Strycharz, 1968). Since the blood group glycolipids possess several more terminal sugars than the Lewis[a] glycolipid, it is possible that the fucose-containing glycolipids that resemble Lewis[a] antigen arise by the loss of enzymes that function late in the production of A and B antigens, both of which are now thought to be pure glycolipids, not glycoproteins (Uhlenbruck, 1967). Furthermore, there seems to be a close correlation between the transformed cell phenotype and the exposure on the cell surface of Forssman antigen (GL-5) (Fogel and Sachs, 1964; O'Neill, 1968; Robertson and Black, 1969; Burger, 1971b). This compound, like Lewis[a] antigen, lacks many of the terminal sugars possessed by the more complicated higher gangliosides of normal cells. It may well be

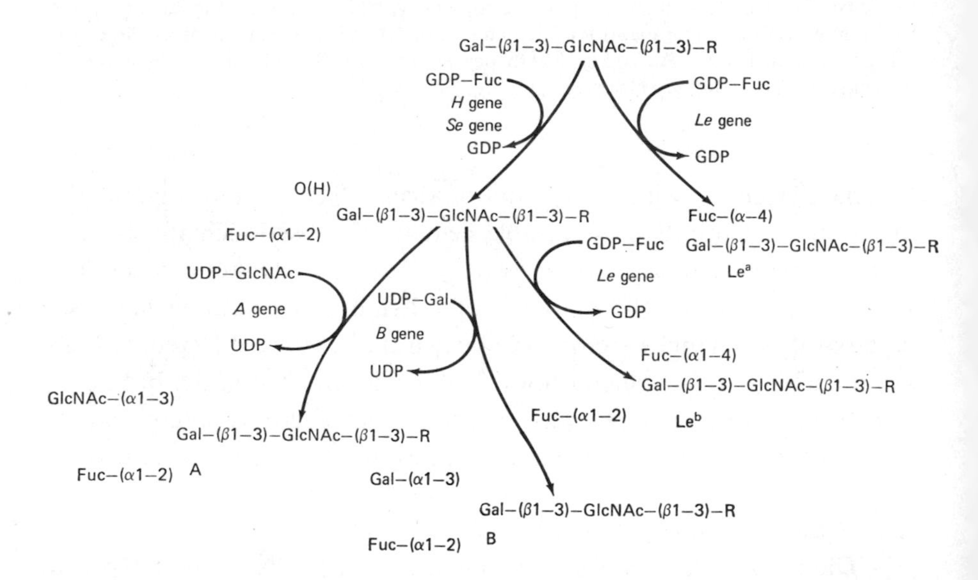

Figure 3.21. Biosynthesis of structures responsible for ABO and Lewis blood types in man. Revised from Kobata et al., *Biochem. Biophys. Res. Commun.* 32, 272, 1968.

that the glycolipids of transformed and untransformed cells always share a common core (Figure 3.21), but in transformed cells specific terminal sugars, such as NANA or galactosamine, are not added.

Masking of Membrane Glycolipids in Normal Cells

Many membrane glycolipids are effectively buried in the cell surface and do not react with antisera directed against them. For example, the most abundant glycolipid of the human erythrocyte membrane is GL-4 (globoside); it constitutes some 80 percent of the total glycolipids. Yet anti-globoside antiserum completely fails to bind to erythrocyte membranes. In contrast, antiserum directed against the much less common A and B blood group glycoproteins binds very strongly to intact red cells. The failure of the globosides to react with antisera may result from shielding by specific erythrocyte proteins, for when erythrocytes are treated with trypsin, they become competent to bind large quantities of anti-globoside antibody. A similar situation has been found by Uhlenbruck et al. (1970) using an invertebrate agglutinin that is specific for *N*-acetylgalactosamine, the terminal sugar of both GL-4 (globoside) and GL-5 (the Forssman antigen).

Globosides, however, are not masked in fetal erythrocytes, a fact which suggests that sometime during ontogenetic development they become cryptic (Hakomori, 1969). They are also much more available for antiserum binding in transformed PyBHK or SV3T3 cells (Hakomori et al., 1968). Most importantly, while trypsin treatment increases globoside availability in normal 3T3 or BHK cells, it has no effect on transformed cells, a finding which strongly parallels the finding of Burger (1969) and Inbar and Sachs (1969a) that trypsin treatment leads to increased agglutination of normal cells by wheat germ agglutinin and con A.

Proteins of Surface Membranes of Cultured Cells

Electrophoresis through SDS-polyacrylamide gels of the proteins found in the surface membranes of L-cells (mouse fibroblasts) and BHK cells reveals the presence in both cell types of more than twenty prominent components with molecular weights ranging from $\sim$15,000 to

~230,000 daltons (Gahmberg, 1971; Podoula et al., 1972). Several of these proteins are selectively iodinated when intact cells are treated with lactoperoxidase, indicating that at least some of their tyrosine and histidine residues are exposed to the exterior surface. In particular the protein(s) with a molecular weight of about 230,000 daltons appears to be very accessible to iodination when intact cells are exposed to this enzyme, which does not penetrate intact cells. This protein is presumably located at the outside surface of the membrane. Most of the other membrane proteins are, by contrast, iodinated only if purified membrane fragments are exposed to lactoperoxidase and so are presumably located for the most part on the inner surface of the cell membrane.

Patterns of Surface Glycoproteins

The idea that the transformation of a normal cell to a cancer cell may be accompanied by a change in the amounts or the types (or both) of the glycoproteins at the cell's surface has never lacked supporters (see reviews by Cook, 1968; Winzler, 1970; Roseman, 1970; Kraemer, 1971). For it is generally believed that carbohydrate groups play a crucial role in the process of cell-cell recognition and, being much larger, the glycoproteins have a greater intrinsic molecular specificity than the glycolipids. It has often been speculated, therefore, that the presence or absence of one or more glycoproteins might distinguish the surface of a cancer cell from the surface of its normal counterpart. And the recent discoveries of Burger and Sachs and their respective colleagues that carbohydrate groups are involved in the binding of agglutinins to cell surfaces (see below) is consistent with this speculation and has given renewed impetus to investigations of membrane glycoproteins.

The above-mentioned obstacles to purifying cell membranes, however, make the analysis of membrane glycoproteins extremely difficult. None the less, we do know for certain that, in sharp contrast to erythrocytes, whose surfaces contain only one major glycoprotein, other types of cells probably have a variety of different glycoproteins on their surfaces (Gahmberg, 1971; Glossmann and Neville, 1971).

For example, Gates and Morrison (1972) resolved by SDS-polyacrylamide gel electrophoresis four glycoproteins and at least ten other proteins in extracts of the surface membrane of mouse Ehrlich ascites tumour cells. The molecular weights of these four glycoproteins were estimated to be about 180,000; 130,000; 100,000 and 50,000 daltons but these numbers are only very approximate. Similarly Allan and Crumpton (1972) suggest that seven of the twenty or so proteins that can be resolved by SDS gel electrophoresis of extracts of plasma membranes of human thymocytes are glycosylated. One or more of these glycosylated proteins carries the HLA antigenic determinants. Human HLA transplantation antigens and corresponding murine H2 antigens have been purified by several groups (Mann et al., 1968; Davies et al., 1968; Sanderson, 1968; Hammerling et al., 1971; Sanderson et al., 1972; Yamane et al., 1972); the HLA antigenic activity resides in glycopeptide fragments (molecular weight ~45,000) that are released by limited proteolytic (papain, ficin and bromelin) digestion of surface membranes.

Immunoglobulin molecules must also be common components of the surfaces of lymphocytes (Raff et al., 1970; Pernis et al., 1970; Coombs et al., 1970; Rabellino et al., 1971; Diener and Paetkau, 1972; Cone et al., 1972). The extent to which immunoglobulins are glycosylated varies, γM immunoglobulins having more carbohydrate residues than γG immunoglobulins. Five different oligosaccharide side groups are attached to a species of γM immunoglobulin isolated from a patient with Walderstrom macroglobulinemia. Each side group is located at one or more specific sites along the heavy chain (Shimizu et al., 1971). Although each carbohydrate side chain has a specific structure (Hickman et al., 1972), they all seem to be derived from a common core structure consisting of

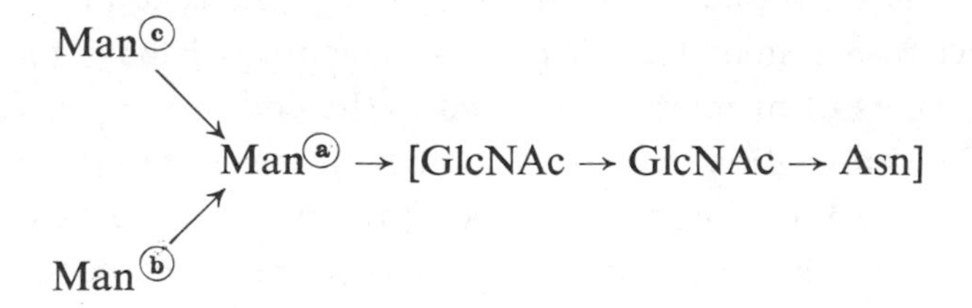

One side group contains two to four additional mannose groups while two others have outer branches containing sialic acid and Gal → GlcNAc sequences, as well as a fucose residue linked to one of the two

GlcNAc residues in the core. Conceivably the surrounding amino acid sequences influence the activity of the various glycosyltransferases, and so the same set of enzymes could be involved in the synthesis of oligosaccharides positioned at many different sites along a polypeptide chain.

Difficulties in Comparing the Glycoproteins of Normal vs. Transformed Cells

Unfortunately we still have no way of telling what degenerative changes occur to most membrane proteins as we disrupt cells and isolate membrane fractions. As a result many of the differences so far reported between the glycoproteins of membranes isolated from transformed and untransformed cells may be artifacts quite irrelevant to the putative changes which accompany transformation. In particular great care has to be taken to minimize the action of hydrolytic enzymes during the separation of membrane fractions, something which is easier said than done, not least because many cells grow only when they are firmly attached to glass or plastic surfaces and chelating agents, such as EDTA, or proteolytic enzymes, such as trypsin, have to be used to detach them.

Over the past several years the methods developed by Warren and Glick (see 1969 and 1971 reviews) to isolate large fragments of plasma membranes have been widely used. The cells are broken open in a homogenizer in the presence of either fluorescein mercuric acid or zinc chloride. These reagents stabilize the resulting membrane fragments during the several centrifugations through sucrose gradients, conceivably by inhibiting the action of many of the hydrolytic enzymes released when the cells are broken. But since no proof exists that all these degradative enzymes are inhibited, there is no reason to believe that the isolated fragments of membrane obtained at the end of the procedure have the same composition as the surface membrane of intact cells.

Indeed we know that procedures involving exposure of cells to even very dilute concentrations of trypsin result in the loss of significant amounts of cell surface. For example, exposure to 0.25% trypsin for 10 minutes at 37°C can remove 30 to 80 percent of the surface carbohydrate components (galactose, mannose, fucose, glucosamine and NANA) without changing the viability of the cells (Shen and Ginsburg,

1968; Onodera and Sheinin, 1970); only components lying outside the lipid bilayer are thus affected. Most likely these carbohydrate groups are derived from glycoprotein molecules that are sensitive to trypsin. And by analogy with recent work on erythrocyte membranes, it is likely that mild trypsin treatment usually digests only that portion of a membrane protein which is external to the lipid bilayer.

Depending on the extent of the trypsin digestion and on the protein, the molecular weights of the released fragments range from <3000 to $>100,000$ daltons. For example, after mild digestion with trypsin, mouse mammary adenocarcinoma ascites cells (Codington et al., 1972) liberate two very large glycopeptides with molecular weights of 100,000 and 250,000 daltons; 65 percent of the amino acids in these fragments are serine and threonine residues, and about 70 percent of their mass is carbohydrate, with galactose, *N*-acetylgalactosamine, *N*-acetylglucosamine and sialic acid present in proportions to 4:2:1:1.

Fucose-containing Glycopeptides in Viral Transformants

The enzymic removal of glycopeptides from the surfaces of intact, viable cells need not always be disadvantageous. It can also be turned to advantage and exploited to compare the external surfaces of intact transformed and untransformed cells. Warren and his colleagues, for example (Buck et al., 1970, 1971a, b), have shown that surface glycopeptides removed by trypsin from BHK cells transformed by RSV are larger than the corresponding molecules removed from untransformed BHK cells. Experiments involving the digestion of these glycopeptides with neuraminidase reveal that those obtained from transformed cells have more sialic acid residues than those from untransformed cells (Warren et al., 1972b).

One interesting experiment suggests that these differences in the amount of sialic acid are related to transformation (Warren et al., 1972a). When chick fibroblasts transformed by the temperature-sensitive mutant strain of RSV called *t*-5 are grown at 35°C, they have the morphology of typical RSV transformants; the cells are rounded and refractile and they grow in foci. When these cells are shifted from 35 to 41°C, their morphology changes to that of untransformed chick fibroblasts; they become spindle-shaped and grow in oriented arrays.

Growth at 41°C also alters the concentration of the larger fucose-containing glycopeptides; more can be isolated from these cells grown at 35°C than from these cells grown at 41°C. In other words, the amount of fucose-containing glycopeptide that could be released by trypsin was dependent upon the phenotype of the cells. Similar results were found with BHK cells transformed by polyoma virus and with Balb/c-3T3 cells transformed by mouse sarcoma virus.

These increases in the amount of sialic acid in surface glycoproteins of transformed cells may be due to increases in the activity of sialic acid transferase (sialyltransferase). It has been claimed that in chick cells transformed by RSV the specific activity of sialyltransferase is 2.5–11 times that in untransformed chick cells (Warren et al., 1972b). Likewise, SV3T3 cells and Py3T3 cells have been reported to have greater sialyltransferase activities than 3T3 cells (Bosmann et al., 1968). On the other hand, Grimes (1970) and Cumar et al. (1970) have reported that sialyltransferase activities in transformed cells are lower than in untransformed cells. And Fishman et al. (1972) stated recently that the only consistent change in a glycosyltransferase activity that they could observe in cells transformed either by polyoma virus or SV40 was a greatly reduced activity of UDP-GalNAc: hemotoside *N*-acetyl-galactosaminyltransferase. They failed to detect reproducible changes in sialyltransferase activity in transformants. Conceivably more than one type of sialyltransferase exists, and the different laboratories have measured different enzymes.

When galactosamine instead of fucose was used as a marker for membrane glycopeptides, somewhat conflicting results were obtained by the different groups that compared normal and transformed cells. Meezan et al. (1969) labeled 3T3 and SV3T3 cells with glucosamine and compared the glycoproteins of various membrane fractions by chromatography on Sephadex columns in the presence of SDS. This method separated membrane proteins largely on the basis of molecular weight. Two peaks of material were found which occurred in greater amounts in SV3T3 cells than in 3T3 cells, while two other peaks were found in larger quantities in the untransformed cells. These differences were found in all the membrane fractions examined (mitochondrial, nuclear and endoplasmic reticulum). By contrast, Sakiyama and Burge (1972) found no differences in the acrylamide gel electrophoresis patterns of fractions of smooth membrane from SV3T3 cells and 3T3 cells labeled with glucosamine. However, gel electrophoresis of the glycoproteins "excreted" into the medium by 3T3 cells revealed a prominent glycoprotein with a molecular weight of 150,000 daltons that was not found in the medium in which SV3T3 cells had grown.

Sheinin and Onodera (1972) examined the galactosamine-containing glycopeptides released by trypsin from intact 3T3 cells. They isolated five main glycopeptides from 3T3 cells, two of which were missing in trypsinates of 3T3 cells transformed by SV40. Nothing so far is known about the chemical composition of these molecules. Subsequently, Sheinin and Onodera (1972) used electrophoresis in SDS gels to examine the glycoproteins of the cell surface that remain after exposure to trypsin. Though the patterns of glycoproteins from normal and transformed cells differed, almost equally large differences were seen between the surface glycoproteins of several different clones of SV3T3 cells. Whether or not any of these changes are related to the primary transformation event is not clear.

Glycoproteins Released from Transformed Cells

The resolution of these conflicting observations may come from further investigation of an observation made by Chiarugi and Urbano (1972). They found that PyBHK cells released more glycosylated macromolecular material into their surrounding medium than did BHK cells. Many of these released molecules had molecular weights of about 150,000 daltons and they may be the BHK counterparts of the molecules that Sakiyama and Burge (1972) found to be preferentially "excreted" from SV3T3 cells.

One can speculate that the increased "excretion" of glycopeptides from transformed cells is brought about by proteases that are secreted into the culture medium, and that one important difference between transformed and untransformed cells is that the former liberate more proteases than the latter. By placing restrictions on the specificity of these putative proteases, this hypothesis can be elaborated to explain several of the observations we have discussed above. This possibility that changes in the composition of the cell membrane after transformation might result from the action of extracellular proteases secreted by the transformed cells rather than from blocks in the biosynthetic pathways leading to the cell surface molecules deserves thorough investigation.

INTERACTION OF CELL SURFACES WITH LECTINS

Lectins (or agglutinins) are molecules, usually isolated from plant seeds, that have the ability to agglutinate a variety of animal cells, particularly erythrocytes (Renkonen, 1948; Makela, 1951; Boyd and Shapleigh, 1954; Boyd et al., 1961; Boyd, 1963). A number of lectins have been chemically characterized and Table 3.5 lists some of their properties.

In the past plant agglutinins have usually been referred to as phytohemagglutinins or phytoagglutinins. But since cell-agglutinating proteins also occur in organisms other than plants, for example the snail, the horseshoe crab and certain fish, the term lectins proposed by Boyd (1963) is now used to encompass all agglutinating proteins.

Given lectins often possess other interesting properties. Some are specific for certain human blood groups (ABO and MN) and so have been used in blood typing as well as more general investigations of the molecular basis of blood group specificity. Even more important is the mitogenic capacity of certain lectins. They can stimulate the transformation of lymphocytes from small resting cells into large blast-like cells which undergo mitotic division. For this reason they are being used to study the biochemical events that characterize the conversion of a quiescent cell into an actively growing one.

Many lectins cause both normal and transformed cells to agglutinate, but at least six (from wheat germ, soybeans, lentils, great northern beans, jack beans and castor beans) preferentially agglutinate cells from spontaneous tumours (Ambrose et al., 1961; Aub et al., 1963) as well as cells transformed by chemicals or viruses (Burger and Goldberg, 1967; Inbar and Sachs, 1969a). The two known lectins which best discriminate between normal and transformed cells are concanavalin A, the agglutinin of jack beans (Sumner and Howell, 1936; Agrawal and

Goldstein, 1967; Olson and Liener, 1967a; Inbar and Sachs, 1969a; Ben-Bassat et al., 1970), and wheat germ agglutinin (Aub et al., 1963, 1965; Burger and Goldberg, 1967; Burger, 1969; Nagata and Burger, 1972). Both have been purified to homogeneity and are now being intensively studied as a means of probing the differences between the membranes of normal cells and transformed cells.

In spite of the vast amount of work now being done with lectins, their role in nature has not yet been established. Numerous speculations have been made, however, as to how they may function. Among the suggestions: They act as antibiotics against soil bacteria; they protect against fungal attack by inhibiting fungal polysaccharases; that because of their capacity to bind to specific sugars, they function in sugar transport; they serve as attachment points for glycoprotein enzymes in multienzyme complexes; they function to control cell division and germination in plants. Real evidence for or against any one of these ideas is still not available and so it is conceivable that the biological functioning of lectins is unrelated to the roles that they have in laboratory experiments.

Structures of Concanavalin A and Wheat Germ Agglutinin

Concanavalin A (con A) is a metalloprotein, which below pH 6 exists as a dimer but above pH 7 assumes a tetrametic form (Kalb and Lustig, 1968), composed of four identical subunits, each with a molecular weight of 27,000 daltons (Olson and Liener, 1967b; Wang et al., 1971). Early X-ray crystallographic studies at 4 Å resolution showed that the tetrameric form of the molecule, which has a molecular weight of 102,000 daltons, consists of two ellipsoidal dimers joined at right angles along a common dyad axis to form a "pseudotetrahedral" molecule (Quiocho et al., 1971; Hardman et al., 1971). The two subunits joined along the axis are more firmly associated than the other two subunits. The amino acid composition of concanavalin A shows no unusual features except for the lack of half-cystine residues (Olson and Liener, 1967a; Agrawal and Goldstein, 1968a) and, unlike most lectins, it is not a glycoprotein. Each subunit contains two metal binding sites, the first directed toward divalent transition metals (except Cu^{++}

Table 3.5. Chemical and Biological Properties of Highly Purified Lectins

Source		Molecular weight	Subunits	Carbohydrates		Mitogenic activity	Specificity		No. of binding sites	Metal requirement
				Percentage	Major constituents		Human blood type	Sugar		
Higher plants: Leguminosae										
Canavalia enisformis (jack bean)		55,000	2	0		+		α-D-Man	2	+
Dolichos biflorus (horse gram)		140,000		3.8	GlcN, Man	−	A	α-D-GalNAc		
Glycine max (soybean)		110,000		5	D-GlcNAc, D-Man	−		D-GalNAc	2	+
*Lens culinaris** (common lentil)		42,000–69,000	2	2	GlcN, Glc	+		α-D-Man	2	+
Lotus tetragonolobus	I	120,000		9.4	GlcN, Hexose		H(O)	α-L-Fuc	4	
	II	58,000		4.8	GlcN, Hexose		H(O)	α-L-Fuc	2	
	III	117,000		9.2	GlcN, Hexose		H(O)	α-L-Fuc	4	
Phaseolus lunatus† (lima bean)	I	269,000	8	4	GlcN, Man, Fuc		A			+
	II	138,000	4	4	GlcN, Man, Fuc		A			+
Phaseolus vulgaris (black kidney bean)		128,000		5.7	Hexosamine Man, Xyl					
Phaseolus vulgaris‡ (red kidney bean)	I	138,000	8	8.9	GlcN, Man	+				
	II	98,000–138,000	4	4.1	GlcN, Man	+		D-GalNAc		
Phaseolus vulgaris (yellow wax bean)				10.4	GlcN, Man, Glc, Ara					

Pisum sativum (garden pea)		53,000		(0.3)	(Glc)		D-Man		+
Robinia pseudoacacia (black locust)		90,000		10.7	GlcN, Man, Fuc, Xyl				
Ulex europeus (gorse)	I	170,000		5.2	GlcN, Man	H(O)	L-Fuc		
	II			21.7	GlcN, Man, Gal, Ara	H(O)	$(\text{D-GlcNAc})_2$		
Other high plants									
Ricinus communis (castor bean)		98,000					D-Gal		
Solanum tuberosum (potato)		(20,000)		5.2	Ara		$(\text{D-GlcNAc})_2$		
Triticum vulgaris (wheat)		26,000		4.5			$(\text{D-GlcNAc})_2$		
Lower plants									
Agaricus campestris (meadow mushroom)		64,000	4	4		+			
Invertebrates									
Helix pomatia (vineyard snail)		100,000		7.3	Gal, Man (hexosamine)	A	α-D-GalNAc	6	
Limulus polyphemus (horseshoe crab)		400,000	18				Sialic acid		+

From Sharon and Lis, 1972.

Roman numerals indicate different lectins isolated from the same plant.

* Also known as *Lens esculenta*.

† Also known as *Phaseolus limensis*.

‡ Two (or more) lectins with distinct activities appear to be present in the red kidney bean, an erythroagglutinin (I) and a leucoagglutinin (II).

and Hg^{++}) and the second toward Ca^{++} (Agrawal and Goldstein, 1968b; Kalb and Levitzki, 1968; Kalb and Lustig, 1968). Agglutination by concanavalin A can only be accomplished when both metal binding sites are filled.

More recent X-ray analysis, together with amino acid sequence analysis, has revealed the exact three-dimensional conformation of concanavalin A (Edelman et al., 1972). About half of the 238 amino acids which make up each chain are arranged in two anti-parallel, pleated-sheet (β) structures with most of the non-β-structure residues having essentially random coil configurations; the final resulting subunit is a globular protein of overall dimensions 42 Å × 40 Å × 39 Å. The saccharide binding site, deduced (Becker et al., 1971) by studying differences in electron density following binding of specific saccharides, is approximately 20 Å from the metal binding sites and some 15 Å from a still different site which binds myoinositol (Hardman and Ainsworth, 1972).

Wheat germ agglutinin (WGA) apparently consists of a single polypeptide chain with a molecular weight of approximately 22,000 daltons (Nagata and Burger, 1972; LeVine et al., 1972). Analysis of the amino acid composition of this lectin shows that it is extremely rich in half-cystine and lysine residues; these two amino acids together constitute almost 40 percent of the molecule. A number of sugar residues are covalently attached to each molecule; glucose (3–5 moles/mole of agglutinin), xylose (1–2 moles/mole of agglutinin) and hexosamine (1–2 moles/mole of agglutinin) are the major carbohydrate components. Unlike concanavalin A, wheat germ agglutinin does not appear to require metal ions for its agglutinating activity.

The amino acid compositions of concanavalin A and wheat germ agglutinin are so different that it is almost certain that the two proteins are not phylogenetically related. It is, therefore, likely that the lectins, as a group of proteins, are related only by their abilities to agglutinate cells, a property that arises from their ability to bind tightly to carbohydrate groups on the cell surface.

The sugar(s) to which a given lectin binds can be revealed either by studying inhibition of agglutination with sugar haptens or by directly measuring the affinity of binding of specific sugars to the various lectins (Goldstein et al., 1965). Such studies reveal that the sugar(s) bound by each lectin is usually highly specific (Table 3.5), and that lectin-mediated cell agglutination is inhibited by the presence of the same sugar hapten(s). However, some lectins bind to cell surface carbohydrate groups that are not directly involved in agglutination. For example, wheat germ agglutinin binds *N*-acetylneuraminic acid (sialic acid), which is abundant on the surfaces of both normal and transformed cell membranes, and yet this sugar is not an inhibitor of cell agglutination (Greenaway and LeVine, 1973).

All lectins are assumed to be di- or multivalent with respect to carbohydrate binding. One of the phytohemagglutinins of the lentil contains two binding sites for D-mannose (Stein et al., 1971); each of the subunits (molecular weight 27,000 daltons) of concanavalin A contains a single binding site for α-methylglucose (So and Goldstein, 1968), so that the tetrameric molecule is tetravalent for carbohydrate binding. Finally, wheat germ agglutinin contains several binding sites (depending on the hapten concentration) for *N*-acetylglucosamine (Greenaway and LeVine, 1973).

Brief treatment of con A with limiting amounts of trypsin or chymotrypsin yields molecules which still specifically bind α-methylglucose but which fail to cause cells to agglutinate. Burger and Noonan (1970) referred to such enzymically treated molecules as "monovalent" con A. They believe that the con A fragments failed to agglutinate cells because they could attach to only one cellular receptor. Subsequent studies (Cunningham et al., 1972), however, revealed the situation to be more complex than first imagined. Both trypsin and chymotrypsin split the 27,000 dalton con A polypeptide into small pieces and it is likely that the "monovalent" con A molecules used by Burger and Noonan were a collection of fragments of different sizes. But nevertheless, this mixture is very useful because it binds to cells without causing aggregation.

Cellular Lectin Receptors

Attempts to isolate the specific membrane molecules that bind lectins are now being made. Membrane proteins solubilized by sodium deoxycholate bind con A. The con A receptor proteins can therefore be separated using affinity chromotography with Sepharose to which con A has been attached (Allan et al., 1972). When pig lymphocyte plasma membranes solubilized by deoxycholate are passed through such columns, most of the membrane glycoproteins are preferentially bound. Some of the bound glycoproteins can be eluted from the column by the addition of the con A-specific hapten methyl α-D-glucopyranoside to yield material enriched some 450-fold for con A binding activity. When analyzed by SDS-polyacrylamide gel electrophoresis, this material displays considerable heterogeneity; it contains two glycoproteins with apparent molecular weights of 27,000 and 33,000 daltons and three glycoproteins about the size of the immunoglobulins. The biological activity of the eluted material was measured by its ability to inhibit con A-induced transformation of lymphocytes (see below). The eluted glycoprotein material had a biological activity similar to that of unseparated mixtures of lipid-free membrane proteins. It is therefore most unlikely that the glycolipids are involved in con A binding; probably the specific receptor(s) of con A is a membrane glycoprotein(s), and by exploiting affinity chromatography and other techniques, it should be possible to obtain this receptor in pure form.

The WGA receptor has also been only partially purified (Burger, 1968; Jansons and Burger, 1971). Several different isolation procedures yield from cell membranes glycoprotein fractions that specifically inhibit WGA-induced cell agglutination but do not inhibit agglutination induced by either con A or phytohemagglutinin; this glycoprotein material does, however, partially inhibit agglutination mediated by the lectin from *Lens culinaris*. Differences between the glycopeptide receptors for con A and WGA have also been found by Walborg (Wray and Walborg, 1971; Smith and Walborg, 1972), who examined the mixture of sialic acid-containing glycopeptides released from Novikoff tumour cells by digestion with papain. When this material was further digested with pronase, it yielded several small glycopeptides with molecular weights of 2000 to 3300 daltons. The larger of these fragments preferentially bound to WGA, while the smaller bound to con A. Because "monovalent" con A (see below) binds to cells and inhibits cell agglutination by normal (divalent) con A but does not inhibit agglutination by WGA, and because WGA and con A do not compete for the same binding site (Ozanne and Sambrook, 1971a), it seems that there are distinct receptors for these two lectins (Jansons and Burger, 1971).

Phytohemagglutinin (PHA), a glycoprotein from the red kidney bean, induces dormant human and other lymphocytes to undergo mitosis (Nowell, 1960), and the receptor on human red cells which specifically binds the lectin has been exhaustively characterized. Kornfeld and Kornfeld (1969) found that treatment with trypsin released from human erythrocytes a soluble glycopeptide that bound PHA, prevented erythrocyte agglutination by PHA, and blocked the mitogenic activity of PHA. The most highly purified samples of this material have a molecular weight of 2000 daltons and it is believed to have the following structure:

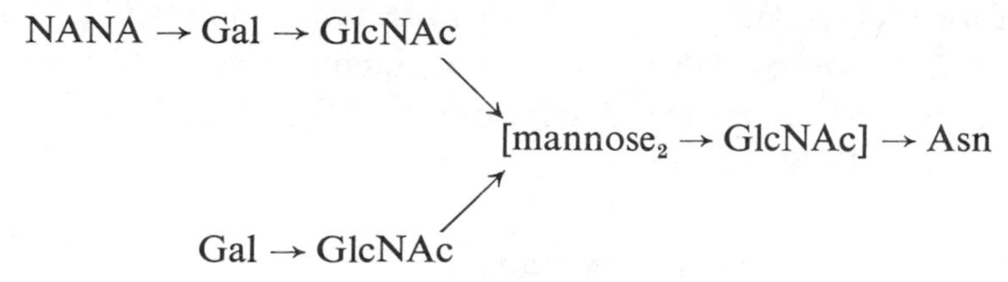

Removal of the various sugars with enzymes reveals that the determinant sugar is the galactose residue that is attached to the terminal *N*-acetylneuraminic acid.

Differential Agglutination of Transformed Cells

Almost all sorts of cells can be agglutinated by high concentrations of con A or WGA, but cells derived from tumours agglutinate at much lower concentrations of the lectins than do cells from similar normal tissue (Ambrose et al., 1961; Aub et al., 1963; Inbar and Sachs, 1969a). However, the concentrations of lectins required to agglutinate normal cells from different tissues varies; for example, some normal cells agglutinate at lectin concentrations which usually cause only tumour cells to clump (Liske and Franks, 1968; Kapeller and Doljanski, 1972). Moreover, several sorts of tumours yield cells which have been reported not to clump even in high concentrations of WGA or con A (Liske and Franks, 1968; Gantt et al., 1969). Also, the age of the animal from which the cells are taken alters the lectin-mediated agglutination of the cells in some cases. Cells from some embryonic tissues are more readily agglutinated by WGA and con A than are cells from the corresponding tissues of the adult (Moscona, 1971; Weiser, 1972). It is, however, impossible to determine how much of this variability is artifactual and stems from differences in the treatment of the cells (e.g., exposure to trypsin) as they are isolated from tissues.

Many of the problems inherent to studies with lectins of freshly isolated cells are avoided if cultured cells are used, and lectins can be used to study changes in the surface of cells after transformation in vitro. Untransformed 3T3 cells, BHK cells and cells of secondary cultures of human, mouse, hamster or chick embryo fibroblasts are less readily agglutinated by con A or WGA than are cells which have transformed spontaneously or which have been transformed in vitro by chemicals or by DNA and RNA tumour viruses (Burger and Goldberg, 1967; Inbar and Sachs, 1969a; Burger and Martin, 1972; Kapeller and Doljanski, 1972).

Correlation with Saturation Density

Differences in the lectin-mediated agglutinability of untransformed and transformed cells usually correlate with differences in their saturation densities. Pollack and Burger (1969) investigated several derivatives of 3T3 cells which had different saturation densities. They found cells with high saturation densities were agglutinated by lower concentrations of WGA than were cells with low saturation densities (Figure 3.22). Furthermore, variants of transformed cells, selected for the property of increased growth control, were less agglutinable than the parental transformed cells. Consonant with this observation, variants of SV3T3 cells, which were selected for resistance to agglutination by con A, grew to lower saturation densities than the parental transformed cells (Ozanne and Sambrook, 1971b; Culp and Black, 1972).

Further evidence that increased lectin agglutinability is associated with a loss of growth control comes from experiments with a temperature-sensitive mutant of SV3T3. At 31°C the mutant cells attained high saturation densities and were easily agglutinated by WGA. But at

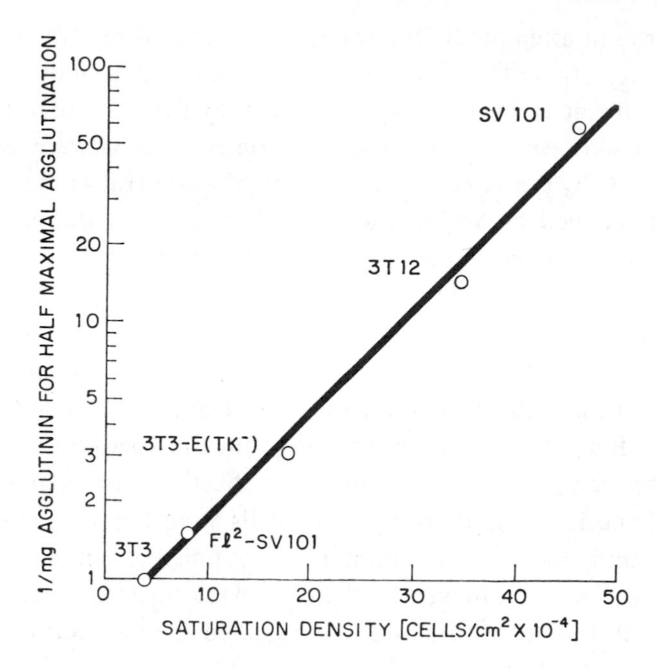

Figure 3.22. Agglutinability of cell lines derived from 3T3 mouse fibroblasts. SV101 = SV40-transformed 3T3 cell line; 3T12 = spontaneously transformed 3T3 cell line; 3T3-E(TK⁻) = thymidine kinaseless line which had lost some density-dependent inhibition of growth; Fl²-SV101 = variant isolated from transformed SV101 with FdU. Courtesy of R. Pollack and M. Burger.

39°C these cells grew to a lower saturation density and were no longer agglutinable by low concentrations of WGA or con A (Renger and Basilico, 1972). Since wild-type SV40 could be rescued from most of these mutant cells, presumably it was not an altered virus but a cellular change which increased the growth control of the cells and rendered them non-agglutinable.

Equally pertinent may be the observation that transformed cells that revert to a normal morphology when grown in the presence of Bu_2-cAMP (Hsie and Puck, 1971; Johnson et al., 1971) simultaneously become less agglutinable by lectins. Sheppard (1971) showed that Py3T3 cells exposed to Bu_2-cAMP assumed the morphology of 3T3 cells, had increased growth control and also became less agglutinable by WGA or con A.

The recent attempts by Inbar et al. (1972a) to directly correlate the degree of agglutinability of a cell line with its ability to induce tumours are very difficult to evaluate. Previously, they found that transformed cells which were easily agglutinated by con A at 24°C were not agglutinated at 4°C by the same concentrations of lectin (Inbar et al., 1971a). Other lectins such as WGA and soybean agglutinin did not display this temperature dependence. Since the cells bound equal amounts of con A at both temperatures, it was speculated that some metabolic change at a second site had to occur before the cells could be agglutinated by con A. Subsequently Inbar et al. (1972a) observed that transformed cells subcultured after trypsinization were not agglutinated by con A during the first 24 hours after replating, but they became agglutinable after longer periods in culture. Again this effect was not seen with other lectins. If the trypsinized transformed cells were harvested during the 24 hours and injected into animals, they failed to induce tumours. Tumours did arise, however, if the cells were injected after they had been several days in culture and had regained their agglutinability.

These latter results were taken as further evidence for a second site on the surface of transformed cells and on trypsinized normal cells that is necessary for agglutination mediated by con A and that plays a vital role in the formation of a tumour. Much more direct evidence, however, will have to be presented before this conclusion can be seriously considered. Conceivably, the failure of the newly trypsinized transformed cells to form tumours merely reflects the fact that the cells, unless given a 24-hour recovery period after trypsinization, quickly die after injection in a host animal.

A Direct Role for Viral Genes

Several lines of evidence suggest that when a cell has been transformed by a virus, the continued functioning of one or more of the viral genes is involved in maintaining the surface alteration that renders the cells readily agglutinable. Suggestive evidence comes from experiments with 3T3 cells abortively transformed by SV40 or by polyoma virus. Such cells became temporarily agglutinable by WGA or con A if they were allowed to undergo a round of division after infection (Inbar and Sachs, 1969a; Benjamin and Burger, 1970; Ben-Bassat et al., 1970;

Eckhart et al., 1971). When, after further cell division, the progeny cells regained a normal morphology, they lost their enhanced agglutinability. Even more direct evidence comes from experiments with BHK cells grown and transformed at the permissive temperature by the temperature-sensitive mutant *ts*-3 (see Chapter 7) of polyoma virus. At the permissive temperature of 31°C, cells transformed by *ts*-3 are easily agglutinated by WGA or con A, but after a shift to the nonpermissive temperature of 38°C, the cells behave more like untransformed cells and become less agglutinable by WGA or con A (Eckhart et al., 1971). These results imply that continued expression of viral gene(s) may be required to maintain the cells in the agglutinable state.

At first there was the suspicion that cells transformed by some RNA tumour viruses might not show any significant increase in agglutinability (Moore and Temin, 1971). But recently, several groups have shown that chicken embryo fibroblasts (CEF) transformed by RSV (Kapeller and Doljanski, 1972; Burger and Martin, 1972; Lehman and Sheppard, 1972) are more agglutinable than normal CEF. The most definitive experiments are those of Burger and Martin, who used CEF transformed by a temperature-sensitive mutant of RSV (*t*-5) which is analogous to the *ts*-3 mutant of polyoma virus. Cells transformed by RSV (*t*-5) and grown at the permissive temperature (35°C) were agglutinated by much lower concentrations of WGA than were required to agglutinate cells of the same transformant clone that had been grown at 41°C. This result proves that in these transformants an RSV gene product is required continuously to maintain enhanced susceptibility to agglutination by WGA.

Surface Changes Induced during Lytic Infections

Cells lytically infected by either SV40 or polyoma virus also become agglutinable if cellular DNA synthesis is allowed to occur (Sheppard et al., 1971; Benjamin and Burger, 1970; Eckhart et al., 1971). Detectable changes in agglutinability begin about 18 hours after infection and are completed by 24 hours.

The situation with the adenoviruses is less clear. Several lines of hamster cells transformed by adenovirus type 5 have been shown to agglutinate more readily than untransformed hamster cells (e.g., BHK cells and NIL cells) after the addition of con A or WGA (Cline and Livingston, 1971). The weakly oncogenic human adenovirus type 2 also induces increased agglutinability late in lytic infections provided that DNA synthesis simultaneously occurs. But in marked contrast, cells lytically infected with "wild-type" adenovirus type 12, a highly oncogenic virus, are not readily agglutinated by lectins even when cellular DNA synthesis occurs. To further confuse the matter, cells lytically infected by two non-transforming mutants of adeno-12 are rendered more readily agglutinable by con A (Salzberg and Raskas, 1972).

The relationship between lytic infection, transformation and agglutinability remains obscure. It is clear, however, that infection with some nononcogenic viruses can increase the agglutinability of cells. By 2 hours after infection with vaccinia virus, rabbit kidney cells become easily agglutinable by con A (Zarling and Tevethia, 1971). This change occurs several hours before the synthesis of vaccinia virus DNA commences, but if cycloheximide, an inhibitor of protein synthesis, is present during the first several hours of infection, the addition of con A does not cause the cells to agglutinate. This suggests that an enzyme coded by the virus is involved in the surface alteration. Changes in the surface of BHK cells infected with Newcastle disease virus (NDV) start some 5 hours after infection, and progeny virus do not appear until at least 8 hours after infection (Poste and Reeve, 1972). Because NDV infection eventually leads to cell fusion, controls were done which showed that agglutination induced by con A 4 hours after infection can be reversed by the specific hapten methyl-D-glucopyranoside, while *N*-acetylglucosamine reversed agglutination induced by WGA.

The cell surface changes that we now measure by increased agglutinability thus may somehow promote the multiplication of many viruses. There are hints that high concentrations of cAMP may inhibit infection by SV40 (J. Robb, personal communication 1972). A gene which, somehow, lowers intracellular concentrations of cAMP might therefore confer a selective advantage on a virus. Since adenyl cyclase, the enzyme which makes cAMP, is found largely, if not exclusively, on the plasma membrane, one can speculate that the surface changes that accompany lytic infection and can be detected with lectins may be a device to modify the outer surface of a host cell so as to prevent the functioning of adenyl cyclase.

In summary, cells from most tumours are more readily agglutinated by con A or WGA than are cells from the corresponding normal tissues. Cells transformed in vitro are always more readily agglutinated than the untransformed parental cells. The degree of agglutinability of a cell line correlates with its degree of growth control. Finally, in cells transformed by RNA or DNA tumour viruses, a viral gene product seems to be required to maintain enhanced susceptibility to agglutination.

Agglutination of Normal Cells after Exposure to Proteases

The differences between the susceptibility of untransformed and transformed cells to lectin-induced agglutination disappear if the normal cells are briefly exposed to a proteolytic enzyme such as trypsin (Burger, 1969; Inbar and Sachs, 1969a; Sela et al., 1970; Nicolson and Blaustein, 1972). For several hours after exposure to trypsin, a variety of untransformed cell types (e.g., BHK, 3T3, and chicken embryo fibroblasts) agglutinate as if they were transformed cells. At the same time, such trypsinized cells, if they are part of a confluent non-dividing monolayer, escape from growth control, and most of these cells go through one further round of division (Burger, 1969; Sefton and Rubin, 1970) (Figure 3.23). The cells divide only once perhaps because the surface change caused by trypsin is only transitory; cells making proteins repair their surfaces and become refractory to lectin-induced agglutination within 6 hours after the removal of the trypsin.

How trypsin alters cell surfaces so as to render the cells more readily agglutinable is not clear. Initially it was thought (Burger, 1969) that trypsin exposed cryptic binding sites for lectins that normally are masked in untransformed cells. But now that we are no longer sure that transformed cells bind more molecules of a lectin than do untransformed cells (see below), alternative possibilities must be considered. Nicolson (1971, 1972), for example, suggests the primary effect of trypsin is to increase the fluidity of the surface membrane by splitting off many surface glycopeptides.

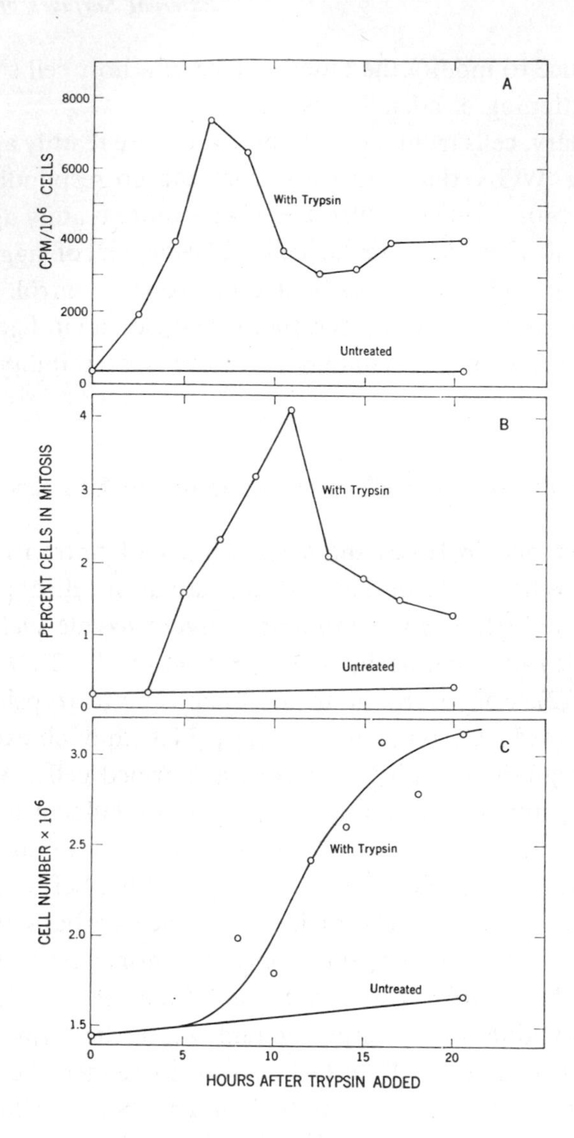

Figure 3.23. **(A)** Rate of incorporation of [^{3}H] thymidine, **(B)** mitosis, and **(C)** increase in cell number in the first 24 hours following the addition of trypsin. Chick embryo cells (1×10^6) from trypsinized primary cultures were added to 60-mm plastic petri dishes in medium 199 plus 2% tryptose phosphate broth and 2% chicken serum. After 40 hours of incubation at 38°C the cultures were fully confluent and 3 μg/ml of crystalline trypsin (Sigma) was added to the medium. At interals the cultures were labeled with 0.25 μCi/ml of [^{3}H] thymidine for 1 hour and the radioactivity insoluble in 5% trichloroacetic acid was counted. Reproduced with permission from Sefton and Rubin, *Nature* 227, 843, 1970.

Why the surface changes induced by trypsin not only lead to agglutination by lectins but also result in loss of normal growth control is equally obscure. Burger and his colleagues (Fox et al., 1971), who found that cells engaged in mitosis appeared to bind more WGA labeled with fluorescein isothiocyanate than cells in G_1, S and G_2, suggested that unmasking of cryptic binding sites was necessary to initiate a new round of cell division, and that contact inhibition was a result of the failure of closely packed normal cells to expose these cryptic sites. In proposing this theory, Burger et al. were strongly influenced by their previous observation (Burger and Noonan, 1970) that the addition of "monovalent" con A (see above) to transformed Py3T3 cells restores the cell's growth control. At low cell densities Py3T3 cells multiply in the presence of "monovalent" con A at the rate at which they multiply in its absence, but when Py3T3 cells have formed a confluent monolayer in the presence of "monovalent" con A, they behave like untransformed 3T3 cells and cease multiplying. Because "monovalent" con A molecules should not form agglutinating cross-bridges, Burger and Noonan suggested that these molecules provided a coating on the surface of the transformed cells which mimics the effect of the molecules that coat the surface of untransformed cells (Figure 3.24). The cessation of division of Py3T3 cells in the presence of "monovalent" con A was not the result of damage caused to the cells (see below), for when the monovalent agglutinin is removed by a hapten competition, the transformed cells started to multiply again and

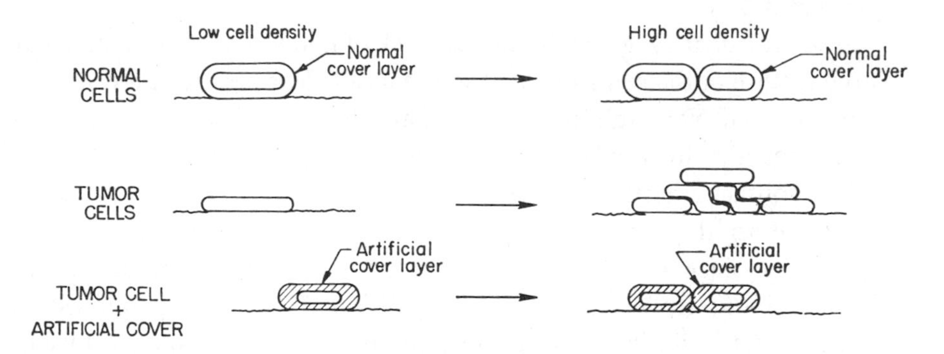

Figure 3.24. A simple model for the action of "monovalent" agglutinin. Reproduced with permission from Burger and Noonan, *Nature* 228, 515, 1970.

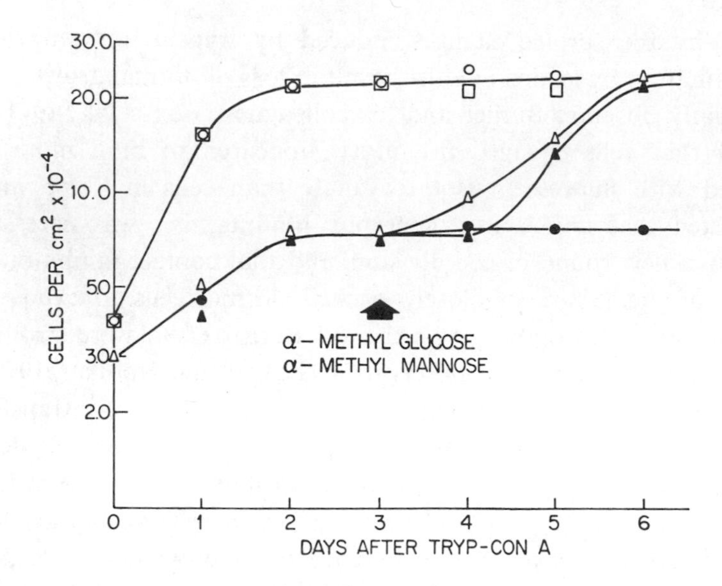

Figure 3.25. Restoration of density-dependent inhibition of growth by covering agglutinin receptor sites. Subsequent removal of the cover again abolishes inhibition of growth. (○—○) Py3T3 cells, control; (□—□) Py3T3 cells + 50 μg trypsinized con A preincubated for 12 hours with 10^{-2} M α-methylglucoside; (△—△) Py3T3 cells + 50 μg trysinized con A + 10^{-2} M α-methylglucoside or α-methylmannoside added at day 3; (▲—▲) Py3T3 cells + 50 μg trypsinized con A + 10^{-3} M α-methylglucoside or α-methylmannoside added at day 3; (●—●) Py3T3 cells + 50 μg trypsinized con A. Reproduced with permission from Burger and Noonan, *Nature* 228, 514, 1970.

achieved the high saturation density characteristic of Py3T3 cells (Figure 3.25).

Unfortunately, given the current lack of knowledge of the molecular architecture of the cell surface, it is not at all clear how to test critically Burger's and Noonan's hypothesis. Thus, Sheppard's (1972a) observation that the intracellular concentrations of cAMP transiently fall when contact-inhibited 3T3 cells are exposed to trypsin may be more germane to the question of how trypsin induces cells to divide—especially since the serum "growth factors" (Todaro et al., 1965; Temin, 1966; Holley and Kiernan, 1968) and insulin (Temin, 1967), which like trypsin induce rounds of cell division, also induce transient falls in intracellular concentrations of cAMP (Sheppard, 1972a). It is tempting to suggest (see reviews by Sheppard, 1972b; Willingham et al., 1972) that trypsin, the serum factors and insulin all induce cell division by lowering the intracellular concentrations of cAMP. Supporting this hypothesis are new experiments which show that insulin directly inhibits the activity of membrane-bound adenyl cyclase (Illiano and Cuatrecasas, 1972).

Equally important may be the fact that the concentration of cAMP in transformed cell lines is significantly lower than that in their untransformed counterparts. Measurements of the steady state amounts of cAMP in untransformed, transformed and revertant 3T3 cell lines revealed that transformed cells possess less than one-half the amount of cAMP present in normal (revertant) density-dependent cells (Sheppard, 1972a). To test further whether these differences related to transformation per se, or whether they represented secondary changes, the concentrations of cAMP in chicken cells infected by RSV were determined at various intervals after infection until transformation was completed (usually between 72–96 hours) (Sheppard and Lehman, 1972). The concentration of cAMP started to decrease around 72 hours after infection, coincidentally with the production of progeny viruses, increased glucose transport, and increased agglutinability.

One can also speculate that cAMP may mediate the rounds of cell division which are induced by exposing confluent non-dividing cultures of chick embryo fibroblasts to neuraminidase (Vaheri et al., 1972). For the removal of sialic acid residues from glycosylated molecules at the cell surface might inhibit adenyl cyclase, which is located in the plasma membrane, and therefore lower the intracellular concentrations of cAMP.

Excess Proteases (Glycosidases) in Transformed Cells?

The fact that untransformed cells after exposure to trypsin (a protease) or neuraminidase (a glycosidase) transiently assume some of the phenotypic characteristics of transformed cells raises the possibility that, at least in part, the cancer cell phenotype stems from the autodigestion of cell surface material. One might speculate that after transformation cells either make and secrete hydrolytic enzymes not present in normal cells, or that transformation causes cells to release hydrolytic enzymes at a faster rate than before transformation.

Preliminary experiments designed to test this hypothesis give mildly encouraging results. First, transformed cells seem to possess more extracellular protease activity than their normal counterparts (Schnebli, 1972). And second, addition of several trypsin and chyrotrypsin inhibitors—e.g., *N*-α-tosyl-L-arginine methylester (TAME); *N*-α-tosyl-L-lysyl-chloromethane (TLCK); *N*-α-tosyl-L-phenylalanyl-chloromethane (TPCK)—to transformed cells causes them to cease growth at saturation densities characteristic of untransformed cells (Schnebli and Burger, 1972). By contrast, the same amounts of these inhibitors have essentially no effect on the growth of untransformed cells. But since only a fourfold greater amount of inhibitors strongly inhibits normal cell growth, the possibility that the primary effects of these protease inhibitors on transformed cells is also one of "non-specific" toxicity has not yet been excluded.

Interestingly, several of the inhibitors (TAME, TLCK and TPCK) used by Schnebli and Burger had previously been used by Troll et al. (1970) to block phorbol ester-promoted cocarcinogenesis in mouse skin. Skin tumours normally are induced by the single application of a primary carcinogen like dimethylbenzanthrene, followed by repeated application of a "promoting" substance like phorbol ester (the active principle of croton oil) (Berenblum and Shubik, 1947). When, however, protease inhibitors are present during the promoting period, no tumours arise. One way to interpret these findings is to suggest that newly arising cancer cells, induced by the action of the primary carcinogen, have only a marginal ability to escape from normal growth controls. But in the presence of an abundance of proteases they may be able to form a visible tumour; conversely, if proteases are inhibited, multiplication of the tumour cells may be restricted.

While the general suspicion has existed for some time that many primary carcinogens act either by causing somatic mutations or by activating latent tumour viruses, the way promoters (cocarcinogens) work has been a total mystery. The finding that several antiproteases block promoters from functioning suggests for the first time a mechanism for their action. Promoters may increase the supply of proteases, conceivably by directly causing their release from lysosomes (Allison and Mallucci, 1964; Weissman et al., 1968). Or they may act more indirectly by causing an inflammatory response marked by acute influxes of hydrolase-rich phagocytic cells.

Independent support for the idea that proteases are involved in the cancerous phenotype may come from further characterization of the growth stimulating factor released from RSV-transformed cells (Rubin, 1970). This factor has properties suggestive of proteases and further experiments may soon identify the active component. Furthermore there is an abundance of peptidases in the periphery of many tumours where stromal and vascular stimulation occurs (Sylven and Malmgroen, 1957), and so the vascularization factor recently found in tumours by Folkman et al. (1971) may turn out to be a protease.

The normal regulation of growth may also in part be controlled by such enzymes. Salivary gland growth is stimulated by proteases (Ershoff and Bajawa, 1963) and "growth factors" that induced growth of skin, epithelial, mesenchymal and nerve tissue have been found to contain peptidase and esterase activities (Jones and Ashwood-Smith, 1970; Attardi et al., 1967; Greene et al., 1968).

Amounts of Lectins Bound by Cells

It was initially suspected that the differential agglutination of transformed cells by con A and WGA was a consequence of differential affinities of the two sorts of cells for these lectins. It was argued that transformed cells were more readily agglutinated because they bound more lectin than untransformed cells. Furthermore, because untransformed cells were rendered as agglutinable as transformed cells by exposure to proteases, Burger (1969, 1970) argued that lectin binding sites were not synthesized "de novo" after transformation, but that they were "cryptic" or masked in untransformed cells. Transformation, he suggested, led to the exposure of the cryptic binding sites either because some coating material was removed from the surface or because cell surface components were reorganized such that the binding sites became available to interaction with the lectins. Early experiments designed to directly test this hypothesis tended to support it. For example, Fox et al. (1971), who applied fluorescein-tagged WGA to growing cells, reported that Py3T3 cells showed bright surface fluorescence when seen through the UV microscope, whereas 3T3 cells, under identical conditions, hardly fluoresced unless they were in mitosis or early G_1. Likewise, when Shoham and Sachs (1972) added low concentrations of fluorescein-labeled con A to transformed interphase hamster cells, they observed more fluorescing cells than in equivalent experiments with normal interphase hamster cells. Quantitation of these observations, however, has proved very difficult and they do not prove that the transformed cells bound more lectin than the untransformed cells. For the observed differences in the brightness of the fluorescence could result from differences in the distribution rather than the numbers of the binding sites on the two cell surfaces (see below).

In the first attempt to measure the number of con A molecules bound by untransformed and transformed cells, Inbar and Sachs (1969b) substituted a radioisotope of Ni, ^{63}Ni, for one of the divalent transition metals required by con A and found that transformed 3T3 and BHK cells bound 2 to 3 times more of the label than did untransformed cells. The difference, however, could be increased to 6- to 10-fold if the data were expressed as the number of con A molecules bound per unit of surface area because they noted that transformed cells were smaller than the corresponding untransformed cells.

Several other groups subsequently reported conflicting findings after discovering ^{63}Ni to be an unsatisfactory label. Using ^{63}Ni they could achieve neither saturation bindings nor adequate hapten inhibition of binding, possibly because the ^{63}Ni was not covalently linked to con A. When instead they labeled con A with ^{125}I or ^{3}H acetyl groups, they found that cells which were readily agglutinated did not bind significantly more lectin molecules than cells which were not agglutinated (Arndt-Jovin and Berg, 1971; Cline and Livingston, 1971; Ozanne and Sambrook, 1971a). For example, both untransformed 3T3 cells and 3T3 cells transformed by SV40 bound 10^7 molecules of con A or 5×10^7 molecules of WGA. Furthermore, 3T3 cells rendered agglutinable by exposure to trypsin did not bind more molecules of WGA or con A than 3T3 cells not exposed to trypsin. Neither were differences observed between the binding of con A or WGA to untransformed BHK cells, to BHK cells transformed by polyoma virus, or to BHK cells transformed by the temperature-sensitive mutant of polyoma virus *ts3* (see Chapter 7) and grown at either 31 or 38°C. In each case about 3×10^7 molecules of con A or WGA bound to the cells irrespective of their agglutinability. And even if the number of molecules of lectin bound was expressed per unit of surface area or per milligram of protein, only twofold variations between transformed and untransformed cells were generated. This group of experiments suggested, therefore, that agglutinability may result from something other than an exposure of new binding sites.

Noonan and Burger (personal communication 1972), however, do not agree with this conclusion, for when they use [^{3}H]-acetylated con A and a different set of assay conditions (incubation at 4°C for several minutes instead of 15 minutes at 37°C), they consistently find 3- to 5-fold more con A molecules bind per unit area of surface to transformed 3T3 cells than to untransformed 3T3 cells. Furthermore, using the Renger and Basilico (1972) line of temperature-sensitive SV3T3 cells, they detect changes in the amount of con A bound to the same cells as the cell phenotype changes. Cells of this mutant line at 32°C have the growth properties of typical transformed 3T3 cells and agglutinate in low concentrations of WGA or con A, but at 39°C they behave like 3T3 cells. When grown at 32°C these cells bind 5-fold more con A molecules than they bind when they are grown at 39°C.

The binding assay of Noonan and Burger reveals only 10–20 percent of the number of binding sites that are revealed by the assays used by other groups. It is possible, therefore, that Burger and Noonan may be measuring only a subset of the con A binding sites measured by others, and this subset may include sites which are, in fact, exposed only after transformation and which are not detected by less discriminatory assays.

But whatever the case may be, both transformed and untransformed cells bind large numbers of lectin molecules, and whether or not the differences in the very large numbers of molecules of lectins bound to the two classes of cells are sufficient to account for the observed differences between their agglutinability is an open question. Other changes may have to occur on the surfaces of transformed cells, in addition to any exposure of new binding sites, to enhance their agglutinability.

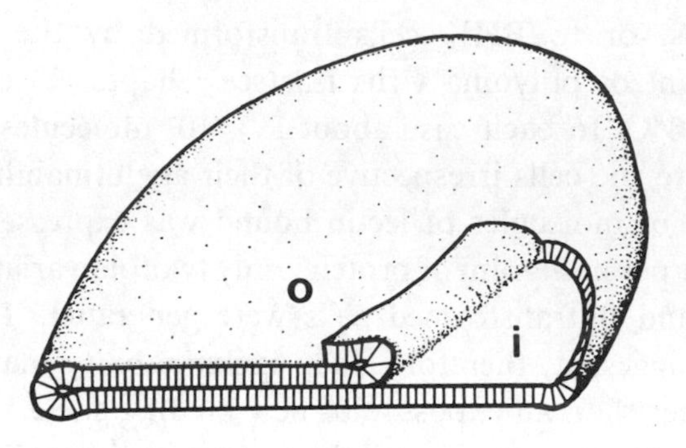

Figure 3.26. Diagrammatic cross-section through a mounted erythrocyte ghost. The upper membrane has torn back, revealing both the inner (*i*) and outer (*o*) surfaces of the membrane. Reproduced with permission from Nicolson and Singer, *Proc. Nat. Acad. Sci.* 68, 942, 1971.

Clusters of Lectin Binding Sites

A most promising new technique for examining in the electron microscope some facets of the molecular architecture of the cell surface has been developed by Nicolson and Singer (1971). They conjugated ferritin to con A molecules and then exposed erythrocytes to this tagged con A. They then lysed the red cells on an air–water interface; the plasma membranes of the cells were often flattened and spread by surface tension forces and they could be picked off the water surface onto electron microscope grids (Figure 3.26). In the electron microscope the distribution of the ferritin-labeled con A molecules could be discerned (Figure 3.27) and their numbers counted.

Nicolson (1971) soon modified this technique to examine membranes of other cells. In particular, he studied the distribution of lectin binding sites on the surfaces of transformed and normal cells. He found that the con A binding sites on SV3T3 cells were arranged on the outer

---

Figure 3.27. (**a**) A mounted rabbit erythrocyte ghost stained with a solution of ferritin-conjugated concanavalin A containing 100 mM D-galactose. The inner (*i*) and outer (*o*) surfaces of the membrane are distinguishable by their absorbances in the micrograph. The insert shows the entire ghost, and the brackets indicate the area magnified in the figure. Bar equals 0.2 μm; insert bar equals 1 μm.

(**b**) Same as in (**a**), except that the staining solution of ferritin-conjugated concanavalin A contained 100 mM sucrose, an inhibitor of concanavalin A.

(**c**) Same as in (**a**), except that the mounted rabbit erythrocyte ghost was stained with ferritin-conjugated anti-(human) spectrin (tecktins). Reproduced with permission from Nicolson and Singer, *Proc. Nat. Acad. Sci.* 68, 942, 1971.

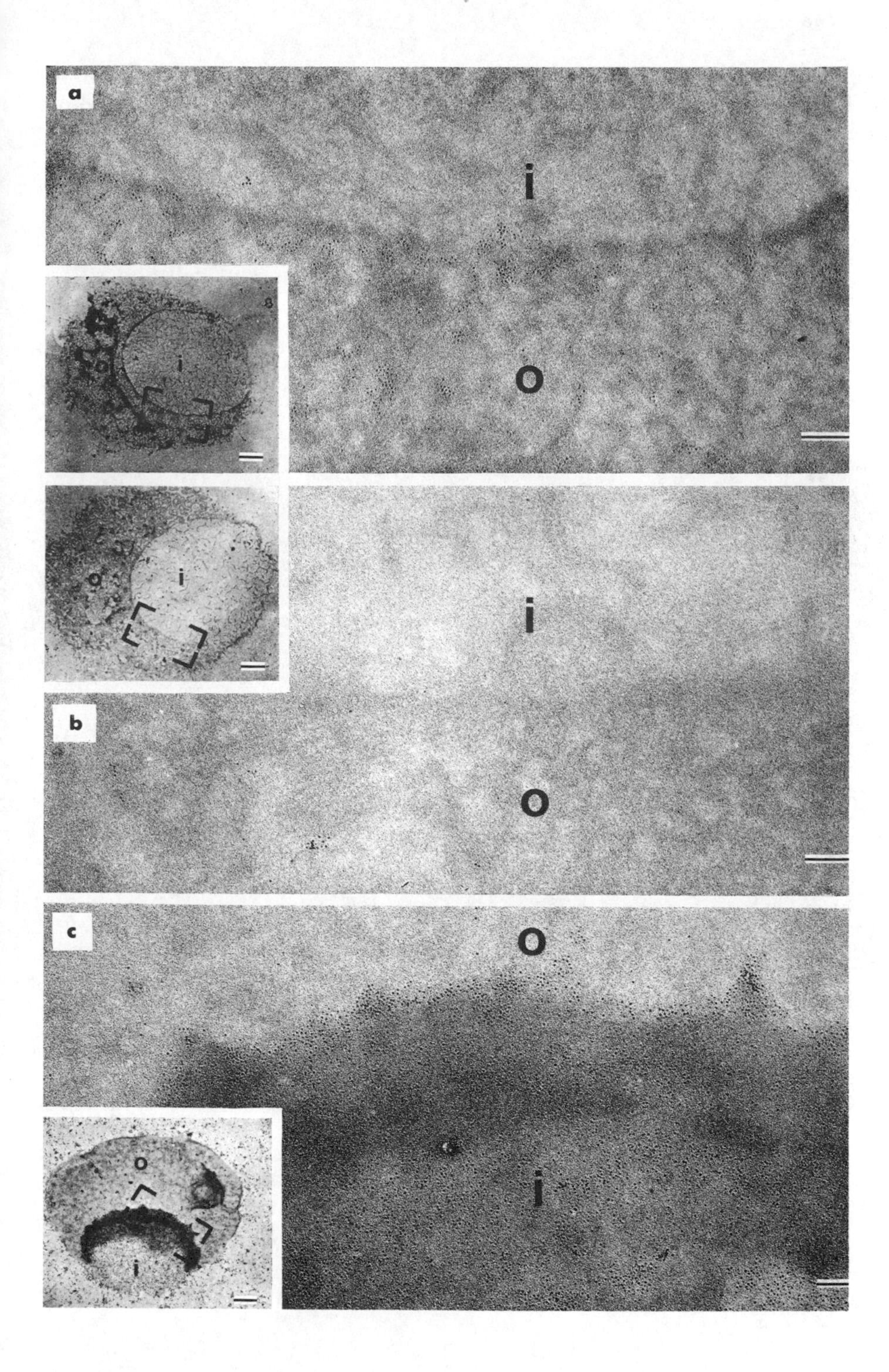
a
i
o
b
i
o
c
o
i

surface of the cell membrane in random clusters, surrounded by areas essentially devoid of binding sites (Figure 3.28). By contrast the binding sites were evenly distributed over the surface membranes of 3T3 cells. He also showed that trypsinized 3T3 cells possessed binding sites clustered in a pattern resembling that seen on transformed cells (Nicolson, 1972). Most importantly, he found approximately the same number of binding sites on the surface of each sort of cell and so he postulated that it was a topological reorganization of the con A binding sites, rather than an increase in their number, that enabled lectins to agglutinate transformed cells, perhaps by some bridging mechanism

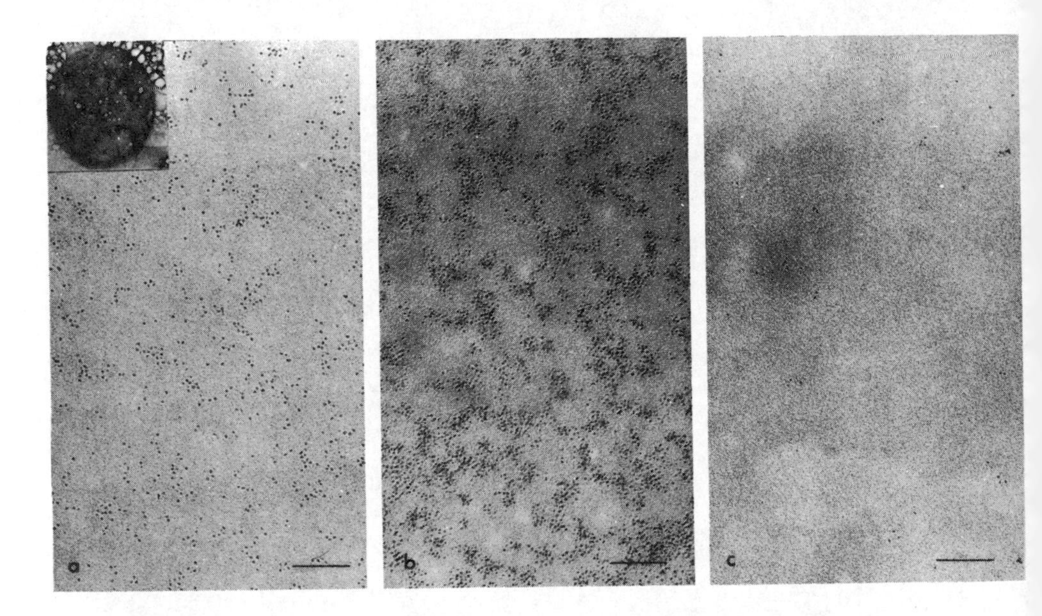

Figure 3.28. (**a**) The outer cell surface of a normal 3T3 mouse fibroblast that has been lysed at an air–water interface and picked up on a coated electron microscope grid. The flattened cell membrane (inset, low magnification) was directly stained with a ferritin-conjugated concanavalin A solution, washed, and then air-dried. The ferritin appears bound to the membrane mainly in a dispersed distribution. Magnification 61,500 ×, bar equals 0.1 μ; inset magnification 1500 ×.

(**b**) The same as in (**a**) except that an SV40-transformed 3T3 cell was stained with ferritin-conjugated concanavalin A in a parallel experiment. The ferritin-agglutinins are now present in a more clustered state on the membrane surface.

(**c**) A control experiment. The same as in (**b**) except that 0.2 M sucrose, an inhibitor of concanavalin A, was present in the staining solution. Reproduced with permission from Nicolson, *Nature New Biol*. 233, 244, 1971.

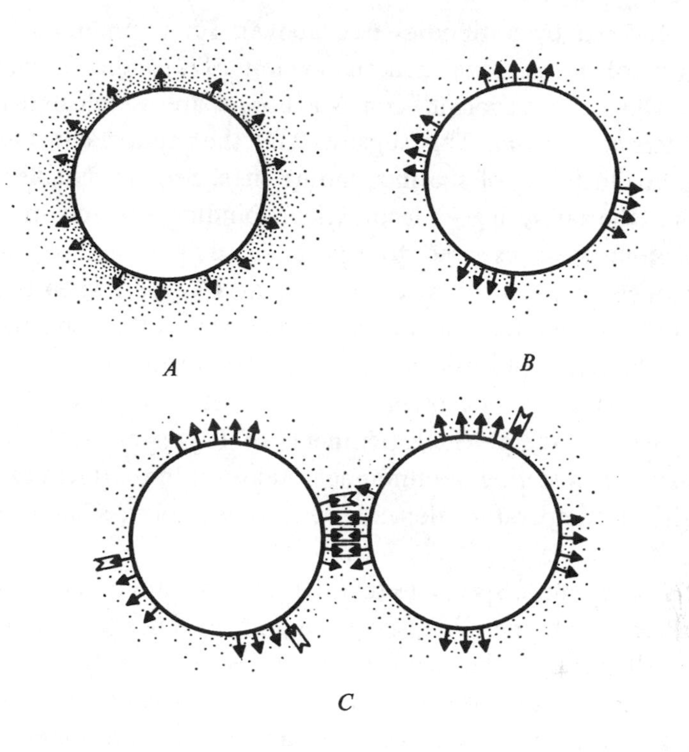

Figure 3.29. A mechanism to explain the difference in cell agglutination by plant agglutinins after brief treatment with proteolytic enzymes. (**A**) The agglutinin binding sites are normally present in a dispersed random distribution that does not favour cell agglutination at low agglutinin concentrations. (**B**) After treatment with proteolytic enzymes, clustering of agglutinin binding sites occurs due to local changes in charge distribution on the cell surface, thereby increasing cell agglutination (**C**) by lowering cell repulsion and allowing multiple agglutinin cross-bridges to form between adjacent cells. Reproduced with permission from Nicolson, *Nature New Biol.*, 1972.

(Figure 3.29). A similar clustering of con A binding sites on the surfaces of BHK cells transformed by polyoma virus has also been reported (Martinez-Palomo et al., 1972).

Whether clusters of con A binding sites exist as such in the transformed or trypsinized cells, or whether they are the result of the binding of the lectin, has not yet been clearly established. Two recent observations suggest that con A may induce the clustering of its specific binding sites perhaps in a way analogous to the clustering of cell surface antigens that is induced by antibodies (see above). First, preliminary experiments (Nicolson, personal communication 1972) indicate that surface sites to which "monovalent" con A is bound are not clustered but are distributed at random. This suggests that the multivalent con A may induce the clustering of specific binding sites. Second, Nicolson (1972) finds that the clustering of lectin (ricin) binding sites on the surfaces of red blood cells exposed to trypsin is temperature dependent; it occurs much more slowly when cells are at 4°C than when they are at 37°C. This observation he interprets in terms of a temperature-dependent diffusion process which randomly brings con A receptors sufficiently close to each other for multivalent con A to form cross-bridges between them. The experiments of Taylor et al. (1971), which show that the clustering of immunoglobulins on the surfaces of lymphocytes also is temperature dependent, lend support to Nicolson's suggestion.

It is tempting to speculate that the fundamental difference between the surface membrane of a cancer cell and the surface membrane of a normal cell is simply that the former is a less viscous, two-dimensional fluid. There may be fewer restraints on the diffusion of molecules in the surface membranes of cancer cells than in the surface membranes of normal cells, and after transformation there may be fewer restraints on which molecules can lie next to which molecules in the cell surface. Perhaps in the membranes of normal cells interactions between glycopeptides and other membrane components stabilize the membrane and impose some "order" on its constituents. If transformation leads to the loss (or modification) of glycopeptides, these constraints would be changed and may even be lost altogether. As a result, the transformed cells may be rendered insensitive to contact inhibition of division and to other factors which regulate cell division, and they may lose their sense of address and begin metastasizing.

Differential Killing of Transformed Cells

Growing cultures of normal and transformed cells respond differently to added con A. At certain concentrations of con A transformed cells round up, detach from the surface of the dish, and do not grow when replated; by contrast, the normal cells appear to be unaffected by con A at these concentrations. The ability of con A to kill transformed cells seems to be related to its ability to cause its specific receptors to cluster. "Monovalent" con A prepared by digesting con A with trypsin is not toxic for transformed cells.

The differential toxicity of con A for transformed and untransformed cells has been used by several groups to isolate con A-resistant cells from populations of 3T3 cells transformed by SV40. The clones of con A-resistant cells were not agglutinated by con A or WGA and they grew to saturation densities characteristic of untransformed 3T3 cells (Ozanne and Sambrook, 1971b; Culp and Black, 1972).

At sublethal concentrations con A affects the transport of metabolites across the cell membrane. Several groups (Inbar et al., 1971a, b; Isselbacher, 1972; Ben-Bassat et al., 1971) have shown that both con A and WGA inhibit the active transport of several hexoses and amino acids into both normal and transformed cells, but transformed cells are more susceptible to this inhibition than untransformed cells. Con A has also been shown to inhibit phagocytosis by polymorphomonuclear leucocytes (Berlin, 1972) as well as the movement of cells in culture (Friberg et al., 1971); the migration of easily agglutinated cells seems to be more susceptible to inhibition by con A than is the migration of nonagglutinable cells.

Attempts to Use Lectins as Anti-Tumour Agents

Since con A is more toxic to transformed cells than to normal cells, Sachs and his collaborators (Shoham et al., 1970; Inbar et al., 1972b) have attempted to use it as an anti-tumour agent. In their first experiments (Shoham et al., 1970) con A was injected into an area of a hamster that had previously been injected with sufficient transformed cells to induce tumours. Often the animals receiving con A did not develop tumours or the tumours appeared later than in control animals; this effect was inhibited by the injection of methylglucose with the con A, so presumably, con A had to be able to interact with the cells. Subsequently in similar experiments they measured the effect of con A on the appearance of ascites tumours, following the injection of YAC lymphoma cells into mice (Inbar et al., 1972b). Again addition of con A reduced the incidence of tumours, but in both sets of experiments the concentrations of con A that had to be used to obtain protection from the transplanted tumour cells were so large that some of the animals died of con A poisoning. Smaller doses of con A did not kill the mice but neither did they protect them significantly from the tumours.

Shier (1971) has tried in another way to exploit the fact that cells bind lectins. He made synthetic WGA binding sites by attaching short chains of *N*-acetylglucosamine residues to polypeptide backbones. He then used these synthetic receptor molecules to immunize animals prior to the injection of enough tumour cells to induce a tumour in an unimmunized animal. The tumours took slightly longer to appear in the immunized animal than in the control animals that had been immunized with the polypeptide alone. Why such a procedure protects the animals is obscure since WGA interacts with both normal and transformed cells.

Stimulation of Lymphocyte Division

We are just learning how lectins such as PHA and con A stimulate dormant lymphocytes to divide. Large numbers (about 10^6) of con A receptors are present on the surface of the average-sized lymphocytes (Edelman and Millette, 1971), and transformed lymphocytes begin to appear some 24 hours to 30 hours after addition of con A. At the same time insulin binding sites begin to appear (Cuatrecases, 1971; Krug et al., 1972). As no insulin binding sites are present on the surface of the dormant small lymphocytes when they are isolated from human blood, it is very tempting to believe that their appearance is connected with morphological transformation into a blastlike cell and the initiation of cell division. This de novo appearance of insulin receptors, however, is not the first event associated with transformation, since activation of cellular RNA and protein systems occurs within the first 24 hours following addition of con A. Conceivably the initial transforming event results from an insulin-like effect of con A itself since the concentration of cAMP is decreased in lymphocytes exposed to the lectin PHA (Smith et al., 1971a) while compounds that enhance the concentration of lymphocyte cAMP (aminophylline, isoproterenol, prostaglandins) inhibit PHA-induced transformation (Smith et al., 1971b).

Why con A binding should inhibit adenyl cyclase activity is not yet known. Possibly it relates to its effect on the mobility of membrane proteins (Yahara and Edelman, 1972). Prior binding of con A to a lymphocyte prevents the formation of "caps" induced by the addition of appropriate anti-Ig antibodies, suggesting that when large numbers of con A molecules are bound to their receptors, the free diffusion of surface antigen-antibody complexes is prevented. This effect is completely reversible; the addition of the hapten α-methyl-D-mannoside quickly leads to patch (cap) formation. The con A does not apparently prevent capping of antigen-antibody complexes by inhibiting the binding of antibody to surface antigens; neither does the con A cause its own receptor molecules on the surfaces of lymphocytes to cluster.

Literature Cited

ABERCROMBIE, M., J. E. M. HEAYSMAN and S. M. PEGRUM. 1970a. The locomotion of fibroblasts in culture. I. Movements of the leading edge. Exp. Cell Res. *59:* 393.

———. 1970b. The locomotion of fibroblasts in culture. II. "Ruffling." Exp. Cell Res. *60:* 437.

———. 1970c. The locomotion of fibroblasts in culture. III. Movements of particles on the dorsal surface of the leading lamella. Exp. Cell Res. *62:* 389.

———. 1971. The locomotion of fibroblasts in culture. IV. Electron microscopy of the leading lamella. Exp. Cell Res. *67:* 359.

ADELSTEIN, R. S. and M. A. CONTI. 1973. The characterization of contractile proteins from platelets and fibroblasts. Cold Spring Harbor Symp. Quant. Biol. *37:* 599.

ADELSTEIN, R. S., T. D. POLLARD and W. M. KUEHL. 1971. Isolation and characterization of myosin and two myosin fragments from human blood platelets. Proc. Nat. Acad. Sci. *68:* 2703.

AGRAWAL, B. B. L. and I. J. GOLDSTEIN. 1967. Protein-carbohydrate interaction. VI. Isolation of concanavalin A by specific adsorption on cross-linked dextran gels. Biochim. Biophys. Acta *147:* 262.

———. 1968a. Protein-carbohydrate interaction. VII. Physical and chemical studies on concanavalin A, the hemagglutinin of the jack bean. Arch. Biochem. Biophys. *124:* 218.

———. 1968b. Protein-carbohydrate interaction. XV. The role of bivalent cations in concanavalin A—polysaccharide interaction. Canadian J. Biochem. *46:* 1147.

ALLAN, D. and M. J. CRUMPTON. 1972. Isolation and composition of human thymocyte plasma membrane. Biochim. Biophys. Acta *274:* 22.

ALLAN, D., J. AUGER and M. J. CRUMPTON. 1972. Glycoprotein receptors for concanavalin A isolated from pig lymphocyte plasma membrane by affinity chromatography in sodium deoxycholate. Nature New Biol. *236:* 23.

ALLISON, A. C. and L. MALLUCCI. 1964. Uptake of hydrocarbon carcinogens by lysosomes. Nature *203:* 1024.

ALLISON, A. C., P. DAVIES and S. DE PETRIS. 1971. Role of contractile microfilaments in macrophage movement and endocytosis. Nature New Biol. *232:* 153.

AMBROSE, E. J., J. A. DUDGEON, D. M. EASTY and G. C. EASTY. 1961. The inhibition of tumour growth by enzymes in tissue culture. Exp. Cell Res. *24:* 220.

ANDERSON, C. W. and R. F. GESTELAND. 1972. Pattern of protein synthesis in monkey cells infected by Simian Virus 40. J. Virol. *9:* 758.

ARNDT-JOVIN, D. J. and P. BERG. 1971. Quantitative binding of ^{125}I-concanavalin A to normal and transformed cells. J. Virol. *8:* 716.

ATTARDI, D. G., M. J. SCHLEISINGER and S. SCHLESINGER. 1967. Submaxillary gland of mouse: properties of a purified protein affecting muscle tissue in vitro. Science *156:* 1253.

AUB, J. C., C. TIESLAU and A. LANKESTER. 1963. Reaction of normal and tumor cell surfaces to enzymes. I. Wheat-germ lipase and associated mucopolysaccharides. Proc. Nat. Acad. Sci. *50:* 613.

AUB, J. C., B. H. SANFORD and M. N. COTE. 1965. Studies on reactivity of tumor and normal cells to a wheat germ agglutinin. Proc. Nat. Acad. Sci. *54:* 396.

BECKER, J. W., G. N. REEKE and G. M. EDELMAN. 1971. Location of the saccharide binding site of concanavalin A. J. Biol. Chem. *246:* 6123.

BEN-BASSAT, H., M. INBAR and L. SACHS. 1970. Requirement for cell replication after SV40 infection for a structural change of the cell surface membrane. Virology *40:* 854.

———. 1971. Changes in the structural organization of the surface membrane in malignant cell transformation. J. Membrane Biol. *6:* 183.

BENDER, W. W., H. GARAN and H. C. BERG. 1971. Proteins of the human erythrocyte membrane as modified by pronase. J. Mol. Biol. *58:* 783.

BENJAMIN, T. L. and M. M. BURGER. 1970. Absence of a cell membrane alteration function in non-transforming mutants of polyoma virus. Proc. Nat. Acad. Sci. *67:* 929.

BENNETT, G. and C. P. LEBLOND. 1970. Formation of cell coat material for the whole surface of columnar cells in the rat small intestine, as visualized by radioautography with L-fucose-^{3}H. J. Cell Biol. *46:* 409.

BERENBLUM, I. and P. A. SHUBIK. 1947. A new, quantitative approach to the study of the stages of chemical carcinogenesis in the mouse's skin. Brit. J. Cancer *1:* 379.

BERG, H. C. 1969. Sulfanic acid diazonium salt: A label for the outside of the human erythrocyte membrane. Biochim. Biophys. Acta *183:* 65.

BERLIN, R. D. 1972. Effect of concanavalin A on phagocytosis. Nature New Biol. *235:* 44.

BETTEX-GALLAND, M. and E. F. LÜSCHER. 1965. Thrombosthenin, the contractile protein from blood platelets and its relation to other contractile proteins. In *Advances in protein chemistry* (ed. C. B. Anfinsen et al.) Vol. 20, p. 1. Academic Press, New York.

BLASIE, J. K., C. R. WORTHINGTON and M. M. DEWEY. 1969. Molecular localization of frog retinal receptor photopigment by electron microscopy and low-angle X-ray diffraction. J. Mol. Biol. *39:* 407.

BLAUROCK, A. E. 1971. Structure of the nerve myelin membrane: Proof of the low-resolution profile. J. Mol. Biol. *56:* 35.

BLAUROCK, A. E. and M. H. F. WILKINS. 1969. Structure of frog photoreceptor membranes. Nature New Biol. *223:* 906.

BOOYSE, F. M., T. P. HOVEKE, D. ZSCHOCKE and M. E. RAFELSON. 1971. Human platelet myosin. Isolation and properties. J. Biol. Chem. *246:* 4291.

BOSMANN, H. B. 1971. Platelet adhesiveness and aggregation: the collagen: glycosyl, polypeptide: N-acetylgalactosaminyl and glycoprotein: galactosyltransferases of human platelets. Biochem. Biophys. Res. Commun. *43:* 1118.

BOSMANN, H. B., A. HAGOPIAN and E. H. EYLAR. 1968. Membrane glycoprotein biosynthesis: changes in levels of glycosyl transferases in fibroblasts transformed by oncogenic viruses. J. Cell. Physiol. *72:* 81.

Bouchilloux, S., O. Chabaud, M. Michel-Béchet, M. Ferrand and A. M. Athouël-Haon. 1970. Differential localization in thyroid microsomal subfractions of a mannosyltransferase, two *N*-acetyl-glucosaminyl-transferases and a galactosyltransferase. Biochem. Biophys. Res. Commun. *40:* 314.

Boyd, W. C. 1963. The lectins: Their present status. Vox Sanguinis (Basel) *8:* 1.

Boyd, W. C. and E. Shapleigh. 1954. Specific agglutinating activity of plant agglutinins (lectins). Science *119:* 419.

Boyd, W. C., E. Waszczenko-Facharczenko and S. Goldwasser. 1961. List of plants tested for hemagglutinating activity. Transfusion *1:* 374.

Branton, D. 1966. Fracture faces of frozen membranes. Proc. Nat. Acad. Sci. *55:* 1048.

———. 1969. Membrane structure. Ann. Rev. Plant Physiol. *20:* 209.

Bretscher, M. S. 1971a. Human erythrocyte membranes: Specific labeling of surface proteins. J. Mol. Biol. *58:* 775.

———. 1971b. Major protein which spans the human erythrocyte membrane. J. Mol. Biol. *59:* 351.

———. 1971c. Major human erythrocyte glycoprotein spans the cell membrane. Nature New Biol. *231:* 229.

———. 1972. Asymmetrical lipid bilayer structure for biological membranes. Nature New Biol. *236:* 11.

———. 1973. Membrane structure: Some general principles. Science. In press.

Brown, P. K. 1972. Rhodopsin rotates in the visual receptor membrane. Nature New Biol. *236:* 35.

Buck, C. A., M. C. Glick and L. Warren. 1970. A comparative study of glycoproteins from the surface of control and Rous sarcoma virus transformed hamster cells. Biochemistry *9:* 4567.

———. 1971a. Glycopeptides from the surface of control and virus-transformed cells. Science *172:* 169.

———. 1971b. Effect of growth on the glycoproteins from the surface of control and Rous sarcoma virus transformed hamster cells. Biochemistry *10:* 2176.

Burger, M. M. 1968. Isolation of a receptor complex for a tumour specific agglutinin from the neoplastic cell surface. Nature *219:* 499.

———. 1969. A difference in the architecture of the surface membrane of normal and virally transformed cells. Proc. Nat. Acad. Sci. *62:* 994.

———. 1970. Proteolytic enzymes initiating cell division and escape from contact inhibition of growth. Nature *227:* 170.

———. 1971a. Cell surfaces in neoplastic transformation. In *Current topics in cellular regulation* (ed. B. L. Horecker and E. R. Stadtman) Vol. 3, p. 135. Academic Press, New York.

———. 1971b. Forssman antigen exposed on surface membrane after viral transformation. Nature *231:* 125.

Burger, M. M. and A. R. Goldberg. 1967. Identification of a tumor-specific determinant on neoplastic cell surfaces. Proc. Nat. Acad. Sci. *57:* 359.

Burger, M. M. and G. S. Martin. 1972. Agglutination of cells transformed by Rous sarcoma virus by wheat germ agglutinin and concanavalin A. Nature New Biol. *237:* 9.

Burger, M. and K. D. Noonan. 1970. Restoration of normal growth by covering of agglutinin sites on tumour cell surfaces. Nature *228:* 512.

Carter, S. B. 1967. Effects of cytochalasins on mammalian cells. Nature *213:* 261.

Caspar, D. L. D. and D. A. Kirschner. 1971. Myelin membrane structure at 10 Å resolution. Nature New Biol. *231:* 46.

Chiarugi, V. P. and P. Urbano. 1972. Electrophoretic analysis of membrane glycoproteins in normal and polyoma virus transformed BHK21 cells. J. Gen. Virol. *14:* 133.

CLARKE, M. 1971. Isolation and characterization of a water-soluble protein from bovine erythrocyte membranes. Biochem. Biophys. Res. Commun. *45:* 1063.

CLARKE, M. and J. GRIFFITH. 1972. Isolation and properties of a protein from human erythrocyte membranes. Fed. Proc. 1091: 412 (Abstr.).

CLINE, M. J. and D. C. LIVINGSTON. 1971. Binding of ^{3}H-concanavalin A by normal and transformed cells. Nature New Biol. *232:* 155.

CODINGTON, J. F., B. H. SANFORD and R. W. JEANLOZ. 1972. Glycoprotein coat of the TA3 cell. Isolation and partial characterization of a sialic acid containing glycoprotein fraction. Biochemistry *11:* 2559.

CONE, R. A. 1972. Rotational diffusion of rhodopsin in the visual receptor membrane. Nature New Biol. *236:* 39.

CONE, R. E., J. SPRENT and J. J. MARCHALONIS. 1972. Antigen-binding specificity of isolated cell-surface immunoglobulin from thymus cells activated to histocompatibility antigens. Proc. Nat. Acad. Sci. *69:* 2556.

COOK, G. M. W. 1968. Glycoproteins in membranes. Biol. Rev. Cambridge Phil. Soc. *43:* 363.

COOMBS, R. R. A., B. W. CURNER, C. A. JANEWAY, JR., A. B. WILSON, P. G. H. GELL and A. S. KELUS. 1970. Immunoglobulin determinants on the lymphocytes of normal rabbits. I. Demonstration by the mixed antiglobulin reaction of determinants recognized by anti-γ, anti-μ, anti-Fab and anti-allotype sera, anti-As^4, anti-As^6. Immunology *18:* 417.

CUATRECASAS, P. 1971. Perturbation of the insulin receptor of isolated fat cells with proteolytic enzymes. Direct measurement of insulin-receptor interactions. J. Biol. Chem. *246:* 6522.

CULP, L. A. and P. H. BLACK. 1972. Contact-inhibited revertant cell lines isolated from Simian Virus 40-transformed cells. III. Concanavalin A-selected revertant cells. J. Virol. *9:* 611.

CULP, L. A., W. J. GRIMES and P. H. BLACK. 1971. Contact-inhibited revertant cell lines isolated from SV40-transformed cells. I. Biologic, virologic and chemical properties. J. Cell Biol. *50:* 682.

CUMAR, F. A., R. O. BRADY, E. H. KOLODNY, V. W. MCFARLAND and P. T. MORA. 1970. Enzymatic block in the synthesis of gangliosides in DNA virus-transformed tumorigenic mouse cell lines. Proc. Nat. Acad. Sci. *67:* 757.

CUNNINGHAM, B. A., J. L. WANG, M. N. PFLUMM and G. M. EDELMAN. 1972. Isolation and proteolytic cleavage of the intact subunit of concanavalin A. Biochemistry *2:* 3233.

DALLNER, G., P. SIEKEVITZ and G. E. PALADE. 1966. Biogenesis of endoplasma reticulum membranes. I. Structure and chemical differentiation in developing rat hepatocyte. J. Cell Biol. *30:* 73.

DANIELLI, J. F. and H. DAVSON. 1935. A contribution to the theory of permeability of thin films. J. Cell. Comp. Physiol. *5:* 495.

———. 1943. The structure of the plasma membrane. In *Permeability of natural membranes*, p. 60. Cambridge University Press.

DAVIES, D. A. L., J. COLOMBANI, D. C. VIZA and J. DAUSSET. 1968. Human HL-A transplantation antigens: Separation of molecules carrying different immunological specificities determined by a single genotype. Biochem. Biophys. Res. Commun. *33:* 88.

DAVIS, W. C. 1972. H-2 antigen on cell membranes: An explanation for the alteration of distribution by indirect labeling techniques. Science *175:* 1006.

DEFENDI, V. and G. GASIC. 1963. A surface mucopolysaccharide of polyoma virus transformed cells. J. Cell. Comp. Physiol. *62:* 23.

DEN, H., A. M. SCHULTZ, M. BASAU and S. ROSEMAN. 1971. Glycosyltransferase activities in normal and polyoma-transformed BHK cells. J. Biol. Chem. *246:* 2721.

DEWEY, M. M. and L. BARR. 1970. Some considerations about the structure of cellular membranes. In *Current topics in membranes and transport*, p. 1. Academic Press, New York.

DIENER, E. and V. H. PAETKAU. 1972. Antigen recognition: Early surface-receptor phenomena induced by binding of a tritium-labeled antigen. Proc. Nat. Acad. Sci. *69:* 2364.

EBASHI, S., M. ENDO and I. OHTSUKI. 1969. Control of muscle contraction. Quart. Rev. Biophys. *2:* 4.

ECKHART, W., R. DULBECCO and M. M. BURGER. 1971. Temperature-dependent surface changes in cells infected or transformed by a thermosensitive mutant of polyoma virus. Proc. Nat. Acad. Sci. *68:* 283.

EDELMAN, G. M. and C. F. MILLETTE. 1971. Molecular probes of spermatozoan structures. Proc. Nat. Acad. Sci. *68:* 2436.

EDELMAN, G. M., B. A. CUNNINGHAM, G. N. REEKE, JR., J. W. BECKER, M. J. WAXDAL and J. L. WANG. 1972. The covalent and three-dimensional structure of concanvalin A. Proc. Nat. Acad. Sci. *69:* 2580.

EDIDIN, M. and A. WEISS. 1972. Antigen cap formation in cultured fibroblasts: A reflection of membrane fluidity and of cell motility. Proc. Nat. Acad. Sci. *69:* 2456.

ERSHOFF, B. H. and G. S. BAJAWA. 1963. Submaxillary gland hypertrophy in rats fed proteolytic enzymes. Proc. Soc. Exp. Biol. Med. *113:* 879.

FAIRBANKS, G., T. L. STECK and D. F. H. WALLACH. 1971. Electrophoretic analysis of the major polypeptides of the human erythrocyte membrane. Biochemistry *10:* 2606.

FINE, R. E. and D. BRAY. 1971. Actin in growing nerve cells. Nature New Biol. *234:* 115.

FINEAN, J. B. 1961. *Chemical ultrastructure in living tissues*. Thomas, Springfield, Illinois.

FINEAN, J. B. and R. E. BURGE. 1963. The determination of the Fourier transform of the myelin layer from a study of swelling phenomena. J. Mol. Biol. *7:* 672.

FISHMAN, P. H., V. W. MCFARLAND, P. T. MORA and R. O. BRADY. 1972. Ganglioside biosynthesis in mouse cells: Glycosyltransferase activities in normal and virally-transformed lines. Biochem. Biophys. Res. Commun. *48:* 48.

FOGEL, M. and L. SACHS. 1962. Induction and repression of antigenic material (Forssman type) in relation to cell organization. Develop. Biol. *10:* 411.

FOLKMAN, J., E. MERLER, C. ABERNATHY and G. WILLIAMS. 1971. Isolation of a tumor factor responsible for angiogenesis. J. Exp. Med. *133:* 275.

FOX, T. O., J. R. SHEPPARD and M. M. BURGER. 1971. Cyclic membrane changes in animal cells: Transformed cells permanently display a surface architecture detected in normal cells only during mitosis. Proc. Nat. Acad. Sci. *68:* 244.

FRIBERG, S., JR., A. J. COCHRAN and S. H. GOLUB. 1971. Assessment of concanavalin A reactivity by inhibition of tumour cell migration. Nature New Biol. *232:* 121.

FRICKE, H. 1925. The electric capacity of suspensions with special reference to blood. J Gen. Physiol. *9:* 137.

FRYE, L. D. and M. EDIDIN. 1970. The rapid intermixing of cell surface antigens after formation of mouse-human heterokaryons. J. Cell Sci. *7:* 319.

FURTHMAYR, H. and R. TIMPL. 1970. Immunochemical studies on structural proteins of the red cell membrane. Eur. J. Biochem. *15:* 301.

GAHMBERG, C. G. 1971. Proteins and glycoproteins of hamster kidney fibroblast (BHK21) plasma membranes and endoplasmic reticulum. Biochim. Biophys. Acta *249:* 81.

GANOZA, M. C. and C. A. WILLIAMS. 1969. In vitro synthesis of different categories of specific protein by membrane-bound and free ribosomes. Proc. Nat. Acad. Sci. *63:* 1370.

Gantt, R. R., J. R. Martin and V. J. Evans. 1969. Agglutination of in vitro cultured neoplastic and non-neoplastic cell lines by a wheat germ agglutinin. J. Nat. Cancer Inst. *42:* 369.

Gates, R. E. and M. Morrison. 1972. Investigation of the plasma membrane of tumor cells. Fed. Proc. *1088:* 412 (Abstr.).

Geren, B. 1954. The formation from the Schwann cell surface of myelin in the peripheral nerves of chick embryos. Exp. Cell Res. *7:* 558.

Ginsburg, V. and A. Kobata. 1971. Structure and function of surface components of mammalian cells. In *Structure and function of biological membranes* (ed. L. I. Rothfield) p. 439. Academic Press, New York.

Glossmann, H. and D. M. Neville, Jr. 1971. Glycoproteins of cell surfaces. A comparative study of three different cell surfaces of the rat. J. Biol. Chem. *246:* 6339.

Goldman, R. D. 1971. The role of three cytoplasmic fibers in BHK-21 cell motility. I. Microtubules and the effects of colchicine. J. Cell Biol. *51:* 752.

———. 1972. The effects of cytochalasin B on the microfilaments of baby hamster kidney (BHK-21) cells. J. Cell Biol. *52:* 246.

Goldman, R. D. and D. M. Knipe. 1973. Functions of cytoplasmic fibers in non-muscle cell motility. Cold Spring Harbor Symp. Quant. Biol. *37:* 523.

Goldstein, I. J., C. E. Hollerman and J. M. Merrick. 1965. Protein-carbohydrate interaction. I. The interaction of polysaccharides with concanavalin A. Biochim. Biophys. Acta *97:* 68.

Gorter, E. and F. Grendel. 1925. On biomolecular layers of lipoids in the chromocytes of the blood. J. Exp. Med. *41:* 439.

Greenaway, P. and D. LeVine. 1973. The binding of *N*-acetylneuraminic acid by wheat-germ agglutinin. Nature New Biol. *241:* 191.

Greene, L. A., E. M. Shooter and S. Varcn. 1968. Enzymatic activities of mouse nerve growth factor and its subunits. Proc. Nat. Acad. Sci. *69:* 1383.

Grimes, W. J. 1970. Sialic acid transferases and sialic acid levels in normal and transformed cells. Biochemistry *9:* 5083.

Guidotti, G. 1972. Membrane proteins. Annu. Rev. Biochem. *41:* 731.

Hakomori, S. 1969. Differential reactivities of fetal and adult human erythrocytes. Vox Sanguinis (Basel) *16:* 478.

———. 1970. Cell-density dependent changes of glycolipids in fibroblasts and loss of this response in the transformed cells. Proc. Nat. Acad. Sci. *67:* 1741.

———. 1971. Glycolipid changes associated with malignant transformation. In *The dynamic structure of cell membranes* (ed. D. F. Hölzl Wallach and H. Fischer) p. 65. Springer-Verlag, New York.

Hakomori, S. and R. W. Jeanloz. 1964. Isolation of a glycolipid containing fucose, galactose, glucose, and glucosamine from human cancerous tissue. J. Biol. Chem. *239:* 3606.

Hakomori, S. and W. T. Murakami. 1968. Glycolipids of hamster fibroblasts and derived malignant-transformed cell lines. Proc. Nat. Acad. Sci. *59:* 254.

Hakomori, S. and B. D. Strycharz. 1968. Investigations on cellular blood-group substances. I. Isolation and chemical composition of blood-group ABH and Le^b isoantigens of sphingoglycolipid nature. Biochemistry *7:* 1279.

Hakomori, S., J. Koscielak, K. J. Block and R. W. Jeanloz. 1967. Immunologic relationship between blood group substances and a fucose-containing glycolipid of human adenocarcinoma. J. Immunol. *98:* 31.

Hakomori, S., C. Teather and H. Andrews. 1968. Organizational differences of cell surface "hematoside" in normal and virally transformed cells. Biochem. Biophys. Res. Commun. *33:* 563.

HAKOMORI, S., B. SIDDIQUI, Y.-T. LI, S.-C. LI and C. G. HELLERQVIST. 1971a. Anomeric structures of globoside and ceramide trihexoside of human erythrocytes and hamster fibroblasts. J. Biol. Chem. *246:* 2271.

HAKOMORI, S., T. SAITO and P. K. VOGT. 1971b. Transformation by Rous sarcoma virus: Effects on cellular glycolipids. Virology *44:* 609.

HAMMERLING, U., D. A. L. DAVIES and A. J. MANSTONE. 1971. Transplantation antigens in a high molecular weight from mouse H-2 antigens. Immunochemistry *8:* 7.

HARDMAN, K. D. and C. F. AINSWORTH. 1972. Myo-inosita binding site of concanavalin A. Nature New Biol. *237:* 54.

HARDMAN, K. D., M. K. WOOD, M. SCHIFFER, A. B. EDMUNDSON and C. F. AINSWORTH. 1971. Structure of concanavalin A at 4.25 Å resolution. Proc. Nat. Acad. Sci. *68:* 1393.

HEITZMANN, H. 1972. Rhodopsin is the predominant protein of rod outer segment membranes. Nature New Biol. *235:* 114.

HICKMAN, S., R. KORNFELD, C. K. OSTERLAND and S. KORNFELD. 1972. The structure of the glycopeptides of a human γM-immunoglobulin. J. Biol. Chem. *247:* 2156.

HOFFMAN-BERLING, H. 1964. *Primitive motile systems in cell biology* (ed. R. D. Allen and N. Kamiya) p. 365. Academic Press, New York.

HOLLEY, R. W. and J. A. KIERNAN. 1968. "Contact inhibition" of cell division in 3T3 cells. Proc. Nat. Acad. Sci. *60:* 300.

HOOGEVEEN, J., R. JULIANO, J. COLEMAN and A. ROTHSTEIN. 1970. Water-soluble proteins of the human red cell membrane. J. Membrane Biol. *3:* 156.

HSIE, A. W. and T. T. PUCK. 1971. Morphological transformation of Chinese hamster cells by dibutyryl adenosine cyclic 3′:5′-monophosphate and testosterone. Proc. Nat. Acad. Sci. *68:* 358.

HUBBELL, W. L. and H. M. MCCONNELL. 1968. Spin-label studies of the excitable membranes of nerve and muscle. Proc. Nat. Acad. Sci. *61:* 12.

HULLA, F. W. and W. B. GRATZER. 1972. Association of high-molecular weight proteins in the red cell membrane. FEBS Letters *25:* 275.

ILLIANO, G. and P. CUATRECASAS. 1972. Modulation of adenylate cyclase activity in liver and fat cell membranes by insulin. Science *175:* 906.

INBAR, M. and L. SACHS. 1969a. Interaction of the carbohydrate-binding protein concanavalin A with normal and transformed cells. Proc. Nat. Acad. Sci. *63:* 1418.

———. 1969b. Structural differences in sites on the surface membrane of normal and transformed cells. Nature *223:* 710.

INBAR, M., H. BEN-BASSAT and L. SACHS. 1971a. A specific metabolic activity on the surface membrane in malignant cell-transformation. Proc. Nat. Acad. Sci. *68:* 2748.

———. 1971b. Location of amino acid and carbohydrate transport sites in the surface membrane of normal and transformed mammalian cells. J. Membrane Biol. *6:* 195.

———. 1972a. Membrane changes associated with malignancy. Nature New Biol. *236:* 3.

———. 1972b. Inhibition of ascites tumor development by concanavalin A. Int. J. Cancer *9:* 143.

INGRAM, V. M. 1969. A side view of moving fibroblasts. Nature *222:* 641.

ISSELBACHER, K. J. 1972. Increased uptake of amino acids and 2-deoxy-D-glucose by virus-transformed cells in culture. Proc. Nat. Acad. Sci. *69:* 585.

JAMIESON, J. D. and G. E. PALADE. 1968. Intracellular transport of secretory proteins in the pancreatic exocrine cell. J. Cell Biol. *39:* 58.

JANSONS, V. K. and M. M. BURGER. 1971. Distinct agglutinin receptor molecules on transformed cell surfaces. Fed. Proc. *30:* 692.

JOHNSON, G. S., R. M. FRIEDMAN and I. PASTAN. 1971. Restoration of several morphological characteristics of normal fibroblasts in sarcoma cells treated with adenosine-3':5'-cyclic monophosphate and its derivatives. Proc. Nat. Acad. Sci. *68:* 425.

JONES, R. O. and M. J. ASHWOOD-SMITH. 1970 Some preliminary observations on the biochemical and biological properties of an epithelial growth factor. Exp. Cell Res. *59:* 161.

KALB, A. J. and A. LEVITZKI. 1968. Metal-binding sites of concanavalin A and their role in the binding of α-methyl-D-glucopyranoside. Biochem. J. *109:* 669.

KALB, A. J. and A. LUSTIG. 1968. The molecular weight of concanavalin A. Biochim. Biophys. Acta *168:* 366.

KAPELLER, M. and F. DOLJANSKI. 1972. Agglutination of normal and Rous sarcoma virus-transformed chick embryo cells by concanavalin A and wheat germ agglutinin. Nature New Biol. *235:* 184.

KAY, H. E. M. and D. M. WALLACE. 1961. A and B antigens of tumors arising from urinary epithelium. J. Nat. Cancer Inst. *26:* 1349.

KE, B. 1965. Optical rotatory dispersion of chloroplast-lamallae fragments. Arch. Biochem. Biophys. *112:* 554.

KEITH, A. D., A. S. WAGGONER and O. H. GRIFFITH. 1968. Spin-labeled mitochondrial lipids in *Neurospora crassa*. Proc. Nat. Acad. Sci. *61:* 819.

KIJIMOTO, S. and S. HAKOMORI. 1971. Enhanced glycolipid: α-galactosyltransferase activity in contact-inhibited hamster cells, and loss of this response in polyoma transformants. Biochem. Biophys. Res. Commun. *44:* 557.

———. 1972. Contact-dependent enhancement of net synthesis of Forssman glycolipid antigen and hematoside in NIL cells at the early stage of cell-to-cell contact. FEBS Letters *25:* 38.

KORN, E. D. 1966. Structure of biological membranes. Science *153:* 1491.

KORNBERG, R. D. and H. M. MCCONNELL. 1971a. Lateral diffusion of phospholipids in a vesicle membrane. Proc. Nat. Acad. Sci. *68:* 2564.

———. 1971b. Inside-outside transitions of phospholipids in vesicle membranes. Biochemistry *10:* 111.

KORNFELD, S. 1969. Decreased phytohemagglutinin receptor sites in chronic lymphocytic leukemia. Biochem. Biophys. Acta *192:* 542.

KORNFELD, S. and R. KORNFELD. 1969. Solubilization and partial characterization of a phytohemagglutinin receptor site for human erythrocytes. Proc. Nat. Acad. Sci. *63:* 1439.

KRAEMER, P. M. 1971. Complex carbohydrates of animal cells: Biochemistry and physiology of the cell periphery. In *Biomembranes*, (ed. L. A. Manson) Vol. 1, p. 67. Plenum Press, New York.

KRUG, U., F. KRUG and P. CUATRECASAS. 1972. Emergence of insulin receptors on human lymphocytes during *in vitro* transformation. Proc. Nat. Acad. Sci. *69:* 2604.

LEHMAN, J. M. and J. R. SHEPPARD. 1972. Agglutinability by plant lectins increases after RNA virus transformation. Virology *49:* 339.

LEHNINGER, A. L. 1970. *Biochemistry*. Worth Publishers, New York.

LENARD, J. 1970. Protein and glycolipid components of human erythrocyte membranes. Biochem. J. *9:* 1129.

LENARD, J. and S. J. SINGER. 1966. Protein conformation in cell membrane preparations as studied by optical rotatory dispersion and circular dichroism. Proc. Nat. Acad. Sci. *56:* 1828.

LEVINE, D., M. J. KAPLAN and P. J. GREENAWAY. 1972. The purification and characterization of wheat germ agglutinin. Biochem. J. *129:* 847.

LIN, E. C. C. 1971. The molecular basis of membrane transport systems. In *Structure and function of biological membranes* (ed. L. I. Rothfield) p. 285. Academic Press, New York.

LISKE, R. and D. FRANKS. 1968. Specificity of the agglutinin in extracts of wheat germ. Nature *217:* 860.

MACLENNAN, D. H. 1970. Purification and properties of an adenosine triphosphatase from sarcoplasmic reticulum. J. Biol. Chem. *245:* 4508.

MACLENNAN, D. H. and P. T. S. WONG. 1971. Isolation of a calcium-sequestering protein from sarcoplasmic reticulum. Proc. Nat. Acad. Sci. *68:* 1231.

MACLENNAN, D. H., G. H. ILES, C. C. YIP and P. SEEMAN. 1973. Isolation of sarcoplasmic reticulum proteins. Cold Spring Harbor Symp. Quant. Biol. *37:* 469.

MAKELA, O. 1951. Studies on hemagglutinins of Leguminosae seeds. Amer. Med. Exp. Fenn. *35:* suppl. II.

MANN, D. L., G. N. ROGENTINE, JR. and J. L. FAHEY. 1968. Solubilization of human leucocyte membrane isoantigens. Nature *217:* 1180.

MARCHESI, S. L., E. STEERS, V. T. MARCHESI and T. W. TILLACK. 1970. Physical and chemical properties of a protein isolated from red cell membranes. Biochem. J. *9:* 50.

MARCHESI, V. T. and E. P. ANDREWS. 1971. Glycoproteins: Isolation from cell membranes with lithium diiodosalicylate. Science *174:* 1247.

MARCHESI, V. T. and E. STEERS, JR. 1968. Selective solubilization of a protein component of the red cell membrane. Science *159:* 203.

MARTINEZ-PALOMO, A., R. WICKER and W. BERNHARD. 1972. Ultrastructural detection of concanavalin surface receptors in normal and in polyoma-transformed cells. Int. J. Cancer *9:* 676.

MARTONOSI, A. 1971. The structure and function of sarcoplasmic reticulum membranes. In *Biomembranes* (ed. L. A. Manson) Vol. 1, p. 191. Plenum Press, New York.

MARTONOSI, A. and R. A. HALPIN. 1971. Sarcoplasmic reticulum. X. The protein composition of sarcoplasmic reticulum membranes. Arch. Biochem. Biophys. *144:* 66.

MAZIA, D. and A. RUBY. 1968. Dissolution of erythrocyte membranes in water and comparison of the membrane protein with other structural proteins. Proc. Nat. Acad. Sci. *61:* 1005.

MEEZAN, E., H. C. WU, P. H. BLACK and P. W. ROBBINS. 1969. Comparative studies on the carbohydrate-containing membrane components of normal and virus-transformed mouse fibroblasts. II. Separation of glycoproteins and glycopeptides by Sephadex chromatography. Biochemistry *8:* 2518.

MELCHIOR, D. L., H. J. MOROWITZ, J. M. STURTEVANT and T. Y. TSONG. 1970. Characterization of the plasma membrane of *Mycoplasma laidlawii*. VII. Phase transitions of membrane lipids. Biochim. Biophys. Acta *219:* 114.

MONTAGNIER, L. 1971. Factors controlling the multiplication of untransformed and transformed BHK21 cells under various environmental conditions. In Ciba Symp., *Growth control in cell cultures* (ed. G. E. W. Wolstenholme and J. Knight) p. 33. Churchill Livingston, London.

MOODY, M. F. 1963. X-ray diffraction pattern of nerve myelin: A method for determining the phases. Science *142:* 1173.

MOORE, E. G. and H. M. TEMIN. 1971. Lack of correlation between conversion by RNA tumour viruses and increased agglutinability of cells by concanavalin A and wheat germ agglutinin. Nature *231:* 117.

MORA, P. T., R. O. BRADY and V. W. MCFARLAND. 1969. Gangliosides in DNA virus-transformed and spontaneously transformed tumorigenic mouse-cell lines. Proc. Nat. Acad. Sci. *63:* 1290.

MORA, P. T., F. A. CUMAR and R. O. BRADY. 1971. A common biochemical change in SV40 and polyoma virus transformed mouse cells coupled to control of cell growth in culture. Virology *46:* 60.

MOSCONA, A. A. 1971. Embryonic and neoplastic cell surfaces: Availability of receptors for concanavalin A and wheat germ agglutinin. Science *171:* 905.

NAGATA, Y. and M. M. BURGER. 1972. Wheat germ agglutinin: Isolation and crystallization. J. Biol. Chem. *247:* 2248.

NICOLSON, G. L. 1971. Difference in topology of normal and tumour cell membranes shown by different surface distributions of ferritin-conjugated concanavalin A. Nature New Biol. *233:* 244.

———. 1972. Topography of membrane concanavalin A sites modified by proteolysis. Nature New Biol. *239:* 193.

NICOLSON, G. L. and J. BLAUSTEIN. 1972. The interaction of *Ricinus communis* agglutinin with normal and tumor cell surfaces. Biochim. Biophys. Acta *266:* 543.

NICOLSON, G. L. and S. J. SINGER. 1971. Ferritin-conjugated plant agglutinins as specific saccharide stains for electron microscopy: Application to saccharides bound to cell membranes. Proc. Nat. Acad. Sci. *68:* 942.

NOVIKOFF, A. B. and E. HOLTZMAN. 1970. *Cells and organelles.* Holt, Rinehart, and Winston, New York.

NOWELL, P. C. 1960. Phytohemagglutinin: An initiator of mitosis in cultures of normal human leukocytes. Cancer Res. *20:* 462.

OHTA, N., A. B. PARDEE, B. R. MCAUSLAN and M. M. BURGER. 1968. Sialic acid contents and controls of normal and malignant cells. Biochim. Biophys. Acta *158:* 98.

OLSON, M. O. J. and I. E. LIENER. 1967a. Some physical and chemical properties of concanavalin A, the phytohemagglutinin of the jack bean. Biochemistry *6:* 105.

———. 1967b. The association and dissociation of concanavalin A, the phytohemagglutinin of the jack bean. Biochemistry *6:* 3801.

O'NEILL, C. H. 1968. An association between viral transplantation and Forssman antigen detected by immune adherence in cultured BHK21 cells. J. Cell Sci. *3:* 405.

ONODERA, K. and R. SHEININ. 1970. Macromolecular glucosamine-containing component of the surface of cultivated mouse cells. J. Cell Sci. *7:* 337.

OVERTON, E. 1895. Uber die osmotischen Eigenshaften der lebenden Pflanzen und Tierzelle. Arch. Naturforsch. Ges. (Zurich) *40:* 159.

OZANNE, B. and J. SAMBROOK. 1971a. Binding of radioactively labelled concanavalin A and wheat germ agglutinin to normal and virus-transformed cells. Nature New Biol. *232:* 156.

———. 1971b. Isolation of lines of cells resistant to agglutination by concanavalin A from 3T3 cells transformed by SV40. Lepetit Colloq. Biol. Med. *2:* 248.

PERDUE, J. F., R. KLETZIEN and K. MILLER. 1971. The isolation and characterization of plasma membrane from cultured cells. I. The chemical composition of membrane isolated from uninfected and oncogenic RNA virus-converted chick embryo fibroblasts. Biochim. Biophys. Acta *249:* 419.

PERDUE, J. F., R. KLETZIEN and V. L. WRAY. 1972. The isolation and characterization of plasma membrane from cultured cells. IV. The carbohydrate composition of membranes isolated from oncogenic RNA virus-converted chick embryo fibroblasts. Biochim. Biophys. Acta *266:* 505.

PERNIS, B., L. FORNI and L. AMANTE. 1970. Immunoglobulin spots on the surface of rabbit lymphocytes. J. Exp. Med. *132:* 1001.

PHILLIPS, D. R. and M. MORRISON. 1970. The arrangement of proteins in the human erythrocyte membrane. Biochem. Biophys. Res. Commun. *40:* 284.

———. 1971. Position of glycoprotein polypeptide chain in the human erythrocyte membrane. FEBS Letters *18:* 95.

PINTO DA SILVA, P. and D. BRANTON. 1970. Membrane splitting in freeze-etching: Covalently bound ferritin as a membrane marker. J. Cell Biol. *45:* 598.

PODOULA, J. F., C. S. GREENBERG and M. C. GLICK. 1972. Proteins exposed on the surface of mammalian membranes. Biochemistry *11:* 2616.

POLLACK, R. E. and M. M. BURGER. 1969. Surface-specific characteristics of a contact-inhibited cell line containing the SV40 genome. Proc. Nat. Acad. Sci. *62:* 1074.

POSTE, G. and P. REEVE. 1972. Agglutination of normal cells by plant lectins following infection with nononcogenic viruses. Nature New Biol. *237:* 113.

PUCK, T. T., C. A. WALDREN and A. W. HSIE. 1972. Membrane dynamics in the action of dibutyryl adenosine 3′:5′-cyclic monophosphate and testosterone on mammalian cells. Proc. Nat. Acad. Sci. *69:* 1943.

QUIOCHO, F. A., G. N. REEKE, J. W. BECKER, W. N. LIPSCOMB and G. M. EDELMAN. 1971. The structure of concanavalin A at 4 Å resolution. Proc. Nat. Acad. Sci. *68:* 1853.

RABELLINO, E., S. COLON, H. M. GREY and E. R. UNANUE. 1971. Immunoglobulin on the surface of lymphocytes. I. Distribution and quantitation. J. Exp. Med. *133:* 156.

RAFF, M. C., M. STERNBERG and R. B. TAYLOR. 1970. Immunoglobulin determinants on the surface of mouse lymphoid cells. Nature *225:* 553.

RAMBOURG, A. and C. P. LEBLOND. 1967. Electron microscope observations on the carbohydrate-rich cell coat present at the surface of cells in the rat. J. Cell Biol. *32:* 27.

RAMBOURG, A., W. HERNANDEZ and C. LEBLOND. 1969. Detection of complex carbohydrates in the Golgi apparatus of rat cells. J. Cell Biol. *40:* 395.

RAPPORT, M. M., G. LISELOTTE, V. T. SKIPSKI and N. F. ALONZO. 1959. Immunochemical studies of organ and tumor lipids. Cancer *12:* 438.

REDMAN, C. M. 1969. Biosynthesis of serum proteins and ferritin by free and attached ribosomes of rat liver. J. Biol. Chem. *244:* 4308.

RENGER, H. C. and C. BASILICO. 1972. Mutation causing temperature-sensitive expression of cell transformation by a tumor virus. Proc. Nat. Acad. Sci. *69:* 109.

RENKONEN, K. O. 1948. Studies of hemagglutinins present in the seeds of some representatives of the family Leguminosae. Ann. Med. Exp. Fenn. *26:* 66.

RENKONEN, O., C. G. GAHMBERG, K. SIMONS and L. KÄÄRIÄINEN. 1972. The lipids of the plasma membranes and endoplasmic reticulum from cultured baby hamster kidney cells (BHK21). Biochim. Biophys. Acta *255:* 66.

ROBBINS, P. W. and I. MACPHERSON. 1971. The control of glycolipid synthesis in a cultured hamster cell line. Nature *229:* 569.

ROBERTSON, H. T. and P. H. BLACK. 1969. Changes in surface antigens of SV40-virus transformed cells. Proc. Soc. Exp. Biol. Med. *130:* 363.

ROBERTSON, J. D. 1959. The ultrastructure of cell membranes and their derivatives. Biochem. Soc. Symp. (Cambridge, England) *16:* 3.

———. 1960. The molecular structure and contact relationships of cell membranes. In *Progress in biophysics* (ed. B. Katz and J. A. V. Butler) p. 343. Pergamon Press, New York.

———. 1964. Unit membranes: A review with recent new studies of experimental alterations and a new subunit structure in synaptic membranes. In *Cellular membranes in development* (ed. M. Locke) p. 1. Academic Press, New York.

ROSEMAN, S., 1968. Biosynthesis of glycoproteins, gangliosides and related substances. In *Biochemistry of glycoproteins and related substances* (ed. E. Rossi and E. Stoll). S. Karger, Basel and New York.

———. 1970. The synthesis of complex carbohydrates by multiglycosyltransferase systems and their potential function in intercellular adhesion. In *Chemistry and physics of lipids*, Vol. 5, p. 270. North-Holland, Amsterdam.

ROSENBERG, S. A. and G. GUIDOTTI. 1968. The protein of human erythrocyte membranes. Preparation, solubilization, and partial characterization. J. Biol. Chem. *243:* 1985.

———. 1969. Fractionation of the protein components of human erythrocyte membranes. J. Biol. Chem. *244:* 5118.

ROSENTHAL, A. S., F. M. KREGENOW and H. L. MOSES. 1970. Some characteristics of Ca^{2+}-dependent ATPase activity associated with a group of erythrocyte membrane proteins which form fibrils. Biochim. Biophys. Acta *196:* 254.

ROSSITER, R. J. 1968. Metabolism of phosphatides. In *Metabolic pathways* (ed. D. M. Greenberg) Vol. 2, p. 69. Academic Press, New York.

ROTH, S. and D. WHITE. 1972. Intercellular contact and cell-surface galactosyl transferase activity. Proc. Nat. Acad. Sci. *69:* 485.

ROTH, S., E. J. MCGUIRE and S. ROSEMAN. 1971a. An assay for intercellular adhesive specificity. J. Cell Biol. *51:* 525.

———. 1971b. Evidence for cell-surface glycosyltransferases. Their potential role in cellular recognition. J. Cell Biol. *51:* 536.

RUBIN, H. 1970. Overgrowth stimulating factor released from Rous sarcoma cells. Science *167:* 1271.

SABATINI, D. D., Y. TASHIRO and G. E. PALADE. 1966. On the attachment of ribosomes to microsomal membranes. J. Mol. Biol. *19:* 503.

SAKIYAMA, H. and B. W. BURGE. 1972. Comparative studies of the carbohydrate-containing components of 3T3 and Simian Virus 40 transformed 3T3 mouse fibroblasts. Biochemistry *11:* 1366.

SAKIYAMA, H., S. K. GROSS and P. W. ROBBINS. 1972. Glycolipid synthesis in normal and virus-transformed hamster cell lines. Proc. Nat. Acad. Sci. *69:* 872.

SALZBERG, S. and H. J. RASKAS. 1972. Surface changes of human cells productively infected with human adenoviruses. Virology *48:* 631.

SANDERSON, A. R. 1968. HL-A substances from human spleens. Nature *220:* 192.

SANDERSON, A. R., P. CRESSWELL and K. I. WELSH. 1972. Involvement of carbohydrate in the immunochemical determinant area of HL-A substances. Nature New Biol. *230:* 8.

SATIR, B., C. SCHOOLEY and P. SATIR. 1972. Membrane reorganization during secretion in tetrahymena. Nature *235:* 53.

SCHACHTER, H., I. JABBAL, R. L. HUDGIN and L. PINTERIC. 1970. Intracellular localization of liver sugar nucleotide glycoprotein glycosyltransferase in Golgi-rich fraction. J. Biol. Chem. *245:* 1090.

SCHMITT, F. O., R. S. BEAR and G. L. CLARK. 1935. X-ray diffraction studies on nerve. Radiology *25:* 131.

SCHMITT, F. O., R. S. BEAR and J. PALMER. 1941. X-ray diffraction studies on the structure of the nerve myelia sheath. J. Cell. Comp. Physiol. *18:* 31.

SCHNEBLI, H. P. 1972. A protease-like activity associated with malignant cells. Schweiz. Med. Wschr. *102* (in press).

SCHNEBLI, H. P. and M. M. BURGER. 1972. Selective inhibition of growth of transformed cells by protease inhibitors. Proc. Nat. Acad. Sci. *69:* 3825.

SCHROEDER, T. E. 1970. The contractile ring. I. Fine structure of dividing mammalian (HeLa) cells and the effects of cytochalasin B. Z. Zellforsch. Mikrosk. Anat. *109:* 431.

SEFTON, B. M. and H. RUBIN. 1970. Release from density dependent growth inhibition by proteolytic enzymes. Nature *227:* 843.

SEGREST, J. P., R. L. JACKSON, E. P. ANDREWS and V. T. MARCHESI. 1971. Human erythrocyte membrane glycoprotein: A re-evaluation of the molecular weight as determined by SDS polyacrylamide gel electrophoresis. Biochem. Biophys. Res. Commun. *44:* 390.

SELA, B., H. LIS, N. SHARON and L. SACHS. 1970. Different locations of carbohydrate-containing sites in the surface membrane of normal and transformed mammalian cells. J. Membrane Biol. *3:* 267.

SHAPIRO, A. L., E. VIÑUELA and J. V. MAIZEL, JR. 1967. Molecular weight estimation of polypeptide chains by electrophoresis in SDS-polyacrylamide gel. Biochem. Biophys. Res. Commun. *28:* 815.

SHARON, N. and H. LIS. 1972. Lectins: Cell-agglutinating and sugar-specific proteins. Science *177:* 949.

SHEININ, R. and K. ONODERA. 1972. Studies of the plasma membrane of normal and virus-transformed 3T3 mouse cells. Biochim. Biophys. Acta *274:* 49.

SHEININ, R., K. ONODERA, G. YOGEESWARAN and R. K. MURRAY. 1971. Studies of components of the surface of normal and virus-transformed mouse cells. Le Petit Colloq. Biol. Med. *2:* 274.

SHEN, L. and V. GINSBURG. 1968. Release of sugars from HeLa cells by trypsin. In *Biological properties of the mammalian surface membrane* (ed. L. A. Manson) p. 67. Wistar Inst. Press, Philadelphia.

SHEPPARD, J. R. 1971. Restoration of contact-inhibited growth to transformed cells by dibutyryl adenosine 3′:5′-cyclic monophosphate. Proc. Nat. Acad. Sci. *68:* 1316.

———. 1972a. Difference in the cyclic adenosine 3′:5′-monophosphate levels in normal and transformed cells. Nature New Biol. *236:* 14.

———. 1972b. Cyclic AMP and the growth of transformed cells. In *Membranes and viruses in immunopathology* (ed. S. Day and R. A. Good), Academic Press, New York.

SHEPPARD, J. R. and J. M. LEHMAN. 1972. Viral transformation and cellular cyclic AMP levels. Virology *49:* 339.

SHEPPARD, J., A. J. LEVINE and M. M. BURGER. 1971. Cell-surface changes after infection with oncogenic viruses: Requirements for synthesis of host DNA. Science *172:* 1345.

SHIER, W. T. 1971. Preparation of a "chemical vaccine" against tumor progression. Proc. Nat. Acad. Sci. *68:* 2078.

SHIMIZU, A., F. W. PUTNAM, C. PAUL, J. R. CLAMP and I. JOHNSON. 1971. Structure and role of the five glycopeptides of human IgM immunoglobulins. Nature *231:* 73.

SHOHAM, J. and L. SACHS. 1972. Differences in the binding of fluorescent concanavalin A to the surface membrane of normal and transformed cells. Proc. Nat. Acad. Sci. *69:* 2479.

SHOHAM, J., M. INBAR and L. SACHS. 1970. Differential toxicity on normal and transformed cells in vitro and inhibition of tumour development in vivo by concanavalin A. Nature *227:* 1244.

SIDDIQUI, B. and S. HAKOMORI. 1971. A revised structure for the Forssman glycolipid hapten. J. Biol. Chem. *240:* 5766.

SIEKEVITZ, P. and G. E. PALADE. 1968. A cytochemical study of the pancreas of the guinea pig. V. In vivo incorporation of leucine-1-C^{14} into the chymotrypsinogen of various cell fractions. J. Biophys. Biochem. Cytol. *7:* 619.

SINGER, S. J. and G. L. NICOLSON. 1972. The fluid mosaic model of the structure of cell membranes. Science *175:* 720.

SKOU, J. C. 1965. Enzymatic basis for active transport of Na^+ and K^+ across the cell membrane. Physiol. Rev. *45:* 596.

SMITH, D. F. and E. F. WALBORG, JR. 1972. Purification of wheat germ agglutinin receptor sites of the AS-30D rat ascites hepatoma cell surface. Fed. Proc. *31:* 411.

SMITH, J. W., A. L. STEINER and C. PARKER. 1971a. Human lymphocyte metabolism—Effects of cyclic and noncyclic nucleotides on stimulation by phytohemagglutinin. J. Clin. Invest. *50:* 442.

SMITH, J. W., A. L. STEINER, W. M. NEWBERRY, JR. and C. W. PARKER. 1971b. Cyclic adenosine-3′,5′-monophosphate in human lymphocytes—Alterations after phytohemagglutinin stimulation. J. Clin. Invest. *50:* 432.

So, L. L. and I. J. GOLDSTEIN. 1968. Protein-carbohydrate interaction. XX. On the number of combining sites on concanavalin A, the phytohemagglutinin of the jack bean. Biochim. Biophys. Acta *165:* 398.

SPOONER, B. S., K. M. YAMADA and N. K. WESSELLS. 1971. Microfilaments and cell locomotion. J. Cell Biol. *49:* 595.

STECK, T. L., G. FAIRBANKS and D. F. H. WALLACH. 1971. Disposition of the major proteins in the isolated erythrocyte membrane. Proteolytic dissection. Biochemistry *10:* 2617.

STEIM, J. M., M. E. TOURTELLOTTE, J. C. REINERT, R. N. MCELHANEY and R. L. RADER. 1969. Calorimetric evidence for the liquid-crystalline state of lipids in a biomembrane. Proc. Nat. Acad. Sci. *63:* 104.

STEIN, M. D., I. K. HOWARD and H. J. SAGE. 1971. Studies on a phytohemagglutinin from the lentil. IV. Direct binding studies of *Lens culinaris* hemagglutinin with simple saccharides. Arch. Biochem. Biophys. *146:* 353.

STEIN, W. D. and J. F. DANIELLI. 1956. Structure and function in red cell permeability. Discussions Faraday Soc. *21:* 238.

STOFFEL, W. 1971. Sphingolipids. Annu. Rev. Biochem. *40:* 57.

SUMNER, J. B. and S. F. HOWELL. 1936. The role of divalent metals in the reversible inactivation of jack bean hemagglutinin. J. Biol. Chem. *115:* 583.

SYLVEN, B. and H. MALMGROEN. 1957. Histological distribution of proteinase and peptidase activity in solid tumour transplants. Acta Radiol. Suppl., 154.

TAL, C. 1965. The nature of the cell membrane receptor for the agglutination factor present in the sera of tumour patients and pregnant women. Proc. Nat. Acad. Sci. *54:* 1318.

TAYLOR, R. B., P. H. DUFFUS, M. C. RAFF and S. DE PETRIS. 1971. Redistribution and pinocytosis of lymphocyte surface immunoglobulin molecules induced by anti-immunoglobulin antibody. Nature New Biol. *233:* 225.

TEMIN, H. M. 1966. Studies on carcinogenesis by avian sarcoma viruses. III. The differential effect of serum and polyanions on multiplication of uninfected and converted cells. J. Nat. Cancer Inst. *37:* 167.

———. 1967. Studies on carcinogenesis by avian sarcoma viruses. VI. Differential multiplication of uninfected and of converted cells in response to insulin. J. Cell. Physiol. *69:* 377.

TILNEY, L. G. 1968. Studies on the microtubules in Heliozoa. IV. The effect of colchicine on the formation and maintenance of the axopodia of *Actinosphaerium nucleofilum.* J. Cell Sci. *3:* 549.

TILNEY, L. G. and M. MOOSEKER. 1971. Actin in the brush-border of epithelial cells of the chicken intestine. Proc. Nat. Acad. Sci. *68:* 2611.

TILNEY, L. G. and K. R. PORTER. 1965. Studies on the microtubules in Heliozoa. I. Fine structure of *Actinosphaerium* with particular reference to axial rod structure. Protoplasma *60:* 317.

TODARO, G. J., G. K. LAZAR and H. GREEN. 1965. The initiation of cell division in a contact-inhibited mammalian cell line. J. Cell. Comp. Physiol. *66:* 325.

TOURTELLOTTE, M. E., D. BRANTON and A. KEITH. 1970. Membrane structure: Spin labelling and freeze etching of *Mycoplasma laidlawii*. Proc. Nat. Acad. Sci. *66:* 909.

TRAYER, H. R., Y. NOZAKI, J. A. REYNOLDS and C. TANFORD. 1971. Polypeptide chains from human red blood cell membranes. J. Biol. Chem. *246:* 4485.

TROLL, W., A. KLASSEN and A. JANOFF. 1970. Tumorigenesis in mouse skin: Inhibition by synthetic inhibitors of proteases. Science *169:* 1211.

UHLENBRUCK, G. 1967. Chemie der Blutgrippensubstanzen. Bibl. Haematol. *25:* 1.

UHLENBRUCK, G., U. REIFENBERG and M. HEGGEN. 1970. On the specificity of broad spectrum agglutinins. IV. Invertebrate agglutinins: Current status, conceptions and further observations of the variation of the Hel receptor in pigs. Z. Immunitaetsforsch. Alerg. Klin. Immunol. *139:* 486.

UNANUE, E. R., W. D. PERKINS and M. J. KARNOVSKY. 1972. Endocytosis by lymphocytes of complexes of anti-Ig with membrane-bound Ig[1]. J. Immunol. *108:* 569.

VAGELOS, P. R. 1971. Regulation of fatty acid biosynthesis. In *Current topics in cellular regulation* (ed. B. L. Horecker and E. R. Stadtman) Vol. 4, p. 119. Academic Press, New York.

VAHERI, A., E. RUOSLAHTI and S. NORDLING. 1972. Neuraminidase stimulates division and sugar uptake in density-inhibited cell cultures. Nature New Biol. *238:* 211.

WALLACH, D. F. H. and P. H. ZAHLER. 1966. Protein conformation in cellular membranes. Proc. Nat. Acad. Sci. *56:* 1552.

WANG, J. L., B. A. CUNNINGHAM and G. M. EDELMAN. 1971. Unusual fragments in the subunit structure concanavalin A. Proc. Nat. Acad. Sci. *68:* 1130.

WARREN L. and M. C. GLICK. 1969. Isolation of surface membranes of tissue culture cells. In *Fundamental techniques in virology* (ed. K. Habel and N. P. Salzman) p. 66. Academic Press, New York.

———. 1971. The isolation of the surface membranes of animal cells: A survey. In *Biomembranes* (ed. L. A. Manson) Vol. 1, p. 257. Plenum Press, New York.

WARREN, L., D. CRITCHLEY and I. MACPHERSON. 1972a. Surface glycoproteins and glycolipids of chicken embryo cells transformed by a temperature-sensitive mutant of Rous sarcoma virus. Nature *235:* 275.

WARREN, L., J. P. FUHRER and C. A. BUCK. 1972b. Surface glycoproteins of normal and transformed cells. A difference determined by sialic acid and a growth-dependent sialyl transferase. Proc. Nat. Acad. Sci. *69:* 1838.

WEBER, A. 1966. Energized calcium transport and relaxing factors. Current Topics Bioenergetics *1:* 203.

WEBER, K. and M. OSBORN. 1969. The reliability of molecular weight determinations by dodecyl sulfate-polyacrylamide gel electrophoresis. J. Biol. Chem. *244:* 4406

WEISER, M. M. 1972. Concanavalin A agglutination of intestinal cells from the human fetus. Science *177:* 525.

WEISSMAN, G., W. TROLL, B. L. VAN DUUREN and G. SESSA. 1968. Studies on lysosomes. X. Effects of tumor-promoting agents upon biological and artificial membrane systems. Biochemistry *17:* 2421.

WESSELS, N. K., B. S. SPOONER, J. F. ASH, M. O. BRADLEY, M. S. LUDUENA, E. L. TAYLOR, J. T. WRENN and K. M. YAMADA. 1971. Microfilaments in cellular and developmental processes. Science *171:* 135.

WHITTAM, R. 1967. The molecular mechanism of active transport. In *The neurosciences* (ed. G. C. Qarton et al.) p. 313. Rockefeller Institute, New York.

WILKINS, M. H. F., A. E. BLAUROCK and D. M. ENGELMAN. 1971. Bilayer structure in membranes. Nature New Biol. *230:* 72.

WILLINGHAM, M. C., G. S. JOHNSON and I. PASTAN. 1972. Control of DNA synthesis and mitosis in 3T3 cells by cyclic AMP. Biochem. Biophys. Res. Commun. *48:* 743.

WINZLER, R. J. 1969. In *Red cell membrane: structure and function* (ed. G. A. Jamieson and T. J. Greenwalt) p. 157. Lippincott, Philadelphia.

———. 1970. Carbohydrates in cell surfaces. Int. Rev. Cytol. *29:* 77.

WRAY, V. P. and E. F. WALBORG, JR. 1971. Isolation of tumor cell surface binding sites for concanavalin A and wheat germ agglutinin. Cancer Res. *31:* 2072.

WU, H. C., E. MEEZAN, P. H. BLACK and P. W. ROBBINS. 1969. Comparative studies on the carbohydrate-containing membrane components of normal and virus-transformed mouse fibroblasts. I. Glucosamine-labeling patterns in 3T3, spontaneously transformed 3T3, and SV40-transformed 3T3 cells. Biochemistry *8:* 2509.

YAHARA, I. and G. M. EDELMAN. 1972. Restriction of the mobility of lymphocyte immunoglobulin receptors by concanavalin A. Proc. Nat. Acad. Sci. *69:* 608.

YAMANE, K., A. SHIMADA and S. G. NATHENSON. 1972. Peptide comparison of two histocompatability-2 (H-2b and H-2d) alloantigens. Biochemistry *11:* 2398.

ZAMBRANO, F., M. CELLINO and M. CANESSA-FISCHER. 1971. The molecular organization of nerve membranes. IV. The lipid composition of plasma membranes from squid retinal axons. J. Membrane Biol. *6:* 289.

ZARLING, J. M. and S. T. TEVETHIA. 1971. Expression of concanavalin A binding sites in rabbit kidney cells infected with vaccinia virus. Virology *45:* 313.

Note Added in Proof

Two important papers* have just appeared on the way transformed cells dissolve fibrous clots, an observation first made in 1925 by E. Fischer. This fibrinolytic activity results from the interaction of two separate proteins, one called the "cell factor" specifically released from the tumour cells, the other a component of normal serum. Certain trypsin inhibitors block fibrinolysis, showing that at least one of the two proteins is a latent protease activated by interaction with the second protein, which itself may also be a protease. The "cell factor" interacts best with factors from sera from homologous animals, with, in some cases, the presence of the serum factor being masked by the presence of specific inhibitors. Tests on cells (e.g. 3T3) transformed by several different viruses reveal that the specificity of the "cell factor" resides in the cellular genome, not in the respective tumour virus genome.

The final proteolytic activity is not restricted to fibrinolysis, but also acts upon cell surfaces, conceivably being a key factor in maintaining the distinctive phenotype of the cancer cell. Thus when tumour cells are grown in the presence of inhibitory serum, they resemble normal cells. Correspondingly, normal cells exposed to this protease activity take on a cancerous phenotype. Preliminary experiments reveal that "cell factors" are released from a variety of human tumours as well as from virally transformed cells. Normal cells show little, if any, release of cell factor; nor is any released during the lytic cycle of most viruses, nor during the growth of noncytocidal viruses like AMV.

* Unkeless, J. C., A. Tobia, L. Ossowski, J. P. Quigley, D. B. Rifkin and E. Reich. 1973. An enzymatic function associated with transformation of fibroblasts by oncogenic viruses. I. Chick embryo fibroblast cultures transformed by avian RNA tumor viruses. J. Exp. Med. *137:* 85. Ossowski, L., J. C. Unkeless, A. Tobia, J. P. Quigley, D. B. Rifkin and E. Reich. 1973. II. Mammalian fibroblast cultures transformed by DNA and RNA tumor viruses. J. Exp. Med. *137:* 112.

4

Structure and Composition of Polyoma Virus, SV40 and the Papilloma Viruses

Polyoma virus, SV40 and the papilloma viruses are oncogenic members of the papova group of viruses (*pa*pilloma, *po*lyoma and *va*cuolating viruses, Melnick, 1962). Polyoma virus was first detected as a contaminant in cell-free extracts of Ak mice being used for the transmission of murine leukemia (Gross, 1953a, b). It was first propagated in vitro by Stewart and Eddy and their coworkers, who used mouse embryo cultures (Stewart et al., 1958; and review by Stewart, 1960). The virus was named polyoma (Stewart and Eddy, 1959) because of its ability to induce a wide range of tumours when injected into animals of several species. The cytopathic effects of polyoma virus concomitant with productive infections were described by Eddy and Stewart (1959) and Eddy et al. (1960) and the transformation of mouse and hamster cells in culture was observed in several laboratories at the same period (see review by Gross, 1970).

Simian virus 40 was discovered by Sweet and Hilleman (1960) in cultures of monkey kidney cells of the type being used to produce and to test poliomyelitis vaccines. They initially called this latent simian virus vacuolating virus, because the development of prominent cytoplasmic vesicles characterizes its cytopathic effect during productive infections, but subsequently the nomenclature of Hull et al. (1956) was adopted and vacuolating virus has since been called Simian virus 40 (SV40). It was first shown to be oncogenic in newborn hamsters by Eddy et al. (1961, 1962), and the transformation by SV40 of cells in tissue culture was first reported by Shein and Enders (1962).

The papilloma viruses, which have been isolated from many species and which usually cause benign skin papillomas or warts in their natural hosts, form a very compact subgroup. They have a much longer history than either polyoma virus or SV40 but far less is known about them chiefly because there is no adequate tissue culture system. As long ago as 1894 Variot observed that human warts can be transmitted by inoculation and in 1907 Ciuffo showed that this could be achieved with cell-free filtrates (see review by Gross, 1970). Thereafter, numerous investigators reported the transmission of warts by cell-free filtrates in man, rabbits, cattle, horses and dogs. Shope, in particular, spent many years investigating the biology of the papilloma viruses, especially Shope rabbit papilloma virus (see review by Gross, 1970).

GENERAL DESCRIPTION OF THE PAPOVA VIRUSES

All the papova viruses have a similar chemical composition and structure. They contain only protein and DNA and the DNA molecules are covalently closed, circular superhelices (supercoils). These viruses fall into two natural groups, the polyoma group and the papilloma group. Polyoma virus and SV40 particles are very similar in size and structure (see Table 4.1); probably both contain about 3.6×10^6 daltons of DNA (see Table 4.2), which constitutes about 12 percent of the mass of the particle, enclosed in a spherical protein capsid with icosahedral symmetry (Klug, 1965). Early estimates of the number of capsomers in the capsids of the papova viruses ranged from 42 (Howatson and Crawford, 1963; Horne and Wildy, 1963) to 92 (Mattern et al., 1963). Klug and Finch, however, showed conclusively that the capsids of both the papilloma group (Klug and Finch, 1965; Finch and Klug, 1965) and the polyoma group (Klug, 1965) are skew icosahedral structures with a triangulation number of 7 and 72 capsomers. This structure has been

Table 4.1. STRUCTURAL FEATURES OF THE PAPOVA VIRUSES

	Polyoma Virus	SV40	Papilloma Viruses
Size	45 mμ	45 mμ	55 mμ
Symmetry	icosahedral	icosahedral	icosahedral
Triangulation number (t)	7 laevo	7 dextro	7 laevo (rabbit) or dextro (human)
Number of structural subunits (60 × t)	420	420	420
Number of capsomers	72	72	72
DNA mol wt	3×10^6 (?)	3.6×10^6	5×10^6
DNA content (% w/w)	12.5	12.5	10
Virion mol wt	28×10^6	28×10^6*	47×10^6
Protein composition			
Major capsid polypeptide	47,000 (75% total protein)	45,000 (75% total protein)	?
Minor capsid polypeptides	35,000 23,000	30,000	?
Internal histone-like proteins	3 proteins with mol wt 10,000–20,000		?

The molecular weights of the polypeptides of polyoma virus and SV40 cited in this table are only estimates obtained from the electrophoretic mobilities of the various molecules in gel electrophoresis systems. Rigorous determinations of the molecular weights of these polypeptides have yet to be made.

* Anderer et al. (1967).

confirmed for both polyoma virus and SV40 (Mattern et al., 1967; Anderer et al., 1967).

The papilloma viruses are larger than the viruses of the polyoma group; they contain about 5×10^6 daltons of DNA (Watson and Littlefield, 1960; Crawford, 1965; Kleinschmidt et al., 1965) enclosed in a proportionately larger capsid. But apart from the structure and composition of the virions virtually nothing is known about the molecular biology of the papilloma viruses; there are no satisfactory tissue culture systems for the analysis either of productive infections or of transformation. By contrast, polyoma virus and SV40 have probably been more extensively studied than any other tumour viruses.

POLYOMA VIRUS AND SV40

Polyoma virus and SV40 are very small, although not quite the smallest animal viruses. The particles are about 45 mμ in diameter (Figure 4.1), have a sedimentation coefficient, $S_{20,w}$, of 240 and contain about 3.6×10^6 daltons of DNA. With only about 6000 base pairs the genomes of these two viruses can code only for polypeptides weighing in total about 200,000 daltons. The particles contain only protein and DNA; the absence of any lipid means that the virions are resistant to lipid solvents. They are also resistant to inactivation by heating. Infectious virus has a density of about 1.34 g per ml in caesium or rubidium chloride, whereas empty capsids have a density of about 1.29 g per ml.

The viruses are usually propagated in either primary or secondary cultures of mouse embryos (polyoma virus) or monkey kidney (SV40), or in susceptible cell lines of these species. Each infected cell yields

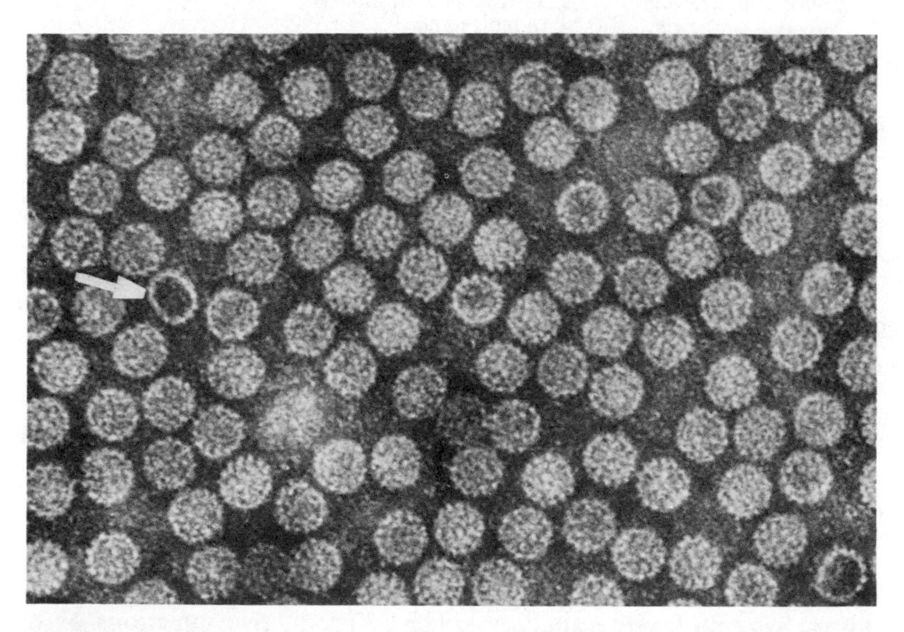

Figure 4.1. Polyoma virus particles stained with 3% phosphotungstic acid. Arrow points to an empty capsid. ×160,000.

about 10^5 physical particles (10^3 infectious particles). This means that 10^6 infected cells in a petri dish yield about 10^{11} particles containing 0.5 μg of DNA and 4 μg of protein. The direct extraction of viral DNA from infected cells (Hirt, 1967) results in somewhat larger yields of viral DNA per cell. The purification procedures for polyoma virus and SV40 (Crawford, 1969b; Black et al., 1964) are comparatively straightforward and include velocity and equilibrium gradient centrifugation. Neuraminidase, which destroys polyoma virus receptor sites, is often used to free virions from cellular debris.

The chief difficulty in preparing large quantities of these viruses is obtaining sufficient quantities of susceptible cells; unfortunately, cells which will grow in suspension and which are also susceptible to these viruses have not so far been selected. Further, many monkey and mouse cell lines which grow in monolayer cultures become resistant to infection by the viruses. Finally, purified polyoma virus and SV40 strongly adsorb to surfaces, especially of such plastics as polycarbonate, and this phenomenon can greatly reduce virus yields.

In the literature there are frequent references to large-plaque and small-plaque polyoma virus and SV40. Both polyoma virus and SV40 when first isolated were found to produce large plaques during lytic infections of permissive cells; subsequently, however, strains which produce comparatively small plaques in these conditions were independently isolated in various laboratories. The small-plaque strains seem to be plaque morphology mutants, but both large- and small-plaque strains are usually considered to be wild-type virus as far as other criteria are concerned. Since their discovery, stocks of both polyoma virus and SV40 have been passed from one laboratory to another and at the same time some research groups have independently obtained fresh isolates of these viruses from the cells of naturally infected animals. As a result there may well be genetic differences between stocks of so-called wild-type virus which are currently propagated in different laboratories. Furthermore when these viruses are propagated at high multiplicities of infection, particles with partially deleted genomes and particles containing host cell DNA covalently bound to the viral DNA accumulate (Tai et al., 1972; Lavi and Winocour, 1972; and see below).

DNAs of Polyoma Virus and SV40

Size and Base Composition

The absolute molecular weight of a DNA molecule can be computed if the base sequence of the DNA is known, but to date no DNA genome has been completely sequenced. The values for the molecular weights of SV40 and polyoma virus DNAs, which are listed in Table 4.2, are, therefore, only estimates which have been normalized against the estimated molecular weights of the genomes of certain coliphages. The considerable variations in the values listed in Table 4.2 have at least two sources. First, the limits of error of the various methods used to measure molecular weights are large. For example, Table 4.3 shows that the contour length of polyoma virus DNA, measured from electron micrographs, varies by at least 15 percent depending upon the way the DNA is mounted onto grids. Second, the size of the DNA in SV40 and polyoma virus particles varies depending upon the way in which

Table 4.2. MOLECULAR WEIGHT OF POLYOMA VIRUS AND SV40 DNA

	Molecular Weight ($\times 10^6$ daltons) Calculated from		
	Sedimentation Velocity	Band Width	Electron Microscopy
Polyoma virus	2.8[a]	2.4[a]	3.1[a]
		2.5[b]	3.7[c]
SV40	3.2[d]	2.7[d]	3.6[e]
	2.33[g]		3.3[f]
			2.9[g]
			2.53[g] (linear)

Estimates of molecular weights are calculated from caesium chloride equilibrium density gradient determinations and have been corrected to the sodium salt of DNA. Values derived from measurements of the length of DNA refer to circular molecules unless specified otherwise; one micron is assumed to correspond to 1.96×10^6 daltons of sodium salt DNA. The low values calculated from electron micrographs (g) are for the free salt of DNA.

Data from: [a] Weil and Vinograd, 1963; [b] Crawford, 1964; [c] Caro, 1965; [d] Crawford and Black, 1964; [e] Tai et al., 1972; [f] Crawford et al., 1966; [g] Anderer et al., 1967.

Table 4.3. Measurement of the Contour Length of Polyoma Virus DNA

Investigator	Protein	Reported length (μ)
Stoeckenius in Weil and Vinograd (1963)	Cytochrome *c*	1.58 ± 0.16
Crawford et al. (1966)	Cytochrome *c*	1.54 ± 0.13
Caro (1965)	Cytochrome *c*	1.90 ± 0.10
Hirt (unpublished data)	Cytochrome *c*	1.85 ± 0.04
Hirt (unpublished data)	Methylated albumin	2.44 ± 0.14
Hirt and Barblan (unpublished data)	No protein	1.56 ± 0.10*

The DNA molecules were prepared according to the procedure of Kleinschmidt et al. (1962).

* Hirt and Barblan adsorbed from solution polyma virus DNA molecules onto films of either cytochrome *c* or methylated albumin which were on carbon-coated electron microscope grids. The DNA molecules were not, therefore, subject to surface tension forces.

the virus is grown. When permissive cells are infected with high multiplicities of these viruses, the progeny particles have on the average shorter genomes than the genomes in virus particles harvested from cells infected at low multiplicities; most investigators have not routinely used recently plaque-purified virus propagated at low multiplicities of infection.

The most recent estimate of the molecular weight of SV40 DNA is 3.6×10^6 daltons (Tai et al., 1972) and it is probably the most accurate yet made for two important reasons. First, only those SV40 genomes which were likely to be nondefective were measured. Second, the lengths of these molecules were compared with the lengths of lambda phage DNA molecules which were on the same electron microscope grid, so that errors stemming from variations in the way the molecules were treated were minimized.

If the molecular weight of polyoma virus DNA is identical to that of SV40 DNA, namely about 3.6×10^6 daltons, both molecules must contain about 6000 base pairs, sufficient to code for about 2000 amino acids or ten proteins of molecular weight 20,000 daltons.

By assaying the 5′-deoxyribonucleotides liberated by the enzymic hydrolysis of polyoma virus labeled with ^{32}P, Smith et al. (1960) estimated the relative amounts of the four nucleotides in polyoma virus

DNA to be dCMP, 23: dAMP, 23: dGMP, 19: dTMP, 25. These figures agree well with the subsequent estimates, based on measurements of the buoyant density and thermal denaturation, that polyoma virus DNA contains 48–49 percent dGMP plus dCMP (Crawford, 1963; Weil, 1963); the buoyant density of SV40 DNA (Crawford and Black, 1964) indicates that it contains 41 percent dGMP plus dCMP. The base composition of SV40 DNA in particular is similar to those of host cells (dGMP + dCMP content, 42–44 percent) (see review by Crawford, 1969a). The nearest neighbour patterns of SV40 DNA and polyoma virus DNA are very similar to each other and fairly similar to the pattern obtained from host cell DNA (see review by Crawford, 1969a). Polyoma virus DNA contains little or no 5-methylcytosine (Winocour, 1967; Kaye and Winocour, 1967), and when the DNA is methylated using *E. coli* enzymes, its infectivity is unchanged (Kaye et al., 1967). There is no homology detectable by hybridization between SV40 DNA and polyoma virus DNA (Winocour, 1965a; Benjamin, 1966), and no homology is detected between host DNA and DNA from plaque-purified virus.

Structure of Polyoma and SV40 DNA

The DNA of polyoma virus and SV40 (Figure 4.2 and Table 4.4) consists of two covalently circular strands (Dulbecco and Vogt, 1963; Weil and Vinograd, 1963; Crawford and Black, 1964) intertwined with each other and base-paired in the usual Watson-Crick fashion; this covalently circular molecule is referred to as component I (Weil and Vinograd, 1963). As there are ten base pairs per turn of the helix, there must be about 500 turns in the double helical DNA of these viruses in normal conditions of temperature and ionic strength. But perhaps for some reason connected with the location and mechanism of replication, polyoma virus and SV40 DNA molecules apparently contain less than 500 turns when the last internucleotide bond is completed. Apparently as a result, the molecules assume a right-handed superhelical configuration (Vinograd et al., 1965). The number of turns in the superhelix (Table 4.4) has been estimated to be 15–20 from studies of the alkaline denaturation of component I DNA of polyoma virus (Vinograd and Lebowitz, 1966) and about 12 from studies of the action of ethidium

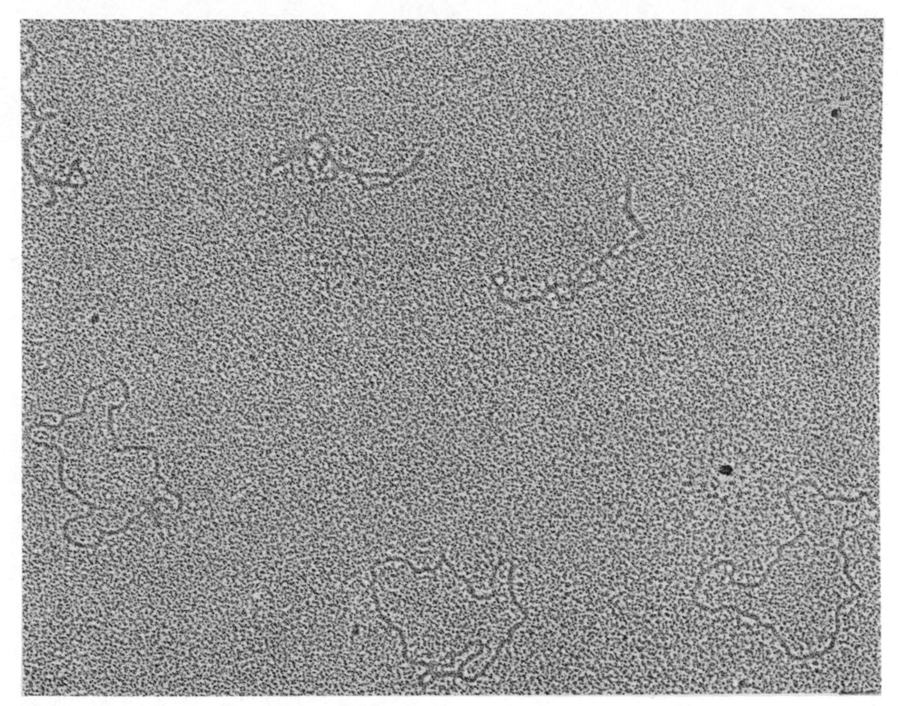

Figure 4.2. Polyoma virus DNA, components I (top) and II. Specimens were rotary shadowed at an angle of 14°. ×120,000.

bromide on the superhelical DNA (Crawford and Waring, 1967a). The mean number of superhelix turns in DNA molecules extracted from SV40 has been estimated to be about 15 from experiments with ethidium bromide and other intercalating drugs (Bauer and Vinograd, 1968). However, the superhelix density of the closed SV40 DNA extracted directly from infected cells is much more heterogeneous and appears to be only three-fourths as large as the superhelix density of the viral DNA (Eason and Vinograd, 1971). The reasons for this are unclear, but one explanation is that the intracellular pool of viral DNA undergoes cycles of nicking and resealing in a heterogeneous environment, resulting in populations of molecules with correspondingly heterogeneous superhelix densities.

Because each of the two complementary strands of the DNA of polyoma virus and SV40 is a covalently bonded circle, the complementary strands cannot be separated (without breaking one of them) by

Table 4.4. Properties of Superhelical DNA from Polyoma Virus and SV40

	Sedimentation Velocity[a]	
	Neutral pH	Alkaline pH
Component I	20 S superhelix	53 S cyclic coils
Component II	16 S relaxed circle	18 S circular single strand 16 S linear single strand

Estimates of Number of Superhelical Turns

Method	Virus	Number of Turns
Alkaline titration	Polyoma	15–20[b]
Intercalation of ethidium bromide and:		
(a) Velocity centrifugation	Polyoma	12[c]
(b) Equilibrium centrifugation	SV40	15[d]

Data from: [a] Vinograd et al., 1965. [b] Vinograd and Lebowitz, 1966. [c] Crawford and Waring, 1967a. [d] Bauer and Vinograd, 1968.

melting conditions such as heating (Weil, 1963) or exposure to alkaline media with a pH of up to 12.5 (Vinograd et al., 1965). Under these conditions most hydrogen bonds in the molecules are broken without the strands separating; therefore the molecules regain their native configuration when cooled or returned to neutral pH. The properties of polyoma virus and SV40 DNAs in extremely alkaline conditions, at a pH above 12.4, have not been investigated. However, these DNAs would be expected to behave in a way similar to the replicative form of ϕX174 (Pouwels et al., 1968). Above pH 12.5 this form of ϕX174 DNA collapses, and when such collapsed molecules are returned to neutrality, they do not renature. Apparently the complementary strands lose all register if all the base pairs are disrupted and they become so entangled that they cannot find their complements when reneutralized. If, however, the collapsed molecules are shifted from pH 12.5 to pH 11.9

so that some base pairs are formed, the molecules completely renature when subsequently returned to neutrality.

Component I DNA loses its superhelical configuration as soon as a single phosphodiester bond is broken in either of the complementary strands because free rotation about the phosphodiester bond opposite the nick is then possible (Vinograd et al., 1965). The nicked DNA, which is called component II, is less compact than the superhelical component I from which it is readily separated by centrifugation; the sedimentation velocity of component I DNA is about 1.3 times that of component II (see Figure 4.3 and Table 4.4). The discovery of the superhelical configuration, therefore, provided a completely satisfactory explanation for two earlier and puzzling observations: first, that DNA extracted from polyoma virus particles can invariably be separated into faster and slower sedimenting components; and second, that treatment with DNase rapidly converts the faster sedimenting to the slower sedimenting component with single-hit kinetics (Dulbecco and Vogt, 1963; Weil and Vinograd, 1963). The component II molecules that are found in

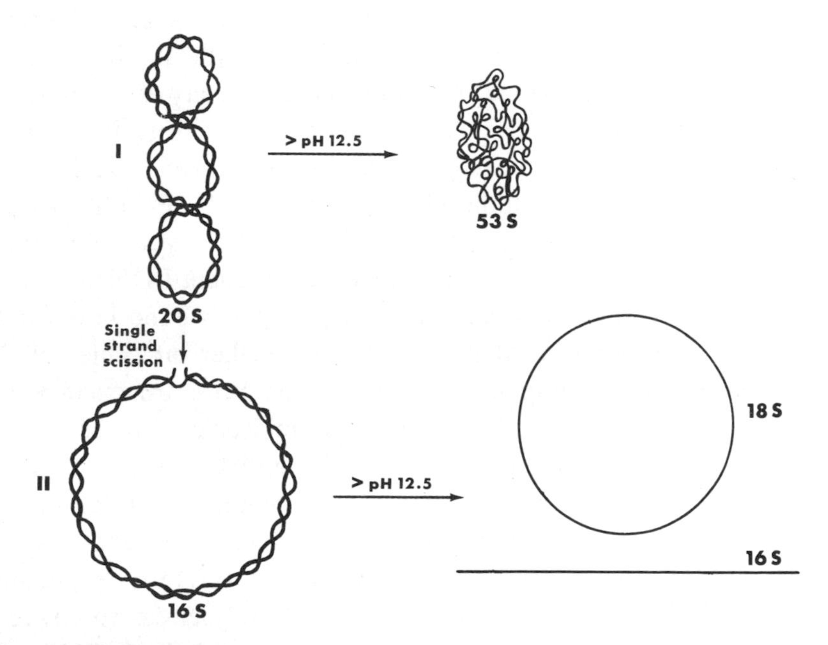

Figure 4.3. Diagrammatic representation of the different configurations of the DNA molecule of polyoma virus and their sedimentation coefficients.

preparations of viral DNA must either come from virus particles containing this form of the DNA or be generated from component I by nucleases during the isolation procedure.

The properties of the DNA of polyoma virus and SV40 which result from the superhelical configuration should not be confused with those which result from the covalent circularity of these molecules. Covalent circularity is a prerequisite for a superhelix but covalently bonded circular DNAs are not necessarily superhelices. The superhelical structure accounts for a slight inherent tendency of component I towards denaturation (see review by Crawford, 1969a), while the shear resistance and comparatively low viscosity of the DNA of polyoma virus and SV40 are a consequence of both the superhelical configuration and the covalent circularity.

Other properties of these DNAs, including the rapid renaturation (Weil, 1963), the failure of the strands to separate on denaturation, and the limited uptake of intercalating dyes are related to the covalent circularity. As each ethidium molecule is intercalated into the double helix, the adjacent base pairs are moved apart and the angle between them is reduced probably by about 12°. Thus the double helix is progressively untwisted and at the same time the density of the molecule is progressively reduced. The saturating amount of drug that can be inserted into a double-helical DNA molecule of given size is much greater if the ends of the molecule can rotate freely. In other words, linear and component II polyoma virus or SV40 DNAs can bind many more ethidium molecules than superhelical component I DNAs, which, because they are covalently closed, rapidly become wound into a tight superhelical conformation that cannot bind further molecules of the drug. Because DNA component II binds much more drug than DNA component I, the density of component II molecules decreases more than the density of component I molecules and so the two types of molecule, after they have bound saturating amounts of ethidium, can readily be separated by centrifugation. Component I DNA always sediments at a higher density than component II DNA in caesium chloride/ethidium bromide gradients. It is certainly easier to envisage the integration of circular, but not necessarily superhelical, DNA into the DNA of the host cell than the integration of a linear DNA. The superhelical structure is not essential for biological activity; component I and component II have comparable infectivities and transforming activities (Crawford et al., 1964), and probably the minimal requirement for full biological activity is one intact, circular, single strand of DNA.

In many preparations of polyoma virus DNA and in some preparations of SV40 DNA, a third component, component III, is resolved by centrifugation. This is linear DNA of the host cell, which, during the process of maturation, is enclosed in viral capsids to form pseudovirions.

Partially Deleted and Substituted Genomes

When SV40 is passed repeatedly and at high multiplicities through African green monkey kidney cells, the yield of plaque-forming particles decreases markedly, whereas the yield of physical particles is reduced only slightly (Uchida et al., 1966). The defective virions produced in this way are heterogeneous in density and most are lighter than infectious particles (Uchida et al., 1968). Their DNA is circular but 10–50 percent shorter than the DNA of infectious particles (Yoshiike, 1968). In other words, these light particles are spontaneous deletion mutants of SV40. Similar light particles also occur in stocks of polyoma virus (Thorne et al., 1968; Blackstein et al., 1969) but they have been less well characterized.

Although most defective SV40 particles are incapable of completing a productive infection without the complementation of an infectious virus, they retain some biological activity. In African green monkey kidney cells, for example, they can induce T antigen (Sauer et al., 1967; Uchida et al., 1968), stimulate cellular DNA synthesis, and even produce short progeny DNA molecules in the absence of fully infectious virus (Kato et al., unpublished data cited by Yoshiike and Furuno, 1969). Moreover Uchida and Watanabe (1968) have shown that the capacity of defective particles to induce tumours in newborn hamsters parallels their capacity to induce T antigen in African green monkey cells. Likewise the capacity of these particles to transform 3T3 cells parallels their capacity to induce T antigen (Uchida and Watanabe, unpublished data cited by Yoshiike and Furuno, 1969). The defective genomes cannot, however, be rescued from transformed cells by fusion with African green monkey kidney cells and viral capsid antigen is not detectable in the heterokaryons.

By selecting and plaque-purifying SV40 particles which are heavier than most defective particles but lighter than most infectious particles, Yoshiike and Furuno (1969) succeeded in isolating two stocks of defective particles which are infectious, albeit yielding between one-hundredth and one-thirtieth the yield obtained from wild-type infections. Apparently these defective particles, the DNA of which is about 14 percent shorter and more heterogeneous than that of wild-type SV40, are able to complement one another and so complete the lytic cycle. The dose response curves for both the particles and their DNA indicate that successful infection is at least a two-hit event. Because at least some defective particles can complement each other, the process of shortening the DNA, which generates these deletion mutants, cannot always span the same site in the SV40 genome.

Between 1966, when Uchida et al. first detected defective SV40 particles, and early 1972, when Lavi and Winocour and Tai et al. published their experiments with stocks of SV40 that had been passaged at high and at low multiplicities of infection, most groups gave surprisingly little attention to the way in which they prepared their stocks of SV40 and to the possibility that large proportions of the particles in their stocks might well be defective. Since February 1972, however, most sensible investigators have scrupulously controlled the multiplicity at which they infect cells when growing SV40 and polyoma virus, for the work of Lavi and Winocour and Tai et al. has finally convinced everyone that the genomes of populations of progeny SV40 particles vary depending on the multiplicity at which the host cells are infected.

In 1969 Aloni et al. claimed that 5 percent of each genome of their strain of SV40, which had been grown on BSC1 cells, was homologous to BSC1 cell DNA. These data met with a skeptical reception; however, Lavi and Winocour (1972) confirmed them when they showed that the extent to which SV40 DNA hybridizes to BSC1 cell DNA depends on the multiplicity of infection that was used when the stocks of virus were grown. They challenged denatured BSC1 cell DNA bound to filters with labeled, sheared, component I DNA from either plaque-purified SV40 or from stocks of SV40 that had been passaged four times and at high multiplicities of infection. The amount of radioactivity retained on the filters after challenge with the DNA from plaque-purified virus was about twice background, but with DNA from repeatedly passaged virus it was twenty times background.

One obvious interpretation of these data is that the SV40 genomes can acquire, presumably by recombination, host DNA sequences, and that repeated passage at high multiplicities favours this event. If this interpretation is correct, the host DNA sequences detected in the SV40 DNA by Lavi and Winocour were probably derived from highly reiterated sequences in the BSC1 cell genome because the hybridization technique they used would not detect unique sequences of cellular DNA. Whether or not any unique sequences of host DNA are integrated into viral DNA and packaged into virus particles is still unknown. But whatever the nature of the host DNA sequences in these high multiplicity SV40 particles, Lavi and Winocour demonstrated, by sequentially hybridizing this viral DNA first to BSC1 cell DNA and then to low multiplicity SV40 DNA, that the host and viral sequences are covalently linked.

Simultaneous with the report of the work of Lavi and Winocour, Tai et al. (1972) reported that the DNA from stocks of SV40 particles that had been passaged at high multiplicities of infection contained significantly greater numbers of molecules shorter than unit length than did the DNA from stocks of SV40 which had been passaged at low multiplicities of infection. Furthermore, when component I DNA from virus grown at high multiplicities was nicked once in one or both strands, denatured and renatured, many heteroduplex molecules with a wide variety of shapes, including molecules with substitution and deletion loops, were formed (see Figure 4.4). By contrast, component I DNA from stocks of virus grown at low multiplicities of infection, after identical treatment, renatured to form almost exclusively (95%) circular duplex molecules which resemble component II SV40 DNA.

Analysis of the shapes and lengths of the various heteroduplex SV40 DNA molecules formed in the first experiment led Tai et al. to two conclusions. First, when SV40 is passaged at high multiplicities of infection, segments of the viral genome are deleted and may be replaced or substituted by nonhomologous DNA. Second, the deleted segment

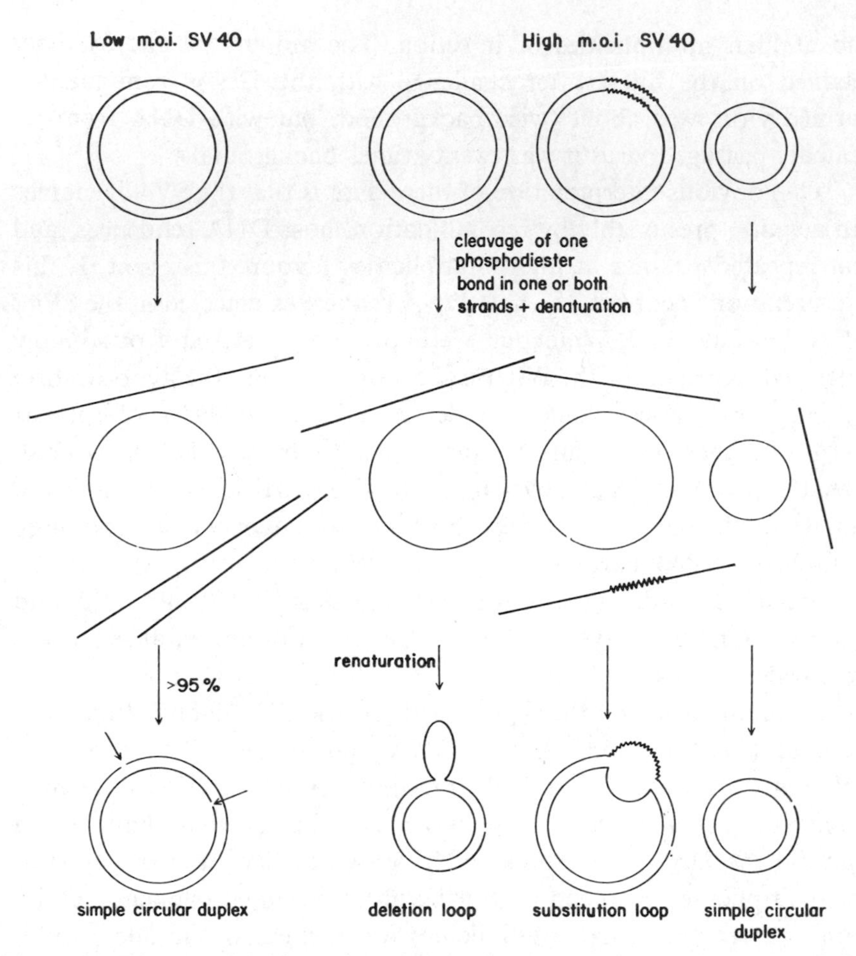

Figure 4.4. Diagram of the formation of heteroduplex SV40 DNA molecules containing substitution and deletion loops during renaturation of single-stranded SV40 DNA from stocks of virus passaged at high multiplicities of infection.

of SV40 DNA is usually longer than the segment of DNA that replaces it. (In substitution type heteroduplexes, on the average 25 percent of the SV40 genome was deleted and replaced by a segment of DNA measuring only 20 percent of the length of the SV40 genome.) In other words, passage of SV40 at high multiplicities of infection leads to the accumulation of particles containing partially deleted, partially substituted genomes.

The origin of the segments of DNA that substitute for deleted segments of the SV40 genome was not established by Tai et al. But in view of Lavi and Winocour's data, it seems most likely that the substitution DNA will prove to be cellular; the other alternative is that it is segments of the SV40 genome which have either been duplicated or inverted.

These findings obviously have important and ramifying implications. For example, they indicate that previous estimates of the molecular weight of SV40 DNA must now be considered unreliable because the stocks of virus that were used probably contained variable proportions of defective particles with partially deleted genomes. Likewise, estimates of the ratio of infectious particles to physical particles in particular stocks of SV40 can have no general significance. And, of course, the presence of quite large segments of host cell DNA in the genomes of some SV40 particles is not without significance for models of SV40 transformation and replication.

Physical Mapping of SV40 DNA

Although the genomes of SV40 and polyoma virus are unlikely to contain more than ten structual genes, progress towards genetic maps of the genomes has been painfully slow because recombination between different strains of virus replicating in the same cell is very hard to detect. Recently, however, several specific chemical features of SV40 DNA have been mapped (see Figure 4.5) by physical techniques including (1) partial denaturation mapping, (2) analysis of fragments of SV40 DNA liberated by digestion with various bacterial restriction endonucleases, and (3) nucleic acid hybridization including analysis in the electron microscope of heteroduplex molecules.

Partial denaturation mapping. The technique of partial denaturation mapping, which was developed for the analysis of lambda phage DNA (Inman, 1967; Inman and Schnös, 1970; Schnös and Inman, 1970), relies on the fact that segments of DNA molecules rich in A–T base pairs denature more readily than segments rich in G–C base pairs. When a linear double-stranded DNA molecule is partially denatured by alkali, regions rich in A–T base pairs separate into single strands, whereas regions rich in G–C base pairs remain double-stranded, and in the presence of formaldehyde the denatured stretches of the molecule

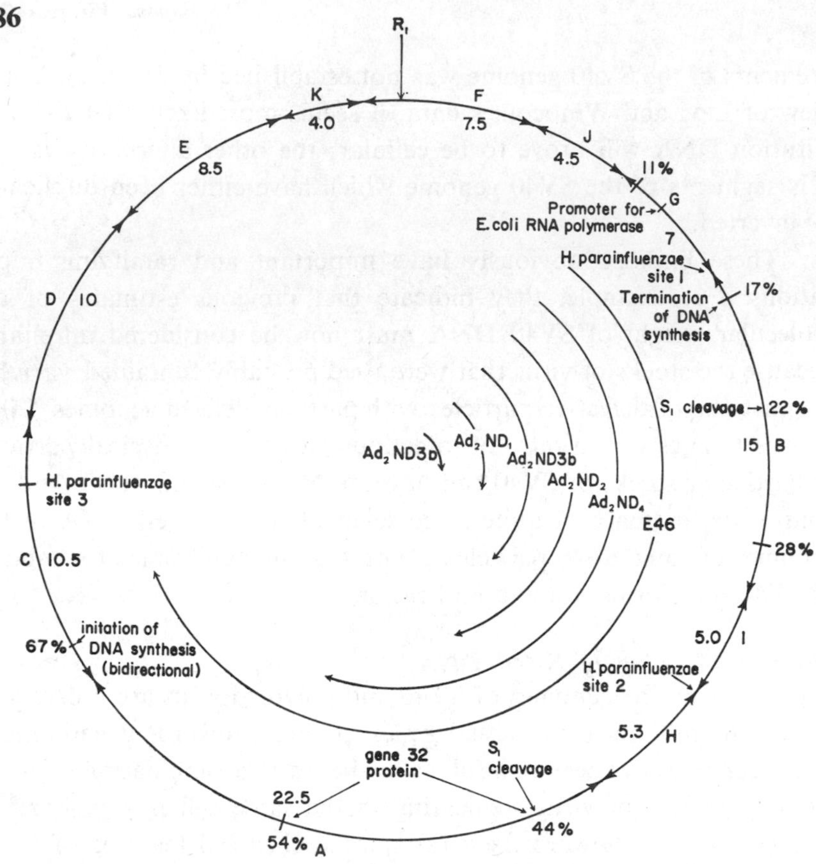

Figure 4.5. A physical map of the SV40 genome showing the sites at which bacterial restriction endonucleases cleave the genome. The eleven cleavage sites of the *H. influenzae* enzyme are indicated by double arrow heads. Also shown are the sites at which replication of SV40 DNA is initiated and terminated, the site of a strong promoter for *E. coli* RNA polymerase, the site of binding to the gene 32 protein of phage T4, and the segments of SV40 DNA that are integrated into the genomes of various hybrid adenoviruses.

are prevented from renaturing. Partially denatured DNA molecules appear in the electron microscope as a series of single-stranded loops or bubbles connected by double-stranded linear segments; the distance of the bubbles from the ends of the molecules can be measured and is characteristic for the particular species of DNA.

Mulder and Delius (1972) used this technique to show that the R_1 restriction endonuclease from *E. coli* carrying a resistance transfer factor cleaved both strands of SV40 component I DNA at a unique site, and they then mapped the A–T rich regions of this DNA with respect to the R_1 cleavage site (see Figure 4.6).

Sites of cleavage by restriction endonucleases. In 1971 Danna and Nathans reported the seminal observation that the restriction endonuclease of *Hemophilus influenzae* cuts both strands of component I SV40 DNA at eleven specific sites to yield eleven specific fragments. Using the products of the partial digestion of component I SV40 DNA by the *Hemophilus influenzae* restriction enzyme, Danna and Nathans (unpublished data) succeeded in ordering the eleven fragments liberated by complete digestion (see Figure 4.5). Subsequently they showed that the restriction enzyme of *Hemophilus parainfluenzae* cleaved SV40 DNA at only three sites, which they have mapped with respect to the eleven cleavage sites of the *Hemophilus influenzae* enzyme (see Figure 4.5).

Danna and Nathan's finding that the restriction enzyme of *Hemophilus influenzae* cleaves SV40 DNA at eleven specific sites spurred on other groups, notably that of Berg, who had been screening the action of various DNases on SV40 DNA. Berg and his colleagues have focused their attention on the R_1 restriction endonuclease.

The restriction enzyme R_1 isolated from *E. coli*, carrying a resistance transfer factor, cuts component I SV40 DNA at a unique site to yield a linear molecule (Morrow and Berg, unpublished data; Mulder and Delius, 1972). This cleavage site has been localized by denaturation mapping of SV40 DNA (Mulder and Delius, 1972) and by hybridizing the linear SV40 DNA, after it has been denatured, to the denatured DNAs of various adeno-SV40 hybrid viruses (Morrow, Berg and Kelly, unpublished data and see Chapter 8). The R_1 cleavage site (see Figure 4.5), is 11 percent of the SV40 genome away from the beginning of the segment of SV40 DNA that is integrated into the adenovirus DNA of the genomes of all the adeno-SV40 hybrid viruses studied so far. The hybridization experiments that revealed the R_1 cleavage site also, of course, revealed the segments of SV40 DNA which are integrated into genomes of the various adeno-SV40 hybrid viruses (Morrow, Berg and Kelly, unpublished data).

The sequence in lambda phage DNA which is recognized and cleaved by the R_1 enzyme has been determined by Goodman and Boyer (unpublished results): the enzyme makes staggered cuts in the

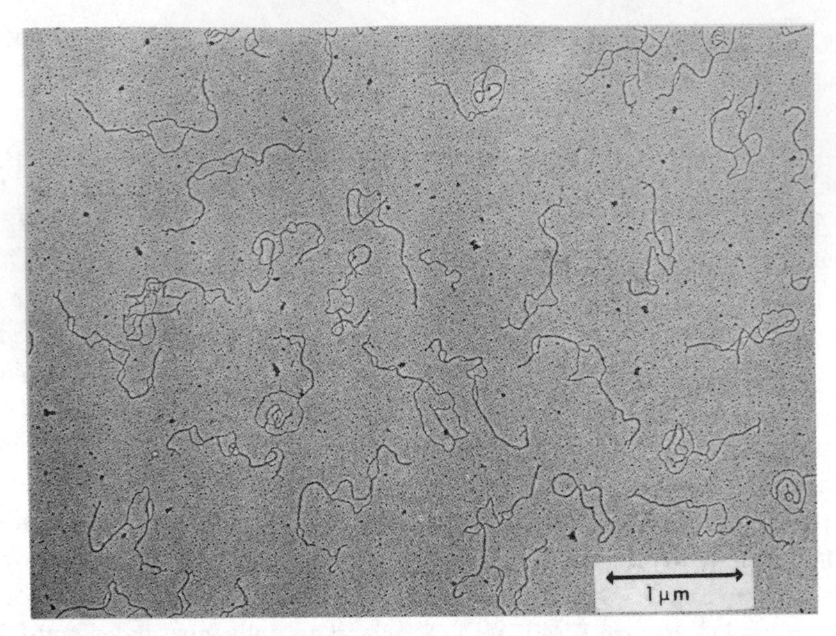

Figure 4.6. *Above:* Electron micrograph of SV40 DNA converted to linear molecules by endonuclease R_1 and partially denatured at low pH (pH 10.9; 60 mM NaOH). Courtesy of H. Delius. *Below:* Histogram of the native regions as derived from maps of the linear SV40 DNA shown above. Reproduced with permission from Mulder and Delius, 1972.

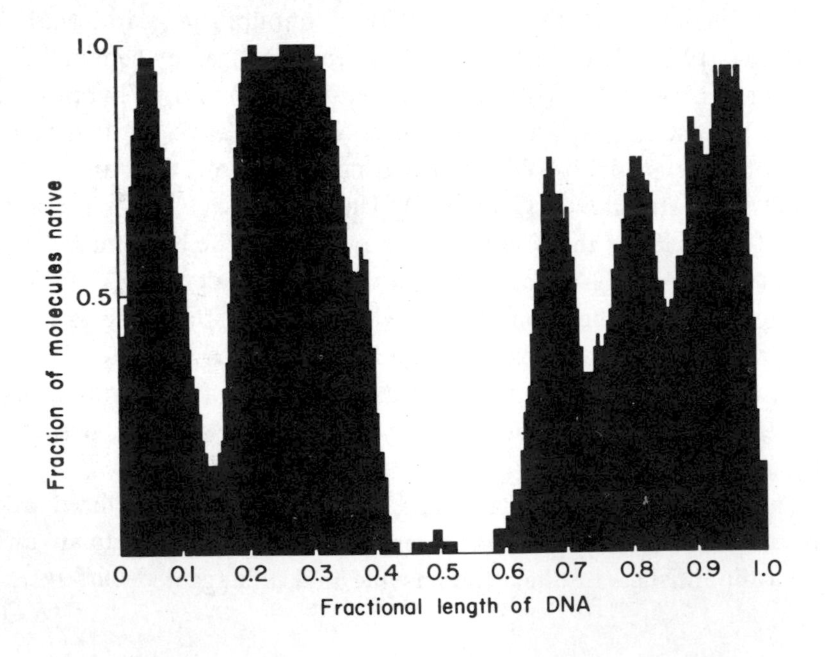

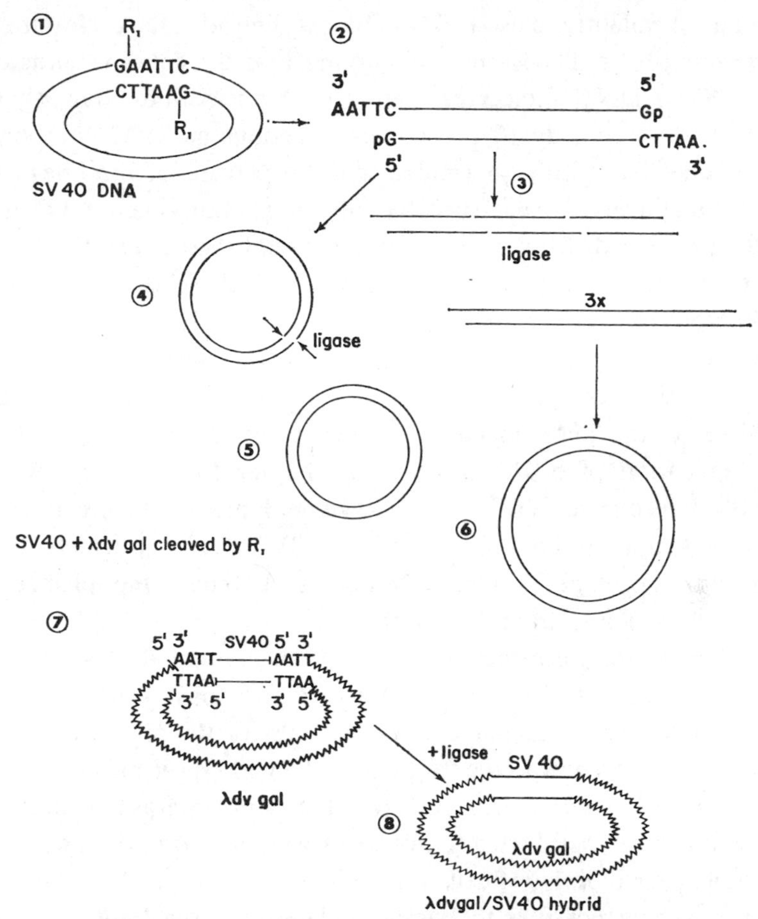

Figure 4.7. Cleavage of SV40 DNA by endonuclease R_1 and synthesis of an SV40/λdv*gal* hybrid DNA. The SV40 DNA is cleaved at a unique site (1) to liberate linear molecules with sticky ends (2). These linear molecules can recircularize or align in linear arrays (3) and (4). In the presence of *E. coli* ligase the molecules can be covalently linked to yield oligomeric or circular monomeric DNA (5, 6, 7). When linear SV40 DNA is mixed with linear λdv*gal* DNA in the presence of ligase, one product is a hybrid SV40/λdv*gal* DNA (7, 8).

two DNA chains so that the linear double-strand DNA it produces has cohesive ends (see Figure 4.7). Presumably this same palindromic hexanucleotide sequence occurs at the R_1 cleavage site in SV40 DNA because Jackson, Symons and Berg (unpublished data), using DNA ligase, have integrated linear SV40 DNA produced by R_1 cleavage into linear λdv*gal* phage DNA, also produced by R_1 cleavage, to yield a circular, covalently closed SV40/λdv*gal* hybrid DNA (Figure 4.7). Moreover Mertz, Davis and Berg found that the cohesive ends of the linear SV40 DNA molecules were 6 ± 2 nucleotides long, that this linear DNA was only 10 percent as infectious as SV40 component I DNA, and that with ligase more than 90 percent of the linear DNA could be converted back to fully infectious component I DNA. According to Beard, Morrow and Berg (unpublished data), the S_1 endonuclease from *Aspergillus oryzae* also specifically cleaves component I SV40 DNA at either of two sites that are located 22 percent and 44 percent of the total genome length from the R_1 cleavage site (see Figure 4.5).

Several other specific sites in the SV40 DNA have been mapped (see Figure 4.5). These include the sites of initiation and termination of the replication of SV40 DNA (see Chapter 5), a strong promoter sequence recognized by *E. coli* RNA polymerase (Weissmann, unpublished data), and a unique site in the SV40 DNA to which the gene 32 protein of phage T4 binds (Mulder and Delius, unpublished data; Morrow and Berg, unpublished data).

That all this information has been accumulated in such a short space of time is eloquent testimony to, amongst other things, the power of restriction endonucleases as tools for the dissection of DNA genomes. And using specific fragments of SV40 DNA and enzymes such as *E. coli* RNA polymerase, *E. coli* DNA polymerase I, and *E. coli* ligase, it is now feasible to attempt to answer many questions concerning the molecular biology of cells infected or transformed by SV40. Moreover it is now possible to begin to determine the base sequence of SV40 DNA either by directly sequencing the SV40 DNA fragments or by sequencing RNA transcripts. Indeed Weissmann et al. (unpublished data) have already indirectly determined the sequence of a stretch of SV40 DNA over 100 base pairs long.

Proteins of Polyoma Virus and SV40 Virions

About 88 percent of each polyoma virion and each SV40 virion is protein; the remaining 12 percent is DNA; there is no detectable lipid or carbohydrate (see Table 4.1). When the total proteins from SV40 particles and polyoma virus particles are separated by electrophoresis

Figure 4.8. Fifteen percent polyacrylamide gel electrophoretograms of (a) SV40 and (b) polyoma virus proteins, and (c) empty polyoma virus capsids. The three smallest polypeptides in both SV40 and polyoma virus particles are believed to be histones; they are not present in empty capsids.

through SDS polyacrylamide gels, several polypeptides are resolved (see Figure 4.8). The three smallest polypeptides in both sorts of virions are basic, have molecular weights in the range of 10,000–20,000 daltons and seem to be derived from histones coded by the host cell genomes. The evidence for this claim, which in part rests on the fact that histones are rich in lysine and arginine but lack tryptophan, is the following: First, when mouse embryo cells are labeled with [^{3}H]lysine and subsequently infected with polyoma virus, these three basic proteins in the progeny virions are more heavily labeled than other proteins of the virions (Frearson and Crawford, 1972). Second, when primary mouse kidney cells are infected with polyoma virus and fed [^{3}H]tryptophan and [^{14}C]arginine, both amino acids are incorporated into the major capsid protein of the progeny virions, but only the [^{14}C]arginine is incorporated into the three small basic polypeptides of the virions (Roblin et al., 1971); it is known that histones do not contain tryptophan. Finally, Hirt (unpublished data) has shown, using [^{35}S]methionine as label, that after tryptic digestion the three small basic polypeptides of both polyoma virus and SV40 particles give the same three distinct fingerprints. Furthermore, three basic polypeptides, with the same electrophoretic mobilities as the three basic polypeptides in the virions, can be extracted from the isolated nuclei of uninfected host cells. The fingerprints of these three cellular basic polypeptides are identical to those of the three small basic polypeptides in the virions. Frearson and Crawford (1972) also found matching tryptic peptides in the fingerprints of the complete histone fraction extracted from mouse embryo cells and the fingerprints of these three virion polypeptides.

The empty capsids that occur in stocks of polyoma virus and SV40 do not contain the three histone-like polypeptides (Frearson and Crawford, 1972; Hirt, unpublished data), which therefore are probably internal proteins associated with the viral DNA. The fact that after virions are disrupted by treatment with alkali some basic proteins stay attached to the DNA, while the major capsid protein is removed (Anderer et al., 1968; Estes et al., 1971), is consonant with this claim. The infectivity of the DNA-internal polypeptide complex is very low but can be increased 1000- to 10,000-fold in the presence of DEAE-dextran, as is the case for pure DNA. Whether alkali degradation of SV40 particles mimics intracellular uncoating of SV40 and whether the internal proteins play a role in modifying transcription of SV40 DNA is uncertain. Huang et al. (1972) have reported data which suggest that SV40 internal proteins may have some regulatory role, but these observations have yet to be confirmed.

The major capsid proteins of SV40 and polyoma virus particles account for about 70 to 80 percent of the total protein in virions. Estimates of the molecular weights of these proteins, based on their electrophoretic mobilities in gels, vary, but it is generally agreed that the major capsid protein of polyoma virus is between 1000 and 2000 daltons larger than that of SV40. Fine et al. (1968), Frearson and Crawford (1972), and Roblin et al. (1971) all estimate that the major capsid protein of polyoma virus has a molecular weight of between 46,000 and 50,000 daltons. This protein, therefore, accounts for almost a quarter of the total coding capacity of the viral genome.

Girard et al. (1970), Estes et al. (1971), Barban and Goor (1971), and Hirt and Gesteland (1971) estimate that the major capsid protein of SV40 has a molecular weight of about 45,000 daltons. Earlier reports that SV40 has two major capsid proteins with molecular weights of 16,900 and 16,400 daltons (Anderer et al., 1967; Schlumberger et al., 1968) are in error. Fingerprint analyses of the major capsid proteins of SV40 and polyoma virus particles labeled with [^{35}S]methionine indicate that these two proteins are not related (Hirt and Gesteland, 1971).

In addition to the major capsid protein, empty polyoma virus capsids and virions contain small amounts of a polypeptide with a molecular weight of about 23,000 daltons, and two polypeptides which have almost identical electrophoretic mobilities and estimated molecular weights of 35,000 daltons. Fingerprint analyses indicate that all these proteins are related (Hirt, unpublished data) but whether or not they are also related to the major capsid protein has yet to be determined. The amino acid composition of SV40 has been determined by Schlumberger et al. (1968) and of polyoma virus by Murakami et al. (1968) and by Kass (1970). Murakami et al. report that the N-terminal amino acid of the major capsid protein of polyoma has a blocked amino group.

Apart from the major capsid protein small amounts of two other proteins can be extracted from SV40 virions and empty capsids. The larger of these has an electrophoretic mobility slightly less than that of the major capsid protein (Girard et al., 1970; Estes et al., 1971) and as Hirt (unpublished data) has shown by fingerprint analysis, this protein is closely related to the major capsid protein. The smaller of the two minor components in SV40 virions and empty capsids has a molecular weight of about 30,000 daltons and yields a fingerprint quite distinct from that of the major capsid protein.

Cuzin et al. (1971) and Kaplan et al. (1972) have detected in extensively purified preparations of polyoma virus and SV40 an endonuclease activity which apparently converts component I DNA to component II DNA; to prove that this activity is intrinsic to the virions and is not the result of some contamination with a cellular enzyme will be a formidable task.

Pseudovirions

The discovery of pseudovirions, particles which contain linear DNA derived from the fragmented host genome enclosed in viral capsids (see Table 4.5), stemmed from a series of reports, based on the results of hybridization experiments, that there was some homology between the base sequences of polyoma DNA and host cell DNA (Axelrod et al., 1964; Winocour, 1965b; Reich et al., 1966). This apparent homology might result from contamination of virus preparations by cell DNA, from the encapsidation of cell DNA in viral capsid protein, or from authentic regions of homology between the viral DNA and the

Table 4.5. OCCURRENCE OF SV40 AND POLYOMA PSEUDOVIRIONS

Cell	Virus	Pseudovirions	Induction of host-cell DNA synthesis	Fragmentation of host genome	Sedimentation coefficient of DNA
Primary					
AGMK (a)	SV40	+	+	+	11–15 S
VERO (b)	SV40	+	+	+	15 S
BSC1 (c)	SV40	−	−	−	
CV1 (c)	SV40	−	+	−	
Primary mouse embryo (d)	Polyoma	+	+	+	14 S

Data from: (a) Levine and Teresky, 1970; (b) Trilling and Axelrod, 1970; (c) Ritzi and Levine, 1970; (d) see review by Winocour, 1969.

host DNA. The second of these explanations proved to be correct (Michel et al., 1967; Winocour, 1968; also see review by Winocour, 1969). During productive infections the host genome is often fragmented, and random (Winocour, 1968) linear pieces of host DNA are enclosed in polyoma capsids. Only size appears to restrict which pieces of host DNA are encapsidated; all the linear host DNA extracted from polyoma pseudovirions sediments at 14 S or less (Michel et al., 1967). The DNA from pseudovirions, which sediments slower than polyoma virus or SV40 component I and II DNAs, is called component III.

The proportion of pseudovirions in preparations of polyoma virus varies markedly depending on the type of cell in which the virus is grown (see review by Crawford, 1969a), and so far it has proved impossible to isolate polyoma virus completely free of pseudovirions, which commonly account for up to 20–40 percent of the particles in a virus preparation (see reviews by Crawford, 1969a; Winocour, 1969).

There is often much less component III DNA in preparations of SV40 DNA than in preparations of polyoma virus DNA (see review by Crawford, 1969a), and recently Levine and his colleagues have shown that production of pseudovirions during SV40 infections depends on the nature of the host cell. When SV40 is grown on primary African green monkey kidney cells or on the VERO line of monkey cells, cellular DNA synthesis is induced; the cellular DNA is fragmented and pseudovirions, which contain linear pieces of host DNA with sedimentation coefficients of between 11 S and 15 S, are produced (see

Table 4.5 and Levine and Teresky, 1970; Trilling and Axelrod, 1970). However, when SV40 is grown on the BSC1 line of African green monkey cells, the virus induces neither the synthesis of cellular DNA nor its fragmentation, and no pseudovirions are formed. When SV40 is grown on another monkey cell line, CV1, although the synthesis of cellular DNA is induced, the cellular DNA is not fragmented and pseudovirions are not produced (see Table 4.5 and Ritzi and Levine, 1970). These experiments indicate that cellular functions are involved in both viral induction of cellular DNA synthesis and the fragmentation of cellular DNA, which is the prerequisite of the formation of pseudovirions.

Apart from the observation that polyoma virus pseudovirions are uncoated in mouse embryo fibroblasts and human embryonic cells (Osterman et al., 1970; Qasba and Aposhian, 1971), and at least some of the pseudovirion DNA reaches the nuclei of infected cells, there is no information about the biological activity of the DNA in pseudovirions. The possibility that pseudovirions might be capable of transducing cellular genes has not been excluded, but at least one attempt to transfer a thymidine kinase gene from primary mouse kidney cells to variant L-cells and BHK cells defective in this enzyme resulted in failure (Widmer, 1971).

It is impossible to attach any significance to most of the reports in the literature of homology between host cell DNA and polyoma virus or SV40 DNAs because the viral DNAs used were, or may have been, contaminated with cellular DNA from pseudovirions (see review by Winocour, 1969). This criticism cannot, however, be leveled at the experiments of Aloni et al. (1969) and Lavi and Winocour (1972) because when SV40 replicates in BSC1, pseudovirions are not produced (Ritzi and Levine, 1970); moreover, they removed linear DNA molecules from their preparations of SV40 DNA before hybridizing it with BSC1 DNA.

THE PAPILLOMA VIRUSES

The physical characterization of all animal viruses has been greatly hampered by the difficulty of obtaining large quantities of purified virus particles. The first viruses to be characterized were those which occur naturally in large quantities and which are not easily inactivated. Shope rabbit papilloma virus exhibits these two characteristics and consequently it was one of the first viruses to be studied by physicochemical techniques. The papillomas of cottontail rabbits, by contrast to the papillomas borne by domestic rabbits, contain large amounts of the Shope virus and simple extraction procedures yield 0.01 mg to 1 mg of virus per gram of papilloma tissue. Not surprisingly, rabbit trappers, notably those in the State of Kansas, have found a brisk market for any cottontail rabbits with warts. Human warts often yield comparable amounts of virus, although the yield is more variable, but, of course, human warts are a good deal harder to obtain.

Papilloma virus particles are between 50 and 55 mμ in diameter and because they contain no lipid they are not inactivated by lipid solvents; they also resist inactivation by heat. The particles have icosahedral symmetry with the 72 capsomers in a skew structure, either left-handed (Shope rabbit papilloma virus) or right-handed (human papilloma virus). The particles have a sedimentation coefficient, $S_{20,W}$, of 300 and a density in caesium or rubidium chloride of 1.34 g per ml.

DNAs of Papilloma Viruses

The papilloma viruses contain about 5×10^6 daltons of DNA (see Table 4.1). Depending on the virus and the methods used, estimates of the molecular weights of the DNAs of papilloma viruses range from 4×10^6 daltons to 5.3×10^6 daltons (see review by Crawford, 1969a). This means the genomes of these viruses contain about 8000–8500 base pairs, sufficient to code for about a dozen proteins with molecular weights of about 25,000 daltons.

The anomalous properties of the papilloma DNAs (Watson and Littlefield, 1960; Crawford, 1964, 1965) were resolved when Vinograd et al. (1965) suggested that covalently circular DNA could assume a superhelical configuration. There are about 20 turns in the superhelix of papilloma component I DNA (Crawford and Waring, 1967b). The sense of the superhelix of human papilloma virus DNA and Shope rabbit papilloma virus DNA is the same even though the capsids of the two viruses are mirror images of each other (Klug and Finch, 1965; Finch and Klug, 1965). Preparations of human papilloma DNA and Shope rabbit papilloma DNA contain, in addition to components I and II, a small percentage (3 percent and 6 percent, respectively) of linear component III DNA, which is presumed to be host DNA from pseudovirions (Crawford, 1969a).

Chemical analysis, thermal denaturation and buoyant density measurements indicate that the DNA of Shope rabbit papilloma virus has a dGMP + dCMP content of 48–49.5 percent (Watson and Littlefield, 1960), whereas human, canine and bovine papilloma DNAs have 41 percent, 43 percent and 45.5 percent dGMP + dCMP (Crawford and Crawford, 1963). The nearest neighbour patterns of human and Shope rabbit papilloma viruses (Subak-Sharpe et al., 1966; Morrison et al., 1967) are similar to each other and to that of host DNA. Human papilloma DNA (Follet and Crawford, 1967) contains seven regions rich in adenine and thymidine. There is no homology detectable by RNA/DNA or DNA/DNA hybridizations between the DNAs of human and Shope rabbit papilloma viruses or between polyoma virus DNA and Shope papilloma virus DNA (see review by Crawford, 1969a).

Proteins of Papilloma Virions

The number of proteins in the virions of papilloma viruses is unknown. In the electron microscope the capsomers of these viruses appear as short hollow cylinders, about 10 mμ long (Howatson and Crawford, 1963), which may be composed of five or six subunits since each has five or six neighbours in the capsid. There are no reliable estimates of the molecular weight(s) of the polypeptide chain(s) which comprise the capsomers. Amino acid analyses of Shope rabbit papilloma virus have been reported (Knight, 1950; Kass and Knight, 1965) and the amino-terminal residues of the virus protein(s) are apparently blocked, as is the case with the major capsid protein of polyoma virus (Murakami et al., 1968).

Biology of the Papilloma Viruses

The papilloma viruses rarely have any cytopathic effect on cells in tissue culture. Human papilloma virus occasionally induces a cytopathic effect in monkey kidney cells (Mendelson and Kligman, 1961) and in human and murine fetal skin cells (Oroszlan and Rich, 1964), but there is no reliable assay based on cytopathogenicity for any papilloma virus. Bovine papilloma virus transforms bovine and murine cells (Black et al., 1963; Thomas et al., 1963; Thomas et al., 1964), as does bovine papilloma DNA (Boiron et al., 1965); but with both intact virus and viral DNA the efficiency of transformation is extremely low. Human papilloma virus also transforms human cells but again at an extremely low efficiency (Noyes, 1965). Presumably the papilloma viruses have a restricted host range that does not include any of the cells tested so far, added to which, as Fenner (1968) suggests, the temperature of incubation (37–39°C) which has been used may be too high to allow these viruses to multiply. The distribution of papilloma viruses in warts certainly indicates that they replicate only in a highly specialized cellular environment. Shope and Hurst (1933) showed that papilloma viruses replicate only in epithelial cells at the site at which they are inoculated. Subsequently Noyes and Mellors (1957) found that viral antigen is present in the nuclei of cells in the superficial, keratohyalin and horny layers of the stratum Malpighii of the skin but is not detectable in the deeper, proliferating layers of the epidermis. Moreover, only keratinized, superficial cells yield infectious virus (Noyes, 1959). Under the electron microscope virus particles can be seen only in the nucleoli in the deepest part of the stratum Malpighii, in the entire nucleus of cells closer to the surface, and throughout the nucleus and cytoplasm of the keratinized cells on the surface of the skin (Stone et al., 1959). Apparently, synthesis of capsid proteins and maturation of progeny particles occur only as epidermal cells become keratinized.

Moreover, Shope papilloma virus behaves differently in domestic and cottontail rabbits. Even the keratinized cells of papillomas induced in domestic rabbits contain very little infectious virus and viral antigen (Shope and Hurst, 1933), whereas the corresponding cells of cottontail rabbits are replete with the virus. The cells of papillomas of domestic rabbits do, however, contain infectious papilloma DNA, as do the cells of squamous cell carcinomas which develop from persistent rabbit papillomas (Ito, 1962).

It seems, therefore, that the papilloma viruses are extremely fastidious and that a transformed cell can, but does not always, become productively infected as the state of differentiation of the cell changes. The proliferating cells in the epidermis are infected and transformed; they are no longer subject to normal growth control. As these cells divide, differentiate by becoming keratinized and migrate to the surface of the epidermis, the virus may apparently be induced to replicate and infectious progeny particles then accumulate.

Literature Cited

Aloni, Y., E. Winocour, L. Sachs and J. Torten. 1969. Hybridization between SV40 DNA and cellular DNAs. J. Mol. Biol. *44:* 333.

Anderer, F. A., H. D. Schlumberger, M. A. Koch, H. Frank and H. J. Eggers. 1967. Structure of simian virus 40: II. Symmetry and components of the virus particles. Virology *32:* 511.

Anderer, F. A., M. A. Koch and H. D. Schlumberger. 1968. Structure of simian virus 40: III. Alkaline degradation of the virus particle. Virology *34:* 452.

Axelrod, D., K. Habel and E. T. Bolton. 1964. Polyoma virus genetic material in a virus-free polyoma-induced tumor. Science *146:* 1466.

Barban, S. and R. Goor. 1971. Structural proteins of SV40. J. Virol. *7:* 198.

Bauer, W. and J. Vinograd. 1968. The interaction of closed circular DNA with intercalative dyes: I. The superhelix density of SV40 DNA in the presence and absence of dye. J. Mol. Biol. *33:* 141.

Benjamin, T. L. 1966. Virus-specific RNA in cells productively infected or transformed by polyoma virus. J. Mol. Biol. *16:* 259.

Black, P. H., J. W. Hartley, W. P. Rowe and R. J. Huebner. 1963. Transformation of bovine tissue culture cells by bovine papilloma virus. Nature *199:* 1016.

Black, P. H., E. M. Crawford and L. V. Crawford. 1964. The purification of simian virus 40. Virology *24:* 381.

Blackstein, M. E., C. P. Stanners and A. J. Farmilo. 1969. Heterogeneity of polyoma virus DNA: Isolation and characterization of non-infectious small supercoiled molecules. J. Mol. Biol. *42:* 301.

Boiron, M., M. Thomas and Ph. Chenaille. 1965. A biological property of deoxyribonucleic acid extracted from bovine papilloma virus. Virology *26:* 150.

Caro, L. G. 1965. The molecular weight of lambda DNA. Virology *25:* 226.

Crawford, L. V. 1963. The physical characteristics of polyoma virus: II. The nucleic acid. Virology *19:* 279.

———. 1964. A study of Shope papilloma virus DNA. J. Mol. Biol. *8:* 489.

———. 1965. A study of human papilloma virus DNA. J. Mol. Biol. *13:* 362.

———. 1969a. Nucleic acids of tumor viruses. Advance. Virus Res. *14:* 89.

———. 1969b. Purification of polyoma virus, p. 75. *In* (K. Habel and N. P. Salzman, ed.) Fundamental techniques in virology. Academic Press, New York.

CRAWFORD, L. V. and P. H. BLACK. 1964. The nucleic acid of simian virus 40. Virology *24:* 388.

CRAWFORD, L. V. and E. M. CRAWFORD. 1963. A comparative study of polyoma and papilloma viruses. Virology *21:* 258.

CRAWFORD, L. V. and M. J. WARING. 1967a. Supercoiling of polyoma virus DNA measured by its interaction with ethidium bromide. J. Mol. Biol. *25:* 23.

——. 1967b. The supercoiling of papilloma virus DNA. J. Gen. Virol. *1:* 387.

CRAWFORD, L., R. DULBECCO, M. FRIED, L. MONTAGNIER and M. STOKER. 1964. Cell transformation by different forms of polyoma virus DNA. Proc. Nat. Acad. Sci *52:* 148.

CRAWFORD, L. V., E. A. C. FOLLETT and E. M. CRAWFORD. 1966. An electron microscopic study of DNA from three tumor viruses. J. de Microscopie *5:* 597.

CUZIN, F., D. BLANQUY and P. ROUGET. 1971. Activité endonucléastique de preparation purifié du virus polyome. Compt. Rend. Acad. Sci. *273:* 2650.

DANNA, K. and D. NATHANS. 1971. Specific cleavage of SV40 DNA by restriction endonuclease of *Hemophilus influenzae*. Proc. Nat. Acad. Sci. *68:* 2913.

DULBECCO, R. and M. VOGT. 1963. Evidence for a ring structure of polyoma virus DNA. Proc. Nat. Acad. Sci. *50:* 236.

EASON, R. and J. VINOGRAD. 1971. Superhelix density heterogeneity of intracellular SV40 DNA. J. Virol. *7:* 1.

EDDY, B. E. and S. E. STEWART. 1959. Characteristics of the SE polyoma virus. Amer. J. Pub. Health *49:* 1486.

EDDY, B. E., G. S. BORMAN, R. L. KIRSCHSTEIN and R. H. TOUCHETTE. 1960. Neoplasms in guinea pigs infected with SE polyoma virus. J. Infect. Diseases *107:* 361.

EDDY, B. E., G. S. BORMAN, W. H. BERKELEY and R. D. YOUNG. 1961. Tumors induced in hamsters by injection of Rhesus monkey kidney cell extracts. Proc. Soc. Exp. Biol. Med. *107:* 191.

EDDY, B. E., G. S. BORMAN, G. E. GRUBBS and R. D. YOUNG. 1962. Identification of the oncogenic substance in Rhesus monkey kidney cell cultures as simian virus 40. Virology *17:* 65.

ESTES, M. K., E. HUANG and J. PAGANO. 1971. Structural polypeptides of SV40. J. Virol. *7:* 635.

FENNER, F. 1968. The biology of animal viruses: Vol. II. The pathogenesis and ecology of viral infections. Academic Press, New York.

FINCH, J. T. and A. KLUG. 1965. The structures of viruses of the papilloma-polyoma type: III. Structure of rabbit papilloma virus. J. Mol. Biol. *13:* 1.

FINE, R., M. MASS and W. T. MURAKAMI. 1968. Protein composition of polyoma virus. J. Mol. Biol. *36:* 167.

FOLLETT, E. A. C. and L. V. CRAWFORD. 1967. Electron microscope study of the denaturation of human papilloma virus DNA: II. The specific location of denatured regions. J. Mol. Biol. *28:* 461.

FREARSON, P. M. and L. V. CRAWFORD. 1972. Polyoma virus basic proteins. J. Gen. Virol. *14:* 141.

GIRARD, M., L. MARTY and F. SUAREZ. 1970. Capsid proteins of simian virus 40. Biochem. Biophys. Res. Commun. *40:* 97.

GROSS, L. 1953a. A filterable agent, recovered from Ak leukaemic extracts, causing salivary gland carcinomas in C3H mice. Proc. Soc. Exp. Biol. Med. *83:* 414.

GROSS, L. 1953b. Neck tumors, or leukaemia, developing in adult C3H mice following innoculation, in early infancy, with filtered (Berkefeld N) or centrifuged (144,000 × *g*), Ak-leukaemic extracts. Cancer *6:* 948.

GROSS, L. 1970. Oncogenic viruses, 2nd ed. Pergamon Press.
HIRT, B. 1967. Selective extraction of polyoma DNA from infected mouse cell cultures. J. Mol. Biol. *26:* 365.
HIRT, B. and R. F. GESTELAND. 1971. Characterization of SV40 and polyoma virus. Lepetit. Colloq. Biol. Med. *2:* 98. North-Holland, Amsterdam.
HORNE, R. W. and P. WILDY. 1963. Virus structure revealed by negative staining. Advance. Virus Res. *10:* 101.
HOWATSON, A. F. and L. V. CRAWFORD. 1963. Direct counting of the capsomers in polyoma and papilloma viruses. Virology *21:* 1.
HUANG, E-S., M. NONOYAMA, and J. S. PAGANO. 1972. Structure and function of the polypeptides in SV40. J. Virol. *9:* 930.
HULL, R. N., J. R. MINNER and J. W. SMITH. 1956. New viral agents recovered from tissue cultures of monkey kidney cells: I. Origin and properties of cytopathic agents S.V. 1, S.V. 2, S.V. 4, S.V. 5, S.V. 6, S.V. 11, S.V. 12 and S.V. 15. Amer. J. Hyg. *63:* 204.
INMAN, R. B. 1967. Denaturation maps of the left and right sides of the lambda DNA molecule determined by electron microscopy. J. Mol. Biol. *28:* 103.
INMAN, R. B. and M. SCHNÖS. 1970. Partial denaturation of thymine- and 5-bromouracil-containing λDNA in alkali. J. Mol. Biol. *49:* 93.
ITO, Y. 1962. Relationship of components of papilloma virus to papilloma and carcinoma cells. Cold Spring Harbor Symp. Quant. Biol. *27:* 387.
KAPLAN, J. C., S. M. WILBERT and P. H. BLACK. 1972. Endonuclease activity associated with purified SV40. J. Virol. *9:* 800.
KASS, S. J. 1970. Chemical studies on polyoma and Shope papilloma viruses. J. Virol. *5:* 381.
KASS, S. J. and C. A. KNIGHT. 1965. Purification and chemical analysis of Shope papilloma virus. Virology *27:* 273.
KAYE, A. M. and E. WINOCOUR. 1967. On the 5-methylcytosine found in the DNA extracted from polyoma virus. J. Mol. Biol. *24:* 475.
KAYE, A. M., B. FRIDLENDER, R. SALOMON and S. BAR-MEIR. 1967. Methylation of DNA *in vitro:* Enzymic activity from different bacterial strains on DNA from various sources. Biochim. Biophys. Acta *142:* 331.
KLEINSCHMIDT, A. K., D. LANG, D. JACHERTS and R. K. ZAHN. 1962. Darstellung der Längenmessungen des gesamten Desoxyribonukleinsäure-Inhaltes von T_2-Bakteriophagen. Biochim. Biophys. Acta *61:* 857.
KLEINSCHMIDT, A. K., S. J. KASS, R. C. WILLIAMS and C. A. KNIGHT. 1965. Cyclic DNA of Shope papilloma virus. J. Mol. Biol. *13:* 749.
KLUG, A. 1965. Structure of viruses of the papilloma-polyoma type: II. Comments on other work. J. Mol. Biol. *11:* 424.
KLUG, A. and J. T. FINCH. 1965. Structure of viruses of the papilloma-polyoma type: I. Human wart virus. J. Mol. Biol. *11:* 403.
KNIGHT, C. A. 1950. Amino acids of the Shope papilloma virus. Proc. Soc. Exp. Biol. Med. *75:* 843.
LAVI, S. and E. WINOCOUR. 1972. Acquisition of sequences homologous to host DNA by closed circular SV40 DNA. J. Virol. *9:* 309.
LEVINE, A. J. and A. K. TERESKY. 1970. Deoxyribonucleic acid replication in simian virus 40-infected cells: II. Detection and characterization of simian virus 40 pseudovirions. J. Virol. *5:* 451.
MATTERN, C. F. T., A. C. ALLISON and W. P. ROWE. 1963. Structure and composition of K virus, and its relation to the "papovavirus" group. Virology *20:* 413.
MATTERN, C. F. T., K. K. TAKEMOTO and A. M. DE LEVA. 1967. Electron microscopic observations on multiple polyoma virus-related particles. Virology *32:* 378.

MELNICK, J. L. 1962. Papova virus group. Science *135:* 1128.
MENDELSON, C. G. and A. M. KLIGMAN. 1961. Isolation of wart virus in tissue culture. Arch. Dermatol. *83:* 559.
MICHEL, M. R., B. HIRT and R. WEIL. 1967. Mouse cellular DNA enclosed in polyoma viral capsids (pseudovirions). Proc. Nat. Acad. Sci. *58:* 1381.
MORRISON, J. M., H. M. KEIR, H. SUBAK-SHARPE and L. V. CRAWFORD. 1967. Nearest neighbour base sequence analysis of the deoxyribonucleic acids of a further three mammalian viruses: Simian virus 40, human papilloma virus and adenovirus type 2. J. Gen. Virol. *1:* 101.
MULDER, C. and H. DELIUS. 1972. Specificity of the break produced by restricting endonuclease R_1 in SV40 DNA as revealed by partial denaturation. Proc. Nat. Acad. Sci. *69:* 3215.
MURAKAMI, W. T., R. FINE, M. R. HARRINGTON and Z. BEN-SASSAN. 1968. Properties and amino acid composition of polyoma virus purified by zonal ultracentrifugation. J. Mol. Biol. *36:* 153.
NOYES, W. F 1959. Studies on the Shope rabbit papilloma virus: II. The location of infective virus in papillomas of the cottontail rabbit, J. Exp. Med. *109:* 423.
———. 1965. Studies on the human wart virus: II. Changes in primary human cell cultures. Virology *25:* 358.
NOYES, W. F. and R. C. MELLORS. 1957. Fluorescent antibody detection of the antigens of the Shope papilloma virus in papillomas of the wild and domestic rabbit. J. Exp. Med. *106:* 555.
OROSZLAN, S. and M. A. RICH. 1964. Human wart virus: *In vitro* cultivation. Science *146:* 531.
OSTERMAN, J. V., A. WADDEL and H. V. APOSHIAN. 1970. DNA and gene therapy: Uncoating of polyoma pseudovirions in mouse embryo cells. Proc. Nat. Acad. Sci. *67:* 37.
POUWELS, P. H., C. M. KNIJNENBURG, J. VAN ROTTERDAM, J. A. COHEN and H. S. JANSZ. 1968. Structure of the replicative form of bacteriophage φX174: VI. Studies on alkali denatured double-stranded φX DNA. J. Mol. Biol. *32:* 169.
QASBA, P. K. and H. V. APOSHIAN. 1971. DNA and gene therapy: Transfer of mouse DNA to human and mouse embryonic cells by polyoma pseudovirions. Proc. Nat. Acad. Sci. *68:* 2345.
REICH, P. R., P. H. BLACK and S. M. WEISSMAN. 1966. Nucleic acid homology studies of SV40 virus-transformed and normal hamster cells. Proc. Nat. Acad. Sci. *56:* 78.
RITZI, E. and A. J. LEVINE. 1970. Deoxyribonucleic acid replication in simian virus 40-infected cells: III. Comparison of simian virus 40 lytic infection in three different monkey kidney cell lines. J. Virol. *5:* 686.
ROBLIN, R., E. HARLE and R. DULBECCO. 1971. Polyoma virus proteins. I. Multiple virion components. Virology *45:* 555.
SAUER, G., H. KOPROWSKI and V. DEFENDI. 1967. The genetic heterogeneity of simian virus 40. Proc. Nat. Acad. Sci. *58:* 599.
SCHLUMBERGER, H. D., F. A. ANDERER and M. A. KOCH. 1968. Structure of the simian virus 40: IV. The polypeptide chains of the virus particle. Virology *36:* 42.
SCHNÖS, M. and R. B. INMAN. 1970. Position of branch points in replicating DNA. J. Mol. Biol. *51:* 61.
SHEIN, H. M. and J. F. ENDERS. 1962. Transformation induced by simian virus 40 in human renal cell cultures: I. Morphology and growth characteristics. Proc. Nat. Acad. Sci. *48:* 1164.
SHOPE, R. E. and E. W. HURST. 1933. Infectious papillomatosis of rabbits. J. Exp. Med. *58:* 607.

SMITH, J. D., G. FREEMAN, M. VOGT and R. DULBECCO. 1960. The nucleic acid of polyoma virus. Virology *12:* 185.

STEWART, S. E. 1960. The polyoma virus. Advance. Virus Res. *7:* 61.

STEWART, S. E. and B. E. EDDY. 1959. Properties of a tumour-inducing virus recovered from mouse neoplasms. *In* (M. Pollard, ed.) Perspectives in virology. Wiley and Sons, New York.

STEWART, S. E., B. E. EDDY and N. G. BORGESE. 1958. Neoplasms in mice inoculated with a tumor agent carried in tissue culture. J. Nat. Cancer Inst. *20:* 1223.

STONE, R. S., R. E. SHOPE and D. H. MOORE. 1959. Electron microscope study of the development of the papilloma virus in the skin of the rabbit. J. Exp. Med. *110:* 543.

SUBAK-SHARPE, H., R. R. BÜRK, L. V. CRAWFORD, J. M. MORRISON, J. HAY and H. M. KEIR. 1966. An approach to evolutionary relationships of mammalian DNA viruses through analysis of the pattern of nearest neighbor base sequence. Cold Spring Harbor Symp. Quant. Biol. *31:* 737.

SWEET, B. H. and M. R. HILLEMAN. 1960. The vacuolating virus, SV40. Proc. Soc. Exp. Biol. Med. *105:* 420.

TAI, H. T., C. A. SMITH, P. A. SHARP and J. VINOGRAD. 1972. Sequence heterogeneity in closed SV40 DNA. J. Virol. *9:* 317.

THOMAS, M., J. P. LEVY, J. TANZER, M. BOIRON and J. BERNARD. 1963. Transformation *in vitro* de cellules de peau de veau embryonnaire sous l'action d'extraits accelulaires de papillomes bovins. Compt. Rend. Acad. Sci. *257:* 2155.

THOMAS, M., M. BOIRON, J. TANZER, J. P. LEVY and J. BERNARD. 1964. *In vitro* transformation of mice cells by bovine papilloma virus. Nature *202:* 709.

THORNE, H. V., J. EVANS and D. WARDEN. 1968. Detection of biologically defective molecules in component I of polyoma virus DNA. Nature *219:* 728.

TRILLING, D. M. and D. AXELROD. 1970. Encapsidation of free host DNA by simian virus 40: A simian virus 40 pseudovirus. Science *168:* 268.

UCHIDA, S. and S. WATANABE. 1968. Tumorigenicity of the antigen-forming defective virions of simian virus 40. Virology *35:* 166.

UCHIDA, S., S. WATANABE and M. KATO. 1966. Incomplete growth of simian virus 40 in African green monkey kidney culture induced by serial undiluted passages. Virology *28:* 135.

UCHIDA, S., K. YOSHIIKE, S. WATANABE and A. FURUNO. 1968. Antigen-forming defective viruses of simian virus 40. Virology *34:* 1.

VINOGRAD, J. and J. LEBOWITZ. 1966. Physical and topological properties of circular DNA. J. Gen. Physiol. *49:* 103.

VINOGRAD, J., J. LEBOWITZ, R. RADLOFF, R. WATSON and P. LAIPIS. 1965. The twisted circular form of polyoma viral DNA. Proc. Nat. Acad. Sci. *53:* 1104.

WATSON, J. D. and J. W. LITTLEFIELD. 1960. Some properties of DNA from Shope papilloma virus. J. Mol. Biol. *2:* 161.

WEIL, R. 1963. The denaturation and the renaturation of DNA of polyoma virus. Proc. Nat. Acad. Sci. *49:* 480.

WEIL, R. and J. VINOGRAD. 1963. The cyclic helix and cyclic coil forms of polyoma viral DNA. Proc. Nat. Acad. Sci. *50:* 730.

WIDMER, C. 1971. Ph.D. thesis, University of Lausanne.

WINOCOUR, E. 1965a. Attempts to detect an integrated polyoma genome by nucleic acid hybridization: I. "Reconstruction" experiments and complementary tests between synthetic polyoma RNA and polyoma tumor DNA. Virology *25:* 276.

———. 1965b. Attempts to detect an integrated polyoma genome by nucleic acid hybridization: II. Complementation between polyoma virus DNA and normal mouse synthetic RNA. Virology *27:* 520.

WINOCOUR, E. 1967. On the apparent homology between DNA from polyoma virus and normal mouse synthetic RNA. Virology *31:* 15.

———. 1968. Further studies on the incorporation of cell DNA into polyoma-related particles. Virology *34:* 571.

———. 1969. Some aspects of the interaction between polyoma virus and cell DNA. Advance. Virus Res. *14:* 153.

YOSHIIKE, K. 1968. Studies on DNA from low-density particles of SV40: I. Heterogeneous defective virions producсd by successive undiluted passages. Virology *34:* 391.

YOSHIIKE, K. and A. FURUNO. 1969. Heterogeneous DNA in simian virus 40. Fed. Proc. *28:* 1899.

5

The Lytic Cycle of Polyoma Virus and SV40

Polyoma virus and SV40 can give rise to either lytic (productive) infections or incomplete (abortive) infections, depending on the type of cell which is infected. During both types of response the virus adsorbs to and penetrates the cell and is uncoated. During productive infections the viral genome, once liberated from its capsid, is transcribed and translated in the permissive environment; T antigen and the synthesis of cellular enzymes involved in DNA metabolism are induced. The viral DNA is replicated and the synthesis of cellular DNA is initiated about 12 hours after infection. Progeny virus particles are assembled from the newly made genomes, viral capsid protein and the other structural proteins of the virion. The infected cells do not divide and by 36 hours after infection, they begin to die, liberating virus. Figure 5.1 shows the time course of these events during the productive infection of primary cultures of monkey kidney cells by SV40 (Hatanaka and Dulbecco, 1966); productive polyoma virus infections of mouse cells (Dulbecco et al., 1965; Sabin and Koch, 1964) are very similar, but progeny polyoma virus particles are usually matured and released sooner than progeny SV40.

By contrast during incomplete infections of nonpermissive cells only part of the viral genome appears to be expressed; cellular enzymes

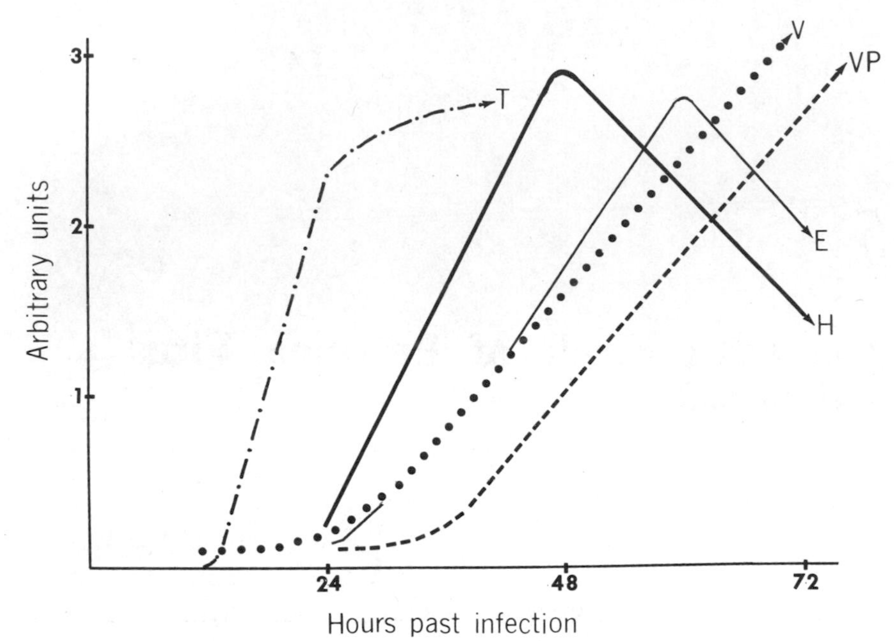

Figure 5.1. Time course of events during productive infection of monkey kidney cells by SV40. T = T antigen; H = host DNA; V = virus DNA; E = enzymes; VP = virus progeny.

involved in DNA synthesis, T antigen and cellular DNA synthesis are all induced, but little or no capsid antigen, viral DNA or progeny virus are made. The cells transiently exhibit some of the characteristics of transformed cells, but many revert to normal after three or four cell divisions while a minority may become stably transformed.

In this chapter we discuss what is known about the events of productive infections.

VIRUS ADSORPTION, PENETRATION AND UNCOATING

Because of the high ratio of physical particles to infectious particles in purified stocks of all animal viruses, including the DNA tumour viruses, it is extremely difficult to follow the early stages of virus infection. The particle to infectivity ratio of polyoma virus, for example, is about 10^2 (Crawford, 1963). Following the fate of labeled virus particles after infection, by measuring radioactivity (Bourgaux, 1964) or tracing unlabeled virus particles by electron microscopy or immunofluorescence (Fraser and Crawford, 1965), unavoidably gives data more relevant to the 99 percent of particles which do not infect than to those few particles which do succeed in infecting the cell concerned. This is a serious reservation, which should always be kept in mind when considering any experiment of this kind.

The first stage of infection by polyoma virus is the reversible physical adsorption of the virus to receptor sites on the cell surface which perhaps contain neuraminic acid; treatment with neuraminidase (receptor-destroying enzyme) prevents the uptake of polyoma virus (Crawford, 1962; Weisberg, 1964; Fried, 1970). Adsorption is strongly dependent on temperature and pH (Crawford, 1962). It also varies with different plaque types; small-plaque polyoma virus is more tightly adsorbed under usual conditions than large-plaque virus. As for most animal viruses, reversible adsorption of polyoma virus is followed by irreversible adsorption and penetration of the cell in pinocytotic vesicles. The virus appears attached to the inner membrane of these vesicles in the cytoplasm; they usually accumulate, with their virus contents, around the nuclear envelope before they become indistinct. It is not known whether polyoma virions are uncoated in the cytoplasm or after the virions have penetrated the nucleus.

Very little is known about the adsorption of SV40. The virus probably attaches to sites on the cell surface which apparently do not contain neuraminic acid since treatment with neuraminidase has no effect on the uptake of the virus. There is some evidence that intact SV40 particles penetrate the nucleus between 30 to 60 minutes after adsorption and are uncoated there (Barbanti-Brodano et al., 1970; Hummeler et al., 1970).

SYNTHESIS OF VIRUS-SPECIFIC RNA

The genetic expression of polyoma virus and SV40 is regulated during productive infection and occurs in at least two distinct phases. The early phase includes the induction of T antigen and of cellular DNA synthesis; the late phase is marked by the onset of replication of viral DNA, includes the synthesis of structural proteins of the virion and culminates in the assembly, maturation and release of progeny particles. These late events can be suppressed by various inhibitors which interfere with the replication of the viral DNA (Butel and Rapp, 1965; Pétursson and Weil, 1968).

Purified polyoma virus and SV40 DNAs are infectious (Crawford et al., 1964). Neither polyoma virus particles nor SV40 particles contain RNA polymerase activity, and the synthesis of early and late species of SV40 RNA in nuclei isolated at late times from productively infected monkey cells is inhibited by α-amanitin. These data all suggest that cellular RNA polymerase II transcribes early and late species of SV40 RNA.

In 1966 Gershon and Sachs reported that the replication of polyoma virus DNA and the synthesis of viral proteins depended on RNA synthesis after infection, and several investigators (Granboulan and Tournier, 1965; Benjamin, 1966; Fried and Pitts, 1968; Oda and Dulbecco, 1968a) reported that productive infections of polyoma virus and SV40 entailed a general stimulation of RNA synthesis in the infected cells. This continued synthesis of cellular RNA after infection complicated early attempts to detect and analyze by hybridization experiments the synthesis of virus-specific RNA. These difficulties notwithstanding, Benjamin in 1966 succeeded for the first time in unequivocally demonstrating the presence of polyoma virus RNA in productively infected mouse kidney cells. Pulse-labeled RNA was extracted from infected cells either before or after the onset of viral DNA synthesis and was hybridized to polyoma virus DNA immobilized on filters. About 0.01 percent of the labeled RNA extracted from cells soon after infection and before the onset of viral DNA synthesis hybridized to the viral DNA. But about 1 percent of the labeled RNA from cells in which the viral DNA was replicating hybridized. Clearly the onset of DNA replication was accompanied by a sharp increase in the amount of viral RNA in the infected cells. This result was quickly confirmed and similar data were obtained from experiments with cells lytically infected by SV40 (Hudson et al., 1970; Aloni et al., 1968; Oda and Dulbecco, 1968b; Sauer and Kidwai, 1968).

Extent of Transcription of Viral DNA Early and Late in Lytic Infections

These results raised several questions; for example: Are the same sequences present in early and late viral RNAs? How much of the viral genome is transcribed during a productive infection? Does the early RNA specify functions essential for the transcription of late RNA?

Aloni et al. (1968), Oda and Dulbecco (1968b) and Sauer and Kidwai (1968) attempted to answer the first of these questions by competition hybridization experiments in which they measured the extent to which early and late SV40 RNAs compete for complementary viral DNA sequences. Their results indicated (1) that populations of early and late viral RNAs are not identical; (2) that about a third of the SV40 DNA sequences transcribed during the complete lytic cycle are present in early RNA; and (3) that late RNA contains all the early RNA sequences as well as sequences not found at early times.

Martin and Axelrod (1969b) then reported that combined early and late SV40 RNAs saturate almost 50 percent of the sequences in the double-stranded viral DNA genome, which means that the equivalent of one strand of the duplex viral DNA is transcribed during the complete lytic cycle. The equivalent of one strand of the polyoma virus genome is transcribed during the complete lytic cycle (Martin and Axelrod, 1969a), but the relative distribution of sequences between early and late RNAs is not known.

The picture that emerged from these investigations was that the equivalent of about one-third of a single strand of SV40 or polyoma virus DNA is transcribed off parental infecting genomes before DNA replication begins and that after DNA replication starts the equivalent of two-thirds of a single strand is additionally transcribed. And the finding that inhibitors of viral DNA replication also uncouple the transcription in vivo of early and late viral RNAs (Carp et al., 1969; Hudson et al., 1970; Sauer, 1971) suggests that the association of DNA synthesis and transcription of late viral RNAs may be more than coincidental.

Transcriptional Control: Strand-Switching

Experimental analyses of the mechanism which controls the early/late transcriptional switch during the lytic cycle of SV40 became feasible when Westphal (1970) and Westphal and Kiehn (1970) found that *E. coli* RNA polymerase assymetrically and completely transcribes one strand of component I SV40 DNA. As Westphal pointed out, this RNA offered investigators a tool with which to separate the two strands of SV40 DNA, for only one of the two strands of denatured, linear SV40 DNA hydridizes to the RNA transcript made by the *E. coli* enzyme.

Lindstrom and Dulbecco (1972) challenged SV40 RNA made in vitro by *E. coli* RNA polymerase with populations of virus-specific RNA molecules extracted from monkey cells either early or late after infection with SV40. Some of the late RNA, but none of the early RNA, hybridized with the in vitro RNA to form RNA/RNA hybrids. This result suggested first, that RNA synthesized early in infection is transcribed from the strand of SV40 DNA that is used as a template by *E. coli* RNA polymerase in vitro; and second, that those RNA sequences present only at late times in infected cells are transcribed from the complementary, antiparallel strand of the SV40 DNA. In other words, a transcriptional strand-switching mechanism operates during the lytic cycle of SV40.

Khoury et al. (1972), Khoury and Martin (1972) and Sambrook et al. (1972) adapted a technique devised by Britten and Kohne (1968) to allow the separation of single-stranded nucleic acid from double-stranded nucleic acid to separate the two strands of SV40 DNA using RNA transcribed in vitro. Having separated the two strands of SV40 DNA they fragmented them to pieces 300–500 bases long and challenged the DNA fragments with early RNA and with late RNA. The pattern of RNA/DNA hybridization they obtained proved that early RNA is transcribed off 30–40 percent of that strand which is used as template by *E. coli* RNA polymerase, while late RNA is transcribed off 60–70 percent of the complementary DNA strand. Since early RNA does not hybridize with late RNA, the segments of the two strands which are used as templates cannot overlap in the duplex DNA. Finally Sambrook et al. (1972) were able to estimate that 10–11 hours after monkey cells are infected with SV40, 0.0017 percent of the total mass of cellular RNA is virus-specific, while at late times (32–48 hours post infection) the mass percent of early viral RNA is at least tenfold greater.

Comparable analyses of the control of transcription of polyoma virus DNA in lytically infected mouse cells have not yet been done.

Lindstrom and Dulbecco (unpublished data) have, however, obtained preliminary evidence of a transcriptional strand-switch during polyoma virus replication. Furthermore they believe that concomitant with viral DNA synthesis some RNA is transcribed off a region of the early RNA template strand that is not used as template before DNA replication starts.

The biological activity of polyoma virus and SV40 RNAs made in permissive cells has yet to be assayed in cell-free systems that support protein synthesis. But there can be no doubting that at least some of these molecules are messages; they can be isolated from polysomes of infected cells (Martin, 1970; Kajioka, 1972) and they contain polyadenylic acid sequences (Weinberg et al., 1972a).

Size of Viral RNAs

Numerous attempts have been made to measure the size of SV40 and polyoma virus RNAs transcribed in vivo (Hudson et al., 1970; Acheson et al., 1971; Kajioka, 1972; Oda and Dulbecco, 1968b; Martin, 1970; Tonegawa et al., 1970; Sokol and Carp, 1971; Weinberg et al., 1972a). RNA molecules isolated from the nuclei of lytically infected cells vary greatly in size as judged by their sedimentation and electrophoretic properties, and many are apparently larger than a single transcript of the entire SV40 or polyoma virus genome. By contrast, viral RNAs extracted from cytoplasmic polysomes are much more homogeneous and sediment between 18 S and 28 S. Presumably these viral RNAs are derived by cleavage maturation from the larger molecules in the nucleus, but without knowing the pool sizes of the substrates of RNA synthesis in mammalian cells and the rates of turnover of viral RNAs in vivo, this precursor-product relationship will be extremely difficult to prove.

If both early and late SV40 RNAs are transcribed in vivo off free DNA molecules, each DNA strand must contain one initiation site and one termination site for transcription, the DNA between these signals acting as template for the respective RNAs. Further, the early RNAs should have molecular weights of about 500,000 daltons and the late

RNAs molecular weights of about 1,000,000 daltons. Weinberg et al. (1972b) have devised methods for purifying viral RNAs made in vivo. By hybridizing these RNAs against separated linear strands of SV40 DNA and examining the RNA/DNA hybrids in the electron microscope, it may be possible to test these suppositions.

Alternatively, of course, viral RNAs may be transcribed, at least during the early stages of the lytic cycle, off viral genomes integrated into cellular DNA. There is evidence suggesting that such integration may occur (see below), and Sambrook et al. (1972) have devised a model, which, assuming an integrated viral genome, accounts for what we now know about the pattern of transcription of SV40 DNA. The idea that viral DNA might integrate during the lytic cycle first arose when Acheson et al. (1971) reported the presence of virus-specific RNA molecules larger than a complete transcript of the polyoma virus genome in the nuclei of lytically infected cells and Weinberg et al. (1972a) reported the same phenomena in monkey cells infected with SV40. If these "giant" RNA molecules do, indeed, contain both host and viral RNA sequences as Rozenblatt and Winocour (unpublished data) believe, it will be hard to escape the conclusion that these molecules are transcripts of a tandem array of host and viral genes. This possibility is, however, far from proven and there are reasons for thinking that all the models of transcriptional control currently under discussion may be considerable oversimplifications.

First, the ratio of noninfectious particles to infectious particles in stocks of virus used to infect permissive cells is very high, and biochemical investigations of the synthesis of viral RNA in infected cells may therefore yield data about the transcription of noninfectious rather than infectious genomes. Second, and more important, Aloni (1972) has recently reported an intriguing set of experiments which suggest that late in the infection of monkey cells by SV40 large tracts of the viral DNA are transcribed symmetrically and that subsequently one or both of the RNA strands are degraded. Aloni, unlike other investigators, exposed the infected cells to very short pulses of [^{3}H]uridine and then extracted the viral RNA immediately or after chasing for various times with cold uridine. He then self-annealed the isolated RNA, removed with ribonuclease all RNA that remained single-stranded, and then analyzed the surviving labeled, double-stranded RNA. Hybridization experiments with single-stranded SV40 DNA revealed that most of the sequences (about 60 percent of the size of a complete viral transcript) in the double-stranded RNA molecules were homologous to SV40 DNA. Aloni's observations suggest that the regulation of expression of SV40 DNA may involve extensive but highly regulated post-transcriptional processing of viral RNA molecules. At present we do not have the slightest clue as to how selective destruction of particular viral RNA molecules might be achieved; neither is it obvious how best to begin to elucidate this mechanism.

SYNTHESIS OF VIRUS-SPECIFIC PROTEIN AND ANTIGENS

Very little is known about the synthesis of polyoma virus or SV40 proteins during productive infections not least because only one protein undoubtedly specified by these viruses, their major capsid protein, has been identified and partially characterized. The capsid proteins of SV40 and polyoma virus weigh about 45,000 daltons and account for about one-quarter of the genome of these viruses; the number and nature of the proteins specified by the other three-quarters of the SV40 and polyoma virus genomes are obscure.

New antigens, either viral gene products, derepressed cellular gene products, or substances arising from the interaction of viral gene products with components of the infected cells can, however, be detected in infected cells by immunochemical techniques. Apart from the V antigen (viral capsids), which is unique to productively infected cells, these include the so-called T antigen, a protein or mixture of proteins which appears in the nuclei of infected nonpermissive as well as permissive cells, the U antigen, which is associated with the nuclear membrane, the transplantation antigen(s) and the surface antigen(s). At least some of the surface antigens of infected cells are probably cellular gene products and they are discussed together with the transplantation antigens in Chapters 3 and 6. There is every reason to believe that the viral capsid protein is coded by the viral genome, and the T antigen and U antigen may also be viral gene products.

Viral Capsid Protein

Capsids can be demonstrated by immunofluorescent staining in the nuclei of permissive cells where progeny virus particles mature and accumulate late in productive infections. There is little doubt that viral capsid protein is synthesized in the cytoplasm, where virus-specific RNA can be detected in polyribosomes, and capsid protein must be made in large amounts. But immunofluorescent staining with antisera made against whole virions has repeatedly failed to reveal capsid protein in the cytoplasm of productively infected cells (Melnick et al., 1964). Presumably this antiserum reacts with assembled capsids but not necessarily with the unassembled subunit proteins, although it is possible that capsid subunits are immunologically reactive but are transported too rapidly to the nucleus to be detected by these methods.

Several attempts have been made to identify, in extracts of monkey cells infected by SV40, polypeptide chains which were not also present in similar extracts of unifected cells in the hope that at least some of the polypeptides in this class would prove to be coded by the SV40 genome (Fischer and Sauer, 1972; Ozer, 1972; Anderson and Gesteland, 1972; Walter et al., 1972). All four groups used SDS polyacrylamide gel electrophoresis to analyze the cell extracts and they all detected two polypeptides unique to the extracts of infected cells, one of which coelectrophoreses with the major capsid protein of SV40 while the other comigrates with the minor (30,000 daltons) capsid protein. Furthermore, Anderson and Gesteland (1972) showed that the former polypeptide yields a fingerprint similar to that of the major capsid protein. The four groups also detected several other polypeptides that are present only in extracts of infected cells, but these polypeptides have not been further characterized; which, if any of them, is specified by the viral genome or gives rise to T or U antigenicites (see below) is unknown.

Ozer and Tegtmeyer (1972) made antisera against purified intact SV40 virions and against SV40 capsid protein obtained by disrupting virions with alkali. Using these antisera, the former specific for viral capsids, the latter able to cross-react with capsids and denatured capsid protein, they attempted to measure the pool of capsid protein and the pool of capsids in monkey cells infected with SV40. They found that the pool of unassembled capsid protein in such cells was small, and pulse-chase experiments suggested that capsid protein was rapidly assembled into empty capsids which then acquired viral DNA and histone-like proteins as they matured into progeny virions.

T Antigen

T antigen appears in the nuclei of productively infected cells, in infected nonpermissive cells, transformed cells and the cells of tumours induced by SV40 or polyoma virus. As yet T antigens have not been chemically analyzed and because they are detected by immunochemical tests, it is impossible to be sure that the T antigens produced in different cells infected with the same virus are chemically identical. The antigenic sites of the T antigens in these different types of cells are, however, sufficiently similar to allow serological cross reactions.

T antigen is detectable by complement fixation (Black et al., 1963; Habel, 1965, 1966; Melnick, 1969) or fluorescent antibody tests (Pope and Rowe, 1964; Kawamura, 1969) in the nuclei of permissive cells as early as 6 to 10 hours after infection (Hoggan et al., 1965) and it continues to accumulate until 24–48 hours after infection. Its appearance precedes the onset of viral DNA synthesis and the appearance of detectable amounts of capsid protein; a positive reaction in a T antigen test is, therefore, one of the first signs that a cell is infected with SV40 or polyoma virus (Rapp et al., 1964).

T antigens are virus-specific and their induction apparently depends on the expression of some part of the viral genome. Four early mutants of polyoma virus, which at the nonpermissive temperature fail to replicate their DNA, also fail to induce T antigen in mouse cells (Oxman et al., 1972). Since the inhibition with cytosine arabinoside of the replication of wild-type polyoma virus DNA in these cells does not prevent the induction of T antigen, it seems that the failure of the early mutants to induce T antigen at the nonpermissive temperature is not an indirect consequence of the failure in DNA replication, but a direct consequence of an alteration in a viral gene which either codes for T antigen or is required for its induction. The same conclusion can be drawn from the observation reported by Yoshiike et al. (1972) that a deletion mutant of SV40, whose genome is 13 percent shorter than the wild-type SV40 genome, fails to induce T antigen, and from the observation that the ability of SV40 to induce T antigen is inactivated by ultraviolet light at the same rate as infectivity (Yamamoto and Shimojo, 1971). However, neither these findings nor the observation that interferon inhibits both transcription of early virus-specific RNA and the amount of T antigen in monkey cells infected with SV40 (Oxman and Levin, 1971) proves that polyoma virus and SV40 DNAs contain the structural gene for their respective T antigens. Probably the only way to decide whether or not T antigen is coded by the viral genome is to show that, in a cell-free system capable of transcribing and translating the entire SV40 or polyoma virus genome, one of the products is identical to T antigen purified from infected cells. At present such an experiment is technically impossible and the several attempts which have been made to purify T antigens from productively infected or transformed cells have not been very successful (Kit et al., 1967a; Potter et al., 1969; Del Villano and Defendi, 1970).

U Antigen

The U antigen, which is detected by immunofluorescent staining at the nuclear membrane but not within the nucleus, is produced early in infections with SV40. It appears to be distinct from T antigen but, like T antigen, U antigen has not been proven to be a product of a viral gene. Nothing is known of its chemistry or function.

In short, we know so little about the synthesis of viral proteins in vivo because there are as yet no methods for selectively inhibiting host protein synthesis and making the detection of viral proteins in infected cells feasible. As a result, attempts are being made to transcribe and translate the polyoma virus genome in vitro.

EXPRESSION OF SV40 AND POLYOMA VIRUS GENOMES IN VITRO

At least in principle, it is possible to determine the number and nature of the proteins specified by the genomes of SV40 and polyoma virus by transcribing the DNA in vitro and translating the virus-specific RNA in a cell-free system. However, at present mammalian RNA polymerases and mammalian cell-free systems capable of the coupled transcription and translation of added DNA are not as efficient or as well characterized as their bacterial counterparts. This has meant that attempts to transcribe or transcribe and translate SV40 or polyoma virus DNAs have involved the use of bacterial enzymes and cell-free systems, and data obtained from experiments with such heterologous systems are open to the criticism that they may be completely irrelevant to what happens in the normal host cell. Notwithstanding this ominous possibility, several groups are currently analyzing the products made by *E. coli* RNA polymerase and *E. coli* systems capable of coupled transcription and translation when primed with SV40 or polyoma virus DNA.

Transcription of Polyoma Virus and SV40 DNA

SV40 and polyoma virus DNAs are efficient templates for *E. coli* RNA polymerase and this enzyme has been widely used for the production of virus-complementary RNA (Winocour, 1965; Westphal and Dulbecco, 1968). The number of *E. coli* RNA polymerase molecules which bind to SV40 DNA depends inversely upon the salt concentration (Crawford et al., 1965; Pettijohn and Kamiya, 1967; Winocour et al., 1971). At optimal reaction conditions the amount of SV40-complementary RNA produced can exceed up to 10-fold the amount of SV40 DNA, from stocks of virus propagated at low multiplicities of infection, which is used as template. In these conditions as many as six RNA chains may be synthesized simultaneously on one template molecule, and some molecules of RNA polymerase may move round the circular template several times to yield RNA molecules several times longer than the circumference of the template (Delius et al., 1973). The RNA transcribed from covalently closed circular SV40 DNA does not self-anneal even after prolonged incubation; it is therefore the product of assymmetric transcription. When an excess of this RNA is incubated with denatured SV40 DNA, the two strands of SV40 DNA can be separated as a DNA:RNA duplex and a single strand of SV40 DNA either by centrifugation in caesium chloride gradients (Westphal, 1970) or hydroxylapatite columns (Khoury et al., 1972; Sambrook et al., 1972).

Because *E. coli* RNA polymerase transcribes SV40 DNA asymmetrically, it has been speculated that this bacterial enzyme recognizes a promoter(s) site that is also recognized by mammalian cell RNA polymerase. Investigators have, therefore, tried to demonstrate a specific promoter site for *E. coli* RNA polymerase on SV40 DNA but the results are conflicting. Delius et al. (1973) for example, after examining in the electron microscope complexes of the DNA, enzyme and RNA product, have reached the conclusion that bacterial RNA polymerase can initiate at many different, nonclustered sites on the superhelical SV40 DNA. However, if SV40 DNA which has been converted into linear molecules by R_1 endonuclease is used as template, they detect a single, strong promoter as well as scattered, weaker initiation sites. The strong promoter is 16 percent of the genome from the R_1 endonuclease cleavage site and the polymerase transcribes from this promoter the short arm of the template (Westphal et al., 1973).

Zain et al. (unpublished data) have analyzed by RNA/DNA hybridization and by fingerprinting the RNA transcribed by *E. coli* RNA polymerase off SV40 DNA and the SV40 DNA moiety of the genome of the adeno-SV40 hybrid virus $Ad2^{+}ND_1$. They conclude that the bacterial enzyme initiates at a specific site on the SV40 DNA, but that it transcribes the DNA in the opposite direction to that claimed by Westphal et al. In short, we still do not know whether *E. coli* RNA polymerase recognizes a promoter site on SV40 DNA that is also recognized by animal cell RNA polymerases in vivo.

Analysis of the transcription of polyoma virus DNA in vitro by *E. coli* RNA polymerase lags behind the analysis of transcription of SV40 DNA in vitro. However, Blangy and Vogt (unpublished data) find that the DNA of stocks of polyoma virus propagated at low multiplicities of infection and the DNA of *ts-a* mutant polyoma virus are transcribed symmetrically by the bacterial enzyme, whereas the DNA of stocks of polyoma virus passaged at high multiplicities of infection is transcribed asymmetrically. And Lindstrom (unpublished data) has used the latter RNA to effect the separation of the strands of polyoma virus DNA.

Analysis of the transcription of SV40 DNA and polyoma virus DNA by mammalian RNA polymerases is still in its infancy, not least because of the difficulties in purifying these mammalian enzymes. RNA polymerases isolated from HeLa and KB cells readily accept SV40 DNA as template (Sugden and Sambrook, 1970; Keller and Goor, 1970). Purified RNA polymerase II from KB cells transcribes both strands of component I SV40 DNA (Sugden, unpublished data). The strand that is completely transcribed in vitro by the *E. coli* enzyme is also completely transcribed by KB RNA polymerase II, but the eukaryotic polymerase in addition transcribes 75 percent of the complementary antiparallel DNA chain. The extent to which KB RNA polymerase II transcribes SV40 DNA depends in part on the state of the DNA template; component II SV40 DNA and linear SV40 DNA are less efficient templates than component I DNA. Why KB RNA polymerase II preferentially uses component I DNA is not known. Component I DNA is also transcribed by RNA polymerases A (I) and B (II) from calf thymus and both enzymes transcribe at least parts of both DNA strands (Mandel and Chambon, 1971).

To date RNA made in vitro off SV40 and polyoma virus duplex DNA has been used to separate the strands of these DNAs. These RNAs promise to be useful tools for the determination of the base sequence of these viral DNAs and eventually perhaps they may be used as messenger to program the synthesis in vitro of viral proteins.

Synthesis of SV40 and Polyoma Virus Protein

The several attempts made to transcribe and translate the DNA of polyoma virus or SV40 in cell-free systems obtained from *E. coli* have met with limited success. SV40 DNA was found to program an *E. coli* cell-free system by Gelfand and Hayashi (1969), Bryan et al. (1969) and Crawford et al. (1971). The polypeptides made in vitro have been analyzed by fingerprinting (Crawford et al., 1971); none of them contained amino acid sequences corresponding to those in the major capsid protein of SV40 particles. Recently Anderson and Gesteland (unpublished data), who find that the polypeptides made when their cell-free systems are programmed with SV40 DNA (from stocks of

SV40 propagated at low multiplicities of infection) have molecular weights of up to 60,000 daltons, analyzed the SV40 RNA that is transcribed in this system and presumably acts as messenger. They find as expected that transcription is almost completely asymmetric and the RNA that finds its way into the polysomes is transcribed from the strand of SV40 DNA that is transcribed early during the lytic cycle. Sequences found in late SV40 RNA made in vivo are not detectable in the polysomal RNA of the cell-free systems. These results may explain why coat protein is not amongst the polypeptide products made in this system, and they also raise hopes that some of the polypeptides made in vitro may be products of genes expressed early in the lytic cycle. Attempts to establish this are in progress as are experiments in which the cell-free system is primed with those fragments of the SV40 genome (obtained by digestion of SV40 DNA with *Hemophilus influenzae* restriction endonuclease) that are known to specify early SV40 RNA.

The DNA of large-plaque strains of polyoma virus is a less efficient primer in the coupled cell-free system than SV40 DNA, but some of the polypeptides made when large-plaque polyoma virus DNA is used migrate on SDS gels with the major capsid protein of polyoma virus (Crawford and Gesteland, 1973). Moreover, tryptic fingerprints of these polypeptides made in vitro and labeled with [^{35}S]methionine are similar to fingerprints of authentic capsid protein. In short, it seems that polyoma virus DNA can be transcribed and translated in an extract of *E. coli* to yield products, some of which may be identical to capsid protein. Consistent with these findings, Blangy and Vogt (unpublished data) have reported that the DNA of some stocks of polyoma virus is transcribed symmetrically in vitro by purified *E. coli* RNA polymerase.

Taken together these various experiments with heterologous cell-free systems capable of coupled transcription and translation encourage hopes that it may eventually be possible to synthesize in vitro many, if not all, of the gene products of SV40 and polyoma virus. Of course, it can always be objected that polypeptides made in a heterologous cell-free system may bear little resemblance to those made in the natural host. However, the fact that polyoma virus apparently primes the synthesis by an *E. coli* extract of capsid protein diminishes doubts about the fidelity of transcription and translation in heterologous systems, and in any case attempts are being made to derive cell-free systems from animal cells which support efficient transcription and translation.

INDUCTION OF HOST ENZYMES

During productive and incomplete infections with SV40 or polyoma virus, the activity of several enzymes involved in the synthesis of DNA increases (see Table 5.1 and reviews by Weil et al., 1967; Kit et al., 1967b; Kit, 1968; Eckhart, 1968). Many of these induced enzymes are

Table 5.1. ENZYMES INDUCED BY POLYOMA VIRUS AND SV40 INFECTION

Enzyme	Relative Increase in Activity after Infection	Virus	Cell Infected	Ref.
DNA polymerase	3–7	Py	P, NP*	a, b, c
	3	SV40	P, NP	d
DNA ligase	4–10	Py	P	e
	4	SV40	P	e
Thymidine kinase	3–7	Py	P	c
	3–6	Py	P	f
	3–55	Py	P	g
	10–15	Py	P, NP	b
	10–20	Py	P	h
	4–15	SV40	P, NP	i, d
dTMP kinase	1–5	Py	P	c
dTDP kinase	3	Py	P	g
		SV40	P, NP	d
Cytidine kinase	3–4	Py	P	c
dCMP deaminase	2–9	Py	P	c
	10–15	Py	P	b
CDP reductase	3–4	Py	P	c
dTMP synthetase	2	Py	P	f
	2	SV40	P	i
Dehydrofolate reductase	3	Py	P	j
	3	SV40	P	j

Modified from Eckhart (1968).

* P = permissive; NP = nonpermissive.

Data compiled from: (a) Dulbecco et al., 1965; (b) Hartwell et al., 1965; (c) Kara and Weil, 1967; (d) Kit et al., 1967d; (e) Sambrook and Shatkin, 1969; (f) Frearson et al., 1965; (g) Kit et al., 1966a; (h) Sheinin, 1966c; (i) Kit et al., 1966b; (j) Frearson et al., 1966.

involved in the synthesis of pyrimidine deoxyribonucleotides, in particular thymidine derivatives. The induction is clearly selective because there is no general induction of cellular enzyme synthesis after infection; the activities of uridine kinase, dUMP phosphatase, dCMP kinase and dAMP kinase, for example, do not increase (Dulbecco et al., 1965; Hartwell et al., 1965; Kit et al., 1965, 1966a). Nothing is known about the mechanism of induction.

There is little doubt that the increased activities of many of the induced enzymes is the result of de novo synthesis because in the presence of inhibitors of protein synthesis induction does not occur (Frearson et al., 1966; Sambrook and Shatkin, 1969; Hartwell et al., 1965), and neither detectable viral nor cellular DNA synthesis is necessary for enzyme induction (Sambrook and Shatkin, 1969; Kara and Weil, 1967).

The genomes of SV40 and polyoma virus are too small to code for these enzymes and also code for the structural proteins of the virions. However, of the ten enzymes known to be induced, only one, thymidine kinase, has been shown conclusively to be specified by the host genome. After polyoma infects mutants of BHK21 and 3T3 cells which lack thymidine kinase (BHK21 TK^- and 3T3 TK^-), thymidine kinase activity is not detected (Littlefield and Basilico, 1966; Basilico et al., 1969). The infection proceeds normally in the absence of thymidine kinase and so this enzyme cannot play an essential role in the infection of cells in tissue culture. The induction of thymidine kinase and other enzymes may, however, be important for infections of animals.

Although thymidine kinase is not coded by the viral genome, there is evidence that the enzyme in some infected cells differs from that in uninfected cells. Thymidine kinase in mouse kidney cells infected with polyoma virus has been reported to be less sensitive to inactivation by heat (Hartwell et al., 1965; Sheinin, 1966c) and to have a lower K_m with thymidine as substrate (Sheinin, 1966c) than the same enzyme in uninfected cells. But, using deoxyuridine as substrate, Kit et al. (1966a) found no difference in the K_m of thymidine kinase in cells infected with polyoma virus and uninfected cells, and Sheinin (1967) found no difference in the thermostability or K_m of thymidine kinase in uninfected rat cells and rat cells incompletely infected with polyoma virus.

Likewise, Kit et al. (1967b) found no difference between the thymidine kinase in uninfected and SV40-transformed mouse kidney cells, but Hatanaka and Dulbecco (1967) found that the thymidine kinase induced as SV40 replicates in monkey kidney cells differs significantly from the thymidine kinase in the uninfected cells. The DNA polymerase induced during productive infections of SV40 also differs significantly from the same enzyme in uninfected cells (Kit et al., 1967c). It appears therefore, that the thymidine kinase, and perhaps other enzymes, induced during productive infections of polyoma virus or SV40 is modified in some way, whereas the enzyme induced during infections of nonpermissive cells and in transformed cells is not detectably different from that in uninfected cells. However, all measurements of the physical properties of the enzymes induced by polyoma virus and SV40 have been made with crude extracts and so their significance is questionable.

INDUCTION OF HOST DNA SYNTHESIS

Polyoma virus and SV40 induce the synthesis of cellular DNA during both productive and incomplete infections. During productive infections the initiation of host DNA synthesis occurs at about the same time as the onset of viral DNA replication. During incomplete infections the time lapse between infection and the start of cellular DNA synthesis is much the same as in productive infections but there is no detectable synthesis of viral DNA.

In most cases DNA synthesis in infected cells has been measured by the uptake of radioactive thymidine into acid-insoluble material. This method is convenient but suffers from two possible disadvantages. First, there may be poor equilibration of the labeled DNA precursor with the internal pools of the cell. Second, the internal pools may change during viral infection.

However, cellular DNA synthesis has also been detected by using several radioactive precursors which follow other pathways into DNA (Winocour et al., 1965) and by colorimetric measurements of an increase in the total amount of DNA per cell after infection (Molteni et al., 1966). Since the net DNA synthesis detected by all three methods is approximately the same, there seems to be no reason to believe that the use of radioactive thymidine leads to any special artifacts. Various procedures can be used to discriminate between host cell and viral DNAs: fractionation on columns of methylated-albumin-kieselguhr (Dulbecco et al., 1965; Sheinin, 1966a), molecular hybridization of nucleic acids (Dulbecco et al., 1965), DNA infectivity determinations and most recently, selective extraction (Hirt, 1967) of virus and host cell DNA (Pétursson and Weil, 1968; Ritzi and Levine, 1970).

Viral induction of cellular DNA synthesis is most easily detected in cultures which have stopped multiplying because they have depleted the growth factors in their medium and/or because they are subject to topographical inhibition (Todaro et al., 1965; Fried and Pitts, 1968; Dulbecco and Stoker, 1970; Dulbecco, 1970). When such resting cultures are infected and control cultures are mock infected and their old medium replaced, virus-induced host DNA synthesis can be readily analyzed (Molteni et al., 1966; Werchau et al., 1966; Sheinin, 1966a, b; Henry et al., 1966). Other methods, such as X-irradiation or treatment with 5-fluorodeoxyuridine before infection, facilitate the detection of induced cellular DNA synthesis (Minowada, 1964; Gershon et al., 1965; Winocour, 1969) at least in part because these treatments lower the level of DNA synthesis in cultures at the time of infection. Although more difficult to detect, viral-induced DNA synthesis has been demonstrated after the infection of growing mouse embryo cells with polyoma virus (Branton and Sheinin, 1968). The induction, however, was only demonstrated after infection with low (Branton and Sheinin, 1968) but not high (Sheinin, 1966a) input multiplicities of virus. As Winocour (1969) points out, the latter observation may be explained by the fact that host DNA synthesis may continue for only a short time in growing cells undergoing rapid infection, and may, therefore, have escaped detection.

Stimulation of DNA synthesis following infection with polyoma virus was first detected by autoradiography (Minowada and Moore, 1963). Subsequently, Dulbecco et al. (1965) reported that (a) stationary mouse kidney cells infected with polyoma virus synthesized about 10-fold more DNA than mock-infected control cultures, (b) most of the DNA made in infected cells is cellular, and (c) DNA synthesis is accompanied by the induction of several enzymes involved in DNA metabolism. Similar findings have since been reported for productive (Weil et al., 1965; Winocour et al., 1965; Gershon et al., 1965; Hatanaka and Dulbecco, 1966; Werchau et al., 1966; Kit et al., 1967c; Fried and Pitts, 1968; Sambrook and Shatkin, 1969) and incomplete (Gershon et al., 1965, 1966; Sheinin, 1966b; Henry et al., 1966) infections by both polyoma virus and SV40.

Characteristics of the Induction

1. The extent of infection and the extent of induction of host DNA synthesis depend on the multiplicity of infection at input multiplicities ranging from 0.5 to 1,000 plaque-forming units per cell (Basilico et al., 1966). In infected cultures DNA and viral capsid proteins are synthesized in the same cells (Vogt et al., 1966; Basilico et al., 1966).

2. Cellular DNA synthesis in stationary cultures is induced synchronously by infection with polyoma virus and SV40 (Dulbecco et al., 1965; Henry et al., 1966; Ritzi and Levine, 1970). Since most resting cells are arrested at the same stage—the postmitotic (G_1) phase of their growth cycle (Todaro et al., 1965; Weil et al., 1967)—the virus releases the block to DNA synthesis, which is controlled by the host, at about the same time after infection in most cells. Adding 5-fluorodeoxyuridine during the early stages of productive infections further synchronizes induction of cellular DNA synthesis (Pétursson and Weil, 1968).

3. Polyoma virus and SV40 are able to induce cellular DNA synthesis in productively infected cells even in the presence of low concentrations of 5-fluorodeoxyuridine (Ben-Porat and Kaplan, 1967) and prolonged treatment of nonpermissive cells with this drug before infection with polyoma virus results in a pronounced synthesis of cellular DNA after infection (Winocour, 1969). This release by infection of the 5-fluorodeoxyuridine block to DNA synthesis is remarkable because uninfected cultures are not stimulated to synthesize DNA when 5-fluorodeoxyuridine is replaced by thymidine or uridine. However, enzyme induction occurs in infected cells in the absence of DNA synthesis (Sambrook and Shatkin, 1969; Kara and Weil, 1967) and it is possible that the elevated levels of thymidilic acid synthetase

formed (see Table 5.1) allow the cells to overcome the 5-fluorodeoxy-uridine block.

4. Substantial amounts of cellular DNA are made (Gershon et al., 1965; Kit et al., 1967c) and most of the newly synthesized DNA contains one parental strand made before infection (Weil et al., 1965; Ben-Porat et al., 1967; Kasamaki et al., 1968). Moreover, the density distribution in caesium chloride equilibrium gradients of the cellular DNA synthesized after infection is the same as that isolated from uninfected cells. The synthesis of histones and nuclear acidic proteins is induced concomitantly with the induction of cellular DNA synthesis (Shimono and Kaplan, 1969; Hancock and Weil, 1969; Winocour and Robbins, 1970; Rovera et al., 1972), and replication of mitochondrial DNA is induced during both productive and abortive infections by SV40 (Levine, 1971). According to Hirai et al. (1971a), two complete rounds of DNA replication may occur before mitosis in Chinese hamster cells infected with SV40 to yield polyploid cells.

Taken together these results indicate that DNA synthesis is induced throughout the entire cell and that replication of the cellular genome is accompanied by synthesis of the protein components of chromatin.

5. The pattern of DNA synthesis induced by polyoma virus differs from the pattern of normal DNA synthesis. In primary mouse kidney cells productively infected with polyoma virus, replication of the light satellite fraction of the mouse genome precedes the replication of the rest of the genome (Smith, 1970). When, however, secondary mouse embryo fibroblasts are induced by serum factor(s) to replicate their DNA, the satellite fraction is replicated after most of the genome has been replicated (Smith, 1971). DNA synthesis in monkey kidney cells infected with SV40 starts in the region of nucleoli (Granboulan and Tournier, 1965); but although this suggests that in these cells also the virus may induce the replication of satellite DNA first, Tobia et al. (1972) have reported that replication of satellite DNA in CV1 cells infected with SV40 occurs late in the S phase as it does in CV1 cells induced to divide by serum.

6. Increased mitotic rates have been reported after the induction of cellular DNA synthesis in resting cells incompletely or productively

infected with polyoma virus (Molteni et al., 1966; Gershon et al., 1966; Winocour and Robbins, 1970). However the proportion of polyploid cells as well as mitotic cells increases in cultures of Chinese hamster cells infected with SV40 (Lehman and Defendi, 1970). Therefore virus-induced DNA synthesis does not inevitably result in mitosis.

7. During productive infections by polyoma virus (Ben-Porat et al., 1966, 1967; Cheevers et al., 1970) or SV40 (Ritzi and Levine, 1970) some cellular DNA is degraded; such degradation is confined to, but is not an inevitable consequence of, lytic infections (Ritzi and Levine, 1970). The cellular DNA suffers both single strand (Cheevers et al., 1970) and double strand breaks (Ben-Porat et al., 1966, 1967; Ritzi and Levine, 1970) which generate the small fragments of host DNA that are encapsidated to form pseudovirions. Cellular nucleases appear to play a crucial role in this degradation because when SV40 productively infects BSC1 and CV1 cells, the cellular DNA is not degraded; but when the same virus productively infects AGMK cells, cellular DNA is fragmented (Ritzi and Levine, 1970). Degradation of cellular DNA may therefore reflect the induction, by the infecting virus, of possibly specific, cellular nucleases.

8. The induction of cellular DNA synthesis during productive infection by SV40 or polyoma virus has been observed in over a dozen different types of cells from several species of animals. To date there is only one example of a cell which supports the lytic growth of either virus without concomitant cellular DNA synthesis; that is, BSC1 cells in stationary cultures which are productively infected with SV40 (Gershon et al., 1966; Ritzi and Levine, 1970). BSC1 cells in stationary cultures do, however, respond to added serum factor(s) by increasing their DNA synthesis. There are several plausible interpretations of this finding; for example:

(a) The induction of cellular DNA synthesis may not be essential for the replication of polyoma virus and SV40. But naively one might speculate that the replication of the DNA of these viruses depends on cellular enzymes; if so, polyoma virus and SV40 would be unable to replicate efficiently, if at all, in most cells in an animal because most cells are not dividing and presumably do not contain the enzymic machinery required for DNA replication. By carrying into the cell a function which induces that cell to replicate its DNA, polyoma virus and SV40 would increase greatly the range of cell types in which they can replicate.

(b) SV40 may in fact induce cellular DNA synthesis in BSC1 cells but the amount of cellular DNA made may be too small to detect. One can speculate that productive infections depend on the replication of only a small, specific part of the cell genome and that the replication of the rest of the cell genome, which occurs in most infected cells, is irrelevant to the replication of these viruses.

(c) The DNA replication function(s) which polyoma virus and SV40 induce in most cell types may already be present in BSC1 cells, but for some reason it may not be sufficient to switch on the synthesis of the DNA of BSC1 cells. Alternatively, this function(s), absent in uninfected cells, may be induced by SV40, but may be necessary but not sufficient to induce synthesis of the BSC1 DNA; whereas in other cells the same induced function(s) may be both necessary and sufficient to induce replication of the entire cell genome.

Until more data is available it is impossible to decide which, if any, of these explanations is correct.

Induction—A Viral Function

The stimulation of cellular DNA synthesis depends on the expression of part, but not all, of the viral genome because:

(a) The ability of polyoma virus (Gershon et al., 1965) or SV40 (Defendi and Jensen, 1967) to induce cellular DNA synthesis is inactivated by nitrous acid or ultraviolet irradiation but at a rate up to 5 times slower than the inactivation of infectivity. The inactivation of both properties, however, follows first-order kinetics. Hence, only part of the viral genome can be involved in inducing cellular DNA synthesis.

On the other hand, Basilico et al. (1966) reported that ^{60}Co-irradiation of polyoma virus inactivates the capacity of the virus to form plaques and to induce cellular DNA synthesis at about equal rates, implying that both functions depend on the integrity of the entire genome. However, X-irradiation induces breaks in both strands of DNA (Laterjet et al., 1967). Such drastic treatment is likely to be lethal for most functions and unlikely to indicate accurately the target size of a particular function.

(b) One temperature-sensitive mutant of polyoma, *ts*-3, does not induce cellular DNA synthesis in 3T3 cells at the nonpermissive temperature.

Mechanism of Induction

The mechanism of induction of cellular DNA synthesis is obscure. It depends on the integrity of at least part of the viral genome and it does not appear to involve the late events of a lytic infection; abortively infected cells are induced to synthesize DNA and in lytic infections induction precedes late events. The induction is now known to be sensitive to interferon (Dulbecco, 1969; Dulbecco and Johnson, 1970; Taylor-Papadimitriou et al., 1971) and presumably, therefore, depends on the translation of viral messenger RNA. The host-range mutants of polyoma virus isolated by Benjamin induce the synthesis of at least some cellular DNA even though wheat germ agglutinin sites on the cell surfaces are not exposed. Therefore the exposure of these sites is not essential for the induced replication of at least part of the cell genome. During productive infections (Dulbecco et al., 1965; Weil et al., 1965; Molteni et al., 1966; Werchau et al., 1966; Gershon et al., 1966) the uptake of thymidine reaches a maximum and then declines. Whether or not this reflects a cessation of induction, changes in the sizes of the pools of DNA precursors, or death of the infected cells is unknown.

In summary, during productive infections by SV40 or polyoma virus, the induction of cellular DNA synthesis has been detected in every cell type which has been examined except BSC1 cells infected with SV40. The significance of this induction for the replication of the viral DNA and the mechanism of induction are unknown, but further investigations of BSC1 cells infected with SV40 may provide answers to these outstanding questions.

INTEGRATION OF VIRAL AND HOST DNAs DURING PRODUCTIVE INFECTIONS

It is generally accepted that SV40 and polyoma virus DNAs are covalently integrated into the DNA of transformed nonpermissive cells (see Chapter 6), but the idea that integration might also occur during the replication of these viruses in permissive cells has only recently been discussed seriously. Unambiguous evidence of integration during the lytic cycle has not yet been obtained but suggestive data have been reported. Three groups of investigators have apparently detected viral DNA sequences associated with the cellular DNA of infected permissive cells.

Hirai and Defendi (1972) infected CV1 cells with SV40 in the presence of cytosine arabinoside, which inhibits viral DNA synthesis. At various times after infection they lysed the cells in alkali and centrifuged the lysate through alkaline sucrose gradients to separate any 53 S component I viral DNA from chromosomal DNA, which sediments at 100 S and above. Immediately after infection most SV40 DNA, assayed by hybridization with ^{3}H-labeled SV40 RNA made in vitro by *E. coli* RNA polymerase, appeared in the 53 S region of the gradient. Twenty hours later most SV40 DNA sedimented with the 100 S cellular DNA. By contrast, 30 hours after infection in the absence of cytosine arabinoside, most of the newly replicated SV40 DNA sedimented at 53 S—free progeny genomes—but the 100 S cellular DNA contained the same concentration of SV40 DNA sequences as the 100 S cellular DNA from cells 30 hours after infection in the presence of cytosine arabinoside. These results suggest that (1) SV40 DNA molecules are integrated into cellular DNA throughout the lytic cycle, and (2) the number of genomes integrated remains fairly constant; it does not increase as progeny genomes accumulate. Hirai et al. (1971b) have also shown that SV40 DNA becomes associated by alkali-stable bonds to the DNA of semipermissive Chinese hamster embryo cells within 20 hours of infection. Babiuk and Hudson (1972) reported similar experiments done with mouse cells infected by polyoma virus in the absence of any inhibition of DNA synthesis. Their results suggest that the parental genomes are efficiently integrated into cellular DNA within 6 hours of infection and that after the onset of viral DNA replication the amount of integrated polyoma virus DNA increases. Finally Ralph and Colter (1972) have reported experiments in which they infected mouse embryo cells, whose DNA had been labeled with BrdU, with polyoma virus and detected some 8–15 copies of the viral

DNA associated with the cellular DNA 14 hours after infection in the absence of any viral DNA synthesis.

Estimates of the precise number of genome equivalents of viral DNA that become associated with the DNA of permissive cells vary greatly, and the source and significance of the variation is unclear. But there is no reason to dispute the essential finding that some viral DNA is probably integrated into the DNA of some lytically infected cells, and this integration must involve a recombination process and therefore endonuclease and ligase activities at the very least.

Two other sets of data can also be construed as evidence for integration during the lytic cycle. First, the results of Tai et al. (1972) and Lavi and Winocour (1972) (see Chapter 4) when taken together indicate that as SV40 genomes replicate in monkey cells some of the SV40 DNA molecules become partially deleted and segments of the host DNA are incorporated into these partially deleted molecules. Obviously such deletions and substitutions might arise if SV40 DNA molecules are first integrated into and then excised from host DNA molecules. Second, virus-specific RNA molecules, longer than one complete transcript of the SV40 genome, can be found in the nuclei of infected monkey cells. Such molecules might arise from the transcription of tandem host and viral genes.

It hardly needs saying, however, that the significance of integration during the lytic cycle remains a matter for speculation; none of the data discussed above establishes that integration is a necessary event during the lytic cycle of SV40 or polyoma virus, and it is possible that viral genomes that become integrated never give rise to infectious progeny. Finally any viral DNA molecule that becomes integrated into cellular DNA must be excised intact if it is to act as a template for the synthesis of infectious progeny genomes.

REPLICATION OF VIRAL DNA

The genomes of SV40 and polyoma virus are replicated in the nuclei of permissive cells (Minowada and Moore, 1963), whereas in incompletely infected cells there is no detectable viral DNA synthesis. Because these two viruses can code for only about eight proteins, the replication of their DNA must be dependent on many host enzymes, including, no doubt, some of those which are induced by infection. The synthesis of viral DNA and cellular DNA is initiated after a population of permissive cells is infected.

In part because of the induction of cellular DNA synthesis, viral DNA never amounts to more than about 20 percent of the total DNA in productively infected cells (Hirt, unpublished data). Moreover, all attempts to inhibit the synthesis of cellular DNA without inhibiting the replication of viral DNA have failed. As a result it proved extremely difficult to analyze the replication of polyoma virus or SV40 DNA until Hirt (1967) devised a method for selectively precipitating most of the cellular DNA which leaves many, if not all, the viral DNA molecules in the supernatant. Using this technique and 5-bromodcoxyuridine as a density label, Hirt proved that polyoma virus DNA is replicated semiconservatively. Since polyoma virus and SV40 DNAs are covalently bonded circular DNA molecules (component I), the two strands can be separated only if at least one phosphodiester bond in either strand is cleaved. Nothing, however, is known about the nuclease(s) which achieves this, the polymerase(s) which synthesizes the progeny strands, or the ligase(s) which reforms covalent circles from linear molecules during the semiconservative replication.

Two models (see Figure 5.2) have been proposed to explain the semiconservative replication of the circular DNAs of *E. coli* and certain coliphages, the Cairns model (Cairns, 1963) and the rolling circle model (Gilbert and Dressler, 1968). The Cairns model predicts that all the DNA strands in the replicating structure are as long as or smaller, but never longer than, the strands in the parental molecule. A further prediction of this model is that the two replicating progeny strands are of equal length. The rolling circle model, on the other hand, predicts that the replicating structure will include linear DNA strands longer than the parental strands and that the two progeny strands are not of the same length.

Several lines of evidence indicate that replicating SV40 DNA molecules have properties consistent with the Cairns model. In 1971 Danna and Nathans discovered that the restriction endonuclease of *Hemophilus influenzae* cuts SV40 DNA at specific sites to yield eleven

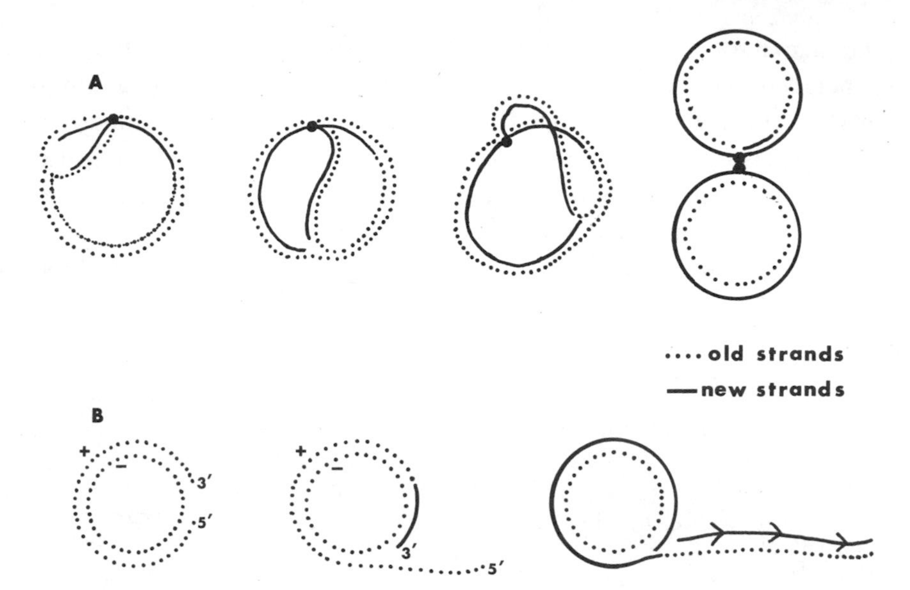

Figure 5.2. Diagrammatic representation of the replication of circular DNA. **A.** The Cairns model, in which the duplication always starts at the same point (12 o'clock) and advances in the same direction (counterclockwise). **B.** The rolling circle. The closed inner circle represents the negative strand serving as a template for elongation of the positive strand, which grows by the addition of nucleotides to its 3′ hydroxyl end. Complementary fragments are synthesized on the growing positive strand. (Redrawn from Cairns, 1963 and Gilbert and Dressler, 1968).

fragments. With this enzyme Nathans and Danna (1972) were able to prove, from the pattern of labeling of the DNA fragments obtained from partially replicated SV40 DNA molecules labeled with [^{3}H]-thymidine, that replication of SV40 DNA starts at a unique origin. Thoren et al. (1972) have reached the same conclusion from their analyses, by nucleic acid hybridization, of the genetic complexity of nascent daughter DNA strands. Subsequently by refining their experimental techniques, Danna and Nathans (unpublished data) established that the unique origin of replication lay in that part of the SV40 DNA molecule which gave rise to fragments A and C (see Chapter 4) and that replication proceeded in both directions from the origin.

Fareed et al. (1972) used the R_1 restriction endonuclease, which cuts SV40 DNA at just one site, to convert partially replicated, circular SV40 DNA molecules into linear molecules. After examining the shapes of these molecules in the electron microscope, they also concluded that replication of SV40 DNA was initiated at a unique origin, which was about one-third of the length of the genome from the R_1 cleavage site (see Figure 5.3). This places the origin of replication in the fragment C obtained by Danna and Nathans. Furthermore, Fareed et al. concluded that two replication forks move in opposite directions, but at the same rate, around the circular SV40 DNA molecule.

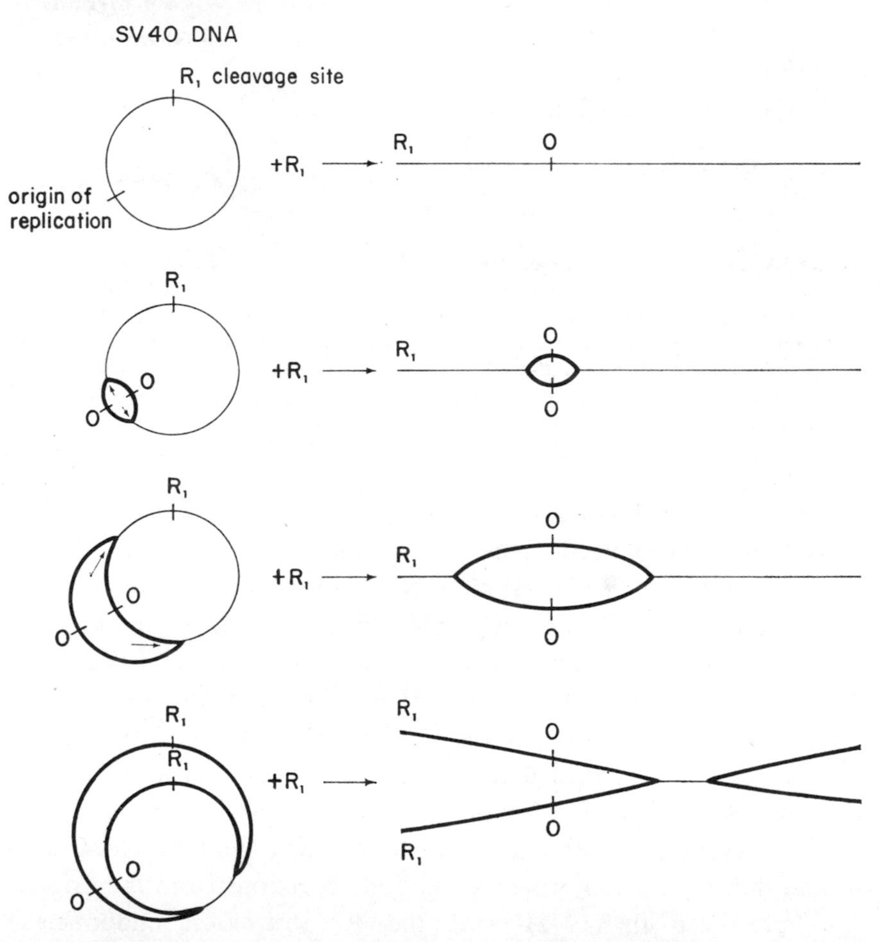

Figure 5.3. Diagrammatic representation of the experiment by Fareed et al. (1972) which established the unique origin of replication of SV40 DNA and that the daughter DNA chains grow at the same rate in both directions from the origin.

When 5-minute pulses of [^{3}H]thymidine were fed to monkey cells infected by SV40, or to mouse cells infected by polyoma virus, replicating viral DNA molecules were isolated that sedimented as a broad peak, faster than component I DNA (Bourgaux et al., 1969; Levine et al., 1970). In the electron microscope the majority of such molecules were seen to have three branches, two branch points and no ends, as predicted by the Cairns model (Sebring et al., 1971; Jaenisch et al., 1971). Furthermore, one of the three branches of most of these molecules was superhelical (see Figure 5.4) and was the unreplicated part of the parental molecule. When these triple-branched molecules were denatured at alkaline pH, the nascent daughter strands, which contained all the [^{3}H]thymidine label, were liberated; these daughter strands were never larger than the parental strands. Both parental strands in the partially replicated molecules were covalently closed circles. The relaxed triple-branched molecules observed by Hirt (1969) in electron microscope preparations of DNA extracted from mouse cells productively infected with polyoma virus contained parental strands

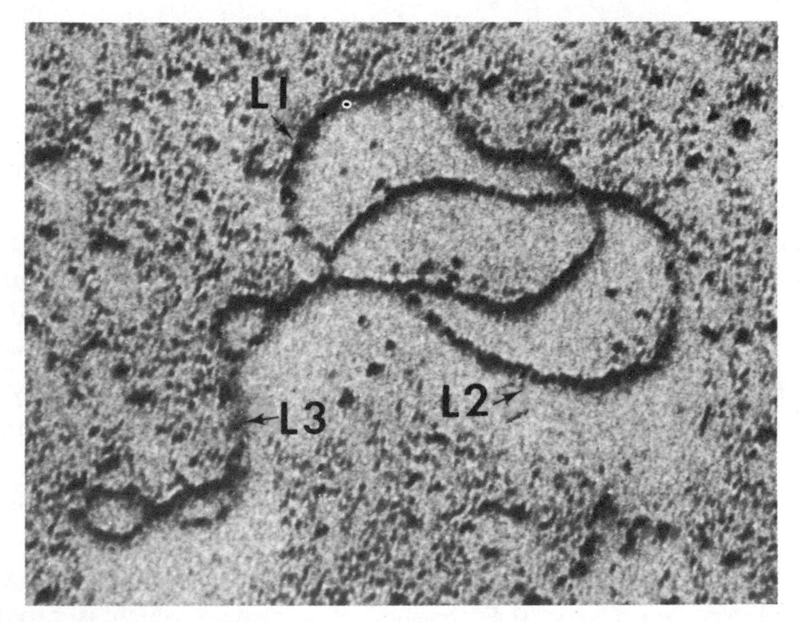

Figure 5.4. Electron micrograph of a partially replicated SV40 DNA molecule showing three branches, one of which (L3) is superhelical, unreplicated parental DNA. Courtesy of N. P. Salzman.

with at least one broken phosphodiester bond. In Hirt's preparations there were triple-branched molecules, one branch of which was superhelical, but he failed to recognize their significance. Widmer (1971) has since shown, however, that most triple-branched molecules of replicating polyoma virus DNA have one superhelical branch.

If the parental strands of replicating SV40 DNA and polyoma virus DNA molecules are not nicked, how is their separation achieved? Champoux and Dulbecco (1972) have shown that extracts of uninfected mouse embryo cells contain an activity which converts component I polyoma virus DNA into a relaxed, circular molecule whose two strands are both covalently closed. A similar activity exists in *E. coli* cells and it has been more thoroughly characterized (Wang, 1971). Apparently both bacterial and mammalian cells contain a "DNA untwisting enzyme" that is capable of untwisting closed, circular, superhelical DNA. This "untwisting" possibly involves first, an endonuclease activity that breaks a phosphodiester bond, and then a ligase activity that promptly remakes the broken bond. During replication of papova virus DNA, separation of the parental strands would be possible if an "untwisting enzyme" repeatedly cut and resealed the parental DNA molecules.

Little is known about the enzymes required to replicate SV40 DNA or polyoma virus DNA. Bourgaux and Bourgaux-Ramoisy (1972) have shown that, in the presence of puromycin, [^{3}H]thymidine from a 30-minute pulse fed to mouse cells infected by polyoma virus is incorporated into component I DNA which has fewer superhelical turns than component I DNA made in the absence of puromycin. After extraction the component I DNA labeled in the presence of the drug is not associated with protein, whereas in the absence of puromycin polyoma virus DNA can be extracted as a DNA-protein complex (Green et al., 1971); and in a similar study White and Eason (1971) found that newly synthesized SV40 DNA can be extracted from cells in the form of a nucleoprotein complex. Bourgaux and Bourgaux-Ramoisy tentatively conclude from these findings that replication of polyoma virus DNA depends on the presence of a protein which is rapidly turned over in infected cells. Kang et al. (1971), who analyzed SV40 DNA replication in monkey cells in the presence and absence of cycloheximide, believe that protein synthesis is required at two different stages in the replication of SV40 DNA: (1) the initiation of DNA synthesis, and (2) the conversion of the 25 S replicating form to 21 S mature viral DNA (finishing step).

A similar conclusion with regard to initiation of DNA synthesis has also been reached by Tegtmeyer (1972) from temperature-shift experiments with three temperature-sensitive mutants of SV40 belonging to one complementation group (see Chapter 7). When AH cells, a line of African green monkey kidney cells, are infected with these mutant strains of SV40 at 33°C, viral DNA is replicated; but when the infected cells are shifted to 41°C, viral DNA replication rapidly ceases. Component I SV40 DNA continues to accumulate for a few minutes, but cell DNA synthesis continues for several hours. Tegtmeyer concludes that a protein is required to initiate replication of SV40 DNA, that this protein rapidly turns over, that it may be coded by an SV40 gene, and finally, that it is not required for the completion of synthesis of an SV40 DNA molecule once initiation has occurred.

Whatever the number and nature of the enzymes that are required to replicate SV40 DNA prove to be, both of the daughter DNA strands are synthesized discontinuously as a series of short DNA chains (analogous to the Okazaki pieces of replicating *E. coli* chromosomes), which are ligated to the nascent daughter molecules. This was shown by Fareed and Salzman (1972), who exposed monkey cells infected with SV40 to short (15–45 second) pulses of [^{3}H]thymidine and then extracted the replicating SV40 DNA molecules. The [^{3}H]thymidine incorporated during pulses at 37°C was found in DNA chains that sediment at about 4 S in alkaline sucrose gradients and, therefore, contain about 75 nucleotides. These 4 S DNA molecules were found associated with SV40 DNA molecules at every stage of replication, and pulse-chase experiments proved that the 4 S DNA was a precursor of both nascent daughter molecules. In other words, both strands of SV40 DNA are replicated discontinuously from start to finish.

Lack of appropriate cell-free systems has hampered all investigations of the biochemistry of DNA replication. One attempt has been made to isolate a cell-free system that supports replication of polyoma virus DNA. Winnacker et al. (1971) extracted nuclei from mouse cells infected by polyoma virus and incubated them with the four deoxynucleoside triphosphates, including [^{3}H]dTTP. Label was incorporated into replicating polyoma virus DNA. Winnacker et al. have yet to show, however, that in this cell-free system new rounds of replication of polyoma virus DNA are initiated. The DNA synthesis they detected might well have been restricted to molecules whose replication had been initiated before the cell nuclei were extracted. It is obvious that we know more about the replication of SV40 DNA than about the replication of polyoma virus DNA. But because of the close parallels between the biologies of these two viruses, we believe that the mechanism of replication of polyoma virus DNA will prove, in all essentials, identical to that of SV40 DNA.

Finally we should mention that during the replication of polyoma virus and SV40 DNA a very small percentage (1–2) of oligomeric progeny molecules are generated. Most oligomeric polyoma virus DNA molecules are circular, superhelical dimers (Mulder and Vogt, unpublished data), but some catenated dimers can also be detected (Meinke and Goldstein, 1971). By contrast, circular and catenated dimers of SV40 DNA occur in about equal proportions and some higher oligomers, up to hexamers, occur (Jaenisch and Levine, 1971; Rush et al., 1971). And Jaenisch and Levine (1972) have reported that in monkey cells infected with SV40 and exposed to cycloheximide, the formation of dimeric and tetrameric DNAs is not impaired, but the number of monomeric SV40 DNA molecules made is inhibited by up to 90 percent; as a result the proportion of oligomeric DNAs increases to about 8 percent of the total SV40 DNA in the cells. Kit and Nakajima (1971) find no increase in oligomeric DNAs when CV1 cells are infected with SV40 in the presence of cycloheximide. Oligomeric SV40 DNAs are infectious (Jaenisch and Levine, 1971). There are several ways in which these oligomers might arise, including recombination, excision of tandem integrated viral genomes, or by some abberation during replication. Experiments establishing the origin of these molecules have yet to be done.

MATURATION AND ASSEMBLY OF PROGENY VIRIONS

We know from immunological and electron microscopic studies that progeny virions are first found in the nuclei of cells productively infected with SV40 (see review by Fenner, 1968; Mayor et al., 1962; Rapp et al., 1965; Granboulan et al., 1963) and polyoma virus (see review by Fenner, 1968; Mattern et al., 1966; Mattern and DeLeva, 1968; Bernhard et al., 1959). But virtually nothing is known about the mechanism of maturation and assembly of progeny virus particles. The functions of the minor protein components of the virions, the role of cellular membrane sites, the contributions of self-assembly mechanisms and the time and way in which polyoma virus or SV40 DNA is packaged are all open questions.

T and V antigens are detected in the nucleus before the appearance of progeny virions (Mayor et al., 1962; Rapp et al., 1965; Granboulan et al., 1963; Mattern et al., 1966), but we assume that all viral proteins are synthesized in the cytoplasm and transported from there to the nucleus. There is no direct evidence as yet to support this assumption, but the transport of some structural proteins of adenoviruses 2 and 5 from the cytoplasm to the nucleus has been observed (Velicer and Ginsberg, 1970; Horwitz et al., 1969). Autoradiographic studies show that the other component of polyoma virus and SV40 particles, the viral DNA, is replicated in the nuclei of infected cells (Minowada and Moore, 1963). Moreover, Champoux (unpublished data) finds that when mouse embryo cells infected with polyoma virus are pulse-labeled late in infection with [^{3}H]thymidine, the labeled polyoma virus replicative intermediates could be recovered exclusively from isolated nuclei.

Because SV40 and polyoma virus are simple structures, it has repeatedly been suggested that self-assembly mechanisms may play a role in the production of progeny particles (Green, 1970; Fenner, 1968; Leberman, 1968); several RNA plant viruses and bacteriophages of comparable complexity are capable of self-assembly. Friedman (unpublished data) has recently shown that purified polyoma virus can be disrupted into particles the size of capsomers by exposure to pH 10.6 in the presence of a reducing agent. Moreover, such capsomers will reassemble in vitro into particles which resemble naturally occurring empty capsids, but the relevance of this observation to the intracellular events during the assembly of polyoma virions remains to be seen.

The assembly mechanism(s), whatever it may be, produces a

variety of aberrant particles. These include polyoma and SV40 pseudovirions and defective particles and several different forms of tubular, elongated structures, some of which contain DNA (Anderer et al., 1967; Mattern and DeLeva, 1968). Similar structures appear during the maturation of other viruses (Chambers et al., 1966; Breedis et al., 1962; Finch and Klug, 1965) and the suggestion, for which there is little evidence, that they may be developmental precursors of mature virions (Bernhard et al., 1959) has been discussed by Kiselev and Klug (1969). Empty capsids free of DNA are also produced, usually in abundance, during SV40 and polyoma virus productive infections (Winocour, 1963; Crawford, 1963; Koch et al., 1967) and several different forms of small spherical particles have also been described.

The significance of the naturally occurring SV40 and polyoma virus capsids which are free of DNA (Crawford et al., 1962; Koch et al., 1967; Murakami et al., 1968; Maizel et al., 1968) is not known. It is possible that they are intermediates in the maturation of virions since such empty capsid particles have been shown to be intermediates in the assembly of polio virus (Jacobson and Baltimore, 1968) and T4 phage (Luftig et al., 1971). The data of Ozer (1972) and Ozer and Tegtmeyer (1972) on the sizes of the pools of capsid protein and capsids in monkey cells infected with SV40 are consistent with this idea, but the alternative possibility, namely, that progeny virions are assembled around progeny DNA molecules, has yet to be rigorously excluded. Nobody has yet observed the reassembly in vitro of infectious polyoma virus or SV40 particles from capsomers, core protein and DNA. Friedmann's (1971) attempt to do this resulted in the reassembly of spherical, noninfectious particles that sediment at 140 S and have greatly reduced DNA to protein ratios.

Literature Cited

Acheson, N. H., E. Buetti, K. Scherrer and R. Weil. 1971. Transcription of the polyoma virus genome: Synthesis and cleavage of giant late polyoma-specific RNA. Proc. Nat. Acad. Sci. *68:* 2231.

Aloni, Y. 1972. Extensive symmetrical transcription of SV40 DNA in virus-yielding cells. Proc. Nat. Acad. Sci. *69:* 2404.

Aloni, Y., E. Winocour and L. Sachs. 1968. Characterization of the SV40-specific RNA in virus-yielding and transformed cells. J. Mol. Biol. *31:* 415.

Anderer, F. A., H. D. Schlumberger, M. A. Koch, H. Frank and H. J. Eggers. 1967.

Structure of simian virus 40: II. Symmetry and components of the virus particles. Virology *32:* 511.

ANDERSON, C. W. and R. F. GESTELAND. 1972. Pattern of protein synthesis in monkey cells infected by SV40. J. Virol. *9:* 758

BABIUK, L. A. and J. B. HUDSON. 1972. Integration of polyoma virus DNA into mammalian genomes. Biochem. Biophys. Res. Comm. *47:* 111.

BARBANTI-BRODANO, G., P. SWETLY and H. KOPROWSKI. 1970. Early events in the infection of permissive cells with simian virus 40: Adsorption, penetration, and uncoating. J. Virol. *6:* 78.

BASILICO, C., G. MARIN and G DI MAYORCA. 1966. Requirement for the integrity of the viral genome for the induction of host DNA synthesis by polyoma virus. Proc. Nat. Acad. Sci. *56:* 208.

BASILICO, C., Y. MATSUYA and H. GREEN. 1969. Origin of the thymidine kinase induced by polyoma virus in productively infected cells. J. Virol. *3:* 140.

BENJAMIN, T. L. 1966. Virus-specific RNA in cells productively infected or transformed by polyoma virus. J. Mol. Biol. *16:* 359.

BEN-PORAT, T. and A. S. KAPLAN. 1967. Correlation between replication and degradation of cellular DNA in polyoma virus-infected cells. Virology *32:* 457.

BEN-PORAT, T., C. COTO and A. S. KAPLAN. 1966. Unstable DNA synthesized by polyoma virus-infected cells. Virology *30:* 74.

BEN-PORAT, T., A. S. KAPLAN and R. W. TENNANT. 1967. Effect of 5-fluorouracil on the multiplication of a virulent virus (pseudorabies) and an oncogenic virus (polyoma). Virology *32:* 445.

BERNHARD, W., H. L. FEBVRE and R. CRAMER. 1959. Mise en évidence au microscope électronique d'un virus dans des cellules infectées *in vitro* par l'agent du polyome. Compt. Rend. Acad. Sci. *249:*483.

BLACK, P. H., W. P. ROWE, H. C. TURNER and R. J. HUEBNER. 1963. A specific complement-fixing antigen present in SV40 tumor and transformed cells. Proc. Nat. Acad. Sci. *50:* 1148.

BOURGAUX, P. 1964. The fate of polyoma virus in hamster, mouse and human cells. Virology *23:* 46.

BOURGAUX, P. and D. BOURGAUX-RAMOISY. 1972. Is a specific protein responsible for the supercoiling of polyoma DNA? Nature New Biol. *235:* 105.

BOURGAUX, P., D. BOURGAUX-RAMOISY and R. DULBECCO. 1969. The replication of the ring-shaped DNA of polyoma virus: I. Identification of the replicative intermediate. Proc. Nat. Acad. Sci. *64:* 701.

BRANTON, P. E. and R. SHEININ. 1968. Control of DNA synthesis in cells infected with polyoma virus. Virology *36:* 652.

BREEDIS, C., L. BERWICK and T. F. ANDERSON. 1962. Fractionation of Shope papilloma virus in cesium chloride density gradients. Virology *17:* 84.

BRITTEN, R. J. and D. E. KOHNE. 1968. Repeated sequences in DNA. Science *161:* 529.

BRYAN, R. N., D. H. GELFAND and M. HAYASHI. 1969. Initiation of SV40 DNA-directed protein synthesis with N-formylmethionine *in vitro*. Nature *224:* 1019.

BUTEL, J. S. and F. RAPP. 1965. The effect of arabinofuranosylcytosine on the growth cycle of SV40. Virology *27:* 490.

CAIRNS, J. 1963. The chromosome of *Escherichia coli*. Cold Spring Harbor Symp. Quant. Biol. *28:* 43.

CARP, R. I., G. SAUER and F. SOKOL. 1969. The effect of actinomycin D on the transcription of simian virus 40 deoxyribonucleic acid. Virology *37:* 214.

CHAMBERS, V. C., S. HSIA and Y. ITO. 1966. Rabbit kidney vacuolating virus: Ultrastructural studies. Virology *29:* 32.

CHAMPOUX, J. J. and R. DULBECCO. 1972. An activity from mammalian cells that untwists superhelical DNA—A possible swivel for DNA replication. Proc. Nat. Acad. Sci. *69:* 143.

CHEEVERS, W. P., P. E. BRANTON and R. SHEININ. 1970. Characterization of abnormal DNA formed in polyoma virus-infected cells. Virology *40:* 768.

CRAWFORD, L. V. 1962. The adsorption of polyoma virus. Virology *18:* 177.

———. 1963. The physical characteristics of polyoma virus: II. The nucleic acid. Virology *19:* 279.

CRAWFORD, L. V. and R. F. GESTELAND. 1973. Synthesis of polyoma proteins *in vitro.* J. Mol. Biol. In press.

CRAWFORD, L. V., E. M. CRAWFORD and D. H. WATSON. 1962. The physical characterisation of polyoma virus: I. Two types of particles. Virology *18:* 170.

CRAWFORD, L., R. DULBECCO, M. FRIED, L. MONTAGNIER and M. STOKER. 1964. Cell transformation by different forms of polyoma virus DNA. Proc. Nat. Acad. Sci. *52:* 148.

CRAWFORD, L. V., E. M. CRAWFORD, J. P. RICHARDSON and H. S. SLAYTER. 1965. The binding of RNA polymerase to polyoma and papilloma DNA. J. Mol. Biol. *14:* 593.

CRAWFORD, L., R. GESTELAND, G. RUBIN and B. HIRT. 1971. The use of mammalian DNAs to direct protein synthesis in extracts from *E. coli*. Lepetit Colloq. Biol. Med. *2:* 104.

DANNA, K. and D. NATHANS. 1971. Specific cleavage of SV40 DNA by restriction endonuclease of *Hemophilus influenzae*. Proc. Nat. Acad. Sci. *68:* 2913.

DEFENDI, V. and F. JENSEN. 1967. Oncogenicity by DNA tumor viruses: Enhancement after ultraviolet and cobalt-60 radiations. Science *157:* 703.

DELIUS, H., H. WESTPHAL and N. AXELROD. 1973. Length measurements of RNA synthesized *in vitro* by *E. coli* RNA polymerase. J. Mol. Biol. In press.

DEL VILLANO, B. and V. DEFENDI. 1970. Preparation and use of an immunosorbent to study the molecular composition of the SV40 T-antigen from hamster tumor cells. Bacteriol. Proc. Abstr. *222:* 188.

DULBECCO, R. 1969. Cell transformation by viruses: Two minute viruses are powerful tools for analyzing the mechanism of cancer. Science *166:* 962.

———. 1970. Topoinhibition and serum requirement of transformed and untransformed cells. Nature *227:* 802.

DULBECCO, R. and T. JOHNSON. 1970. Interferon-sensitivity of the enhanced incorporation of thymidine into cellular DNA induced by polyoma virus. Virology *42:* 368.

DULBECCO, R. and M. G. P. STOKER. 1970. Conditions determining initiation of DNA synthesis in 3T3 cells. Proc. Nat. Acad. Sci. *66:* 204.

DULBECCO, R., L. H. HARTWELL and M. VOGT. 1965. Induction of cellular DNA synthesis by polyoma virus. Proc. Nat. Acad. Sci. *53:* 403.

ECKHART, W. 1968. Transformation of animal cells by oncogenic DNA viruses. Physiol. Rev. *48:* 513.

FAREED, G. C. and N. P. SALZMAN. 1972. Intermediate in SV40 DNA chain growth. Nature New Biol. *238:* 274.

FAREED, G. C., C. F. GARON and N. P. SALZMAN. 1972. Origin and direction of SV40 DNA replication. J. Virol. *10:* 484.

FENNER, F. 1968. The biology of animal viruses, vol. I. and II. Academic Press, New York.

FINCH, J. T. and A. KLUG. 1965. The structure of viruses of the papilloma-polyoma type: III. Structure of rabbit papilloma virus (with an appendix on the topography of contrast in negative-staining for electron-microscopy). J. Mol. Biol. *13:* 1.

FISCHER, H. and G. SAUER. 1972. Identification of virus-induced proteins in cells productively infected with SV40. J. Virol. *9:* 1.

FRASER, K. B. and E. M. CRAWFORD. 1965. Immunofluorescent and electron-microscopic studies of polyoma virus in transformation reactions with BHK21 cells. Exp. Mol. Pathol. *4:* 51.

FREARSON, P. M., S. KIT and D. R. DUBBS. 1965. Deoxythymidylate synthetase and deoxythymidine kinase activities of virus-infected animal cells. Cancer Res. *25:* 737.

———. 1966. Induction of dehydrofolate reductase activity by SV40 and polyoma virus. Cancer Res. *26:* 1653.

FRIED, M. A. 1970. Characterization of a temperature-sensitive mutant of polyoma virus. 1970. Virology *40:* 605.

FRIED, M. and J. D. PITTS. 1968. Replication of polyoma virus DNA: I. A resting cell system for biochemical studies on polyoma virus. Virology *34:* 761.

FRIEDMANN, T. 1971. *In vitro* reassembly of shell-like particles from disrupted polyoma virus. Proc. Nat. Acad. Sci. *68:* 2574.

GELFAND, D. H. and M. HAYASHI. 1969. DNA-dependent RNA-directed protein synthesis *in vitro:* II. Synthesis of ϕX-174 coat protein component. Proc. Nat. Acad. Sci. *63:* 135.

GERSHON, D. and L. SACHS. 1966. The early synthesis of RNA in polyoma virus development. Virology *29:* 44.

GERSHON, D., P. HAUSEN, L. SACHS and E. WINOCOUR. 1965. On the mechanism of polyoma virus-induced synthesis of cellular DNA. Proc. Nat. Acad. Sci. *54:* 1584.

GERSHON, D., L. SACHS and E. WINOCOUR. 1966. The induction of cellular DNA synthesis by simian virus 40 in contact-inhibited and in X-irradiated cells. Proc. Nat. Acad. Sci. *56:* 918.

GILBERT, W. and D. DRESSLER. 1968. DNA replication: The rolling circle model. Cold Spring Harbor Symp. Quant. Biol. *33:* 473.

GRANBOULAN, N. and P. TOURNIER. 1965. Horaire et localisation de la synthèse des nucléiques pendant la phase d'éclipse du virus SV40. Inst. Pasteur Ann. *109:* 837.

GRANBOULAN, N., P. TOURNIER, R. WICKER and W. BERNHARD. 1963. An electron microscope study of the development of SV40 virus. J. Cell Biol. *17:* 423.

GREEN, M. 1970. Oncogenic viruses. Annu. Rev. Biochem. *39:* 701.

GREEN, M., H. MILLER and S. HENDLER. 1971. Isolation of a polyoma nucleoprotein complex from infected mouse cell cultures. Proc. Nat. Acad. Sci. *68:* 1032.

HABEL, K. 1965. Specific complement-fixing antigens in polyoma tumors and transformed cells. Virology *25:* 55.

———. 1966. Virus tumor antigens: Specific fingerprints? Cancer Res. *26:* 2018.

HANCOCK, R. and R. WEIL. 1969 Biochemical evidence for induction by polyoma virus of replication of the chromosomes of mouse kidney cell. Proc. Nat. Acad. Sci. *63:* 1144.

HARTWELL, L., M. VOGT and R. DULBECCO. 1965. Induction of cellular DNA synthesis by polyoma: II. Increase in the rate of enzyme synthesis after infection with polyoma virus in mouse embryo kidney cells. Virology *27:* 262.

HATANAKA, M. and R. DULBECCO. 1966. Induction of DNA synthesis by SV40. Proc. Nat. Acad. Sci. *56:* 736.

———. 1967. SV40-specific thymidine kinase. Proc. Nat Acad. Sci. *58:* 1888.

HENRY, P., P. H. BLACK, M. N. OXMAN and S. M. WEISSMAN. 1966. Stimulation of DNA synthesis in mouse cell line 3T3 by simian virus 40. Proc. Nat. Acad. Sci. *56:* 1170.

HIRAI, K. and V. DEFENDI. 1972. Integration of SV40 DNA into the DNA of permissive monkey cells. J. Virol. *9:* 705.

HIRAI, K., J. LEHMAN and V. DEFENDI. 1971a. Reinitiation within one cell cycle of the DNA synthesis induced by SV40. J. Virol. *8:* 828.

———. 1971b. Integration of SV40 DNA into the DNA of primary infected Chinese hamster cells. J. Virol. *8:* 708.

HIRT, B. 1967. Selective extraction of polyoma DNA from infected mouse cell cultures. J. Mol. Biol. *26:* 365.

———. 1969. Replicating molecules of polyoma virus DNA. J. Mol. Biol. *40:* 141.

HOGGAN, M. D., W. P. ROWE, P. H. BLACK and R. J. HUEBNER. 1965. Production of tumor specific antigens by oncogenic viruses during acute cytolytic infections. Proc. Nat. Acad. Sci. *53:* 12.

HORWITZ, M. S., M. D. SCHARFF and J. V. MAIZEL. 1969. Synthesis and assembly of adenovirus 2: I. Polypeptide synthesis, assembly of capsomers, and morphogenesis of the virion. Virology *39:* 682.

HUDSON, J., D. GOLDSTEIN and R. WEIL. 1970. A study on the transcription of the polyoma viral genome. Proc. Nat. Acad. Sci. *65:* 226.

HUMMELER, K., N. TOMASSINI and F. SOKOL. 1970. Morphological aspects of the uptake of simian virus 40 by permissive cells. J. Virol. *6:* 87.

JACOBSON, M. F. and D. BALTIMORE. 1968. Morphogenesis of poliovirus: I. Association of the viral RNA with coat protein. J. Mol. Biol. *33:* 369.

JAENISCH, R. and A. LEVINE. 1971. DNA replication in SV40-infected cells. V. Circular and catenated oligomers of SV40 DNA. Virology *44:* 480.

———. 1972. DNA replication in SV40-infected cells. VI. The effect of cycloheximide on the formation of SV40 oligomeric DNA. Virology *48:* 373.

JAENISCH, R., A. MAYER and A. LEVINE. 1971. Replicating SV40 molecules containing closed circular template DNA strands. Nature New Biol. *233*: 72.

KAJIOKA, R. 1972. Some studies on the status of polyoma-specified RNA after transcription. Virology *48*: 284.

KANG, H. S., T. B. ESHBACH, D. A. WHITE and A. J. LEVINE. 1971. DNA replication in SV40-infected cells. IV. Two different requirements for protein synthesis during SV40 DNA replication. J. Virol. *7:* 112.

KARA, J. and R. WEIL. 1967. Specific activation of the DNA synthesizing apparatus in contact inhibited cells by polyoma virus. Proc. Nat. Acad. Sci. *57:* 63.

KASAMAKI, A., T. BEN-PORAT and A. S. KAPLAN. 1968. Polyoma virus-induced release of inhibition of cellular DNA synthesis caused by iododeoxyuridine. Nature *217:* 756.

KAWAMURA, A. 1969. Fluorescent antibody techniques and their applications. University Tokyo Press, Tokyo. University Park Press, Baltimore, Md. and Manchester, England.

KELLER, W. and R. GOOR. 1970. Mammalian RNA polymerase: Structural and functional properties. Cold Spring Harbor Symp. Quant. Biol. *35:* 671.

KHOURY, G. and M. A. MARTIN. 1972. Comparison of SV40 DNA transcription *in vivo* and *in vitro*. Nature New Biol. *238:* 4.

KHOURY, G., J. C. BRYNE and M. A. MARTIN. 1972. Patterns of SV40 DNA transcription after acute infection of permissive and nonpermissive cells. Proc. Nat. Acad. Sci. *69:* 1925.

KISELEV, N. A. and A. KLUG. 1969. The structure of viruses of the papilloma-polyoma type: V. Tubular variants built of pentamers. J. Mol. Biol. *40:* 155.

KIT, S. 1968. Viral induced enzymes and viral carcinogenesis. Advance. Cancer Res. *11:* 73.

KIT, S. and K. NAKAJIMA. 1971. Analysis of molecular forms of SV40 DNA synthesized in cycloheximide-treated cell cultures. J. Virol. *7:* 87.

KIT, S., P. M. FREARSON and D. R. DUBBS. 1965. Enzyme-induction in polyoma-infected mouse embryo cells. Fed. Proc. *24:* 596.

Kit, S., D. R. Dubbs and P. M. Frearson. 1966a. Enzymes of nucleic acid metabolism in cells infected with polyoma virus. Cancer Res. *26:* 638.

Kit, S., D. R. Dubbs, P. M. Frearson and J. L. Melnick. 1966b. Enzyme induction in SV40-infected green monkey kidney cultures. Virology *29:* 69.

Kit, S., J. L. Melnick, M. Anken, D. R. Dubbs, R. A. de Torres and T. Kitahara. 1967a. Non-identity of some simian virus 40-induced enzymes with tumor antigen. J. Virol. *1:* 694.

Kit, S., J. L. Melnick, D. R. Dubbs, L. J. Piekarski and R. A. de Torres. 1967b. Virus-directed host response. *In* (M. Pollard, ed.) Perspectives in virology. Academic Press, New York.

Kit, S., R. A. de Torres, D. R. Dubbs and M. L. Salvi. 1967c. Induction of cellular deoxyribonucleic acid synthesis by simian virus 40. J. Virol. *1:* 738.

Kit, S., L. J. Piekarski and D. R. Dubbs. 1967d. DNA polymerase induced by simian virus 40. J. Gen. Virol. *1:* 163.

Koch, M. A., H. J. Eggers, F. A. Anderer, H. D. Schlumberger and H. Frank. 1967. Structure of simian virus 40: I. Purification and physical characterization of the virus particle. Virology *32:* 503.

Laterjet, R., R. Cramer and L. Montagnier. 1967. Inactivation by UV-, X-, and γ-radiations of the infecting and transforming capacities of polyoma virus. Virology *33:* 104.

Lavi, S. and E. Winocour. 1972. Acquisition of sequences of homologous to host DNA by closed circular SV40 DNA. J. Virol. *9:* 309.

Leberman, R. 1968. The disaggregation and assembly of simple viruses, p. 183. In (L. V. Crawford and M. G. P. Stoker, ed.) The molecular biology of viruses. Cambridge University Press, Cambridge.

Lehman, J. M. and V. Defendi. 1970. Changes in DNA synthesis regulation in Chinese hamster cells infected with SV40. J. Virol. *6:* 738.

Levine, A. J. 1971. Induction of mitochondrial DNA synthesis in monkey cells infected by SV40 and (or) treated with calf serum. Proc. Nat. Acad. Sci. *68:* 717.

Levine, A. J., H. S. Kang and F. E. Billheimer. 1970. DNA replication in SV40 infected cells: I. Analysis of replicating SV40 DNA. J. Mol. Biol. *50:* 549.

Lindstrom, D. M. and R. Dulbecco. 1972. Strand orientation of SV40 transcription in productively infected cells. Proc. Nat. Acad. Sci. *69:* 1517.

Littlefield, J. W. and C. Basilico. 1966. Infection of thymidine kinase deficient BHK cells with polyoma virus. Nature *211:* 250.

Luftig, R. B., W. B. Wood and R. Okinawa. 1971. Bacteriophage T4 head morphogenesis. On the nature of gene 49—defective heads and their role as intermediates. J. Mol. Biol. *57:* 555.

Maizel, J. V., D. O. White and M. D. Scharff. 1968. The polypeptides of adenovirus: II. Soluble proteins, cores, top components and the structure of the virion. Virology *36:* 126.

Mandel, J. L. and P. Chambon. 1971. Contrôle positif de la transcription chez les eucaryotes. C. R. Soc. Biol. (Paris) *165:* 509.

Martin, M. A. 1970. Characteristics of SV40 DNA transcription during lytic infection, abortive infection and in transformed mouse cells. Cold Spring Harbor Symp. Quant. Biol. *35:* 833.

Martin, M. A. and D. Axelrod. 1969a. Polyoma virus gene activity during lytic infection and in transformed animal cells. Science *164:* 68.

———. 1969b. SV40 gene activity during lytic infection and in a series of SV40 transformed mouse cells. Proc. Nat. Acad. Sci. *64:* 1203.

MATTERN, C. F. T. and A. M. DELEVA. 1968. Observations of polyoma virus filaments. Virology *36:* 683.

MATTERN, C. F. T., K. K. TAKEMOTO and W. A. DANIEL. 1966. Replication of polyoma virus in mouse embryo cells: Electron microscopic observations. Virology *30:* 242.

MAYOR, H. D., S. E. STINEBAUGH, R. M. JAMISON, L. E. JORDAN and J. L. MELNICK. 1962. Immunofluorescent, cytochemical and microcytological studies on the growth of the simian vacuolating virus (SV40) in tissue cultures. Exp. Mol. Pathol. *1:* 397.

MEINKE, W. and D. A. GOLDSTEIN. 1971. Studies on the structure and formation of polyoma DNA replicative intermediates. J. Mol. Biol. *61:* 543.

MELNICK, J. L. 1969. Analytical serology of animal viruses. In (J. B. G. Kwapinski, ed.) Analytical serology of microorganisms, Vol. I. Interscience, New York.

MELNICK, J. L., S. E. STINEBAUGH and F. RAPP. 1964. Incomplete simian papova virus SV40: Formation of non-infectious viral antigen in the presence of fluorouracil. J. Exp. Med. *119:* 313.

MINOWADA, J. 1964. Effect of X-irradiation on DNA synthesis in polyoma virus-infected cultures. Exp. Cell Res. *33:* 161.

MINOWADA, J. and G. E. MOORE. 1963. DNA synthesis in X-irradiated cultures infected with polyoma virus. Exp. Cell Res. *29:* 31.

MOLTENI, P., V. DE SIMONE, E. GROSSO, P. A. BIANCHI and E. POLLI. 1966. The incorporation of thymidine into deoxyribonucleic acid in mouse fibroblasts infected with polyoma virus. J. Biochem. *98:* 78.

MURAKAMI, W. T., R. FINE, M. R. HARRINGTON and Z. BEN-SASSAN. 1968. Properties and amino acid composition of polyoma virus purified by zonal ultracentrifugation. J. Mol. Biol. *36:* 153.

NATHANS, D. and K. J. DANNA. 1972. Specific origin of SV40 DNA replication. Nature New Biol. *236:* 200.

ODA, K. and R. DULBECCO. 1968a. Induction of cellular RNA synthesis in BSC-1 cells infected by SV40. Virology *35:* 439.

———. 1968b. Regulation of transcription of the SV40 DNA in productively infected and in transformed cells. Proc. Nat. Acad. Sci. *60:* 525.

OXMAN, M. N. and M. J. LEVIN. 1971. Interferon and transcription of early virus-specific RNA in cells infected with SV40. Proc. Nat. Acad. Sci. *68:* 299.

OXMAN, M. N., K. K. TAKEMOTO and W. ECKHART. 1972. Polyoma T antigen synthesis by temperature-senisitve mutants of polyoma virus. Virology *49:* 675.

OZER, H. L. 1972. Synthesis and assembly of SV40. I. Differential synthesis of intact virions and empty shells. J. Virol. *9:* 41.

OZER, H. L. and P. TEGTMEYER. 1972. Synthesis and assembly of SV40. II. Synthesis of the major capsid protein and its incorporation into viral particles. J. Virol. *9:* 52.

PETTIJOHN, D. and T. KAMIYA. 1967. Interaction of RNA polymerase with polyoma DNA. J. Mol. Biol. *29:* 275.

PÉTURSSON, G. and R. WEIL. 1968. A study on the mechanism of polyoma-induced activation of the cellular DNA-synthesizing apparatus: Synchronization by FdU of virus-induced DNA synthesis. Arch. Ges. Virusforsch. *24:* 1.

POPE, J. H. and W. P. ROWE. 1964. Detection of specific antigen in SV40 transformed cells by immunofluorescence. J. Exp. Med. *120:* 121.

POTTER, C. W., B. C. MCLAUGHLIN and J. S. OXFORD. 1969. Simian virus 40-induced T and tumor antigens. J. Virol. *4:* 574.

RALPH, R. K. and J. S. COLTER. 1972. Evidence for the integration of polyoma virus DNA in a lytic system. Virology *48:* 49.

RAPP, F. T., T. KITAHARA, J. S. BUTEL and J. L. MELNICK. 1964. The synthesis of SV40 tumor antigen during replication of SV40. Proc. Nat. Acad. Sci. *52:* 1138.

RAPP, F., J. S. BUTEL, L. A. FELDMAN, T. KITAHARA and J. L. MELNICK. 1965. Differential effects of inhibitors on the steps leading to the formation of SV40 tumor and virus antigens. J. Exp. Med. *121:* 935.

RITZI, E. and A. J. LEVINE. 1970. Deoxyribonucleic acid replication in simian virus 40-infected cells: III. Comparison of simian virus 40 lytic infection in three different monkey kidney cell lines. J. Virol. *5:* 686.

ROVERA, G., R. BASERGA and V. DEFENDI. 1972. Early increase in nuclear acidic protein synthesis after SV40 infection. Nature New Biol. *237:* 240.

RUSH, M. G., R. EASON and J. VINOGRAD. 1971. Identification and properties of complex forms of SV40 DNA isolated from SV40-infected African green monkey (BSC1) cells. Biochim. Biophys. Acta *228:* 585.

SABIN, A. and M. A. KOCH. 1964. Source of genetic information for specific complement-fixing antigen in SV40 induced tumors. Proc. Nat. Acad. Sci. *52:* 1131.

SAMBROOK, J. F. and A. S. SHATKIN. 1969. Polynucleotide ligase activity in cells infected with polyoma SV40 or vaccinia. J. Virol. *4:* 719.

SAMBROOK, J., P. A. SHARP and W. KELLER. 1972. Transcription of SV40. I. Separation of the strands of SV40 DNA and hybridization of the separated strands to RNA extracted from lytically infected and transformed cells. J. Mol. Biol. *70:* 57.

SAUER, G. 1971. Apparent differences in transcriptional control in cells productively infected and transformed by SV40. Nature New Biol. *231:* 135.

SAUER, G. and J. R. KIDWAI. 1968. The transcription of the SV40 genome in productively infected and in transformed cells. Proc. Nat. Acad. Sci. *61:* 1256.

SEBRING, E. D., T. J. KELLY, M. M. THOREN and N. P. SALZMAN. 1971. Structure of replicating SV40 DNA molecules. J. Virol. *8:* 479.

SHEININ, R. 1966a. Deoxyribonucleic acid synthesis in cells replicating polyoma virus. Virology *28:* 621.

———. 1966b. DNA synthesis in rat embryo cells infected with polyoma virus. Virology *29:* 167.

———. 1966c. Studies on the thymidine kinase activity of mouse embryo cells infected with polyoma virus. Virology *28:* 47.

———. 1967. Deoxyribonucleic acid synthesis in cells infected with polyoma virus. *In* (J. S. Colter and W. Paranchych, ed.) The molecular biology of viruses. Academic Press, New York.

SHIMONO, H. and A. S. KAPLAN. 1969. Correlation between the synthesis of DNA and histones in polyoma virus-infected mouse embryo cells. Virology *37:* 690.

SMITH, B. J. 1970. Light satellite-band DNA in mouse cells infected with polyoma virus. J. Mol. Biol. *47:* 101.

———. 1971. Light satellite-band replication in mouse cells. Lepetit Colloq. Biol. Med. *2:* 220.

SOKOL, F. and R. I. CARP. 1971. Molecular size of SV40-specific RNA synthesized in productively infected cells. J. Gen. Virol. *11:* 177.

SUGDEN, B. and J. SAMBROOK. 1970. RNA polymerase from HeLa cells. Cold Spring Harbor Symp. Quant. Biol. *35:* 663.

TAI, H. T., C. A. SMITH, P. A. SHARP and J. VINOGRAD. 1972. Sequence heterogeneity in closed SV40 DNA. J. Virol. *9:* 317.

TAYLOR-PAPADIMITRIOU, J., M. STOKER and P. RIDDLE. 1971. Further manifestations of abortive transformation of BHK21 cells by polyoma virus. Int. J. Cancer *7:* 269.

TEGTMEYER, P. 1972. SV40 DNA synthesis: The viral replicon. J. Virol. *10:* 591.

THOREN, M. M., E. D. SEBRING and N. P. SALZMAN. 1972. Specific initiation site for SV40 DNA replication. J. Virol. *10:* 462.

TOBIA, A. M., E. H. BROWN, R. J. PARKER, C. L. SCHILDKRAUT and J. J. MAIO. 1972

DNA replication in synchronized cultured mammalian cells. IV. Replication of African green monkey component α and bulk DNA. Biochim. Biophys. Acta *277:* 256.

TODARO, G. J., G. K. LAZAR and H. GREEN. 1965. The initiation of cell division in a contact-inhibited mammalian cell line. J. Cell. Comp. Physiol. *66:* 325.

TONEGAWA, S., G. WALTER, A. BERNARDINI and R. DULBECCO. 1970. Transcription of the SV40 genome in transformed cells and during lytic infection. Cold Spring Harbor Symp. Quant. Biol. *35:* 823.

VELICER, L. F. and H. S. GINSBERG. 1970. Synthesis, transport and morphogenesis of type 5 adenovirus capsid proteins. J. Virol. *5:* 338.

VOGT, M., R. DULBECCO and B. SMITH. 1966. Induction of cellular DNA synthesis by polyoma virus: III. Induction in productively infected cells. Proc. Nat. Acad. Sci. *55:* 956.

WALTER, G., R. ROBLIN and R. DULBECCO. 1972. Protein synthesis in SV40-infected monkey cells. Proc. Nat. Acad. Sci. *69:* 921.

WANG, J. C. 1971. Interaction between DNA and an *E. coli* protein ω. J. Mol. Biol. *55:* 523.

WEIL, R., M. R. MICHEL and G. K. RUSCHMANN. 1965. Induction of cellular DNA synthesis by polyoma virus. Proc. Nat. Acad. Sci. *53:* 1468.

WEIL, R., G. PÉTURSSON, J. KARA and H. DIGGELMANN. 1967. The interaction of polyoma virus with the genetic apparatus of host cells, p. 593. *In* (J. S. Colter and W. Paranchych, ed.) The molecular biology of viruses. Academic Press, New York.

WEINBERG, R. A., Z. BEN-ISHAI and J. E. NEWBOLD. 1972a. PolyA associated with SV40 messenger RNA. Nature New Biol. *238:* 111.

WEINBERG, R. A., S. O. WARNAAR and E. WINOCOUR. 1972b. Isolation and characterization of SV40 RNA. J. Virol. *10:* 193.

WEISBERG, R. A. 1964. The morphology and transplantability of mouse embryo cells transformed *in vitro* by polyoma virus. Virology *23:* 553.

WERCHAU, H., H. WESTPHAL, G. MAASS and R. HASS. 1966. Untersuchungen über den Nucleinsaurestoffwechsel von Affennierengewebe-Kulturzellen nach Infektion mit SV40: II. Einflub der Infektion auf den DNS-Stoffwechsel der Zelle. Arch. Ges. Virusforsch. *19:* 351.

WESTPHAL, H. 1970. SV40 DNA strand selection by *Escherichia coli* RNA polymerase. J. Mol. Biol. *50:* 407.

———. 1971. Transcription of superhelical and relaxed circular SV40 DNA by *E. coli* RNA polymerase in the presence of rifampicin. Lepetit Colloq. Biol. Med. *2:* 77.

WESTPHAL, H. and R. DULBECCO. 1968. Viral DNA in polyoma- and SV40-transformed cell lines. Proc. Nat. Acad. Sci. *59:* 1158.

WESTPHAL, H. and E. D. KIEHN. 1970. The in vitro product of SV40 DNA transcription and its specific hybridization with DNA of SV40-transformed cells. Cold Spring Harbor Symp. Quant. Biol. *35:* 819.

WESTPHAL, H., H. DELIUS and C. MULDER. 1973. Visualization of SV40 *in vitro* transcription complexes. Lepetit Colloq. Biol. Med. *4:* In Press.

WHITE, M. and R. EASON. 1971. Nucleoprotein complexes in SV40-infected cells. J. Virol. *8:* 363.

WIDMER, C. 1971. Ph.D. thesis, University of Lausanne.

WINNACKER, G. L., G. MAGNUSSAN and P. REICHARD. 1971. Synthesis of polyoma DNA by isolated nuclei. Biochem. Biophys. Res. Comm. *44:* 952.

WINOCOUR, E. 1963. Purification of polyoma virus. Virology *19:* 158.

———. 1965. Attempts to detect an integrated polyoma genome by nuclei acid hybridization. I. "Reconstruction" experiments and complementarity tests between synthetic polyoma RNA and polyoma tumor DNA. Virology *25*: 276.

WINOCOUR, E. 1969. Some aspects of the interaction between polyoma virus and cell DNA. Advance. Virol. Res. *14:* 153.

WINOCOUR, E. and E. ROBBINS. 1970. Histone synthesis in polyoma and SV40-infected cells. Virology *40:* 307.

WINOCOUR, E., A. M. KAYE and V. STOLLAR. 1965. Synthesis and transmethylation of DNA in polyoma-infected cultures. Virology *27:* 156.

WINOCOUR, E., R. A. WEINBERG and M. HERZBERG. 1971. SV40 DNA transcription *in vivo* and *in vitro*. Lepetit Colloq. Biol. Med. *2:* 63.

YAMAMOTO, H. and H. SHIMOJO. 1971. Inactivation of T antigen-forming capacities of SV40 and adenovirus 12 by ultraviolet irradiation. J. Virol. *7:* 419.

YOSHIIKE, K., A. FURUNO and K. SUZUKI. 1972. Denaturation maps of complete and defective SV40 DNA molecules. J. Mol. Biol. *70:* 415.

6

Transformation by Polyoma Virus and SV40

Cell division is controlled in multicellular organisms. Tumours grow because cancer cells continue to multiply in conditions which regulate the multiplication of normal cells. Polyoma virus and SV40 can effect this change in the growth control of a cell and these viruses cause tumours when they are injected into susceptible animals. Transformation of cells growing in vitro by these viruses is believed to be an analogous process to the induction of tumours in animals because cells transformed in vitro by polyoma virus and SV40 acquire a set of properties (see Table 6.1), some of which, including increased malignancy, are characteristic of tumour cells. It is well to remember, however, that the tumours which polyoma virus and SV40 induce when they are injected in large amounts into susceptible rodents do not usually metastasize; neither does the histology of such induced tumours closely resemble the histology of spontaneous malignant tumours. This is also true of the tumours which develop when cells transformed in vitro by these viruses are injected into host animals. In other words, transformation of cultivated fibroblasts by these viruses provides a model system in which we can study in a quantitative way at least some of the events that lead to the seemingly spontaneous development of malignant tumours in animals and man.

Cell lines are maintained in culture by the periodic replacement of the medium and serum. Populations of untransformed cells divide while they remain sparse, but as the cell density and the number of cell to cell contacts increases, the growth rate of the culture dramatically decreases; the cells of some lines stop dividing altogether once they form a confluent monolayer. If such untransformed cells are infected

Table 6.1. PROPERTIES OF SV40- OR POLYOMA VIRUS-TRANSFORMED CELLS

Growth
High or indefinite saturation density*
Different, usually reduced, serum requirement*
Growth in agar or Methocel suspension*
Tumour formation upon injection into susceptible animals
Not susceptible to contact inhibition of movement
Growth in a less oriented manner*
Growth on monolayers of normal cells*
Surface
Increased agglutinability by plant lectins
Changes in composition of glycoproteins and glycolipids
Tight junctions missing
Fetal antigens revealed
Virus-specific transplantation antigen
Different staining properties
Increased rate of transport of nutrients
Evidence of Virus
Virus-specific antigens detectable
Virus DNA sequences detected
Virus messenger RNA present
In some cases virus can be rescued

Transformed cells show many, if not all, of these properties which are not shared by untransformed parental cells. Several of these properties (*) have formed the basis of selection procedures for isolating transformants.

with tumour viruses, some of the cells may be transformed; the transformants no longer respond to all of the controls which regulate the multiplication of untransformed cells in culture. As a result the transformants continue to multiply in conditions which severely reduce the rate of multiplication of untransformed cells.

The differences between transformed cells and untransformed cells can be classified into three inter-related groups, concerned with (a) changes in patterns of cell growth, (b) changes at the cell surface, and (c) changes in the intracellular population of macromolecules resulting from the presence of the viral genome (see Table 6.1).

Transformed cells can be isolated from untransformed cells in mixed populations by manipulating the conditions of culture so that only cells possessing one or more of the properties listed in Table 6.1 can survive and/or multiply. A transformed cell always exhibits several, but seldom exhibits the complete set, of the properties listed in Table 6.1. For example, a cell selected because it can grow in low concentrations of serum may not be able to grow to a high or indefinite saturation density (Smith et al., 1971), and a cell selected because it can grow on top of a monolayer of parental untransformed cells may not be able to grow suspended in agar. More importantly, transformants isolated by different procedures are not equally tumourigenic when transplanted into susceptible animals, but increased tumourigenicity does appear to be correlated with increased saturation density (Aaronson and Todaro, 1968).

Clearly, each of the changes listed in Table 6.1 cannot be caused by a different virus-coded protein because polyoma virus and SV40 contain only a small amount of genetic information (see Chapter 4). It follows that most of the observed alterations in transformed cells must be either pleiotropic or indirect responses of the cell to the virus.

Changes in Growth Patterns

When most types of untransformed cells are plated in vitro, they will grow to a certain density and then either stop dividing or divide at a greatly reduced rate. In the same conditions the corresponding transformed cells continue to multiply and may reach saturation densities 10–25 times greater than those of the untransformed cells, when nutritional factors probably become limiting. Until recently it was thought that the growth of cells in tissue culture is controlled solely by "contact inhibition" (Stoker and Rubin, 1967), the individual cells in a culture responding to the close proximity of other cells by ceasing to multiply. It is now clear, however, that although cell with cell contact plays a role in inhibiting cell division (Dulbecco, 1970; Burger and Noonan, 1970)

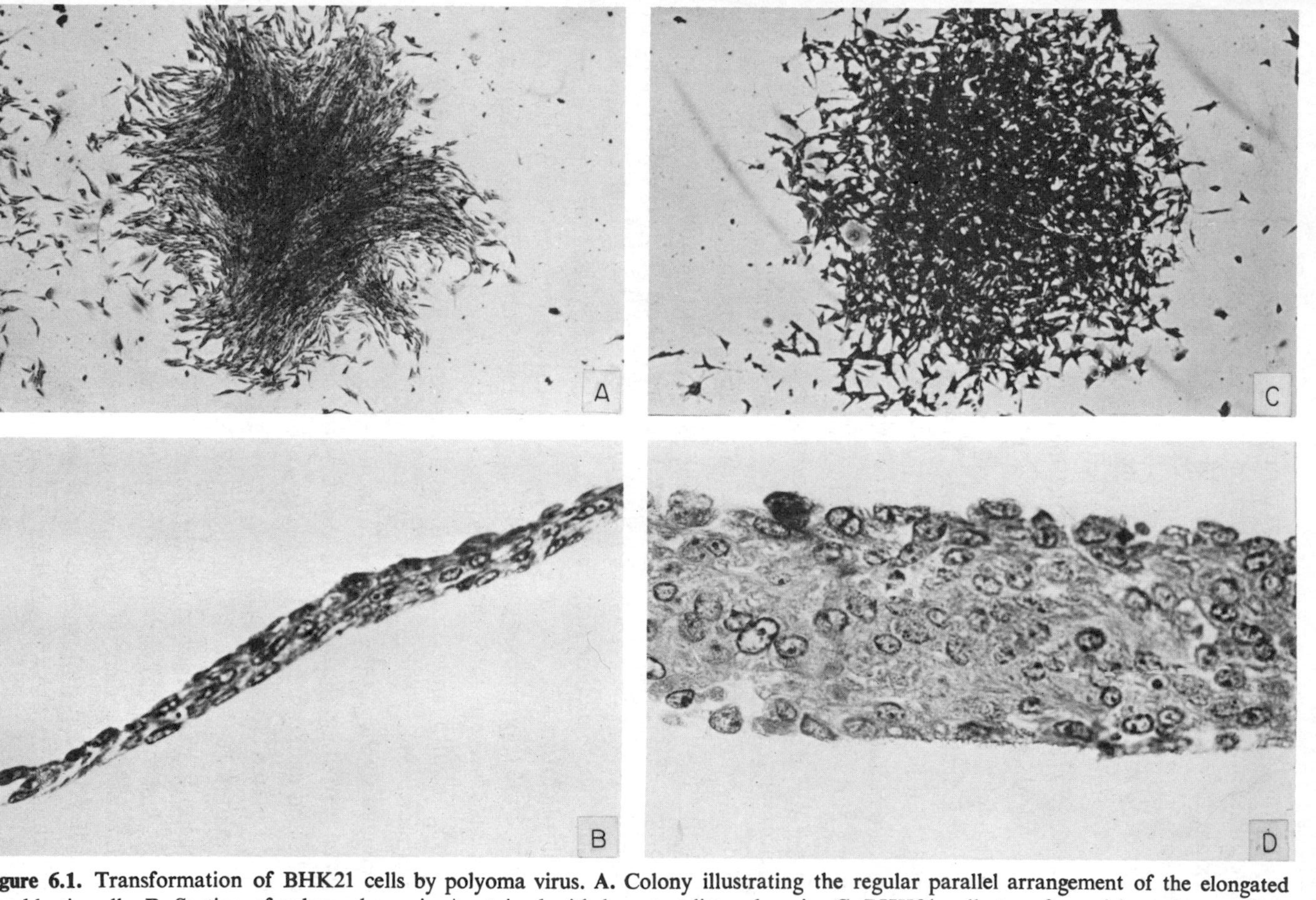

Figure 6.1. Transformation of BHK21 cells by polyoma virus. **A.** Colony illustrating the regular parallel arrangement of the elongated fibroblastic cells. **B.** Section of colony shown in A, stained with hematoxylin and eosin. **C.** BHK21 cells transformed by polyoma virus, illustrating the random orientation of more rounded cells. **D.** Section of colony shown in C, stained with hematoxylin and eosin. Courtesy of M. G. P. Stoker.

as well as cell movement (Abercrombie and Heaysman, 1954), some, as yet unidentified, factor(s) in serum also has a part in regulating the extent to which cells in culture multiply. The addition of a fresh dose of serum to a non-dividing monolayer of untransformed cells results in further rounds of division so that the cells pile up on top of one another (see Figure 6.1 and Kruse and Miedema, 1965; Temin, 1967; Todaro et al., 1967; Holley and Kiernan, 1968).

Viral transformation changes and usually reduces a cell's serum dependence; transformed cells require much less serum factor(s) than untransformed cells to initiate their division cycle (Temin, 1967; Bürk, 1967; Holley and Kiernan, 1968; Eagle et al., 1970; Dulbecco, 1970; Clarke et al., 1970; Paul et al., 1971). What serum factors are, how they work and whether or not they act in vivo remain unanswered questions. Attempts to fractionate the various survival and growth-promoting factors in calf serum (Paul et al., 1971; Lipton et al., 1971, 1972; Holley and Kiernan, 1971) confirm that untransformed and transformed cells in culture differ in their response to serum factors (Holley and Kiernan, 1968; Clarke et al., 1970). They also show that serum contains several substances which can be loosely classified together as growth factors; these include factors which are required by cells for survival, for multiplication and for migration. Furthermore we now know that at least some transformed cells may not only grow in the absence of serum, but also secrete into their medium substances which can satisfy the serum requirement of other cells (Austin et al., 1971; Alfred and Pumper, 1960; Rubin, 1966, Stoker et al., 1971). Shodell (1972) has detected in the serum-free medium, in which cells of a mouse L-cell line have grown, material which satisfies the serum requirement of BHK cells and causes these cells to proliferate in the absence of serum. He has fractionated this material into two active fractions, one of which stimulates BHK cells to synthesize DNA, while the other, which is a heat-labile macromolecule, stimulates the cells subsequently to divide.

Even though the mechanism of action of growth factors is unknown, their discovery clearly represents a significant advance, which gives reason to hope that we may someday be able to decipher the signals which guide cells through their growth cycles.

Altered Cell Surfaces

The striking changes in the properties of a cell's surface that accompany transformation have been discussed at length in Chapter 3. Suffice it here to reiterate that these changes include the appearance of new antigens, as well as the uncovering or renewed synthesis of fetal antigens, and the uncovering or redistribution of receptor molecules that bind lectins. Experiments with hamster cells transformed by the *ts3* mutant of polyoma virus prove that some of these surface alterations are controlled by a viral gene (see Chapter 7).

Since only the outer surface of a cell is in direct contact with the cell's environment, it seems reasonable to anticipate that those changes that occur at the cell surface on transformation may, at least in part, be the cause of the patterns of growth and social behaviour which distinguish transformed from untransformed cells. Although this belief motivates much of the current research into the physiology, chemistry and structure of cell surface membranes, no one has yet proven that any of the known changes that occur at the cell surface on transformation actually cause cells to have different growth properties and different social behaviour.

Assays for Transformation

In order to define how transformants differ from their normal counterparts, it is clearly essential to have available cloned lines of both sorts of cells. Polyoma virus and SV40 transform cells of many different species. Whether any given type of cell becomes transformed or whether it supports a lytic infection depends chiefly upon its genetic properties (see Table 6.2); other parameters, such as the multiplicity of infection and the physiological state of the cells, are of comparatively minor importance. Cells of the natural host species of these two viruses, mouse cells and monkey cells respectively, are fully permissive; in such permissive cells the genome of the virus is fully expressed, the viral DNA is replicated, progeny virus particles are released and the cell is killed. By contrast, completely nonpermissive cells survive an

Table 6.2. Response of Cells of Different Species to Infection by SV40 and Polyoma Virus

Species	Infection by SV40	Infection by Polyoma Virus
Human	Semipermissive	NT*
Mouse	Nonpermissive	Permissive
Monkey	Permissive	NT
Rat	Non- or semipermissive	Semipermissive
Guinea pig	Non- or semipermissive	NT
Rabbit	Non- or semipermissive	NT
Cow	Non- or semipermissive	NT
Hamster	Semipermissive	Semipermissive

* NT = not tested.

infection and progeny virus particles are never released, but some viral genes are expressed at least transiently.

Between these two extremes lie cells which are semipermissive for polyoma virus or SV40. When populations of such semipermissive cells are infected, some of the cells support the replication of the virus and die whereas others provide a nonpermissive environment and so survive the infection. We do not, at present, know what causes a cell to be nonpermissive or permissive. However, Basilico et al. (1970) showed that the yield of polyoma virus particles obtained from infected mouse (permissive)/hamster (nonpermissive) hybrid cells was positively correlated with the number of murine chromosomes in these hybrids. This experiment strongly suggests that mouse cells are permissive for polyoma virus because some mouse chromosomal gene specifies a factor(s) essential for replication of the virus. If this is the case, it seems reasonable to suggest that nonpermissive cells lack the ability to make such a factor(s).

Even though nonpermissive cells do not yield virus after infection, some virus genes are expressed. For example, 3T3 cells are totally nonpermissive for SV40, but after they are infected by this virus, they acquire SV40-specific T antigen. Furthermore, after high multiplicity infection by SV40, 3T3 cells replicate their DNA and go through one or more rounds of mitosis (Smith et al., 1971) and both early and late species of SV40 RNAs are synthesized (Khoury et al., 1972). The SV40 DNA, however, is never replicated, and neither SV40 capsid (V) antigen nor progeny particles can ever be detected. These findings suggest that the factor lacking in nonpermissive cells acts selectively at some post-transcriptional level, perhaps during the processing or transport from the nucleus of viral RNAs, or at the level of translation.

Presumably semipermissive cells for one reason or another do not continuously supply adequate amounts of the factor(s) required for replication of these viruses. We can further speculate that at some low frequency, and at random, individual nonpermissive cells in a population make enough of the factor(s) to support replication of the virus and as a result, if they are infected, eventually die. It now seems clear that populations of BHK cells are semipermissive rather than absolutely nonpermissive for polyoma virus because (1) Fraser and Gharpure (1963) found that a small proportion of the cells in populations of BHK cells infected with polyoma virus stained with antibody against polyoma virus; (2) Bourgaux (1964) showed that some BHK cells infected with polyoma virus incorporated [^{32}P]orthophosphate into material which cosedimented with polyoma virus; (3) Folk (1973) has shown that as many as one percent of the cells in some clones of BHK cells transformed by polyoma virus yield progeny virus when they are chilled from 39°C to 31°C.

Human cells are the classic example of cells semipermissive for SV40. Butel et al. (1972) for example, in a valuable review of transformation by SV40, cite the following: One transforming unit of SV40 assayed on 3T3 cells corresponds to 10^3 infectious units assayed in monkey cells and to 10^5 physical particles. Concentrations of SV40 which transform ten percent of a population of 3T3 cells transform only 0.03 percent of a population of human diploid cells. And spontaneous production of small amounts of infectious virus can be detected in populations of SV40-transformed human cells (Dubbs and Kit, 1971). Syrian hamster cells are also semipermissive for SV40; transformants can be obtained which yield SV40 on fusion with monkey cells and which can be induced to yield SV40 after exposure to agents such as mitomycin C (Burns and Black, 1968, 1969), bromodeoxyuridine, caffeine and ultraviolet or X-irradiation (Rothschild and Black, 1970) and cycloheximide, or after amino acid deprivation (Kaplan et al., 1972).

Some rabbit cells (Black and Rowe, 1963), rat cells (Diderholm et al., 1966), bovine cells (Diderholm et al., 1965) and some guinea pig cells (Diderholm et al., 1966) can be transformed by SV40. Although these systems have not been thoroughly characterized, it seems likely that all these sorts and species of cells are semipermissive for SV40.

In summary, the outcome of infection by wild-type SV40 and polyoma virus depends first and foremost on the genetic properties (and usually, therefore, species) of the host cell. Permissive cells are killed; nonpermissive cells survive, but they may become stably transformed and acquire a subset of the set of properties listed in Table 6.1. Clearly, if permissive cells are infected with defective mutant viruses, which lack functions essential for replication, they may survive and be transformed. So as a general rule it is safe to say that permissive cells are transformed by defective viruses, whereas nonpermissive cells are transformed by wild-type viruses. But exceptions to those rules are known.

Cells transformed by polyoma virus or SV40 can be obtained by four methods, three of which (methods 2, 3 and 4 below) are selective while the other (method 1 below) is nonselective. The three selective methods select for different characters but they, and the nonselective method, all yield cells which usually grow to high saturation densities, contain viral nucleic acids and have permanently exposed agglutinin receptor sites. The four methods can be summarized as follows:

1. Confluent monolayers or cells in suspension are exposed to the virus and then plated at low cell densities. This method imposes no selection because both the untransformed and the transformed cells form colonies which can be distinguished by their density and morphology (Stoker and Macpherson, 1961; Todaro and Green, 1964a). This method was originally devised for obtaining BHK cells transformed by polyoma virus but it has since become the procedure most commonly used for obtaining 3T3 cells transformed by SV40.

2. Cells in suspension are infected with the virus and then plated in 0.33% agar or 1.2% Methocel. Transformed cells are able to form

large colonies whereas untransformed cells are unable to multiply. This technique has superseded the first method for the selection of BHK or NIL-2 cells transformed by polyoma virus; for these two virus-cell combinations either method detects transformants with the same efficiency but the second method is much simpler (Macpherson and Montagnier, 1964).

3. Subconfluent monolayers of cells are exposed to the virus; after 2–3 weeks without replating, transformed clones can be selected as dense clones of cells growing out from the monolayer of untransformed cells.

4. Confluent cultures of resting cells are infected with virus and plated at low concentrations in medium lacking serum growth factor(s). Many cells are induced by virus infection to multiply for a few generations (abortive transformation), but they do not survive transfer into fresh medium lacking serum factor(s). The cells which are stably transformed by the virus acquire the ability to multiply continuously in factor-free medium. Most of these cells show most if not all of the properties typical of virus-transformed cells (see Table 6.1). A minority of transformants selected in this way, however, while retaining the ability to multiply in medium lacking serum, grows only to low saturation densities in complete medium. Growth in serum-free medium, therefore, allows the selection of a class of viral transformants which would not be selected in the other assays. To date this method has been used only with Balb/c-3T3 cells infected with SV40 (Smith et al., 1970).

Parameters of the Transformation Event

In addition to the genetic properties of the cell, the multiplicity of infection, the genetic properties of the virus and the physiological state of the cells can affect the frequency of transformation.

Multiplicity of Infection

Transformation is an inefficient process. The frequency of transformation by the DNA tumour viruses varies greatly, but usually 10^4–10^5 infectious units (that is, 10^6–10^7 particles) are needed per transforming event (Stoker and Abel, 1962; Macpherson and Montagnier,

1964; Todaro and Green, 1966a). However, at low input multiplicities the number of cells transformed is directly proportional to the multiplicity of infection (Stoker and Abel, 1962; Macpherson and Montagnier, 1964; Todaro and Green, 1966a), which suggests that a single virus particle or DNA molecule is sufficient to initiate transformation.

With polyoma virus and BHK cells only a maximum of about 5 percent of the cells in a population ever become transformed at very high input multiplicities of virus. This low frequency of transformation is not the result of genetic inhomogeneity in the cell population since freshly isolated subclones show the same behaviour (Macpherson and Stoker, 1962; Black, 1964), although the genotype of a BHK cell has some influence on the probability of its being transformed by polyoma virus (Stoker and Smith, 1964). In the SV40-3T3 system up to 40 percent of the cells can be transformed (Todaro and Green, 1966a) if very high multiplicities of virus (10^6 infectious particles per cell) are used.

Physiological State of the Cells

There is no evidence that cells have to be in any particular physiological state in order to be transformed; experiments involving the infection of synchronized populations of cells have revealed only small differences in the rates of transformation of cells infected at different stages of their growth cycle (Basilico and Marin, 1966). Cells in log phase, however, are more susceptible to transformation than resting cells (Todaro and Green, 1966b), and cells surviving X-irradiation (Stoker, 1964) or exposure to the base analogues 5-bromodeoxyuridine or 5-iododeoxyuridine (Todaro and Green, 1964b) are transformed at increased rates. Furthermore, at least in the polyoma virus-BHK cell system, the probability of transformation may be affected by the pH and the concentration of Mg^{++} (Stoker and Abel, 1962; Kisch and Fraser, 1964).

Genetic Properties of the Host Cell

Human cells obtained from some persons with a high risk of developing cancer are ten- to fiftyfold more susceptible to transformation in vitro by SV40 than cells obtained from normal persons. Cells with increased transformation rates have been obtained from a family with a history of multiple cases of sarcoma, from patients with Fanconi's anemia, from heterozygous relatives of patients with Fanconi's anemia, and from patients with Down's syndrome (Todaro et al., 1966; Todaro and Martin, 1967; Potter et al., 1970). Moreover, Mukerjee et al. (1970) have recently found that cells from people suffering from Klinefelter's syndrome are also very susceptible to transformation by SV40. All these syndromes are characterized not only by a high incidence of chromosomal abnormalities, but also by a high incidence of cancer. (Skin fibroblasts from patients with primary immunodeficiencies, who also have a greater than normal risk of developing cancer, are not, however, more susceptible to transformation by SV40 than skin fibroblasts from healthy persons [Kersey et al., 1972].) The basis of the increased susceptibility to transformation is unknown; it seems to be concerned with an event after the adsorption of the virus, but before T antigen becomes detectable (Aaronson and Todaro, 1968). Aaronson (1970) has shown that the differences in transformation rates among human cell strains infected with intact SV40 particles are eliminated when SV40 DNA is used as transforming agent. Apparently, normal human cells are resistant to transformation by SV40 because of some block in either the penetration or uncoating of the virus, and this block, which is lacking in susceptible cells, seems to be specific for SV40. For example, there is no such block against the SV40 genome in adeno-SV40 hybrid virus particles.

Genetic Properties of the Virus

Not all the viral genes required for the replication of polyoma virus and SV40 are essential for transformation. At least some types of defective genomes which are either deleted (Uchida and Watanabe, cited in Yoshiike and Furuno, 1969) or functionally deficient (Basilico and di Mayorca, 1965; Benjamin, 1965; Latarjet et al., 1967; Altstein et al., 1967; Aaronson, 1970) can transform. It seems likely that both functionally complete (wild-type) and certain defective genomes can transform nonpermissive cells, but only defective genomes transform permissive cells.

EVENTS DURING TRANSFORMATION

Abortive Transformation of Nonpermissive Cells

Abortive transformation is the transient acquisition by a nonpermissive cell, which has been infected with polyoma virus or SV40, of the capacity to divide in restrictive conditions which inhibit the division of uninfected cells. For a few cell generations abortively and stably transformed cells have the same growth properties, but the abortively transformed cells rapidly resume the pattern of growth of uninfected, untransformed cells. The cells which are destined to be stably transformed emerge from the larger population of abortive transformants (Fox and Levine, 1971).

Abortive transformation was first observed in cultures of BHK cells infected with polyoma virus (Stoker, 1968) and subsequently in 3T3 cells infected with SV40 (Smith et al., 1970, 1971, 1972). Different methods are used to detect abortive transformation by SV40 and polyoma virus.

Stoker (1968) infected BHK21 cells in suspension with polyoma virus and plated them in Methocel. Whereas uninfected cells did not divide, many of the infected cells multiplied for about five or six generations and then stopped dividing. When these clones were picked and examined, they were normal by the two criteria used, the orientation of cell growth and the ability to grow in suspension. A small proportion of the infected cells went on to be stably transformed.

Abortive transformation by SV40 has been demonstrated using a selective medium which contains low concentrations of serum growth factor(s) (Smith et al., 1970, 1971). This medium allows 3T3 cells transformed by SV40 (SV3T3) to grow, but it restricts the multiplication of Balb/c-3T3 cells. After Balb/c-3T3 cells, growing in this medium, were infected with SV40, the cells were induced to synthesize DNA and to divide; most of the colonies tested were not stably transformed and grew only to low saturation density even in complete medium. One new type of transformant was isolated which grew well but only to low saturation densities in medium depleted of serum factor; its phenotype is very similar to the "flat" revertants of SV3T3 isolated by Pollack et al. (1968).

In short, in conditions of infection by polyoma virus and SV40 which cause a small proportion of cells to transform, a much larger proportion of the cells escape from growth control for a few generations. The mechanism of this change is unknown. However, there is little doubt that abortive transformation depends on virus gene function, because it is blocked by interferon (Dulbecco and Johnson, 1970; Taylor-Papadimitriou and Stoker, 1971) and it is prevented by UV irradiation of the virus (Smith et al., 1971). Stoker (1968) has speculated that abortively transformed cells contain viral DNA in a plasmid state and that stably transformed cells contain integrated viral DNA.

Biochemical Changes

Infection of 3T3 cells with SV40 results in a 3- to 8-fold increase in the specific activity of several enzymes concerned with DNA synthesis (for example, thymidine kinase, DNA polymerase, dCMP deaminase and dTMP kinase [Kit et al., 1967a]), and in this respect infections of nonpermissive cells resemble productive infections (see Chapter 5).

Since no differences have been detected between these induced enzymes and those present in uninfected cells (Hatanaka and Dulbecco, 1967; Kit, 1966), it seems probable that all of the enzymes induced in infected nonpermissive cells and stably transformed cells are coded by the cell. The mechanism of this enzyme induction is obscure; neither DNA synthesis nor cell division is required (Kit, 1966; Kit et al., 1966). Because, however, protein synthesis is necessary (Frearson et al., 1966), it seems likely that the enzymes are synthesized de novo.

Cellular DNA synthesis is induced both in 3T3 cells infected with SV40 and in nonpermissive, rat embryo cells infected with polyoma virus (Gershon et al., 1965; Sheinin, 1966; Gershon et al., 1966; Henry et al., 1966; Kit et al., 1967b; May et al., 1971). Attempts to detect viral DNA synthesis and capsid proteins have been unsuccessful (Todaro and Green, 1966a,b; Gershon et al., 1966; Henry et al., 1966). However, viral RNA can be detected in abortively infected cells (Khoury et al., 1972). Using the separated strands of SV40 DNA, Khoury et al.

estimate that 38 percent of the sequences of early strand and 62 percent of those of the late strand of SV40 DNA are transcribed into stable RNA species. These values, which are indistinguishable from those obtained when late lytic RNA is hybridized to the separated strands (see Chapter 5), are most unexpected in view of the absence of late viral functions in abortively infected cells. The result suggests that the block to the expression of late SV40 functions in 3T3 cells occurs at a step after transcription and either during processing or translation of the viral RNA.

Virus-specific Antigens

In conditions where 30 percent of 3T3 cells infected with SV40 will eventually be transformed, almost all of the cells contain a new antigen, the SV40-specific T antigen (see Chapter 5), in the nucleus a few hours after infection (Black, 1966). Gradually, and concomitant with cell division, T antigen disappears from those cells which do not become stably transformed (Oxman and Black, 1966) whereas it remains in those cells which become stably transformed. Likewise a T antigen specific for polyoma virus is synthesized during infections of hamster and rat embryo cells by polyoma virus (Fogel et al., 1967). New surface antigen(s), detected by immunofluorescence, appears on the surface of cells infected with polyoma virus (Irlin, 1967 and see Chapter 3); this antigen(s) is different from T antigen, seems to appear earlier than T antigen, and is probably specified by the cell (Häyry and Defendi, 1970).

Surface Changes

After infection with SV40, 3T3 cells become agglutinable by concanavalin A (Ben-Bassat et al., 1970; Sheppard et al., 1971). Infected cultures of dividing cells do not become agglutinable until after at least one cell division has occurred (Ben-Bassat et al., 1970), and after two or three further divisions the cell surface returns to normal. By contrast, resting cultures of nonpermissive cells become agglutinable about three days after infection with SV40 (Sheppard et al., 1971).

Genetic Changes after Abortive Transformation

Even though abortive transformation is a transient phenomenon, it can result in permanent genetic changes. Smith et al. (1972) have shown that at least some clones of 3T3 cells that had been abortively transformed at some distant time in their past still carry several copies of viral DNA. Three clones of 3T3 cells were examined which had been stimulated to divide as a result of infection with SV40. After the cells had resumed a normal morphology and were indistinguishable phenotypically from uninfected 3T3 cells, the DNA was extracted from them and was hybridized to ^{32}P-labeled SV40 DNA. Two of the three cell lines examined were found to contain between 3 and 10 copies of the viral DNA per diploid cell. The state of this viral DNA is unknown, but two lines of evidence suggest that it may not be as firmly associated with the cellular DNA as is the integrated viral DNA in stable SV40 transformants. First, three subclones derived from one of the cell lines each contained different amounts of SV40 DNA (4.8, 6.2, and 10 copies of viral DNA per diploid cell). Second, it was difficult to obtain reproducible results from different DNA preparations extracted from the same cell line. How the viral DNA is maintained and the mechanism by which expression of viral functions is suppressed as the cell returns to its untransformed state are unknown.

Fixation of Transformation

At least one round of cell division is necessary to "fix" the transformed state. Todaro and Green (1966b) infected 3T3 cells with SV40 and seeded the cells at different concentrations so that they underwent varying numbers of divisions before reaching their saturation density. They found that at least one cell division is required for the transformed state to become irreversibly fixed; infected cells which did not undergo one division after infection were not transformed. Using interferon the fixation event in the SV40-3T3 system has been further localized. Todaro and Green (1967) synchronized cultures of 3T3 cells, infected them with SV40 at different stages of the cell cycle and added interferon to block the expression of functions coded by the virus. They found that interferon added before the S period prevented transformation, but there was no effect if interferon was added after the S period. Interferon also blocks abortive transformation (Dulbecco and Johnson, 1970; Taylor-Papadimitriou and Stoker, 1971). The mechanism by which interferon prevents transformation is not known, but the fact that after one cell division transformation is insensitive to interferon

(Todaro and Green, 1966b) suggests that the fixation of the transformed state occurs either in the S or G_2 phases of the cell cycle (Basilico and Marin, 1966). The nature of the fixation event is unknown.

VIRUS GENES IN STABLY TRANSFORMED CELLS

Rescue of SV40 and Polyoma Virus

Once transformed cell lines have been established by cloning cells through antiserum, infectious virus cannot be isolated from the cells (Dulbecco and Vogt, 1960). However, it is known that at least one complete viral genome is present in many cells that are transformed by SV40 but are free of detectable virus because SV40 can be recovered from the transformed cells by one of two methods:

(I) *Cocultivation with permissive cells.* When transformed cells are mixed with cells permissive to infection by the virus, SV40 can often be found in the culture a few days later (Gerber and Kirchstein, 1962). Contact between the transformed cells and the indicator cells is necessary (Gerber, 1966), and the yield of virus is increased and appears earlier if the two cells are fused artificially into heterokaryons with inactivated Sendai virus (Gerber, 1966; Koprowski et al., 1967; Watkins and Dulbecco, 1967; Tournier et al., 1967). It is not known whether, in the absence of Sendai virus, induction of SV40 depends on the rare spontaneous fusion of transformed cells with indicator cells. Usually less than 10 percent of any population of transformed cells is induced by fusion (Tournier et al., 1967; Watkins and Dulbecco, 1967; Koprowski et al., 1967), but pretreatment of transformed cells with base analogues before making heterokaryons increases the percentage of heterokaryons which yield virus (Watkins, 1970); this small proportion is not simply the result of genetic inhomogeneity because subclones of the transformed lines all yield virus in fusion experiments, albeit at varying frequencies (Watkins and Dulbecco, 1967). The mechanism of virus rescue is not understood. The permissive cells may supply a factor(s) necessary for multiplication of the virus which might act at any stage of the expression of the viral genome or affect the replication of viral DNA; alternatively, these putative factors may perform more specific functions (for example, the excision of viral DNA from cell DNA, or the supply of some factor regulating transcription). According to Wever et al. (1970) when SV40 virus is rescued from transformed hamster cells by fusing them with African green monkey kidney cells (CV1), viral capsids can first be detected in the nucleus of the transformed hamster cell in the heterokaryon. Rescue does not, therefore, appear to depend on the transfer of viral DNA from the transformed to the permissive cell nucleus, but the small number of transformed cells induced at any one time makes further biochemical analysis very difficult. Some exceptional lines of cells transformed by SV40 exist in which virus production is induced not only when these cells are fused with permissive cells, but also when the cells are treated with mitomycin C (Burns and Black, 1969; Gerber, 1964), proflavin or hydrogen peroxide (Gerber, 1964). The reason for this unusual behaviour is not known.

Not all lines of cells transformed by SV40 yield virus when fused with permissive cells. When, however, cells of a pair of transformed lines, neither of which yield virus when fused alone with permissive cells, are simultaneously fused to a permissive cell (generating a troika), wild-type virus can be rescued (Knowles et al., 1968). It is supposed that the original, uninducible transformed cells contain defective viral genomes; presumably these defective genomes are induced by fusion with permissive cells and they then complement and recombine in the permissive environment to yield wild-type virus.

So far it has proved impossible to rescue virus from some SV40-transformed cell lines even though viral DNA and SV40-specific T antigen can be detected (Melnick et al., 1964). The reason for this is not known, but it does not appear to be related either to the amount of SV40 DNA in the transformed cells or to the extent of transcription of viral DNA (Ozanne et al., 1973).

By contrast with most SV40-transformed cells, cells transformed by polyoma virus do not usually yield virus after fusion with permissive cells (Watkins and Dulbecco, 1967). Three exceptions have been noted:

(a) Polyoma virus can be rescued by fusion from a line of rat embryo cells transformed by the large-plaque strain of polyoma virus (Fogel and Sachs, 1969). Ninety-two out of 100 clonal isolates of this line show a small amount of spontaneous virus production in about one in 10^4 cells. The spontaneous yield is increased 100- to 200-fold by X-irradiation (Fogel and Sachs, 1969) or UV-irradiation (Fogel and Sachs, 1970) and 700- to 2000-fold by treatment with mitomycin C (Fogel and Sachs, 1970).

(b) Virus can be rescued from BHK cells transformed by the *ts-a* and *ts-c* temperature-sensitive mutants of polyoma virus (Summers and Vogt, 1971; Folk, 1973; Fried, unpublished data). The possible reasons for this result are discussed in Chapter 7.

(c) Virus can be rescued occasionally from a few lines of BHK cells transformed by wild-type polyoma virus (Folk, 1973). Why these few lines of cells transformed by polyoma virus should behave differently from the majority is still a matter for speculation. One possibility is that BHK cells which are commonly used for transformation by polyoma virus are partly permissive, so that transformation is most often performed by defective virus mutants which cannot multiply and therefore cannot be rescued from the stable transformants. It may be that those PyBHK cell lines from which wild-type polyoma virus can be rescued arise as a consequence of infection of a small class of BHK cells, which for unknown reasons are truly nonpermissive.

II. *Extraction of infectious SV40 DNA from transformed cells.* When DNA extracted from virus-free SV40-transformed hamster, mouse or monkey cells is introduced into permissive simian cells, infectious SV40 is produced by the recipient cells (Boyd and Butel, 1972). Large amounts of high molecular weight DNA are necessary and DEAE-dextran is required to promote its uptake—two obligatory conditions which probably explain the failure of previous workers to achieve virus rescue. The DNA transfer method seems to be a more efficient method of assaying for the presence of intact viral genomes in transformed cells than cell fusion because infectious SV40 DNA can be recovered from some cell lines that fail to yield virus by heterokaryon formation.

The mechanism by which the DNA transfer method succeeds in rescuing SV40 virus is obscure. Presumably the first stage involves excision of the viral sequences from adjacent cellular DNA. Whether this step involves specific cleavage enzymes or whether it occurs as a consequence of random nicking of the transformed cell DNA is not known, but the efficiency of the process argues strongly for the first possibility. Presumably once the viral DNA is cut loose from the shackles that bind it to its integration site, it is free to undergo a cycle of replication in the permissive cell (see Chapter 5).

Viral DNA in Transformed Cells

It is clear that SV40 and polyoma virus DNA must persist in those lines of transformed cells from which virus can be rescued. However, direct physical evidence of the presence of viral DNA, both in these cells and in the lines of transformed cells which do not yield virus on cell fusion, has been very hard to obtain chiefly because one molecule of viral DNA weighs only about one-two-millionths of the weight of the DNA of a cell. Such minute amounts of viral DNA have, however, recently been detected by two methods, both involving nucleic acid hybridization.

DNA/RNA Hybridization

Westphal and Dulbecco (1968), Sambrook et al. (1968), Tai and O'Brien (1969), Westphal and Kiehn (1970), Hirai and Defendi (1971) and Levine et al. (1970) have all attempted to detect viral DNA in transformed cells by measuring the hybridization between virus-specific RNA and DNA from transformed cells. Highly radioactive, virus-specific RNA was synthesized in vitro with *E. coli* RNA polymerase using highly purified, component I viral DNA (see Chapters 4 and 5) as template. The RNA was hybridized with DNA extracted from transformed and untransformed cells, utilizing the Gillespie and Spiegelman technique (1965). Whereas about 10^{-3} of the input RNA hybridized with DNA from untransformed cells, the amount of RNA which hybridized with DNA from transformed cells was at least 3- to 8-fold greater. This increased hybridization is virus-specific; DNA from cells transformed by polyoma virus showed increased hybridization only with polyoma virus-specific RNA and not with SV40-specific RNA, and vice-versa. With reconstruction experiments Westphal and Dulbecco (1968) were able to show that there were multiple genome equivalents of viral DNA in each cell, ranging from about five copies in some 3T3 cells transformed by polyoma virus to about 60 copies in one line of hamster cells transformed by SV40. Tai and O'Brien (1969) found about 60 copies of SV40 DNA in transformed hamster cells while Levine et al. (1970) found between two and nine copies of SV40 DNA per transformed cell.

Although there is little doubt that DNA/RNA hybridization specifically detects viral sequences in transformed cells, recent work suggests that the technique seriously overestimates the number of viral DNA copies in each transformed cell. Under certain conditions, SV40 DNA/RNA hybrids are selectively lost from nitrocellulose filters (Haas et al., 1972) so that the reconstruction experiments that calibrate the technique used to measure the number of viral genomes in transformed cells suffer from a systematic error. When the error is corrected the number of virus genome equivalents per transformed cell seems to be about 5-fold lower than the original estimates (Westphal, unpublished results) and in good agreement with results obtained by the second hybridization method.

DNA/DNA Hybridization

This method, developed by Gelb et al. (1971), takes advantage of the fact established by Britten and Kohne (1968) and Wetmur and Davidson (1968) that the rate of reannealing in solution of any sequence of denatured DNA is proportional to its concentration. It is possible therefore to determine directly the amount of viral DNA in the genomes of transformed cells by following the rate of reannealing of small amounts of viral DNA in the presence of DNA from transformed cells. Gelb et al. (1971) took highly radioactive denatured SV40 DNA and allowed it to reanneal in the presence of denatured DNA from a variety of uninfected, untransformed and SV40-transformed cell lines. The rate of reannealing of the labeled viral DNA was faster in the presence of transformed cell DNA than in the presence of DNA from untransformed cells. Gelb et al. were able to calculate that the SV40-transformed cell lines that they examined contained between one and three copies of SV40 DNA. Other cell lines examined since then have been found to contain up to nine copies of viral DNA (Ozanne et al., 1973).

Surprisingly, Gelb et al. (1971) also reported the presence of about 0.5 copies of SV40 DNA in untransformed cells. Although the reason for this finding is still not certain, it now seems most likely that the highly radioactive SV40 DNA that was used as a probe in the hybridization test contained some host DNA sequences (Tai et al., 1972; Lavi and Winocour, 1972).

Viral DNA Integrated into Cell DNA

The state of the viral DNA in transformed cells was examined by Sambrook et al. (1968) using the DNA/RNA hybridization technique. They found that the ability of cellular DNA to hybridize with SV40-specific RNA remained associated with the high molecular weight fraction after zonal and equilibrium sedimentation in alkaline gradients, in caesium chloride gradients containing ethidium bromide and during Hirt (1967) extractions of DNA from transformed cells. From these results they concluded that SV40 DNA is integrated into the cellular DNA at least in the line of SV3T3 cells examined. The same conclusion has been reached by Hirai and Defendi (1971) for a line of Chinese hamster cells transformed by SV40 and by Green et al. (1970) for cells transformed by adenovirus 2. It is not known whether the viral DNAs are integrated individually or in tandem in one chromosome or many.

Weiss and her colleagues (Weiss et al., 1968; Weiss, 1970) exploited the fact that human chromosomes are shed preferentially from hybrids made by fusing normal mouse cells with SV40-transformed human cells in an attempt to determine the number of human chromosomes into which viral DNA is integrated during transformation. Such hybrids lose SV40 T antigen only after they have lost virtually all the human chromosomes. At first sight this result suggests that there are many SV40 genomes integrated into many human chromosomes, but an equally plausible alternative cannot be ruled out. The selection pressures which lead to the loss of human chromosomes are not known. The SV40 genome may be integrated in only one or a few human chromosomes which, for other reasons, are retained in hybrid cells until most of the human chromosome complement has been shed. An unambiguous answer to this question is unlikely to come from this cytogenetic approach but will probably depend on the isolation of individual chromosomes which can be chemically analyzed for the presence of viral DNA, or on hybridization experiments using viral DNA or RNA of extremely high specific activity (Gall and Pardue, 1969).

Transcription of Viral DNA in Transformed Cells

Virus-specific RNA has been found in cells transformed by polyoma virus (Benjamin, 1966; Martin and Axelrod, 1969a) and by SV40 (Aloni et al., 1968; Oda and Dulbecco, 1968; Sauer and Kidwai, 1968; Martin and Axelrod, 1969b; Tonegawa et al., 1970; Darnell et al., 1970; Wall and Darnell, 1971; Sauer, 1971; Martin, 1970; Ozanne et al., 1973; Khoury et al., unpublished data). Three types of experiments have been performed aimed at defining the sequences of SV40 DNA that are transcribed in transformed cells.

Hybridization competition experiments have been done by Aloni et al. (1968), Sauer and Kidwai (1968), Oda and Dulbecco (1968), Tonegawa et al. (1970) and Sauer (1971). All four groups find that, in hybridization tests with viral DNA, the viral RNA extracted from transformed cells can compete with about 30–40 percent of the RNA found late in lytic infection. This result implies that the transcription of viral DNA in transformed cells is controlled, and it is compatible with the observation that late viral functions (synthesis of free viral DNA and structural proteins) are not expressed.

Saturation hybridization experiments have been done by Martin and Axelrod (1969b) who grew permissive cells for several days in ^{32}P, in otherwise phosphate-free medium, before infecting with SV40. At various times after infection, they isolated RNA and hybridized it to SV40 DNA on nitrocellulose filters. They assumed that the long labeling period had ensured steady-state labeling, so that virus-specific RNA would have the same specific activity as the total RNA in the cells. After estimating the specific activity of the RNA, they showed that virus-specific RNA present late in infection saturated 50 percent (equivalent to one strand) of SV40 DNA (see Chapter 5). Using a similar labeling regimen with transformed cell lines, they showed that an excess of steady-state, labeled RNAs would saturate from 15–50 percent of the sequences in denatured SV40 DNA.

Martin and Axelrod, on the basis of these results, suggested that all of the virus-specific RNA sequences present late in lytic infection may also be present in some lines of transformed cells, so that control of expression of viral genes is at a post-transcriptional level.

Hybridization against separated strands of SV40 DNA. Clearly the competition hybridization and the saturation hybridization gave different results, each with their own set of implications. This unsatisfactory state of affairs, which lasted for several years, was recently resolved when the separated strands of SV40 DNA became available (Khoury et al., 1972; Sambrook et al., 1972).

Ozanne et al. (1973) and Khoury et al. (unpublished data) have examined the percentage of the early and late strands of SV40 DNA that is transcribed in different transformed cells. The results of the two groups are in close agreement and can be summarized as follows:

1. Whereas none of the control cells contained RNA that hybridized to SV40 DNA, the RNA extracted from all of the cell lines transformed by SV40 showed virus specific sequences.
2. In different transformants, different percentages of the virus genome are transcribed. However, there does not seem to be any correlation between the extent of transcription and the number of virus genomes per cell or the ease with which they can be rescued.
3. Many of the transformed cell lines examined contain more RNA sequences complementary to the early strand (-ve) of SV40 DNA (see Chapter 5) than are ever expressed during lytic infection. In productively infected cells not more than 30 percent of the sequences of the early (-ve) strand appear in stable species of RNA. However, in transformed cells the percentage ranges from a low of about 30 percent to a high of over 80 percent. This result means that at least some of the viral sequences in some transformed cells are "anti-late," complementary to the RNA sequences made late in the lytic cycle. Presumably this RNA is non-informational so that it is slightly surprising to find it present in stable species. The reason for the different transcriptional pattern shown by different lines of transformed cells is unknown. However, one possibility is that the amount of transcription of the early strand in transformants is a reflection of the site of integration of the viral DNA. This can be stringently tested by mapping the orientation of integration of viral DNA in cell lines which show different amounts of transcription.

4. A few lines of transformed cells contain RNA transcribed from a small segment (4–20 percent) of the late strand of SV40 DNA. We do not know yet, however, whether this RNA contains sequences also found late in infected permissive cells rather than "anti-early" sequences.
5. In some, but not all revertants (cells which have resumed a normal phenotype) the concentration of viral RNA sequences is at least 5-fold less than in other transformed cell lines, and it is impossible to add enough RNA from these cells to reach saturation in the hybridization reaction. Whether this reduction in intracellular SV40 RNA concentration is a cause or a consequence of the cells' resuming a normal phenotype remains to be established.

From these results it seems likely that expression of the transformed phenotype requires the presence of RNA coded by the "early" strand of SV40 DNA. Although this RNA may be necessary for transformation, it is by no means sufficient because concanavalin A-resistant cells that exhibit a revertant phenotype contain high concentrations of RNA transcribed from the early strand of SV40 DNA.

Size of Virus-specific RNA in Transformed Cells

The size of SV40-specific RNA in transformed cells has been examined by Tonegawa et al. (1970), Darnell et al. (1970), Lindberg and Darnell (1970) and Wall and Darnell (1971). Both groups find in the cell nucleus virus-specific RNA molecules which sediment at 60–70 S and are longer than transcripts of the entire viral genome. Smaller virus-specific RNA molecules, sedimenting at about 20 S, occur in the cytoplasm. The nuclear RNA, which is probably a precursor of the cytoplasmic RNA (Tonegawa et al., 1970), has been resolved by acrylamide gel electrophoresis into several discrete peaks ranging in molecular weight from 1.2×10^6 to 3.0×10^6 daltons. The most plausible explanation for the existence of these giant molecules is that they result from initiation of transcription at a cell promoter, with read-through and termination within the integrated viral DNA or vice-versa (Wall and Darnell, 1971). Some, if not all, of the RNA molecules containing SV40 sequences are covalently linked to poly-adenylic acid residues (Weinberg et al., 1972). All these data are consistent with the following model (see Figure 6.2): In transformed

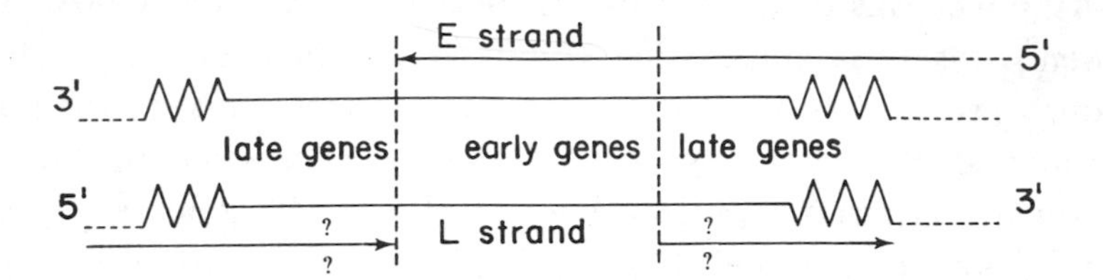

Figure 6.2. Diagram of a model of the transcription of integrated SV40 DNA. This model envisages that when SV40 DNA integrates into cell DNA the recombination event involves a breakage in the SV40 late genes, leaving intact the SV40 early genes. Transcription may occur from host promoters through host genes to viral terminators or vice-versa.

cells viral DNA is integrated in such a way that the functions necessary to maintain transformation are preserved intact—probably with a break somewhere in the late genes. Among different cell lines the position of the break point in the late genes may differ. This model explains the occurrence of large RNA transcripts, which contain both host and virus-specific sequences, and it preserves the integrity of the early genes including those responsible for maintaining transformation. Clearly, however, it is an oversimplification and does not take into account any post-transcriptional control events, either at the level of RNA processing or translation.

VIRUS-SPECIFIC PROTEINS IN TRANSFORMED CELLS

No proteins indisputably coded by the virus have been isolated from transformed cells in spite of many attempts. However, virus-specific antigens—the T (tumour) antigen located in the nucleus, the U antigen located on the nuclear membrane, and the surface antigens—are presented in transformed cells.

T Antigen

A new virus-specific antigen appears in the nuclei of cells transformed by polyoma virus, SV40 and the adenoviruses; this antigen is called the tumour antigen (T antigen, neoantigen, or the induced complement-fixing antigen, ICFA). T antigen also appears early after the infection of nonpermissive cells, and in productively infected cells it persists throughout the lytic cycle of the virus (see Chapter 5).

T antigen was discovered by Huebner and coworkers (1963, 1965) in adenovirus 12 tumours. They found that extracts of adenovirus 12 tumours gave a positive complement fixation reaction with the serum from animals bearing adenovirus 12 tumours. The same reaction was subsequently reported for cells transformed by SV40 (Black et al., 1963) and polyoma virus (Habel, 1965, 1966). The reaction occurs with cells transformed in vitro or in vivo. Apart from complement fixation, T antigen can be detected by either direct or indirect immunofluorescence tests, which have the advantage of revealing the cellular location of T antigen.

Properties of T Antigen

1. T antigen is a protein or mixture of proteins. Its appearance after infection is blocked by interferon (Oxman and Black, 1966), by puromycin and by cycloheximide (Kit et al., 1966; Gilden and Carp, 1966). The T antigen of SV40 has been partially purified; it is a heat-labile protein but virtually nothing is known of its chemistry, and estimates of its molecular weight range from 70,000 daltons to 300,000 daltons (Gilden et al., 1965; Kit et al., 1967c; Potter et al., 1969; Del Villano and Defendi, 1970).

2. T antigen is virus-specific. Cells transformed by polyoma virus react only with sera from animals bearing tumours induced by polyoma virus and not with sera from animals bearing tumours induced by SV40, and vice-versa (Black et al., 1963; Habel, 1965, 1966). Moreover, when polyoma virus infects 3T3 cells transformed by SV40, the polyoma T antigen is induced in cells which already have the SV40 T antigen (Todaro et al., 1965).

3. T antigen is found in the nucleus of cells transformed or infected by SV40 and polyoma virus (Black et al., 1963; Habel 1965, 1966). However, hamster cells transformed by the defective SV40 in stocks of PARA (adenovirus-SV40 hybrids) contain SV40-specific T antigen in the cytoplasm and not in the nucleus (Butel et al., 1969; Duff et al., 1970; Richardson and Butel, 1971).

4. T antigen does not appear to be part of the viral capsid. Anti-virion antibody does not react with tumour cells or transformed cells and anti-T antibody does not neutralize the infectivity of the virus

(Black et al., 1963; Rapp et al., 1964c; Pope and Rowe, 1964; Sabin and Koch, 1964).

5. T antigen is produced in all infected cells, irrespective of whether they are permissive or nonpermissive, and in transformed cells. In productively infected cells T antigen appears very early, from 6 to 8 hours after infection and before detectable amounts of viral DNA or capsids are made (Hatanaka and Dulbecco, 1966; Rapp et al., 1964a; Sabin and Koch, 1964; Black, 1966 and see Chapter 5).

6. Synthesis of T antigen is not blocked by inhibitors of DNA synthesis such as cytosine arabinoside or 5-fluorodeoxyuridine (Sabin and Koch, 1964; Rapp et al., 1964b; Rowe et al., 1965). Synthesis of T antigen is not, therefore, dependent on the replication of viral DNA.

7. The function of T antigen is unknown. Since most, if not all, transformed cells contain T antigen, its detection provides an easy method for determining whether or not a cell is transformed by a particular virus. It is not known whether T antigen is required for transformation or for the maintenance of the transformed state. We do know, however, that the synthesis of T antigen, even if necessary, is not sufficient for transformation. When 3T3 cells are infected with SV40, the number of cells in which T antigen is induced greatly exceeds the number of cells which become stably transformed (Black, 1966).

8. Although T antigens, found in infected and transformed cells, are virus-specific, there is no evidence that they are coded by the virus. Synthesis of T antigen appears to depend on the transcription and translation of some viral genes but not on the expression of the entire viral genome; three types of physically defective SV40 genomes all induce T antigen: (a) circular DNAs about 85 percent the mass of the complete SV40 genome (Uchida et al., 1968), (b) linear viral DNA (Winocour, private communication), and (c) the defective SV40 DNA in SV40-adenovirus hybrids (Huebner et al., 1964; and see Chapter 8).

In short, T antigen could be a virus-coded protein, a derepressed cell protein or a protein produced by the interaction of viral and cellular proteins.

U Antigen

The U antigen, which appears to be induced by SV40, is distinct from the T antigen; it is located at the nuclear membrane whereas T antigen is present throughout the nucleus (Lewis et al., 1969), and antisera which react with U antigen without cross-reacting with T antigen have been obtained (Lewis and Rowe, 1971). Moreover, U antigen is not inactivated by heating cells at 56°C whereas T antigen is inactivated at this temperature.

U antigen was first detected in human embryonic kidney cells, African green monkey kidney cells and rat embryo cells infected with the adeno-SV40 hybrid virus $Ad2^+ND_1$ (Lewis et al., 1969 and see Chapter 8). Although most sera from hamsters bearing tumours induced by SV40 reacted with the U antigen induced by $AD2^+ND_1$ in HEK cells, some hamster sera were found which failed to react with U antigen but reacted normally with T antigen (Lewis et al., 1969). Subsequently it was found that hamsters with tumours induced by SV40 often develop antibodies against T antigen long before antibodies against U antigen are formed. Eventually, however, the hamsters which initially do not contain antibodies against U antigen develop these antibodies (Rowe, personal comminication). In HEK and AGMK cells infected with $Ad2^+ND_1$, U antigen is first detectable at about the time that viral DNA synthesis is initiated; when DNA synthesis is blocked with 5-fluorodeoxyuridine or cytosine arabinoside, U antigen cannot be detected (Lewis et al., 1969). By contrast, in cells infected with wild-type SV40, U antigen appears early after infection and its appearance is not blocked by 5-fluorodeoxyuridine and cytosine arabinoside (Lewis and Rowe, 1971).

Sera specific for SV40 capsid (V antigen) from hamsters hyperimmune to the virus do not react with U antigen induced by $Ad2^+ND_1$; neither do sera from hamsters bearing tumours induced by adenoviruses 7, 12 or 18, polyoma virus, Rous sarcoma virus or sera against a variety of viruses react with U antigen (Crumpacker et al., 1971). Nothing is known about the nature or function of U antigen.

Virus-specific Transplantation Antigens

Transformed cells acquire new antigens on their surfaces which are detectable by immunological and biological tests (see Chapter 3).

The virus-specific transplantation antigen on cells transformed by polyoma virus was discovered by Habel (1961) and Sjögren and coworkers (1961). The corresponding antigen induced by SV40 was discovered in 1963 (Khera et al., 1963; Habel and Eddy, 1963; Koch and Sabin, 1963; Defendi, 1963). These workers observed that animals immunized with the virus were resistant to subsequent challenges with transplantable tumours produced by that virus. The tumour rejection mechanism appears to be mediated by sensitized immune lymphocytes and is highly specific; there is no cross reaction between cells transformed in vitro with SV40 or polyoma virus or between cells from tumours induced by these two viruses (Sjögren, 1965). Animals can be immunized with injections of X-irradiated tumour cells or with injections of cells transformed in culture. It is presumed, therefore, that injections of virus induce immunity because they transform some cells which, because they carry the virus-specific transplantation antigen, stimulate an immune response leading to their rejection and the rejection of any other cells bearing this transplantation antigen.

The same transplantation antigen is present in cells of different species transformed by the same virus (Sjögren, 1965). This was demonstrated by the rejection of transplantable tumours within a species if the animals had previously been immunized with transformed cells or tumour cells produced by the same virus but from a different species.

Two pieces of evidence suggest that TSTA is present on the surface of transformed cells. First, the rejection of tumour cells by immune animals depends on contact between sensitized lymphocytes and TSTA (Coggin et al., 1967). Second, newborn hamsters injected with cell membranes from SV40-induced tumour cells become immunologically tolerant to SV40 TSTA (Tevethia and Rapp, 1966), and adult hamsters can be immunized against SV40 tumour cells by injections of membranes from SV40 tumour cells (Coggin et al., 1969). SV40 TSTA is also apparently synthesized as the virus replicates in monkey cells, and its appearance is blocked by inhibitors of protein synthesis and actinomycin D but not by inhibitors of DNA synthesis (Girardi and Defendi, 1970).

The only unequivocal assay for the transplantation antigen is the tumour rejection test. Other methods such as agglutination tests, cytotoxicity tests and plant agglutinin tests also detect surface antigens peculiar to transformed cells, but the antigens detected with these tests are not, or not only, the transplantation antigen (Häyry and Defendi,

1970; Tevethia et al., 1968). Although the transplantation antigen is virus-specific, it is not known whether the antigen is coded by the virus, is a derepressed cellular protein, or is a cryptic component of the cell membrane of untransformed cells which is exposed by transformation. Whether individual cells transformed by SV40 or polyoma virus have unique antigens, as do cells transformed by chemical carcinogens, in addition to a common virus-specific transplantation antigen(s), remains to be seen.

Surface Antigen

A surface antigen not detectable on the surface of normal cells can be detected on the surface of transformed cells by immunofluorescence, mixed hemagglutination tests (Häyry and Defendi, 1968, 1969, 1970; Metzgar and Oleinick, 1968), cytotoxic antibodies (Hellström and Sjögren, 1965; Tevethia and Rapp, 1965) and colony inhibition tests (Hellström and Sjögren, 1965, 1967). The chemical identity of the surface antigen(s) is unknown but it is not coded by the virus (Häyry and Defendi, 1970) since it is exposed on the surface of normal cells after mild treatment with trypsin or chymotrypsin. The surface antigen is not, however, species-specific, for infection with a particular virus results in the exposure of cross-reacting antigens on the surfaces of cells of different species. Furthermore, spontaneously transformed cells not treated with trypsin, or cells transformed by polyoma virus, adenoviruses or Rous sarcoma virus, do not cross react with antisera against cells transformed by SV40.

The surface antigen(s) is an anomaly, at least the surface antigen studied by Defendi and Häyry; it is specified by the cell, not by the virus, but it is virus-specific and not species-specific. These properties of the surface antigen should be borne in mind when considering claims that the absence of species-specificity is a strong argument that a substance or function is viral rather than cellular.

The surface antigen is probably not responsible for the binding of wheat germ agglutinin because the agglutinin receptor site is not virus-specific. The nature of the surface antigen of SV40-transformed cells and the relationship between S antigen and TSTA is extensively discussed by Butel and her colleagues (1972) in their review article. However, it seems that the most straightforward interpretation of these confusing data is that infection with each particular virus exposes a particular preexisting antigen; trypsin, by contrast, exposes the complete set of these antigens.

TRANSFORMED PERMISSIVE CELLS

As we have already said, as a general rule permissive cells are transformed by defective viruses. Such transformants are not easy to isolate without manipulating the experimental conditions in ways that either increase the proportion of defective virus particles in the stock used or prevent the spread of progeny wild-type virus throughout the culture (Fernandes and Moorhead, 1965; Jensen and Koprowski, 1969; Rapp and Trulock, 1970; Barbanti-Brodano et al., 1970; Sauer and Hahn, 1970; Defendi, 1968; Shiroki and Shimojo, 1971). And because transformation of permissive cells is a rare event, it cannot be quantitated.

Using a stock of SV40 that had been heavily irradiated with ultraviolet light, Shiroki and Shimojo (1971) isolated lines of African green monkey kidney cells transformed by SV40 (SV-AGMK). These cells had the following properties: (1) They contained SV40 T antigen and SV40 RNA; (2) they did not contain capsid (V) antigen; (3) they did not yield SV40 when fused with uninfected AGMK cells, but they were susceptible to superinfection by SV40 and were killed as a consequence of it. This set of properties is exactly that expected of permissive cells transformed by a defective virus.

Knowles et al. (1968) also transformed AGMK cells with SV40 without taking any special measures to ensure a high proportion of defectives in the stock of SV40 they used. Their SV-AGMK transformants had curious properties. When cells from each of these lines were fused with fresh AGMK cells, SV40 was not rescued. When, however, two transformants from different lines were fused with an AGMK cell, SV40 was often rescued. Knowles et al. argued that this indicated that cells of each transformant line contained defective SV40 and that each line was transformed by a different defective SV40 genome. Rescue of virus by fusion of two different transformants with a fresh AGMK cell was, they suggested, the result of complementation of defective SV40 genomes within the heterokaryon. Curiously, however, if pairs of transformants of different lines were fused together, virus was not rescued, even though the transformants, and presumably therefore the heterokaryons, were susceptible to superinfection by wild-type SV40; and the virus rescued by fusion of two transformants with one fresh AGMK cell was apparently wild-type SV40.

Clearly, if, in populations of permissive cells, nonpermissive variants arise which can no longer support virus replication, these cells should be susceptible to transformation by wild-type virus. Shiroki and Shimojo (1971) have apparently isolated SV40 transformants of such variant, nonpermissive AGMK cells. These transformed cells contain SV40 T antigen and SV40 RNA, at least some yield SV40 virus when fused with normal AGMK cells, and as expected they cannot be superinfected by wild-type SV40.

Finally Koprowski et al. (1967) and Swetly et al. (1969) have reported isolating clones of AGMK cells transformed by SV40 that can be superinfected and killed by wild-type SV40 and that yield wild-type SV40 upon fusion with uninfected AGMK cells. Sauer and Kidwai (1968) reported competition hybridization experiments which indicated that about 80 percent of the sequences present in SV40 RNA made late in the lytic cycle were also present in the SV40-specific RNA made in these transformants. After the transformed cells had been passed for 30 months, however, only 35 percent of these sequences could still be detected (Sauer, 1971). Whether this change reflects loss of segments of the SV40 DNA during passage of these cells is unknown; attempts to rescue the viral genome from stocks of cells passaged for 30 months have not been made.

In summary, we can tentatively identify three classes of SV40-transformed monkey cells: cells transformed by defective virus, variant nonpermissive cells transformed by wild-type virus, and finally, a rare class of permissive cells apparently transformed by wild-type virus.

Lines of mouse cells transformed by polyoma virus have been isolated on several occasions (Vogt and Dulbecco, 1960; Hellström et al., 1962; Todaro and Green, 1965; Benjamin, 1970). Many of these cells resist superinfection by polyoma virus particles but can be superinfected by polyoma virus DNA, and the transforming viral genome cannot be rescued by fusion (Watkins and Dulbecco, 1967; Watkins, 1971). These properties strongly suggest that such transformants arise when permissive cells are infected by defective polyoma virus particles, and most stocks of polyoma virus and SV40 contain defective particles (see Chapter 4).

Virus-specific Repressors

The parallels between the biologies of SV40 and polyoma virus and lysogenic bacteriophages such as lambda may well be more apparent than real, but they have nevertheless led to searches for SV40-specific and polyoma virus-specific repressors akin to the c_1 repressor gene product of lambda. If, for example, the mechanism of transformation of permissive monkey cells by SV40 involved a diffusible repressor protein that functioned in a way similar to lambda repressor, a clone of such cells should, after transformation, be resistant to superinfection by SV40 particles and SV40 DNA because they contain the putative repressor. And on the face of things such cells might be expected not to yield SV40 after fusion with uninfected monkey cells. As we have mentioned, several lines of monkey cells transformed by SV40 have been isolated. Some of these cells do not apparently yield SV40 on fusion with uninfected monkey cells and some resist superinfection by SV40 virus and by SV40 DNA (Shiroki and Shimojo, 1971 and Butel et al., 1971b). One might, therefore, speculate that these cells contain a diffusible, SV40-specific repressor protein but there is a more plausible explanation. These SV40-transformed AGMK cells were not isolated from infected clones of cells known to be permissive for SV40 but from infected uncloned populations of cells, at least some of which were permissive for SV40. The transformants could therefore have arisen from rare variant monkey cells in the mass population, which had spontaneously lost the ability to support the replication of SV40 and were in fact nonpermissive.

But in 1968 Cassingena and his colleagues and followers (Cassingena and Tournier, 1968; Cassingena et al., 1969a, b; Suarez et al.,

1971) claimed that crude extracts of SV40- and polyoma virus-transformed cells contained specific repressor proteins because in the presence of polylysine or polyornithine the extracts inhibited plaque formation by these viruses. These reports have not been substantiated; indeed a compelling body of evidence has been marshalled together, all of which argues against the existence of specific SV40 and polyoma virus repressors which act in the manner of the lambda repressor. For example, several lines of SV40-transformed monkey cells can be superinfected by SV40 particles or, if not by SV40 particles, by SV40 DNA (Barbanti-Brodano et al., 1970; Butel et al., 1971a; Jensen and Koprowski, 1969; Rapp and Trulock, 1970; Swetly et al., 1969). Such cells should not exist if SV40 genomes in transformed cells specify an effective immunity repressor.

Basilico's group reached the same conclusion from a different sort of experiment. They had shown that hybrid cells obtained by fusing BHK cells with 3T3 cells are permissive for polyoma virus so long as the hybrids retain mouse chromosomes; this result indicates that BHK cells are probably nonpermissive for polyoma virus because they lack a factor required for virus replication that is present in 3T3 cells (Basilico and Wang, 1971). Using the same techniques they produced hybrids of polyoma-transformed BHK cells and 3T3 cells and showed that these cells could be superinfected by small plaque polyoma virus and were killed as the polyoma virus replicated. This result indicates that BHK cells transformed by polyoma virus do not contain a diffusible immunity repressor which specifically inhibits the replication of superinfecting polyoma virus.

In short, claims that SV40 and polyoma virus specify a repressor analogous to that of phage lambda have not been substantiated, but as Sambrook (1972) has emphasized none of these experiments have any bearing on the possibility that other sorts of repressors exist in cells transformed by SV40 or polyoma virus.

DOUBLE TRANSFORMATION OF CELLS

Polyoma virus and SV40 have similar biologies, but they are biochemically quite distinct viruses. It is not surprising, therefore, that an appropriate cell transformed by one of these viruses can be subsequently transformed by the other. Todaro and Green (1965), using colony morphology as a marker, detected the transformation by SV40 of 3T3 cells already transformed by polyoma virus, and these double transformants contained both SV40-specific and polyoma virus-specific T antigens. And Takemoto and Habel (1966) showed that cells of hamster tumours induced by SV40 could be transformed by polyoma virus to yield cells with the specific T antigens and tumour-specific transplantation antigens of both viruses.

Cells can also be doubly transformed by SV40 and a human adenovirus. Some hamster cells transformed in vitro or in vivo by stocks of PARA (defective SV40-adenovirus hybrids, see Chapter 8) contain both SV40-specific and adenovirus 7-specific T antigens (Rapp et al., 1966, 1969; Butel et al., 1971b; Richardson and Butel, 1971; Duff and Rapp, 1970) and both SV40-specific and adenovirus 7-specific RNAs (Levin et al., 1969). Furthermore, some cells transformed by stocks of PARA transcapsidated by adenovirus 12 (see Chapter 8) contain SV40-specific RNA, adenovirus 7-specific RNA and adenovirus 12-specific RNA.

Cells transformed by SV40 can be doubly transformed by the RNA tumour virus, murine sarcoma virus (Renger, 1972), and rat cells infected but not transformed by Rauscher leukemia virus have an enhanced susceptibility to transformation by SV40 (Rhim et al., 1971). All these data clearly indicate that different DNA-transforming viruses and DNA- and RNA-transforming viruses do not exclude one another by competition, perhaps because the DNA of different transforming viruses is integrated at different sites in the host genome.

REVERSION OF TRANSFORMATION

A cell's loss of growth control can be reversible. For example, by applying selection pressures to cells transformed by viruses, it is possible to obtain cell lines which exhibit growth control comparable to that of untransformed cells (Table 6.3). Such revertants have been selected by the following procedures:

1. *Negative selection of transformed cells at high density with*

Table 6.3. PROPERTIES OF REVERTANTS OF SV40- AND POLYOMA VIRUS-TRANSFORMED CELLS

Reversion System			Concomitant Reversion to				Evidence of Presence of Transforming Virus				Retransformed by	
Property selected against	Transformed line	Killing agent	Low density	Anchorage dependence	Increased serum requirement	Loss of lectin sites	Virus rescuable	DNA	RNA	T antigen	Same virus	Other virus
Growth to high density	SV3T3	FdU		+	–	+	10^{-6}*	+	+	+	–	MSV
	Py3T3	FdU		ND	ND	+	NP	ND	ND	+	–	ND
	SV3T3	Colchicine		ND	+	ND	10^{-3}	ND	+	+	–	ND
	PyBHK	Glutaraldehyde		+	ND	+	NP	ND	ND	+	–	ND
Growth at 39°C	SV3T3	FdU		ND	+	+	10^{-3}	ND	ND	+	–	MSV
Growth in agar suspension	PyBHK	BrdU-light	+		+	ND	NP	ND	ND	+	–	MSV
Growth in 1% serum	SV3T3	BrdU-light	+	ND		ND	10^{-6}	ND	+	+	–	ND
Exposure of lectin sites	SV3T3	Con A	+	ND	–		10^{-6}	+	+	+	–	–
Chromosomes carrying viral genome	PyBHK	Thioguanine	+	+	ND	+	NP	ND	ND	–	+	ND

NP = not possible with transformed parent. ND = not determined.
These data are drawn from references cited in the text and Pollack and Vogel (personal communication).
* Fraction of revertant cells yielding virus on fusion with permissive cells.

5-fluorodeoxyuridine. The drug 5-fluorodeoxyuridine prevents the synthesis of thymidylic acid and kills dividing cells. When cultures of transformed cells growing at a high cell density are exposed to 5-fluorodeoxyuridine, all the cells which are dividing are killed. The survivors are the cells which had stopped growing before the 5-fluorodeoxyuridine treatment, presumably because they maintained partial growth control and susceptibility to contact inhibition despite being transformed (Pollack et al., 1968; Culp and Black, 1972; McNutt et al., 1971). The "flat" revertants isolated by this procedure by Pollack et al. (1968) are less sensitive to agglutination by concanavalin A than SV3T3 cells (Pollack and Burger, 1969); they have more chromosomes than the parental SV3T3 cells (Pollack et al., 1970); they are less tumourigenic than SV3T3 cells (Pollack and Teebor, 1969); they contain the same amount of SV40 DNA per diploid quantity of cell DNA as the parental SV3T3 cells; finally, although these revertants contain some SV40 RNA, it is present in lower concentrations than in SV3T3 cells; and because of the difficulties in obtaining useful amounts of SV40-specific RNA from the "flat" revertants, the pattern of transcription of the viral DNA has yet to be compared with that in SV3T3 cells (Ozanne et al., 1973).

Revertant cells which grow to low saturation densities have also been isolated using colchicine as a killing agent (Vogel and Pollack, unpublished data) and by seeding transformed cells on monolayers of untransformed cells fixed with glutaraldehyde (Rabinowitz and Sachs, 1968). The rationale of this last selection procedure is obscure; it is not obvious why monolayers of glutaraldehyde-fixed untransformed cells should select for or induce revertant cells. However, about 3 percent of the transformed cells plated formed colonies and about 90 percent of these clones which grew up on fixed monolayers seemed to be revertants.

2. *Selection of anchorage-dependent PyBHK revertants with bromodeoxyuridine.* Dividing cells incorporate the thymidine analog BrdU and can be killed by irradiation with blue light. This method was used to kill polyoma-transformed BHK cells in suspension culture and hence to recover variants of transformed cells which do not grow in suspension (Wyke, 1971a,b).

3. *Selection for serum dependence.* 3T3 cells grow poorly in medium containing 1% serum instead of the usual 10% serum, whereas their transformed derivatives grow in 1% serum (Holley and Kiernan, 1968; Smith et al., 1971). When 3T3 cells transformed by SV40 were grown in 1% serum in the presence of BrdU and subsequently exposed to blue light, most of the cells were killed (Vogel and Pollack, unpublished data). The survivors were revertants which not only required high concentrations (10%) of serum to grow, but also grew to low saturation densities.

4. *Selection for resistance to concanavalin A.* Concanavalin A is more toxic for SV3T3 cells than for 3T3 cells (see Chapter 3). When populations of SV3T3 cells are cultured for a few generations in the presence of concanavalin A, most cells are killed. The survivors have the following properties: (1) They are neither killed nor agglutinated by concentrations of concanavalin A which kill or agglutinate SV3T3 cells; (2) the survivors grow to low saturation densities in media containing 1% serum and in media containing 10% serum; (3) the survivors contain the same amount of SV40 DNA per diploid quantity of cell DNA as SV3T3 cells, and the extent and pattern of transcription of the viral DNA is the same in the survivors as in SV3T3 cells (Ozanne and Sambrook, 1971; Ozanne et al., 1973; Culp and Black, 1972).

5. *Selection of hybrid transformed lines with reduced chromosome complements.* Marin and Littlefield (1968) fused cells of two strains of BHK21, one of which was resistant to 5-bromodeoxyuridine while the other was resistant to 6-thioguanine. The resulting hybrids, which were able to incorporate both drugs and were therefore sensitive to both, were transformed with polyoma virus. The transformed and sensitive hybrids were then selected for resistance to 6-thioguanine. The development of such resistance implies the loss, through aberrant mitoses, of the chromosome(s) specifying the pathway which determines sensitivity to thioguanine. Marin and Littlefield argued that by selecting for the loss of chromosomes in this way, the chromosomes carrying the polyoma genomes might be lost concomitant with the loss of the chromosomes determining sensitivity to the drug. Two of Marin and Littlefield's surviving clones proved to be revertants (Marin and Macpherson, 1969) that grew to low saturation densities and could not grow suspended in

Methocel. The revertants selected by the technique of Marin and Littlefield differ from those selected by the other methods; Marin and Littlefield's revertants have lost the viral genomes along with some cell chromosomes because they do not contain T antigen, whereas all the revertants selected by other methods still express some virus-specific characters, for example they contain T antigen and virus-specific RNA (Ozanne et al., 1973).

6. *Passage of transformed cells at high dilution* (Rabinowitz and Sachs, 1969a,b). It seems that nearly all cells of the line transformed by polyoma virus with which Rabinowitz and Sachs worked can be induced to revert transiently to normal phenotype by plating at high dilutions. Why this happens is not clear.

7. *Selection of a temperature-sensitive revertant SV3T3.* Using fluorodeoxyuridine as a killing agent for dividing cells, Renger and Basilico (1972) selected from populations of SV3T3 cells those which grow at 39°C to the low saturation density of untransformed 3T3 cells. Like SV3T3 cells these temperature-sensitive revertant/transformants form colonies when plated on top of monolayers of 3T3 cells at 32°C, but at 39°C the temperature-sensitive cells, but not the SV3T3 cells, lose this ability. The temperature-sensitive cells contain T antigen at both temperatures and they require less serum than 3T3 at both 32°C and 39°C. The number of chromosomes per cell at 32°C and 39°C is 68 and 67, respectively. And what is most interesting—SV40 rescued from these cells appears to be wild type. The simplest interpretation of these data is that these cells carry a temperature-sensitive mutation in a cell gene, the product of which is essential for the maintenance of transformation.

The phenotypic changes which allow the selection of revertants from populations of transformants could stem either from a mutation in the transforming viral DNA or from a mutation or some other change in the cell genome. If SV3T3 revertants contain mutated SV40 DNA, the SV40 virus that can be rescued from the revertants by fusion with monkey cells should be mutant rather than wild type. Complete SV40 can be rescued from some, but not all, revertants of SV40-transformed cells, and to date, the rescued virus has always proved to be wild type in its transforming ability and infectivity (Pollack et al., 1968; Ozanne and Sambrook, 1971; Culp et al., 1971; Culp and Black, 1972; Renger and Basilico, 1972). This suggests that at least in these revertants from which SV40 can be rescued, reversion is the result of some change in the cell genome rather than some change in the viral genome. Unfortunately however, interpretation of the rescue experiments cannot be rigorous because the revertants, as well as the parental transformants, contain more than one genome equivalent of SV40 DNA. It is possible, therefore, that the genome which is rescued is not the genome controlling transformation and reversion. Because of technical limitations it is not at present possible to resolve this ambiguity, but it would be interesting to know the properties of any SV40 that could be rescued from revertants, particularly the temperature-sensitive revertants of Renger and Basilico (1972), by infecting monkey cells with high molecular weight DNA from revertant cells (Boyd and Butel, 1972).

If transformed cells revert because of some change in the cell genome, we might expect that it would be impossible to retransform the revertants with the same virus that they already harbour, and for the revertants that have so far been examined this is the case (Pollack, unpublished data; Wyke, 1971a,b; Ozanne and Sambrook, 1971). The only exception is that reported by Marin and Littlefield (1968); the revertants they isolated by selecting for chromosome loss could be retransformed by polyoma virus. Why revertant cells which have not lost chromosomes cannot be retransformed by the virus they harbour is unknown. The block to retransformation might be trivial, a change at the cell surface which blocks virus receptor sites for example, or it might be more subtle. But whatever the mechanism of the blockage, this facet of the phenotype of revertants is predicted by the hypothesis that reversion stems from a change in the cell genome. And the fact that many revertant cells have more chromosomes (and DNA, Vogel et al., unpublished data) than the parental transformed cells (Pollack et al., 1970; Culp et al., 1971; Culp and Black, 1972; Scher and Nelson-Rees, 1971; Mondal et al., 1971; Ozanne, unpublished data; Vogel and Pollack, unpublished data) is also consistent with this hypothesis.

Indeed, based on the results of analysis of the chromosomal compositions of parental transformants, revertants and so called re-revertants (revertant cells which have resumed the transformed cell phenotype),

Rabinowitz and Sachs (1970), Pollack et al. (1970) and Hitosumachi et al. (1971) have suggested that as the chromosome composition of transformed cells alters, and therefore the dosage of particular genes alters, the phenotype of the cell changes. They envisage that the changes in a cell's behaviour which we label transformation, reversion and re-reversion, stem from interactions between a transforming viral gene function and various cell functions, the expression of which depends at least in part upon the cell gene dosage.

It is also pertinent to this discussion that reversion of the transformed cell phenotype is not restricted to SV40 and polyoma virus transformants. Revertants have been isolated from populations of cells transformed by chemical carcinogens and by RNA tumour viruses (Macpherson, 1965; Mondal et al., 1971; Fischinger et al., 1972; Nomura et al., 1972), and presumably the reversion of cells transformed by chemicals does not stem from mutations of a viral genome.

In short, although we do not yet understand the underlying biochemical mechanisms of reversion, further analysis of the phenomenon should shed new light on the mechanism of transformation and of tumourigenicity, for most revertant cells are less tumourigenic than their parental transformed cells (Pollack and Teebor, 1969; Mondal et al., 1971).

THE VIRAL GENOME AND MAINTENANCE OF TRANSFORMATION

Cells transformed by SV40 or polyoma virus contain at least part of the viral genome integrated into one or more of their chromosomes. The mere integration of all or part of one or more viral genomes into the cell genome is not, however, sufficient to cause a cell to acquire the pattern of growth and the social behaviour which define transformed cells. Both revertant cells and abortively transformed cells may contain all or part of the viral genome, and some of this viral DNA may be transcribed without the cell's being transformed. Obviously these findings raise the question, Is the continued presence and expression of the transforming virus genome at all necessary to maintain the transformed cell phenotype? The properties of those types of revertants obtained by selecting for chromosome loss suggest that reversion can occur because the cells have lost the chromosomes that carry the viral genome(s), but the only firm evidence that viral gene products are required continuously to keep a cell transformed comes from experiments with a mutant strain of polyoma, which is discussed in the next chapter (Dulbecco and Eckhart, 1970).

Literature Cited

AARONSON, S. A. 1970. Susceptibility of human cell strains to transformation by simian virus 40 and simian virus 40 deoxyribonucleic acid. J. Virol. *6:* 470.

AARONSON, S. A. and G. J. TODARO. 1968. SV40 T antigen induction and transformation in human fibroblast cell strains. Virology *36:* 254.

ABERCROMBIE, M. and J. E. M. HEAYSMAN. 1954. Observations on the social behaviour of cells in tissue culture: II. "Monolayering" of fibroblasts. Exp. Cell Res. *6:* 293.

ALFRED, L. J. and R. W. PUMPER. 1960. Biological synthesis of a growth factor for mammalian cells in tissue culture. Proc. Soc. Exp. Biol. Med. *103:* 688.

ALONI, Y., E. WINOCOUR and L. SACHS. 1968. Characterization of the simian virus 40-specific RNA in virus-yielding and transformed cells. J. Mol. Biol. *31:* 415.

ALTSTEIN, A. D., G. I. DEICHMAN, O. F. SARYCHEVA, N. N. DODONOVA, E. M. TSETLIN and N. N. VASSILIEVA. 1967. Oncogenic and transforming activity of hydroxylamine-inactivated SV40 virus. Virology *33:* 746.

AUSTIN, P. E., E. A. MCCULLOCH and J. E. TILL. 1971. Characterization of the factor in L-cell conditioned medium capable of stimulating colony formation by mouse marrow cells in culture. J. Cell. Physiol. *77:* 121.

BARBANTI-BRODANO, G., P. SWETLY and H. KOPROWSKI. 1970. Early events in the infection of permissive cells with SV40: Adsorption, penetration and uncoating. J. Virol. *6:* 78.

BASILICO, C. and G. DI MAYORCA. 1965. Radiation target size of the lytic and the transforming ability of a polyoma virus. Proc. Nat. Acad. Sci. *54:* 125.

BASILICO, C. and G. MARIN. 1966. Susceptibility of cells in different stages of the mitotic cycle to transformation by polyoma virus. Virology *28:* 429.

BASILICO, C. and R. WANG. 1971. Susceptibility to superinfection of hybrids between polyoma "transformed" BHK and "normal" 3T3 cells. Nature New Biol. *230:* 105.

BASILICO, C., Y. MATSUYA and H. GREEN. 1970. The interaction of polyoma virus with mouse-hamster somatic hybrid cells. Virology *41:* 295.

BEN-BASSAT, H., M. INBAR and L. SACHS. 1970. Requirement for cell replication after SV40 infection for a structural change of the cell surface membrane. Virology *40:* 854.

BENJAMIN, T. L. 1965. Relative target sizes for the inactivation of the transforming and reproductive abilities of polyoma virus. Proc. Nat. Acad. Sci. *54:* 121.

———. 1966. Virus-specific RNA in cells productively infected or transformed by polyoma virus. J. Mol. Biol. *16:* 359.

———. 1970. Host range mutants of polyoma virus. Proc. Nat. Acad. Sci. *67:* 394.

BLACK, P. H. 1964. Studies on the genetic susceptibility of cells to polyoma virus transformation. Virology *24:* 179.

———. 1966. Transformation of mouse cell line 3T3 by SV40: Dose response relationship and correlation with SV40 tumor antigen production. Virology *28:* 760.

BLACK, P. H. and W. P. ROWE. 1963. SV40-induced proliferation of tissue culture cells of rabbit, mouse and porcine origin. Proc. Soc. Exp. Biol. Med. *114:* 721.

BLACK, P. H., W. P. ROWE, H. C. TURNER and R. J. HUEBNER. 1963. A specific complement-fixing antigen present in SV40 tumor and transformed cells. Proc. Nat. Acad. Sci. *50:* 1148.

BOURGAUX, P. 1964. Multiplication of polyoma virus in cells of a continuous hamster line susceptible to transformation. Virology *24:* 120.

BOYD, V. A. L. and J. S. BUTEL. 1972. Demonstration of infectious DNA in transformed cells. I. Recovery of SV40 from yielder and nonyielder transformed cells. J. Virol. *10:* 399.

BRITTEN, R. J. and D. E. KOHNE. 1968. Repeated sequences in DNA. Science *161:* 529.

BURGER, M. and K. D. NOONAN. 1970. Restoration of normal growth by covering of agglutinin sites on tumour cell surfaces. Nature *228:* 512.

BÜRK, R. R. 1967. The detection and extraction of anomin, a growth inhibitor from non-tumor cells. Wistar Symp. Monogr. *7:* 39. Wistar Institute Press, Philadelphia.

BURNS, W. H. and P. H. BLACK. 1968. Analysis of SV40-induced transformation of hamster kidney tissue in vitro. V. Variability of virus recovery from cell clones inducible with mitomycin C and cell fusion. J. Virol. *2:* 600.

———. 1969. Analysis of SV40-induced transformation of hamster kidney tissue *in vitro:* VI. Characteristics of mitomycin C in induction. Virology *39:* 625.

BUTEL, J. S., M. J. GUENTZEL and F. RAPP. 1969. Variants of defective simian papovavirus 40 (PARA) characterized by cytoplasmic localization of simian papovavirus 40 tumour antigen. J. Virol. *4:* 632.

BUTEL, J. S., L. S. RICHARDSON and J. L. MELNICK. 1971a. Variation in properties of SV40-transformed simian cell lines detected by superinfection with SV40 and human adenoviruses. Virology *46:* 844.

BUTEL, J. S., S. S. TEVETHIA and M. NACHTIGAL. 1971b. Malignant transformation in vitro by "non-oncogenic" variants of defective SV40 (PARA). J. Immunol. *106:* 969.

BUTEL, J. S., S. S. TEVETHIA and J. L. MELNICK. 1972. Oncogenicity and cell transformation by papovavirus SV40: The role of the viral genome. Advan. Cancer Res. *15:* 1. Academic Press, New York.

CASSINGENA, R. and P. TOURNIER. 1968. Mise dan évidence d'un "répresseur" spécifique dans les cellules d'especes différentes transformées par le virus SV40. C. R. Acad. Sci. (Paris) *267:* 2251.

CASSINGENA, R., P. TOURNIER, S. ESTRADE and M. F. BOURALI. 1969a. Blocage de l'action du "répresseur" du virus SV40 par un facteur constitutif des cellules permissives pour ce virus. C. R. Acad. Sci. (Paris) *269:* 261.

CASSINGENA, R., P. TOURNIER, E. MAY, S. ESTRADE and M. F. BOURALI. 1969b. Synthese du "répresseur" du virus SV40 dan l'infection productive et abortive. C. R. Acad. Sci. (Paris) *268:* 2834.

CLARKE, G. D., M. G. P. STOKER, A. LUDLOW and M. THORNTON. 1970. Requirement of serum for DNA synthesis in BHK21 cells: Effects of density, suspension and virus transformation. Nature *227:* 798.

COGGIN, J. H., V. M. LARSON and M. R. HILLEMAN. 1967. Immunologic responses in hamsters to homologous tumour antigens measured in vivo and in vitro. Proc. Soc. Exp. Biol. Med. *124:* 1295.

COGGIN, J. H., L. H. ELROD, K. R. AMBROSE and N. G. ANDERSON. 1969. Induction of tumor-specific transplantation immunity in hamsters with cell fractions from adenovirus and SV40 tumor cells. Proc. Soc. Exp. Biol. Med. *132:* 328.

CRUMPACKER, C. S., P. H. HENRY, T. KAKEFUDA, M. J. LEVIN, A. M. LEWIS, JR. and W. P. ROWE. 1971. Studies of non-defective Ad2-SV40 hybrid viruses: III. Base composition, molecular weight and conformation of the Ad.$2^{+}ND_1$ genome. J. Virol. *7:* 352.

CULP, L. and P. BLACK. 1972. Contact-inhibited revertant cell lines isolated from SV40-transformed cells. III. Concanavalin A-selected revertant cells. J. Virol. *9:* 611.

CULP, L., W. GRIMES and P. BLACK. 1971. Contact-inhibited revertant cell lines isolated from SV40-transformed cells. I. Biologic, virologic and chemical properties. J. Cell Biol. *50:* 682.

DARNELL, J. E., G. N. PAGOULATOS, U. LINDBERG and R. BALINT. 1970. Studies on the relationship of mRNA to heterogeneous nuclear RNA in mammalian cells. Cold Spring Harbor Symp. Quant. Biol. *35:* 555.

DEFENDI, V. 1963. Effect of SV40 virus immunization on growth of transplantable SV40 and polyoma virus tumors in hamsters. Proc. Soc. Exp. Biol. Med. *113:* 12.

———. 1968. Studies with virally induced transplantation antigens. Transplantation *6:* 642.

DEL VILLANO, B. and V. DEFENDI. 1970. The preparation and use of an immonosorbent to study the molecular composition of the SV40 T antigen from hamster tumor cells. Bacteriol. Proc. Abstr. *222:* 188.

DIDERHOLM, H., R. BERG and T. WESSLÉN. 1966. Transformation of rat and guinea pig cells *in vitro* by SV40, and the transplantability of the transformed cells. Int. J. Cancer *1:* 139.

DIDERHOLM, H., B. STENKVIST, J. PONTÉN and T. WESSLÉN. 1965. Transformation of bovine cells *in vitro* after inoculation of simian virus 40 or its nucleic acid. Exp. Cell Res. *37:* 452.

DUBBS, D. R. and S. KIT. 1971. Spontaneous virus production by clonal lines of SV40-transformed cells and effects of superinfection by DNA from mutant SV40 strains. J. Virol. *8:* 430.

DUFF, R. and F. RAPP. 1970. Quantitative characteristics of the transformation of hamster cells by PARA (defective SV40)-adenovirus 7. J. Virol. *5:* 568.

DUFF, R., F. RAPP and J. S. BUTEL. 1970. Transformation of hamster cells by variants of PARA-adenovirus 7 able to induce SV40 tumour antigen in the cytoplasm. Virology *42:* 273.

DULBECCO, R. 1970. Topoinhibition and serum requirement of transformed and untransformed cells. Nature *227:* 802.

DULBECCO, R. and W. ECKHART. 1970. Temperature-dependent properties of cells transformed by a thermosensitive mutant of polyoma virus. Proc. Nat. Acad. Sci. *67:* 1775.

DULBECCO, R. and T. JOHNSON. 1970. Interferon-sensitivity of the enhanced incorporation of thymidine into cellular DNA induced by polyoma virus. Virology *42:* 368.

DULBECCO, R. and M. VOGT. 1960. Significance of continued virus production in tissue cultures rendered neoplastic by polyoma virus. Proc. Nat. Acad. Sci. *46:* 1617.

EAGLE, H., G. E. FOLEY, H. KOPROWSKI, H. LAZARUS, E. M. LEVINE and R. A. ADAMS. 1970. Maximum population density, inhibition by normal cells, serum requirement, growth in soft agar, and xenogeneic transplantability: Growth characteristics of virus-transformed cells. J. Exp. Med. *131:* 863.

FERNANDES, M. V. and P. S. MOORHEAD. 1965. Transformation of African green monkey kidney cultures infected with SV40. Tex. Rep. Biol. Med. *23:* 242.

FISCHINGER, P., S. NOMURA, P. PEEBLES, D. HAAPALA and R. BASSIN. 1972. Reversion of MSV-transformed mouse cells: Variants without a rescuable sarcoma virus. Science *176:* 1033.

FOGEL, M. and L. SACHS. 1969. The activation of virus synthesis in polyoma transformed cells. Virology *37:* 327.

FOGEL, M. and L. SACHS. 1970. Induction of virus synthesis in polyoma transformed cells by ultraviolet light and mitomycin C. Virology *40:* 174.

FOGEL, M., R. GILDEN and V. DEFENDI. 1967. Polyoma virus-induced "complement-fixing antigen" in tumors and infected cells as detected by immunofluorescence (31920). Proc. Soc. Exp. Biol. Med. *124:* 1047.

Folk, W. R. 1973. Induction of virus synthesis in polyoma transformed BHK-21 cells. J. Virol. In press.

Fox, T. O. and A. J. Levine. 1971. Relationship between virus-induced cellular DNA synthesis and transformation by SV40. J. Virol. *7:* 473.

Fraser, K. B. and M. Gharpure. 1963. Immunofluorescent tracing of polyoma virus in transformation experiments with BHK-21 cells. Virology *18:* 505.

Frearson, P. M., S. Kit and D. R. Dubbs. 1966. Induction of dihyrofolate reductase activity by SV40 and polyoma virus. Cancer Res. *26:* 1653.

Gall, J. G. and M. L. Pardue. 1969. Formation and detection of RNA–DNA hybrid molecules in cytological preparations. Proc. Nat. Acad. Sci. *63:* 378.

Gelb, L. D., D. E. Kohne and M. A. Martin. 1971. Quantitation of SV40 sequences in African green monkey, mouse and virus-transformed cell genomes. J. Mol. Biol. *57:* 129.

Gerber, P. 1964. Virogenic hamster tumor cells: Induction of virus synthesis. Science *145:* 833.

———. 1966. Studies on the transfer of subviral infectivity from SV40-induced hamster tumor cells to indicator cells. Virology *28:* 501.

Gerber, P. and R. L. Kirschstein. 1962. SV40-induced ependymomas in newborn hamsters: I. Virus-tumor relationships. Virology *18:* 582.

Gershon, D., P. Hausen, L. Sachs and E. Winocour. 1965. On the mechanism of polyoma virus-induced synthesis of cellular DNA. Proc. Nat. Acad. Sci. *54:* 1584.

Gershon, D., L. Sachs and E. Winocour. 1966. The induction of cellular DNA synthesis by simian virus 40 in contact-inhibited and in X-irradiated cells. Proc. Nat. Acad. Sci. *56:* 918.

Gilden, R. V. and R. I. Carp. 1966. Effects of cycloheximide and puromycin on synthesis of simian virus 40T antigen in green monkey kidney cell. J. Bacteriol. *91:* 1295.

Gilden, R. V., R. I. Carp, F. Taguchi and V. Defendi. 1965. The nature and localization of the SV40-induced complement-fixing antigen. Proc. Nat. Acad. Sci. *53:* 684.

Gillespie, D. and S. Spiegelman. 1965. A quantitative assay for DNA-RNA hybrids with DNA immobilized on a membrane. J. Mol. Biol. *12:* 829.

Girardi, A. J. and V. Defendi. 1970. Induction of SV40 transplantation antigen (TrAg) during the lytic cycle. Virology *42:* 688.

Green, M., J. T. Parsons, M. Piña, K. Fujinaga, H. Caffier and I. Landgraf-Leurs. 1970. Transcription of adenovirus genes in productively infected and in transformed cells. Cold Spring Harbor Symp. Quant. Biol. *35:* 803.

Haas, M., M. Vogt and R. Dulbecco. 1972. Loss of SV40 DNA-RNA hybrids from nitrocellulose membranes: Implications for the study of virus-host DNA interactions. Proc. Nat. Acad. Sci. *69:* 2160.

Habel, K. 1961. Resistance of polyoma virus immune animals to transplanted polyoma tumors (26453). Proc. Soc. Exp. Biol. Med. *106:* 722.

———. 1965. Specific complement-fixing antigens in polyoma tumors and transformed cells. Virology *25:* 55.

———. 1966. Virus tumor antigens: Specific fingerprints? Cancer Res. *26:* 2018.

Habel, K. and B. E. Eddy. 1963. Specificity of resistance to tumor challenge of polyoma and SV40 virus-immune hamsters (28259). Proc. Soc. Exp. Biol. Med. *113:* 1.

Hatanaka, M. and R. Dulbecco. 1966. Induction of DNA synthesis by SV40. Proc. Nat. Acad. Sci. *56:* 736.

———. 1967. SV40-specific thymidine kinase. Proc. Nat. Acad. Sci. *58:* 1888.

Häyry, P. and V. Defendi. 1968. Use of mixed haemagglutination technique in detection of virus-induced antigen(s) on SV40-transformed cell surface. Virology *36:* 317.

HÄYRY, P. and V. DEFENDI. 1969. Demonstration of specific antigen(s) on the surface of SV40-transformed cells using the mixed haemagglutination technique. Transplantation Proc. *1:* 119.

———. 1970. Surface antigen(s) of SV40-transformed tumor cells. Virology *41:* 22.

HELLSTRÖM, I. and H. O. SJÖGREN. 1965. Demonstration of H-2 isoantigens and polyoma specific tumor antigens by measuring colony formation *in vitro*. Exp. Cell Res. *40:* 212.

———. 1967. *In vitro* demonstration of humoral and cell-bound immunity against common specific transplantation antigen(s) of adenovirus 12-induced mouse and hamster tumors. J. Exp. Med. *125:* 1105.

HELLSTRÖM, I., K. E. HELLSTRÖM and H. O. SJÖGREN. 1962. Further studies on superinfection of polyoma-induced mouse tumors with polyoma virus in vitro. Virology *16:* 282.

HENRY, P., P. H. BLACK, M. N. OXMAN and S. M. WEISSMAN. 1966. Stimulation of DNA synthesis in mouse cell line 3T3 by simian virus 40. Proc. Nat. Acad. Sci. *56:* 1170.

HIRAI, K. and V. DEFENDI. 1971. Homology between SV40 DNA and DNA of normal and SV40-transformed Chinese hamster cells. Biochem. Biophys. Res. Comm. *42:* 714.

HIRT, B. 1967. Selective extraction of polyoma DNA from infected mouse cell cultures. J. Mol. Biol. *26:* 365.

HITOTSUMACHI, S., A. RABINOWITZ and L. SACHS. 1971. Chromosomal control of reversion in transformed cells. Nature *231:* 511.

HOLLEY, R. W. and J. A. KIERNAN. 1968. Contact inhibition of cell division in 3T3 cells. Proc. Nat. Acad. Sci. *60:* 300.

———. 1971. Studies of serum factors required by 3T3 and SV3T3 cells. In *Growth control in cell cultures* (CIBA Found. Symp.), ed. G. E. W. Wolstenholme and J. Knight. Churchill/Livingstone, London.

HUEBNER, R. J., W. P. ROWE, H. C. TURNER and W. T. LANE. 1963. Specific adenovirus complement-fixing antigens in virus-free hamster and rat tumors. Proc. Nat. Acad. Sci. *50:* 379.

HUEBNER, R. J., R. M. CHANOCK, B. A. RUBIN and M. J. CASEY. 1964. Induction by adenovirus type 7 of tumors in hamsters having the antigenic characteristics of SV40 virus. Proc. Nat. Acad. Sci. *52:* 1333.

HUEBNER, R. J., M. J. CASEY, R. M. CHANOCK and K. SCHELL. 1965. Tumours induced in hamsters by a strain of adenovirus type 3: Sharing of tumor antigens and "neoantigens" with those produced by adenovirus type 7 tumors. Proc. Nat. Acad. Sci. *54:* 381.

IRLIN, I. S. 1967. Immunofluorescent demonstration of a specific surface antigen in cells infected or transformed by polyoma virus. Virology *32:* 725.

JENSEN, F. C. and H. KOPROWSKI. 1969. Absence of repressor in SV40-induced transformed cells. Virology *37:* 687.

KAPLAN, J. C., S. M. WILBERT and P. H. BLACK. 1972. Analysis of SV40-induced transformation of hamster kidney tissue in vitro. VIII. Induction of infectious SV40 from virogenic transformed hamster cells by amino acid deprivation or cycloheximide treatment. J. Virol. *9:* 448.

KERSEY, J. H., R. A. GATTI, R. A. GOOD, S. A. AARONSON and G. J. TODARO. 1972. Susceptibility of cells from patients with primary immunodeficiency diseases to transformation by SV40. Proc. Nat. Acad. Sci. *69:* 980.

KHERA, K. S., A. ASHKENAZI, F. RAPP and J. L. MELNICK. 1963. Immunity in hamsters to cells transformed *in vitro* and *in vivo* by SV40. Tests for antigenic relationship among the papovaviruses. J. Immunol. *91:* 604.

KHOURY, G., J. C. BYRNE and M. A. MARTIN. 1972. Patterns of SV40 DNA transcription after acute infection of permissive and nonpermissive cells. Proc. Nat. Acad. Sci. *69:* 1925.

KISCH, A. L. and K. B. FRASER. 1964. The effect of pH on transformation of BHK21 cells by polyoma virus: I. Relationship between transformation rate and synthesis of viral antigen. Virology *24:* 186.

KIT, S. 1966. Induction of enzymes of DNA metabolism by simian virus 40. *In* (Y. Ito, ed.) Subviral carcinogenesis: 1st Int. Symp. Tumor Viruses. Nogoya, Japan.

KIT, S., D. R. DUBBS, P. M. FREARSON and J. L. MELNICK. 1966. Enzyme induction in SV40-infected green monkey kidney cultures. Virology *29:* 69.

KIT, S., L. J. PIEKARSKI and D. R. DUBBS. 1967a. DNA polymerase induced by simian virus 40. J. Gen. Virol *1:* 163.

KIT, S., R. A. DE TORRES, D. R. DUBBS and M. L. SALVI. 1967b. Induction of cellular deoxyribonuclei acid synthesis by simian virus 40. J. Virol. *1:* 738.

KIT, S., J. L. MELNICK, M. ANKEN, D. R. DUBBS, R. A. DE TORRES and T. KITAHARA. 1967c. Non-identity of some simian virus 40-induced enzymes with tumor antigen. J. Virol. *1:* 684.

KNOWLES, B. B., F. C. JENSEN, Z. S. STEPLEWSKI and H. KOPROWSKI. 1968. Rescue of infectious SV40 after fusion between different SV40-transformed cells. Proc. Nat. Acad. Sci. *61:* 42.

KOCH, M. A. and A. B. SABIN. 1963. Specificity of virus-induced resistence to transplantation of polyoma and SV40 tumors in adult hamsters. Proc. Soc. Exp. Biol. Med. *113:* 4.

KOPROWSKI, H., F. C. JENSEN and Z. S. STEPLEWSKI. 1967. Activation of production of infectious tumor virus SV40 in heterokaryon cultures. Proc. Nat. Acad. Sci. *58:* 127.

KRUSE, P. F. and E. MIEDEMA. 1965. Production and characterisation of multiple-layered populations of animal cells. J. Cell Biol. *27:* 273.

LATARJET, R., R. CRAMER and L. MONTAGNIER. 1967. Inactivation by UV-, X-, and γ-radiations, of the infecting and transforming capacities of polyoma virus. Virology *33:* 104.

LAVI, S. and E. WINOCOUR. 1972. Acquisition of sequences of homologous to host DNA by closed circular SV40 DNA. J. Virol. *9:* 309.

LEVIN, M. J., P. H. BLACK, S. L. COGHILL, C. B. DIXON and P. H. HENRY. 1969. *In vitro* transformation by the adenovirus-SV40 hybrid viruses. V. Virus-specific ribonucleic acid in cell lines transformed by the adenovirus 2-SV40 and adenovirus 12-SV40 transcapsidant hybrid viruses. J. Virol. *4:* 704.

LEVINE, A. S., M. N. OXMAN, P. H. HENRY, M. J. LEVIN, G. T. DIAMANDOPOULOS and J. F. ENDERS. 1970. Virus-specific deoxyribonucleic acid in simian virus 40-exposed hamster cells: Correlation with S and T antigens. J. Virol. *6:* 199.

LEWIS, A. M. and W. P. ROWE. 1971. Studies on non-defective Ad2-SV40 hybrid viruses: I. A newly characterized SV40 antigen induced by the $Ad2^+$ ND_1 genome. J. Virol. *7:* 189.

LEWIS, A. M., M. J. LEVIN, W. H. WEISE, C. S. CRUMPACKER and P. H. HENRY. 1969. A non-defective (competent) adenovirus-SV40 hybrid isolated from the Ad2-SV40 hybrid population. Proc. Nat. Acad. Sci. *63:* 1128.

LINDBERG, U. and J. E. DARNELL. 1970. SV40-specific RNA in the nucleus and polyribosomes of transformed cells. Proc. Nat. Acad. Sci. *65:* 1089.

LIPTON, A., I. KLINGER, D. PAUL and R. W. HOLLEY. 1971. Migration of mouse 3T3 fibroblasts in response to a serum factor. Proc. Nat. Acad. Sci. *68:* 2799.

LIPTON, A., D. PAUL, M. HENAHAN, I. KLINGER and R. W. HOLLEY. 1972. Serum requirements for survival of 3T3 and SV3T3 fibroblasts. Exp. Cell Res. *74:* 466.

MACPHERSON, I. 1965. Reversion in hamster cells transformed by Rous sarcoma virus. Science *143:* 1731.

MACPHERSON, I. and L. MONTAGNIER. 1964. Agar suspension culture for the selective assay of cells transformed by polyoma virus. Virology *23:* 291.

MACPHERSON, I. A. and M. G. P. STOKER. 1962. Polyoma transformation of hamster cell clones: An investigation of genetic factors affecting cell competence. Virology *16:* 147.

MARIN, G. and J. W. LITTLEFIELD. 1968. Selection of morphologically normal cell lines from polyoma-transformed BHK21/13 hamster fibroblasts. J. Virol. *2:* 69.

MARIN, G. and I. MACPHERSON. 1969. Reversion in polyoma-transformed cells: Retransformation, induced antigens and tumorigenicity. J. Virol. *3:* 146.

MARTIN, M. 1970. Characteristics of SV40 DNA transcription during lytic infection, abortive infection and in transformed mouse cells. Cold Spring Harbor Symp. Quant. Biol. *35:* 833.

MARTIN, M. A. and D. AXELROD. 1969a. Polyoma virus gene activity during lytic infection and in transformed animal cells. Science *164:* 68.

———. 1969b. SV40 gene activity during lytic infection and in a series of SV40 transformed mouse cells. Proc. Nat. Acad. Sci. *64:* 1203.

MAY, E., P. MAY and R. WEIL. 1971. Analysis of the events leading to SV40-induced chromosome replication and mitosis in primary mouse kidney cell cultures. Proc. Nat. Acad. Sci. *68:* 1208.

MCNUTT, N., L. CULP and P. BLACK. 1971. Contact-inhibited revertant cell lines isolated from SV40-transformed cells. II. Ultrastructural study. J. Cell Biol. *50:* 691.

MELNICK, J. L., K. S. KHERA and F. RAPP. 1964. Papova virus SV40: Failure to isolate infectious virus from transformed hamster cells synthesizing SV40-induced antigens. Virology *23:* 430.

METZGAR, R. S. and S. R. OLEINICK. 1968. The study of normal and malignant cell antigens by mixed agglutination. Cancer Res. *28:* 1366.

MONDAL, S., M. EMBLETON, H. MARQUARDT and C. HEIDELBERGER. 1971. Productions of variants of decreased malignancy and antigenicity from clones transformed *in vitro* by methylcholanthrene. Int. J. Cancer *8:* 410.

MUKERJEE, D., J. BOWEN and D. E. ANDERSON. 1970. Simian papovavirus 40 transformation of cells from cancer patient with XY/XXY mosaic Klinefelter's syndrome. Cancer Res. *30:* 1769.

NOMURA, C., P. FISCHINGER, C. MATTERN, P. PEEBLES, R. BASSIN and R. FRIEDMAN. 1972. Revertants of mouse cells transformed by murine sarcoma virus. J. Virol. *50:* 51.

ODA, K. and R. DULBECCO. 1968. Regulation of transcription of the SV40 DNA in productively infected and in transformed cells. Proc. Nat. Acad. Sci. *60:* 525.

OXMAN, M. and P. H. BLACK. 1966. Inhibition of SV40 T antigen formation by interferon. Proc. Nat. Acad. Sci. *55:* 1133.

OZANNE, B. and J. SAMBROOK. 1971. Isolation of lines of cells resistant to agglutination by concanavalin A from 3T3 cells transformed by SV40. Lepetit Colloq. Biol. Med. *2:* 248.

OZANNE, B., A. VOGEL, P. SHARP, W. KELLER and J. SAMBROOK. 1973. Transcription of SV40 DNA sequences in different transformed cell lines. Lepetit Colloq. Biol. Med. *4.* In press.

PAUL, D., A. LIPTON and I. KLINGER. 1971. Serum factor requirements of normal and SV40-transformed 3T3 mouse fibroblasts. Proc. Nat. Acad. Sci. *68:* 645.

POLLACK, R. E. and M. M. BURGER. 1969. Surface-specific characteristics of a contact-inhibited cell line containing the SV40 viral genome. Proc. Nat. Acad. Sci. *62:* 1074.

POLLACK, R. E. and G. W. TEEBOR. 1969. Relationship of contact inhibition to tumor transplantability, morphology and growth rate. Cancer Res. *29:* 1770.

POLLACK, R. E., H. GREEN and G. J. TODARO. 1968. Growth control in cultured cells: Selection of sub-lines with increased sensitivity to contact inhibition and decreased tumor-producing ability. Proc. Nat. Acad. Sci. *60:* 126.

POLLACK, R., S. WOLMAN and A. VOGEL. 1970. Reversion of virus-transformed cell lines: hyperploidy accompanies retention of viral genes. Nature *228:* 938.

POPE, J. H. and W. P. ROWE. 1964. Detection of specific antigen in SV40-transformed cells by immunofluorescence. J. Exp. Med. *120:* 121.

POTTER, C. W., B. C. MCLAUGHLIN and J. S. OXFORD. 1969. Simian virus 40-induced T and tumor antigens. J. Virol. *4:* 574.

POTTER, C. W., A. M. POTTER and J. S. OXFORD. 1970. Comparison of transformation and T antigen induction in human cell lines. J. Virol. *5:* 293.

RABINOWITZ, Z. and L. SACHS. 1968. Reversion of properties in cells transformed by polyoma virus. Nature *220:* 1203.

———. 1969a. The formation of variants with a reversion of properties of transformed cells: II. *In vitro* formation of variants from polyoma-transformed cells. Virology *38:* 343.

———. 1969b. The formation of variants with a reversion of properties of transformed cells: I. Variants from polyoma-transformed cells grown *in vivo*. Virology *38:* 336.

———. 1970. Control of the reversion properties in transformed cells. Nature *225:* 136.

RAPP, F. and S. C. TRULOCK. 1970. Susceptibility to superinfection of simian cells transformed by SV40. Virology *40:* 961.

RAPP, F., T. KITAHARA, J. S. BUTEL and J. L. MELNICK. 1964a. Synthesis of SV40 tumor antigen during replication of simian papova virus (SV40). Proc. Nat. Acad. Sci. *52:* 1138.

RAPP, F., J. L. MELNICK and T. KITAHARA. 1964b. Tumor and virus antigens of simian virus 40: Differential inhibition of synthesis by cytosine arabinoside. Science *147:* 625.

RAPP, F., J. S. BUTEL and J. L. MELNICK. 1964c. Virus-induced intranuclear antigen in cells transformed by papovavirus SV40. Proc. Soc. Exp. Biol. Med. *116:* 1131.

RAPP, F., J. S. BUTEL, S. S. TEVETHIA, M. KATZ and J. L. MELNICK. 1966. Antigenic analysis of tumors and sera from animals inoculated with PARA-adenovirus populations. J. Immunol. *97:* 833.

RAPP, F., S. PAULUZZI and J. S. BUTEL. 1969. Variation in properties of plaque progeny of PARA (defective simian papovavirus 40)-adenovirus 7. J. Virol. *4:* 626.

RENGER, H. 1972. Retransformation of temperature-sensitive SV40 at non-permissive temperature. Nature New Biol. *240:* 19.

RENGER, H. and C. BASILICO. 1972. Mutation causing temperature-sensitive expression of cell transformation by a tumor virus. Proc. Nat. Acad. Sci. *69:* 109.

RHIM, J. S., C. GREENAWALT, K. K. TAKEMOTO and R. J. HUEBNER. 1971. Increased transformation efficiency of SV40 in rat embryo cells infected with Rauscher leukaemia virus. Nature New Biol. *230:* 81.

RICHARDSON, L. S. and J. S. BUTEL. 1971. Properties of transformed hamster cells containing SV40 tumor antigen in the cytoplasm. Int. J. Cancer *7:* 75.

ROTHSCHILD, H. and P. H. BLACK. 1970. Analysis of SV40-induced transformation of hamster kidney tissue in vitro. VII. Induction of SV40 virus from transformed hamster cell clones by various agents. Virology *42:* 251.

ROWE, W. P., S. G. BAUM, W. E. PUGH and M. D. HOGGAN. 1965. Studies of adenovirus SV40 hybrid viruses: I. Assay system and further evidence for hybridization J. Exp. Med. *122:* 943.

RUBIN, H. 1966. A substance in conditioned medium which enhances the growth of small numbers of chick embryo cells. Exp. Cell Res. *41:* 138.

SABIN, A. B. and M. A. KOCH. 1964. Source of genetic information for specific complement-fixing antigens in SV40 virus-induced tumors. Proc. Nat. Acad. Sci. *52:* 1131.

SAMBROOK, J. 1972. Transformation by polyoma virus and SV40. Advan. Cancer Res. *16:* 141. Academic Press, New York.

SAMBROOK, J., H. WESTPHAL, P. R. SRINIVASAN and R. DULBECCO. 1968. The integrated state of DNA in SV40-transformed cells. Proc. Nat. Acad. Sci. *60:* 1288.

SAMBROOK, J., P. A. SHARP and W. KELLER. 1972. Transcription of SV40. I. Separation of the strands of SV40 DNA and hybridization of the separated strands to RNA extracted from lytically infected and transformed cells. J. Mol. Biol. *70:* 57.

SAUER, G. 1971. Apparent differences in transcriptional control in cells productively infected and transformed by SV40. Nature New Biol. *231:* 135.

SAUER, G. and E. C. HAHN. 1970. Interaction of SV40 with SV40-transformed and non-transformed monkey kidney cells. Z. Krebsforsch. *74:* 40.

SAUER, G. and J. R. KIDWAI. 1968. The transcription of the SV40 genome in productively infected and transformed cells. Proc. Nat. Acad. Sci. *61:* 1256.

SCHER, C. and W. A. NELSON-REES. 1971. Direct isolation and characterization of "flat" SV40-transformed cells. Nature New Biol. *233:* 263.

SHEININ, R. 1966. DNA synthesis in rat embryo cells infected with polyoma virus. Virology *29:* 167.

SHEPPARD, J. R., A. J. LEVINE and M. M. BURGER. 1971. Cell-surface changes after infection with oncogenic viruses: Requirement for synthesis of host DNA. Science *172:* 1345.

SHIROKI, K. and H. SHIMOJO. 1971. Transformation of green monkey kidney cells by SV40 genome: The establishment of transformed cell lines and the replication of human adenoviruses and SV40 in transformed cells. Virology *45:* 163.

SHODELL, M. 1972. Environmental stimuli in the progression of BHK/21 cells through the cell cycle. Proc. Nat. Acad. Sci. *69:* 1455.

SJÖGREN, H. O. 1965. Transplantation methods as a tool for detection of tumor-specific antigens. Prog. Exp. Tumor Res. *6:* 289.

SJÖGREN, H. O., I. HELLSTRÖM and G. KLEIN. 1961. Transplantation of polyoma virus-induced tumors in mice. Cancer Res. *21:* 329.

SMITH, H. S., C. D. SCHER and G. J. TODARO. 1970. Abortive transformation of BALB/3T3 by simian virus 40. Bacteriol. Proc. Abstr. *217:* 187.

———. 1971. Induction of cell division in medium lacking serum growth factor by SV40. Virology *44:* 359.

SMITH, H. S., L. D. GELB and M. A. MARTIN. 1972. Detection and quantitation of SV40 genetic material in abortively transformed BALB/3T3 clones. Proc. Nat. Acad. Sci. *69:* 152.

STOKER, M. 1964. Further studies on radiation-induced sensitivity of hamster cells to transformation by polyoma virus. Virology *24:* 123.

———. 1968. Abortive transformation by polyoma virus. Nature *218:* 234.

STOKER, M. G. P. and P. ABEL. 1962. Conditions affecting transformation by polyoma virus. Cold Spring Harbor Symp. Quant. Biol. *27:* 375.

STOKER, M. G. P. and I. A. MACPHERSON. 1961. Studies on the transformation of hamster cells by polyoma virus *in vitro*. Virology *14:* 359.

STOKER, M. G. P. and H. RUBIN. 1967. Density dependent inhibition of cell growth in culture. Nature *215:* 171.

STOKER, M. and A. SMITH. 1964. Characteristics of normal cells in mixed clones arising after delayed transformation by polyoma virus. Virology *24:* 175.

STOKER, M., G. D. CLARKE and M. THORNTON. 1971. Density dependent stimulation of thymidine incorporation in BHK21 cells by active material released from the same cells. J. Cell. Physiol. *78:* 345.

SUAREZ, H., G. SONENSHEIN, R. CASSINGENA and P. TOURNIER. 1971. Study on the mode of inhibition by SV40 "repressor" of productive SV40 infection. Lepetit Colloq. Biol. Med. *2:* 1.

SUMMERS, J. and M. VOGT. 1971. Recovery of virus from polyoma-transformed BHK21. Lepetit Colloq. Biol. Med. *2:* 306.

SWETLY, P., G. BRODANO, B. KNOWLES and H. KOPROWSKI. 1969. Response of SV40-transformed cell lines and cell hybrids to superinfection with SV40 and its DNA. J. Virol. *4:* 348.

TAI, H. T. and R. L. O'BRIEN. 1969. Multiplicity of viral genomes in an SV40-transformed hamster cell line. Virology *38:* 698.

TAI, H. T., C. A. SMITH, P. A. SHARP and J. VINOGRAD. 1972. Sequence heterogeneity in closed SV40 DNA. J. Virol. *9:* 317.

TAKEMOTO, K. K. and K. HABEL. 1966. Hamster tumor cells doubly transformed by SV40 and polyoma viruses. Virology *30:* 20.

TAYLOR-PAPADIMITRIOU, J. and M. G. P. STOKER. 1971. Effect of interferon on some aspects of transformation by polyoma virus. Nature New Biol. *230:* 114.

TEMIN, H. M. 1967. Control by factors in serum of multiplication of uninfected cells and cells infected and converted by avian sarcoma viruses. Wistar Inst. Symp. Monogr. *7:* 103. Wistar Institute Press, Philadelphia.

TEVETHIA, S. S. and F. RAPP. 1965. Demonstration of new surface antigens in cells transformed by papovavirus SV40 by cytotoxic tests. Proc. Soc. Exp. Biol. Med. *120:* 455.

———. 1966. Prevention and interruption of SV40 induced transplantation immunity with tumor cell extracts. Proc. Soc. Exp. Biol. Med. *123:* 612.

TEVETHIA, S. S., G. TH. DIAMANDOPOULOS, F. RAPP and J. F. ENDERS. 1968. Lack of relationship between virus-specific and transplantation antigens in hamster cells transformed by simian papovavirus SV40. J. Immunol. *101:* 1192.

TODARO, G. J. and H. GREEN. 1964a. An assay for cellular transformation by SV40. Virology *23:* 117.

———. 1964b. Enhancement by thymidine analogs of susceptibility of cells to transformation by SV40. Virology *24:* 393.

———. 1965. Successive transformations of an established cell line by polyoma virus and SV40. Science *147:* 513.

———. 1966a. High frequency of SV40 transformation of mouse cell line 3T3. Virology *28:* 756.

———. 1966b. Cell growth and the initiation of transformation by SV40. Proc. Nat. Acad. Sci. *55:* 302.

———. 1967. Simian virus 40 transformation and the period of cellular deoxyribonucleic acid synthesis. J. Virol. *1:* 115.

TODARO, G. J. and G. M. MARTIN. 1967. Increased susceptibility of Down's syndrome fibroblasts to transformation by SV40 (31974). Proc. Soc. Exp. Biol. Med. *124:* 1232.

TODARO, G. J., K. HABEL and H. GREEN. 1965. Antigenic and cultural properties of cells doubly transformed by polyoma virus and SV40. Virology *27:* 179.

TODARO, G. J., H. GREEN and M. R. SWIFT. 1966. Susceptibility of human diploid fibroblast strains to transformation by SV40 virus. Science *153:* 1252.

TODARO, G., Y. MATSUYA, S. BLOOM, A. ROBBINS and H. GREEN. 1967. Stimulation of RNA synthesis and cell division in resting cells by a factor present in serum. Wistar Inst. Symp. Monogr. *7:* 87. Wistar Institute Press, Philadelphia.

TONEGAWA, S., G. WALTER, A. BERNARDINI and R. DULBECCO. 1970. Transcription of the SV40 genome in transformed cells and during lytic infection. Cold Spring Harbor Symp. Quant. Biol. *35:* 823.

TOURNIER, P., R. CASSINGENA, R. WICKERT, J. COPPEY and H. SUAREZ. 1967. Étude de mécanisme de l'induction chez des cellules de hamster Syrien transformées par le virus SV40: I. Propriétés d'une lignée cellulaire clonale. Int. J. Cancer *2:* 117.

UCHIDA, S., K. YOSHIIKE, S. WATANABE and A. FURUNO. 1968. Antigen-forming defective viruses of simian virus 40. Virology *34:* 1.

VOGT, M. and R. DULBECCO. 1960. Virus-cell interaction with a tumor-producing virus. Proc. Nat. Acad. Sci. *46:* 365.

WALL, R. and J. E. DARNELL. 1971. Presence of cell and virus-specific sequences in the same molecules of nuclear RNA from virus transformed cells. Nature New Biol. *232:* 73.

WATKINS, J. F. 1970. In *Defectivité, demasquage, et stimulation des virus oncogenes*, pp. 135–138. CNRS, Paris.

———. 1971. Cell fusion in the study of tumor cells. Int. Rev. Exp. Pathol. *10:* 115.

WATKINS, J. F. and R. DULBECCO. 1967. Production of SV40 virus in heterokaryons of transformed and susceptible cells. Proc. Nat. Acad. Sci. *58:* 1396.

WEINBERG, R. A., Z. BEN-ISHAI and J. E. NEWBOLD. 1972. PolyA associated with SV40 messenger RNA. Nature New Biol. *238:* 111.

WEISS, M. C. 1970. Further studies on loss of T-antigen from somatic hybrids between mouse cells and SV40-transformed human cells. Proc. Nat. Acad. Sci. *66:* 79.

WEISS, M. C., B. EPHRUSSI and L. J. SCALETTA. 1968. Loss of T-antigen from somatic hybrids between mouse cells and SV40-transformed human cells. Proc. Nat. Acad. Sci. *59:* 1132.

WESTPHAL, H. and R. DULBECCO. 1968. Viral DNA in polyoma and SV40-transformed cell lines. Proc. Nat. Acad. Sci. *59:* 1158.

WESTPHAL, H. and E. D. KIEHN. 1970. The in vitro product of SV40 DNA transcription and its specific hybridization with DNA of SV40-transformed cells. Cold Spring Harbor Symp. Quant. Biol. *35:* 819.

WETMUR, J. G. and N. DAVIDSON. 1968. Kinetics of renaturation of DNA. J. Mol. Biol. *31:* 649.

WEVER, G. H., S. KIT and D. R. DUBBS. 1970. Initial site of synthesis of virus during rescue of simian virus 40 from heterokaryons of simian virus 40-transformed and susceptible cells. J. Virol. *5:* 578.

WYKE, J. 1971a. A method of isolating cells incapable of multiplication in suspension culture. Exp. Cell Res. *66:* 203.

———. 1971b. Phenotypic variation and its control in polyoma-transformed BHK21 cells. Exp. Cell Res. *66:* 209.

YOSHIIKE, K. and A. FURUNO. 1969. Heterogeneous DNA of simian virus 40. Fed. Proc. *28:* 1899.

7

Genetics of Polyoma Virus and SV40

The idea that the properties of transformed cells can be influenced by the expression of viral genes stems from the observations that viral DNA can become part of the genetic material of transformed cells and that virus-specific products persist in transformed cells. By isolating mutant viruses with defective functions, it should be possible to associate a particular viral gene product with particular changes which occur on transformation. To this end temperature-sensitive mutants of polyoma virus and recently SV40, and host-range mutants of polyoma virus have been selected.

The isolation of temperature-sensitive mutants of polyoma virus was first reported in 1965 by Fried, but for reasons that are obscure attempts to isolate comparable mutants of SV40 failed for several years. During the past two years, however, several groups have managed to obtain useful SV40 mutants. Not surprisingly, today we know more about the polyoma virus mutants than the SV40 mutants simply because the former have been studied for a considerably longer time. But this state of affairs is changing quite rapidly. SV40 seems to be more convenient to handle in the laboratory than polyoma virus and it can be readily rescued from transformed cells; as a result we know far more about SV40 than polyoma virus, with the exception of the genetics of these viruses, and this exception seems destined to be short lived.

TEMPERATURE-SENSITIVE MUTANTS OF POLYOMA VIRUS

Temperature-sensitive mutants of polyoma virus have been isolated after treating large-plaque polyoma virus with nitrous acid (Fried, 1965a; Vogt, unpublished data; di Mayorca et al., 1969) or hydroxylamine (di Mayorca et al., 1969). The viruses which survived exposure to the mutagen were plated at 31°C on secondary mouse embryo cells or 3T3 cells. Individual plaques, formed after 4 weeks, were then plated at both the permissive temperature, 31°C, and the nonpermissive temperature, 38.5°C. From 2–5 percent of these plaque isolates were found to be temperature-sensitive (*ts*) mutants. About 200 *ts* mutants have so far been isolated, but few of them have been examined in detail. Many show a leakiness which is dependent on the input multiplicity at the nonpermissive temperature. For example, the mutant *ts-a*, isolated by Fried (1965a), produces about 300- to 1,000-fold less virus than the wild type when grown at 38.5°C in cells infected with multiplicities of less than 0.1 plaque-forming unit per cell, but produces yields as large as those of the wild type at input multiplicities of 100 plaque-forming units per cell (Fried, 1970). Multiplicity-dependent leakiness hampers complementation tests which have to be carried out at low multiplicity. Representative results of such a test are shown in Table 7.1.

Table 7.1. Complementation of Polyoma Temperature-sensitive Mutants

Infecting virus	Virus yield (p.f.u./ml) 38°C 40–48 hours	31°C 72–96 hours
ts10	10^3–10^4	10^7–10^8
ts-a	2×10^5	10^7–10^8
ts-a × *ts10*	10^7	10^7–5×10^7
wild-type	10^8	10^8

From Eckhart (unpublished results).

Assignment of Complementation Groups

Complementation tests reveal three (or perhaps four) classes of *ts* mutants (di Mayorca et al., 1969; Eckhart, 1969, 1971a). Mutants in the same class do not complement each other, but complement members of the other classes. On the basis of physiological tests the mutants can be divided into five classes. Table 7.2 shows the pooled results from both genetic and biochemical tests.

Mutants of class I and class IV, which are separable by complementation tests, are defective in late viral functions that are not required for viral DNA synthesis; at least some of these mutants do not synthesize virion (V) antigen at the nonpermissive temperature (di Mayorca et al., 1969). Class I mutants are probably mutated in a gene which specifies a structural protein of the virus particle because at least one mutant of this class makes particles at the permissive temperature

Table 7.2. POLYOMA VIRUS FUNCTIONS DEFINED BY TEMPERATURE-SENSITIVE AND HOST-RANGE MUTANTS AT NONPERMISSIVE CONDITIONS

	Temperature-sensitive Mutants					Host-range Mutants
	I *ts10*	II *ts-a*	III P155	IV	V *ts3*	NG18
Viral DNA synthesis	+	−	−	+	−	ND
Transformation assayed by growth in agar	+	−	+	+	+	−
T antigen induction	+	−	−	ND	ND	+
Host DNA synthesis induction	+	+	+	+	− (in Balb/c-3T3 cells only)	+ (in Swiss 3T3 cells and BHK cells)
Exposure of wheat germ agglutinin site	ND	+	ND	ND	−	−
Abortive transformation	ND	+	ND	ND	ND	−
Capsid antigen	−	−	ND	−	−	−

ND = not determined; + = function expressed in nonpermissive conditions.

which are more labile than wild-type virus when heated to 68°C (Eckhart, 1969).

Mutants of class II and class III are closely related because they are not separable by complementation tests. They can, however, be distinguished by their transforming ability; class II mutants do not transform cells at the nonpermissive temperature whereas mutant P155, the only class III mutant identified so far, does transform cells at the nonpermissive temperature. Mutant P155 is, however, leaky (Eckhart, 1969; Oxman et al., 1972) and high multiplicities of virus are used in transformation assays. P155 may, therefore, be a class II mutant which, because it is leaky, is able at high multiplicities of infection to transform cells even at the nonpermissive temperature.

During infections at the nonpermissive temperature, the mutants of both class II and class III fail to induce detectable amounts of T antigen, when assayed by indirect immunofluorescent staining (Oxman et al., 1972), but it is not known whether the mutants also fail to induce detectable amounts of T antigen when assayed by complement fixation. Neither is it known if the T antigen made by mutants at the permissive temperature has different physical properties from the T antigen made by wild-type polyoma virus. Because of the failure of the mutants of class II and class III to induce, at nonpermissive temperatures, T antigen detectable by immunofluorescent staining, it seems unlikely that these are mutants of a structural gene for T antigen; if they were, one would expect to find at least some mutants producing immunologically cross-reacting material. It seems more likely that class II and class III mutants are defective in a function which controls the induction of T antigen. Recent experiments by Eckhart (unpublished data), which were modeled on those done by Tegtmeyer (1972) with one of his temperature-sensitive mutants of SV40 (see below), indicate that one function of the product of the polyoma virus gene(s) defined by the *ts-a* class II mutants is the initiation of rounds of replication of the viral DNA in permissive cells. Because T antigen is induced in permissive cells long before viral DNA replication is initiated, it is hard to avoid the conclusion either that the *ts-a* class II mutant is a double mutant or that the product of the *ts-a* gene acts in two ways at two different times during the lytic cycle.

Transformation by Class II (*ts-a*) Mutants

Mutants of class II, which includes *ts-a*, are defective in a function which is required for the establishment of transformation, since BHK21 cells infected at the nonpermissive temperature are rarely transformed by *ts-a*. However, the function which is defective in class II mutants is not required to maintain the transformed-cell phenotype. BHK21 can be transformed by these mutants at the permissive temperature and the transformants, *ts-a*-BHK, retain their transformed character when they are grown at the nonpermissive temperature (Fried, 1965b; Eckhart, 1969; di Mayorca et al., 1969). Because they are defective in a function required to establish transformation, class II mutants, in particular *ts-a*, have been the subject of many experiments and much speculation.

Vogt (1970) isolated several lines of 3T3 cells transformed by *ts-a*. 3T3 cells are permissive for wild-type polyoma virus and are usually killed by it, but the *ts-a* mutant does not kill all the cells at 38.5°C and some of the surviving cells are transformed. The transformed cells, called *ts-a*-3T3, multiply at high temperature even though small amounts of infectious virus can be recovered from them. However, when *ts-a*-3T3 cells are shifted to 31°C, virus multiplication is induced in a considerable fraction of the cells at a rate similar to that observed in a normal infection of 3T3 cells at 31°C and the induced cells are killed.

Vogt (1970) and Cuzin et al. (1970) used the *ts-a*-3T3 line to try to elucidate the nature of the function controlled by the gene(s) containing the *ts-a* mutation. To determine whether the *ts-a* function is needed continuously or only transiently during the lytic cycle, *ts-a*-3T3 cells growing at 38.5°C were shifted to 31°C for varying lengths of time before being returned to the high temperature. No increase in the yield of progeny *ts-a* particles was detected unless the cells had been kept for 8 hours or longer at 31°C; progeny virus then appeared after about 36 hours further incubation at 38.5°C. The yield of virus increased with longer exposure to the low temperature. Two explanations for this dependence of the yield on the length of incubation at the permissive temperature are possible. Perhaps all cells are "induced" immediately after the shift to 31°C, and the number of virus particles in each cell increases only as long as the cells remain at 31°C. Alternatively, the probability of induction for individual cells is proportional to the time at the low temperature, but once a cell acquires the capacity to produce one or a few virions, it goes on to give a full yield regardless of temperature.

The kinetics of viral DNA synthesis after brief shifts to lower temperature provide further evidence that the *ts-a* function is needed for the initiation of DNA synthesis but not for its continuation. Pulse-labeled viral DNA, isolated from cells by the Hirt (1967) procedure, was first detectable 24 hours after the shift to 31°C. If, however, cells were induced for 24 hours at low temperature and then further incubated at the high temperature in the presence of labeled DNA precursor, DNA synthesis was observed to continue at nearly the same rate as when the cells remained at 31°C. How can this result be reconciled with Eckhart's experiments (see above), which indicate that the *ts-a* gene product is required specifically for the initiation of each round of viral DNA replication? Because *ts-a* shows a multiplicity-dependent leakiness (see above), it is reasonable to suggest that sufficient *ts-a* genomes are replicated during 24 hours at the permissive temperature (31°C) to permit leakage through the temperature-sensitive blockage when the cells are returned to 38.5°C.

Formation of Oligomeric ts-a DNA

The viral DNA synthesized by *ts-a*-3T3 cells when they are shifted from 38.5°C to 31°C takes several forms in contrast to the viral DNA produced in 3T3 cells productively infected at 31°C with *ts-a* or wild-type polyoma virus. In such productively infected cells component I and component II DNA (molecular weight 3×10^6 daltons, see Chapter 4) account for 98 percent of the viral DNA made. In induced *ts-a*-3T3 cells, however, in addition to monomeric viral DNA, large amounts of viral DNAs, which sediment in neutral sucrose gradients at 1.32 and 1.54 times the velocity of the monomeric DNA, are formed (Cuzin et al., 1970). These two sedimentation velocities are those expected of circular DNA molecules, respectively two and three times the length of the monomeric form of polyoma virus DNA. Four lines of evidence indicate that these molecules must be covalently closed, circular dimers or trimers of the viral genome: (1) They sediment faster than monomeric circles both in neutral and alkaline velocity gradients. (2) In caesium chloride gradients in the presence of ethidium bromide, they band at the same position as polyoma component I DNA. (3) Under the electron microscope they appear as circles with contour lengths two and three times that of polyoma DNA extracted from virions. (4) The monomer, dimer and trimer species of viral DNA from induced *ts-a*-3T3 cells were found to be equally infective on a molecular basis. Dimers and trimers, however, were never observed in mature virus particles isolated from induced *ts-a*-3T3 cells, presumably because they are too large to be packaged in polyoma virus capsids.

The relative proportions of oligomers and monomers produced varies in different subclones of *ts-a*-3T3. In the original *ts-a*-3T3 line there were more monomers than dimers and more dimers than trimers after induction. However, in other subclones the dimers were often the most common form, and circular DNA, which had a length 1.6 times that of the monomer, was consistently isolated from one clone. These molecules must contain at least one viral genome since they are infectious, although at a lower specific activity than the dimers (Mulder and Vogt, unpublished results); but it is now believed that the various oligomeric polyoma virus DNAs also contain host DNA sequences (Blangy and Vogt, unpublished data).

Within any one subclone of *ts-a*-3T3, the relative amounts of DNA present in monomers, dimers and trimers did not vary either with the length of the pulse of [^{3}H]thymidine or with the time after the temperature shift at which the pulse of thymidine was given. Neither could label be chased from one DNA form to another; these results suggest that the oligomeric forms of *ts-a* DNA are probably not intermediates in the replication of the monomeric DNA (Cuzin et al., 1970).

Rescue of ts-a from Transformed Cells

So far we have discussed some of the properties of permissive mouse 3T3 cells transformed by the *ts-a* mutant of polyoma virus. But, of course, *ts-a*, like wild-type polyoma virus, also transforms non-permissive cells including hamster BHK cells at low temperature, and the cells retain their transformed phenotype when subsequently shifted to high temperature. The transformed cells are free of virus, but

Summers and Vogt (1971) have shown that *ts-a* can be rescued from *ts-a*-BHK cells by fusion with permissive cells. Until recently it seemed that *ts-a* mutant polyoma virus was unique in this respect, but Fried (unpublished data) has shown that the *ts-c* mutant (a late mutant) can be rescued from transformed hamster cells by fusion with mouse cells, and Folk (1973) has shown that *ts-a* can be induced to replicate merely by chilling some clones of *ts-a*–BHK transformants from 39°C to 32°C.

The Group V Mutant *ts3*

Induction of cellular DNA synthesis occurs normally after the infection of 3T3 cells, Balb/c- 3T3 cells or of baby mouse kidney cells with mutants of classes I through IV (di Mayorca et al., 1969; Fried, 1970; Eckhart, 1971b). However, mutant *ts3* does not induce cellular DNA synthesis in Balb/c-3T3 cells (Dulbecco and Eckhart, 1970) and *ts3* also fails to induce the exposure of the wheat germ agglutinin site on the cell surface when it infects Balb/c-3T3 cells at the nonpermissive temperature (Eckhart et al., 1971). Wild-type virus induces this change within 24–30 hours after the infection of monolayers of Swiss and Balb/c-3T3 cells (Eckhart et al., 1971; Benjamin and Burger, 1970). DNA synthesis is required for the agglutinin site to be exposed because 5-fluorodeoxyuridine and hydroxyurea prevent both DNA synthesis and the exposure. Cellular DNA synthesis rather than viral DNA synthesis is required, because when Balb/c-3T3 cells are infected by another *ts* mutant of polyoma virus, which is unable to replicate its own DNA but can induce cellular DNA synthesis at the nonpermissive temperature, the wheat germ agglutinin site is exposed at the nonpermissive temperature (Eckhart et al., 1971).

The connection between the *ts3* gene product and the cell surface change has been further explored in the following experiments. When BHK cells, transformed by *ts3* at 31°C and subsequently grown at 31°C, are shifted to 38.5°C, they lose some of the characters of transformed cells—the wheat germ agglutinin site is masked 12–24 hours after the temperature shift, the cells growing in dense cultures assume a somewhat more normal morphology, and they behave more like normal cells than transformed cells in the topoinhibition test of Dulbecco (1970). After the shift down to 31°C, they reassume the full transformed-cell phenotype within 24 hours (Eckhart et al., 1971). However, BHK cells transformed by *ts3* retain the ability to grow in suspension in agar at 38.5°C; this means, of course, that not all the facets of the transformed-cell phenotype are controlled by the temperature selective function of *ts3*; it also means that *ts3*-BHK transformants can be selected at 38.5°C by virtue of their ability to grow in agar.

Ts3 is clearly a crucially important mutant. The temperature-sensitive properties of BHK cells transformed by *ts3* constitute the only direct evidence we have that products coded by polyoma virus, and by analogy SV40, are needed continuously for the maintenance of at least some of the phenotypic characteristics of transformed cells. In other words, the chief tenet of tumour virology, namely, that tumour viruses are not hit-and-run carcinogens like chemical carcinogens, rests on the properties of *ts3* and on the properties of some temperature-sensitive mutants of Rous sarcoma virus (see Chapter 11). Obviously much effort is currently being spent trying to elucidate further the properties of this mutant. The results have not been without their surprises, for recently Eckhart (unpublished data) found that when mouse cells at the nonpermissive temperature are infected with *ts3* DNA, instead of *ts3* virus particles, a lytic cycle ensues, the cells are killed and progeny virus are produced. In other words, in permissive cells infections initiated by *ts3* DNA are not temperature sensitive, but those initiated by *ts3* virus particles are. This property of the *ts3* mutant is shared by one of the recently isolated mutants of SV40.

TEMPERATURE-SENSITIVE MUTANTS OF SV40

Since 1970 several groups have reported the isolation and partial characterization of SV40 temperature-sensitive mutants (Tegtmeyer et al., 1970; Kit et al., 1970; Tegtmeyer and Ozer, 1971; Tegtmeyer, 1972; Robb and Martin, 1972; Kimura and Dulbecco, 1972), which fall into three complementation and four physiological classes, and for which a uniform nomenclature has been proposed (Robb et al., 1972b).

There are two classes of SV40 late-function mutants that correspond in some ways to polyoma virus mutants of classes I and IV (see Table 7.2), and although Tegtmeyer and Ozer (1971) failed to detect complementation between their mutants in these two classes, which they distinguished by physiological criteria, Kimura and Dulbecco (1972), working with apparently similar mutants, have observed complementation. At the nonpermissive temperature, mutants in both these classes synthesize viral DNA in permissive cells and transform nonpermissive cells (Tegtmeyer and Ozer, 1971; Kimura and Dulbecco, 1972). One class of SV40 late mutants also makes V antigen and empty virus particles, whereas the other class produces some capsid antigens but fails to form V antigen and empty virus particles (Kit et al., 1970; Tegtmeyer and Ozer, 1971; Ozer and Tegtmeyer, 1972; Kimura and Dulbecco, 1972). Finally the virions of mutants in both these classes are thermolabile compared to wild-type SV40 virions (Tegtmeyer and Ozer, 1971; Kimura and Dulbecco, 1972; Tegtmeyer, 1972). This discovery is important, of course, because it implies that both the genes defined by these two classes of mutants specify virion proteins.

A third class of SV40 mutants has properties that are very similar to the class II (Table 7.2) *ts-a* group of mutants of polyoma virus (Tegtmeyer and Ozer, 1971; Tegtmeyer, 1972). At the nonpermissive temperature these mutants do not transform mouse cells, and neither mutant virus particles nor DNA extracted from them induces T antigen. Mutant DNA replication in permissive monkey cells is also temperature sensitive. Using density labels coupled with temperature shifts Tegtmeyer (1972) has obtained compelling evidence that the function mutated in these class II mutants is essential for the initiation of each round of replication of viral DNA in permissive cells but is not essential for elongation of daughter DNA chains once they have been initiated. Eckhart (unpublished data) believes that the product of the *ts-a* gene of polyoma virus also functions as an initiator of DNA synthesis.

SV101 (Robb and Martin, 1972) is the sole representative of the fourth class of SV40 mutants; its properties are reminiscent of some of those of the *ts3* (class V) mutant of polyoma virus. When monkey cells at the nonpermissive temperature are infected with SV101 virions, absorption and penetration occur but T, U and V antigens are not induced; cell DNA synthesis is not induced and viral DNA is not replicated. Furthermore, using virions, Robb and Martin failed to detect complementation between SV101 and any of their other SV40 mutants. When, however, monkey cells at the nonpermissive temperature are infected with SV101 DNA, SV40 T, U and V antigens are induced and temperature-sensitive SV101 progeny particles are liberated.

SV101 particles absorb to and penetrate nonpermissive 3T3 cells at 33°C and at 38°C; the induction of T antigen at 38°C is markedly inhibited, but at both temperatures some of the cells are abortively transformed and the percentage of infected cells that emerges as stable transformants is the same (Robb et al., 1972a). Moreover the percentage of cells infected with SV101, at 33°C or 38°C, that becomes stably transformed is the same as the percentage of cells infected by wild-type SV40 that is stably transformed. Finally, virus rescued by fusion from 3T3 cells transformed by SV101 is temperature sensitive.

Function of the *ts-a* Gene

Pooling the data obtained by various groups, we can say that the class II mutants (Table 7.2) of polyoma virus and SV40 have the following properties:

1. In permissive cells the induction of T antigen and the initiation of replication of the viral DNA are both temperature sensitive.
2. Permissive cells infected at the permissive temperature and then shifted to the nonpermissive temperature may be transformed by these mutants.
3. When transformed permissive cells are shifted back to the permissive temperature, mutant virus particles as well as large proportions of oligomeric viral DNA molecules are produced.
4. In permissive cells infection with mutant DNA is as temperature sensitive as infection with mutant virus particles.
5. In nonpermissive cells the mutated gene function is required for the establishment, but not the maintenance, of transformation.
6. At the nonpermissive temperature virus can be rescued from transformed nonpermissive cells by fusion with permissive cells.

It is impossible from these data alone to reach any satisfactory conclusion as to the function of the protein specified by the gene defined by the class II mutants of polyoma virus and SV40, and it seems likely that short of isolating and purifying this protein we shall remain more or less ignorant of its function or functions. The minimal statement that can be made is that this protein, which is not a component of the virion, almost certainly acts on the viral DNA because the *ts-a* function initiates DNA replication in permissive cells, thereby preventing them from becoming transformed, and it is necessary for the establishment of transformation of nonpermissive cells. We can speculate that it might melt or separate the strands of the covalently circular viral DNA and so promote initiation of replication and recombination between host and viral chromosomes; alternatively, it might be a site-specific endonuclease or it might have some other activity.

Defective Function(s) of *ts3* and SV101

Ts3 and SV101 have three important properties in common: (1) The virions are temperature sensitive in permissive cells but their DNA is not. (2) When virions are used, these mutants are not complemented by mutants of any other class. (3) The virions transform nonpermissive cells at both permissive and nonpermissive temperatures. But the lesion or lesions in the *ts3* genome also renders temperature sensitive a viral function(s) required for the maintenance of two characteristics of *ts3*-transformed BHK cells.

These findings are difficult to interpret in any satisfactory way. We can speculate that both *ts3* and SV101 virions cannot be uncoated in their respective permissive cells at the nonpermissive temperature; this would explain why these mutant virions are temperature sensitive, why they fail to complement any other mutant and why naked mutant DNA, uncoated by the experimenter, is not temperature sensitive. This hypothesis implies that the mutation is *ts3* and SV101 genomes is in a late gene but that the product of this late gene, a component of the virion, is involved in a very early event during infection, namely uncoating. Since uncoating presumably involves interaction between the virion and host cell enzymes, the fact that *ts3* and SV101 particles are uncoated at both permissive and nonpermissive temperatures in their respective nonpermissive cells can be accounted for simply by postulating that permissive and nonpermissive cells contain different enzymes responsible for uncoating. The great snag to this hypothesis is that it totally fails to account for the most interesting and important property of *ts3*, the temperature sensitivity of two of the phenotypic traits of *ts3*-BHK transformants. How a gene which specifies a virion protein involved in uncoating can also regulate facets of the phenotype of transformed cells is, to say the least, far from obvious. Perhaps *ts3* is a double mutant or perhaps the hypothesis that the *ts3* protein is involved in uncoating is illusory. After all, the only evidence which supports the uncoating-deficiency hypothesis is the observation that *ts3* and SV101 DNAs are not temperature sensitive in permissive cells. Infection with DNA is very inefficient and therefore large amounts have to be used, and we have no idea whether or not the controlled expression of the viral genome is maintained when naked DNA, rather than a virion, infects cells. Such uncertainties mean that all interpretations of these experiments will remain dubious until we have isolated the gene products specified by the mutated *ts3* genome.

In short, studies of temperature-sensitive mutants of polyoma virus and SV40 have established that two genes probably code for virion proteins, and that there may well be two or three genes which function before replication of the viral DNA begins. But the precise nature of the proteins these genes specify remains obscure.

HOST-RANGE MUTANTS OF POLYOMA VIRUS

Four polyoma mutants (or possibly four isolates of the same mutant), which do not grow in normal cells but are still able to grow in permissive cells previously transformed by polyoma virus, have been isolated by Benjamin (1970). It seems likely that the mutants can grow in permissive, transformed cells because they are complemented by the transforming viral genome(s) or some cellular gene(s) expressed as a consequence of transformation by polyoma virus. The mutants are defective in some function(s) expressed in the transformed host which is required for lytic growth; and since they are unable to transform normal rat or hamster fibroblasts, the mutated function may also be required for transformation.

Benjamin has used a line of 3T3 cells transformed by polyoma virus called 3T3-Py3. These cells contain polyoma T antigen and polyoma-specific RNA; they are productively infected by wild-type polyoma virus and yield as many progeny virus as untransformed 3T3 cells. Benjamin mutagenized the small-plaque strain of polyoma virus with nitrosoguanidine or hydroxylamine and plated the surviving virus on 3T3-Py3. Four mutants were isolated which form plaques on 3T3-Py3 but not on 3T3 or mouse embryo cells. The average burst size of these mutants is less than 1 percent that of wild-type virus growing in 3T3 cells, but in 3T3-Py3 cells the burst sizes of wild-type virus and the mutants were equal. Viral DNA isolated from the mutants has the same host range as the mutant virions; the failure of the mutants to grow in 3T3 cells is not therefore the result of a failure to adsorb or uncoat in these cells. None of the four mutants complement each other in 3T3 cells and none of them transform BHK or rat embryo cells (Benjamin, 1970). They induce cellular DNA synthesis in resting 3T3 monolayers (Benjamin and Burger, 1970), but, in contrast to wild-type virus, they do not induce the exposure of the wheat germ agglutinin site. It seems, therefore, that in productively infected cells, although host DNA synthesis is required for exposure of the agglutinin site, it is not sufficient to bring about that exposure, and another function, perhaps coded by the virus, is necessary.

The host-range mutants have not yet been assigned to any of the complementation groups defined by the *ts* mutants; however, it may well prove to be the case that Benjamin's mutants are absolute defectives in the *ts3* function. Two pieces of evidence support this conjecture: (a) the host-range mutants and *ts3* seem to be involved in the exposure of the wheat germ agglutinin site, and (b) the host-range mutants are complemented by a function expressed in transformed cells, and we know from the temperature shift experiments with cells transformed by *ts3* (Eckhart et al., 1971) that the *ts3* function is present in transformed cells. The fact that NG18 (the best characterized host-range mutant) induces cellular DNA synthesis in Swiss 3T3 cells, whereas *ts3* does not induce DNA synthesis in Balb/c-3T3 cells, does not necessarily mean that the *ts3* mutation and the NG18 mutation are not in the same gene. *Ts3* may be a double mutant with the second mutation in the function required to induce cellular DNA synthesis; alternatively, the differences in response may be caused by intrinsic differences between Swiss and Balb/c cells.

RECOMBINATION

Attempts have been made to demonstrate genetic recombination between temperature-sensitive mutants of polyoma virus (Ishikawa and di Mayorca, 1971). These experiments suggest that recombination may occur at a low frequency (for example, about 0.24% with one pair of mutants [Ishikawa and di Mayorca, 1971]) after mixed infection by two different mutants.

Dubbs and Kit (1970) have also apparently detected recombination between plaque-morphology variants of SV40. They transformed mouse cells with a mixed stock of SV40 carrying two distinct plaque-morphology markers (fuzzy plaque and small clear plaque). They then fused the transformed mouse cells with permissive monkey cells and determined the plaque morphology of the rescued SV40 virus. In addition to the two parental plaque types, they detected a third plaque type, large clear plaques. Subsequently Dubbs et al. (1972) isolated from monkey cells infected with a mixed stock of fuzzy plaque and small clear plaque oligomeric SV40 DNA molecules. Eleven percent of the plaques produced when this DNA was used to infect monkey cells had the large clear plaque morphology. Furthermore the large clear plaque strain isolated by both these methods bred true. These various experiments have established that recombination can occur between pairs of polyoma virus and pairs of SV40 genomes. But the frequency of recombination is not high enough to allow recombination analysis of the genomes of these viruses.

Literature Cited

Benjamin, T. L. 1970. Host-range mutants of polyoma virus. Proc. Nat. Acad. Sci. *67:* 394.

Benjamin, T. L. and M. M. Burger. 1970. Absence of a cell membrane alteration function in non-transforming mutants of polyoma virus. Proc. Nat. Acad. Sci. *67:* 929.

Cuzin, F., M. Vogt, M. Dieckmann and P. Berg. 1970. Induction of virus multiplication in 3T3 cells transformed by a thermosensitive mutant of polyoma virus: II. Formation of oligomeric polyoma DNA molecules. J. Mol. Biol. *47:* 317.

di Mayorca, G., J. Callender, G. Marin and R. Giordano. 1969. Temperature-sensitive mutants of polyoma virus. Virology *38:* 126.

Dubbs, D. R. and S. Kit. 1970. Isolation of double lysogens from 3T3 cells transformed by plaque morphology mutants of SV40. Proc. Nat. Acad. Sci. *65:* 536.

Dubbs, D. R., S. Kit, R. Jaenisch and A. J. Levine. 1972. Isolation of SV40 recombinants from cells infected with oligomeric forms of SV40 DNA. J. Virol. *9:* 717.

Dulbecco, R., 1970. Topoinhibition and serum requirement of transformed and untransformed cells. Nature *227:* 802.

Dulbecco, R. and W. Eckhart. 1970. Temperature-dependent properties of cells transformed by a thermosensitive mutant of polyoma virus. Proc. Nat. Acad. Sci. *67:* 1775.

Eckhart, W. 1969. Complementation and transformation by temperature-sensitive mutants of polyoma virus. Virology *38:* 120.

———. 1971a. Polyoma gene functions required for cell transformation. In CIBA Found. Symp., *Strategy of the Viral Genome*, p. 267. Churchill/Livingstone, London.

———. 1971b. Induced cellular DNA synthesis by early and late temperature-sensitive mutants of polyoma virus. Proc. Roy. Soc. London B *177:* 59.

Eckhart, W., R. Dulbecco and M. Burger. 1971. Temperature-dependent surface changes in cells infected or transformed by a thermosensitive mutant of polyoma virus. Proc. Nat. Acad. Sci. *68:* 283.

Folk, W. R. 1973. Induction of virus synthesis in polyoma-transformed BHK-21 cells. J. Virol. In press.

Fried, M. 1965a. Isolation of temperature-sensitive mutants of polyoma virus. Virology *25:* 669.

———. 1965b. Cell-transforming ability of a temperature-sensitive mutants of polyoma virus. Proc. Nat. Acad. Sci. *53:* 486.

———. 1970. Characterization of a temperature-sensitive mutant of polyoma virus. Virology *40:* 605.

Hirt, B. 1967. Selective extraction of polyoma DNA from infected mouse cell cultures. J. Mol. Biol. *26:* 365.

Ishikawa, A. and G. di Mayorca. 1971. Recombination between two temperature-sensitive mutants of polyoma virus. Lepetit Colloq. Biol. Med. *2:* 294. North-Holland, Amsterdam.

Kimura, G. and R. Dulbecco. 1972. Isolation and characterization of temperature-sensitive mutants of simian virus 40. Virology *49:* 394.

Kit, S., S. Tokuno, K. Nakajima, D. Trkula and D. R. Dubbs. 1970. Temperature-sensitive simian virus 40 mutant defective in late function. J. Virol. *6:* 286.

Oxman, M. N., K. K. Takemoto and W. Eckhart. 1972. Polyoma T antigen synthesis by temperature-sensitive mutatants of polyoma virus. Virology *49:* 675.

Ozer, H. L. and P. Tegtmeyer. 1972. Synthesis and assembly of simian virus 40. II. Synthesis of the major capsid protein and its incorporation into viral particles. J. Virol. *9:* 52.

Robb, J. A. and R. G. Martin. 1972. Genetic analysis of simian virus 40. III. Characterization of a temperature-sensitive mutant blocked at an early stage of productive infection in monkey cells. J. Virol. *9:* 956.

Robb, J. A., H. S. Smith and C. D. Scher. 1972a. Genetic analysis of simian virus 40. IV. Inhibited transformation of BALB 3T3 cells by a temperature-sensitive early mutant. J. Virol. *9:* 969.

Robb, J. A., P. Tegtmeyer, R. G. Martin and S. Kit. 1972b. Proposal for a uniform nomenclature for simian virus 40 mutants. J. Virol. *9:* 562.

Summers, J. and M. Vogt. 1971. Recovery of virus from polyoma-transformed BHK21. Lepetit Colloq. Biol. Med. *2:* 306.

Tegtmeyer, P. 1972. SV40 DNA synthesis: The viral replicon. J. Virol. *10:* 591.

Tegtmeyer, P. and H. L. Ozer. 1971. Temperature-sensitive mutants of SV40: Infection of permissive cells. J. Virol. *8:* 516.

Tegtmeyer, P., C. Dohan, Jr. and C. Reznikoff. 1970. Inactivating and mutagenic effects of nitrosoguanidine on simian virus 40. Proc. Nat. Acad. Sci. *66:* 745.

Vogt, M. 1970. Induction of virus multiplication in 3T3 cells transformed by a thermosensitive mutant of polyoma virus: I. Isolation and characterization of Ts-a-3T3 cells. J. Mol. Biol. *47:* 307.

8

The Adenoviruses

The adenoviruses, which were discovered in 1953 (Rowe et al., 1953; Hilleman and Werner, 1954), seem largely to cause respiratory diseases in their natural hosts, but many have oncogenic potential for rodents. This was discovered by Trentin et al. (1962), who found that human adenovirus 12 induces tumours in newborn hamsters. Since then human adenovirus 12 has been shown to cause tumours in newborn rats (Huebner et al., 1963), the African rodent *Mastomys natalensis* (Rabson et al., 1964) and some strains of mice (Rabson et al., 1964; Yabe et al., 1964). At the same time the list of adenoviruses with oncogenic potential has grown rapidly; in addition to those human adenoviruses that possess oncogenic potential, some adenoviruses of simian (Hull et al., 1965), bovine (Darbyshire, 1966) and avian (Sarma et al., 1965) origin have been shown to induce tumours in newborn rodents.

In vitro transformation by adenoviruses was first demonstrated by McBride and Wiener (1964), who infected cultures of newborn hamster kidney cells with human adenovirus 12 and detected transformed cells about 8 to 10 weeks later. Subsequently Freeman et al. (1967) reported that the transformation by human adenovirus 12 of rat embryo fibroblasts, grown in a medium containing only low concentrations of Ca^{++}, is more rapid and reproducible.

The adenoviruses, described formally as non-enveloped, icosahedral viruses containing linear double-stranded DNA (Pereira et al., 1963), are much larger and more complex than the papova viruses. Each particle weighs about 175×10^6 daltons and contains 12–14 percent ($20–25 \times 10^6$ daltons) DNA; the genome is therefore large enough to code for 25–50 average-sized polypeptides. The virion shows some morphological differentiation and is composed of at least nine species of polypeptide chains. In spite of this increased complexity, the papova

Table 8.1. ADENOVIRUSES

Natural Host Species	Serological Subgroup*	Serological Types
Human	I II III	31
Simian		19
Bovine		3(−5)
Canine		2
Murine		2
Avian†		10
Opossum		1

Adapted from Wilner (1969).

* The human adenoviruses can be classified into subgroups on the basis of serological tests. Subgroup I contains types 3, 7, 11, 14, 16, 20, 21, 25, 28; subgroup II contains types 8, 9, 10, 13, 15, 17, 19, 22, 23, 24, 26, 27, 29, 30; subgroup III contains types 1, 2, 4, 5, 6, 12, 18, 31.

† The avian adenoviruses are morphologically similar to other adenoviruses but lack the common group-specific antigen. One avian serotype (CELO) has been studied in detail (Laver et al., 1971). This virus has a larger DNA (30×10^6 daltons) and a different polypeptide composition than human adenoviruses.

viruses and the adenoviruses have much in common:

1. Both replicate and are assembled in the nuclei of permissive cells.

2. Both cause tumours in newborn rodents and transform cells in tissue culture.

3. No infectious virus can be found in tumours or cells transformed by either of the two groups of viruses. In both cases, however, the viral genome persists, since viral DNA, viral RNA and virus-specific antigens can be detected in the cells. Adenoviruses, however, have not so far been rescued from transformed cells by fusion with permissive cells.

More than 50 different adenoviruses have been isolated from a variety of animal species (Table 8.1) and all except the avian adenoviruses share a group-specific antigen. Three subgroups of human adenoviruses have been distinguished on the basis of their ability to agglutinate rhesus monkey and rat erythrocytes (Rosen, 1960). Each subgroup contains a number of serotypes that are recognized because of type-specific antigens in their capsid (Table 8.1). These antigens can be identified by hemagglutination inhibition or neutralization assays.

MORPHOLOGY AND POLYPEPTIDE COMPOSITION OF THE VIRION

Electron micrographs show that the adenoviruses are about 80 mμ in diameter and have 20 equilateral triangular faces (see Figure 8.1) (Horne et al., 1959; Valentine and Pereira, 1965; Norrby, 1966). The capsid comprises 252 capsomers, 240 of which have six neighbours and are called hexons, while 12 sit at the vertices, have five neighbours and are called pentons. Each penton consists of a base anchored in the capsid, from which projects a fibre consisting of a rod with a 4-mμ knob at the end (Valentine and Pereira, 1965; Pettersson et al., 1968). The length of the fibre varies among the different serological subgroups (Rosen, 1960). Viruses of subgroup I have the shortest fibres, about 10 mμ long, while those of subgroup III are the longest, some 25–30 mμ (Norrby, 1968, 1969b) (Figure 8.1). Pentons may be selectively removed from the adenovirus type 2 capsid after dialysis against distilled water (Laver et al., 1969) or Tris-maleate buffer at pH 6.0–6.5 (Prage et al., 1970). The penton-less virions are stable in the cold, but their DNA becomes sensitive to DNase.

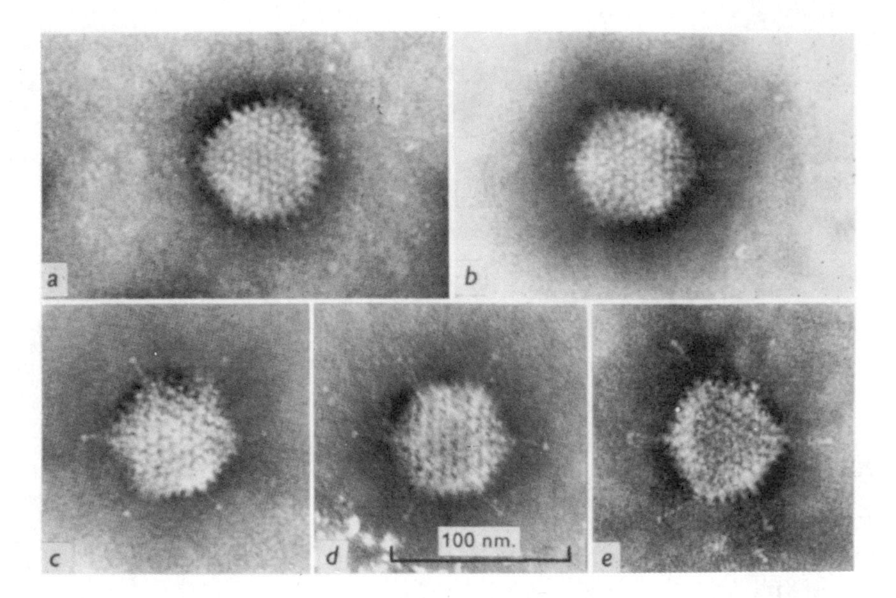

Figure 8.1. Adenovirus particles of different human serotypes showing the icosahedral array of capsomers and fibres of different lengths. **a,** Type 3 (subgroup I); **b,** type 15 (subgroup II); **c,** type 4 (subgroup III); **d,** type 6 (subgroup III); **e,** type 2 (subgroup III). Reprinted with permission from Norrby, *J. Gen. Virol.* 5, 224, 1969.

The 180 hexons which form the triangular faces of the adenovirus icosahedron have slightly different properties than the hexons which surround the pentons (Figure 8.2); after degradation of the particle with SDS, urea or pyridine (Smith et al., 1965; Maizel et al., 1968a; Prage et al., 1970), the hexons of the triangular faces are released as symmetrical aggregates of nine hexons and they are associated with one or two small polypeptides (Maizel et al., 1968b; Everitt et al., 1973).

Electron microscopic studies in thin sections of adenoviruses have revealed a dense core which contains DNA and stains heavily with uranyl acetate (Epstein, 1959; Epstein et al., 1960; Bernhard et al., 1961). This core has been isolated from virions by a variety of treatments including exposure to acetone (Laver et al., 1967, 1968), 5 M urea (Maizel et al., 1968b), 10% pyridine or repeated freezing and thawing (Prage et al., 1968, 1970). The core consists of the viral DNA associated with two polypeptides. One of these, "the major core protein," has a molecular weight of 17,000 daltons and is rich in arginine (21%) and

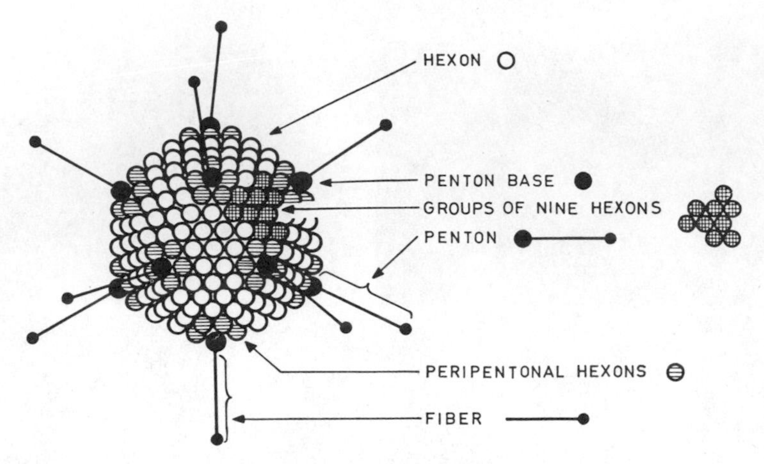

Figure 8.2. Schematic drawing showing the location of various components in the adenovirus capsid. Reprinted with permission from Philipson and Pettersson, *Advances in Tumor Virus Research*, vol. 18. Academic Press, 1972.

alanine (18%) and thus resembles the arginine-rich histones (Laver, 1970; Prage and Pettersson, 1971; Russell et al., 1971). It differs from cellular histones because it contains tryptophan and because it is precipitated by virus-specific antisera (Prage and Pettersson, 1971) and, unlike the core proteins of SV40 and polyoma virus particles, there is no evidence that this protein is derived from the host. The second core protein has a molecular weight of about 45,000 daltons and is moderately rich in arginine (Laver, 1970).

The polypeptide composition of adenoviruses has been studied by SDS-polyacrylamide gel electrophoresis (Maizel et al., 1968a,b; Pereira and Skehel, 1971; Everitt et al., 1973). Human adenovirus type 2, which has been extensively studied, may contain as many as 13 polypeptides (Everitt et al., 1973). Some of these polypeptides may be derived from the major capsid components by proteolytic degradation (Pereira and Skehel, 1971) and only nine or ten have so far been shown to be antigenically and topologically distinct (polypeptides II–X) (Figure 8.3). Five of these are integral parts of hexons (II), penton bases (III), fibres (IV) and the core (V and VII). The origin of the remaining polypeptides is not yet unambiguously established but some tentative locations have been assigned to most of them. Polypeptide VI

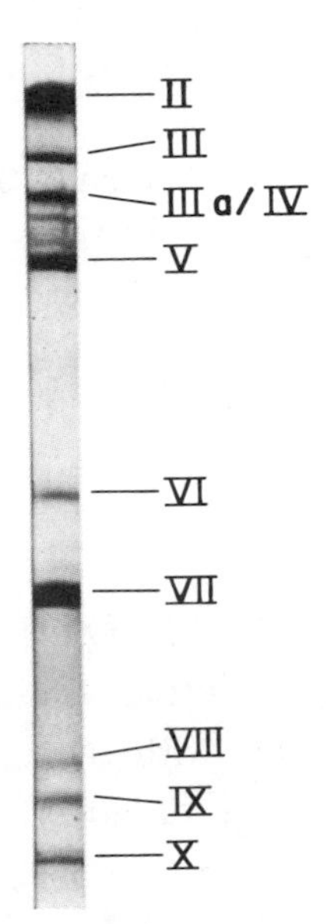

Figure 8.3. The polypeptides of adenovirus type 2 resolved by SDS-polyacrylamide gel electrophoresis. The anode is toward the bottom. (Modified from Everitt et al., 1973).

seems to be associated with all hexons in the capsid, while polypeptide IX and possibly polypeptide VIII are located adjacent to those hexons which form the triangular faces of the particle (Everitt et al., 1973; Maizel et al., 1968b). Polypeptide "IIIa", which is difficult to resolve from the polypeptide of the fibre (IV), appears to be derived from the peripentonal region (Everitt et al., 1973); the origin of the smallest polypeptide "X" is not yet established. The polypeptides which so far have been resolved by SDS gel electrophoresis account for approximately 30 percent of the coding capacity of the adenovirus genome.

The Structural Proteins

Unlike most other viruses, capsid proteins of the adenoviruses are soluble under non-denaturing conditions. This has made possible the detailed biochemical and immunological studies of these proteins, which have perhaps been more extensively analyzed than proteins from any other group of animal viruses. During lytic adenovirus infection the major structural components, hexons and pentons are produced in considerable excess, and it has been estimated that only 5–10 percent of the viral proteins made actually assemble into mature virions (Green, 1962). As a result hexons, pentons and fibres can conveniently be purified from infected cells and their amino acid compositions have been determined (Pettersson et al., 1967, 1968; Boulanger et al., 1969; Pettersson and Höglund, 1969).

Hexons, Fibres and Pentons

The hexons have a molecular weight of about 350,000 daltons (Franklin et al., 1971) (Table 8.2); low angle X-ray diffraction studies have shown that hexons resemble prolate ellipsoids with the dimensions 9×11 mμ (Lindquist, Philipson and Pettersson, unpublished data). On SDS-polyacrylamide gels hexons migrate as a single band, whose molecular weight has been estimated to be 120,000 daltons (Maizel et al., 1968b; Horwitz et al., 1970). The hexons from human adenoviruses of serotypes 2 and 5 have been crystallized (Pereira et al., 1968; Franklin et al., 1971; Cornick et al., 1971). Hexon crystals are of the cubic type and their space group is designated $P2_13$ (Franklin et al., 1971; Cornick et al., 1971). Such crystals are characterized by having 12 asymmetrical units per cell, and since the unit cell contains four hexons, there must be three crystallographic asymmetrical units per hexon. These units may correspond to the 120,000-daltons band revealed when hexons with a molecular weight of 350,000 daltons are subjected to SDS-polyacrylamide gel electrophoresis. X-ray diffraction analysis to a resolution of 10 Å has recently been reported by Franklin and coworkers (1972). Although the hexons are structurally simple, they are immunologically complex. All hexons contain one type-specific determinant known as ε (Kohler, 1965; Norrby, 1969a; Pettersson, 1971) and the type specificity is reflected by slight differences in the amino acid composition of hexons from different serotypes (Pettersson, 1971). In addition all hexons, except those from the avian adenoviruses, carry group-specific determinants which show a different degree of relatedness between members of the immunological subgroups (Table 8.1).

Table 8.2. Properties of Components of Adenoviruses

	Hexon	Penton Base	Fibre	Major Core Protein
Antigenic specificity	Group + type	Group + subgroup + intersubgroup	Type + subgroup + intersubgroup	Group + type
Number per capsid	240	12	12	~1000
Morphology	9 × 11 mμ ellipsoid[a]	8 mμ sphere[d]	10–25 × 2 mμ rod	
Mol wt (daltons)	310,000–360,000[b]	400,000–515,000[d] (base + fibre)	200,000[e] (serotype 2)	
Polypeptide size* (daltons)	120,000[c]; presumably 3/hexon	70,000[c]; presumably 5/penton base	60–65,000[c]; presumably 3/fibre	17,000
Biological activity		Cell-detaching factor Endonuclease Hemagglutination	Inhibition of cell macromolecular synthesis Hemagglutination	

Five additional polypeptides (IIIa, VI, VIII, IX and X) have been identified on SDS-polyacrylamide gels (Everitt et al., 1973).

* Estimated from SDS-polyacrylamide gel electrophoresis.

[a] Lindquist et al. (unpublished data). [b] Franklin et al. (1971). [c] Maizel et al. (1968b). [d] Pettersson and Höglund (1969); Wadell (1970). [e] Sundquist et al. (1973). [f] Prage and Pettersson (1971); Laver (1970).

The fibre has a complicated morphology; a 40-mμ sphere is attached to a rodlike structure, the length of which differs for members of the three immunological subgroups (see Figure 8.1) (Norrby, 1968, 1969b). Viruses in subgroup I have the shortest fibres; those in subgroup III have the longest fibres and the length of the fibre positively correlates with antigenic complexity.

Fibres from human adenoviruses of serotype 2 in subgroup III have a molecular weight of about 200,000 daltons, and polyacrylamide gel electrophoresis in SDS reveals one polypeptide chain with a molecular weight of 60,000–65,000 daltons (Sundquist et al., 1973) (Table 8.2). Thus, there appear to be three of these polypeptides in each fibre of viruses of serotypes belonging to the immunological subgroup III; it is not yet known, however, if all three peptides of the fibre have identical primary structures. All fibres carry type-specific antigenic determinants in their spherical part (Norrby et al., 1969), and the long fibres from subgroups II and III carry, in addition, antigenic determinants which are shared with other related serotypes (for detailed information, see Norrby, 1968, 1969b). The fibres attach to specific receptors on KB and HeLa cells (Philipson et al., 1968). They are also involved in the interaction between adenoviruses and the various kinds of erythrocytes which these viruses agglutinate. It has been observed that, when KB cells are exposed to fibres, synthesis of macromolecules is turned off, and in vitro fibres inhibit DNA and RNA polymerases. It is conceivable, therefore, that the decline in synthesis of protein and DNA which occurs late after infection is caused by an accumulation of this structural protein (Levine and Ginsberg, 1967, 1968; Ginsberg et al., 1967).

The fibre is attached to the penton base by non-covalent bonds (Norrby and Skaaret, 1967; Pettersson and Höglund, 1969) that can be disrupted by 2.5 M guanidine-HCl or 8% pyridine. The molecular weight of the entire penton (base plus fibre) is in the range 400,000–515,000 (Pettersson and Höglund, 1969; Wadell, 1970), and SDS-polyacrylamide gel electrophoresis has revealed that the base is composed of identical polypeptides with a molecular weight of about 70,000 daltons (Table 8.2). The base, which is very sensitive to proteolytic degradation by trypsin, contains antigenic determinants which are group and subgroup reactive (Wadell and Norrby, 1969; Pettersson and Höglund, 1969).

A number of structural units have been found to be involved in hemagglutination caused by adenoviruses. Norrby and coworkers have shown that agglutination of red cells can be caused by aggregates of 12 pentons ("dodecons"), dimers of pentons and fibres, and intact virions (see reviews by Norrby, 1968, 1969b). Monomers of pentons and fibres also agglutinate erythrocytes in the presence of heterologous antibodies which link pentons and fibres into dimeric units.

Enzyme Activities

No DNA polymerase activity has so far been identified in preparations of purified virions. An endonuclease activity, which cuts adenovirus DNA (31 S) into fragments sedimenting at about 18 S, has been isolated from cells infected with adenoviruses and from purified preparations of adenovirus type 2 and 12 (Burlingham and Doerfler, 1972). Pentons purified from infected KB cells have been found to carry a similar activity, although it has yet to be established that this activity is a property of the penton rather than of some associated peptide (Burlingham et al., 1971; Doerfler, personal communication). The penton endonuclease preferentially cleaves DNA in regions rich in G + C because it is inhibited to a greater extent by dG:dC polymers than by dA:dT polymers. No physiological function has yet been assigned to this endonucleolytic activity. Whole pentons or isolated penton bases also cause clumping and detachment of monolayer cells (Pereira, 1958; Everett and Ginsberg, 1958; Rowe et al., 1958; Pettersson and Höglund, 1969) ("the early cytopathic effect of adenoviruses"), but the biological significance of this effect is not understood.

THE GENOME OF ADENOVIRUSES

The adenoviruses contain linear duplex DNA with molecular weights in the range 20×10^6–25×10^6 daltons (Green et al., 1967a; Van der Eb et al., 1969). The highly tumourigenic serotypes (types 12, 18 and 31) have a slightly smaller genome than the nononcogenic serotypes (Green et al., 1967a). On the basis of reciprocal DNA/DNA or DNA/RNA hybridization, the human adenoviruses have been divided into three groups, groups A, B and C (Table 8.3). The members of any one group are more closely related to each other by their base sequence than to members of any other group, and each group has a characteristic content of G + C. It has been noted that among the human adenoviruses there is a correlation between oncogenicity and the G + C content of the DNA; the DNAs of the highly oncogenic serotypes have the lowest G + C content (48–49 percent). The simian adenoviruses do not, however, follow this rule; for instance, the highly oncogenic simian adenovirus SA7 has a G + C content of around 60 percent (Piña and Green, 1968; Goodheart, 1971). Although the DNAs of some adenoviruses and their host cells have very similar base compositions, nearest neighbour analyses show that the DNA of adenovirus 2 differs markedly from that of its host cell (Morrison et al., 1967). DNA molecules that have been extracted from serotypes 2 and 12 are not circularly permuted since they have unique denaturation maps (Doerfler and Kleinschmidt, 1970; Doerfler et al., 1972). Digestion with exonuclease III does not generate circular forms of adenovirus DNA and thus the adenovirus DNA is not terminally redundant (Green et al.,

Table 8.3. PROPERTIES OF HUMAN ADENOVIRUSES

Group	Members	Oncogenicity	% DNA	DNA % G + C	DNA/mRNA† % Homology
A	Ad 12, 18, 31	Highly oncogenic*	11.6–12.5	48–49[a]	30–60[b]
B	Ad 3, 7, 11, 14, 16, 21	Weakly oncogenic* (except Ad 11)	12.5–13.7	49–52[a]	40–100[b]
C	Ad 1, 2, 5, 6	Nononcogenic, but transform rat embryo cells[c]	12.5–13.7	57–59[a]	90–100[b]

Adapted from Green (1970).

* Highly oncogenic adenoviruses induce tumours in a large proportion of newborn hamsters within two months after injection with a purified virus; weakly oncogenic adenoviruses induce tumours in a small proportion of animals after 4–18 months.

† Hybridization of virus-specific RNA from transformed cells with viral DNA from other group members, given as percent homologous hybridization.

[a] Piña and Green (1965 and unpublished data). [b] Fujinaga and Green (1966, 1967a, b); Fujinaga et al. (1969). [c] Freeman et al. (1967).

1967a). On the other hand, it has been observed that each of the two strands of denatured adenovirus DNA is able to form single-stranded circles (Wolfson and Dressler, 1972; Garon et al., 1972). The circle formation is caused by terminal complementary sequences which are present on both strands.

The strands of adenovirus DNA have been separated by copolymer binding using IG or UG copolymers (Kubinski and Rose, 1967; Landgraf-Leurs and Green, 1971; Patch et al., 1972). Both strands bind the copolymers and separation is based on quantitative differences in binding capacity between the two strands. The two halves of adenovirus type 2 DNA differ in G + C content and they have been separated on CsCl gradients after degradation by controlled shearing (Kimes and Green, 1970; Doerfler and Kleinschmidt, 1970).

The DNA of adenovirus type 2 has been cleaved into six unique fragments by restriction endonuclease $R.R_1$ from *E. coli* carrying the drug-resistance transfer factor RTF-1 (Pettersson et al., 1973). All six fragments have been separated by gel electrophoresis and they will presumably be useful for mapping specific functions on the adenovirus chromosome.

The DNA of simian adenovirus 7 has been reported to be infectious and to cause tumours (Burnett and Harrington, 1968a,b), and Nicolson and McAllister (1972) find that human adenovirus 1 DNA can be used to infect human embryo kidney cells, albeit very inefficiently.

THE LYTIC CYCLE

Adsorption and Penetration

The lytic cycle of the adenoviruses has been studied chiefly using adenovirus type 2 and KB cells in Spinner culture (Green, 1962, 1965, 1966, 1970). The time course of infection is shown in Figure 8.4. Similar results have been obtained with other types of adenoviruses, but the growth cycle of the highly oncogenic group is a little longer (Mak and Green, 1968).

Adsorption and penetration have been studied by electron microscopy (Dales, 1962; Morgan et al., 1969; Chardonnet and Dales, 1970a,b)

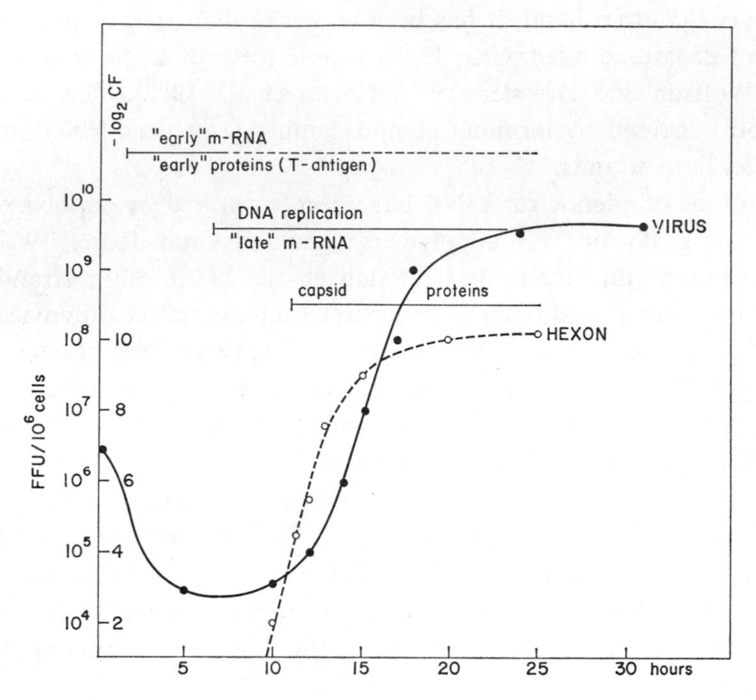

Figure 8.4. Growth curve of type 2 adenovirus in suspension cultures of KB cells. (●——●) Intracellular virus measured as fluorescent focus-forming units/10^6 cells. (○——○) Hexon antigen monitored by complement fixation. (Modified from Philipson and Pettersson, *Advances in Tumor Virus Research*, vol. 18. Academic Press, 1972).

and by following the fate of radioactively labeled virus (Philipson et al., 1968; Lawrence and Ginsberg, 1967; Philipson, 1967; Lonberg-Holm and Philipson, 1969; Sussenbach, 1967). The ratio of particles to plaque-forming units of the adenoviruses is high (estimates range from 11:1 to 2300:1; Green et al., 1967b), and because one can never be sure that the small number of infectious virions follows the same biochemical path as the noninfectious majority, any results obtained by following the fate of input particles are open to question. Given that caveat, it seems likely that the virions attach to specific receptors on the plasma membrane (Philipson et al., 1968). The particles enter the cell by direct penetration (Morgan et al., 1969) so that within 15 minutes the attached particles are turned into partially uncoated virions which appear to lack pentons and some of the peripentonal hexons (Sussenbach, 1967;

Lonberg-Holm and Philipson, 1969). These structures are later converted into membrane-bound cores which enter the nucleus where the DNA becomes exposed and transcription begins (Lonberg-Holm and Philipson, 1969). The initial stages of the virus uncoating take place in the presence of inhibitors of protein synthesis and are probably brought about by preexisting enzymes (Lawrence and Ginsberg, 1967). The time course for the uncoating events is highly dependent on the multiplicity of infection; high multiplicities give a more rapid and synchronous infection and under optimal conditions uncoating takes approximately two hours (Lonberg-Holm and Philipson, 1969).

Early Events of the Lytic Cycle

Synthesis of Early Viral RNA

Since adenovirus particles do not seem to contain any RNA polymerase, it is likely that at least early viral RNA is synthesized by a cell enzyme. Several lines of evidence (Ledinko, 1971; Price and Penman, 1972a; Wallace and Kates, 1972; Chardonnet et al., 1972) indicate that viral messenger RNA at both early and late times is transcribed by an RNA polymerase which is inhibited by the cyclic peptide α-amanitin and thus resembles RNA polymerase II of mammalian cells (Roeder and Rutter, 1970). A segment of polyadenylic acid is added to the viral transcripts both early and late after infection (Philipson et al., 1971); this poly(A) segment is 150–250 nucleotides long and is probably added to the RNA after transcription since poly(A) does not hybridize to adenovirus DNA (Philipson et al., 1971). The poly(A) sequences may be required for proper cleavage and transport of the viral RNA because in the presence of analogues to adenosine (cordycepin, toyocamycin) much less viral messenger RNA can be detected on the cytoplasmic polysomes and high molecular weight RNA accumulates in the nucleus (Philipson et al., 1971; McGuire et al., 1972). Since nearly all viral RNA, both in the nucleus and in the cytoplasm, contains poly(A), it is possible to purify viral messenger RNA by affinity chromatography on columns of polyuridylic acid linked to agarose beads (Lindberg et al., 1972). Viral messenger RNA is assayed by hybridization of pulse-labeled RNA with viral DNA (Rose et al., 1965; Bello and Ginsberg, 1969; Thomas and Green, 1969).

We still do not have a complete picture of the pattern and control of transcription of adenovirus DNA in permissive cells, but the following relevant observations have been made. (1) The RNA sequences transcribed prior to the onset of DNA replication ("early RNA") differ from those transcribed late in the infectious cycle ("late RNA"). (2) Ten to twenty percent of the genome (the complete genome is defined as sequences corresponding to one strand of adenovirus DNA) of adenovirus type 2 is transcribed early after infection (Fujinaga and Green, 1970). This early RNA can be detected as early as 1–2 hours after infection, but at these very early times the viral RNA constitutes only a minute fraction of the total RNA synthesized in the cells (Wall et al., 1972). (3) Some sequences of early RNA are apparently not synthesized late after infection (Lucas and Ginsberg, 1971), and some sequences are synthesized at an enhanced rate in the presence of inhibitors of protein synthesis (Parsons and Green, 1971). (4) Viral RNA from the nucleus shows broad distribution in size on sucrose gradients and molecules as large as 10^7 daltons have been observed (Wall et al., 1972). (5) By contrast, polysomal RNA from the early stages of infection sediments between 15 S and 30 S, with broad peaks at 15–17 S and 20–24 S and a minor peak at 26–28 S (Parsons and Green, 1971; Lindberg et al., 1972). (6) Competition hybridization experiments between unlabeled polysomal RNA and radioactively labeled nuclear RNA show that about 25 percent of the sequences present in nuclear RNA may never reach the cytoplasm (Wall et al., 1972).

Synthesis of Viral Protein

Although the adenoviruses are assembled in the nucleus, their proteins are synthesized in the cytoplasm because adenovirus-specific messenger RNA is found associated with cytoplasmic polysomes (Thomas and Green, 1966; Velicer and Ginsberg, 1968); however, very little is known about the early proteins coded by the virus. Most workers have used complement fixation or immunofluorescence to detect new antigens early after virus infection. The most prominent of these appears to be T antigen, which, like the T antigen induced by papova viruses, reacts with sera from tumour-bearing hamsters (Rouse and Schlesinger, 1967; Russell et al., 1967; Pope and Rowe, 1964; Shimojo et al., 1967). But unlike the T antigen of the papova viruses, adenovirus T antigen is found in both the cytoplasm and the nucleus. The induction of T antigen does not depend on DNA synthesis, for it is detected in the presence of 5-fluorodeoxyuridine (Gilead and Ginsberg, 1965) or cytosine arabinoside (Feldman and Rapp, 1966). Another antigen known as P antigen is also detected early after infection (Russell and Knight, 1967). This antigen is probably a mixture of T antigen and one of the core proteins (Russell and Skehel, 1972).

The early proteins have also been studied by SDS-polyacrylamide gel electrophoresis of infected cell extracts after labeling with [^{35}S]-methionine of high specific activity. By this method at least five prominent bands, which are absent in uninfected cells, have been observed early after infection (Russell and Skehel, 1972). There is, however, no direct evidence that these polypeptides or the T antigen are coded by the viral genome.

Induced Enzyme Synthesis

In exponentially growing cells there is little increase in any of those enzyme activities which are induced after infection with polyoma virus or SV40. In slower growing cells, however, such as human embryo kidney or monkey kidney cells, up to 20-fold increases in thymidine kinase and a minor increase in DNA polymerase activities have been reported (Takahashi et al., 1966; Ledinko, 1967; Kit et al., 1967; Bresnick and Rapp, 1968; Ogino and Takahashi, 1969).

Synthesis of Viral DNA

The DNAs of group C adenoviruses (see Table 8.3) have a considerably higher content of G + C than mammalian DNA and thus viral and host cell DNA can be separated by equilibrium centrifugation in CsCl or by chromatography on methylated albumin kieselguhr. Synthesis of viral DNA in KB cells infected with serotypes 2 or 5 begins in the nucleus 6–7 hours after infection. The maximum rate of synthesis is reached about 13 hours after infection and at this time the rate of synthesis of host cell DNA has declined so that 90 percent or more of the newly synthesized DNA is viral (Ginsberg et al., 1967; Pettersson, 1973). No information is yet available about the nature of the enzymes involved in the synthesis of adenovirus DNA.

The replicating adenovirus DNA has a higher buoyant density than mature viral DNA probably because of an increased content of single-stranded DNA (Sussenbach et al., 1972; Pettersson, 1973). Pulse-labeling for short times with radioactive thymidine reveals that the isotope is incorporated into DNA which bands 5–10 mg/ml heavier than completed adenovirus DNA (Pettersson, 1973). This difference in buoyant density is eliminated after digestion with a single-strand specific nuclease and it has been estimated that the replicating DNA contains 25 percent single-stranded DNA (Pettersson, 1973). The intermediates of replication contain no DNA that is longer than one complete strand of adenovirus DNA (Horwitz, 1971). Electron microscopy of adenovirus DNA, synthesized in isolated nuclei, has shown that the two strands of adenovirus DNA may be replicated asynchronously so that structures arise which consist of duplex DNA with tails of single-stranded DNA (Sussenbach et al., 1972). It has also been reported that during replication of adenovirus type 12 DNA three distinct classes of fragments are synthesized in addition to unit-length DNA (Sussenbach and van der Vliet, 1972). Approximately 20 percent of all viral DNA synthesized during infection is incorporated into mature virions (Green, 1962b).

Late Events of the Lytic Cycle

Synthesis of Viral RNA

After the onset of viral DNA replication the transcriptional pattern is altered so that, for example, 80–100 percent of the viral genome is transcribed in cells infected with type 2 adenoviruses (Fujinaga et al., 1968). There is presently no clue as to how this change is accomplished and several explanations are conceivable: (1) The virus might code for an RNA polymerase which, unlike host polymerases, is able to synthesize late RNA. (2) The host-cell polymerase might, after infection, become modified so that it recognizes promotors for late RNA. (3) The basic proteins that are associated with the adenovirus DNA might restrict transcription during the early phase after infection but not at late times. (4) There might be a change in the physical state of the viral template. The viral DNA may, for instance, become integrated early after infection and thereby gain access to host promotors and terminators.

Late RNA in the nucleus consists of a heterogeneous population of RNA molecules, the majority of which have sedimentation coefficients in the range 10–43 S (Parsons et al., 1971); but molecules as large as 80 S have been observed and the large species (RNA $>$ 45 S) contain sequences which are not present on the polysomes (Wall et al., 1972). The polysomal RNA has been fractionated by sucrose gradient centrifugation and gel electrophoresis, and a heterogeneous population of RNAs has been resolved in the size range 0.3–2.0 $\times$ 10^6 (12–30 S), with prominent peaks at 0.95 $\times$ 10^6 (22 S) and 1.6 $\times$ 10^6 daltons (26 S) (Parsons et al., 1971; Lindberg et al., 1972; Bhaduri et al., 1972). Late after infection about 95 percent of the messenger RNA present on polysomes is specified by the virus (Lindberg et al., 1972), but only 30 percent of the viral RNA in the cytoplasm is associated with the polysomes (Raskas and Okubo, 1971). Since the viral messenger RNA isolated from the polysomes is smaller than the virus-specific RNA in the nucleus, it appears that cleavage occurs before transport of the messenger RNA to the cytoplasm (Parsons and Green, 1971; Wall et al., 1972; McGuire et al., 1972). The processing of adenovirus RNA has been studied by Raskas and coworkers (Raskas, 1970; Brunner and Raskas, 1972). They have shown that isolated nuclei from adenovirus-infected cells release virus-specific RNA after incubation in the presence of ATP. Before release the RNA is cleaved to the same size as RNA found on polysomes in infected cells (Brunner and Raskas, 1972).

A virus-specific 5.5 S RNA (VA RNA) is synthesized in abundance late after infection but its function is unknown (Ohe et al., 1969; Ohe, 1972). This RNA consists of 156 nucleotides and its entire sequence has been determined; there appears to be one copy of VA RNA per viral genome, and VA RNAs from cells infected with different serotypes show some minor sequence differences (Ohe and Weissman, 1970; Ohe, 1972). These RNAs have a high degree of secondary structure and their base sequences would allow many regions to form base-paired loops (Ohe and Weissman, 1970). Unlike the adenovirus messenger RNA, VA RNA is apparently not transcribed by the α-amanitin-sensitive RNA polymerase since isolated nuclei continue to synthesize VA RNA in the presence of this drug (Price and Penman, 1972b). VA RNA also differs from other RNAs in that it is not derived from a large precursor molecule (Price and Penman, 1972b).

Host Macromolecular Synthesis

KB cells infected with adenovirus do not divide, and during the replication cycle of the virus most of the macromolecular synthesis of the host cell is gradually shut down. Synthesis of cellular DNA begins to decline 6–8 hours after infection, and by 10–13 hours about 90 percent of the newly synthesized DNA is viral (Ginsberg et al., 1967; Pettersson, 1973). Late in infection the synthesis of ribosomal RNA decreases to about 20 percent of that in uninfected cells (Raskas et al., 1970), while the synthesis of cell proteins declines after 16 hours. Even at late times, however, the synthesis of cellular transfer RNA and messenger RNA continues (Ginsberg et al., 1967; Mak and Green, 1968). No new species of transfer RNAs specified by adenoviruses have been recognized (Raska et al., 1970; Kline et al., 1972).

Late Proteins and Virus Assembly

In cells infected with adenovirus 5, newly synthesized hexon antigen and fibre antigen can be detected about 11 hours after infection, and 2 or 3 hours later maturation of progeny virus begins (Russell et al., 1967; Russell and Skehel, 1972). Most of the proteins synthesized in the cell during the late stages of infection are coded by the virus and concomitant viral DNA synthesis is required for their continued synthesis. Whereas hexons, pentons and fibres are synthesized in large excess, the synthesis of the core proteins is less excessive (White et al., 1969). Only 5–10 percent of the viral proteins synthesized during the infectious cycle become incorporated into mature virions (Green, 1962; White et al., 1969). The viral proteins are synthesized on 200 S polyribosomes and they are released, assembled, and transported to the nucleus a few minutes after synthesis (Velicer and Ginsberg, 1970; Horwitz et al., 1970). There is presently no evidence that any of the proteins in the virion are produced by a cleavage of large precursor molecules (Horwitz et al., 1969; White et al., 1969).

The virus is assembled in the nucleus where large quantities of viral proteins accumulate late after infection. Thin sections of infected cells display typical large crystalline structures ("paracrystals"; Morgan et al., 1957), which appear to be reservoirs of structural proteins (Boulanger et al., 1970), in particular of the major core protein (Marusyk et al., 1972).

Adenovirus maturation requires an exogenous supply of arginine (Rouse and Schlesinger, 1967; Russell and Becker, 1968). If this amino acid is omitted from the culture medium, the yield of virus is reduced by several orders of magnitude. Small amounts of viral DNA and capsid proteins, including those which are rich in arginine, are synthesized during arginine starvation but the synthesized proteins fail to assemble into mature particles (Everitt et al., 1971; Rouse and Schlesinger, 1972). All RNA sequences present under normal conditions appear to be transcribed in the absence of arginine (Raska et al., 1972) and presently it is not known what causes this specific requirement. It has, however, been reported that viral proteins synthesized in arginine-deficient medium assemble in vitro in the presence of extracts from infected cells that have been maintained in normal medium (Winters and Russell, 1971); it may therefore be possible to elucidate the role of arginine in maturation.

Empty shells and incomplete particles which contain reduced amounts of DNA are synthesized in cells infected with certain serotypes (Mak, 1971; Prage et al., 1972). These particles appear to lack the two core proteins (polypeptides V and VII) which are replaced by another peptide, larger than the major core protein (Prage et al., 1972).

Incomplete Infection

Whether an adenovirus infection is lytic, incomplete or transforming depends on the type of adenovirus and the type of cell involved, as well as the state of growth of the cells. Some of the different responses which have been described are shown in Table 8.4. There seems to be no meaningful pattern to these observations, and currently we

Table 8.4. RESPONSES OF DIFFERENT TYPES OF CELLS TO INFECTION BY ADENOVIRUSES

Cell Type	Adenovirus Serotype	Response
KB or HeLa	All human	Permissive
AGMK	Most human	Semipermissive*
Hamster embryo, BHK21, NIL-2	Ad 2	Permissive
	Ad 12	Nonpermissive†

* T antigen, viral DNA and RNA synthesis normal; synthesis of capsid proteins reduced; virus replication enhanced by simultaneous infection with SV40.

† Normal synthesis of T antigen; no synthesis of viral DNA, late messenger RNA or capsid proteins.

have no knowledge of the factor(s) which causes infections of different cells with the same virus to have such dissimilar consequences. Human adenoviruses, for example serotype 2, are unable to replicate efficiently in African green monkey kidney (AGMK) cells. A small amount of virus is synthesized but simultaneous infection with SV40 enhances the yield of adenovirus approximately 1000-fold (Rabson et al., 1964; Friedman et al., 1970). Other agents like simian adenoviruses (Naegele and Rapp, 1967; Altstein and Dodonova, 1968), adeno-SV40 hybrid viruses (Rowe and Baum, 1965) and the unidentified agent MAC also act as helpers (Butel and Rapp, 1967). The helper does not increase the number of cells susceptible to infection, but rather enhances the yield from each infected cell. During unenhanced infection, the virus seems to penetrate the cells and synthesis of T antigen and viral DNA occurs (Feldman et al., 1966; Friedman et al., 1970). The majority of the RNA sequences which are synthesized in AGMK cells in the presence of a helper are also made in the absence of helper (Baum et al., 1968; Fox and Baum, 1972), but the translation of late messenger RNA seems to be impaired because little capsid protein can be detected (Friedman et al., 1970; Baum et al., 1972). SV40 is unable to replicate its DNA during mixed infection, and it is likely that an early SV40 function facilitates translation of the late adenovirus messenger RNA (Friedman et al., 1970). The gene(s) which is responsible for this function is apparently present in the adeno-SV40 hybrid Ad2$^+$ND$_I$ (see below), although this virus contains only about 15 percent of the SV40 genome.

In hamster cells adenovirus 2 goes through a normal lytic cycle, whereas replication of serotype 12 is even more restricted than it is in AGMK cells. Adenovirus 12 enters BHK21 cells and induces T antigen synthesis and at least part of the early genes are transcribed (Strohl et al., 1967; Raska and Strohl, 1972). The DNA of serotype 12 is, however, unable to replicate in these cells even in mixed infections with adenovirus type 2 (Doerfler, 1969; Doerfler and Lundholm, 1970). No capsid proteins are detected and the incoming genome becomes fragmented into 18 S segments, possibly by the endonuclease that is associated with the pentons (Burlingham and Doerfler, 1971; Burlingham et al., 1971). Some viral genomes appear to integrate into the host chromosomes (Doerfler, 1968, 1970) and a small population of BHK cells becomes transformed (Strohl et al., 1970). The majority of the cells die, however, perhaps because of extensive chromosome fragmentation (Strohl, 1969a,b, 1972). In BHK cells arrested in the G_1 phase, adenovirus 12 induces cellular DNA synthesis (Strohl, 1969a,b). Most enzymes involved in DNA replication are activated (Zimmerman et al., 1970) but cyclic AMP blocks this stimulation and also suppresses T antigen synthesis (Zimmerman and Raskas, 1972).

ADENO-SV40 HYBRID VIRUSES

The PARA Adenovirus 7 (E46+)

The adeno-SV40 hybrid virions consist of recombinant DNA containing all or part of the SV40 genome and part or all of the adenovirus genome enclosed in adenovirus capsids. They were originally isolated either from stocks of adenoviruses which had been adapted to grow in rhesus monkey kidney cells for vaccine production (types 1–5 and 7) (Hartley et al., 1956), or from propagation of adenovirus and SV40 together in African green monkey kidney cultures (type 12) (Schell et al., 1966).

The first hybrid to be discovered was isolated from the twenty-eighth passage of the adenovirus 7 vaccine strain, L.L., in monkey kidney tissue cultures. This strain, isolated and propagated continuously in rhesus monkey kidney cells from 1955–1961 for a total of 22 passages, was then found to be contaminated with SV40. After two consecutive passages in the presence of SV40 antiserum the L. L. strain was freed of infectious SV40 virions. After the 28th passage the L. L. strain of adenovirus 7 had the following unusual properties (Huebner et al., 1964; Rowe and Baum, 1964; Rapp et al., 1964):

1. It productively infected monkey cells.
2. It induced tumours in hamsters which closely resembled those induced by SV40 and contained SV40 T antigen.
3. Anti-adenovirus antiserum destroyed oncogenicity but anti-SV40 antiserum had no effect.
4. Heating for 10 minutes at 56°C rendered the virus incapable of inducing SV40 T antigen. This treatment inactivates adenovirus 7 but has little or no effect on SV40.
5. During productive infection of monkey cells, no SV40 viral antigen or SV40 virions were produced.
6. All the particles had the morphology of adenovirions.

This strain of virus, which has been designated E46+ or PARA* adenovirus type 7 contains particles with SV40 and adenovirus genetic material enclosed in adenovirus capsids. Subsequent studies (Rowe and Baum, 1965) have shown the PARA adenovirus 7 strain contains two types of viruses, wild-type adenovirus type 7 and virus containing hybrid genomes of SV40 and adenovirus 7. The two DNAs present in the hybrid genome are covalently linked since, although the DNAs of SV40 and adenovirus 7 have different buoyant densities, the two moieties of hybrid DNA band at the same density after equilibrium centrifugation in alkaline CsCl (Baum et al., 1966). The amount of SV40 DNA in the hybrid genome has been determined by electron microscopy of heteroduplex molecules formed between adenovirus 7 DNA and hybrid DNA. Kelly and Rose (1971) have shown that the hybrid DNA contains an insertion corresponding to 75 percent of the complete SV40 DNA, beginning at 0.05 fractional lengths from one of the ends of the hybrid DNA. Approximately 10 percent of the original

* PARA = particles aiding replication of adenovirus.

adenovirus 7 DNA seems to be deleted and Kelly and Rose (1971) propose that the hybrid DNA was generated by two recombination events; one led to the insertion of a complete SV40 genome into the chromosome of adenovirus type 7, and a second intramolecular event deleted some SV40 DNA and some adenovirus DNA from the recombinant. Both genomes in the hybrid virus are defective. SV40 T and U antigens, but no SV40 capsid proteins, are synthesized in cells productively infected with PARA adenovirus 7. On AGMK cells the PARA adenovirus 7 strain plaques with two-hit kinetics (Figure 8.5). The hybrid carries the helper function which is required for growth of human adenoviruses on AGMK cells, and the plaques originate from cells which were simultaneously infected with the hybrid and wild-type adenovirus 7. The progeny of such plaques consists of a mixture of adenovirus 7 and hybrid virus. On human cells the PARA adenovirus

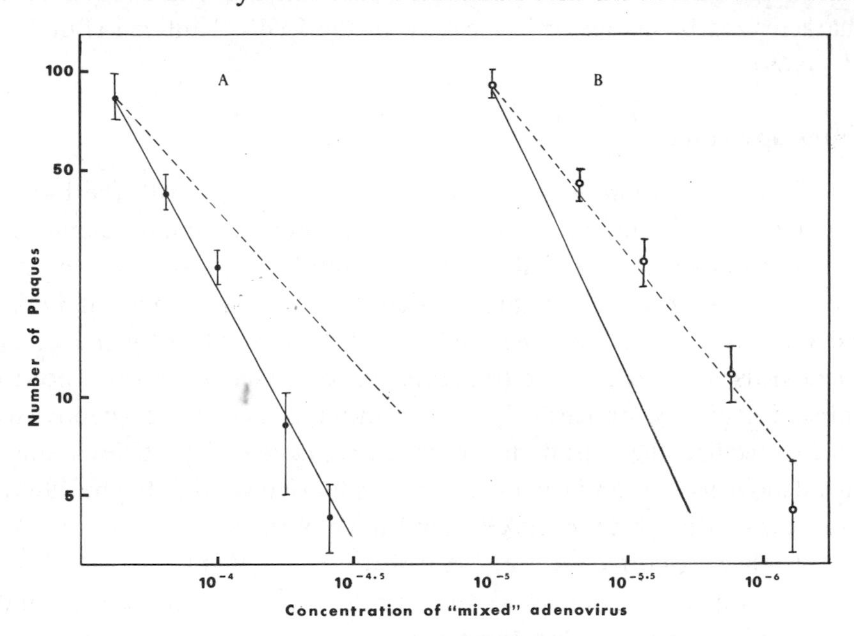

Figure 8.5. Results of plating adeno-SV40 hybrid viruses; mixed population of wild-type adenovirus 7 and PARA adenovirus 7 (Ad7-SV40) on African green monkey kidney cells **(A)** alone, and **(B)** in presence of high concentration of adenovirus type 7. Broken line = one-hit curve; i.e., one particle of dilute inoculum produces a plaque. Solid line = two-hit curve; i.e., plaque production occurs only when a cell is simultaneously infected with hybrid and wild-type particles. (Redrawn from Fenner, *The Biology of Animal Viruses*, Academic Press, 1968).

7 strain plaques with one-hit kinetics and at low multiplicities of infection the hybrid is lost. Since the hybrid genome is defective, it is unable to replicate and all plaques on human cells originate from cells which were infected with wild-type virions. The original strain of PARA adenovirus 7 was shown to be highly oncogenic when inoculated subcutaneously into newborn hamsters (Huebner et al., 1964). The tumour-bearing animals produced antibodies to SV40 T antigen and usually also to the T antigen of adenovirus type 7. In vitro transformation of hamster cells and human skin fibroblasts has been described (Black and Todaro, 1965) and transformation occurs with one-hit kinetics (Duff and Rapp, 1970). The transformed cells contain SV40 T antigen but usually not adenovirus T antigen (Diamond, 1967). Variants of the original hybrid population have been isolated; some of these exhibit different degrees of oncogenicity (Rapp et al., 1969) and others are characterized by cytoplasmic location of the SV40 T antigen (Butel et al., 1969).

Transcapsidation

Transcapsidation occurs when cells are coinfected with the E46+ population and an excess of another adenovirus immunologically distinct from adenovirus 7 (Rapp et al., 1965; Rowe, 1965). Among the progeny of such infections are particles containing the hybrid DNA (adenovirus 7 and SV40) enclosed in a capsid specified by the other adenovirus (Figure 8.6). The transcapsidated particle is sensitive not to antisera against adenovirus 7, but to antisera against the adenovirus which specified the capsid; however, transcapsidated particles induce an adenovirus 7 T antigen in infected cells (Rowe and Pugh, 1966). Transcapsidation between PARA and adenoviruses 1, 2, 5, 6, and 12 has also been reported (Rowe, 1965; Rapp et al., 1968).

Transcapsidated adeno-SV40 hybrid viruses are genetically and phenotypically mixed DNA tumour viruses.

Adeno 2-SV40 Hybrid Viruses

The generation of an adeno-SV40 hybrid virus is a rare event, but after the discovery of PARA adenovirus 7, several other of these viruses were found in stocks of various adenoviruses which had been

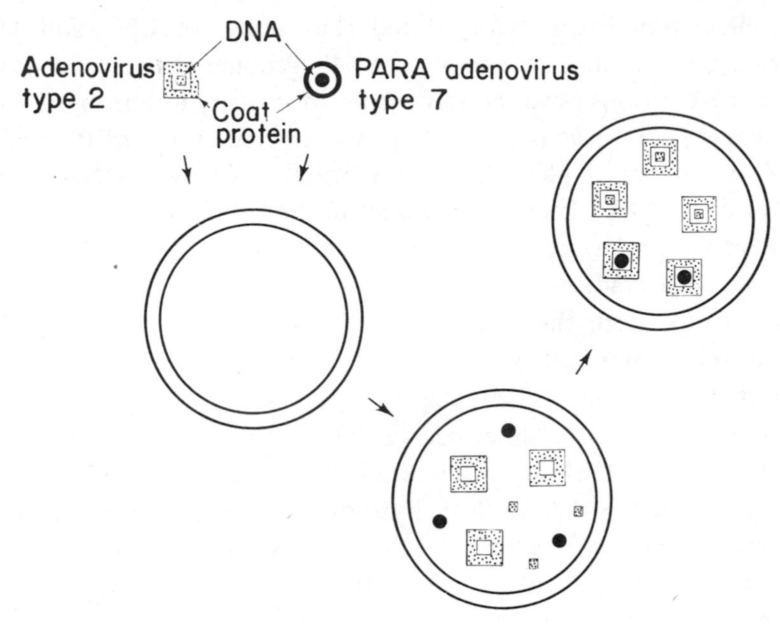

Figure 8.6. Diagrammatic representation of the phenomenon of transcapsidation. Simian cells were simultaneously infected with adenovirus type 2 and PARA adenovirus 7. The defective DNA of PARA adenovirus 7 replicates, but no adenovirus type 7 capsids are synthesized. Consequently the progeny genome of PARA adenovirus 7 becomes encased in an adenovirus type 2 capsid. (Redrawn from Rapp, *The Molecular Biology of Viruses*, p. 273. Cambridge University Press, 1968).

"adapted" to grow in monkey cells. A few adeno-SV40 hybrids have been described in which either the adenovirus or the SV40 genomes are not defective, by contrast to the adeno 7-SV40 hybrid, in which both genomes are defective. Such viruses have been isolated from a strain of adenovirus type 2 which was propagated in monkey cells. This strain, known as $Ad2^{++}$, contains wild-type adenovirus type 2, complete SV40 virions and a mixed population of hybrids between SV40 and adenovirus type 2 (Lewis et al., 1969). Several genetically stable hybrids have been segregated from this original strain after plaquing on AGMK cells. Two of these, known as $Ad2^{+}$HEY and $Ad2^{+}$LEY, yield on AGMK cells complete SV40 virus in addition to wild-type adenovirus 2 and hybrid virus (Lewis and Rowe, 1970). Both the LEY and HEY strains are mixtures of hybrid virus and wild-type adenovirus type 2. The yield of SV40 from the LEY strain is 0.01–0.1 percent of that from the HEY strain. It has been shown that the progeny of SV40 virus in both cases originates from hybrid viruses and thus both the LEY and HEY hybrids must contain a complete SV40 genome (Lewis and Rowe, 1970). LEY virions have the same buoyant density as adenovirus type 2 whereas HEY virus is 4 mg/ml lighter (Wiese et al., 1970), and its DNA is 3 mg/ml lighter than non-hybrid adenovirus 2 DNA (Crumpacker et al., 1970). The two genomes in the hybrid DNA of HEY are covalently linked since they are not separated after centrifugation in alkaline CsCl gradients or in alkaline sucrose gradients (Crumpacker et al., 1970). Both the HEY and LEY strains plaque with one-hit kinetics on human cells (only wild-type virus can replicate) and with two-hit kinetics on AGMK cells. Thus they resemble the PARA adenovirus 7 strain but differ because of their ability to yield infectious SV40 virus. The HEY strain gives SV40 plaques with one-hit kinetics (Lewis and Rowe, 1970). Both hybrids contain defective adenovirus genomes and the replication of the hybrid virus requires a mixed infection with wild-type adenovirus 2. It is not yet known why the HEY strain gives higher yields of SV40 than the LEY strain though both contain complete SV40 genomes; possibly an adenovirus function has been deleted from HEY which suppresses replication of SV40 virus in AGMK cells simultaneously infected with SV40 and adenovirus.

Another important group of hybrid viruses has been derived from the original $Ad2^{++}$ stock; in 1969 Lewis and coworkers isolated an unusual hybrid virus, designated $Ad2^{+}ND_1$, from HEK cells infected with pool B55, a derivative $Ad2^{++}$ stock. The $Ad2^{+}ND_1$ hybrid plaques with one-hit kinetics and almost equal titres on both human and AGMK cells (Lewis et al., 1969). It contains a nondefective adenovirus type 2 genome with a minor insertion of SV40 DNA and, unlike the defective hybrids, it remains stable after replication in human cell lines. The inserted fragment of SV40 DNA apparently contains the helper function which enables human adenoviruses to grow on AGMK cells. The two genomes in $Ad2^{+}ND_1$ are covalently linked (Levin et al., 1971). Electron microscopy of heteroduplex molecules of adenovirus 2 and $Ad2^{+}ND_1$ DNA has shown that the hybrid DNA contains an insertion corresponding to 18 percent of the complete SV40 DNA, and approximately 1.3×10^6 daltons of adenovirus DNA has been deleted from the hybrid DNA (Kelly and Lewis, unpublished data). The only detectable SV40 function in cells lytically infected with $Ad2^+ND_1$ is the induction of U antigen. This antigen resembles the SV40 T antigen; it is detected by antisera from hamsters carrying SV40-induced tumours, but differs from the T antigen in its heat stability. SV40-specific RNA has been detected in cells lytically infected with $Ad2^+ND_1$ (Oxman et al., 1971). RNA/DNA competition hybridization studies have shown that this RNA contains part of the sequences which are transcribed from the SV40 DNA early after lytic infection (Oxman et al., 1971).

In addition to $Ad2^+ND_1$ a number of other nondefective hybrids have been isolated (Kelly and Lewis, personal communication). The biological properties of these viruses is summarized in Table 8.5 (Lewis et al. and Lewis and Rowe, unpublished data). Differences in the biological properties of these hybrid viruses stem from differences in the fraction of the adenovirus 2 genome that is deleted and the fraction of the SV40 genome that is inserted. The amount of SV40 DNA inserted in each hybrid genome has been estimated by hybridizing SV40 RNA transcribed in vitro by *E. coli* RNA polymerase to the DNAs of the various hybrid viruses (Henry et al., unpublished data). Kelly and

Table 8.5. PROPERTIES OF ADENO-SV40 HYBRID VIRUSES

Hybrid	% SV40 Genome in Hybrid	SV40 Functions			Growth in AGMK
		T Antigen	U Antigen	TSTA[a]	
PARA adenovirus 7 (E46+)	75[b]	+	+	+	+
$Ad2^{++}HEY$	at least 100[c]	+	+	+	+
$Ad2^{++}LEY$	at least 100[c]	+	+	−	+
$Ad2^+ND_1$	see Table 8.6	−	+	−	+
$Ad2^+ND_2$	see Table 8.6	−	+	+	+
$Ad2^+ND_3$	see Table 8.6	−	−	−	−
$Ad2^+ND_4$	see Table 8.6	+	+	+	+
$Ad2^+ND_5$	see Table 8.6	−	−	−	−

[a] SV40-specific transplantation antigen.
[b] Estimated by electron microscopy (Kelly and Rose, 1971).
[c] Estimated by hybridization of in vitro synthesized SV40 c-RNA to hybrid DNA (Levine, personal communication).

Table 8.6. SIZE OF SV40 DNA IN NONDEFECTIVE ADENO-SV40 HYBRIDS AND MAP OF INDUCING FUNCTIONS OF SV40-SPECIFIC ANTIGENS

	SV40 DNA Segment		SV40 Antigen-inducing	% Ad2 Genome
	Daltons*	% Genome†	Functions‡	Deleted
$Ad2^+ND_1$	2.4×10^5	18	U	5.4
$Ad2^+ND_2$	6.2×10^5	32	U TSTA	6.1
$Ad2^+ND_3$	4.8×10^4	7		5.3
$Ad2^+ND_4$	8.4×10^5	43	U TSTA T	4.5
$Ad2^+ND_5$	5.3×10^5	28		7.1

* Amount of SV40 DNA estimated by hybridizing in vitro synthesized SV40-specific c-RNA with hybrid virus DNA.

† Estimated by heteroduplex mapping.

‡ Bar represents length of inserted SV40 DNA.

Lewis (unpublished data) have used the heteroduplex mapping technique to map the position of the SV40 DNA in the genomes of the hybrid viruses (Figure 8.7). Their conclusions can be summarized as follows: (1) Each nondefective hybrid contains a single substitution (SV40 DNA inserted, adenovirus 2 DNA deleted). (2) In all the hybrids the SV40 DNA sequences begin at the same position in the adenovirus 2 DNA molecule, namely, 14 percent from one end of the adenovirus DNA molecule. (3) The SV40 DNAs inserted are completely overlapping (see Table 8.6). (4) The corresponding adenovirus 2 DNA deletions form an overlapping series, all starting 14 percent of the genome length from one end of the adenovirus 2 DNA molecule.

Analyses of the SV40-specific RNA molecules made in VERO cells infected with these nondefective hybrids (Levine et al., unpublished data) indicate that the SV40 RNA species induced by a given hybrid virus are completely represented in the RNA species induced by all those hybrid viruses which contain more SV40 DNA. Moreover, the SV40 DNA sequences specified by $Ad2^+ND_4$ include all the RNA sequences transcribed early during the infection of monkey cells with wild-type SV40. Since the $Ad2^+ND_4$ genome contains about 40 percent of the wild-type SV40 genome, we can conclude that (1) the SV40 early genes comprise not more than 40 percent of the total SV40

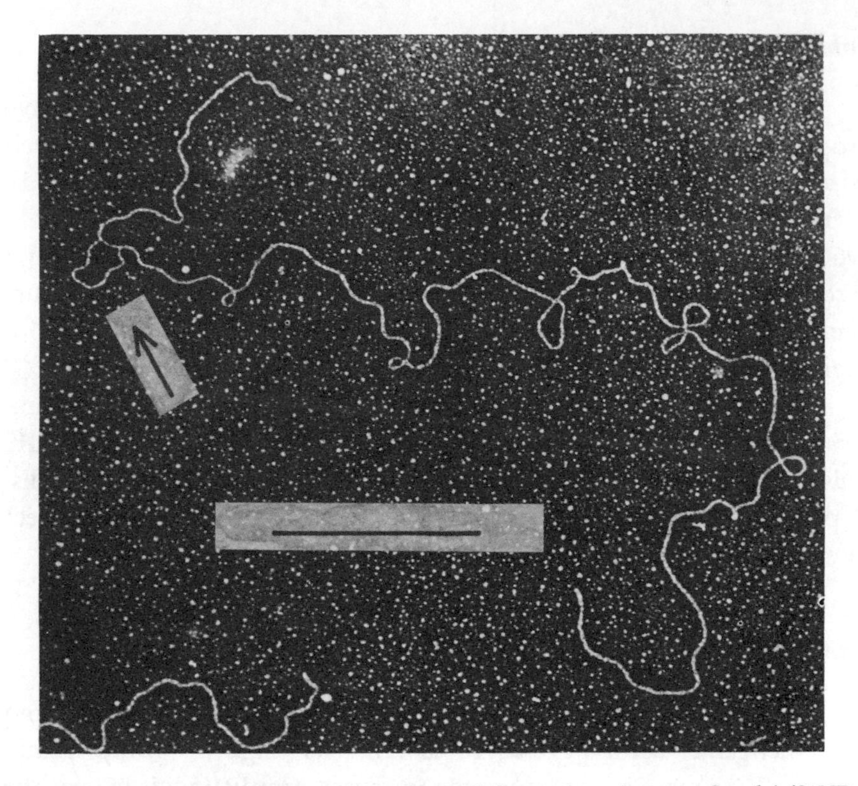

Figure 8.7. Heteroduplex molecule between DNA from adenovirus type 2 and Ad2^{+}ND$_{3}$. A region of non-matched DNA (arrow) is seen close to one of the ends of the heteroduplex molecule. Photograph courtesy of Dr. T. Kelly.

genome—this agrees well with estimates obtained from experiments with monkey cells infected with wild-type SV40 (see Chapters 5 and 6); (2) all the SV40 early genes are contiguous; (3) the induction of T, U and TSTA antigens are all early SV40 functions. Finally, by comparing the SV40-specific antigens induced by the various nondefective hybrid viruses, a self-consistent map of the genes regulating the induction of these functions can be constructed (see Table 8.6). The only anomaly is that Ad2^{+}ND$_{5}$, which contains all the SV40 DNA sequences in Ad2^{+}ND$_{1}$, does not induce SV40-specific U antigen. Ad2^{+}ND$_{5}$ carries the largest Ad2 DNA deletion and it is possible that information required for the expression of SV40 U antigen is specified by part of the Ad2 genome which is deleted in Ad2^{+}ND$_{5}$.

Other Adeno-SV40 Hybrid Viruses

Other adeno-SV40 hybrids which yield SV40 progeny and have properties somewhat comparable to Ad2^{++} have been described, such as adeno 4-SV40 (Easton and Hiatt, 1965; Beardmore et al., 1965), adeno 5-SV40 (Lewis et al., 1966), and adeno 12-SV40 (Schell et al., 1966). Adeno 12-SV40 hybrid virus has a considerably increased and accelerated oncogenicity for newborn hamsters; the resulting tumours phenotypically resemble those induced by adenoviruses but have both adenovirus 12 and SV40 T antigens.

One other adeno-SV40 hybrid virus, adeno 3-SV40 (Lewis et al., 1966) has been described. This virus induces SV40 T antigen in AGMK cells but does not produce SV40 plaques when plated on these cells. It presumably contains a defective SV40 genome and in this respect resembles PARA adenovirus 7.

TRANSFORMATION BY THE ADENOVIRUSES

The human adenoviruses are divided into "highly," "weakly" and "nononcogenic" serotypes (see Table 8.3). The "highly" (group A) and "weakly" (group B) oncogenic adenoviruses are distinguished by the frequency with which hamsters develop tumours after inoculation with virus. There is, however, presently no evidence that the tumours, once established, are fundamentally different. Adenoviruses belonging to group C transform rat embryo cells in vitro (Freeman et al., 1967; McAllister et al., 1969a) but are unable to induce tumours in vivo. Another group of adenoviruses (group "D") has been recognized; viruses in this group are not oncogenic in rodents but they transform rat and hamster cells in vitro (McAllister et al., 1969b). Viruses of groups C and D differ because they induce serologically different T antigens. It is not yet known what determines the range of oncogenicity of the various types of adenoviruses. It could be a reflection of the degree to which the cells of hamsters and other rodents are permissive for the virus of each particular group. We might assume, by analogy with the papova viruses, that the more nonpermissive the cell is, the more likely it is to be transformed. Alternatively, we might argue that some types of adenoviruses possess "tumourigenic cistrons" and some do not, or that the viral genes responsible for tumour formation have different degrees of expression in the different types of adenoviruses. Any acceptable theory must account for the observation that all the human adenoviruses of groups A and B cause tumours in animals and transform rat and hamster embryo cells in vitro, while the human adenoviruses of group C transform cells in vitro but do not induce tumours (McAllister et al., 1969a). Cell transformation with adenoviruses requires a large number of infectious units (10^4–10^6 plaque-forming units/focus-forming unit) and the frequency of transformation is low. Adenovirus-transformed cells are selected in much the same way as polyoma- and SV40-transformed cells, although they have not been examined as extensively as SV3T3 cells. Cells transformed by adenoviruses have a characteristic morphology, and BHK21 cells which have been: ransformed with adenovirus type 12 or polyoma virus are distinguishable (Strohl et al., 1967). Like SV3T3 cells, adenovirus-transformed cells form disoriented arrays, grow to higher saturation densities than the untransformed cells and have an altered surface membrane (see Chapters 3 and 6). No infectious adenovirus is detectable in transformed cells, but sera from tumour-bearing animals occasionally contain antibodies to capsid proteins (Huebner et al., 1964b). The production of infectious virus has never been induced either by making heterokaryons of transformed cells and permissive cells or by treating the transformed cells with physical or chemical agents that sometimes induce the production of SV40 and polyoma virus in transformed cells, which, before induction, were free of virus (Gerber, 1966; Burns and Black, 1969; Fogel and Sachs, 1969, 1970). Several lines of evidence indicate, however, that adenovirus genes persist in transformed cells.

Viral DNA Sequences in Transformed Cells

Viral DNA has been detected in adenovirus-transformed cells by the method of Westphal and Dulbecco (1968). RNA, transcribed in vitro (c-RNA) from adenovirus DNA by the *E. coli* RNA polymerase, has been shown to hybridize specifically to DNA extracted from cells transformed by the homologous adenovirus (Green et al., 1970). The amount of viral DNA in adenovirus-transformed cells has been measured by two different methods. Green and coworkers (1970) hybridized radioactive RNA transcribed in vitro to filters containing known amounts of DNA from transformed cells and reported that rat cells transformed by adenovirus type 2 contain between 14 and 37 copies of viral DNA per cell. Hamster cells transformed by adenovirus types 12 and 7 contain 22 to 97 copies of viral DNA as estimated by the same method. Pettersson and Sambrook (1973) measured the amounts of viral DNA in rat cells transformed by adenovirus type 2 by following the rate of renaturation of ^{32}P-labeled viral DNA in the presence and absence of DNA from transformed cells. One copy of viral DNA per diploid quantity of cell DNA was detected by this method. It has been observed that the filter-binding method used by Green et al. (1970) may give inaccurate results because this method has to be calibrated by reconstruction experiments (Haas et al., 1972), and it seems likely that rat cells transformed by adenovirus type 2 contain very few copies of viral DNA per cell. Although no unambiguous experimental evidence is yet available, it appears likely that the viral DNA is integrated into the DNA of adenovirus-transformed cells.

Viral RNA Sequences in Transformed Cells

Virus-specific RNA in adenovirus-transformed cells was first reported by Fujinaga and Green (1966). These investigators were able to hybridize radioactive polysomal RNA from cells transformed by serotype 12 to filters containing adenovirus type 12 DNA. About 2 percent of the messenger RNA free in the cytoplasm or in the polysomes of cells transformed by adenovirus 12 is viral (Fujinaga and Green, 1966). This is considerably more than would be expected from the estimated amount of viral DNA in each transformed cell and indicates either that the viral DNA is preferentially transcribed or that the viral messenger RNA is unusually stable. This bias becomes even more exaggerated when it is realized that only part of the viral genome is transcribed in adenovirus-transformed cells. Hybridization experiments have shown that early RNA from cells lytically infected with adenovirus type 2 efficiently blocks the binding of RNA from rat cells transformed by adenovirus type 2 to filters containing adenovirus 2 DNA (Fujinaga and Green, 1970; Green et al., 1970). It has been estimated that approximately 50 percent of the sequences transcribed in infected permissive cells at early times is also transcribed in cells transformed by adenovirus type 2; in other words, 5–10 percent of the genome or 1–5 genes are expressed in these cells, and Green et al. claim that the RNA sequences present in the transformed rat cells are absent late after lytic infection with adenovirus type 2. Hamster cells transformed by adenovirus type 7 show a slightly different pattern of transcription; about 50 percent of the adenovirus genome is expressed in the transformed cells; all the sequences expressed in the transformed cells are present early in lytic infection, and some of these sequences are also found late in the lytic cycle (Green et al., 1970). Messenger RNA has been isolated from cells transformed by groups A, B and C adenoviruses and has been compared by base sequence analysis and hybridization to homologous and heterologous DNAs (Fujinaga et al., 1969; Fujinaga and Green, 1967a,b, 1968). These studies have shown that messenger RNA from transformed cells hybridizes only to DNAs from serotypes belonging to the same group, even though DNAs from members of the different groups share between 10 and 30 percent of their base sequences. Base composition analysis has shown that the viral messenger RNAs from cells transformed by adenoviruses belonging to different groups all have a similar content of G + C (47–51 percent), although the DNAs from members of the different groups show a great variation in their base composition (Fujinaga and Green, 1968). This information taken together suggests that in adenovirus-transformed cells only a minor part of the genome is expressed, and different sequences of RNA are transcribed in cells transformed by adenoviruses of the three groups A, B and C.

Nuclear and polysomal RNA have been isolated from cells transformed by adenovirus type 2. Analysis by gel electrophoresis shows that the virus-specific RNA in the nucleus comprises a heterogeneous population of molecules, with sedimentation coefficients in the range 10–35 S, while virus-specific RNA in the cytoplasm sediments at 10–25 S (Green et al., 1970). Since only 5–10 percent of the viral genome is expressed in cells transformed by adenovirus type 2, this information suggests that the large virus-specific RNA molecules may contain both viral and cellular sequences. Such molecules could arise as a result of transcriptional read-through of the integrated viral genome. Consistent with this, Tseui et al. (1972) have shown that isolated viral messenger RNA from cells transformed by adenovirus 2 or 7 hybridizes both to cellular and viral DNA.

Antigens in Transformed Cells

Adenovirus-specific T antigen is present in cells transformed by human adenoviruses of all three groups (Huebner et al., 1963; Pope and Rowe, 1964; Huebner, 1967); it also appears early in lytic infections (Hoggan et al., 1965). T antigen is usually detected by antisera obtained from animals which carry adenovirus-induced tumours. Cells transformed with adenoviruses from different groups have serologically distinct T antigens, and there is no antigenic cross reaction with T antigen found in cells infected with or transformed by the papova viruses. There seems to be disagreement about the properties of purified T antigen (Tavitian et al., 1967; Gilead and Ginsberg, 1968a,b; Tockstein et al., 1968). Gilead and Ginsberg purified adenovirus 12 T antigen from cells transformed by and from cells lytically infected with this virus. One protein with a sedimentation coefficient of 2.6 S was observed in their purified preparations of T antigen. On the other hand, Tockstein et al., who reported a method for partial purification of adenovirus T antigens, found four antigens with sedimentation coefficients of 5.9, 3.9, 3.1 and 2.2 S. No biological activity has so far been identified in any preparation of T antigen (Tockstein et al., 1968).

A tumour-specific transplantation antigen (TSTA), responsible for transplantation rejection, appears on the surface of cells transformed by adenoviruses (Sjögren et al., 1967), but virtually nothing is known about its structure. There is no direct evidence that either T antigen or TSTA is coded by the viral genome.

MUTANTS OF ADENOVIRUSES

The first adenovirus mutants were described by Takemori and coworkers, who isolated a number of cytocidal mutants of adenovirus type 12 (Takemori et al., 1968). These mutants were recovered from unmutagenized stocks of virus at a frequency of 0.01 percent, but ultraviolet irradiation increased the mutation frequency about fivefold. The cytocidal mutants can be distinguished from wild-type virus because they cause a different kind of cytopathic effect; on HEK cells mutant virus produces large, clear plaques by contrast to wild-type adenovirus type 12, which produces small, fuzzy-edged plaques. The mutants are less oncogenic than wild type for newborn hamsters and lack the ability to transform hamster cells in vitro (Takemori et al., 1968). The cytocidal mutants are heterogeneous and some of them are unable to replicate in a subline (KB1) of KB cells (Takemori et al., 1969). Recombination is readily detectable among these mutants, and a number of crosses yield recombinants which produce wild-type plaques and which have regained their oncogenicity (Takemori, 1972).

Temperature-sensitive mutants of adenovirus type 12 were first described by Lundholm and Doerfler (1971), who isolated ten such mutants. A more extensive investigation has been reported by Shiroki et al. (1972), who isolated 88 temperature-sensitive mutants. These replicate at 31°C but not at 38°C, and they were isolated after mutagenesis with hydroxylamine, nitrosoguanidine, bromodeoxyuridine or ultraviolet irradiation. Thirty-four of these mutants have been analyzed further and were shown to fall into 13 complementation groups. Mutants in four of these groups (A–D) are unable to synthesize hexons, penton bases and fibres at the nonpermissive temperature. Mutants in three complementation groups (E–G) are each defective in production of two capsid polypeptides, whereas mutants in complementation groups H, I and J are unable to produce either hexons (H), fibres (I) or penton bases (J). Members of the remaining three complementation groups (K–M) produce capsid proteins in normal amounts but are still unable to replicate at the nonpermissive temperature. None of these mutants are defective in synthesis of viral DNA and all mutants tested so far induce T antigen synthesis and are oncogenic for hamsters. Suzuki and Shimojo (1971) have isolated one temperature-sensitive mutant of the "highly oncogenic" serotype 31 which is able to induce T antigen but fails to replicate its DNA at the nonpermissive temperature.

Temperature-sensitive mutants of serotype 5 have been described by Williams and coworkers (1971) and by Ensinger and Ginsberg

(1972). Williams et al. (1971) isolated mutants of serotype 5 after mutagenesis with nitrosoguanidine, hydroxylamine or bromodeoxyuridine. The mutation frequencies ranged between 0.6 and 9.6 percent and most mutants obtained showed very little leakiness. Complementation analysis has so far revealed ten complementation groups of adenovirus type 5 mutants (Williams and Ustacelebi, 1971). Four mutants have been analyzed in recombination experiments and the percentage of wild-type recombinants in the crosses ranged from 0.5–7.6 percent (Williams and Ustacelebi, 1971). Two mutants of adenovirus type 5, which complement each other, fail to induce interferon in chick embryo cells at the nonpermissive temperature (Ustacelebi and Williams, 1972). Ensinger and Ginsberg (1972) isolated eight temperature-sensitive mutants of serotype 5. Their reversion frequencies were 10^{-5} or less, and these mutants have been separated into three complementation groups. Recombination was observed at frequencies between 0.1 and 15 percent. One of their mutants forms a separate complementation group and has been shown to be defective in synthesis of viral DNA. Four other mutants form a second complementation group and these are all unable to synthesize hexons. The third complementation group contains mutants which replicate their DNA and synthesize hexons, pentons and fibres and thus are defective in some undetermined function (Ensinger and Ginsberg, 1972).

In addition to these mutants of human adenoviruses, Isibachi (1971) has isolated 49 mutants of an avian adenovirus ("CELO"), some of which synthesize structural proteins but fail to transport them from the cytoplasm to the nucleus for viral assembly (Isibachi, 1970).

In summary, temperature-sensitive adenoviruses have only recently become available; the mutants isolated so far have only been partially characterized and apparently none of them affect functions involved in transformation.

ADENOVIRUS-ASSOCIATED VIRUS

Many preparations of adenoviruses contain a second very small DNA virus called adenovirus-associated virus (AAV) (Atchison et al.,

1965; Melnick et al., 1965; Mayor et al., 1965; Hoggan, 1965). Four serological types of AAV have been recognized; all are antigenically distinct from the adenoviruses (Hoggan et al., 1966; Parks et al., 1967) and there is no cross-hybridization between the DNAs of adenovirus and AAV (Rose et al., 1966). Adenovirus-associated viruses are defective and multiply only in cells infected with adenovirus or herpesvirus (Blacklow et al., 1970). They depress the yield of adenoviruses, but they do not appear to play any role in transformation or tumour induction by adenoviruses. The AAV virus contains 1.6×10^6 daltons of single-stranded DNA and strands of both plus and minus polarity are found in separate particles (Mayor et al., 1969; Rose et al., 1969). The capsid contains three polypeptides with molecular weights of 66,000, 80,000 and 92,000 daltons (Johnson et al., 1971).

Literature Cited

ALTSTEIN, A. D. and N. N. DODONOVA. 1968. Interaction between human and simian adenoviruses in simian cells: Complementation, phenotypic mixing and formation of monkey cell "adapted" virions. Virology *35:* 248.

ATCHISON, R. W., B. C. CASTO and W. McD. HAMMON. 1965. Adenovirus-associated defective virus particles. Science *149:* 754.

BAUM, S. G., P. R. REICH, R. J. HUEBNER, W. P. ROWE and S. M. WEISSMAN. 1966. Biophysical evidence for linkage of adenovirus and SV40 DNAs in adenovirus 7-SV40 hybrid particles. Proc. Nat. Acad. Sci. *56:* 1509.

BAUM, S. G., W. H. WIESE and P. R. REICH. 1968. Studies on the mechanism of enhancement of adenovirus 7 infection in African green monkey cells by simian virus 40: Formation of adenovirus-specific RNA. Virology *34:* 373.

BAUM, S., M. HORWITZ and J. MAIZEL, JR. 1972. Studies on the mechanism of enhancement of human adenovirus infection in monkey cells by simian virus 40. J. Virol. *10:* 211.

BEARDMORE, W. B., M. J. HAVLICK, A. SERAFINI and I. W. MCLEAN, JR. 1965. Interrelationship of adenovirus (type 4) and papovavirus (SV-40) in monkey kidney cell cultures. J. Immunol. *95:* 422.

BELLO, L. J. and H. S. GINSBERG. 1969. Relationship between deoxyribonucleic acid-like ribonucleic acid synthesis and inhibition of host protein synthesis in type 5 adenovirus-infected KB cells. J. Virol. *3:* 106.

BERNHARD, W., N. GRANBOULAN, G. BARSKI and P. TURNER. 1961. Essais de cytochemie ultrastructurale. Digestion de virus sur coupes ultrafines. Compt. Rend. Acad. Sci. *252:* 202.

BHADURI, S., H. RASKAS and M. GREEN. 1972. A procedure for the preparation of milligram quantities of adenovirus messenger RNA. J. Virol. *10:* 1126.

BLACK, P. H. and G. J. TODARO. 1965. *In vitro* transformation of hamster and human cells with the adeno 7-SV40 hybrid virus. Proc. Nat. Acad. Sci. *54:* 374.

BLACKLOW, N. R., M. D. HOGGAN and M. S. MCCLANAHAN. 1970. Adenovirus-associated viruses: Enhancement by human herpesviruses. Proc. Soc. Exp. Biol. Med. *134:* 952.

BOULANGER, P. A., P. FLAMENCOURT and G. BISERTE. 1969. Isolation and comparative chemical study of structural proteins of the adenoviruses 2 and 5: Hexon and fiber antigen. Europe. J. Biochem. *10:* 116.

BOULANGER, P., G. TORPIER and G. BISERTE. 1970. Investigations on intranuclear paracrystalline inclusions induced by adenovirus 5 in KB-cells. J. Gen. Virol. *6:* 329.

BRESNICK, E. and F. RAPP. 1968. Thymidine kinase activity in cells abortively and productively infected with human adenoviruses. Virology *34:* 799.

BRUNNER, M. and H. RASKAS. 1972. Processing of adenovirus RNA before release from isolated nuclei. Proc. Nat. Acad. Sci. *69:* 3101.

BURLINGHAM, B. and W. DOERFLER. 1971. Three size classes of adenovirus intranuclear deoxyribonucleic acid. J. Virol. *7:* 707.

———. 1972. An endonuclease in cells infected with adenovirus and associated with adenovirions. Virology *48:* 1.

BURLINGHAM, B., W. DOERFLER, U. PETTERSSON and L. PHILIPSON. 1971. Adenovirus endonuclease: Association with the penton of adenovirus type 2. J. Mol. Biol. *60:* 45.

BURNETT, J. P. and J. A. HARRINGTON. 1968a. Simian adenovirus SA7 DNA: Chemical, physical and biological studies. Proc. Nat. Acad. Sci. *60:* 1023.

———. 1968b. Infectivity associated with simian adenovirus type SA7 DNA. Nature *220:* 1245.

BURNS, W. H. and P. H. BLACK. 1969. Analysis of SV40-induced transformation of hamster kidney tissue *in vitro:* VI. Characteristics of mitomycin C induction. Virology *39:* 625.

BUTEL, J. S. and F. RAPP. 1967. Complementation between a defective monkey cell-adapting component and human adenovirus in simian cells. Virology *31:* 573.

BUTEL, J. S., M. GUNTZEL and F. RAPP. 1969. Variants of defective simian papovavirus 40 (PARA) characterized by cytoplasmic localization of the simian papovavirus 40 tumor antigen. J. Virol. *4:* 632.

CHARDONNET, Y. and S. DALES. 1970a. Early events in the interaction of adenovirus with HeLa cells. I. Penetration of type 5 and intracellular release of the DNA genome. Virology *40:* 462.

———. 1970b. Early events in the interaction of adenovirus with HeLa cells. II. Comparative observation on the penetration of type 1, 5, 7, 12. Virology *40:* 478.

CHARDONNET, Y., L. GAZZOLO and B. POGO. 1972. Effect of α-amanitin on adenovirus 5 multiplication. Virology *48:* 300.

CORNICK, G., P. B. SIGLER and H. S. GINSBERG. 1971. Characterization of crystals of adenovirus type 5 hexon. J. Mol. Biol. *57:* 397.

CRUMPACKER, C. S., M. J. LEVIN, W. H. WIESE, A. M. LEWIS, JR. and W. P. ROWE. 1970. The adenovirus type 2-simian virus 40 hybrid population: Evidence for a hybrid deoxyribonucleic acid molecule and the absence of adenovirus-encapsidated circular simian virus 40 deoxyribonucleic acid. J. Virol. *6:* 788.

DALES, S. 1962. An electron microscope study of the early association between two mammalian viruses and their hosts. J. Cell Biol. *13:* 303.

DARBYSHIRE, J. H. 1966. Oncogenicity of bovine adenovirus type 3 in hamsters. Nature *211:* 102.

DIAMOND, L. 1967. Transformation of SV40-resistant hamster cells with an adenovirus 7-SV40 hybrid. J. Virol. *1:* 1109.

DOERFLER, W. 1968. The fate of the DNA of adenovirus type 12 in baby hamster kidney cells. Proc. Nat. Acad. Sci. *60:* 636.

———. 1969. Non-productive infection of baby hamster kidney cells (BHK 21) with adenovirus type 12. Virology *38:* 587.

———. 1970. Integration of the DNA of adenovirus type 12 into the DNA of baby hamster kidney cells. J. Virol. *6:* 652.

DOERFLER, W. and A. K. KLEINSCHMIDT. 1970. Denaturation pattern of the DNA of adenovirus type 2 as determined by electron microscopy. J. Mol. Biol. *50:* 579.

DOERFLER, W. and U. LUNDHOLM. 1970. Absence of replication of the DNA of adenovirus type 12 in BHK21 cells. Virology *40:* 754.

DOERFLER, W., W. HILLMAN and A. K. KLEINSCHMIDT. 1972. The DNA of adenovirus type 12 and its denaturation pattern. Virology *47:* 507.

DUFF, R. and F. RAPP. 1970. Quantitative characteristics of the transformation of hamster cells with PARA (defective SV40)-adenovirus 7. J. Virol. *5:* 568.

EASTON, J. M. and C. W. HIATT. 1965. Possible incorporation of SV40 genome within capsid proteins of adenovirus 4. Proc. Nat. Acad. Sci. *54:* 1100.

ENSINGER, M. and H. GINSBERG. 1972. Selection and preliminary characterization of temperature-sensitive mutants of type 5 adenovirus. J. Virol. *10:* 328.

EPSTEIN, J. 1959. Observation on the fine structure of type 5 adenovirus. J. Biochem. Biophys. Cytol. *6:* 523.

EPSTEIN, M. A., S. J. HOLT and A. K. POWELL. 1960. The fine structure and composition of type 5 adenovirus: An integrated electron microscopical and cytochemical study. Brit. J. Exp. Pathol. *41:* 567.

EVERETT, S. F. and H. S. GINSBERG. 1958. A toxin-like material separable from type 5 adenovirus particles. Virology *6:* 770.

EVERITT, E., B. SUNDQUIST and L. PHILIPSON. 1971. Mechanism of arginine requirement for adenovirus synthesis. J. Virol. *8:* 742.

EVERITT, E., B. SUNDQUIST, U. PETTERSSON and L. PHILIPSON. 1973. Structural proteins of adenoviruses. X. Isolation and topography of low molecular weight antigens from the virion of adenovirus type 2. Virology. In press.

FELDMAN, L. A. and F. RAPP. 1966. Inhibition of adenovirus replication by 1-β-D-arabinofuranosylcytosine. Proc. Soc. Exp. Biol. Med. *122:* 243.

FELDMAN, L. A., J. S. BUTEL and F. RAPP. 1966. Interaction of a simian papovavirus and adenovirus: I. Induction of adenovirus tumor antigen during abortive infection of simian cells. J. Bacteriol. *91:* 813.

FENNER, F. 1968. The biology of animal viruses. Academic Press, New York.

FOGEL, M. and L. SACHS. 1969. The activation of virus synthesis in polyoma-transformed cells. Virology *37:* 327.

———. 1970. Induction of virus synthesis in polyoma-transformed cells by ultraviolet light and mitomycin C. Virology *40:* 174.

FOX, R. and S. BAUM. 1972. Synthesis of viral ribonucleic acid during restricted adenovirus infection. J. Virol. *10:* 220.

FRANKLIN, R. M., U. PETTERSSON, K. ÅKERVALL, B. STRANDBERG and L. PHILIPSON. 1971. Structural proteins of adenoviruses: V. On the size and stucture of the adenovirus type 2 hexon. J. Mol. Biol. *57: 383.*

FRANKLIN, R. M., S. C. HARRISON, U. PETTERSSON, C. I. BRANDÉN, P. E. WERNER and L. PHILIPSON. 1972. Structural studies on the adenovirus hexon. Cold Spring Harbor Symp. Quant. Biol. *36:* 503.

FREEMAN, A. E., P. H. BLACK, R. WOLFORD and R. J. HUEBNER. 1967. Adenovirus type 12-rat embryo transformation system. J. Virol. *1:* 362.

FRIEDMAN, M. P., M. J. LYONS and H. S. GINSBERG. 1970. Biochemical consequences of type 2 adenovirus and SV40 double infection of African green monkey kidney cells. J. Virol. *5:* 586.

FUJINAGA, K. and M. GREEN. 1966. The mechanism of viral carcinogenesis by DNA mammalian viruses: Viral-specific RNA in polyribosomes of adenovirus tumor and transformed cells. Proc. Nat. Acad. Sci. *55:* 1567.

———. 1967a. Mechanism of viral carcinogenesis by deoxyribonucleic acid mammalian viruses: IV. Related virus-specific ribonucleic acids in tumor cells induced by "highly" oncogenic adenovirus type 12, 18, and 31. J. Virol. *1:* 576.

———. 1967b. Mechanism of viral carcinogenesis by DNA mammalian viruses: II. Viral-specific RNA in tumor cells induced by "weakly" oncogenic human adenoviruses. Proc. Nat. Acad. Sci. *57:* 806.

———. 1968. Mechanism of viral carcinogenesis by DNA mammalian viruses. V. Properties of purified viral-specific RNA from human adenovirus-induced tumor cells. J. Mol. Biol. *31:* 63.

———. 1970. Mechanism of viral carcinogenesis by DNA mammalian viruses: VII. Viral genes transcribed in adenovirus type 2 infected and transformed cells. Proc. Nat. Acad. Sci. *65:* 375.

Fujinaga, K., S. Mak and M. Green. 1968. A method for determining the fraction of the viral genome transcribed during infection and its application to adenovirus-infected cells. Proc. Nat. Acad. Sci. *60:* 959.

Fujinaga, K., M. Piña and M. Green. 1969. The mechanism of viral carcinogenesis by DNA mammalian viruses: VI. A new class of virus-specific RNA molecules in cells transformed by group C human adenoviruses. Proc. Nat. Acad. Sci. *64:* 255.

Garon, C. F., K. Berry and J. Rose. 1972. A unique form of terminal redundancy in adenovirus DNA molecules. Proc. Nat. Acad. Sci. *69:* 2391.

Gerber, P. 1966. Studies on the transfer of subviral infectivity from SV40-induced hamster tumor cells to indicator cells. Virology *28:* 501.

Gilead, Z. and H. S. Ginsberg. 1965. Characterization of a tumorlike antigen in type 12 and type 18 adenovirus-infected cells. J. Bacteriol. *90:* 120.

———. 1968a. Characterization of the tumorlike (T) antigen induced by type 12 adenovirus. I. Purification of the antigen from infected KB cells and a hamster cell line. J. Virol. *2:* 7.

———. 1968b. Characterization of a tumorlike (T) antigen induced by type 12 adenovirus. II. Physical and chemical properties. J. Virol. *2:* 15.

Ginsberg, H. S. 1969. Biochemistry of adenovirus infection. *In* (H. B. Levy, ed.) The biochemistry of viruses. Dekker, New York.

Ginsberg, H. S., L. J. Bello and A. J. Levine. 1967. Control of biosynthesis of host macromolecules in cells infected with adenovirus, p. 547. *In* (J. S. Colter and W. Paranchych, ed.) The molecular biology of viruses. Academic Press, New York.

Goodheart, C. 1971. DNA density of oncogenic and non-oncogenic simian adenoviruses. Virology *44:* 645.

Green, M. 1962a. Biochemical studies on adenovirus multiplication. Virology *18:* 601.

———. 1962b. Studies on the biosynthesis of viral DNA. Cold Spring Harbor Symp. Quant. Biol. *27:* 219.

———. 1965. Chemistry and structure of animal virus particles. Amer. J. Med. *38:* 651.

———. 1966. Biosynthetic modifications induced by DNA animal viruses. Annu. Rev. Microbiol. *20:* 189.

———. 1970. Oncogenic viruses. Annu. Rev. Biochem. *39:* 701.

Green, M. and M. Piña. 1964. Biochemical studies on adenovirus multiplication. VI. Properties of highly purified tumorigenic human adenoviruses and their DNA's. Proc. Nat. Acad. Sci. *51:* 1251.

Green, M., M. Piña, R. Kimes, P. Wensink, L. MacHattie and C. A. Thomas, Jr. 1967a. Adenovirus DNA: I. Molecular weight and conformation. Proc. Nat. Acad. Sci. *57:* 1302.

Green, M., M. Pina and R. C. Kimes. 1967b. Biochemical studies on adenovirus multiplication: XII. Plaquing efficiencies of purified human adenoviruses. Virology *31:* 562.

GREEN, M., J. T. PARSONS, M. PIÑA, K. FUJINAGA, H. CAFFIER and I. LANDGRAF-LEURS. 1970. Transcription of adenovirus genes in productively infected and in transformed cells. Cold Spring Harbor Symp. Quant. Biol. *35:* 803.

HAAS, M., M. VOGT and R. DULBECCO. 1972. Loss of SV40 DNA-RNA hybrids from nitrocellulose membranes. Implications for the study of virus-host DNA interactions. Proc. Nat. Acad. Sci. *69:* 2160.

HARTLEY, J. W., R. J. HUEBNER and W. P. ROWE. 1956. Serial propagation of adenovirus (APC) in monkey kidney tissue cultures (22577). Proc. Soc. Exp. Biol. Med. *92:* 667.

HILLEMAN, M. R. and J. H. WERNER. 1954. Recovery of new agent from patients with acute respiratory illness. Proc. Soc. Exp. Biol. Med. *85:* 183.

HOGGAN, M. D. 1965. Presence of small virus-like particles in various adenovirus types 2, 5, 7 and 12 preparations. Fed. Proc. *24:* 248.

HOGGAN, M. D., W. P. ROWE, P. H. BLACK and R. J. HUEBNER. 1965. Production of "tumor-specific" antigens by oncogenic viruses during acute cytolytic infections. Proc. Nat. Acad. Sci. *53:* 12.

HOGGAN, M. D., N. R. BLACKLOW and W. P. ROWE. 1966. Studies of small DNA viruses found in various adenovirus preparations: Physical, biological and immunological characteristics. Proc. Nat. Acad. Sci. *55:* 1467.

HORNE, R. W., S. BRENNER, A. P. WATERSON and P. WILDY. 1959. The icosahedral form of an adenovirus. J. Mol. Biol. *1:* 84.

HORWITZ, M. 1971. Intermediates in the synthesis of type 2 adenovirus deoxyribonucleic acid. J. Virol. *8:* 675.

HORWITZ, M., M. SCHARFF and J. MAIZEL, JR. 1969. Synthesis and assembly of adenovirus 2: II. Polypeptide synthesis, assembly of capsomers and morphogenesis of the virion. Virology *39:* 682.

HORWITZ, M. S., J. V. MAIZEL and M. D. SCHARFF. 1970. Molecular weight of adenovirus type 2 hexon polypeptide. J. Virol. *6:* 569.

HUEBNER, R. J. 1967. Adenovirus-directed tumor and T antigens. *In* (M. Pollard, ed.) Perspectives in virology. Academic Press, New York.

HUEBNER, R. J., W. P. ROWE, H. C. TURNER and W. T. LANE. 1963. Specific adenovirus complement-fixing antigens in virus-free hamster and rat tumors. Proc. Nat. Acad. Sci. *50:* 379.

HUEBNER, R. J., R. M. CHANOCK, B. A. RUBIN and M. J. CASEY. 1964a. Induction by adenovirus type 7 of tumors in hamsters having the antigenic characteristics of SV40 virus. Proc. Nat. Acad. Sci. *52:* 1333.

HUEBNER, R. J., H. G. PEREIRA, A. C. ALLISON, A. C. HOLLINSHEAD and H. C. TURNER. 1964b. Production of type specific C antigen in virus-free hamster tumor cells induced by adenovirus type 12. Proc. Nat. Acad. Sci. *51:* 432.

HULL, R. N., I. S. JOHNSON, C. G. CULBERTSON, C. B. REIMER and H. F. WRIGHT. 1965. Oncogenicity of the simian adenovirus. Science *150:* 1044.

ISIBACHI, M. 1970. Retention of viral antigen in the cytoplasm of cells infected with temperature sensitive mutants of an avian adenovirus. Proc. Nat. Acad. Sci. *65:* 304.

———. 1971. Temperature-sensitive conditional lethal of an avian adenovirus (CELO). I. Isolation and characterization. Virology *45:* 42.

JOHNSON, B. F., H. L. OZER and M. D. HOGGAN. 1971. Structural proteins of adenovirus associated virus type 3. J. Virol. *8:* 860.

KELLY, T. and J. ROSE. 1971. Simian virus 40 integration site in an adenovirus 7-SV40 hybrid DNA molecule. Proc. Nat. Acad. Sci. *68:* 1037.

KIMES, R. and M. GREEN. 1970. Adenovirus DNA. II. Separation of molecular halves of adenovirus type 2. J. Mol. Biol. *50:* 203.

KIT, S., L. J. PIEKARSKI, D. R. DUBBS, R. A. DE TORRES and M. ANKEN. 1967. Enzyme induction in green monkey kidney cultures infected with simian adenovirus. J. Virol. *1:* 10.

KLINE, L., S. WEISSMAN and D. SOLL. 1972. Investigation on adenovirus directed 4S RNA. Virology *48:* 291.

KOHLER, K. 1965. Reinigung und charakterisierung zweier proteins des adenovirus type 2. Z. Naturforsch. *20b:* 747.

KUBINSKI, H. and J. A. ROSE. 1967. Regions containing repeating base-pairs in DNA from some oncogenic and non-oncogenic animal viruses. Proc. Nat. Acad. Sci. *57:* 1720.

LANDGRAF-LEURS, M. and M. GREEN. 1971. Adenovirus DNA. III. Separation of the complementary strands of adenovirus types 2, 7, and 12 DNA molecules. J. Mol. Biol. *60:* 185.

LAVER, W. G. 1970. Isolation of an arginine-rich protein from particles of adenovirus type 2. Virology *41:* 397.

LAVER, W. G., J. R. SURIANO and M. GREEN. 1967. Adenovirus proteins. J. Virol. *1:* 723.

LAVER, W. G., H. G. PEREIRA, W. C. RUSSELL, and R. C. VALENTINE. 1968. Isolation of an internal component from adenovirus type 5. J. Mol. Biol. *37:* 379.

LAVER, W. G., N. G. WRIGLEY, and H. G. PEREIRA. 1969. Removal of pentons from particles of adenovirus type 2. Virology *38:* 599.

LAVER, W. G., H. B. YOUNGHUSBAND and N. G. WRIGLEY. 1971. Purification and properties of chick embryo lethal orphan virus (an avian adenovirus). Virology *45:* 598.

LAWRENCE, W. C. and H. S. GINSBERG. 1967. Intracellular uncoating of type 5 adenovirus deoxyribonucleic acid. J. Virol. *1:* 851.

LEDINKO, N. 1967. Stimulation of DNA synthesis and thymidine kinase activity in human embryonic kidney cells infected by adenovirus 2 or 12. Cancer Res. *27:* 1459.

———. 1971. Inhibition by α-amanitin of adenovirus 12 replication in human embryonic kidney cells and of adenovirus transformation of hamster cells. Nature New Biol. *233:* 247.

LEVIN, M. J., C. S. CRUMPACKER, A. M. LEWIS, JR., M. N. OXMAN, P. H. HENRY and W. P. ROWE. 1971. Studies of non-defective AD2-SV40 hybrid viruses: II. Relationship of adenovirus 2 and SV40 DNAs in the AD.2^{+}ND$_1$ genome. J. Virol. *7:* 343.

LEVINE, A. J. and H. S. GINSBERG. 1967. Mechanism by which fiber antigen inhibits multiplication of type 5 adenovirus. J. Virol. *1:* 747.

———. 1968. Role of adenovirus structural proteins in the cessation of host-cell biosynthetic functions. J. Virol. *2:* 430.

LEWIS, A. M., JR. and W. P. ROWE. 1970. Isolation of two plaque variants from the adenovirus type 2-simian virus 40 hybrid population which differ in their efficiency in yielding simian virus 40. J. Virol. *5:* 413.

LEWIS, A. M., JR., S. G. BAUM, K. O. PRIGGE and W. P. ROWE. 1966. Occurrence of adenovirus-SV40 hybrids among monkey kidney cell adapted strains of adenovirus. Proc. Soc. Exp. Biol. Med. *122:* 214.

LEWIS, A. M., JR., M. J. LEVIN, W. H. WIESE, C. S. CRUMPACKER and P. H. HENRY. 1969. A non-defective (competent) adenovirus-SV40 hybrid isolated from the AD2-SV40 hybrid population. Proc. Nat. Acad. Sci. *63:* 1128.

LINDBERG, U., T. PERSSON and L. PHILIPSON. 1972. Isolation and characterization of adenovirus messenger RNA in productive infection. J. Virol. *10:* 909.

LONBERG-HOLM, K. and L. PHILIPSON. 1969. Early events of virus-cell interaction in a adenovirus system. J. Virol. *4:* 323.

LUCAS, J. J. and H. S. GINSBERG. 1971. Synthesis of virus-specific ribonucleic acid in KB cells infected with type 2 adenovirus. J. Virol. *8:* 203.

LUNDHOLM, U. and W. DOERFLER. 1971. Temperature-sensitive mutants of adenovirus type 12. Virology *45:* 827.

MAIZEL, J., JR., D. WHITE and M. SCHARFF. 1968a. The polypeptides of adenovirus: II. Soluble proteins, cores, top components and the structure of the virion. Virology *36:* 126.

———. 1968b. The polypeptides of adenovirus: I. Evidence of multiple protein components in the virion and a comparison of types 2, 7 and 12. Virology *36:* 115.

MAK, S. 1971. Defective virions in human adenovirus type 12. J. Virol. *7:* 426.

MAK, S. and M. GREEN. 1968. Biochemical studies on adenovirus multiplication. J. Virol. *2:* 1055.

MARUSYK, R., E. NORRBY and H. MARUSYK. 1972. The relationship of adenovirus-induced paracrystalline structures to the virus core protein(s). J. Gen. Virol. *14:* 261.

MAYOR, H. D., R. M. JAMISON, L. E. JORDAN and J. L. MELNICK. 1965. Structure and composition of a small particle prepared from a simian adenovirus. J. Bacteriol. *90:* 235.

MAYOR, H. D., L. JORDAN and M. ITO. 1969. Deoxyribonucleic acid of adeno-associated satellite virus. J. Virol. *4:* 191.

MCALLISTER, R. M., M. O. NICHOLSON, A. M. LEWIS, I. MACPHERSON and R. J. HUEBNER. 1969a. Transformation of rat embryo cells by adenovirus type 1. J. Gen. Virol. *4:* 29.

MCALLISTER, R. M., M. O. NICOLSON, G. REED, J. KERN, R. V. GILDEN and R. J. HUEBNER. 1969b. Transformation of rodent cells by adenovirus 10 and other group D adenoviruses. J. Nat. Cancer Inst. *43:* 917.

MCBRIDE, W. D. and A. WIENER. 1964. *In vitro* transformation of hamster kidney cells by human adenovirus type 12. Proc. Soc. Exp. Biol. Med. *115:* 870.

MCGUIRE, P. M., C. SWART and L. D. HODGE. 1972. Adenovirus messenger RNA in mammalian cells: Failure of polyribosome association in the absence of nuclear cleavage. Proc. Nat. Acad. Sci. *69:* 1578.

MELNICK, J. L., H. D. MAYOR, K. O. SMITH and F. RAPP. 1965. Association of 20-millimicron particles with adenovirus. J. Bacteriol. *90:* 271.

MORGAN, C., G. GOODMAN, H. ROSE, C. HOWE and J. HUANG. 1957. Electron microscopic and histochemical studies of an unusual crystalline protein occurring in cells infected by type 5 adenovirus. J. Biophys. Biochem. Cytol. *3:* 505.

MORGAN, C., H. S. ROSENKRANZ and B. MEDNIS. 1969. Structure and development of viruses as observed in the electron microscope. X. Entry and uncoating of adenoviruses. J. Virol. *4:* 777.

MORRISON, J. M., H. M. KEIR, H. SUBAK-SHARPE and L. V. CRAWFORD. 1967. Nearest neighbor base sequence analysis of the deoxyribonucleic acids of a further three mammalian viruses: simian virus 40, human papilloma virus and adenovirus type 2 J. Gen. Virol. *1:* 101.

NAEGELE, R. F. and F. RAPP. 1967. Enhancement of the replication of human adenoviruses in simian cells by simian adenovirus SV15. J. Virol. *1:* 838.

NICOLSON, M. O. and R. M. MCALLISTER. 1972. Infectivity of human adenovirus-1 DNA. Virology *48:* 14.

NORRBY, E. 1966. The relationship between the soluble antigens and the virion of adenovirus type 3: I. Morphological characteristics. Virology *28:* 236.

———. 1968. Biological significance of structural adenovirus components. Current Topics Microbiol. Immunol. *43:* 1.

———. 1969a. The relationship between soluble antigens and the virion of adenovirus type 3. IV. Immunological characteristics. Virology *37:* 565.

———. 1969b. The structural and functional diversity of adenovirus capsid components. J. Gen. Virol. *5:* 221.

Norrby, E. and P. Skaaret. 1967. The relationship between soluble antigens and the virion of adenovirus type 3. III. Immunological identification of fiber antigen and isolated vertex capsomer antigen. Virology *32:* 489.

Norrby, E., H. Marusyk and M. L. Hammarskjold. 1969. The relationship between the soluble antigens and the virion of adenovirus type 3. V. Identification of antigen specificities available at the surface of virions. Virology *38:* 477.

Ogino, T. and M. Takahashi. 1969. Altered properties of thymidine kinase induced in hamster kidney cells by adenovirus type 5 and 12. Biken J. *12:* 17.

Ohe, K. 1972. Virus-coded origin of low molecular weight RNA from KB cells infected with adenovirus type 2. Virology *47:* 726.

Ohe, K. and S. M. Weissman. 1970. Nucleotide sequence of an RNA from cells infected with adenovirus type 2. Science *167:* 879.

Ohe, K., S. M. Weissman and N. R. Cooke. 1969. Studies on the origin of a low molecular weight ribonucleic acid from human cells infected with adenoviruses. J. Biol. Chem. *244:* 5320.

Oxman, M., A. Levine, C. Crumpacker, M. Levin, P. Henry and A. Lewis. 1971. Studies on non-defective adenovirus 2-SV40 hybrid viruses. IV. Characterization of the SV40 ribonucleic acid species induced by wild type SV40 and by the non-defective hybrid virus $Ad2^{+}ND_1$. J. Virol. *8:* 215.

Parks, W. P., M. Green, M. Piña and J. L. Melnick. 1967. Physico-chemical characterization of adeno-associated satellite virus type 4 and its nucleic acid. J. Virol. *1:* 980.

Parsons, J. T. and M. Green. 1971. Biochemical studies on adenovirus multiplication. XVIII. Resolution of early virus specific RNA species in adeno 2-infected and transformed cells. Virology *45:* 154.

Parsons, J. T., J. Gardner and M. Green. 1971. Studies on adenovirus multiplication. XIX. Resolution of late viral RNA species in the nucleus and cytoplasm. Proc. Nat. Acad. Sci. *68:* 557.

Patch, C., A. M. Lewis and A. S. Levine. 1972. Evidence for a transcription control region of SV40 in the adenovirus 2-SV40 hybrid $Ad2^{+}ND_1$. Proc. Nat. Acad. Sci. *69:* 3375.

Pereira, H. G. and J. J. Skehel. 1971. Spontaneous and tryptic degradation of virus particles and structural components of adenoviruses. J. Gen. Virol. *12:* 13.

Pereira, H. G., R. J. Huebner, H. S. Ginsberg and J. van der Veen. 1963. A short description of the adenovirus group. Virology *20:* 613.

Pereira, H. G., R. C. Valentine and W. C. Russell. 1968. Crystallization of an adenovirus protein (the hexon). Nature *219:* 946.

Pettersson, U. 1971. Structural proteins of adenoviruses: VI. On the antigenic determinants of the hexon. Virology *43: 123.*

———. 1973. Isolation and characterization of replicating deoxyribonucleic acid in KB-cells infected with adenovirus type 2. Submitted J. Virol.

Pettersson, U. and S. Höglund. 1969. Structural proteins of adenoviruses: III. Purification and characterization of the adenovirus type 2 penton antigen. Virology *39:* 90.

Pettersson, U. and J. Sambrook. 1973. The amount of viral DNA in cells transformed by adenovirus type 2. J. Mol. Biol. *73:* 125.

Pettersson, U., L. Philipson and S. Höglund. 1967. Structural proteins of adenoviruses: I. Purification and characterization of the adeno-virus type 2 hexon antigen. Virology *33:* 575.

———. 1968. Structural proteins of adenoviruses: II. Purification and characterization of adenovirus type 2 fiber antigen. Virology *35:* 204.

Pettersson, U., C. Mulder, H. Delius and P. Sharp. 1973. Cleavage of adenovirus type 2 DNA into six unique fragments with endonuclease R.R_1. Proc. Nat. Acad. Sci. *70:* 200.

PHILIPSON, L. 1967. Attachment and eclipse of adenovirus. J. Virol. *1:* 868.

PHILIPSON, L. and U. PETTERSSON. 1972. Structure and function of virion proteins of adenoviruses. Advance. Tumor Virus Res., vol. 18.

PHILIPSON, L., K. LONBERG-HOLM and U. PETTERSSON. 1968. Virus receptor interaction in an adenovirus system. J. Virol. *2:* 1064.

PHILIPSON, L., R. WALL, G. GLICKMAN and J. E. DARNELL. 1971. Addition of polyadenylate sequences to virus-specific RNA during adenovirus replication. Proc. Nat. Acad. Sci. *68:* 2806.

PINA, M. and M. GREEN. 1965. Biochemical studies on adenovirus multiplication: IX. Chemical and base composition analysis of 28 human adenoviruses. Proc. Nat. Acad. Sci. *54:* 547.

———. 1968. Base composition of the DNA of oncogenic simian adenovirus SA 7 and homology with human adenovirus DNAs. Virology *36:* 321.

———. 1969. Biochemical studies on adenovirus multiplication: XIV. Macromolecule and enzyme synthesis in cells replicating oncogenic and nononcogenic human adenovirus. Virology *38:* 573.

POPE, J. H. and W. P. ROWE. 1964. Immunofluorescent studies of adenovirus 12 tumors and of cells transformed or infected by adenoviruses. J. Exp. Med. *120:* 577.

PRAGE, L. and U. PETTERSSON. 1971. Structural proteins of adenoviruses. VII. Purification and properties of an arginine-rich core protein from adenovirus type 2 and type 3. Virology *45:* 364.

PRAGE, L., U. PETTERSSON and L. PHILIPSON. 1968. Internal basic proteins in adenovirus. Virology *36:* 508.

PRAGE, L., U. PETTERSSON, S. HÖGLUND, K. LONBERG-HOLM and L. PHILIPSON. 1970. Structural proteins of adenoviruses: IV. Sequential degradation of the adenovirus type 2 virion. Virology *42:* 341.

PRAGE, L., S. HOGLUND and L. PHILIPSON. 1972. Structural proteins of adenoviruses. VIII. Characterization of incomplete particles of adenovirus type 3. Virology *49:* 745.

PRICE, R. and S. PENMAN. 1972a. Transcription of the adenovirus genome by an α-amanitine-sensitive ribonucleic acid polymerase in HeLa cells. J. Virol. *9:* 621.

———. 1972b. A distinct RNA polymerase activity, synthesizing 5.5S, 5S and 4S RNA in nuclei from adenovirus 2-infected HeLa cells. J. Mol. Biol. *70:* 435.

RABSON, A. S., R. L. KIRSCHSTEIN and F. J. PAUL. 1964. Tumours produced by adenovirus 12 in *Mastomys* and mice. J. Nat. Cancer Inst. *32:* 87.

RAPP, F. 1968. Dependence and complementation among animal viruses containing deoxyribonucleic acid, p. 273. *In* (L. V. Crawford and M. G. P. Stoker, ed.) The molecular biology of viruses. Cambridge University Press, Cambridge.

RAPP, F., J. L. MELNICK, J. S. BUTEL and T. KITAHARA. 1964. The incorporation of SV40 genetic material into adenovirus 7 as measured by intranuclear synthesis of SV40 tumor antigen. Proc. Nat. Acad. Sci. *52:* 1348.

RAPP, F., J. S. BUTEL and J. L. MELNICK. 1965. SV40-adenovirus "hybrid" populations: Transfer of SV40 determinants from one type of adenovirus to another. Proc. Nat. Acad. Sci. *54:* 717.

RAPP, F., M. JERKOFSKY, J. L. MELNICK and B. LEVY. 1968. Variation in the oncogenic potential of human adenoviruses carrying a defective SV40 genome (PARA). J. Exp. Med. *127:* 77.

RASKA, K. and W. A. STROHL. 1972. The response of BHK21 cells to infection with type 12 adenovirus. VI. Synthesis of virus specific RNA. Virology *47:* 734.

RASKA, K., D. FROHWIRTH and R. W. SCHLESINGER. 1970. Transfer ribonucleic acid in KB cells infected with adenovirus type 2. J. Virol. *5:* 464.

RASKA, K., L. PRAGE and R. W. SCHLESINGER. 1972. The effects of arginine starvation on macromolecular synthesis in infection with type 2 adenovirus. II. Synthesis of virus-specific RNA and DNA. Virology *48:* 472.

RASKAS, H. 1971. Release of adenovirus messenger RNA from isolated nuclei. Nature New Biol. *233:* 134.

RASKAS, H. J. and C. K. OKUBO. 1971. Transcription of viral RNA in KB cells infected with adenovirus type 2. J. Cell Biol. *49:* 438.

RASKAS, H. J., D. C. THOMAS and M. GREEN. 1970. Biochemical studies on adenovirus multiplication. XVII. Ribosome synthesis in uninfected and infected KB cells. Virology *40:* 893.

REICH, P. R., S. G. BAUM, J. A. ROSE, W. P. ROWE and S. M. WEISSMAN. 1966. Nucleic acid homology studies of adenovirus type 7-SV40 interactions. Proc. Nat. Acad. Sci. *55:* 336.

ROEDER, R. G. and W. J. RUTTER. 1970. Specific nucleolar and nucleoplasmic RNA polymerases. Proc. Nat. Acad. Sci. *65:* 675.

ROSE, J. A., P. R. REICH and S. M. WEISSMAN. 1965. RNA production in adenovirus-infected KB cells. Virology *27:* 571.

ROSE, J. A., M. D. HOGGAN and A. J. SHATKIN. 1966. Nucleic acid from an adeno-associated virus: Chemical and physical studies. Proc. Nat. Acad. Sci. *56:* 86.

ROSE, J., K. J. BERNS, D. HOGGAN and F. J. KOCZOT. 1969. Evidence for single-stranded adenovirus associated virus genome: Formation of a DNA density hybrid on release of viral DNA. Proc. Nat. Acad. Sci. *64:* 863.

ROSEN, L. 1960. Hemagglutination-inhibition technique for typing adenovirus. Amer. J. Hyg. *71:* 120.

ROUSE, H. C. and R. W. SCHLESINGER. 1967. An arginine-dependent step in the maturation of type 2 adenovirus. Virology *33:* 513.

———. 1972. The effects of arginine starvation on macromolecular synthesis in infection with type 2 adenovirus. I. Synthesis and utilization of structural proteins. Virology *48:* 463.

ROWE, W. P. 1965. Studies of adenovirus-SV40 hybrid viruses: III. Transfer of SV40 gene between adenovirus types. Proc. Nat. Acad. Sci. *54:* 711.

ROWE, W. P. and S. G. BAUM. 1964. Evidence for a possible genetic hybrid between adenovirus type 7 and SV40 viruses. Proc. Nat. Acad. Sci. *52:* 1340.

———. 1965. Studies of adenovirus-SV40 hybrid viruses: II. Defectiveness of the hybrid particles. J. Exp. Med. *122:* 955.

ROWE, W. P. and W. E. PUGH. 1966. Studies of an adenovirus-SV40 hybrid virus: V. Evidence for linkage between adenovirus and SV40 genetic materials. Proc. Nat. Acad. Sci. *55:* 1126.

ROWE, W. P., R. J. HUEBNER, L. K. GILLMORE, R. H. PARROTT and T. G. WARD. 1953. Isolation of a cytogenic agent from human adenoids undergoing spontaneous degeneration in tissue culture. Proc. Soc. Exp. Biol. Med. *84:* 570.

ROWE, W. P., J. W. HARTLEY, B. ROIZMAN and H. B. LEVEY. 1958. Characterization of a factor formed in the course of adenovirus infection of tissue cultures causing detachment of cells from glass. J. Exp. Med. *108:* 713.

RUSSELL, W. C. and Y. BECKER. 1968. A maturation factor for adenovirus. Virology *35:* 18.

RUSSELL, W. C. and B. KNIGHT. 1967. Evidence for a new antigen within the adenovirus. capsid. J. Gen. Virol. *1:* 523.

RUSSELL, W. C. and J. J. SKEHEL. 1972. The polypeptides of adenovirus-infected cells. J. Gen. Virol. *15:* 45.

RUSSELL, W. C., K. HAYASHI, P. J. SANDERSON and H. G. PEREIRA. 1967. Adenovirus antigens—A study of their properties and sequential development in infection. J. Gen. Virol. *1:* 495.

RUSSELL, W. C., K. MCINTOSH and J. J. SKEHEL. 1971. The preparation and properties of adenovirus cores. J. Gen. Virol. *11:* 35.

SARMA, P. S., R. J. HUEBNER and W. T. LANE. 1965. Induction of tumors in hamsters with an avian adenovirus (CELO). Science *149:* 1108.

SCHELL, K., W. T. LANE, M. J. CASEY and R. J. HUEBNER. 1966. Potentiation of oncogenicity of adenovirus type 12 grown in African green monkey kidney cell cultures preinfected with SV40 virus: Persistence of both T antigens in the tumors and evidence for possible hybridization. Proc. Nat. Acad. Sci. *55:* 81.

SCHLESINGER, R. W. 1969. Adenoviruses: The nature of the virion and of controlling factors in productive or abortive infection and tumorigenesis. Advance. Virus Res. *14:* 1.

SHIMOJO, H., H. YAMAMOTO and C. ABE. 1967. Differentiation of adenovirus 12 antigens in cultured cells. Virology *31:* 748.

SHIROKI, K., J. IRISAWA and H. SHIMOJO. 1972. Isolation and preliminary characterization of temperature-sensitive mutants of adenovirus 12. Virology *49:* 1.

SJÖGREN, H. O., J. MINOWADA and J. ANKERST. 1967. Specific transplantation antigens of mouse sarcomas induced by adenovirus type 12. J. Exp. Med. *125:* 689.

SMITH, K. O., W. D. GEHLE and M. D. TROUSDALE. 1965. Architecture of the adenovirus capsid. J. Bacteriol. *90:* 254.

STROHL, W. A. 1969a. The response of BHK21 cells to infection with type 12 adenovirus: I. Cell killing and T antigen synthesis as correlated viral genome functions. Virology *39:* 642.

———. 1969b. The response of BHK21 cells to infection with type 12 adenovirus: II. Relationship of virus-stimulated DNA synthesis to other viral functions. Virology *39:* 653.

———. 1972. Alterations in hamster cell regulatory mechanisms resulting from abortive infection with an oncogenic adenovirus. Advance. Exp. Tumor Virus Res., vol. 18.

STROHL, W. A., A. S. RABSON and H. ROUSE. 1967. Adenovirus tumorigenesis: Role of the viral genome in determining tumor morphology. Science *156:* 1631.

STROHL, W. A., H. ROUSE, K. TEETS and R. W. SCHLESINGER. 1970. The response of BHK21 cells to infection with type 12 adenovirus. III. Transformation and restricted replication of superinfecting type 2 adenovirus. Arch ges. Virusforsch. *31:* 93.

SUNDQUIST, B., U. PETTERSSON and L. PHILIPSON. 1973. Structural proteins of adenoviruses. IX. Molecular weight and subunit composition of the adenovirus type 2 fiber. Virology. *51:* 252.

SUSSENBACH, J. S. 1967. Early events in the infection process of adenovirus type 5 in HeLa cells. Virology *33:* 567.

SUSSENBACH, J. S. and P. VAN DER VLIET. 1972. Characterization of adenovirus DNA in cells infected with adenovirus type 12. Virology *49:* 224.

SUSSENBACH, J. S., P. C. VAN DER VLIET, D. J. ELLENS and H. S. JANSZ. 1972. Linear intermediates in the replication of adenovirus DNA. Nature New Biol. *239:* 47.

SUZUKI, E. and H. SHIMOJO. 1971. A temperature-sensitive mutant of adenovirus type 3 defective in viral deoxyribonucleic acid replication. Virology *43:* 488.

TAKAHASHI, M., S. VEDA and T. OGINO. 1966. Enhancement of thymidine kinase activity of human embryonic kidney cells and newborn hamster kidney cells by infection with human adenovirus types 5 and 12. Virology *30:* 742.

TAKEMORI, N. 1972. Genetic studies with tumorigenic adenoviruses. III. Recombination in adenovirus type 12. Virology *47:* 157.

TAKEMORI, N., J. L. RIGGS and C. ALDRICH. 1968. Genetic studies with tumorigenic adenoviruses. I. Isolation of cytocidal (*cyt*) mutants of adenovirus type 12. Virology *36:* 575.

TAKEMORI, N., J. L. RIGGS and C. ALDRICH. 1969. Genetic studies with tumorigenic adenoviruses. II. Heterogeneity of *cyt* mutants of adenovirus type 12. Virology *38:* 8.

TAVITIAN, A., J. PERIES, J. CHUAT and M. BOIRON. 1967. Estimation of the molecular weight of adenovirus 12 tumor CF antigen by rate-zonal centrifugation. Virology *31:* 719.

THOMAS, D. C. and M. GREEN. 1966. Biochemical studies on adenovirus multiplication: XI. Evidence of a cytoplasmic site for the synthesis of viral-coded proteins. Proc. Nat. Acad. Sci. *56:* 243.

———. 1969. Biochemical studies on adenovirus multiplication: XV. Transcription of the adenovirus type 2 genome during productive infection. Virology *39:* 205.

TOCKSTEIN, G., H. PLASA, M. PIÑA and M. GREEN. 1968. A simple purification procedure for adenovirus type 12 T and tumor antigens and some of their properties. Virology *36:* 377.

TRENTIN, J. J., Y. YABE and G. TAYLOR. 1962. The quest for human cancer viruses. Science *137:* 835.

TSEUI, D., K. FUJINAGA and M. GREEN. 1972. The mechanism of viral carcinogenesis by DNA mammalian viruses: RNA transcripts containing viral and highly reiterated cellular base sequences in adenovirus-transformed cells. Proc. Nat. Acad. Sci. *69:* 427.

USTACELEBI, S. and J. F. WILLIAMS. 1972. Temperature-sensitive mutants of adenovirus defective in interferon induction at non-permissive temperature. Nature *235:* 52.

VALENTINE, R. C. and H. G. PEREIRA. 1965. Antigens and structure of the adenovirus. J. Mol. Biol. *13:* 13.

VAN DER EB, A. J., L. W. KESTEREN and E. F. J. VAN BRUGGEN. 1969. Structural properties of adenovirus DNAs. Biochem. Biophys. Acta *182:* 530.

VELICER, L. and H. GINSBERG. 1970. Synthesis, transport and morphogenesis of type 5 adenovirus capsid proteins. J. Virol. *5:* 338.

WADELL, G. 1970. Structural and biological properties of capsid components of human adenoviruses. Ph.D. thesis, Karolinska Institute, Stockholm.

WADELL, G. and E. NORRBY. 1969. Immunological and other biological characteristics of pentons of human adenoviruses. J. Virol. *4:* 671.

WALL, R., L. PHILIPSON and J. E. DARNELL. 1972. Processing of adenovirus specific nuclear RNA during virus replication. Virology *50:* 27.

WALLACE, R. D. and J. KATES. 1972. On the state of the adenovirus 2 DNA in the nucleus and its mode of transcription. Studies with viral protein complexes and isolated nuclei. J. Virol. *9:* 627.

WESTPHAL, H. and R. DULBECCO. 1968. Viral DNA in polyoma and SV40 transformed cell lines. Proc. Nat. Acad. Sci. *59:* 1158.

WHITE, D. O., M. D. SCHARFF and J. V. MAIZEL, JR. 1969. The polypeptides of adenoviruses. III. Synthesis in infected cells. Virology *38:* 395.

WIESE, W. H., A. M. LEWIS, JR. and W. P. ROWE. 1970. Equilibrium density gradient studies on simian virus 40-yielding variants of the adenovirus type 2-simian virus 40 hybrid population. J. Virol. *5:* 421.

WILLIAMS, J. F. and S. USTACELEBI. 1971. Complementation and recombination with temperature-sensitive mutants of adenovirus type 5. J. Gen. Virol. *13:* 345.

WILLIAMS, J. F., M. GHARPURE, S. USTACELEBI and S. MCDONALD. 1971. Isolation of temperature-sensitive mutants of adenovirus type 5. J. Gen. Virol. *11:* 95.

WILNER, B. I. 1969. A classification of the major groups of human and other animal viruses, p. 120–132. Burgess, Minneapolis, Minn.

WINTERS, W. D. and W. C. RUSSELL. 1971. Studies on assembly of adenovirus *in vitro*. J. Gen. Virol. *10:* 181.

WOLFSON, J. and D. DRESSLER. 1972. Adenovirus 2-DNA contains an inverted terminal repetition. Proc. Nat. Acad. Sci. *69:* 3054.

YABE, Y., L. SAMPER, E. BRYAN, G. TAYLOR and J. J. TRENTIN. 1964. Oncogenic effect of human adenovirus type 12 in mice. Science *143:* 46.

ZIMMERMAN, J. E. and K. RASKA. 1972. Inhibition of adenovirus type 12 induced DNA synthesis in G1-arrested BHK21 cells by dibutyryl adenosine cyclic 3′,5′-monophosphate. Nature New Biol. *239:* 165.

ZIMMERMAN, J., K. RASKA and W. A. STROHL. 1970. The response of BHK21 cells to infection with type 12 adenovirus. IV. Activation of DNA-synthesizing apparatus. Virology *42:* 1147.

9

Herpesviruses

In the first chapter of this book we discussed some of the historical evidence which associates herpesviruses with cancer in five sorts of animals: frogs, domestic fowl, guinea pigs, monkeys and man. In this chapter we discuss the structure and taxonomy of herpesviruses, events during productive infections, and recent information about the association between herpesviruses and tumours.

CLASSIFICATION

The nomenclature and classification of herpesviruses is in a state of disorder (Roizman and de-Thé, 1972). Herpesviruses have been named after the host species (equine herpesviruses), after the disease they cause (herpes simplex, pseudorabies, Marek's disease herpesvirus, etc.), or after their discoverers (Epstein-Barr virus). The herpesviruses have commonly been classified into two groups, depending upon the extent to which infectious virus is released from infected cells (Melnick et al., 1964). Viruses have been assigned to group A if cell-free fluids or extracts readily produce infection in other cells, or to group B if intact cells are required to transmit viral infectivity. However, recent studies suggest that whether or not viral infectivity is strongly cell-associated depends on the cell in which the virus is grown rather than on the virus (Nazerian and Witter, 1970). Classification of herpesviruses into either group A or group B is therefore meaningless.

THE VIRION: STRUCTURE AND COMPOSITION

Herpesviruses are defined as large viruses that have a DNA core, a capsid exhibiting on its surface 162 capsomers arranged in the form of an icosahedron (see Figure 9.1), and a lipid-glycoprotein envelope.

Herpesvirus DNA

Herpesvirus DNA is linear and double-stranded (Kieff et al., 1971). Studies of the DNA of various herpesviruses (see Tables 9.1 and 9.2)

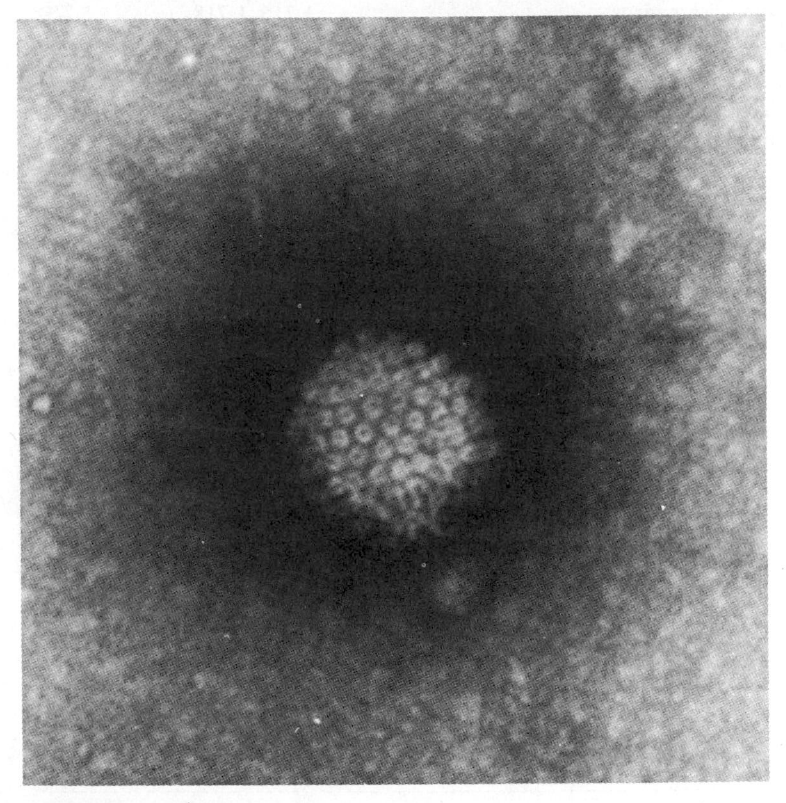

Figure 9.1. High resolution electron micrograph of a negatively stained herpesvirus nucleocapsid which shows the capsomers arranged to form an icosahedron.

Table 9.1 BASE COMPOSITION OF HERPESVIRUSES

	DNA Buoyant Density (g/cm^3)	%G + C (moles)
Infectious bovine rhinotracheitis	1.731	72
Infectious postular vulvovaginitis	1.731	72
Pseudorabies	1.731	72
Herpes simplex subtype 1	1.727	68
Herpes simplex subtype 2	1.729	70
Herpesvirus T	1.726	68
Herpesvirus saimiri	1.729	70
Equine herpesviruses	1.716–1.717	56–58
Human, mouse, vervet cytomegalovirus	1.716–1.717	56–58
Epstein-Barr virus	1.716–1.718	56–58
Marek's disease virus	1.705	45–47
Cat herpesvirus	1.705	45–47
Infectious laryngotracheitis	1.704	45–47
Frog (Lucké) herpesvirus	1.703	43

Data from Roizman and Spear 1970.

Table 9.2 PROPERTIES OF HSV-1 AND HSV-2 DNAS

	HSV-1	HSV-2
Size ($\times 10^6$ daltons)	99 ± 5	99 ± 5
Guanine + cytosine (moles percent)	67	69
Configuration	linear and double-stranded	linear and double-stranded
Alkali-labile bonds	yes, at unique sites	yes, at unique sites
Repetitive sequences ($\times 10^6$ daltons)	<5	16
Percent DNA sequences in common	47	47
Percent matching of base pairs of common sequences	85	85

Data from Kieff et al. 1971; Frenkel and Roizman 1972a; Roizman and Frenkel 1973.

establish the following: (1) The base compositions of herpesvirus DNAs vary considerably, from 45 guanine + cytosine moles percent to 74 guanine + cytosine moles percent (Table 9.1). There is no apparent relationship between the base composition of these DNAs and any particular biological property (Roizman, 1972). Unusual bases in herpesvirus DNAs have not been reported. (2) The size of the DNAs of herpesviruses varies from 92×10^6 daltons (equine abortion virus) to approximately 103×10^6 daltons (Marek's disease herpesvirus); such differences cannot be ascribed entirely to experimental error (Bachenheimer et al., 1972). (3) DNA extracted from herpes simplex 1 and 2 and from herpesviruses associated with Marek's disease and Burkitt's lymphoma fragments on sedimentation through alkaline sucrose density gradients (Lee et al., 1971; Kieff et al., 1971; Nonoyama and Pagano, 1972). HSV-1 DNA, extracted from infected cell nuclei during and after viral DNA synthesis, contains numerous alkali-labile bonds which largely disappear with time and are therefore repaired and/or ligated. The DNA extracted from virions retains some alkali-labile bonds at nonrandom and apparently unique sites. In alkaline sucrose gradients HSV-1 DNA breaks up into one large, single strand, with a molecular weight of 49.5 million daltons, and six smaller fragments weighing 39, 33, 27, 21, 15 and 10 million daltons (Frenkel and Roizman, 1972a). DNA-RNA hybridizations indicate that the fragments weighing 21×10^6 and 27×10^6 daltons together have the same sequence as those weighing 49.5×10^6, while the other four fragments apparently arise from the complementary strand. These interruptions in the DNA may play a role in the transcriptional program of herpes simplex virus (Frenkel and Roizman, unpublished data). (4) Examination in the electron microscope of thin sections digested with various enzymes has indicated the presence of DNA in the core of the herpes virion (Epstein, 1962; Zambernard and Vatter, 1966). Recent studies indicate that the DNA in the core is spooled around a cylindrical mass and takes the form of a toroid (Furlong et al., 1972). (5) Herpes simplex virions contain spermine in the capsid and spermidine in the envelope. The amount of spermine present in the capsid is sufficient to neutralize 50 percent of the phosphate in DNA (Gibson and Roizman, 1971).

The Nucleocapsid of Herpesvirus Particles

Unenveloped nucleocapsids, which have a diameter of 95–105 nm, can be obtained either from the nuclei of cells supporting virus replication or by exposing virus particles isolated from the cytoplasm, late in the infectious cycle, to nonionic detergents that remove the outer envelope. The capsids from the nuclei of infected cells appear in two forms, particles devoid of DNA (form A) and particles containing DNA (form B) (Figure 9.2), which can be separated by centrifugation in sucrose gradients (Gibson and Roizman, 1972). Capsids obtained by stripping the envelope from virions maturing in the cytoplasm are designated form C.

Form A capsids may be precursors to form B capsids, and form B capsids are precursors to form C capsids. Form A capsids contain four proteins (No. 5, 19, 23 and 24). Form B capsids contain, in addition, proteins 21 and 22A, while form C capsids contain proteins 1–5, 19, 21, 23 and 24; protein 22, which is probably derived from 22A, although also associated with the capsid, is removed by detergents (Gibson and Roizman, 1972; Spear and Roizman, 1972).

Apart from this information about the polypeptides in nucleocapsids, the only other data about the structure of these particles come

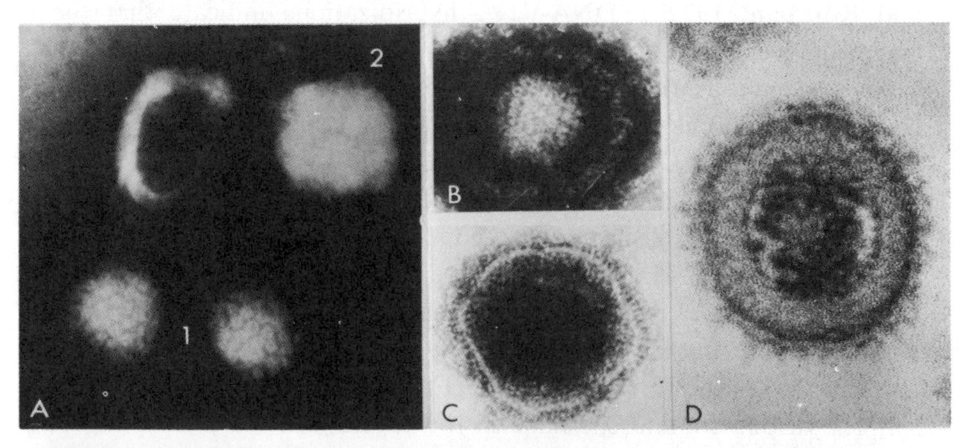

Figure 9.2. **A.** Herpes simplex nucleocapsids and virions negatively stained with silicotungstic acid: (1) nucleocapsids and (2) intact virions, which are impermeable to stain. **B, C** and **D** are enveloped nucleocapsids which have been permeated by stain to reveal details of the structure of the envelope. Courtesy B. Roizman and P. G. Spear.

from electron microscopic studies of negatively stained virions and positively stained thin sections of infected cells. Negative staining reveals capsomers on the outer surface of the capsid, and from their distribution it has been deduced that 162 capsomers are arranged as an icosahedron with a 2:3:5 axis symmetry (Wildy et al., 1960). The predominant element, the hexamer, is a prism approximately 19.5×10^2 nm in longitudinal cross section, with a channel 4 nm in diameter running the length of the prism. The hexamer and the channel have been seen in two views. In capsids impermeable to negative stain, the outline of the hexamer is seen in cross section, face on. Preparations permeated with negative stain reveal the length and width of the hexamer and the channel running its length. Conclusive evidence relating specific proteins with the outer capsids and with the DNA core is lacking. Preliminary evidence, however, based on the amounts of viral proteins in the capsid and the volumes of its components suggests that protein No. 5 is in the outer capsid and protein No. 21 is in the core.

After staining virions with lead citrate, Roizman and his colleagues have shown (Figure 9.3) that the DNA-containing core within the capsid has a toroidal or doughnut shape and that a plug of protein may pass through the centre of this toroid. Thin sections of herpesvirus particles reveal capsids in three forms (Figure 9.3): (1) particles with an electron-opaque core surrounded by an electron-translucent layer and an opaque shell—these are B or C form capsids seen with the toroidal DNA core sideways on; (2) particles with an electron-translucent center surrounded by electron-opaque, electron-translucent and electron-opaque shells—these are B or C capsids seen with the toroidal DNA core face on; (3) particles with a wide, electron-translucent center surrounded by a single electron-opaque shell—these are empty A-form capsids.

The Envelope of Herpesvirus Particles

We can summarize what little is known about the envelope of herpesvirus particles as follows: (1) An intact envelope is not permeable to negative stain (Watson, 1968; Spear and Roizman, 1972). (2) The envelope of intact virions forms a tight sheath around the capsid; often

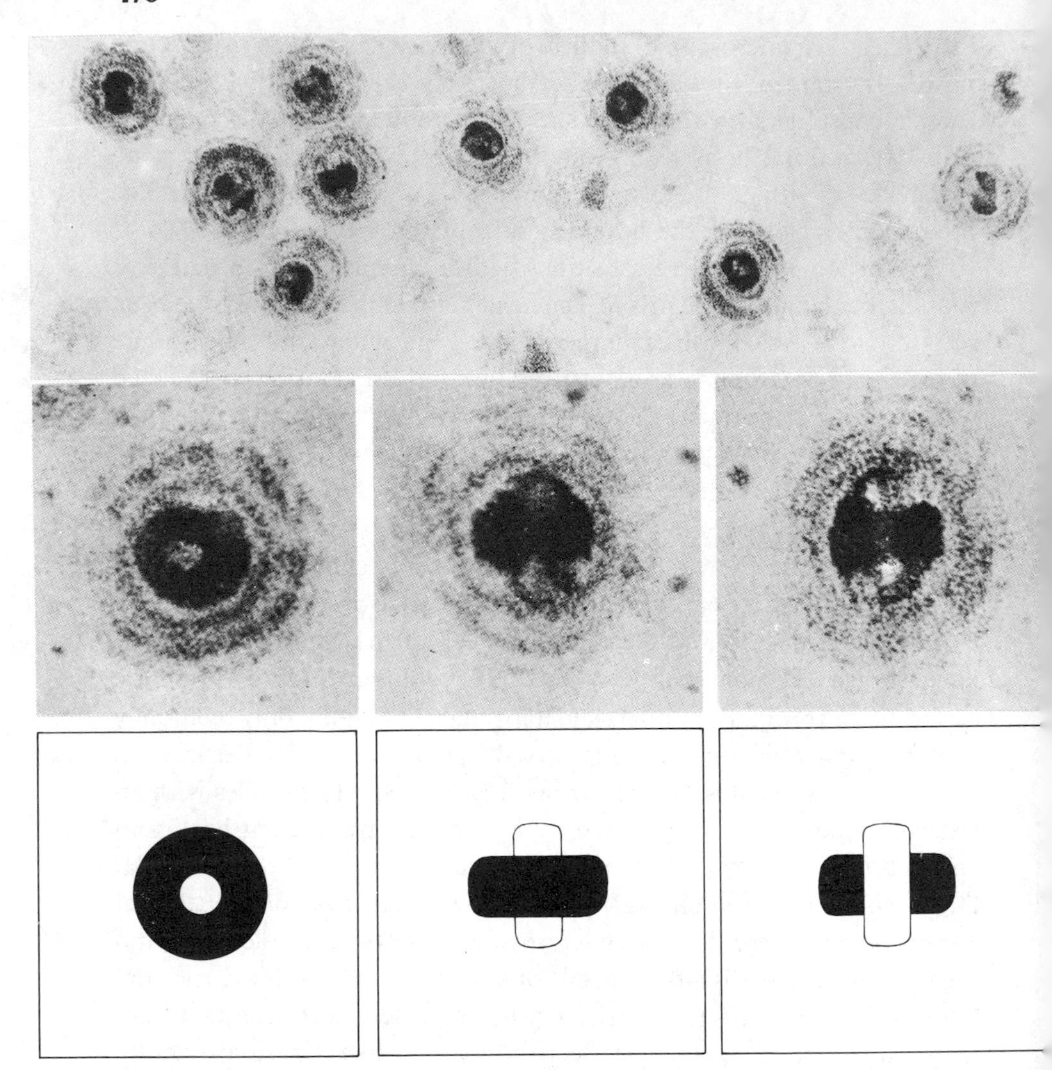

Figure 9.3. Electron micrographs of thin sections of herpes virions stained with lead citrate. These micrographs reveal the toroidal structure of the DNA core and the central plug of protein. From Furlong et al., 1972.

a portion of the envclope appears to form a characteristic "tail." (3) Thin sections of enveloped nucleocapsids indicate that the capsid acquires, as a consequence of envelopment, an electron-opaque (outer) and electron-translucent (inner) shell. The electron-translucent shell accumulates stain and becomes more electron-dense than the outer membrane in preparations made permeable to negative stains. Electron microscopic (Figure 9.2) as well as biochemical studies are compatible with the hypothesis that the envelope consists of inner and outer layers, corresponding to the electron-translucent and opaque layers, respectively. (4) Analysis of the most purified virus preparations obtained to date indicates the presence of lipids and 11 glycoproteins in the envelope (proteins 7, 8, 11, 12, 13, 14, 15, 16, 17, 18, 19) (Spear and Roizman, 1972; Heine et al., 1972; Savage et al., 1972). (5) Viruses which differ immunologically and which cause diverse changes in the social behaviour of the cells they infect contain qualitatively and quantitatively different envelope glycoproteins (Keller et al., 1970).

THE GROWTH CYCLE OF HERPESVIRUSES

Requirements for Infection of Cells

There is general agreement that naked nucleocapsids are not infectious, whereas intact virions (enveloped nucleocapsids) are infectious. Furthermore, exposure of the virus particles to lipid solvents or to lipases destroys infectivity. However, there is evidence that an intact envelope is not required for infectivity; possibly fragments of envelope adhering to the nucleocapsid are sufficient to render it infectious (Spring and Roizman, 1968; Roizman et al., 1969). It has been claimed that the naked nucleocapsid does not adsorb to cell surfaces; in any event it is less stable than the enveloped nucleocapsid. One report (Lando and Ryhiner, 1969) and some preliminary data (Sheldrick, personal communication) indicate that naked double-stranded DNA and possibly single-stranded DNA is infectious. It seems likely that not all of the DNA is essential for infectivity since a deletion mutant capable of multiplying, but lacking 4×10^6 daltons of DNA, has been reported (Bachenheimer et al., 1972).

Adsorption, Penetration and Uncoating

Herpesviruses adsorb to the cell membrane but do not penetrate at low temperatures. Once the virus has adsorbed and the cell is brought to the appropriate temperature, penetration is very rapid (Huang and Wagner, 1964). The nature of the cellular receptors is not known, but sulfated polyanions, both natural and synthetic, compete with cellular receptors for adsorption of virus to cells (Nahmias et al., 1964). Nearly all mammalian cell species tested to date appear to have receptors for adsorption of herpes simplex virus. On the other hand, the distribution of receptors for the herpesvirus associated with human lymphoreticulo-proliferative diseases (Epstein-Barr virus or EBV) is far more restricted; only some lymphocytes have this receptor.

The mechanism of penetration is not known. Thin sections of cells fixed shortly after exposure to virus show what seems to be two modes of penetration (Hummeler et al., 1969). Some particles apparently enter the cell following fusion of the viral envelope with the plasma membrane (Morgan et al., 1968). Other particles are apparently taken into the cell by a mechanism akin to pinocytosis (Dales and Silverberg, 1969). We do not know which of these routes into the cell leads to infection, and both might be artifacts. After the entry of the virus into the cell, the envelope and outer protein capsid are removed, leaving a DNA-protein complex (Sydiskis, personal communication) which only partially protects the DNA from degradation by nucleases (Hochberg and Becker, 1968). The DNA-protein complex then enters the nucleus.

Biosynthesis of the Virus in Productive Infection

The sequence of events leading to the multiplication of herpes simplex virus 1 (HSV-1) and HSV-2 has been analyzed in some detail (Figure 9.4). The steps are transcription of the DNA, synthesis of structural components, assembly of the capsid, and envelopment.

Transcription of Herpesvirus DNA

The following five aspects of the transcription of herpesvirus DNA have been analyzed: (a) the size of the RNA transcripts, (b) the control of transcription, (c) post-transcriptional processing of the RNA, (d) the nature of the enzymes involved in the synthesis of viral RNA and (e) the functions specified by the viral RNA. Almost all the information

Table 9.3. TRANSCRIPTIONAL PROGRAMS OF HSV-1 AND HSV-2 IN PRODUCTIVELY INFECTED HE_p-2 CELLS

Features	HSV-1	HSV-2
Early (*2 hr post infection*)		
Percent DNA transcribed	44	21
Number of RNA classes differing in abundance	2	1
Percent DNA specifying abundant class	14	21
Percent DNA specifying scarce class	30	
...		
Percent DNA transcribed in presence of cycloheximide	44	45
Late (*8 hr post infection*)		
Percent DNA transcribed	48	50
Number of RNA classes differing in abundance	2	2
Percent DNA specifying abundant class	19	31
Percent DNA specifying scarce class	28	19

Data from Roizman and Frenkel 1973.

about these processes comes from experiments with two viruses, HSV-1 and HSV-2 (see Table 9.3).

Approximately 48–50 percent of the DNA of HSV-1 and HSV-2 serves as template for synthesis of RNA. The DNA is transcribed in the nucleus and at least a fraction of the transcripts have a molecular weight much larger than RNA molecules with similar sequences appearing in the polyribosomes (Wagner and Roizman, 1969b; Roizman et al., 1970). These findings imply the existence of a precursor RNA which is processed to yield smaller molecules that function as messengers.

Kinetic hybridization studies of labeled viral DNA with unlabeled excess RNA made early and late in infection have revealed that transcription is regulated in two ways. First, the fraction of the DNA sequences that is transcribed differs at early and late times; second, the number of copies of transcripts of particular sequences, that is, the abundance of transcripts, varies during the infection (Frenkel and Roizman, 1972b; Roizman and Frenkel, 1973). Two hours post infection, before the onset of viral DNA replication, 44 percent of the sequences in HSV-1 DNA and 21 percent of the sequences in HSV-2 DNA are transcribed (see Table 9.3). Eight hours after infection, by which time infectious progeny have appeared, about 50 percent of the sequences in both HSV-1 and HSV-2 DNAs are transcribed. These late RNAs include all the sequences present in RNA at early times. At both early and late times during infections with both these viruses, some RNA sequences are much more abundant than others. For example, in cells infected for 2 hours with HSV-1, RNAs transcribed from 14 percent of the DNA sequences are 140-fold more abundant than RNAs transcribed from 30 percent of the DNA; whereas at late times transcripts from 19 percent of the DNA are 40-fold more abundant than transcripts from 28 percent of the DNA (see Table 9.3). It is also noteworthy that all but about 5 percent of the HSV-1 DNA that is transcribed at late times is also transcribed early. By contrast at early times during HSV-2 infections, only about 21 percent of the DNA sequences are transcribed. These sequences are also transcribed 8 hours after infection, but by then another 31 percent of the DNA is being transcribed to yield a more abundant class of RNA molecules.

RNA transcribed from HSV-1 and HSV-2 DNAs during productive infections is apparently processed in two ways. First, it seems to be cleaved to yield smaller messenger RNA molecules. Second, poly(A) is added after transcription (Bachenheimer and Roizman, 1972), and current experiments (Silverstein et al., unpublished data) indicate that only the abundant species are adenylated. The adenylated species appear in the cytoplasm after a 20- to 25-minute delay.

The source of the enzymes involved in viral RNA metabolism is not clear. However, the entire genome is transcribed in cells exposed at the time of infection and for 6 hours thereafter to cycloheximide at concentrations which completely inhibit protein synthesis (Silverstein et al., unpublished data). This observation indicates either that the virus particles carry with them RNA polymerase or that a stable cell RNA polymerase transcribes the viral DNA. If, as has been reported, viral DNA is infectious, the latter interpretation must be correct. Cellular factors may also be responsible for regulating the abundance of viral RNAs and the extent of their adenylation.

It is perhaps no coincidence that estimates of the amount of viral DNA required to code for the proteins of HSV-1 virus agree well with the amount of viral DNA which at late times acts as template for synthesis of the abundant class of RNA. Also 2 hours after infection with HSV-1, all but four of the species of virion proteins have been synthesized. The amount of DNA required to code for the four missing proteins is estimated to be about 5 percent of the HSV-1 DNA, and late RNA contains transcripts of 5 percent more of the DNA than does early RNA (Frenkel and Roizman, 1972b).

Synthesis of Structural Components

The structural components of the virion are viral DNA, viral proteins, polyamines and lipids. Not much is known about modifications of polyamine and lipid synthesis in infected cells, but it appears almost certain that labeled host lipids and polyamines do become structural components of the virion, whereas host proteins labeled in the same fashion do not.

In cells infected with HSV-1 viral structural proteins are synthesized early in infection, as noted above. The proteins are made on both free and membrane-bound polyribosomes (Sydiskis and Roizman, 1966, 1968). A large fraction of the structural proteins of the virions, and presumably most, if not all, nonstructural proteins, migrates into the nucleus (Olshevsky et al., 1967; Spear and Roizman, 1968; Ben-Porat et al., 1969). The exceptions are the virion envelope proteins, which bind to cell membranes and are glycosylated in situ (Spear and Roizman, 1970).

Herpes simplex 1 virions contain at least 24 structural proteins (Spear and Roizman, 1972). The exact number and function of proteins coded by the viral DNA but not part of the virus particles is unknown. However at least two enzymes in infected cells (Keir, 1968) have been clearly identified as virus coded. These are thymidine kinase and DNA polymerase; and there are reasonably good grounds to suspect that the exonuclease found characteristically only in infected cells is also coded by the virus (Keir, 1968).

Little is known about viral DNA synthesis, which takes place in the nucleus in a space characteristically occupied by a Feulgen-positive Cowdry type A inclusion body (Roizman, 1969). Unlike viral RNA synthesis, the synthesis of viral DNA requires protein synthesis (Roizman and Roane, 1964).

Assembly of Nucleocapsids

The structural proteins entering the nucleus aggregate with the viral DNA to form a capsid. The assembly of the capsid does not appear to be random; rather it is associated with structures, undefined in origin and composition, which appear to aggregate near the nuclear membrane (Schwartz and Roizman, 1969b; Roizman, 1969). The sequence of assembly is not clear. Form A capsids are either an aberrant by-product or become form B nucleocapsids by acquiring DNA and proteins No. 21 and 22A. Protein 22A is then apparently cleaved of a fragment, and native protein 22 remains with the capsid. Proteins 1–4 are added to the capsid at some point during this step. It is this structure which becomes enveloped at the inner lamella of the nuclear membrane. Extensive studies (Frenkel and Roizman, 1972b; Gibson and Roizman, 1972) indicate that structural proteins 5–24, including all membrane glycoproteins, are produced early in infection, whereas proteins 1–4 are made late. Possibly, late synthesis of proteins 1–4 is a regulatory step to prevent premature envelopment of capsids.

The process of assembly is not efficient. Only a small fraction of the DNA and protein aggregates to form virions. The structural components of some herpesviruses form aberrant aggregates resembling microtubules in both the nucleus and cytoplasm (Couch and Nahmias, 1969; Schwartz and Roizman, 1969b).

Envelopment of Nucleocapsids

There is some controversy regarding the site and mechanism of envelopment of the herpesvirus nucleocapsid. Our knowledge of the process of envelopment may be summarized as follows:

1. Early in infection most strains of herpesviruses become enveloped at the inner lamella of the nuclear membrane (Darlington and Moss, 1969; Schwartz and Roizman, 1969a,b). Electron microscopic studies indicate that in nuclei of infected cells, capsids with a DNA core (form B) are dispersed in paracrystalline aggregates or surrounded by some, as yet unidentified, globular material adhering to the nuclear membrane (Furlong and Roizman, unpublished data). At the site of attachment the nuclear membrane becomes thickened (due to the adhesion of the globular material) and surrounds the capsid; envelopment occurs as the particle buds through the membrane to the

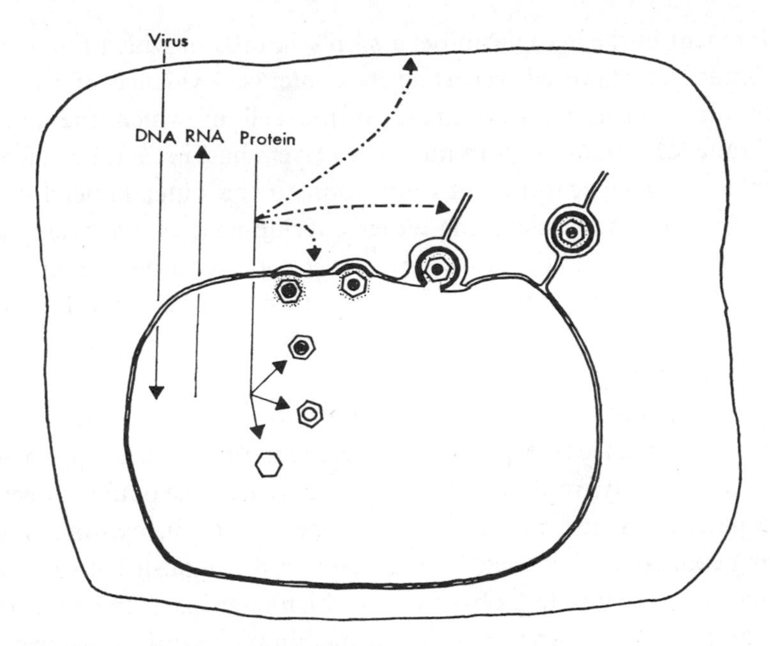

Figure 9.4. Schematic diagram of the multiplication of herpes simplex virus subtype 1 in HE_p-2 cells. The figure shows that after virus enters the cell, the DNA of the virus migrates to the nucleus. Viral RNA is made in the nucleus and migrates into the cytoplasm. The proteins are made in the cytoplasm. Some of the proteins remain in the cytoplasm and become bound to membranes lining the cytoplasm. Some of the proteins, however, enter the nucleus, aggregate around DNA to form particles which become enveloped, and accumulate either in the space between the lamellae of the nuclear membrane or in the endoplasmic reticulum.

perinuclear space (see Figure 9.4). At least in cells infected with herpes simplex, empty capsids rarely become enveloped. Envelopment of empty capsids of the virus associated with Burkitt's lymphoma (Toplin and Schidlovsky, 1966) is, however, more common.

2. In cells infected with some herpesviruses, envelopment seems to occur not only at the nuclear membrane but also at the Golgi membranes, the endoplasmic reticulum, and even at the plasma membrane (Darlington and Moss, 1969; Schwartz and Roizman, 1969a,b). The paramount question is how the nucleocapsids undergoing envelopment in the cytoplasm get there. One school holds that the nucleocapsids become enveloped by the inner lamella of the nuclear membrane, unenveloped by the outer lamella and re-enveloped by cytoplasmic membranes (Stackpole, 1969). The second school holds that

envelopment in the cytoplasm occurs only in cells in which the nuclear membrane has ruptured, releasing its contents. Evidence of ruptured membranes coinciding with areas of the cell in which the nuclear membrane folds upon itself in numerous layers has, in fact, been found. Cytoplasmic nucleocapsids, in opposition to the outer lamella of the nuclear membrane, undergoing either envelopment or unenvelopment have also been found. However, the basic question remains unanswered and our uncertainty reflects not so much the versatility of the herpesviruses as the difficulty in understanding the flow of events from electron micrographs.

3. A pertinent question concerning envelopment is whether the virus acquires its envelope from a segment of modified preexisting membrane or only from membrane synthesized de novo after infection. Since patches of new membrane could be formed in continuity with existing cellular membranes, it is difficult to distinguish between these alternatives. One puzzling observation which might give credence to the hypothesis that envelopes arise from membranes synthesized de novo is that infrequently herpesviruses appear to be enveloped inside the nucleus with membranes which are apparently not continuous with or related to any cellular membrane (Heine et al., 1971).

4. The enveloped particles of herpes simplex virus released from infected cells are generally highly infectious; particle to infectious titre ratios of 10:1 have been reported, but this is by no means the case for most herpesviruses. In the laboratory at least some herpesviruses, such as herpes zoster, cytomegaloviruses, and the virus associated with Burkitt's lymphoma are readily transmitted from culture to culture as infected cells and only with great difficulty as cell-free filtrates. Because of this, these viruses have been classified as "cell-associated" (Melnick et al., 1964), in contrast to herpes simplex, pseudorabies, and a few other herpesviruses that are readily obtained as infectious, cell-free filtrates. However, the term "cell-associated herpesvirus" restates the observation without explaining it and is an anachronism.

Considering the widespread occurrence of varicella and infectious mononucleosis, the production of infectious virions in nature must be very efficient indeed. It has been suggested that the lack of infectivity of varicella and possibly of other "cell-associated" viruses grown in cultured cells might result from the degradation of the virus particle by cellular enzymes (Cook and Stevens, 1968). However, Cook and Stevens's electron micrographs could just as readily be interpreted as indicating that the virus made in these cells is incomplete or defective. Far more likely is the explanation that some herpesviruses require for completion of the biosynthesis of their envelope enzymes found only in differentiated cells which do not grow in culture. An example is Marek's disease herpesvirus, which is "cell-associated" in chick embryo (Biggs, 1968; Churchill, 1968) but appears to multiply very efficiently and to produce infectious progeny in feather follicle cells (Nazerian and Witter, 1970). In follicle cells the envelope of the virus appears to undergo extensive structural modification in cytoplasmic inclusions (Nazerian and Witter, 1970).

Egress

In cells infected with HSV-1, enveloped particles accumulate in the space between the inner and outer lamellae of the nuclear membrane and in the endoplasmic reticulum (Darlington and Moss, 1969; Schwartz and Roizman, 1969a). The exact process by which the virus emerges from the infected cell is not clear. Some claim that portions of the endoplasmic reticulum containing the virus particles form vesicles and transport the particles to the extracellular fluid (Morgan et al., 1959). Others have reported that in infected cells the endoplasmic reticulum connects the perinuclear space with extracellular fluid and serves as a means of egress of the virus from the infected cell (Schwartz and Roizman, 1969a). Regardless of which mechanism actually prevails, it seems clear that enveloped virus is sequestered in a compartment of the infected cell and does not come in contact with the cytoplasm. One rationalization of this observation is that exposure to cytoplasmic enzymes inactivates the virus.

The Fate of the Infected Cell

An invariant consequence of productive infection is structural and functional alterations in the host. The alterations occur in two stages.

The first occurs early in infection and is characterized by complete cessation of host DNA synthesis (Ben-Porat and Kaplan, 1965; Roizman and Roane, 1964) and protein synthesis (Sydiskis and Roizman, 1966, 1968; Ben-Porat et al., 1971) and by a very drastic modification of host RNA metabolism (Roizman et al., 1970). Inhibition of host protein and DNA synthesis begins almost from the time of entrance of the virus into the infected cell and is complete 3–5 hours after infection. The inhibition of host DNA synthesis coincides with aggregation and displacement of the chromosomes to the nuclear membrane. Viral DNA is made in a space free of host DNA. The inhibition of host protein synthesis coincides with the disaggregation of polyribosomes (Sydiskis and Roizman, 1968). The mechanisms of inhibition of host DNA and protein synthesis are not known.

The modification of host RNA synthesis occurs at several levels (Roizman et al., 1970). First, there is a selective reduction in host RNA synthesis. Synthesis of ribosomal precursor RNA is reduced by as much as 70 percent, compared with that of uninfected cells, while synthesis of 4S RNA is inhibited to a very much lesser extent. Second, the host RNA that continues to be made is not processed properly. For example, normally the 45S ribosomal RNA is methylated and then cleaved in a series of steps to yield mature 18S and 28S ribosomal RNAs. In infected cells the small amount of 45S RNA that is made is methylated, but then it is degraded instead of being cleaved into the mature 18S and 28S RNAs. Third, the small amounts of cellular messenger RNA that are synthesized, processed and transported to the cytoplasm do not enter the polyribosomes to direct host protein synthesis. How cellular messengers are selectively excluded from polyribosomes is obscure.

The changes that occur in cell membranes after infection are more apparent and in some respects more interesting. They manifest themselves in four ways. First, the shape of the cell becomes highly distorted. The nucleus in particular acquires bizarre protrusions and indentations. Second, there is an increased leakage of macromolecules from the infected cells (Wagner and Roizman, 1969a; Kamiya et al., 1965). Third, new antigenic determinants can, by various procedures, be detected on the surfaces of infected cells and also in purified membrane preparations from these cells (Roane and Roizman, 1964; Roizman and Spear,

1971). The presence of new antigens was also demonstrated in cytolytic tests employing antibody to viral antigens and complement (Roizman and Spring, 1967) and with ferritin-conjugated antibody by electron microscopy (Nii et al., 1968) and by immunofluorescence (Klein et al., 1968). Several lines of evidence indicate that at least some of these antigenic determinants are also present on the surface of the envelope of the virus (Roizman and Spring, 1967; Pearson et al., 1970). Fourth, cells after infection clump to form loose or tight aggregates, and in some cases infected cells may even fuse to form heterokaryons. Such alterations in the social behaviour of infected cells are determined by the genome of the virus infecting the cell (Roizman, 1962; Ejercito et al., 1968). Furthermore, several sorts of evidence indicate that the alterations in the social behaviour of cells after infection, as well as the changes in their immunological specificity, are caused by the binding of virus-specific glycoproteins to the cell surface and intracellular membranes (Keller et al., 1970; Roizman, 1971). The glycoproteins are the only virus-specific proteins that bind to cell membranes, and the binding is selective because in cells infected with some strains of herpesviruses, the glycoproteins that bind to the cytoplasmic membrane differ from those that bind to the nuclear membrane (Spear et al., 1970; Keller et al., 1970; Roizman, 1971; Roizman and Spear, 1971).

HERPESVIRUSES AND CANCER

Permissive cells productively infected with herpesvirus are killed as the virus replicates (Roizman, 1972). Comparatively little is known about the molecular biology of incomplete infections of nonpermissive cells, which have not as yet been studied to any significant extent in cell culture systems. Several herpesviruses have, however, been associated with tumours in various species, including man, and such associations, if they are causal, presumably involve incomplete infections.

Table 9.4 lists the various sorts of herpesviruses that to date have been associated with tumours, those tumours in question, and animal hosts and culture systems in which the oncogenicity of these viruses can be studied. Of course the crucial question is that of causality: Are the various viruses listed in Table 9.4 the cause of the particular tumours

Table 9.4. ASSOCIATION OF HERPESVIRUSES WITH TUMOURS

Virus	Natural Host	Clinical Disease in Natural Host		Experimental Host Expressing Malignant Property of Virus	
		Nonmalignant	Malignant	Animal	Culture System
Herpes simplex 1[a]	Man	Several	Squamous cell carcinoma of the lip (rare)		
Herpes simplex 2[b]	Man	Several	Cervical, possibly other genital tumours	Mice, hamster (very rare)	BHK cells (transformed by UV-irradiated virus)
Human herpes-viruses associated with lymphoreticulo-proliferative diseases*[c]	Man, chimpanzee (?)	Infectious mononucleosis	Burkitt's lymphoma, postnasal carcinoma	None	Human, primate lymphoblasts
Frog herpesvirus[d]	Frog		Lucké adenocarcinoma	None	None
Herpesvirus saimiri[e]	Squirrel monkey			Marmoset, ring-tail and African green monkey (lymphoma); owl monkey (leukemia)	Permanent lympho-blastoid lines
Herpesvirus ateles[f]	Spider monkey			Owl monkey, marmoset (lymphoma)	
Marek's disease virus	Chickens	Degenerative lesions in feather follicle epithelium, bursa of Fabricus, thymus	Neurolymphomatosis, lymphoid tumours of gonads, kidney, lung, heart, spleen, liver, skin	Turkey, pheasant, quail	None
Guinea pig herpesvirus[g]	Guinea pig strain 2		Lymphocytic leukemia	Hybrid strains, day-old Hartley strain guinea pig	Lymphocytes (increased multiplication)
Cottontail rabbit herpesvirus[h]	Cottontail rabbit	Transient lymphocytic infiltration of liver and kidney	Abnormal lymphocytes in peripheral blood, liver, kidney; occasional death	None	

* Epstein-Barr virus.

Data from: [a] Wyburn-Mason 1957; Kvasnicka 1963, 1964, 1965. [b] Nahmias et al. 1970; Rawls et al. 1969; Aurelian et al. 1970; Duff and Rapp 1971. [c] Klein 1971; Henle et al. 1968; Niederman et al. 1968; de-Thé et al. 1969a,b; Levy et al. 1971. [d] Granoff 1972. [e] Melendez et al. 1972; Deinhardt 1973. [f] Melendez et al. 1972. [g] Hsiung and Kaplow 1969. [h] Hinze 1971a,b.

Table 9.5. Nature of Relationship between Herpesviruses and Associated Tumours

Virus	Origin of Tumour	Expression of Viral Genome in Tumours	Technique for Isolation of Virus	Causal Relationship: Is tumour result of infection?	Causal Relationship: Possible factors modulating outcome of infection
Herpes simplex 2[a]	?	Viral antigens in exfoliated cells, viral DNA and RNA	Not isolated	?	Sexual activity, poor hygiene
Human herpes-viruses associated with lymphoreticulo-proliferative diseases*[b]	Uniclonal	Membrane antigens	Cultivation of tumour cells	Probably; cell lines lacking viral genome not found	Chronic (lymphocyte-depleted) malaria
Frog herpesvirus[c]	?	Summer: viral RNA Winter: infectious virus	Incubation of tumour-bearing frogs or explants at low temperature	Probably	Environmental conditions, host genotype
Herpesvirus saimiri[d]	Multiclonal	No virus or antigens	Prolonged cultivation of tumour cells alone or cocultivation with susceptible cells	Yes	Host genotype
Herpesvirus ateles[e]	?	?			Host genotype
Marek's disease virus[f]	May be multiclonal	None (rare antigen-containing cells produce incomplete virus)	Cultivation of tumour cells	Yes	Host genotype
Guinea pig herpesvirus[g]	?	None	Cultivation of tumour cells	Yes	Host genotype
Cottontail rabbit herpesvirus[h]	?	?	Cultivation of kidney cells from infected animals	?	

* Epstein-Barr virus.

Data from: [a] Royston and Aurelian 1970; Rotkin 1973. [b] Fialkow et al. 1970; Klein 1971; Kafuko and Burkitt 1970. [c] Mizell et al. 1969a,b; Rafferty 1972. [d] Deinhardt 1973. [e] Melendez et al. 1972. [f] Nazerian 1973. [g] Hsiung and Kaplow 1969; Hsiung personal communication. [h] Hinze 1971a,b and personal communication.

or are they simply passengers in the tumour cells? Table 9.5 is a summary of data relevant to this question.

There can be no doubt that Marek's disease of chickens is caused by the so-called Marek's disease virus (MDV) because attenuated strains of MDV (Biggs et al., 1970, 1972) and a herpesvirus of turkeys (HVT) (Eidson et al., 1972) can be used to vaccinate chickens against Marek's disease. Likewise herpesvirus saimiri, isolated from a degenerating primary culture of squirrel monkey kidney cells (Melendez et al., 1968), causes lymphomas when inoculated in several species of monkeys, for example marmosets and owl monkeys (Melendez et al., 1970, 1972). Likewise guinea pig herpesvirus (isolated from cultures of infected lymphocytes) causes lymphocytic leukemia when injected into guinea pigs.

A causal association between the other viruses listed in Table 9.5 and tumours has not yet been firmly established. Extracts of frog kidney tumours that contain frog herpesvirus particles do induce kidney tumours in tadpoles, but purified virus preparations have yet to be obtained.

The human herpesvirus associated with lymphoreticulo-proliferative diseases (EBV) fulfills several of the criteria expected of a tumour virus.

1. Even though virus cannot always be recovered from Burkitt's lymphoma cells by cultivation or cocultivation with susceptible cells, cells from virtually every biopsy of Burkitt's lymphoma and all cell lines established from tumour tissue that have been analyzed to date have proved to contain at least one copy, and sometimes over a hundred copies, of EBV DNA per cell.

2. Antigens that can be detected on the surfaces of Burkitt's lymphoma cells are also present on the envelopes of EBV particles.

3. The epidemiologies of Burkitt's lymphoma and EBV infections suggest that the tumour might develop in people suffering simultaneous lymphocyte depletion and EBV infection, but although these data may be compelling, they are not conclusive. We do know, however, that Burkitt's lymphoma is a clonal disease; and if EBV is involved, it must cause the disease by transforming a cell which then proliferates, rather than by replicating in cells and recruiting new cells to the tumour by infection.

Finally, what is the relationship between HSV-2 and cervical carcinoma? Apart from epidemiological data which show that women with cervical carcinoma usually have high titres of anti-HSV-2 antibodies, only one biopsy of cervical carcinoma tissue has rigorously been screened for evidence of HSV-2 DNA and/or RNA. Cells of that tumour were found to contain one copy of about 40 percent of the HSV-2 genome and RNA transcripts corresponding to about 5 percent of the sequences in the viral DNA (Frenkel et al., 1972). This result is certainly consistent with the idea that HSV-2 causes cervical carcinoma, but by no means proves it.

Obviously we shall not be able to understand at the molecular level the interactions between oncogenic herpesviruses and their various host cells until we have lines of cells in culture derived from biopsies of tumours induced by the viruses and established lines of untransformed cells that can be transformed in vitro by these viruses. The progress towards these goals that has been made recently is summarized in Table 9.6.

Stable lines of tumour cells have not yet been derived from biopsies of tissue from chicks with Marek's disease or from frogs with Lucké adenocarcinoma. Lines of cells harbouring herpesvirus saimiri, the guinea pig herpesvirus and the cottontail rabbit herpesvirus have been obtained recently, but as yet they have not been thoroughly characterized; all that we know about them is that they yield virus and, as expected, produce viral antigens. By contrast a few lines of human cells derived from cervical tumours, and many lines of cells derived from biopsies of Burkitt's lymphoma and from biopsies of postnasal carcinomas, have been established over the past several years and quite extensively studied.

Cervical Tumour Cell Lines

If HSV-2 causes cervical carcinomas, we might expect lines of cells established from these tumours to contain biochemical traces of infection by HSV-2. The most familiar line of cultivated cervical tumour cells is, of course, the HeLa cell line; but these cells have been in culture for so long and have been so widely dispersed that the

Table 9.6. RELATIONSHIP OF VIRUSES WITH NEOPLASTIC CELLS GROWN IN VITRO

Virus	Origin of Cells	Oncogenicity	Special Properties of Cells	Expression of Viral Genome in Cells on Cocultivation		Interaction of Viral and Host Genomes
				Spontaneously	After Induction	
Herpes simplex 2[a]	Cervical tumours	Not tested	?	None	One report of virus induction at alkaline pH	?
	BHK cells transformed by UV-irradiated virus[b]	Produce metastatic tumours	Induce neutralizing antibody in hamsters	Viral antigens, occasionally capsids; infectious virus in human cells; antibody in tumour-bearing hamsters		?
Human herpesviruses associated with lymphoreticulo-proliferative diseases*[c]	Lymphoblasts from tumours, infectious mono; transformed cells	Not tested	Make immunoglobulin chains of unknown specificity	Varies: antigen, traces of early antigen, noninfectious and infectious particles	Arginine deprivation or halogenated pyrimidines increase fraction of cells with viral genome	Noncovalent linkage?
Herpesvirus saimiri[d]	Lymphoblasts from tumours	Yes (recipient cells become malignant)	?	Virus, antigen on continuous cultivation	?	?
Marek's disease and Frog herpesviruses	Not available					
Guinea pig herpesvirus[e]	Leukemic animals	Yes (cells and virus)	?	Viral particles and antigens	?	?
Cottontail rabbit herpesvirus[f]	Lymphocytes	?	?	Viral antigens	?	?

* Epstein-Barr virus.

Data from: [a] Aurelian et al. 1971. [b] Duff and Rapp 1971; Kieff and Rapp personal communication; Rapp and Duff 1972. [c] Klein 1971, 1972; Klein et al. 1972; Gerber et al. 1969; Pope et al. 1968; Henle and Henle 1968; Hampar et al. 1971, 1972a,b; Nonoyama and Pagano 1972. [d] Deinhardt 1973. [e] Hsiung personal communication. [f] Hinze personal communication.

phenotype of most cells called HeLa today bears little, if any, resemblance to the phenotype of the HeLa line originally isolated by G. Gey (see Chapter 2). Except for an unconfirmed report by McKenna et al. (1966), HeLa cells do not appear to contain HSV-2 particles. A report (Aurelian et al., 1971) that a line of human cervical tumour cells established by Melendez produces HSV-2 when exposed to alkaline media has not been confirmed and is not reproducible in the laboratory from which the report emanated. Of greater interest to those seeking to prove that HSV-2 is oncogenic is the work of Duff and Rapp. Following the protocol of Munyon et al. (1971), who infected L-cells lacking thymidine kinase with heavily irradiated HSV-1 and detected in the cells virus-specific thymidine kinase, Duff and Rapp (1971) exposed BHK cells to irradiated HSV-2. At least some of these infected BHK cells were "transformed" in that their morphology changed and the infected cells proved to induce highly metastatic tumours when inoculated into hamsters. These "transformed" cells do not contain any C-type RNA virus particles (Rapp et al., 1972), but as they were cultivated a fraction of the cells degenerated and liberated what appeared to be defective HSV-2 that grew slowly even in human cells highly susceptible to HSV-2 infections (Duff and Rapp, personal communication). This release of "defective" HSV-2 by some of these cells may explain why hamsters inoculated with these cells develop HSV-2-neutralizing antibodies. Unfortunately, however, it is possible that the cells Rapp and Duff used spontaneously transformed and that the HSV-2 infection was coincidental but irrelevant. But, no doubt, this system will attract considerable attention and we may soon know whether or not HSV-2 carries transforming ability.

Human Lymphoblastic Cell Lines

Lines of lymphoblastic cells derived from patients with infectious mononucleosis, Burkitt's lymphoma and postnasal carcinoma, or from cultures of peripheral lymphocytes transformed in vitro by EBV, have been intensively studied. The chief conclusion of these investigations is that all lines of human lymphoblastoid cells, regardless of their origin, contain EBV DNA, which can be detected by nucleic acid hybridization. The extent to which the resident EBV genomes are expressed differs greatly, however, among different lines. In some lines of human lymphoblastoid cells viral membrane antigen and virus particles, differing in the extent of their maturation, are regularly found in a small percentage (1–10%) of the cells in any population (see review by Klein, 1971); the number of such virus-producing cells can in some cases be increased by manipulations such as brief arginine starvation (Henle and Henle, 1968) or brief exposure to halogenated pyrimidines or hydroxyurea (Gerber, 1972; Hampar et al., 1971, 1972a,b), and experiments with cloned cells have established that the cells that are not producing virus in such populations do contain the EBV genome.

Other lines of lymphoblastoid cells, so called nonproducing lines, do not produce either virus or viral membrane antigens, but all of them produce a virus-specific complement-fixing antigen (Klein, 1971 and personal communication). Moreover, cells of some nonproducing lines, for example, the well-known Raji line, are induced to produce viral antigens and replicating EBV DNA when exposed to halogenated pyrimidines (Gerber, 1972; Hampar et al., 1972a). Production of viral antigens and replicating viral DNA cannot, however, be induced in cells of some nonproducing lines and some of these noninducible, nonproducing cells can readily be superinfected with EBV (Klein, personal communication). On the other hand some noninducible, nonproducing cells, for reasons that are obscure, cannot be superinfected.

As we have said, the cells of all the lymphoblastoid lines that have so far been tested contain EBV DNA (Zur-Hausen and Schulte-Holthausen, 1970; Nonoyama and Pagano, 1971); the amount of viral DNA per cell varies widely (from 1 or 2 to more than 100 EBV genome equivalents per cell). However, it is not yet known if noninducible, nonproducing cells contain only a fraction of the EBV genome; neither is the state of the viral DNA in most lymphoblastoid lines known. According to Nonoyoma and Pagano (1972), who used neutral and alkaline gradients to analyze the viral DNA in Raji cells, the EBV DNA in these cells is not associated by alkali-stable bonds with the cell DNA. However, Raji cells contain comparatively large numbers of EBV genomes (Zur-Hausen and Schulte-Holthausen, 1970; Nonoyoma and Pagano, 1971) and while some may be free in the cell, it is hard to exclude the possibility that one or two viral genomes are integrated in a chromosome(s). Furthermore EBV DNA, like HSV DNA, contains alkali-labile bonds. In short, the question of the location of EBV DNA in lymphoblastoid cell lines remains open, but it is pertinent that the cells of one cervical tumour contain at least part of the HSV-2 DNA integrated into host cell DNA (see above). Finally, cells of many stable lymphoblastoid lines have one other property in common: they synthesize immunoglobulins of unknown specificity and significance.

In summary, we know very little about the biology, at the cellular and molecular levels, of the interactions between herpesviruses and their nonpermissive or semi-permissive host cells. What little we do know suggests that the genomes of these viruses, some of which can transform cells in vitro, may establish stable associations with host cell genomes in the same way that SV40 and polyoma virus DNAs stably associate with cell genomes, by integration. Whether or not the products of herpesvirus genomes are required continuously to keep a cell transformed, as is the case with cells transformed by the other DNA and RNA tumour viruses, has yet to be established. But regardless of the molecular mechanisms involved, at least some herpesviruses are oncogenic either for animals of their natural host species or for animals of species maintained in the laboratory.

Literature Cited

AURELIAN, L., I. ROYSTON and H. F. DAVIS. 1970. Antibody to genital herpes simplex virus: Association with cervical atypia and carcinoma *in situ*. J. Nat. Cancer Inst. *45:* 455.

AURELIAN, L., J. D. STRANDBERG, L. V. MELENDEZ and L. A. JOHNSON. 1971. Herpesvirus type 2 isolated from cervical tumor cells grown in tissue culture. Science *174:* 704.

BACHENHEIMER, S. L. and B. ROIZMAN. 1972. RNA synthesis in cells infected with herpes simplex virus. VI. Polyadenylic acid sequences in viral mRNA. J. Virol. *10:* 875.

BACHENHEIMER, S. L., E. D. KIEFF, L. F. LEE and B. ROIZMAN. 1972. Comparative studies of DNAs of Marek's disease and herpes simplex viruses. In *Oncogenesis and Herpesviruses*, p. 74. Int. Agency Res. Cancer, Lyon.

BEN-PORAT, T. and A. S. KAPLAN. 1965. Mechanism of inhibition of cellular DNA synthesis by pseudorabies virus. Virology *25:* 22.

BEN-PORAT, T., H. SHIMONO and A. S. KAPLAN. 1969. Synthesis of proteins in cells infected with herpesvirus. II. Flow of structural viral proteins from cytoplasm to nucleus. Virology *37:* 56.

BEN-PORAT, T., T. RAKUSANOVA and A. S. KAPLAN. 1971. Early functions of the genome of herpesvirus. II. Inhibition of the formation of cell-specific polysomes. Virology *46:* 890.

Biggs, P. M. 1968. Marek's disease: Current state of knowledge. Current Topics Microbiol. Immunol. *43:* 93.

Biggs, P. M., L. N. Payne, B. S. Milne, A. E. Churchill, R. C. Chubb, D. G. Powell and A. H. Harris. 1970. Field trials with an attenuated cell-associated vaccine for Marek's disease. Vet. Rec. *87:* 704.

Biggs, P. M., C. A. W. Jackson, R. A. Bell, F. M. Lancaster and B. S. Milne. 1972. A vaccination study with an attenuated Marek's disease virus. In *Oncogenesis and Herpesviruses*, p. 139. Int. Agency Res. Cancer, Lyon.

Churchill, A. E. 1968. Herpes-type virus isolated in cell culture from tumors of chickens with Marek's disease. I. Studies on cell culture. J. Nat. Cancer Inst. *41:* 939.

Cook, M. L. and J. G. Stevens. 1968. Labile coat: Reason for non-infectious cell-free varicella-zoster virus in culture. J. Virol. *2:* 1458.

Couch, E. F. and A. J. Nahmias. 1969. Filamentous structures of type 2 herpesvirus hominis infection of the chorioallantoic membrane. J. Virol. *3:* 228.

Dales, S. and H. Silverberg. 1969. Viropexis of herpes simplex virus by HeLa cells. Virology *37:* 475.

Darlington, R. W. and H. L. Moss. 1969. The envelope of herpesvirus. Prog. Med. Virol. *11:* 16.

Deinhardt, F. 1973. Herpes saimiri. In *Herpesviruses*, ed. A. S. Kaplan. Academic Press, New York. In press.

de-Thé, G., J. C. Ambrosioni, H. C. Ho and H. C. Kwan. 1969a. Presence of herpes-type virions in Chinese nasopharyngeal tumor cultured *in vitro*. Proc. Amer. Ass. Cancer Res. *10:* 19.

———. 1969b. Lymphoblastoid transformation and presence of herpes-type viral particles in a Chinese nasopharyngeal tumour cultured *in vitro*. Nature *221:* 770.

Duff, R. and F. Rapp. 1971. Properties of hamster embryo fibroblasts transformed *in vitro* after exposure to ultraviolet-irradiated herpes simplex virus type 2. J. Virol. *8:* 469.

Eidson, C. S., S. H. Kleven and D. P. Anderson. 1972. Vaccination against Marek's disease. In *Oncogenesis and Herpesviruses*, p. 147. Int. Agency Res. Cancer, Lyon.

Ejercito, P. M., E. D. Kieff and B. Roizman. 1968. Characterization of herpes simplex virus strains differing in their effect on social behavior of infected cells. J. Gen. Virol. *3:* 357.

Epstein, M. A. 1962. Observations on the fine structure of mature herpes simplex virus and on the composition of its nucleoid. J. Exp. Med. *115:* 1.

Fialkow, P. J., G. Klein, S. M. Gartler and P. Clifford. 1970. Clonal origin for individual Burkitt tumours. Lancet *1:* 384.

Frenkel, N. and B. Roizman. 1972a. Separation of the herpesvirus DNA on sedimentation in alkaline gradients. J. Virol. *10:* 565.

———. 1972b. RNA synthesis in cells infected with herpes simplex virus: Control of transcription and of RNA abundance. Proc. Nat. Acad. Sci. *69:* 2654.

Frenkel, N., B. Roizman, E. Cassai and A. Nahmias. 1972. A herpes simplex 2 DNA fragment and its transcription in human cervical cancer tissue. Proc. Nat. Acad. Sci. *69:* 3784.

Furlong, D., H. Swift and B. Roizman. 1972. Arrangement of herpesvirus DNA in the core. J. Virol. *10:* 1071.

Gerber, P. 1972. Activation of Epstein-Barr virus by 5-bromodeoxyuridine in "virus-free" human cells. Proc. Nat. Acad. Sci. *69:* 83.

Gerber, P., J. Whang-Peng and J. H. Monroe. 1969. Transformation and chromosome changes induced by Epstein-Barr virus in normal human leukocyte cultures. Proc. Nat. Acad. Sci. *63:* 740.

GIBSON, W. and B. ROIZMAN. 1971. Compartmentalization of spermine in herpes simplex virion. Proc. Nat. Acad. Sci. *68:* 2818.

———. 1972. Proteins specified by herpes simplex virus. VIII. Characterization and composition of multiple capsid forms of subtypes 1 and 2. J. Virol. *10:* 1044.

GRANOFF, A. 1972. Lucké tumour-associated viruses. In *Oncogenesis and Herpesviruses*, p. 171. Int. Agency Res. Cancer, Lyon.

HAMPAR, B., J. G. DERGE, L. M. MARTOS and J. L. WALKER. 1971. Persistence of a repressed (switched off) Epstein-Barr genome in Burkitt lymphoma cells made resistant to 5-bromodeoxyuridine. Proc. Nat. Acad. Sci. *68:* 3185.

———. 1972a. Synthesis of Epstein-Barr virus after activation of the viral genome in a "virus-negative" human lymphoblastoid cell (Raji) made resistant to 5-bromodeoxyuridine. Proc. Nat. Acad. Sci. *69:* 78.

HAMPAR, B., J. G. DERGE, L. M. MARTOS, M. A. TAGAMETS and M. A. BURROUGHS. 1972b. Sequence of spontaneous Epstein-Barr virus activation and selective DNA synthesis in activated cells in the presence of hydroxyurea. Proc. Nat. Acad. Sci. *69:* 2589.

HEINE, J. W., P. G. SPEAR and B. ROIZMAN. 1972. The proteins specified by herpes simplex virus. VI. Viral proteins in the plasma membrane. J. Virol. *9:* 431.

HEINE, U., D. V. ABLASHI and G. R. ARMSTRONG. 1971. Morphological studies on herpesvirus saimiri in subhuman and human cell cultures. Cancer Res. *31:* 1019.

HENLE, G., W. HENLE and V. DIEHL. 1968. Relation of Burkitt's tumor-associated herpes-type virus to infectious mononucleosis. Proc. Nat. Acad. Sci. *59:* 94.

HENLE, W. and G. HENLE. 1968. Effect of arginine-deficient media on the herpes-type virus associated with cultured Burkitt (lymphoma) tumor cells (human). J. Virol. *2:* 182.

HINZE, H. C. 1971a. New member of the herpesvirus group isolated from wild cottontail rabbits. Infection and Immunity *3:* 350.

———. 1971b. Induction of lymphoid hyperplasia and lymphoma-like disease in rabbits by herpesvirus sylvilagus. Int. J. Cancer *8:* 514.

HOCHBERG, E. and Y. BECKER. 1968. Adsorption, penetration and uncoating of herpes simplex virus. J. Gen. Virol. *2:* 231.

HSIUNG, G. D. and L. S. KAPLOW. 1969. Herpes-line virus isolated from spontaneously degenerated tissue culture derived from leukaemia-susceptible guinea pigs. J. Virol. *3:* 355.

HUANG, A. S. and R. R. WAGNER. 1964. Penetration of herpes simplex virus into human epidermoid cells. Proc. Soc. Exp. Biol. Med. *116:* 863.

HUMMELER, K., N. TOMASSIAN and B. ZAJAC. 1969. Early events in herpes simplex virus infection: A radioautographic study. J. Virol. *4:* 67.

KAFUKO, G. W. and D. P. BURKITT. 1970. Burkitt's lymphoma and malaria. Int. J. Cancer *6:* 1.

KAMIYA, T., T. BEN-PORAT and A. S. KAPLAN. 1965. Control of certain aspects of the infective process by progeny viral DNA. Virology *26:* 577.

KEIR, H. M. 1968. Virus-induced enzymes in mammalian cells infected with DNA viruses. In *Molecular Biology of Viruses* (18th Symp. Soc. Gen. Microbiol.), p. 67. Cambridge Univ. Press.

KELLER, J. M., P. G. SPEAR and B. ROIZMAN. 1970. The proteins specified by herpes simplex virus. III. Viruses differing in their effects on the social behaviour of infected cells specify different membrane glycoproteins. Proc. Nat. Acad. Sci. *65:* 865.

KIEFF, E. D., S. L. BACHENHEIMER and B. ROIZMAN. 1971. Size, composition and structure of the DNA of subtypes 1 and 2 herpes simplex virus. J. Virol. *8:* 125.

KLEIN, E., R. VAN FURTH, B. JOHANSSON, I. ERNBERG and P. CLIFFORD. 1972. Immunoglobulin synthesis as cellular marker of malignant lymphoid cells. In *Oncogenesis and Herpesviruses*, p. 253. Int. Agency Res. Cancer, Lyon.

KLEIN, G. 1971. Immunological aspects of Burkitt's lymphoma. Advanc. Immunol., vol. 14, p. 187. Academic Press, New York.

———. 1972. EBV-associated membrane antigens. In *Oncogenesis and Herpesviruses*, p. 295. Int. Agency Res. Cancer, Lyon.

KLEIN, G., G. PEARSON, J. S. NADKARNI, J. J. NADKARNI, E. KLEIN, G. HENLE, W. HENLE and P. CLIFFORD. 1968. Relation between Epstein-Barr viral and cell membrane immunofluorescence of Burkitt tumor cells. I. Dependence of cell membrane immunofluorescence on presence of EB virus. J. Exp. Med. *128:* 1011.

KVASNICKA, A. 1963. Relationship between herpes simplex and lip carcinoma. II. Antiherpetic antibodies in patients with lip cancer. Neoplasma (Bratislava) *10:* 82.

———. 1964. Relationship between herpes simplex and lip carcinoma. III. Neoplasma (Bratislava) *10:* 199.

———. 1965. Relationship between herpes simplex and lip carcinoma. IV. Selected cases. Neoplasma (Bratislava) *12:* 61.

LANDO, D. and M-L. RYHINER. 1969. Pouvoir infectieux du DNA d'herpesvirus hominis en culture cellulaire. Compt. Rend. Acad. Sci. (Paris) *269:* 527.

LEE, L., E. D. KIEFF, S. L. BACHENHEIMER, B. ROIZMAN, P. G. SPEAR, B. R. BURMESTER and K. NAZERIAN. 1971. The size and composition of Marek's disease virus DNA. J. Virol. *7:* 289.

LEVY, J. A., S. B. LEVY, Y. HIRSHAUT, G. KAFUKO and A. PRINCE. 1971. Presence of EBV antibodies in sera from wild chimpanzees. Nature *233:* 559.

MCKENNA, J. M., F. E. DAVIS, J. E. PRIER and B. KLEGER. 1966. Induction of neoantigen (G) in human amnion ("Wish") cells by herpesvirus A. Nature *212:* 1602.

MELENDEZ, L. V., M. D. DANIEL, R. D. HUNT and F. G. GARCIA. 1968. An apparently new herpesvirus from primary kidney cultures of the squirrel monkey (*Saimiri sciureus*). Lab. Anim. Care *18:* 374.

MELENDEZ, L. V., M. D. DANIEL, R. D. HUNT, C. E. O. FRASER, F. G. GARCIA, N. W. KING and M. E. WILLIAMSON. 1970. Herpesvirus saimiri. V. Further evidence to consider this virus as the etiological agent of a lethal disease in primates which resemble a malignant lymphoma. J. Nat. Cancer Inst. *44:* 1175.

MELENDEZ, L. V., R. D. HUNT, M. D. DANIEL, C. E. O. FRASER, H. H. BARAHONA, F. G. GARCIA and N. W. KING. 1972. Lymphoma viruses of monkeys: Herpesvirus saimiri and herpesvirus ateles, the first oncogenic herpesviruses of primates. In *Oncogenesis and Herpesviruses*, p. 451. Int. Agency Res. Cancer, Lyon.

MELNICK, J. L., M. MIDULLA, I. WIMBERLY, J. G. BARRERA-ORO and B. M. LEVY. 1964. A new member of the herpesvirus group isolated from South American marmosets. J. Immunol. *92:* 595.

MIZELL, M., I. TOPLIN and J. J. ISAACS. 1969a. Tumor induction in developing frog kidneys by a zonal centrifuge-purified fraction of the frog herpes-type virus. Science *165:* 1134.

MIZELL, M., C. W. STACKPOLE and J. J. ISAACS. 1969b. Herpestype virus latency in the Lucké tumor. In *Biology of Amphibian Tumors*, p. 337. Springer-Verlag, Heidelberg.

MORGAN, C., H. M. ROSE, M. HOLDEN and E. P. JONES. 1959. Electron microscopic observations on the development of herpes simplex virus. J. Exp. Med. *110:* 643.

MORGAN, C., H. M. ROSE and B. MEDNIS. 1968. Electron microscopy of herpes simplex virus. I. Entry. J. Virol. *2:* 507.

MUNYON, W., E. KRAISELBURD, D. DAVIS and J. MANN. 1971. Transfer of thymidine kinase to thymidine kinase-less L cells by infection with ultraviolet-irradiated herpes simplex virus. J. Virol. *7:* 813.

NAHMIAS, A. J., S. KIBRICK and P. BERNFIELD. 1964. Effect of synthetic and biological polyanions on herpes simplex virus. Proc. Soc. Exp. Biol. Med. *115:* 993.

NAHMIAS, A. J., Z. M. NAIB, W. E. JOSEY, F. A. MURPHY and C. F. LUCE. 1970. Sarcomas after inoculation of newborn hamsters with herpesvirus hominis type 2 strains. Proc. Soc. Exp. Biol. Med. *134:* 1065.

NAZERIAN, K. 1973. Marek's disease: A neoplastic disease caused by a herpesvirus. Advanc. Cancer Res. In press.

NAZERIAN, K. and R. L. WITTER. 1970. Cell-free transmission and *in vivo* replication of Marek's disease virus. J. Virol. *5:* 388.

NIEDERMAN, J., R. W. MCCOLLUM, G. HENLE and W. HENLE. 1968. Infectious mononucleosis: Clinical manifestations in relation to EB virus antibodies. J. Amer. Med. Ass. *203:* 205.

NII, S., C. MORGAN, H. M. ROSE and K. C. HSU. 1968. Electron microscopy of herpes simplex virus. IV. Studies with ferritin-conjugated antibodies. J. Virol. *2:* 1172.

NONOYAMA, M. and J. S. PAGANO. 1971. Detection of Epstein-Barr viral genome in nonproductive cells. Nature New Biol. *233:* 103.

———. 1972. Separation of Epstein-Barr virus DNA from large chromosomal DNA in non-virus-producing cells. Nature New Biol. *238:* 169.

OLSHEVSKY, U., J. LEVITT and Y. BECKER. 1967. Studies on the synthesis of herpes simplex virions. Virology *33:* 323.

PEARSON, G., F. DENEY, G. KLEIN, G. HENLE and W. HENLE. 1970. Correlation between antibodies to Epstein-Barr virus-induced membrane antigens and neutralization of EBV infectivity. J. Nat. Cancer Inst. *45:* 989.

POPE, J. H., M. K. HORNE and W. SCOTT. 1968. Transformation of foetal human leukocytes *in vitro* by filtrates of a human leukaemic cell line containing a herpes-like virus. Int. J. Cancer. *3:* 857.

RAFFERTY, K. A., JR. 1972. Pathology of amphibian renal carcinoma. In *Oncogenesis and Herpesviruses*, p. 159. Int. Agency Res. Cancer, Lyon.

RAPP, F. and R. DUFF. 1972. Transformation of hamster cells after infection by inactivated herpes simplex virus type 2. In *Oncogenesis and Herpesviruses*, p. 447. Int. Agency Res. Cancer, Lyon.

RAPP, F., R. CONNER, R. GLASER and R. DUFF. 1972. Absence of leukosis virus markers in hamster cells transformed by herpes simplex virus type 2. J. Virol. *9:* 1059.

RAWLS, W. E., W. TOMPKINS and J. L. MELNICK. 1969. The association of herpes type 2 and carcinoma of the cervix. Amer. J. Epidemiol. *89:* 547.

ROANE, P. R., JR. and B. ROIZMAN. 1964. Studies of the determinant antigens of viable cells. II. Demonstration of altered antigenic reactivity of HEp-2 cells infected with herpes simplex virus. Virology *22:* 1.

ROIZMAN, B. 1962. Polykaryocytosis. Cold Spring Harbor Symp. Quant. Biol. *27:* 327.

———. 1969. The herpesviruses. A biochemical definition of the group. Current Topics Microbiol. Immunol. *49:* 1.

———. 1971. Herpesviruses, membranes and social behaviour of infected cells. Proc. 3rd Int. Symp. Appl. Med. Virol., p. 37. Warren Green, St. Louis, Mo.

———. 1972. Biochemical features of herpesvirus-infected cells, particularly as they relate to their potential oncogenicity. In *Oncogenesis and Herpesviruses*, p. 1. Int. Agency Res. Cancer, Lyon.

ROIZMAN, B. and G. DE-THÉ. 1972. The nomenclature and classification of herpesviruses: A proposal. Bull. World Health Org. 46, p. 547.

ROIZMAN, B. and N. FRENKEL. 1973. Herpes simplex virus DNA: Its transcription and state in productive infection and in human cervical cancer tissue. Cancer Res. In press.

ROIZMAN, B. and P. R. ROANE, JR. 1964. The multiplication of herpes simplex virus. II. The relation between protein synthesis and the duplication of viral DNA in infected HEp-2 cells. Virology *22:* 262.

Roizman, B. and P. G. Spear. 1970. Herpesviruses: Current information on the composition and structure. In *Comparative Virology*, p. 186. Academic Press, New York.

———. 1971. The role of herpesvirus glycoproteins in the modification of membranes of infected cells. Proc. Miami Winter Symp. *2:* 435. North-Holland, Amsterdam.

Roizman, B. and S. B. Spring. 1967. Alteration in immunologic specificity of cells infected with cytolytic viruses. Proc. Conf. Cross Reacting Antigens and Neoantigens, p. 85. Williams & Wilkins, Baltimore, Md.

Roizman, B., S. B. Spring and J. Schwartz. 1969. The herpesvirion and its precursors made in productively and in abortively infected cells. Symp. Viral Defectiveness, Fed. Proc. *28:* 1890.

Roizman, B., S. L. Bachenheimer, E. K. Wagner and T. Savage. 1970. Synthesis and transport of RNA in herpesvirus-infected mammalian cells. Cold Spring Harbor Symp. Quant. Biol: *35:* 753.

Rotkin, I. D. 1973. Epidemiology of cervical cancer. Cancer Res. In press.

Royston, I. and L. Aurelian. 1970. Immunofluorescent detection of herpesvirus antigens in exfoliated cells from human cervical carcinoma. Proc. Nat. Acad. Sci. *67:* 204.

Savage, T., B. Roizman and J. W. Heine. 1972. The proteins specified by herpes simplex virus. VII. Immunologic specificity of the glycoproteins of subtypes 1 and 2. J. Gen. Virol. *17:* 31.

Schwartz, J. and B. Roizman. 1969a. Concerning the egress of herpes simplex virus from infected cells. Electron microscope observations. Virology *38:* 42.

———. 1969b. Similarities and differences in the development of laboratory strains and freshly isolated strains of herpes simplex virus in HEp-2 cells: Electron microscopy. J. Virol. *4:* 879.

Spear, P. G. and B. Roizman. 1968. The proteins specified by herpes simplex virus. I. Time of synthesis, transfer into nuclei, and proteins made in productively infected cells. Virology *36:* 545.

———. 1970. The proteins specified by herpes simplex virus. IV. The site of glycosylation and accumulation of viral membrane proteins. Proc. Nat. Acad. Sci. *66:* 730.

———. 1972. The proteins specified by herpes simplex virus. V. Purification of structural proteins of the herpesvirion. J. Virol. *9:* 143.

Spear, P. G., J. M. Keller and B. Roizman. 1970. The proteins specified by herpes simplex virus. II. Viral glycoproteins associated with cellular membranes. J. Virol. *5:* 123.

Spring, S. B. and B. Roizman. 1968. Herpes simplex virus products in productive and abortive infection. III. Differentiation of infectious virus derived from nucleus and cytoplasm with respect to stability and size. J. Virol. *2:* 979.

Stackpole, C. W. 1969. Herpes-type virus of the frog renal adenocarcinoma. I. Virus development in tumor transplants maintained at low temperature. J. Virol. *4:* 75.

Sydiskis, R. J. and B. Roizman. 1966. Polysomes and protein synthesis in cells infected with a DNA virus. Science *153:* 76.

———. 1968. The sedimentation profiles of cytoplasmic polyribosomes in mammalian cells productively and abortively infected with herpes simplex virus. Virology *34:* 562.

Toplin, I. and G. Schidlovsky. 1966. Partial purification and electron microscopy of virus in the EB-3 cell line derived from a Burkitt lymphoma. Science *152:* 1084.

Wagner, E. K. and B. Roizman. 1969a. RNA synthesis in cells infected with herpes simplex virus. I. The patterns of RNA synthesis in productively infected cells. J. Virol. *4:* 36.

———. 1969b. RNA synthesis in cells infected with herpes simplex virus. II. Evidence that a class of viral mRNA is derived from a high molecular weight precursor synthesized in the nucleus. Proc. Nat. Acad. Sci. *64:* 626.

WATSON, D. H. 1968. The structure of animal viruses in relation to their biological functions. In *Molecular Biology of Viruses* (18th Symp. Soc. Gen. Microbiol.), p. 207. Cambridge Univ. Press.

WILDY, P., W. C. RUSSELL and R. W. HORNE. 1960. The morphology of herpesviruses. Virology *12:* 204.

WYBURN-MASON, R. 1957. Malignant change following herpes simplex. Brit. Med. J. *2:* 615.

ZAMBERNARD, J. and A. E. VATTER. 1966. The fine structural cytochemistry of virus particles found in renal tumors of leopard frogs. I. An enzymatic study of the viral nucleoid. Virology *28:* 318.

ZUR-HAUSEN, H. and H. SCHULTE-HOLTHAUSEN. 1970. Presence of EB virus nucleic acid homology in a "virus-free" line of Burkitt tumour cells. Nature *227:* 245.

10

The RNA Tumour Viruses: Morphology, Composition and Classification

RNA tumour virology has a surprisingly long history, which has been discussed in Chapter 1 and exhaustively chronicled by Gross (1970). Readers interested in the details of the history of the subject cannot do better than read Gross's book, *Oncogenic Viruses.*

In 1908 and 1909 Ellerman and Bang reported the transmission of erythromyeloblastic leukemia of chickens by cell-free filtrates of leukemia cells. Their experiments, which were the first transmission of a truly neoplastic disease by cell-free filtrates, led initially not to a search in other species for infectious agents which might cause leukemia, but to the classification of chicken leukemia apart from the leukemias of other species. The first transmission of a solid tumour by cell-free filtrates was reported three years later by Rous (1911), who isolated several filtrates capable of inducing a variety of sarcomas in chickens. In spite of independent confirmation of these results (Fujinami and Inamoto, 1914), the scientific and medical communities were by and large reluctant to concede the fundamental importance of these two sets of discoveries that proved that viruses can cause malignant diseases, and it was not until 1966 that Rous was awarded the Nobel prize for his work.

During the nineteen-twenties and thirties, however, Rous sarcoma viruses and avian leukemia viruses attracted the attention of increasing numbers of scientists, and the viruses and the diseases they cause were further characterized. At the same time a handful of biologists began systematically to inbreed mice, and work on inbred strains led to the discovery of murine tumour viruses. Using the inbred strain A, Bittner (1936) obtained the first evidence suggesting that a mouse mammary gland carcinoma is transmitted from mother to offspring by a factor in milk. Subsequent experiments by Bittner, Andervont and others showed that weanling mice, of a strain which normally has only a low incidence of mammary carcinoma, develop the disease at a high incidence if the weanlings are foster-nursed by a mouse of a strain which has a high incidence of the disease. Then Bittner (1942) and later Andervont and Bryan (1944) proved that the factor in milk which causes the disease is a filterable virus; this virus has since been characterized (see below).

The availability of inbred strains of mice also made possible the identification of the murine leukemia and sarcoma viruses. Several strains of mice, in which up to 90 percent of the animals spontaneously develop leukemia, have been bred; perhaps the most familiar of these strains are C58 Black and the albino strain Ak. However, numerous attempts to transmit murine leukemia by cell-free filtrates of leukemia cells from these strains failed until in 1946 Zilber reported the induction of tumours with cell-free extracts from chemically induced tumours and in 1951, Gross successfully transmitted leukemia by inoculating C3H mice with extracts of Ak leukemia cells.

Since the pioneering work of Gross, several strains of murine leukemia virus, including the Friend, Moloney, Rauscher and Graffi strains, have been isolated, and murine sarcoma viruses have also been identified (Harvey, 1964; Moloney, 1966; Kirsten and Mayer, 1967).

Leukemia and/or sarcoma viruses have also been isolated from cats, rats, hamsters and a woolly monkey; furthermore there is suggestive evidence for the existence of guinea pig and bovine leukemia viruses, and RNA viruses with biophysical, biochemical and immunological properties similar to those of the leukemia and sarcoma viruses of lower mammals have recently been isolated from both a gibbon and one species of snake (see below). As a result, even though some of these new viruses have not yet been shown to be able to cause tumours, many RNA tumour virologists believe that all species of mammals, birds and reptiles, indeed probably all vertebrates, are hosts to RNA viruses that may induce sarcomas and leukemias.

PATHOLOGY OF RNA TUMOUR VIRUSES

It is impossible to review in a small space the many lesions induced in animals by RNA tumour viruses. However, it is useful to survey briefly the oncogenic potential of these agents and the reader is referred to Gross (1970) for a detailed account of individual viruses.

Neoplasms are conveniently classified according to their tissue of origin (where this can be identified) and their behaviour, whether benign or malignant. Malignant tumours of epithelial origin (carcinomas) are responsible for most human cancer deaths, whilst connective tissue tumours (sarcomas) are far less common. Neoplasms of the hemopoietic and reticulo-endothelial systems form another group, the nomenclature of which is confused. They may be manifest as a leukemic condition in which the blood contains large numbers of circulating tumour cells, or they may be aleukemic, with the neoplasm consisting of solid masses of tumour cells in various organs. In man the solid tumours are often called sarcomas, contrasting them with leukemias of the equivalent cell type (lymphosarcoma/lymphatic leukemia; reticulum cell sarcoma/monocytic leukemia). In animals such as fowl and cattle the various leukemic and aleukemic hemopoietic neoplasms are referred to collectively as leukoses. When the tumour cells resemble stem cell types, the condition is often called blastoma or blastosis. Thus, the erythromyeloblastosis of fowls involves the proliferation of blast-like cells of both the erythrocyte and leukocyte lines.

RNA tumour viruses have been chiefly implicated in tumours of the connective tissue and hemopoietic and reticulo-endothelial systems of various species. Since carcinomas are the commonest human malignancies, it has been argued that such viruses are probably not very important in the natural history of human cancer. In the light of present knowledge this is a valid observation, but it is not true when species other than man are considered. Leukoses, mostly or entirely virus induced, are the commonest malignancies in fowl and cattle, and they are also very common in the cat. Moreover, in young humans sarcomas and leukemias are commoner than carcinomas and it can be argued that the high incidence of human carcinomas reflects the greater relative age achieved by man. Other species that also often survive into old age, such as dogs and horses, also show a high incidence of carcinomas. In addition, some RNA tumour viruses can also induce carcinomas. A renal adenocarcinoma has been observed in cases of fowl leukosis, and the viral origin of mouse mammary carcinoma, whose analogue, breast cancer, is a common neoplasm of women, has long been known. As we improve our techniques for identifying RNA tumour viruses, these agents may yet be revealed in a wider range of neoplasms of man and animals.

MORPHOLOGY OF THE RNA TUMOUR VIRUSES

The structure of the RNA tumour viruses has been repeatedly investigated by electron microscopists during the past decade (see reviews by Vigier, 1970; Gross, 1970); they have used successively thin sectioning techniques, negative staining of intact particles and most recently negative staining followed by freeze drying and freeze etching (Nermut et al., 1972).

The earliest electron microscopic studies established that most RNA tumour virus particles are roughly spherical structures about 100 mμ in diameter (Sharp et al., 1952; Bernhard et al., 1958), comprising a core or nucleoid enclosed in an outer envelope made of a unit membrane, from the outer surfaces of which spikes project (Eckert et al., 1963; de-Thé et al., 1964). The projecting spikes are particularly prominent on the surfaces of the avian viruses; they are seen much less frequently on the surfaces of sectioned or negatively stained murine leukemia or sarcoma viruses. We now know (Rifkin and Compans, 1971) that these spikes are made of the glycoproteins (see below). Furthermore, electron microscopic cytochemical investigations have shown that the envelopes of at least some RNA tumour virus particles, for example particles of avian myeloblastosis virus, murine leukemia virus and mouse mammary tumour virus, may contain an ATPase activity (Beard, 1963; de-Thé, 1964, 1966; El-Fiky et al., 1970). This enzyme is, however, probably of host origin and acquired as the virus particles become enveloped as they bud from the surfaces of infected cells, because avian myeloblastosis virus grown in myeloblasts has an envelope-associated ATPase but the same strain of virus grown in fibroblasts does not (de-Thé, 1964).

A, B and C-type Particles

By virtue of their morphology as revealed in the electron microscope, RNA tumour viruses can be classified into three categories, so-called A-type, B-type and C-type particles (see reviews by Bernhard, 1958, 1960; Sarkar et al., 1972). In thin sections mature C-type particles can be seen to have a centrally located spherical nucleoid, which is enclosed in a unit membrane envelope. All known sarcoma and leukemia virus particles have this morphology, as do several recently discovered viruses which can be classified with the RNA tumour viruses because of their morphology and biochemistry, even though their oncogenic potential is yet to be proven.

Micrographs of thin sections reveal that mature B-type particles characteristically have a solid spherical nucleoid that is eccentrically located with respect to the envelope of the virion (see review by Sarkar et al., 1972). This eccentric nucleoid is enclosed within an inner membrane, which in turn is enclosed within a unit membrane envelope. The outer surface of the envelope of B-type particles is always covered with projecting spikes. The only well-characterized RNA tumour viruses which have this B-type morphology are the murine mammary tumour viruses.

Both B-type and C-type RNA tumour virus particles bud from the surfaces of infected host cells and numerous electron microscopic investigations have been made of this process since the advent of techniques for thin sectioning cells (see reviews by Gross, 1970; Sarkar et al., 1972). In cells releasing B-type particles (see Figure 10.1) toroidal nucleoids enclosed in a membrane envelope develop in the

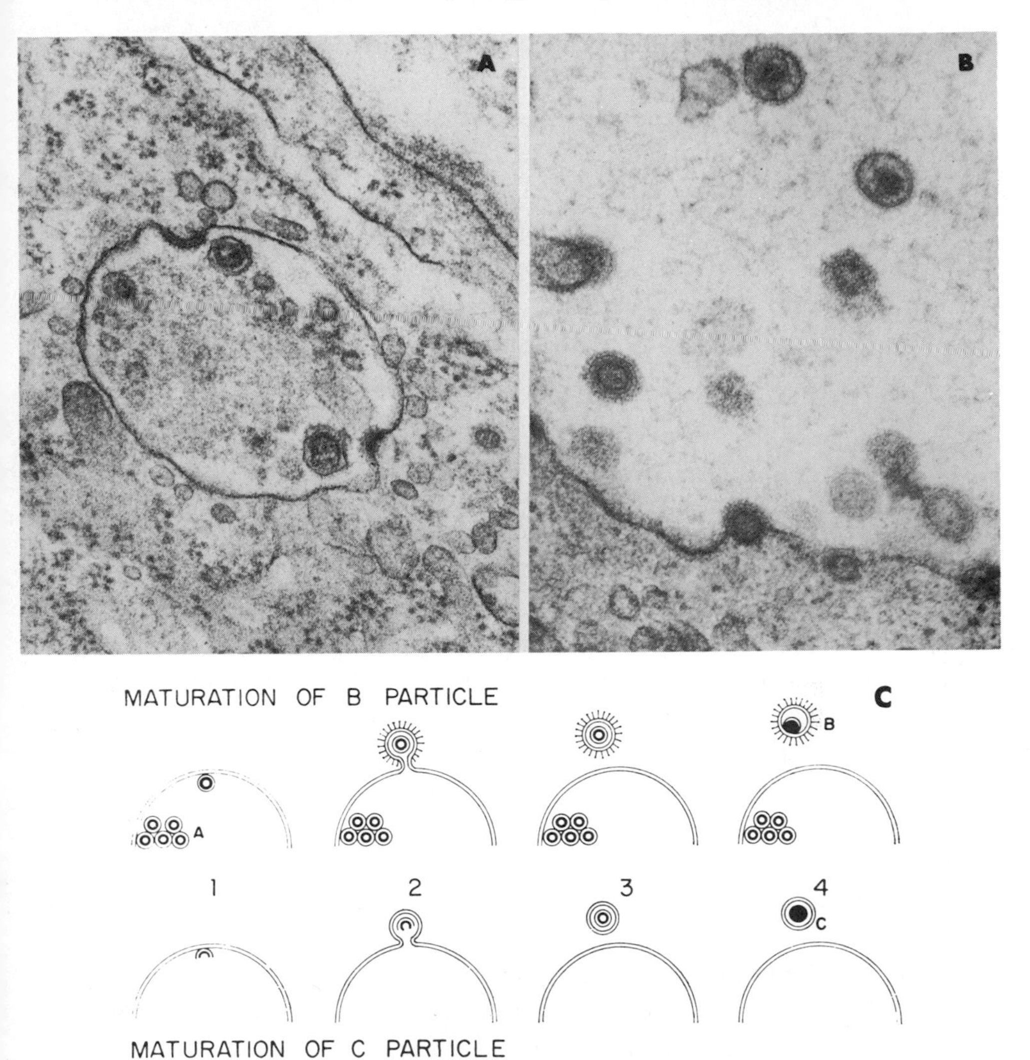

Figure 10.1. A, C-type particles budding into a vacuole (×64,000). **B,** Immature, mature and budding B-type particles (×74,000). **C,** Diagram of the maturation of C- and B-type particles (reproduced by permission from Sarkar et al., 1972).

cytoplasm. These then move to the cell surface from which they bud, acquiring in the process an outer unit membrane envelope, the outer surface of which is covered with regularly arranged spikes (Imai et al., 1966; Tanaka and Moore, 1967; Moore et al., 1971; Gay et al., 1970). Immature intracytoplasmic B-type particles with their toroidal nucleoids enclosed in a membrane are called A-type particles. Their maturation is completed after budding and the acquisition of an envelope has occurred and involves the conversion of the toroidal nucleoid into the spherical and eccentrically located nucleoid of the mature B-type particle (see review by Sarkar et al., 1972).

The budding of the C-type particles is quite different. The nucleocapsids of C-type particles first appear as incomplete toroids, in other words crescent-shaped structures, immediately below the cell surface, and as the virion buds from the surface the nucleoid matures into a complete toroid surrounded by an inner membrane, which in turn is surrounded by a unit membrane envelope. Such particles are classified as A-type particles because of their toroidal nucleoid; their further maturation involves the conversion of the toroidal nucleoid into a centrally placed spherical nucleoid (Sarkar et al., 1972). Any RNA tumour virus particles which have toroidal nucleoids are classified as A-type particles. As we have discussed above, however, the subsequent maturation of A-type particles differs, depending on whether or not they are precursors of B-type or C-type particles. Furthermore, it is not at all clear that all A-type particles mature into particles with spherical nucleoids.

Because, with the exception of the murine mammary carcinoma viruses, all mature RNA tumour virus particles have a C-type morphology, leukemia and sarcoma viruses are often referred to in the literature simply as C-type particles.

Detailed Morphology of C-type Particles

This classification of RNA tumour virus particles into three categories, A-type, B-type and C-type particles, is based on gross morphological characteristics. Until recently the architecture and symmetry of the core of RNA tumour virus particles remained obscure, chiefly because negative staining procedures that had revealed the internal structures of other enveloped viruses, such as influenza viruses and Newcastle disease virus, singularly failed reproducibly to reveal the internal structure of the RNA tumour viruses. For example, while some investigators reported seeing helical, presumably ribonucleoprotein structures in the nucleoids of these viruses after negative staining (Vogt, 1965a; de-Thé and O'Connor, 1966; Sarkar and Moore, 1968; Luftig and Kilham, 1971) and proposed general models of the structure of the nucleoids of these viruses (Sarkar and Moore, 1968; Thomas et al., 1969; Nowinski et al., 1970), others failed to detect such helical structures (Almeida et al., 1967; Hageman et al., 1972).

The precise architecture of the ribonucleoprotein nucleoid of these viruses remains unknown, but recently Nermut et al. (1972), who studied murine leukemia virus particles by the new techniques of freeze etching and negative staining followed by freeze drying, saw in the electron microscope a set of previously undetected structures. Their new information, which has led to the model structure shown in Figure 10.2, can be summarized as follows: (1) The envelope of the virion consists of a unit membrane, the outer surface of which is lined with knob-like structures presumably analogous to the spikes on the surfaces of avian C-type particles. (2) Immediately within the envelope is a core shell consisting of an outer capsid with icosahedral symmetry made of subunits with a hexagonal structure. These subunits presumably contain some of the major internal proteins of these particles (see below). (3) The inner surface of the envelope membrane is very closely associated with this capsid. (4) Immediately within the capsid layer of the core shell lies a core membrane. (5) Within the capsid and core membrane lies the ribonucleoprotein nucleoid, which is a spherical filamentous structure that may possess helical symmetry. As Nermut et al. (1972) point out, several of these structures are similar to structures seen by previous investigators, for example de-Thé and O'Connor (1966) and Padgett and Levine (1966).

No one has yet investigated the structure of avian or other mammalian C-type virus particles by freeze etching but the observations Nermut et al. have made will presumably prove to be of general significance.

CHEMICAL COMPOSITION OF THE RNA TUMOUR VIRUSES

All RNA tumour viruses have very similar overall chemical compositions. They comprise about 60–70% protein, 20–30% lipid, 2% carbohydrate and about 1% RNA (see review by Beard, 1963). In addition, these viruses contain a very small amount of DNA (Levinson

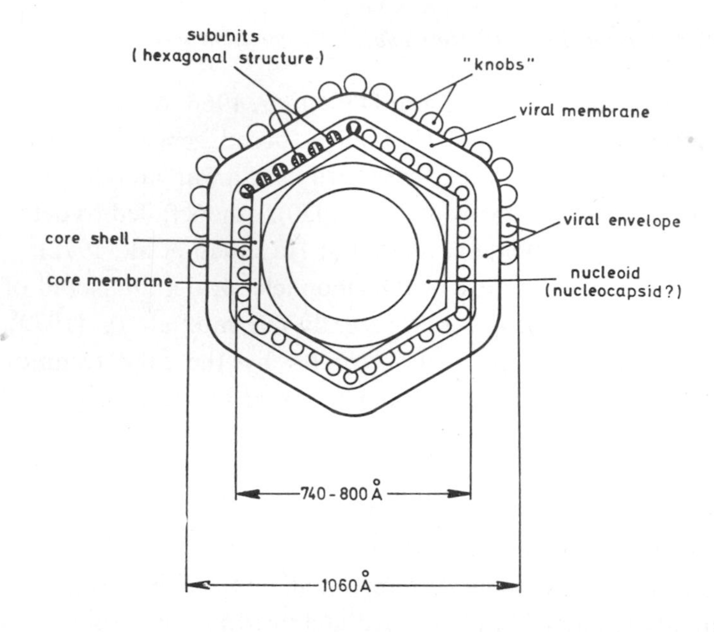

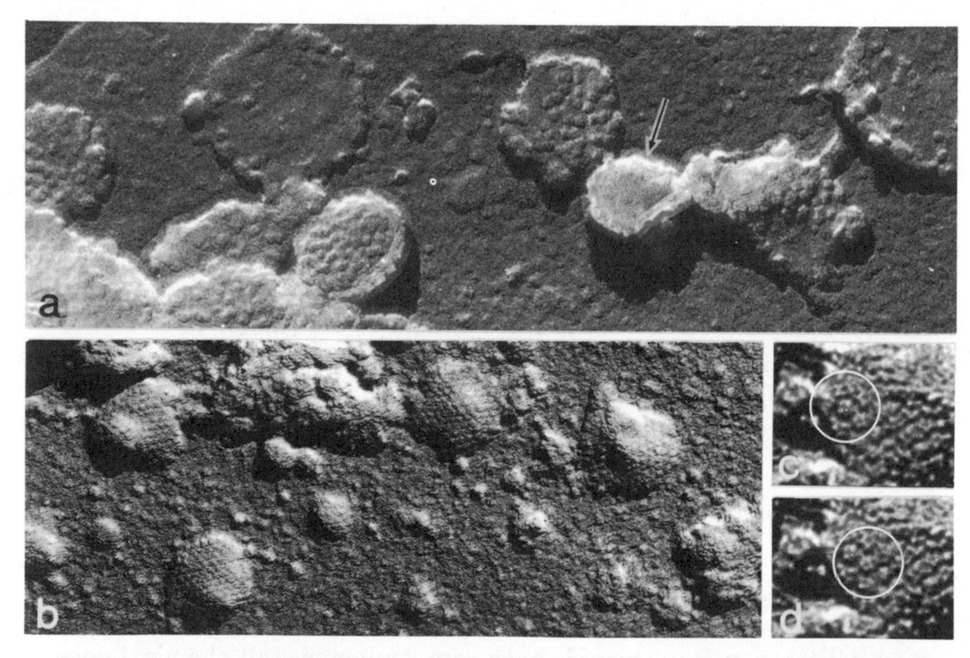

Figure 10.2. *Top:* Diagram of the structure of a murine leukemia virus particle. *Below:* Electron micrographs of Friend virus particles **(a)** and cores **(b)** after treatment with Tween-ether, freeze drying and shadowing. In **(a)** one particle (arrow) is full and knobs cannot be seen. **(c)** A penton (circled) found on the corner of a core. **(d)** A neighbouring hexon. **(a, b)** ×118,800; **(c, d)** ×315,000. Reproduced by permission from Nermut et al., 1972.

et al., 1970; Bishop et al., 1970a,b) which may well be of cellular origin (Varmus et al., 1971).

The Viral RNAs

When RNA extracted with phenol from RNA tumour viruses is sedimented through a neutral sucrose gradient, several components are resolved. The two major components sediment at 60–70S and 4–5S, but in addition there are lesser amounts of 28S, 18S and 7S RNAs (Robinson et al., 1965; Robinson and Baluda, 1965; Duesberg and Robinson, 1966; Watson, 1971; and reviews by Duesberg, 1970; Temin, 1971); all these RNAs are single-stranded and susceptible to digestion by pancreatic ribonuclease. The amounts of the various RNA species vary from preparation to preparation but the ratio of 60–70S RNA to 4–5S RNA is about unity.

The 60–70S RNA

The 60–70S RNA, the largest species in B- and C-type particles, is believed to be the viral genome, but there is no direct evidence that this is the case because 60–70S RNA is not infectious. Claims that infectious RNA can be obtained from RNA tumour viruses have not proven to be reproducible (see review by Temin, 1971), and the possibility that at least some of the 60–70S RNA is derived from the host has not been rigorously excluded. Using the relationship between sedimentation constant and molecular weight of single-stranded RNA derived for the RNA of tobacco mosaic virus (Spirin, 1963), the molecular weight of the 60–70S RNA in tumour viruses is estimated to be $10\text{–}12 \times 10^6$ daltons. Consistent with this Granboulan et al. (1966), who examined the RNAs of avian myeloblastosis virus particles in the electron microscope, saw extended RNA molecules some 8.3 μ long, which they estimated had a molecular weight of about 10^7 daltons. Kakefuda and Bader (1969), on the other hand, reported that the RNAs of murine leukemia virus particles and Rous sarcoma virus particles appear in the electron microscope as extended chains up to 14 μ long, and some of the structures they saw consisted of two closely associated RNA chains.

After 60–70S RNA is irreversibly denatured either by heating or by treatment with DMSO, it sediments in neutral sucrose gradients at about 35S (Duesberg, 1968; Bader and Steck, 1969; Erikson, 1969; Montagnier et al., 1969; Manning et al., 1972) and the 35S RNA has a greater electrophoretic mobility than the undenatured 60–70S RNA Duesberg and Vogt, 1970; Duesberg et al., 1971). It seems, therefore, that on denaturation the 60–70S RNA disaggregates into subunits of lower molecular weight. The alternative possibility, namely that the 60–70S RNA undergoes an extreme conformational change in denaturing conditions, is unlikely (see reviews by Bader et al., 1970; Duesberg, 1970).

Duesberg and Vogt (1970) reported that when 35S RNA from uncloned stocks of Rous sarcoma virus particles is subjected to electrophoresis through SDS-polyacrylamide gels, two classes of molecules differing in their electrophoretic mobilities are resolved; these have been called *a* and *b* subunits. By contrast, when the 35S RNAs from avian leukemia virus particles are analyzed in this way, only *b* subunits are found. Furthermore, 35S RNA from a stock of a Rous sarcoma virus which has lost the ability to transform chick embryo fibroblasts (Toyoshima et al., 1970) apparently consists only of *b* subunits. Since Rous sarcoma virus transforms chick fibroblasts but avian leukemia virus does not transform these cells even though the virus replicates in them, Duesberg suggested (1) that the 60–70S RNA of both these viruses consists of an aggregate of three or four single-stranded RNA subunits, each weighing about $3–4 \times 10^6$ daltons, and (2) that the *a* subunits, apparently unique to the sarcoma viruses, might carry the information necessary for the transformation of fibroblasts. The formal possibility that the appearance of *a* subunits is a consequence of the replication of sarcoma viruses in transformed cells was ruled out by Martin and Duesberg (1972). They transformed fibroblasts by infecting them at the permissive temperature with a mutant strain of Rous sarcoma virus temperature-sensitive for transformation (Martin, 1970) (see Chapter 13). Then by raising the temperature they caused the cells to lose many of the phenotypic characteristics of transformants. Progeny Rous sarcoma virus particles were harvested at this nonpermissive temperature and their RNA was analyzed by gel electrophoresis. Both *a* and *b* subunits were detected, which indicates that the production of *a* subunits is not dependent upon replication of a sarcoma virus in a cell with a transformed phenotype.

All these experiments were performed with uncloned stocks of virus. More recently, Duesberg et al. (1973) have shown that there are no, or very few, *b* subunits in the denatured 60–70S RNA extracted from cloned Rous sarcoma virus particles derived from clonal populations of infected transformed cells. This finding indicates that the genome of a Rous sarcoma virus particle capable of transformation may be made up of three or four *a* subunits with molecular weights of about 3–4 $\times$ 10^6 and identical electrophoretic mobilities. One way to account for both *a* and *b* subunits in stocks of Rous sarcoma virus which propagate by spreading infection is to suggest that such stocks are mixtures of transforming particles, the genomes of which are three or four *a* subunits, and spontaneous deletion mutants (Vogt, 1971a; Martin and Duesberg, 1972), which are unable to transform and have a genome made of three or four *b* subunits. Furthermore, one can speculate that the *b* subunits arise by the partial deletion of *a* subunits. Unfortunately, it is difficult to repeat these experiments with cloned avian leukemia viruses because avian leukemia viruses do not transform chick fibroblasts growing in vitro and as a result it is difficult to select clones of cells infected by single particles of avian leukemia virus. We do not know, therefore, whether the 70S RNA of cloned avian leukemia viruses, like the 70S RNA from uncloned stocks of these viruses, lacks *a* subunits.

It is important to remember that, even if the *a* and *b* subunits are different molecules, a physically homogeneous population of RNA molecules need not have identical base sequences. In other words, Duesberg's data, which indicates the 60–70S RNA molecules of cloned Rous sarcoma virus particles are made of three or four *a* subunits with identical sedimentation coefficients and electrophoretic mobilities, do not prove that the genomes of these viruses are triploid or tetraploid, although, of course, this is a possibility (see Chapter 13).

The evidence we have discussed so far, which indicates that the 60–70S RNA of avian, murine and other C-type RNA tumour viruses consists of an aggregate of subunits with sedimentation coefficients of about 35S and a molecular weight of about 3 $\times$ 10^6, can be summarized as follows: (1) The 60–70S RNA on denaturation sediments at 35S, and (2) the 35S RNAs of avian sarcoma and avian leukemia viruses have different electrophoretic mobilities. Clearly, this evidence is not conclusive but two recent sets of observations also support the segmented genome theory. First, free 30-40S RNA molecules were isolated directly from feline leukemia virus particles (O. Jarrett et al., 1971). Avian tumour virus particles do not usually contain free 30–40S RNA; however, Cheung et al. (1972) found that RSV particles harvested every 5 minutes contained heterogeneous RNA, most of which was 50–60S but some was smaller, whereas virus harvested at hourly intervals contained 60–70S RNA. Duesberg et al. (1973) following up this finding found that the RNA obtained from avian sarcoma virus particles harvested within a few minutes of their release from infected cells sediments at 30–40S, not 60–70S. If, however, these freshly harvested particles are incubated in buffered saline at 40°C for 3 minutes, their RNA on extraction sediments at 60–70S. It seems likely, therefore, that the 30–40S RNA is a precursor of the 60–70S RNA and the conversion of the former to the latter may involve association with 4S RNA molecules (see below). The second set of experiments which argue strongly that the 60–70S genomic RNA of tumour viruses is segmented involves analysis of recombinant virus particles. When a chick fibroblast is infected and transformed by two Rous sarcoma viruses of different subgroups, the progeny virus particles include a very high proportion of recombinants (Vogt, 1971b; Kawai and Hanafusa, 1972b) (see Chapter 13). The most plausible way to explain this high frequency of recombination is to suggest that the 60–70S RNA molecules that enter progeny virions are formed from a single pool of subunit RNA molecules, and in mixed infections half of the molecules in this pool are specified by one of the parental genomes and half by the other parental genome; reassortment of subunits, which might be expected to occur at a high frequency, would of course lead to recombinant 60–70S RNA molecules. The influenza viruses, whose genomes are collections of several single-stranded RNA subunits, offer a precedent for the recombination by reassortment of segmented RNA genomes.

We do not yet know if freshly budded murine C-type viruses contain exclusively 30–40S RNA; neither have the 30–40S RNAs obtained by denaturing 60–70S RNA from these murine viruses been analyzed by gel electrophoresis.

The base composition of those 60–70S RNAs of RNA tumour viruses that have been analyzed show no peculiarities and little can be said about the base sequences of these RNAs. We do know, however, that there is little homology between the sequences of murine leukemia viruses and murine sarcoma viruses (Stephenson and Aaronson, 1971) or murine mammary tumour viruses (Axel et al., 1972a,b), but there is considerable homology between avian sarcoma and leukemia viruses. Furthermore, it has been reported (Lai and Duesberg, 1972; Gillespie et al., 1972) that the 60–70S RNAs of both Rous sarcoma virus and Rauscher mouse leukemia virus contain adenylic acid-rich sequences, which probably consist of a series of tracts of about 10 to 40 adenylic acid residues separated by single guanylic, uridylic or cytidylic acid residues (Horst et al., 1972).

Viral 4–5S and Other RNAs

At least some of the 4–5S RNA molecules in RNA tumour virus particles appear to be host transfer RNAs (Bauer, 1966; Bonar et al., 1967); these RNAs are methylated to the same extent as host-cell transfer RNAs (Erikson, 1969); they can be charged with several species of amino acids (Travnicek, 1968); they hybridize to the DNA of normal and cancer cells to the same extent and can compete with cellular transfer RNAs for the same binding sites (Baluda and Nayak, 1970). Presumably these transfer RNAs of the host cell are incorporated into the virus particles as they are assembled in the cytoplasm, and the incorporation appears to be selective rather than random since in avian myeloblastosis virus the ratios of the fourteen species of tRNA in the virus particles differ greatly from the ratios of the same species in the host cell (Travnicek, 1968; Erikson and Erikson, 1970). Furthermore, different viruses probably contain different proportions of various transfer RNAs, for Bishop et al. (1970a) have reported that the base composition of 4S RNAs from avian myeloblastosis virus particles and Bryan high titre Rous sarcoma virus particles are different. As Temin (1971) has pointed out, the role of the tRNAs in the life cycle of the RNA tumour viruses is obscure and they may simply be contaminants picked up during the maturation of progeny virions. It is also possible that some of the 4–5S RNA found in some preparations of RNA extracted from tumour virus particles contains degradation products of larger RNA species.

When native 60–70S RNA from avian myeloblastosis virus particles, Rous sarcoma virus particles or murine sarcoma and leukemia virus particles, which has been separated from the 4–5S RNA, is exposed to DMSO or is heated, the molecules dissociate into subunits that sediment at about 35S, but in addition, denaturation releases smaller RNA chains sedimenting at 4S (Erikson and Erikson, 1971). This 4S RNA represents about 2.5–3 percent of the total RNA in the undenatured 60–70S aggregates, which means that each 35S subunit must have 4–5 molecules of associated 4S RNA. The 4S RNA molecules that are released by denaturation contain some of the rare bases found in transfer RNAs and therefore they may be related to transfer RNA. Since they are detected only after the 60–70S RNA is denatured, it seems likely that these 4S RNA molecules are associated with the 35S subunits of the 60–70S aggregates by hydrogen bonds, and it is possible that they act as primers for the synthesis of DNA by RNA-dependent DNA polymerase (see Chapter 11) and as linkers holding together the 35S RNA subunits.

RNA tumour virus particles contain, in addition to 60–70S and 4–5S RNAs, RNA molecules sedimenting at 28S, 18S and 7S (Bishop et al., 1970a,b). The 28S and 18S RNAs are similar in size to host ribosomal RNAs, and according to Imai et al. (1966) and Gay et al. (1970), at least some of these viruses contain host ribosomes which presumably are trapped in maturing virions. Such ribosomes could, of course, be the source of the 28S and 18S RNA but that has yet to be proven. The function and origin of the 7S RNA in these viruses is totally obscure, although it is found in cells that are not producing C-type particles (Erikson et al., 1973).

Finally, at least some RNA tumour viruses contain very small amounts of DNA, which sediments at about 7S (Biswal et al., 1971; Levinson et al., 1970; Riman and Beaudreau, 1970; Rokutanda et al., 1970). Hybridization experiments indicate that this DNA is cellular in origin (Varmus et al., 1971) but what function it has and whether or not it is essential for infectivity remains to be seen.

Tumour Virus Lipids

The envelope of RNA tumour virus particles is derived from the cell surface membrane and most, if not all, of the lipids in tumour virus particles are located in the unit membrane envelope of the virion. Bonar and Beard (1959) and Rao et al. (1966) have reported that avian and murine RNA tumour viruses contain phospholipids. More recently, Quigly et al. (1971) have analyzed the phospholipid content of chick embryo fibroblasts and their plasma membranes, before and after infection with the Schmidt-Ruppin strain of RSV, and the phospholipid composition of the progeny virions (see Table 10.1). Infection with this virus does not significantly change the overall phospholipid content of the cells, and there are only small differences between the phospholipid compositions of plasma membrane fractions from uninfected and infected cells.

The phospholipid composition of the Rous sarcoma virions, however, differs significantly from that of the plasma membrane of

Table 10.1. Percent Distribution of Phospholipids

Spot on thin-layer chromatography	Total cell phospholipids		Plasma membrane phospholipids		Rous virus phospholipids
	Normal	Rous	Normal	Rous	
Phosphatidyl-ethanolamine	37	39	24	31	31
Phosphatidyl-choline	40	43	39	36	24
Sphingomyelin	11	9	23	25	38
Phosphatidyl-serine -inositol	8	8	12	9	9
Other phosphate positive spots	<3	<3	<1	<1	<1

Average values from 2–5 experiments.

infected cells (see Table 10.1). Apparently the Rous sarcoma viruses are budded from areas of the plasma membrane exceptionally rich in sphingomyelin and poor in phosphatidylcholine. Studies of antigens on the surfaces of infected cells also indicate that the RNA tumour viruses are budded from particular areas of the cell surface. Virus-specific Gross antigen and cell-specific H2 antigen are localized in patches on the surface of murine cells infected with murine leukemia viruses of the Gross subgroup (Aoki et al., 1970). Progeny viruses are budded from areas of the surface that do not contain either of these antigens, which, as a result, are not present in the envelopes of the virions.

Tumour Virus Carbohydrates and Proteins

As far as is known none of the carbohydrate in RNA tumour virus particles occurs as free carbohydrate; it is found in glycoproteins and other envelope components. The proteins which constitute 60–70% of the mass of RNA tumour virus particles have been analyzed using three techniques, SDS-polyacrylamide gel electrophoresis, gel filtration chromatography, and by immunological methods, notably immunodiffusion. These methods resolve the structural proteins of virions that are present in large amounts. Various enzymes present in these viruses, including RNA-dependent DNA polymerase, are not detected in these analytical systems. The comparatively simple data that have been accumulated about the structural proteins of various C-type RNA viruses have been made needlessly confusing by the different nomenclatures that various investigators have adopted. Tables 10.2 and 10.3 correlate these nomenclatures and we have adopted that of Fleissner (1971) for the avian virus proteins and that of Nowinski et al. (1972a) for the murine virus proteins.

Proteins of Avian C-type Virus Particles

Avian RNA tumour virus particles contain two species of glycoproteins (see Table 10.4) called m1 and m2, the carbohydrate moieties of which contain glucosamine, galactose and fructose (Duesberg et al., 1970). On gel filtration in guanidine hydrochloride, m1 behaves as if its molecular weight were about 100,000 daltons; but when subjected to

Table 10.2. COMPARISON OF NOMENCLATURES FOR PROTEINS OF AVIAN TUMOUR VIRUSES

	Fleissner (1970)	Duesberg et al. (1970)	Bolognesi and Bauer (1970)	Allen et al. (1970)	Hung et al. (1971)
Envelope glyco-proteins	m1	I	G[c]		1, 3[e]
	m2	II	G[c]		2
Internal proteins	gs1	RSV3	4	gsa[d]	4
	gs2	(RSV3)[a]	3		5, 6[f]
	gs3	RSV1[b]	1[b]	gsb[d]	8[b]
	gs4	RSV2	2		7
	p5	(RSV1)[b]	(1)[b]		(8)[b]

[a] The protein corresponding to gs2 appears as a shoulder on RSV3.

[b] Proteins corresponding to gs3 and p5 appear superimposed in PAGE (Fleissner, 1971). The strong gs antigen reported at this mobility must thus be gs3.

[c] Protein G contains carbohydrate and has a molecular weight (ca. 115,000) compatible with a complex of m1 and m2, neither of which appears as distinct monomers in this system.

[d] Identified by molecular weight, gs antigenicity, and amino acid composition.

[e] Protein 1 contains carbohydrate and may represent an aggregate of protein 3 (m1), since m1 aggregates strongly even in 6 M guanidine hydrochloride (Fleissner, 1971). For additional evidence that the smaller glycoprotein forms residual aggregates in SDS during polyacrylamide gel electrophoresis, see Fig. 5 of Duesberg et al. (1970) and Fig. 7 of Hung et al. (1971).

[f] Protein 6 as isolated in 4 M urea probably represents an aggregated form of protein 5 with a different pI. See Bolognesi and Bauer (1970) for evidence that a protein of this size in polyacrylamide gel electrophoresis aggregates strongly in the absence of reducing agents.

polyacrylamide gel electrophoresis in the presence of SDS, it migrates as a polypeptide chain with a molecular weight of about 32,000 daltons. The m1 protein is presumably an oligomer made of subunits weighing 32,000 daltons. By contrast, the other glycoprotein behaves in both systems as a monomeric protein with a molecular weight of about 70,000 daltons. These two proteins, which comprise 10–20 percent of the total virion protein, form the spikes on the outer surface of the envelopes of the virions, from which they can be stripped by proteases without disrupting the envelope. These glycoproteins give rise to the type (subgroup) specific antigenicities of these viruses and they play a part in

Table 10.3. COMPARISON OF NOMENCLATURES FOR PROTEINS OF MAMMALIAN (MURINE) TUMOUR VIRUSES

Nowinski et al. (1972a)	Shanmugam et al. (1972)	Schafer et al. (1972a,b)	Duesberg et al. (1970)	Oroszlan et al. (1971a)	Moroni (1972)
m1	(V)[c]	II_v[e]	I		IV
m2	(VI)[c]	II_{gs}	II		V
gs1 (gs3)[a]	IV	IV, V[a]		3	III
p2[b]	II[b]	1[f]		1[b]	(II)[g]
p3[b]	III[b]	III[f]		2[b]	(II)[g]
p4	I ?[d]	(I)[d]		(I)[d]	I ?[d]

[a] The gs1 and gs3 antigens are on the same molecule. Antigens IV and V of Schafer et al. correspond, respectively, to gs1 and gs3 of Old's group. (Antigen V is also designated gs-interspecies by Schafer's group.)

[b] Protein p2 behaves as though larger than p3 in gel filtration (in 6 M guanidine hydrochloride) and smaller than p3 in polyacrylamide gel electrophoresis in the presence of SDS.

[c] Shanmugam et al. do not explicitly identify these proteins as glycoproteins; in addition they present evidence of a host-cell origin for these species in their virus preparations. Therefore these species are not equivalent to the viral glycoproteins.

[d] Proteins p2 and p4 have essentially identical mobilities in polyacrylamide gel electrophoresis (E. Fleissner; R. Nowinski; unpublished), hence they usually appear superimposed as a single species by this method. It is not clear whether the minor species (I) reported by Moroni and by Shanmugan et al. represents a separation of p4 under their conditions.

[e] The particular glycoproteins carrying these two antigens have not been identified.

[f] Provisional assignments of antigens to polypeptides (W. Schafer, personal communication).

[g] Moroni did not resolve 2 proteins at this position.

controlling the host range of the virus because they are involved in binding virus particles to receptors at the surface of susceptible cells.

The remaining structural proteins (see Tables 10.2 and 10.4) are all located within the envelope of the virion and give rise to the group-specific antigenicity unique to the avian sarcoma and leukemia viruses. One of these proteins, gs1, has been found in all the avian viruses so far tested and may form, together with gs3 and p5, part of the capsid of the core shell (see above) because these three proteins are present in the ratio 1:1:0.5 (Fleissner, 1971). Apparently gs4 is associated with the RNA of these viruses.

Table 10.4. PROTEINS OF AVIAN C-TYPE VIRUSES

Name	Molecular weight* (in daltons) A	B	Molar ratio†	No. of molecules† per virion
m1	<100,000	32,000	0.35	1,000
m2	70,000	70,000	0.15	400
gs1	27,000	25,500	1.00	3,000
gs2	19,000	21,000	0.60	1,800
gs3	15,000	12,000	1.02	3,000
gs4	12,000	14,000	0.67	2,000
p5	10,000	12,000	0.51	1,500

Data from Fleissner (1971).
* Molecular weight estimated from (A) gel filtration in guanidine hydrochloride, (B) SDS-polyacrylamide gel electrophoresis.
† In MC29 (ALV) virions.

Davis and Rueckert (1972) isolated a ribonucleoprotein particle from the Schmidt-Ruppin strain of Rous sarcoma virus. This structure contains 20 percent RNA and five virus polypeptides. All of the 60–70S RNA from the virion is found associated with the particle as well as 15 to 20 molecules of 4S RNA, while the remainder of the 4S RNA is released as the result of the detergent treatment used to isolate the ribonucleoprotein particle. According to Davis and Rueckert the intact SR.RSV virion contains a total of 11 detectable polypeptides; a protein such as the reverse transcriptase probably would not have been observed in these experiments since only a few molecules are found per virion (see Chapter 11).

The total of the molecular weights of all the identifiable proteins is 727,000 daltons. This is about twice the amount of protein coded for by poliovirus. In other words, about 4.8×10^6 daltons of viral RNA would be required to code for all these structural proteins. Finally, none of these proteins correlate with the various enzyme activities present in the virus particles.

Proteins of Murine Leukemia and Sarcoma Viruses

Murine C-type viruses, like their avian counterparts, have two species of glycoproteins containing glucosamine (see Table 10.5) that are located on the outer surface of the virion envelope (Nowinski et al.,

Table 10.5. MOLECULAR WEIGHTS OF PROTEINS OF C-TYPE VIRUSES

Protein	Murine leukemia virus	Hamster sarcoma virus	Protein	Avian leukemia virus
m1	100,000	100,000	m1	100,000
m2	70,000	70,000	m2	70,000
p1 (gs1, gs3)	31,000	31,000	gs1	27,000
			gs2	19,000
p2	15,000	15,000	gs3	15,000
p3	12,000	12,000	gs4	12,000
p4	10,000	10,000	gs5	10,000

Data from Nowinski et al., 1972a; Fleissner, 1971. Molecular weights estimated from guanidine hydrochloride chromatography.

1972a) and probably form the knob-like structures revealed by Nermut et al. (1972). These glycoproteins give rise to the subgroup or type-specific antigenicity of these viruses.

In addition, these murine viruses contain four internal proteins; the largest of these, p1 (also called gs1 or gs3), with a molecular weight of about 30,000 daltons, gives rise to the group-specific antigenicity unique to the murine viruses. This protein, or a protein barely resolvable from p1, or a multimeric form of p1, also gives rise to the interspecies-specific antigenicity (gs3 antigenicity) common to all the mammalian C-type viruses. As Table 10.5 shows, the murine viruses and hamster sarcoma virus lack an internal protein equivalent to the gs2 protein present in the avian viruses.

Proteins of Mouse Mammary Tumour Viruses

Nowinski et al. (1972b) have also analyzed by polyacrylamide gel electrophoresis and gel filtration the proteins of murine B-type particles after labeling the virions in vitro with [^{14}C]iodoacetamide. The molecular weights of the proteins of MMTV particles differ from those of the murine sarcoma-leukemia viruses (see Table 10.6) and there is no immunological cross reaction between proteins of murine B- and C-type particles.

Enzymes in Virions of RNA Tumour Viruses

The virions of RNA tumour viruses contain numerous enzyme activities in addition to ATPase and reverse transcriptase (see Chapter

Table 10.6. Properties of Proteins of the Mouse Mammary Tumour Virus

Protein	Molecular weight[a]	Location in virion	Properties
s1	52,000	nucleoid	major gs antigen of virion; accounts for approximately 25% of total viral protein
s2	23,000	nucleoid	minor gs antigen of virion
s3	23,000	membrane	constituent of viral membrane; common to MTV from all mouse strains
s4		spikes(?)	type-specific antigen, lipid labile, presumed to reside on viral spikes
s5		spikes(?)	

[a] Estimated from polyacrylamide gel electrophoresis.

11). Indeed these viruses contain so many activities that it seems unlikely that many of them are specified by the viral genome or have any important biological role. Most of these enzymes are presumably cellular in origin and are either trapped within virions maturing in the cytoplasm of infected cells or adsorbed to the surface of virions.

The enzymes found to date in purified virions include many involved in nucleic acid synthesis: nucleases, DNA ligase, phosphatases, kinases, RNA methylase and amino acyl RNA synthetases (see Temin and Baltimore, 1972). Also, enzymes of carbohydrate metabolism, such as hexokinase and lactic dehydrogenase, have been detected in virion preparations (Mizutani and Temin, 1971).

CLASSIFICATION AND NOMENCLATURE OF RNA TUMOUR VIRUSES

Largely for historical reasons, the classification and nomenclature of the RNA tumour viruses has become so labyrinthine that it confuses rather than clarifies. In the literature, for example, these viruses are variously described as RNA tumour viruses, leukoviruses, oncornaviruses and C-type RNA viruses: hereafter we shall use only the first of these names.

Unlike the DNA tumour viruses, the replication of RNA tumour viruses need not kill the host cell. The maturation of virus particles by budding from the plasma membrane does not usually cause cytopathic effects; therefore an infected cell can become transformed and proliferate while producing viral progeny. However, as Temin (1962) first reported, virus replication is not necessary for cell transformation; neither is cell transformation necessary for virus replication (Figure 10.3). These phenomena will be discussed in detail in later sections.

Originally the RNA tumour viruses were classified according to the disease which they caused and were named after their discoverers. C-type RNA tumour viruses may be broadly divided into sarcoma and leukemia (leukosis) viruses. The division is not a clear one because some leukemia viruses, such as avian erythroblastosis and lymphoid leukosis viruses, can induce sarcomas when inoculated in high doses intramuscularly. Also leukemia viruses are commonly present in stocks of sarcoma viruses. In general, sarcoma viruses transform fibroblasts in culture, and cell transformation is the criterion by which these viruses are usually assayed. On the other hand, leukemia viruses do not usually transform fibroblasts although they readily infect and replicate in them.

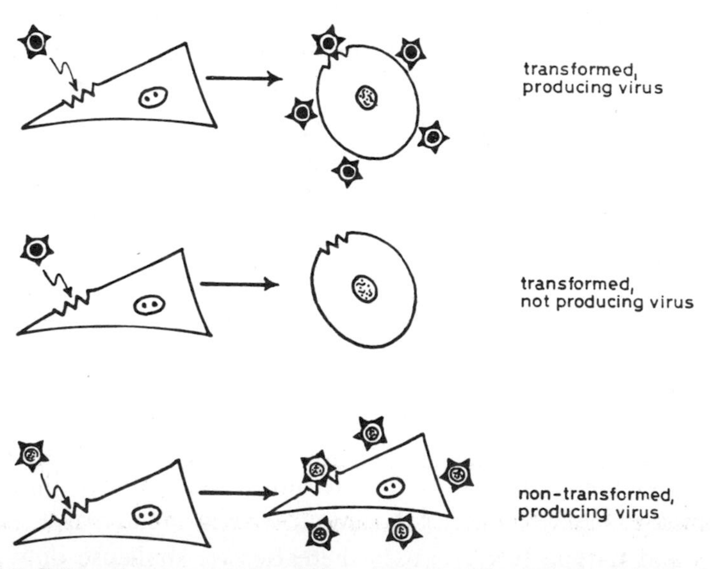

Figure 10.3. Schematic representation of the responses of fibroblasts to infection by RNA tumour viruses.

It is customary to refer to all C-type RNA viruses which fail to transform cultivated fibroblasts as leukemia viruses irrespective of their capacity to induce tumours in vivo and irrespective of the type of tumours they may induce. Certain leukemia viruses may transform other kinds of cells in culture. For instance, chick embryo yolk-sac cells are susceptible to transformation by avian myeloblastosis virus, but are not transformed by avian lymphoid or erythroid leukosis viruses (Baluda and Goetz, 1961; Smith and Moscovici, 1969). The reason why tumour viruses transform only some types of differentiated cells remains a fascinating problem that has received little attention.

Antigenically different viruses can cause the same pathological response in susceptible host animals while antigenically related viruses can cause different diseases. The current classification of the RNA tumour viruses ignores their pathology; instead it is based on three criteria, the host range, the interference pattern, and the antigenicity of the virus. These characteristics can be determined unequivocally by experiments with cultivated cells, and with the exception of the host range of the murine viruses, these characteristics are determined by the chemistry of the envelope of the virus particles.

Because phenotypic mixing frequently occurs between different RNA tumour viruses (see Chapter 13), the envelope of a virion may not have been specified by the genome within that virion. Only if a virus has been cloned can one be certain that the envelope phenotype corresponds to the virion genotype (Vogt, 1967a).

Classification of Avian RNA Tumour Viruses

The leukosis and sarcoma viruses of chickens are often referred to as the avian RNA tumour viruses rather than the chicken RNA tumour viruses. Viruses which have RNA genomes and are associated with tumours have, however, been isolated from other species of birds. For example, reticuloendotheliosis virus of turkeys (Theilen et al., 1966) does not share common antigens with chick leukosis viruses and probably therefore represents a distinct class of avian RNA tumour viruses. Likewise, a virus isolated from anemic ducks (Purchase, personal communication) probably belongs to the same group as the turkey reticuloendotheliosis virus, as may a syncytium-forming virus isolated from chicken tumour tissue (Cook, 1969). But because we know very little about these other avian viruses, we discuss only the leukosis/sarcoma complex of viruses of chickens.

Assays of Avian RNA Tumour Viruses

These viruses have been more thoroughly characterized than any other group of RNA tumour viruses and they serve as a model for the other groups. Analytical studies of avian RNA tumour viruses depended on the development of a quantitative assay. The first reproducible assay for Rous sarcoma virus (RSV) was developed by Keogh (1938), who showed that discrete tumours or "pocks" develop on the chick embryo chorioallantoic membrane following inoculation of a suitable dilution of virus. This technique was refined by Groupé et al. (1957) and by Rubin (1955, 1957), who used it to measure the production of virus by cultivated fibroblasts infected by RSV; the mean number of pocks per chorioallantoic membrane was linearly related to the virus concentration. The chorioallantoic membrane pock assay of RSV is still used to screen the susceptibility of chick embryos to different strains of RSV (Crittenden and Motta, 1969; Payne et al., 1971), but stocks of RSV are nowadays usually assayed by a focus assay involving cultivated chick embryonic fibroblasts (Manaker and Groupé, 1956; Temin and Rubin, 1958; Vogt, 1969).

Temin and Rubin (1958) showed that the number of discrete colonies or "foci" (Figure 10.4) of transformed cells in a culture of fibroblasts infected by RSV is, like the number of chorioallantoic membrane pocks, linearly dependent on the dose of virus inoculated; this indicates that a single transforming RSV particle is sufficient to initiate a focus of transformed cells. Experiments with X-irradiated feeder layers and X-irradiated transformants indicate that a focus grows both by mitosis of the transformed cells and by recruitment of neighbouring cells by secondary infection with progeny RSV (Rubin, 1960a). The capacity of transformed cells to grow in a soft gel (Macpherson and Montagnier, 1964) has also been used as an assay of transformation by RSV (Rubin, 1966; Weiss, 1970; Friis et al., 1971;

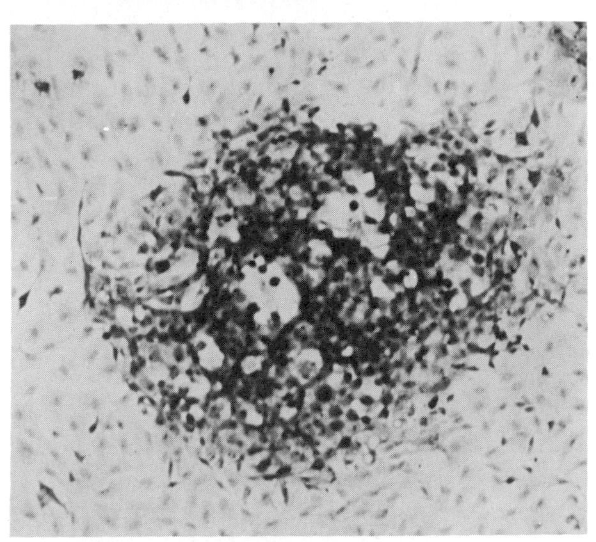

Figure 10.4. Photograph of a focus of quail fibroblasts transformed by BH.RSV(RAV-O). Magnification ×35.

Wyke and Linial, 1973) and this technique is especially useful for the selection of clones of transformed cells.

Avian leukosis viruses that do not transform cultivated fibroblasts are titrated by a variety of methods: (1) detection of group-specific antigens in infected cultures, usually by complement fixation (the "COFAL" test) (Sarma et al., 1964); (2) detection of progeny virus in infected culture fluids by assaying reverse transcriptase activity (see Chapter 11); (3) assay of ability to interfere with sarcoma virus infection (Vogt and Ishizaki, 1966); interference (see below) is a subgroup-specific phenomenon; (4) assay of ability to act as "helper" viruses for defective sarcoma viruses (Hanafusa et al., 1963); (5) fluorescent focus assay for envelope antigens (Vogt, 1964); (6) by plaque assays. Kawai and Hanafusa (1972a) and Wyke and Linial (1973) found that chick cells infected at the nonpermissive temperature with certain temperature-sensitive mutants of RSV failed to take up neutral red (Figure 10.5) after superinfection with nontransforming viruses of subgroups B and D (see below). This cytopathic effect forms the basis of a plaque assay because the number of plaques formed is linearly dependent on the dose of leukosis virus. Dougherty and Rasmussen (1964) and Graf (1972) have also described plaque assays for avian leukosis viruses. Leukosis viruses of two subgroups, B and D (Graf, 1972), infect and kill chick embryonic fibroblasts growing in special conditions and this cytocidal effect can be exploited to assay these viruses. This technique should prove useful for cloning some nontransforming avian RNA tumour viruses.

Methods (1) through (4) are end-point dilution titrations of leukosis viruses, whereas methods (5) and (6) are focal assays.

Envelope Classification of Avian Tumour Viruses

The avian RNA tumour viruses can be classified into five subgroups, A, B, C, D, and E, on the basis of their host range, interference and envelope antigens. These three properties are all determined by the envelope of the virus and the first two reflect the interaction of the viral envelope with specific receptor sites at the cell surface.

Host Range

Vogt and Ishizaki (1965) classified avian tumour viruses into subgroups on the basis of their host range in chickens. The susceptibility of

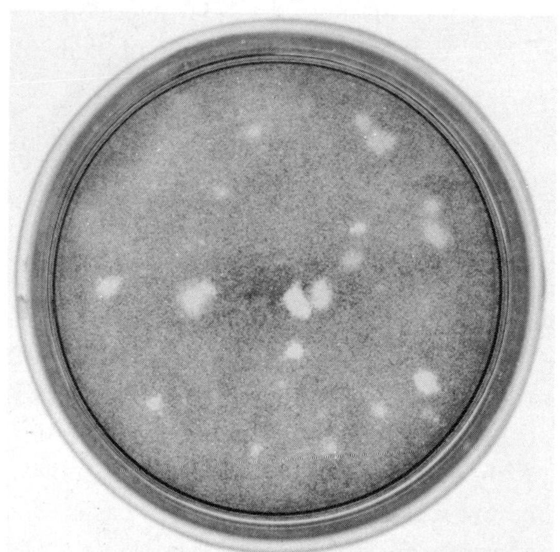

Figure 10.5. **Photograph of plaques produced when chick fibroblasts infected by the** *ts25* mutant of Prague RSV at 41°C are superinfected with avian leukosis virus CZAV. Cells stained with neutral red; magnification ×2.

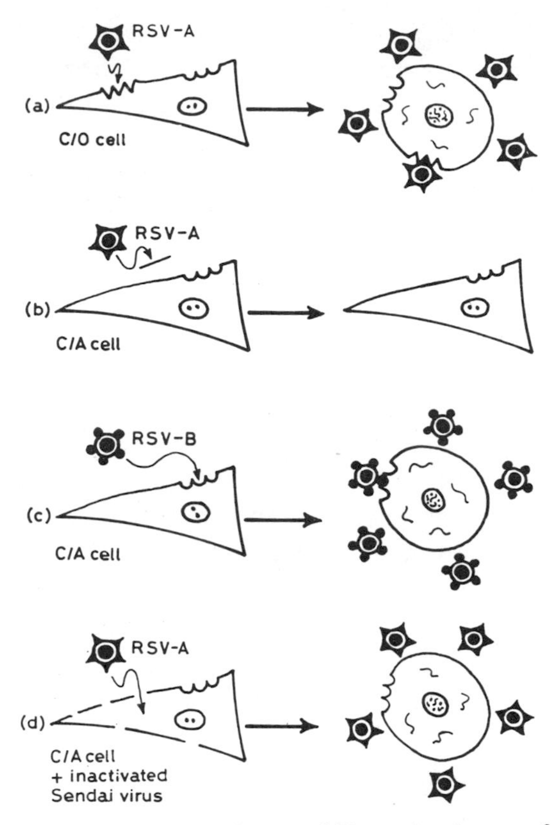

Figure 10.6. Schematic representation of susceptibility and resistance of chick fibroblasts to infection by avian sarcoma viruses (see text).

a cell to infection is expressed at the cellular level and depends on the presence of receptor sites specific for viruses of each subgroup (Figure 10.6a,b,c). If receptors for a particular subgroup are lacking, the cell is resistant to infection by viruses of that subgroup; the efficiency of plating of a virus on resistant cells is 10^{-3} or less than the efficiency of plating on susceptible cells. Virus particles absorb to both susceptible and resistant cells, but they penetrate only susceptible cells (Piraino, 1967; Crittenden, 1968). Resistance may, however, be overcome in two ways: (1) by phenotypic mixing of the virus with a virus of a subgroup to which the cell is susceptible (Hanafusa, 1965; Vogt, 1965b; Crittenden, 1968) or (2) by treating the resistant cells with inactivated Sendai virus or Newcastle disease virus (Robinson et al., 1967; Weiss, 1969a; T.

Hanafusa et al., 1970a), which probably causes fusion between the cellular and viral membranes (Figure 10.6d). These findings suggest that the virus receptors when present are located at the cell surface.

The phenotype of a cell is designated according to the virus subgroups that are excluded (Vogt and Ishizaki, 1965); for example C/AB signifies chick (C) cells resistant to (/) subgroup A and B viruses, while cells susceptible to all known avian tumour viruses are designated C/O (resistant to no subgroup). This designation has certain disadvantages because it identifies resistance, which is a negative attribute. As new subgroups of viruses were identified (Duff and Vogt, 1969; T. Hanafusa et al., 1970b), some cells which were previously designated C/O with respect to viruses of subgroups A and B were found to be resistant to viruses of the new subgroups. For instance, before 1971, unaware of the existence of subgroup E viruses, authors did not distinguish between C/O and C/E cells and many authors still do not do so. In short, C/O should be taken to signify susceptibility only to those subgroups under discussion.

Table 10.7 lists some chick cell phenotypes and their responses to avian tumour viruses of different subgroups. Receptors for the five subgroups are independent except for a relationship between receptors for virus of subgroups B and D. Chick cells which are resistant to subgroup B viruses show a 10- to 100-fold reduction in efficiency of plating of subgroup D viruses (Duff and Vogt, 1969) and exclude subgroup E

Table 10.7. Host Range of Avian Tumour Virus Subgroups in Chicken Cells

Cell type	A	B	C	D	E
C/O	+	+	+	+	+
C/A	−	+	+	+	+
C/E	+	+	+	+	−
C/BE*	+	−	+	±	−
C/AC	−	+	−	+	+

(+) Susceptible; (−) resistant; (±) semi-resistant.

* Resistance to B always also confers resistance to E but not vice versa.

viruses absolutely (Crittenden et al., 1973). However, cells of several other avian species, for example Japanese quail and ring-necked pheasant, are resistant to viruses of subgroup B but are susceptible to subgroup E viruses (Table 10.8).

Table 10.8. HOST RANGE OF AVIAN TUMOUR VIRUSES IN SEVERAL AVIAN SPECIES

	Virus subgroup				
	A	B	C	D	E
Red jungle fowl	+	+	+	+	+, −
Japanese quail	+	−	±	−	+
Ring-necked pheasant	+	−	±	−	+
Bobwhite quail	−	−	−	±	−
Turkey	+	−	+	−	+
Duck	−	−	+	±	NT
Goose	−	−	±	±	−

Symbols as for Table 10.6. NT, not tested.

The susceptibility and resistance of chickens to infection by avian RNA tumour viruses is under genetic control (see Payne et al., 1973). The presence of receptor sites is determined in a simple Mendelian manner by dominant alleles of autosomal loci. Four loci have been described: *tva* and *tvb* (Crittenden et al., 1964; Payne and Biggs, 1964, 1966; Rubin, 1965), *tvc* (Payne and Biggs, 1970) and *tve* (Payne et al., 1971), which control response to viruses of subgroups A, B, C and E respectively. Genetic control of subgroup D receptors has not been studied because chickens which show strong resistance to this subgroup have not yet been found. Alleles for susceptibility and resistance are designated by superscripts, *s* and *r* respectively, for example tva^{s}, tva^{r}, often abbreviated to a^{s}, a^{r}. The *tva* and *tvb* loci segregate independently (Crittenden et al., 1967) but the *tva* and *tvc* loci appear to be linked (Payne and Pani, 1971). Two susceptibility alleles for *tvb* locus, b^{s1} and b^{s2}, have been described (Crittenden et al., 1973). The b^{s1} allele is associated with R_1 erythrocyte antigen (Crittenden et al., 1970). By contrast to susceptibility to viruses of the other subgroups, susceptibility to subgroup E viruses is governed by two unlinked loci (Payne et al.,

1971). There is a receptor gene, *tve*, which may be allelic to the *tvb* locus (Crittenden et al., 1973), with dominant e^s and recessive e^r allelles. But expression of the dominant e^s allele can be blocked by the presence of a dominant, inhibitor gene, I^e. The epistatic influence of I^e on e^s may be a result of the expression of an endogenous virus and it will be discussed further in Chapter 12.

Although a great deal is known about the genetic control of susceptibility of chickens to RNA tumour viruses, the biochemical nature of the viral receptors is unknown. Apart from the association of the b^{s1} allele with the R_1 blood group, the receptors are not known to be antigenic even though they represent highly specific sites at the cell surface. Viruses do not penetrate cells at 4°C (Steck and Rubin, 1966a; Piraino, 1967), which implies that the receptors have a temperature-dependent function.

Interference Patterns

In 1960 Rubin reported that some chick embryos are congenitally infected with a virus that causes susceptible cells to become resistant to infection with RSV. This virus, which he called resistance-inducing factor (RIF), was identified as an avian leukosis virus (ALV) (Rubin, 1960b, 1961). Later, Steck and Rubin (1966a,b) showed that resistance to infection was a result of the leukosis virus blocking the cellular receptor sites so that the challenge RSV could not penetrate. Thus the resistance was neither caused by the induction of interferon nor by contamination with mycoplasma (Pontén and Macpherson, 1966). The ability of leukosis viruses to "interfere" with sarcoma virus infection was exploited as an assay of leukosis virus infection. Hanafusa (1965) and Vogt (1965a,b) found that interference occurred only between viruses which were related by host range and antigenicity, giving rise to a subgroup classification by interference (Vogt and Ishizaki, 1966). Figure 10.7 shows diagrammatically interference by a nontransforming virus of subgroup A. When the cells, preinfected by ALV-A, are challenged with RSV-A or RSV-B, after allowing time for the ALV-A to replicate, receptors for RSV-A are occupied by the ALV-A while receptors for RSV-B remain available. The cells are therefore only infected and transformed by RSV-B.

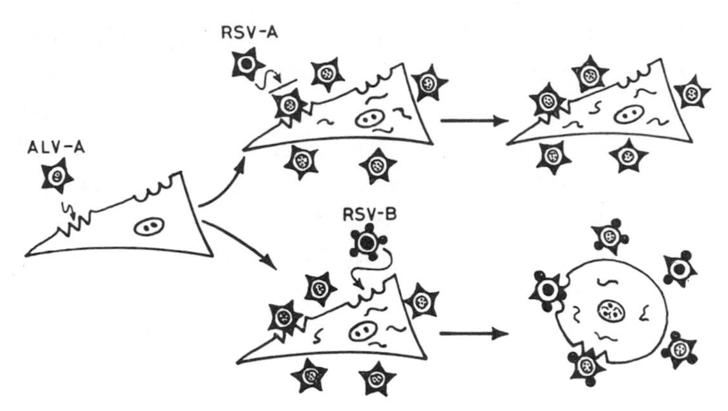

Figure 10.7. Schematic representation of interference of RSV-A infection by preinfection with ALV-A. Note that cells coinfected by ALV-A and RSV-B yield phenotypically mixed progeny.

Table 10.9 summarizes the interference patterns between viruses of the different subgroups. All viruses within a subgroup strongly interfere with one another. In addition, reciprocal interference occurs between viruses of subgroups B and D; the interference of viruses of subgroups B and D with viruses of subgroup E is nonreciprocal. Vogt (1970) has pointed out that this interference pattern is valid only for tests carried out with C/O cells. Only weak interference between subgroup C viruses is observed in cells of certain strains of C/A and C/AB chickens.

When interference tests are carried out between members of different subgroups, enhancement of secondary infection often occurs (Hanafusa,

Table 10.9. Interference Patterns among Avian Tumour Viruses in C/O Cells

Preinfecting ALV subgroup	Subgroup of challenge RSV A	B	C	D	E
A	I	+	+	+	+
B	+	I	+	I	I
C	+	+	I	+	+
D	+	I	+	I	I
E	+	+	+	+	I

I, Interference; (+) susceptible, no interference.

Based on data from Ishizaki and Vogt, 1966; Duff and Vogt, 1969; T. Hanafusa et al., 1970a,b; Vogt and Friis, 1971; Weiss et al., 1971.

1965; Vogt, 1965b; Ishizaki and Shimizu, 1970a). Hanafusa and Hanafusa (1967) found that such virus-induced enhancement is a result of a higher rate of adsorption of the challenge virus to preinfected cells. The same effect is caused by treating cells with polycations, such as DEAE-dextran (Vogt, 1967b; Toyoshima and Vogt, 1969). The polycations apparently attach to the cell surface and increase the rate of adsorption of viruses of subgroups B, C, D and E, but not subgroup A. This probably reflects differences in surface charge between virus particles of subgroup A (positively charged) and those of the other subgroups (negatively charged). It is now customary to treat cells with polycations before infecting them with viruses of subgroups other than A.

Antigenic Cross Reactions

The classification of avian tumour viruses into five subgroups is upheld by immunological studies of the envelope antigens (Ishizaki and Vogt, 1966; Duff and Vogt, 1969; Weiss, 1969b). If antisera against the virus particles are prepared in chickens, the sera react primarily with the envelope antigens rather than with the group-specific antigens in the interior of the virions. The neutralizing antibodies in the antisera apparently bind to the envelope glycoproteins and prevent the latter interacting with the subgroup-specific receptor sites on the surfaces of susceptible cells (Figure 10.8). Thus when antisera are titrated by neutralization tests, they reveal the subgroup specificity of the challenge virus (Ishizaki and Vogt, 1966; Chubb and Biggs, 1968; Weiss and Biggs, 1972). When, however, the same antisera are titrated by immunofluorescence, a more sensitive and specific test than neutralization, type-specific reactions are observed so that different viruses belonging to the same subgroup can be distinguished by virtue of their individual type-specific antigenicities (Vogt, 1970). The subgroup specificity of phenotypically mixed particles revealed by neutralization tests corresponds

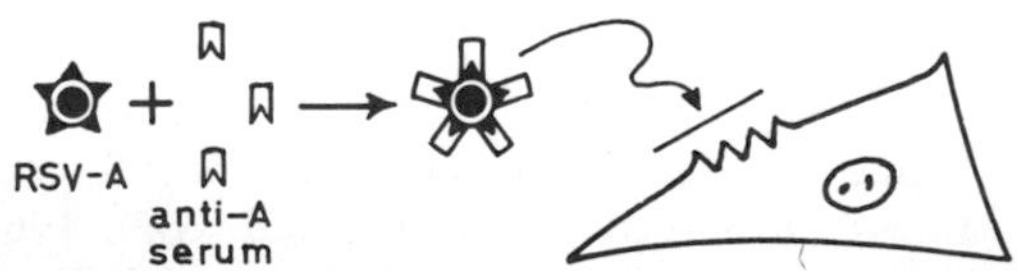

Figure 10.8. Diagram showing neutralization of RSV by antiserum against the envelope antigens of the virion.

to that revealed by tests of the host range of the virus particles (Vogt, 1967a).

Neutralizing antisera raised against subgroup B viruses cross-react with subgroup D viruses and vice versa (Duff and Vogt, 1969; Weiss and Biggs, 1972). Tests of host range and interference also indicate that viruses of subgroups B and D are closely related. With this exception antisera raised against viruses of one subgroup do not cross-react with virus particles of another subgroup; indeed exposure of virus particles to heterologous antisera sometimes enhances the plating efficiency of the treated virus by an unknown mechanism.

Thus, by virtue of their host range, interference and neutralization by antisera, avian tumour viruses may be allocated to five subgroups. Table 10.10 lists most of avian tumour viruses used in experimental studies. The majority of field isolates from the U.S.A., Europe and Japan belong to subgroup A and some to subgroup B. Field isolates of subgroups C and D have not been reported. Subgroup E includes only the viruses endogenous to chick cells, which will be discussed in

Table 10.10. Subgroup Classification of Several Avian Tumour Viruses

	A	B	C	D	E	Not classified
Sarcoma Viruses	SR.RSV-1	SR.RSV-2		SR.RSV-D	SR.RSV-E	BH.RSV
	PR.RSV-A	PR.RSV-B	PR.RSV-C		PR.RSV-E	BS.RSV
	EH.RSV	HA.RSV	B77	CZ.RSV-D		FUSV
	RSV-29		MH-2			
Leukosis Viruses	RAV-1	RAV-2	RAV-7	RAV-50	RAV-0	AMV
	RAV-3	RAV-6	RAV-49	CZAV	RAV-60	MC29
	RAV-4				ILV	Strain R
	RAV-5					
	FAV-1					
	MAV-1	MAV-2				
	RIF-1					
	RPL-12	ES4				

SR, Schmidt-Ruppin strain. PR, Prague strain. EH, Engelbreth-Holm strain. HA, Harris strain. CZ, Carr-Zilber strain. BH, Bryan high titre strain. BS, Bryan standard strain.

RSV, Rous sarcoma virus. RAV, Rous associated virus. B77, Bratislava 77 strain avian sarcoma virus. MH-2, Mill Hill 2 strain avian sarcoma virus. FUSV, Fujinami sarcoma virus. FAV, Fujinami associated virus. CZAV, Carr-Zilber associated virus. RIF, Resistance inducing factor. RPL-12, Lymphoid leukosis isolate 12 of Regional Poultry Laboratory (East Lansing). AMV, Avian myeloblastosis virus. MAV, Myeloblastosis associated virus. ES4, Erythroblastosis virus Strain 4. Strain R, Erythroblastosis virus strain R. MC29, Myelocytoma virus. ILV, Induced leukosis virus.

Chapter 12. The genealogy of some strains of RSV is obscure (Morgan and Traub, 1964) and they may not all be derived from Rous's No. 1 agent. The Prague and Schmidt-Ruppin strains of RSV are represented in several subgroups because of recombination of their subgroup markers with those of leukosis viruses (see Chapter 13). RAV-7 differs from the other viruses of subgroup C, not only antigenically but in its inability to infect species other than chicken (Smida and Smidova, 1971; Vogt, 1970).

Viruses which have not been cloned free of excess associated viruses cannot be assigned with certainty to any subgroup. These include Fujinami sarcoma virus, myelocytoma virus MC29, avian myeloblastosis virus strain BA1-A, and strain R erythroblastosis virus. Some of these viruses may be defective for synthesis of envelope antigens, of which the type viruses are the Bryan high titre and standard strains of RSV.

Bryan Strain RSV

Bryan developed two strains of RSV, the so-called Bryan standard strain (BS.RSV) and the Bryan high titre strain (BH.RSV). Both these strains of viruses have proved to be particularly interesting because they are defective. They are able to transform chick fibroblasts but alone they are unable to replicate infectious progeny.

It had been frequently observed that tumours induced by low doses of Bryan RSV were not infectious, and Temin (1962, 1963) discovered in cultures of chicken fibroblasts infected with low doses of Bryan standard strain RSV transformed (converted) cells that were not producing infectious progeny virus. At the time the existence of these "nonproducer" cells was interpreted as evidence for the idea that the genome of the transforming virus is integrated into that of the host cell, as was Temin's observation that superinfection of "nonproducer" cells by ALV or RSV resulted in the rescue of the Bryan strain RSV.

Also in 1962, Rubin and Vogt isolated a leukosis virus present to excess in their stocks of Bryan high titre RSV; they named this leukosis virus Rous associated virus (RAV) and it is now designated RAV-1. RAV-2 and RAV-3 were subsequently isolated from the same stock of Bryan high titre RSV, and RAV-4, RAV-5 and RAV-7 were isolated from stocks of Bryan standard strain RSV. (RAV-6 is derived from HA.RSV and RAV-49 and RAV-50 from SR.RSV; see Table 10.10).

In order to isolate BH.RSV free of any associated leukosis virus (RAV), Hanafusa et al. (1963) picked single foci from cultures infected with high dilutions of BH.RSV; but they found, as Temin (1962) had found when he picked single foci of cells transformed by BS.RSV, that the foci did not produce infectious virus. If, however, such "non-producer" cells infected by BH.RSV were superinfected with a RAV, infectious BH.RSV was liberated concomitant with the replication of the superinfecting RAV. In other words, production of infectious BH.RSV depended on the presence of the RAV, which acted as a "helper" virus. Other chicken leukosis viruses such as RIF-1 and avian myeloblastosis virus, but not the unrelated paramyxoviruses, were also found to act as "helpers" (Hanafusa, 1964). The envelope properties of infectious BH.RSV replicated in the presence of a helper virus were found to correspond to those of the helper virus, as judged by tests of host range, interference and antigenicity (Hanafusa et al., 1964; Hanafusa, 1965; Vogt, 1965b) (see Figure 10.9).

These various findings led to the suggestion that BH.RSV, and by implication BS.RSV, is unable to produce envelope proteins essential for infectivity and that the role of the helper virus is to donate by phenotypic mixing the missing envelope proteins. The infectious BH.RSV

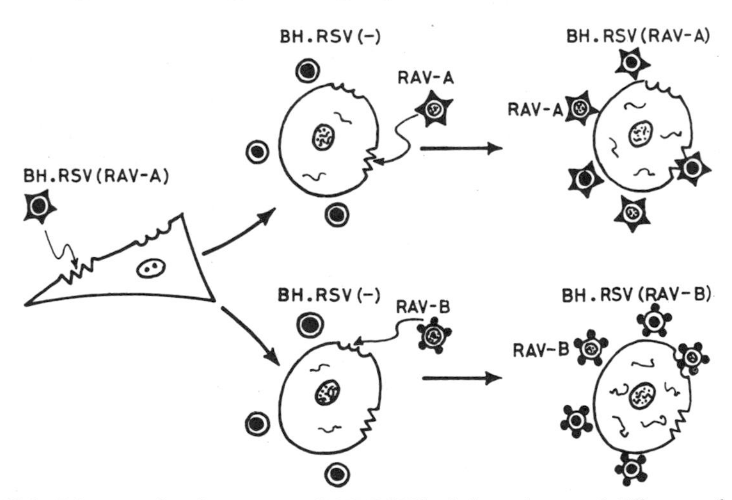

Figure 10.9. Diagram showing rescue of BH.RSV by helper viruses of different subgroups.

particles are called pseudotypes and the helper virus providing the envelope proteins is designated in parentheses, for example RSV-(RAV-1) (Rubin, 1965).

Whilst these experiments elucidating the role of helper viruses were being done, Dougherty and DiStefano (1965), using the electron microscope, discovered that the so-called "nonproducer" cells, infected and transformed by BH.RSV in the absence of a helper virus, do in fact produce C-type particles. These particles from the misnamed "nonproducer" cells proved not only to contain viral RNA (Robinson et al., 1967) but also to be infectious (Vogt, 1967c; Weiss, 1967) and able to replicate without a coinfecting helper virus in some, but not all, C/O cells. (This unusual host range of these particles has led to the recognition of the fifth subgroup of avian RNA tumour viruses, subgroup E [see Chapter 12].)

The virus liberated by "nonproducer" cells not coinfected with a "helper" was named RSV(O) by Vogt (1967c) to denote the absence of any helper. And for a time it seemed that the BH.RSV genome might not after all be defective for replication of infectious progeny. It was argued that the RSV(O) particles simply had an unusual host range. However, the experiments of Weiss (1969a) and T. Hanafusa et al. (1970a,b) revealed that the BH.RSV genome is indeed defective for replication of infectious progeny. They found that infectious RSV(O) particles are only produced in chick cells in which a latent leukosis virus genome is partially expressed; such cells are said to be chick helper factor-positive. Chick cells in which the latent leukosis virus is not expressed (chick helper factor-negative cells) yield noninfectious particles when infected with BH.RSV. These noninfectious particles apparently lack the envelope glycoprotein which confers a subgroup specificity and infectivity (Scheele and Hanafusa, 1971).

These data indicate (1) that the BH.RSV genome is defective and cannot specify an envelope glycoprotein and (2) that the missing envelope glycoprotein can be donated either by an exogenous, coinfecting leukosis virus or by the endogenous leukosis virus(es) present in every chick cell if these latent viruses are being partially expressed. Since all the endogenous viruses belong to subgroup E, infectious BH.RSV produced in the absence of an exogenous helper virus may be designated RSV(E). Noninfectious virus produced in the absence of an exogenous helper in cells not expressing their endogenous viral genome is now designated RSV(−) (see Figure 10.9).

Because BH.RSV is defective for replication but can transform fibroblasts, while leukosis viruses can replicate in fibroblasts but cannot transform these cells, Rubin (1964) suggested that the defectiveness of BH.RSV might somehow be associated with its oncogenicity. This idea was abandoned when it became evident that SR.RSV and other strains of RSV are not defective for replication and can transform fibroblasts (Dougherty and Rasmussen, 1964; Hanafusa, 1964). The idea has, however, recently been resurrected by Weiss (1973) because as far as we know all the mammalian sarcoma viruses resemble BH.RSV in that they are defective for replication but can transform fibroblasts. The nondefective, transforming avian sarcoma viruses may therefore be exceptional (see Chapter 13).

Avian Tumour Viruses in Mammalian Hosts

Infection of rats with avian tumour viruses was first reported by Svet-Moldavsky (1958) and by Zilber and Kryukova (see Zilber, 1964). It was soon confirmed for a variety of other mammalian species (Schmidt-Ruppin, 1959), including primates (Munroe and Windle, 1963). Subsequently, Duff and Vogt (1969) showed that, with the exception of B77 avian sarcoma virus, only subgroup D viruses infect mammals. Hanafusa and Hanafusa (1966) found that phenotypic mixing with RAV-50, a subgroup D virus, conferred infectivity for mammalian cells, albeit at a low efficiency. Altaner and Temin (1970), however, found that the envelope specificity of B77 virus (subgroup C) did not affect its oncogenicity for mammalian cells since phenotypic mixing with RAV-1 (subgroup A) did not affect its plating efficiency on mammalian cells.

With the exception of certain strains of rat and hamster tumour cells infected with B77 virus (Klement and Vesely, 1965; Altaner and Svec, 1966; Šimkovič et al., 1969) mammalian cells transformed by RSV (Svoboda and Hložánek, 1970) appear to be truly nonproducer cells; no C-type particles are released. Mammalian cells remain stably transformed by RSV and synthesize avian gs antigen (Sarma et al., 1964; Bauer and Janda, 1967) although morphological reversion has been observed (Macpherson, 1965). RSV can be rescued by inoculating the transformed mammalian cells into chickens or by cocultivating them with chicken cells (Svoboda, 1960; Šimkovič et al., 1962; Svoboda et al., 1963). Svoboda et al. (1967) and Shevlyaghin et al. (1969) showed that virus is rescued by fusing virogenic mammalian cells with permissive chick cells (Figure 10.10) and in some cases virus synthesis commences in every heterokaryon (Machala et al., 1970). Chick cells other than fibroblasts are inefficient at rescue, but chick fibroblasts need not express chick helper factor to be permissive for rescue of RSV (Svoboda et al., 1971). Finally, mammalian leukemia viruses apparently do not rescue RSV from mammalian cells.

LEUKEMIA AND SARCOMA VIRUSES OF RODENTS

The selection of inbred lines of mice with a high incidence of leukemia (MacDowell and Richter, 1935; Cole and Furth, 1941) led to the discovery of murine leukemia viruses, which have been extensively studied. Murine sarcoma viruses (MSV) have only been isolated in association with murine leukemia viruses. The Harvey and Kirsten strains of murine sarcoma virus were obtained from rats

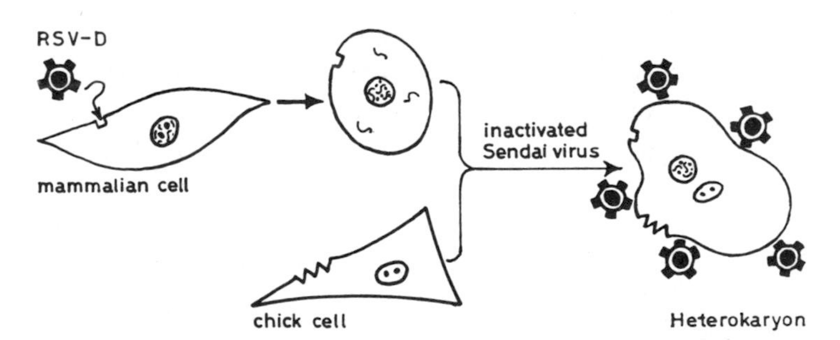

Figure 10.10. Diagram showing transformation of mammalian fibroblast by RSV and rescue of the RSV by fusion of the mammalian cell to a chick cell.

inoculated with stocks of murine leukemia virus. It is possible, therefore, that these two sarcomagenic agents are at least in part derived from a rat virus, in which case murine sarcoma virus may be a misnomer.

By comparison with the murine leukemia viruses, we know very little about the C-type RNA viruses that have been isolated from rats (Chopra et al., 1970; Aaronson, 1971), Syrian hamsters (Graffi et al., 1968; Kelloff et al., 1970a,b) and guinea pigs (cavies) (Jungeblut and Kodza, 1962; Gross, 1970); and although these various viruses are often referred to as leukemia viruses, they have not all been shown to induce leukemia.

The existence of hamster and rat leukemia viruses was surmised before their actual isolation (Bassin et al., 1968; Ting, 1968; Klement et al., 1969a; Perk et al., 1969) because certain hamster and rat cells were found to contain a helper activity for the replication of murine sarcoma virus. Various isolates of Syrian hamster leukemia virus (HaLV) possess the same group-specific (gs1) antigen, which is distinct from the group-specific antigen of the murine viruses (Kelloff et al., 1970b), and HaLV interferes with infection by MSV(HaLV) pseudotypes. MSV(HaLV) pseudotypes have without justification been referred to as hamster sarcoma viruses (HaSV or HSV). Here we shall discuss further only the murine sarcoma and leukemia viruses.

MURINE SARCOMA VIRUS (MSV)

A murine sarcoma virus was first discovered by Harvey (1964), who, after passing Moloney-MLV in rats, obtained a virus stock which induced pleomorphic sarcomas rather than leukemia in Balb/c mice. Subsequently, Moloney (1966) isolated a MSV from rhabdomyosarcomas induced in Balb/c mice by injecting high doses of M-MLV, and Kirsten and Mayer (1967), like Harvey, isolated a MSV from stocks of MLV which had been passaged in rats. Recently, Ball et al. (1973) repeatedly induced sarcomas which liberate MSV by injecting into mice stocks of MLV obtained from murine JLS-V9 cells growing in culture. None of the above sarcoma viruses was obtained from a naturally occurring sarcoma, but Finkel et al. (1966) isolated MSV from an osteosarcoma which arose in a CF1 strain mouse. The different isolates of MSV are usually named after their discoverers, for example, H-MSV: Harvey strain murine sarcoma virus.

The three strains of MSV which have been examined in detail, Harvey, Kirsten and Moloney, are all defective for replication; they can only replicate in the presence of a "helper" MLV to yield MSV-(MLV) pseudotypes (Hartley and Rowe, 1966; Huebner et al., 1966; see Harvey and East, 1971). Certain stocks of MSV that appeared to be helper-independent (O'Connor and Fischinger, 1968) were later shown to be aggregates of MSV and MLV produced during the centrifugation of mixtures of the two viruses (O'Connor and Fischinger, 1969).

Different stocks of MSV show different degrees of defectiveness; one clonal isolate, the nonproducer (NP)MSV, is defective for particle production and induction of gs antigen (Aaronson and Rowe, 1970), while another, the sarcoma-positive leukemia-negative (S+ L−) MSV, induces gs antigen and replicates noninfectious particles (Bassin et al., 1971b). These characteristics are stable even when the MSV is carried by different helper viruses or passed through cells of different lines (Aaronson et al., 1972). This indicates that genetic differences, not subject to recombination with helper viruses, exist between different clones of MSV. The helper functions of MLV include donating glycoproteins of the envelope because MSV pseudotypes are neutralized by antiserum raised against the helper virus (Huebner et al., 1966; Harvey and East, 1971).

Leukemia viruses of other mammalian species will act as helper viruses for defective MSV. For example, hamster leukemia virus rescues MSV from transformed nonproducing Syrian hamster cells to yield pseudotype virus (Kelloff et al., 1970a). Such pseudotypes should be called MSV(HaLV) but they are sometimes called hamster sarcoma virus (Zavada and Macpherson, 1970), even though the sarcomagenic agent is derived from MSV. Likewise rat leukemia virus (RaLV) acts as a helper for MSV; indeed RaLV was identified by this helper activity (Aaronson, 1971). Feline leukemia viruses have also been used as helper viruses for MSV (Fischinger and O'Connor, 1969; Sarma et al., 1970a), and the MSV(FeLV) pseudotypes have been exploited to analyze the biological properties of FeLV (see below).

Focus Formation Assay of MSV

Like the avian sarcoma viruses, MSV can be assayed by focus formation (Hartley and Rowe, 1966; Bassin et al., 1968; Bather et al., 1968; Jainchill et al., 1969). However, when stocks of MSV were assayed on mouse embryonic fibroblasts, a two-hit dose response curve was observed; this was reduced to a one-hit response when the stocks of MSV were titrated in the presence of excess MLV (Hartley and Rowe, 1966). It was concluded that focus formation resulted from dual infection with focus-forming and nonfocus-forming particles. Furthermore, some stocks of MSV which formed foci with a one-hit dose response proved to contain aggregates of MSV and MLV particles (O'Connor and Fischinger, 1968, 1969).

These data seemed to indicate that focus formation by defective MSV differed basically from focus formation by defective avian sarcoma viruses. MSV apparently required a helper virus for the initiation of a focus, whereas defective RSV requires a helper virus for replication of infectious progeny RSV but not for focus formation per se. Subsequent experiments, however, belied this conclusion. Parkman et al. (1970) found that MSV titrated with a one-hit dose response in NRK (rat) cells, forming foci which did not yield MSV particles infectious for mice. An objection, that these NRK rat cells contain a latent rat virus (Aaronson, 1971) or "helper factor," can be raised against these experiments but the results of Aaronson et al. (1970) are unambiguous. They found that MSV titrated in Balb/c cells with a two-hit dose response if the foci were counted 3–5 days after infection; however, if foci were counted 7 days or more after infection, small foci could be seen which developed with a one-hit dose response. We now realize that the foci initiated by dual infection by MSV(MLV) and MLV appear rapidly and become large as a result of recruitment of transformed cells by secondary infection. On the other hand, Balb/3T3 cells infected with solitary MSV(MLV) develop into foci more slowly by mitosis alone. In other words, MSV can transform murine fibroblasts in the absence of a helper virus just as BH.RSV can transform chick fibroblasts in the absence of a helper. Whether or not the transformation event can be assayed depends on whether or not the transformed cells develop into visible foci. Balb/3T3 cells transformed by MSV in the absence of helper viruses do eventually multiply sufficiently to form visible foci. By contrast, cultivated mouse embryo cells infected by MSV in the absence of a helper do not multiply faster than uninfected cells and do not therefore give rise to foci (Hartley and Rowe, 1966). This means that foci of transformed mouse embryo cells always develop by recruitment from cells infected by both MSV and a helper MLV.

Growth in Agar Suspension

Transformation of mouse cells by MSV in the absence of a helper virus can be assayed by plating the infected cells in soft agar (Bassin et al., 1970). The transformed cells are able to multiply to form colonies when cultured in agar suspension, whereas the untransformed cells fail to divide. This method was used by Bassin and his colleagues to isolate S+ L− cells, and hamster cells transformed by MSV have been assayed in this way (Zavada and Macpherson, 1970).

The focus-forming titre of MSV is increased by pretreating the cells to be infected with DEAE dextran (Duc-Nguyen, 1968; Somers and Kirsten, 1968), a procedure which also increases the efficiency of plating of MLV (Rowe et al., 1971). By analogy with the avian tumour viruses (Toyoshima and Vogt, 1969) polycations probably enhance the adsorption of the murine C-type particles.

Summary of the Properties of MSV

The chief properties of all characterized murine sarcoma viruses can be summarized as follows:

1. All MSVs are defective for the replication of infectious progeny particles and as a result all infectious MSV particles are in fact pseudotypes.

2. MLV and other mammalian leukemia viruses can act as helper viruses for the replication of infectious MSV pseudotypes.

3. MSV pseudotypes can induce the formation of foci of transformants by some cultivated cells in the absence of a coinfecting leukemia virus.

4. The formation of foci of mouse embryo cells always depends on double infection by an MSV pseudotype and MLV.

MURINE LEUKEMIA VIRUSES

The leukemogenic virus discovered by Gross (1951) in extracts of tissue from AKR strain mice causes lymphomas, which are typical of the disease prevalent in high leukemia strain mice. Many different isolates of "naturally occurring" leukemia virus have been obtained from mice of these strains as well as from aged or irradiated mice of low leukemic strains (see Gross, 1970). All these viruses are closely related serologically and pathologically to the Gross MLV.

Graffi and his colleagues isolated a second kind of murine leukemia virus, causing mainly myeloid leukemia, from transplantable tumours such as reticulum cell sarcomas, the Ehrlich ascites tumour cells and sarcoma 37 cells (Graffi et al., 1955; Graffi, 1957; Schmidt, 1955). One of these isolates (it is not clear which one) is now known as the Graffi strain MLV. Related leukemia viruses have been isolated from spontaneous tumours (Tennant, 1962) and from sarcoma 37 extracts (Franks et al., 1959; Moloney, 1960). A third kind of MLV was discovered by Friend (1957) after a Swiss mouse, which had been inoculated at birth with an extract of the Ehrlich transplantable tumour, developed an erythroid leukemia. A similar virus was isolated in the same year by Schoolman et al. (1957) from leukemic tissue of a Swiss mouse, and this virus is now known as the Rauscher virus (Rauscher, 1962). Pope (1962) also isolated a virus (WM1-B), which resembles Friend virus, after passaging a lymphoid leukemia virus derived from a leukemic wild mouse.

Pathology of Murine Leukemia Viruses

The pathogenesis of murine leukemia viruses has been summarized by Siegler and Rich (1966). The Gross-type viruses of high leukemic strains of mice, as well as Gross passage A virus, typically give rise to thymic lymphosarcomas. However, if these viruses are inoculated into thymectomized mice (Gross, 1960), myeloid (granulocytic) leukemias often develop instead of lymphosarcomas. Both lymphoid and myeloid tumours have a latent period of several months following inoculation of virus. On the other hand, the erythroblastic leukemias induced by Friend and Rauscher viruses have a latency of only one to three weeks, and adult mice are readily susceptible to the disease (Friend, 1957). The Friend disease is a neoplasm predominantly of the spleen, bone marrow and liver. The target cell is an erythroid precursor and it does not appear to involve the thymus or lymph nodes; neither does it depend on the thymus (Metcalf et al., 1959).

The lymphoid neoplasms found in rats and some strains of mice inoculated with Friend and Rauscher viruses (Mirand and Grace, 1962; Rauscher, 1962; Gross, 1964) are now known to be caused by a helper virus: stocks of the Friend virus comprise mixtures of a defective transforming or "spleen-focus-forming" virus (SFFV) and a non-defective, nontransforming helper virus (Dawson et al., 1968; Fieldsteel et al., 1969). The helper virus may be one of many lymphoid leukemia viruses (LLV), and lymphoid tumours may develop if the erythroleukemia does not kill the host first. Helper virus has been isolated free of Friend virus (SFFV) by end-point dilution (Rowson and Parr, 1970) or by passage in mice resistant to Friend disease (Steeves et al., 1971). The helper viruses appear to be required for the induction of Friend disease but many in fact, like the helper of MSV, merely accelerate the proliferation of neoplastic cells through recruitment by secondary infection.

Some strains of Friend virus induce a polycythemia (Mirand, 1966) in which the Friend cells mature into hemoglobin-synthesizing cells, but most strains of Friend virus give rise to cells which are arrested at an earlier stage of maturation (Friend and Rossi, 1968). Erythropoietin accelerates the onset of Friend disease (Mirand, 1966) but does not appear to be necessary for the differentiation of Friend cells. Certain established lines of cultured Friend cells can be induced to produce hemoglobin by treatment with dimethylsulfoxide (Friend et al., 1971; Ross et al., 1972). This phenomenon is exciting much attention among students of cell differentiation.

Biological Assays of MLV

Assay of MLV in Mice

Lymphoid and myeloid leukemia viruses. In vivo assays of lymphoid and myeloid leukemias (see Gross, 1970) have been superseded by in vitro assays which are much quicker, more quantitative and more economical. But no in vitro assays of MLV have been developed which titrate the oncogenic properties of the virus.

Friend virus. The rapid induction of splenomegaly in adult mice by Friend virus (Friend, 1957; Rowe and Brodsky, 1959) led to the development of a quantitative spleen focus assay (Axelrad and Steeves, 1964; Pluznick and Sachs, 1964). Axelrad and Steeves showed that the number of macroscopic focal lesions in the spleens of C3H or Swiss mice 9 days after infection was linearly related to the dose of virus inoculated. The spleen focus assay has been of great use for the analysis of genetic control of leukemogenesis and of the defectiveness of Friend virus (Lilly and Pincus, 1972), which will be discussed in greater detail below.

Assay of MLV in Culture

Like the avian leukosis viruses, MLV has been assayed by a variety of methods:

1. Detection of viral antigens in infected cultures by complement fixation (Hartley et al., 1965, 1969), immunofluorescence (Pinkel et al., 1966), or radioimmune assay (Scolnick et al., 1972c; Oroszlan et al., 1972a).
2. Detection of progeny virus in infected cultures by assaying reverse transcriptase activity (see Chapter 11).
3. Interference with MSV infection (Sarma et al., 1967).
4. Ability to act as helper viruses for MSV (Huebner et al., 1966; Hartley and Rowe, 1966; Fischinger and O'Connor, 1968).
5. *XC plaque assay.* Klement et al. (1969b) noticed that when mouse embryo cells releasing MSV were mixed with XC cells (a cell line derived from a rat tumour induced by PR-RSV [Svoboda et al., 1963]) in culture, the XC cells formed syncytia. This was developed into a sensitive plaque assay (Rowe et al., 1970) where focal areas of murine

cells supporting replication of MLV are revealed by adding XC cells, which form plaques of syncytia (Figure 10.11). Six days after infection with MLV, the mouse cells are irradiated with ultraviolet light (killing the cells but not the virus) and the XC cells are added. The cultures are fixed and stained 4 days later and the plaques counted. The number of plaques is linearly related to the dilution of MLV. The XC test provides a sensitive assay and a means of cloning MLV.

6. *S+ L− plaque assay*. Bassin et al. (1970) observed that MSV-transformed cells of an S+ L− line suffered a morphological change when superinfected with MLV. The change appeared to complete the transformation of the cells, which formed foci of rounded cells that tended to detach from the culture dish to leave plaques. Bassin et al. (1971a) have refined the technique to provide a quantitative and reliable plaque assay for MLV, giving equivalent titrations to the XC plaque assay. The assay is simpler than the XC test, and may be counted only 5 days after infection. MSV(MLV) is released in the foci produced in S+ L− cultures; therefore, the XC plaque assay is more suitable for cloning MLV.

7. *Focus assay for MLV*. Hackett and Sylvester (1972) recently

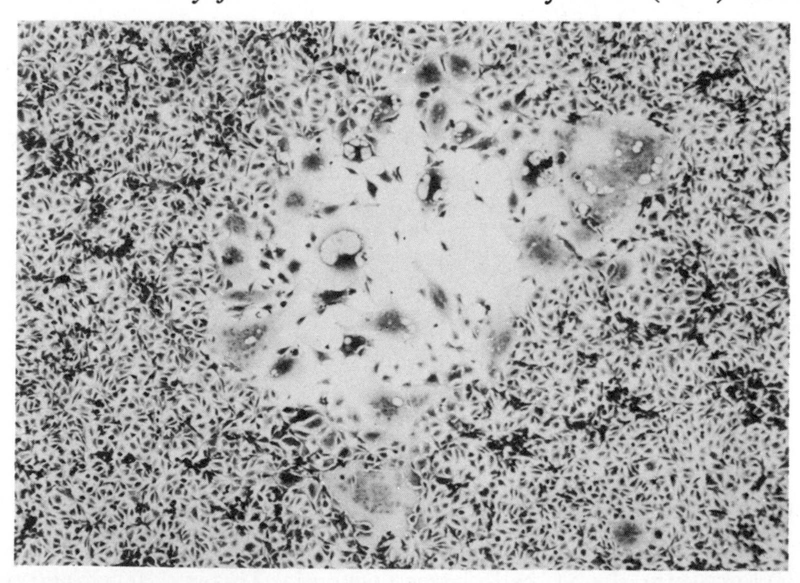

Figure 10.11. Syncytial plaques in an XC cell monolayer induced by MLV released by infected and irradiated mouse cells.

described a subline, UCL-B, of Balb/3T3 cells which is susceptible to transformation by MLV, giving rise to foci. This focus assay is as sensitive as the XC test, but is only suitable for strains of MLV which replicate in Balb/3T3 cells. MSV is not released by the foci but progeny MLV are released.

Classification of MLV

By contrast to the avian RNA tumour viruses, the antigenicity, interference patterns and host range of strains of MLV do not all reflect envelope specificities which allow a tidy classification into consistent subgroups. These three characteristics of MLV are determined independently.

Serological Subgroups of MLV

Serological studies of MLV did not commence with a classification of virus-neutralizing antibodies, but rather with studies of the antigenic properties of the leukemic cells. The murine leukemia viruses have been broadly classified into two serological subgroups, the Gross (G)-AKR subgroup, and the FMR (Friend-Moloney-Rauscher) subgroup, according to the types of antigens associated with the leukemic cells (Old et al., 1964; Aoki et al., 1966). These antigens were detected by cytotoxicity and immunofluorescence tests. It was also found that identical antigens existed in "soluble" form in the plasmas of leukemic mice from which whole virus had been removed by centrifugation (Stück et al., 1964a). A number of cytotoxic antisera also neutralize MLV (Klein and Klein, 1964; Stück et al., 1964a). It seems likely that some of the leukemia antigens represent virion envelope antigens (Aoki et al., 1972).

The naturally occurring leukemias all result in the production of cell-associated and soluble antigens of the G-AKR group (Aoki et al., 1966, 1968; Hartley et al., 1969), whereas FMR antigens are found only in experimentally infected mice.Leukemic cells induced by Gross virus do not absorb FMR antibody, but leukemias induced by Friend, Moloney and Rauscher viruses do weakly absorb Gross antibody (Aoki et al., 1966), probably because stocks of FMR virus are contaminated with Gross virus. Graffi MLV is related to the FMR group (Pasternak, 1969) and the WM1-B strain has not been classified. If leukemic cells induced by one type (for example, Gross) are superinfected with MLV of another type (for example, Rauscher or Graffi), the cells become susceptible to the cytotoxic action of antisera specific to the superinfecting virus (Stück et al., 1964b; Pasternak, 1967). This phenomenon, which resembles phenotypic mixing of viral antigens, is called antigenic conversion. Tumours induced by MSV express the antigens of their helper viruses.

Spontaneous leukemias in mice of the inbred DBA/2 strain do not express G or FMR antigens but do express an antigen which cross-reacts with murine mammary tumour virus antigens (Stück et al., 1964c). Aoki et al. (1966) have suggested that the mammary tumour virus of this strain may have leukemogenic properties.

Interference between MLVs

Sarma et al. (1967) tested interference patterns between Gross, Friend, Moloney and Rauscher MLVs and their MSV pseudotypes. Preinfection of mouse embryo fibroblasts with any of the four strains of MLV induced strong resistance to challenge with all MSV pseudotypes. Therefore, the murine leukemia viruses seem to belong to a single interference group because no distinction was found between the G-AKR and FMR antigenic subgroups.

Host Range of MLV

Hartley et al. (1970) observed that different stains of MLV could be divided into three categories according to their ability to grow on NIH Swiss (N) cells and Balb/c (B) cells. "N-tropic" viruses initiated infection more efficiently on N cells than on B cells, whereas "B-tropic" viruses showed a reciprocal pattern. In other words, N cells are always partially resistant to B-tropic viruses and vice versa. Several strains of MLV which had been passaged in the laboratory for many years, for example Moloney MLV, infected both cell types equally well and were designated "NB-tropic." Hartley et al. (1970) found that the differences in host response to infection with N- or B-tropic viruses were relative, not absolute. The sensitivity of NIH Swiss cells to infection with N-tropic MLV was 100- to 1000-fold greater than that of Balb/c cells, and conversely, the sensitivity of Balb/c cells to infection with B-tropic MLV was 30- to 100-fold greater than that of NIH Swiss cells. The tropism of the viruses was not correlated with their antigenic classification. N- and NB-tropic viruses belong to both G and FMR subgroups, although all B-tropic viruses so far tested belong to the G antigenic subgroup.

Using the XC plaque assay, Pincus et al. (1971a) were able to classify embryonic cells from 12 mouse strains unequivocally as N-type or B-type. Tests with F_1 hybrids of N-type $\times$ B-type mice showed that resistance to infection was dominant for both types of virus. However, NB-type MLV plated as well on F_1 hybrids as on parental N- or B-type cells, indicating that NB-MLV is not a mixture of N- and B-type particles but possesses a distinct phenotype. Backcross studies showed that the host-range patterns were determined by a single Mendelian locus, which proved to be identical to the *Fv-1* locus controlling Friend disease (Pincus et al., 1971b) (see below). Ware and Axelrad (1972) recently showed that resistance and susceptibility operate in cells that are congenic except for the *Fv-1* locus, carrying *Fv-1*^{n} and *Fv-1*^{b} alleles.

N- and B-tropic viruses are genetically stable. However, by forced passage through resistant cells they can be converted to NB-tropic viruses. The N $\rightarrow$ NB adaptation was made in vivo for Friend helper virus (Lilly, 1967), but the B $\rightarrow$ NB adaptation has been accomplished only in vitro (Hartley, Pincus and Rowe, quoted by Lilly and Pincus, 1972). MLV titrates on susceptible cells with single-hit kinetics but on resistant cells with two or more hits. The nature of the resistance is unknown but appears to take place after uncoating of the virion (Hartley et al., 1970). The tropism of MSV replication depends on its helper virus. Yoshikura (1973) has shown that S+L− MSV will infect resistant embryo cells with one-hit kinetics, but the foci will only be revealed by superinfection with an appropriate helper virus. This indicates that the S+L− genome does not carry the host-range marker, or if it does, it is not NB-tropic.

In summary, host-range control of MLV differs from avian viruses in several ways (see Table 10.11): resistance is dominant and reciprocal,

Table 10.11. CONTROL OF HOST RANGE OF AVIAN AND MURINE LEUKEMIA VIRUSES

Avian	Murine
Host range corresponds to envelope antigens	Host range does not correspond to envelope antigens
Susceptibility mediated through cell surface receptor	Susceptibility determined after uncoating of virion
Susceptibility dominant	Resistance dominant
Different susceptibility loci for different subgroups	Both subgroups controlled by one locus
Susceptibility to different subgroups largely independent	Reciprocal resistance between subgroups
Plating efficiency on resistant cells 10^{-5}	Plating efficiency on resistant cells $10^{-1.5}$ to 10^{-3}

but only relative; MLV tropism is not related to the envelope antigens and does not apparently operate at the cell surface receptor level. Whether a DNA provirus forms and integrates in resistant cells remains to be determined. The system is ripe for exploitation by the molecular biologist and the viral geneticist.

The host-range pattern of a few murine viruses does not fit the N-B pattern. One example is the unusual virus carried by NZB mice (Mellors and Huang, 1966; East et al., 1967) which carries murine gs1 antigen and belongs to the G subgroup (Nowinski et al., 1972b). NZB-MLV rescues MSV from transformed hamster cells and plates on hamster and rat cells but does not plate well on N- and B-type mouse cells (Levy and Pincus, 1970).

Genetic Control of Viral Leukemogenesis

Host genetic control of susceptibility to Friend disease was first observed by Odaka and Yamamoto (1962), who studied crosses between a susceptible strain (RF) and a resistant strain (C57BL) of mice. Susceptibility appeared to be dominant in heterozygotes. With the introduction of the spleen focus assay, however, an intermediate response was observed in heterozygotes (DDD × C57BL; DBA × C57BL), which were relatively resistant to focus formation (Axelrad, 1966). Then Lilly (1967) found that the highly inbred mouse strain

Balb/c behaved as a relatively resistant host like the DBA × C57BL heterozygotes. When Lilly subjected the Friend virus to several forced passages through Balb/c mice, he obtained a variant which now showed an equally high-focus titre in the spleens of DBA and Balb/c mice. We now know that the Friend virus was converted from N-tropic to B-tropic and that the relative resistance is that governed by the *Fv-1* host-range gene (Lilly and Pincus, 1972).

A complex picture has now emerged (reviewed by Lilly and Pincus, 1972) involving two host genetic loci and two viral components. Lilly (1970) established that susceptibility was controlled by two independently segregating genes, *Fv-1* and *Fv-2*. The *Fv-2* locus appears to be identical with the *Fv* locus described by Odaka (1969) and is in linkage group II of the mouse genome, whereas the *Fv-1* locus has recently been mapped on linkage group VIII (Rowe et al., unpublished data). Two alleles of the *Fv-2* locus, $Fv\text{-}2^s$ and $Fv\text{-}2^r$, confer susceptibility and resistance to Friend disease, respectively. Unlike the *Fv-1* alleles, resistance at the *Fv-2* locus is recessive, but absolute; it is not related to viral replication and does not affect the plating efficiency of MLV in the XC test (Pincus et al., 1971b).

It is now apparent that Friend leukemia virus stocks consist of two components, named spleen focus-forming virus (SFFV) and lymphoid leukosis virus (LLV) (Dawson et al., 1968). Like MSV, SFFV appears to be defective for replication and LLV acts as a helper virus. Fieldsteel et al. (1969) established a reticulum cell sarcoma induced by Friend virus as a nonproducer cell line in vitro. On superinfection with Moloney MLV or other LLVs, it releases virus which rapidly induces spleen foci in vivo, whereas the helper viruses alone do not. LLV has been freed from SFFV by end-point dilution (Rowson and Parr, 1970) and the host range of the original Friend virus (N-tropic) can be converted to NB-tropic by forced passage through B-type mice (Lilly, 1967; Steeves and Eckner, 1970), or to B-tropic by co-infection with a B-tropic helper virus (Steeves et al., 1971).

The *Fv-2* locus appears to act on SFFV and not on LLV. Mice which are homozygous for $Fv\text{-}2^r$ are resistant to focus formation by SFFV, but are not resistant to replication of LLV or to lymphoid leukemia induced by LLV. Thus the *Fv-2* locus rather specifically controls the transformation of spleen cells by SFFV. Table 10.12 summarizes the properties of LLV and SFFV.

It is not clear whether replication of LLV is necessary for spleen focus formation by SFFV. If double infection with LLV and SFFV were required in order to obtain a spleen focus, a two-hit dose-response curve would be expected, unless there was a great excess of LLV in the Friend virus stock. In mice carrying susceptible *Fv-1* alleles, one-hit dose-response curves are obtained for spleen foci, indicating that the helper virus is not required for focus formation. However, in mice which are resistant at the *Fv-1* locus, a two or more hit response is observed (Steeves et al., 1971), which is reduced to one-hit by adding helper virus of the appropriate tropism. This result suggests that helper virus replication is required for spleen focus formation, and the system requires further study to elucidate the mechanisms involved.

Alleles at loci other than *Fv-1* and *Fv-2* (for example *W* and *Sl*) also affect spleen focus formation, but those loci presumably affect the availability of the target cells as they are known to affect normal hemopoiesis (see Lilly and Pincus, 1972). The $H\text{-}2^{b}$ allele also confers slight resistance to viral leukemogenesis (Lilly, 1968; Tennant and Snell, 1968), and the neighbouring *TLa* and G_{IX} loci on linkage group IX also affect leukemogenesis (see Boyse et al., 1972).

Avian myeloblastosis virus (Moscovici and Zanetti, 1970) and avian erythroblastosis virus strain R (Ishizaki and Shimizu, 1970b) also appear to be comprised of defective transforming particles and helper particles, and the Friend virus story may serve as a useful model for studying other forms of leukemia.

Table 10.12. Friend Virus and Friend Disease in Mice

LLV	SFFV
Causes lymphoid leukemia	Causes erythroid cell transformation
Acts as helper for SFFV replication	Replication defective
Replication controlled by *Fv-1* locus, resistance dominant	Transformation controlled by *Fv-2* locus, resistance recessive

FELINE LEUKEMIA AND SARCOMA VIRUSES

Feline leukemia virus (FeLV) was first described by Jarrett et al. (1964) and subsequently many other independent isolations have been made. Feline leukemia virus can be isolated from about 60 percent of cats with leukemia (O. Jarrett and Laird, unpublished data) and the sera of about 30 percent of cats cross-react with the surface antigens of these viruses (W. Jarrett et al., 1973a). Feline sarcoma viruses have been isolated from several feline fibrosarcomas (Snyder and Theilen, 1969; Gardner et al., 1970). Both feline leukemia and feline sarcoma viruses have the typical C-type virus morphology (Jarrett et al., 1964; Snyder and Theilen, 1969). They contain 60–70S RNA (Jarrett et al., 1971) and an RNA-dependent DNA polymerase (Spiegelman et al., 1970; Hatanaka et al., 1970) which is antigenically related to the corresponding enzyme in murine and hamster C-type particles (Aaronson et al., 1971; Oroszlan et al., 1971b). Feline leukemia and sarcoma viruses have a unique group-specific antigen (Hardy, 1970) and in addition the interspecies-specific antigen common to all mammalian C-type particles (Geering et al., 1970).

Feline sarcoma viruses infect and transform fibroblasts of cats and other species and focus assays have been developed for them (Sarma et al., 1970b, 1971a,b; McDonald et al., 1972). Like avian and murine leukemia viruses, feline leukemia viruses infect and replicate in, but do not transform, fibroblasts (Jarrett et al., 1968). FeLV can be assayed in vitro by measuring the gs antigen (Sarma et al., 1971a) or by measuring interference with FeSV or FeLV pseudotypes of MSV (Sarma and Log, 1971). Plaque assays for FeLV have not yet been described.

FeLV can act as a helper virus for the replication of MSV but, because FeLV does not infect mouse or hamster cells, direct superinfection with FeLV of cells transformed by MSV does not lead to efficient rescue of MSV (FeLV) pseudotypes. These have been obtained by two other methods: (1) Fischinger and O'Connor (1969) showed that cosedimentation of MSV(MLV) and FeLV in the ultracentrifuge yielded aggregates of mixed virus that allowed entry of MSV into feline cells together with the FeLV. The infected cells then yielded MSV(FeLV) pseudotypes possessing the host range of FeLV in addition to progeny FeLV. By cosedimenting MSV(FeLV) with MLV and infecting murine cells with the aggregate virus, MSV(MLV) was obtained again. (2) Sarma et al. (1970a) obtained MSV(FeLV) pseudotypes by cocultivating MSV-transformed hamster cells with feline embryonic fibroblasts infected with FeLV. Presumably the pseudotypes arose from fused feline and hamster cells. The MSV(FeLV) was not infectious for hamster cells, was neutralized by antiserum against FeSV and was subject to interference by FeLV. MSV(FeLV) stocks contain excess FeLV and the MSV genome remains dependent on the FeLV helper virus for replication. MSV(FeLV) pseudotypes have proved useful in the classification of FeLV into subgroups.

Feline leukemia and sarcoma viruses have been divided into three subgroups, A, B and C (Table 10.13), by patterns of viral interference and by host range (Sarma and Log, 1971; O. Jarrett et al., 1973). Interference with FeSV or MSV(FeLV) infection by FeLV occurs only within the subgroups. It was initially thought that all three FeLV subgroups interfered with MSV(FeLV) of subgroup A (Sarma and Log, 1971), but this was probably the result of subgroup B and C viruses in the stock of subgroup A MSV(FeLV) (O. Jarrett et al., 1973). As with the avian leukosis viruses the interference patterns of feline leukemia viruses coincide with the host-range classification: both are determined by the envelope of the virions (O. Jarrett et al., 1973).

Table 10.13. HOST RANGE AND INTERFERENCE PATTERNS OF FELINE LEUKEMIA VIRUSES

	Host range			Interference: Challenge subgroup of MSV(FeLV)		
Subgroup	Feline	Human	Canine	A	B	C
A	+	−	−	I	+	+
B	+	++	+	+	I	+
C	+	+	+	+	+	I

(+) Susceptible, (−) resistant, (I) interference.

It is not known whether there are cats which are genetically resistant to infection by a particular subgroup of FeLV; neither is it likely that this information will be obtained easily because of the lack of sufficiently inbred strains of cats. There is no obvious difference in the susceptibility of different breeds of cats to FeLV (Brodey et al., 1970). However, host-range differences between the subgroups are distinguishable when cells of other species of mammals are infected. Viruses of subgroup A infect only feline cells (O. Jarrett et al., 1972, 1973), whereas those in subgroups B and C infect, in addition to feline cells, canine, bovine and primate cells, including human embryonic fibroblasts (Jarrett et al., 1969; Sarma et al., 1970b). Subgroup B viruses have a greater efficiency of plating on human cells than on feline cells (O. Jarrett et al., 1973). The restricted host range of the subgroup A viruses means, of course, that they are probably safer to use in the laboratory than B and C subgroup viruses.

Most natural isolates of FeLV are mixtures of viruses of subgroups A and B; apparently subgroup C viruses are comparatively rare. Pure subgroup A viruses can be isolated from AB mixtures but propagation of AB mixtures in human fibroblasts, which are resistant to subgroup A viruses, does not select against the subgroup A viruses to yield pure subgroup B viruses (O. Jarrett et al., 1973); subgroup B viruses have not yet been cloned. Either the A genotype is maintained by phenotypic mixing or the AB virus stock represents a single genotype. In the latter case, FeLV restricted to subgroup A may segregate by deletion of the B marker.

The feline sarcoma viruses which grow in human cells (Chang et al., 1970; Sarma et al., 1970b) have been classified as subgroup B (Sarma et al., 1971b,c, 1972). Like FeLV they may prove to be mixtures of viruses of subgroups A and B. FeSV stocks show a one-hit dose response for focus formation, indicating that FeLV is not necessary for cell transformation, but all stocks appear to carry excess FeLV (Sarma et al., 1971b,c; McDonald et al., 1972). Until FeSV has been cloned free of FeLV, one can not judge whether it is defective for viral replication.

Epidemiological studies indicate that virus-associated feline leukemia is transmitted as an infectious disease (Brodey et al., 1970;

Hardy, 1970; W. Jarrett, 1971). Horizontal transmission of FeLV has been demonstrated with experimentally and naturally infected cats (Rickard et al., 1969; W. Jarrett et al., 1973b). Even animals infected as adults can transmit the virus and cause tumours in other adults. These results do not exclude the possibility that congenital transmission of the virus may also occur, but horizontal transmission is probably an important epidemiological factor in feline leukemia.

SEARCH FOR HUMAN C-TYPE VIRUSES

C-type RNA viruses which induce either sarcomas or leukemias have been isolated from chickens, mice and cats and from a woolly monkey (Theilen et al., 1971; Wolfe et al., 1971). Viruses lacking or with an undetermined oncogenic potential but possessing the morphology and biochemical characteristics of sarcoma-leukemia viruses have been isolated from cultivated cells or tumours of Syrian hamsters, rats and guinea pigs and also from a rhesus monkey (Chopra and Mason, 1970; Schlom and Spiegelman, 1971), a gibbon (Kawakami et al., 1972), cattle (Ferrer et al., 1971 and see Gross 1970) and an Asiatic pit viper (Zeigel and Clark, 1969; Gilden et al., 1970). In addition C-type viruses have been induced by exposing cultures of cells from chickens, mice and other species (see Chapter 12) to agents such as 5-iododeoxy-uridine, and it has been postulated that all vertebrates inherit the genetic potential to specify C-type RNA viruses (see Chapters 12 and 13).

These various observations, and in particular the isolation of an oncogenic C-type virus from a primate, a woolly monkey, stimulated extensive searches for human C-type viruses or their components in normal and malignant human tissues. To date, however, there is little to show for all the effort that has been spent. No one has isolated C-type particles directly from biopsies of human tumours; only rarely have electron microscopists seen C-type particles in thin sections of fresh human cells (for example, Dmochowski and Grey, 1957), and unequivocal biochemical evidence of the presence of components of C-type viruses in human cells has yet to be obtained.

C-type particles have, however, been detected in short-term cultures of human tumour cells (Stewart et al., 1964; Dmochowski, 1966;

Dmochowski et al., 1967 and Morton et al., 1969). And Stewart et al. (1972a,b) induced particles resembling C-type viruses by exposing to 5-iododeoxyuridine cultures of cells of stable lines established from a human sarcoma and a human bronchial carcinoma. These viruses have not yet been extensively characterized. If they are not contaminants, they are presumably of human origin; but whether they have any oncogenic potential is another matter.

The liberation of a C-type virus in a stable culture of cells (ESP-1 cells) derived from the pleural effusion of a child with Burkitt's lymphoma was reported in 1971 by Priori et al. However, this virus is probably a murine virus that contaminated the cultures of ESP-1 cells because, although Priori et al. (1971) and Shigematsu et al. (1971) claimed that ESP-1 virus does not contain the group-specific (gs1) antigenicity of murine C-type viruses, Gilden et al. (1971) detected murine gs1 antigenicity in ESP-1 cells. Furthermore, the reverse transcriptase activity in ESP-1 virus particles (Gallo et al., 1971) can be inhibited by antiserum raised against murine reverse transcriptase and to a lesser extent by antiserum raised against feline reverse transcriptase (Scolnick et al., 1972a).

RD-114 virus is another C-type virus liberated by cultivated human cells that has attracted much attention, but its origin is also debatable. McAllister et al. (1971) inoculated kittens in utero with RD cells, a stable line of human sarcoma cells. Prior to this passage in kittens the RD cells did not contain detectable C-type particles, but two RD cell tumours that developed in kittens proved to contain C-type viruses, as did a stable line of cells (RD-114 cells) established from one of these tumours (McAllister et al., 1969). This virus, which contains 60–70S RNA and reverse transcriptase and has the typical C-type morphology (McAllister et al., 1972), might be either a latent human virus activated by passage of RD cells in kittens or it might be a feline virus acquired by the RD cells during their passage in kittens. The available data pertinent to the question of the origin of RD-114 virus can be summarized as follows:

1. RD-114 virus does not carry the group specific (gs1) antigen present in feline leukemia and sarcoma viruses (McAllister et al., 1972; Oroszlan et al., 1972b).

2. The reverse transcriptase in RD-114 virus particles is not immunologically related to the enzyme in feline leukemia virus (Scolnick et al., 1972a), but neither is it immunologically related to the reverse transcriptases in the C-type viruses isolated from a woolly monkey and a gibbon (Scolnick et al., 1972b).

3. The claim that antiserum raised against RD-114 virus reverse transcriptase inhibits the enzyme in primate viruses (Long et al., 1973, cited by Gillespie et al., 1973) has been disputed (Gillespie et al., 1973).

4. The claim that RD-114 RNA specifically hybridizes DNA of human chromosome 14 (Price et al., 1973) is also dubious (Gillespie et al., 1973).

5. According to Gillespie et al. (1973) about 40 percent of the sequences in 60–70S RNA from RD-114 virus particles hybridizes specifically to cat cell DNA and not to DNAs from other mammalian cells, including human cells.

Taken together, these data indicate that RD-114 virus is not closely related to any of the well-characterized feline or primate C-type viruses, while the hybridization data of Gillespie et al. (1973) suggest it is a feline rather than a human virus. Indeed RD-114 virus may well prove to be an endogenous C-type virus of feline cells (Livingston and Todaro, 1973 and see Chapter 12).

In addition to these various attempts to isolate human C-type viruses, attempts have been made to identify either viral RNA or viral reverse transcriptase in human tumour cells. For example, Spiegelman and his colleagues (Hehlmann et al., 1972a,b; Kufe et al., 1972; Gulati et al., 1972) have used a single-stranded DNA transcribed by Rauscher leukemia virus particles as a probe in hybridization experiments for complementary RNA sequences in human leukemic cells, sarcomas and lymphomas. They believe such RNA sequences exist in such human tumour cells but not in healthy human white blood cells or cells from patients with nonmalignant diseases of the blood, such as infectious mononucleosis. However, objections can be raised against these experiments and their interpretation. First, since there is little homology between the RNA sequences of different strains of murine leukemia and sarcoma viruses (Stephenson and Aaronson, 1971), it is not immediately obvious why the RNA of a putative human leukemia virus should be expected to be homologous to the Rauscher strain of MLV. Second, the amount of RNA-DNA hybridization that Spiegelman and his colleagues detect is very small indeed. Such reservations will only be dispelled by independent confirmation of these experiments.

Reverse transcriptase, like 60–70S RNA, appears to be a universal component of RNA tumour viruses, and several attempts have been made to identify in human tumour cells reverse transcriptases with properties identical to those of the reverse transcriptases present in tumour viruses. The results of these experiments, as well as those aimed at identifying reverse transcriptases in other mammalian cells, either uninfected or infected by RNA tumour viruses, have been extensively reviewed by Temin and Baltimore (1972). Suffice it here to say that several enzyme activities capable of using synthetic RNA and natural RNA templates for the synthesis of complementary DNA have been detected in normal, as well as tumour, cells of various species, including man. Whether any of these enzymes are related to the reverse transcriptase in tumour virus particles is, however, still obscure.

In summary, although human cells may well inherit the genetic potential to specify C-type viruses, this potential is usually latent and no one has yet obtained compelling evidence implicating C-type viruses in the etiology of human tumours.

MOUSE MAMMARY TUMOUR VIRUSES

The mouse mammary tumour viruses (MMTV), the first of which was discovered by Bittner (Bittner, 1936; Visscher et al., 1942; Bryan et al., 1942) in the milk of mice of the A and C3H strains (see Chapter 1), are related to the murine sarcoma and leukemia viruses inasmuch as they are enveloped particles containing a 60–70S RNA genome, which in susceptible mice causes tumours. However, five or six strains of MMTV have now been identified and several lines of evidence indicate that none of the strains is closely related to the murine sarcoma and leukemia viruses:

1. MMTV cause carcinomas, tumours of mammary gland epithelial cells which are of epidermal origin, whereas MSV and MLV induce the neoplastic transformation of cells of mesodermal origin.

2. The morphology of MMTV is different from that of MSV and MLV.

3. There is little or no homology between the base sequences of the RNAs of MMTV and MSV or MLV.

4. The MMTVs do not contain the interspecies-specific antigen (gs3) common to all mammalian sarcoma and leukemia viruses so far isolated (Nowinski et al., 1972b).

Identification of MMTV as a B-type Particle

The Bittner agent was identified as a virus because it is transmitted as a filterable agent in the milk of C3H and strain A mice (Bittner, 1942; Andervont and Bryan, 1944). In the mid-fifties, first Dmochowski (1954) and then Bernhard and Bauer (1955) observed B-type particles in sections of mouse mammary tumour tissue, and this raised the obvious possibility that the Bittner milk virus and B-type particles might be one and the same thing. It was, however, then found that mice from high-incidence strains, which, by appropriate fostering, had been freed of the Bittner milk virus and as a result did not develop mammary tumours at an early age, occasionally developed mammary tumours in old age. The cells of these tumours in aged females contained B-type particles identical in morphology to those found in tumours induced by the Bittner milk virus in young mice, but extracts of the tumours from these aged females rarely, if ever, induced tumours when injected into appropriate female recipients (see review by Hageman et al., 1968). This led many investigators to the erroneous conclusion that the B-type particles in mouse mammary tumours were not the etiological agents of those tumours. Hageman and her colleagues (Hageman et al., 1968; Calafat and Hageman, 1968) finally resolved this issue and unequivocally identified the Bittner milk virus, or MMTV-S as it is now called, as a B-type particle by showing that highly purified preparations of B-type particles from mice infected with the Bittner milk virus were highly oncogenic and induced mammary carcinomas in female mice of strains which have only a low incidence of this disease.

The virus that was found in the tumours of aged females freed of MMTV-S was investigated by Boot and Mühlbock (1956), Heston

(1958) and Nandi and DeOme (1965). This virus is not very oncogenic and Nandi and DeOme called it the nodule-inducing virus because it induces nonmalignant nodules in mouse mammary tissue more often than it induces mammary carcinoma. It is now designated MMTV-L (Hageman et al., 1972).

Classification of Mammary Tumour Viruses

Because murine mammary tumour viruses can only be assayed for biological activity in animals (none of the currently available cultured murine cells can be productively infected or transformed by these viruses), comparatively little is known about their biology. For want of other criteria, the various known strains of MMTV are currently classified according to their route of transmission and their pathology in vivo.

Five Strains of MMTV

The virus discovered by Bittner in the milk of C3H mice, which have a high incidence of mammary tumours, used to be called the Bittner agent, but it is now designated as standard MMTV, or MMTV-S. It is very virulent and induces a high incidence of tumours in comparatively young female mice. The other strains discovered subsequently are listed in Table 10.14.

Table 10.14. PROPERTIES OF STRAINS OF MOUSE MAMMARY TUMOUR VIRUSES

	Common name	Natural host	Characteristic	Transmission
MMTV-S	Bittner Virus	C3H	Virulent	Milk
MMTV-P	Mühlbock Virus	GR	Virulent; induces hormone-dependent plaques	Milk, egg, sperm
MMTV-L	Nandi Virus	C3HF	Low oncogenicity	Egg, sperm
MMTV-O	Van Leeuwenhoek Virus	Balb/c	Very low oncogenicity	Egg, sperm
MMTV-X	Timmermans Virus	020	Induced by X-irradiation	Egg, sperm
MMTV-Y		C57BL	Only antigens detected	

Adapted from Hageman et al., 1972.

MMTV-P, which was discovered by Mühlbock (1965) in the GR strain of mice, was initially called the Mühlbock virus (Bentvelzen et al., 1968) but was renamed MMTV-P because the tumours it induces were described as plaques by Foulds (1956). This virus is almost as tumourigenic as MMTV-S. MMTV-L, which was isolated from aged C3H females freed of MMTV-S (Boot and Mühlbock, 1956; Heston, 1958), has a low oncogenic potential. MMTV-O is so named because it was overlooked for many years. Its natural host is the Balb/c strain of mice, and young Balb/c mice rarely develop mammary carcinomas. However, Deringer (1965) found in a population of 17-month-old Balb/c mice a 22 percent incidence of mammary carcinomas, and from the mammary glands of 11- to 15-month-old, retired breeding Balb/c females Hageman et al. (1972) isolated MMTV-O.

MMTV-X is so called because it was detected in mammary tumours induced in mice of the 020 strain that had been X-irradiated and then allowed to drink water containing the carcinogen urethan (Timmermans et al., 1969). This virus could not, however, be detected in the cells of tumours or in cell-free extracts of tumours induced in 020 mice by extreme hormonal stimulation (Timmermans et al., 1969).

Antigenic differences between the various strains of MMTV have been detected, and no doubt these differences will ultimately provide a basis for a subgroup classification of these viruses. What is currently known about the antigens of the various strains of MMTV can be briefly summarized (see Table 10.6 and Hageman et al., 1972). None of these viruses has the interspecies-specific antigen common to all mammalian sarcoma and leukemia viruses and C-type particles. All the strains of MMTV apparently have two group-specific antigens (Blair, 1960, 1963), one in the nucleoid and the other in the envelope and probably the spikes of the envelope. MMTV-P and MMTV-O also have three other antigens, one nucleoid and two envelope antigens, in common with MMTV-S, but the relative proportions of these three antigens in these three strains differ (Hilgers et al., 1971). Finally, MMTV-L has none of the antigens present in MMTV-S other than the two group-specific antigens.

The different strains of MMTV are transmitted from parent to offspring by different routes. While MMTV-S is transmitted as infectious particles in the milk, the other strains are typically transmitted through the gametes from either parent. There is evidence (Bentvelzen, 1972) that MMTV associated with the gametes is not transmitted as an infectious virion but as an integrated "provirus." This concept will be discussed in detail in Chapter 12.

The development of mammary tumours and expression of MMTV is at least in part genetically controlled and is also hormone dependent (see Gross, 1970; Boot et al., 1972; Mühlbock, 1972). Tumours do not develop in male mice of high-incidence strains except when they are experimentally treated with estrogens. Tumours appear most rapidly in force-bred females when litters are removed at birth and the mice are immediately mated again. MMTV is released as mature B-particles by lactating mammary gland epithelium and by the epididymis in males (possibly accounting for some cases of male transmission). Infectious virus is also occasionally "blood borne" where it is associated with red cells and white cells (Nandi et al., 1966 and see Hageman et al., 1972).

Is There a Human Mammary Tumour Virus?

In 1971 Moore et al. reported that certain samples of human milks contain particles which have a morphology essentially identical to that of MMTV-S particles, and they further suggested that the incidence of these particles might be correlated with populations or families having a high incidence or a known history of breast cancer. These observations stimulated several attempts to further characterize human milk particles and to obtain further evidence correlating their occurrence with breast cancer. Schlom et al. (1971, 1972) found that human milk particles band on centrifugation at a density of 1.16–1.91 g/ml, contain 60–70S RNA and reverse transcriptase. Spiegelman and his colleagues (Axel et al., 1972a,b) then attempted to detect in cytoplasmic extracts of human breast cancer tissue RNA complementary to MMTV-S DNA made by reverse transcription (see Chapter 11.) They claim that complementary RNA can be detected specifically in breast cancer tissue but their data have not convinced all of their peers. At the same time after screening many more samples of human milks, Sarkar and Moore (1972) were forced to conclude that most human milk particles seen in the electron microscope cannot be classified as typical B-type particles; indeed a total of only 13 particles with the B-type morphology have been identified in the more than 200 samples of milk that have been screened (Sarkar and Moore, 1972).

These various data, particularly the occurrence of what appears to be typical B-type particles in at least a few samples of human milk, suggest that a human counterpart of MMTV-S may exist. The data currently available, however, fall a long way short of proving that such a virus exists and has a role in the etiology of human breast cancer. Furthermore, to date the search for human mammary carcinoma viruses has been restricted to a search for a milk-borne virus. Most strains of mouse mammary tumour viruses are transmitted through the gametes rather than in milk; whether corresponding human viruses exist is an entirely open question.

Finally, the discovery that cells of a spontaneous mammary carcinoma of an 8-year-old rhesus monkey release a virus, which when mature has the morphology of a C-type particle but which during its intracellular maturation resembles intracellular B-type particles (Chopra and Mason, 1970), raises the possibility that in primates C-type viruses may be associated with mammary carcinoma. This rhesus monkey virus has not yet, however, been shown to possess any oncogenic potential.

Literature Cited

Aaronson, S. A. 1971. Isolation of a rat-tropic helper virus from M-MSV-O. Virology *44:* 29.

Aaronson, S. A. and W. P. Rowe. 1970. Nonproducer clones of murine sarcoma virus transformed BALB/3T3 cells. Virology *42:* 9.

Aaronson, S. A., J. C. Jainchill and G. J. Todaro. 1970. Murine sarcoma virus transformation of BALB/3T3 cells: Lack of dependence on murine leukemia virus. Proc. Nat. Acad. Sci. *66:* 1236.

Aaronson, S. A., W. P. Parks, E. M. Scolnick and G. J. Todaro. 1971. Antibody to the RNA-dependent DNA polymerase of mammalian C-type RNA tumor viruses. Proc. Nat. Acad. Sci. *68:* 920.

Aaronson, S. A., R. H. Bassin and C. Weaver. 1972. Comparison of murine sarcoma viruses in nonproducer and S^+ L^- transformed cells. J. Virol. *9:* 701.

Allen, D. W., P. S. Sarma, H. D. Niall and R. Sauer. 1970. Isolation of a second avian leukosis group-specific antigen (gs-b) from avian myeloblastosis virus. Proc. Nat. Acad. Sci. *67:* 837.

Almeida, J. D., A. P. Waterson and J. A. Drewe. 1967. A morphological comparison of Bittner and influenza viruses. J. Hygiene *65:* 467.

ALTANER, C. and F. SVEC. 1966. Virus production in rat tumors induced by chicken sarcoma virus. J. Nat. Cancer Inst. *37:* 745.

ALTANER, C. and H. M. TEMIN. 1970. Carcinogenesis by RNA sarcoma viruses. XII. A quantitative study of infection of rat cells *in vitro* by avian sarcoma viruses. Virology *40:* 118.

ANDERVONT, H. B. and W. R. BRYAN. 1944. Properties of the mouse mammary-tumor agent. J. Nat. Cancer Inst. *8:* 227.

AOKI, T., L. J. OLD and E. A. BOYSE. 1966. Serological analysis of leukemia antigens of the mouse. Nat. Cancer Inst. Monogr. *22:* 449.

AOKI, T., E. A. BOYSE and L. J. OLD. 1968. Wild-type Gross leukemia virus. I. Soluble antigen (GSA) in the plasma and tissue of infected mice. J. Nat. Cancer Inst. *41:* 89.

AOKI, T., E. A. BOYSE, L. J. OLD, E. DE HARVEN, U. HAMMERLING and H. A. WOOD. 1970. G (Gross) and H-2 cell-surface antigens: Location on Gross leukemia cells by electron microscopy with visually labeled antibody. Proc. Nat. Acad. Sci. *65:* 569.

AOKI, T., R. B. HERBERMAN, P. A. JOHNSON, M. LIU and M. M. STURM. 1972. Wild-type Gross leukemia virus: Classification of soluble antigens (GSA). J. Virol. *10:* 1208.

AXEL, R., J. SCHLOM and S. SPIEGELMAN. 1972a. Evidence for translation of viral-specific RNA in cells of a mouse mammary carcinoma. Proc. Nat. Acad. Sci. *69:* 535.

———. 1972b. Presence in human breast cancer of RNA homologous to mouse mammary tumour virus RNA. Nature *235:* 32.

AXELRAD, A. A. 1966. Genetic control of susceptibility to Friend leukemia virus in mice: Studies with the spleen focus assay method. Nat. Cancer Inst. Monogr. *22:* 619.

AXELRAD, A. A. and R. A. STEEVES. 1964. Assay for Friend leukemia virus: Rapid quantitative method based on enumeration of macroscopic spleen foci in mice. Virology *24:* 513.

BADER, J. P. and T. L. STECK. 1969. Analysis of the ribonucleic acid of murine leukemia virus. J. Virol. *4:* 454.

BADER, J. P., T. L. STECK and T. KAKEFUDA. 1970. The structure of the RNA of RNA-containing tumor viruses. Current Topics Microbiol. Immunol. *51:* 106.

BALL, J. K., D. HARVEY and J. A. MCCARTER. 1973. Evidence for naturally occurring murine sarcoma virus. Nature *241:* 272.

BALUDA, M. A. and I. E. GOETZ. 1961. Morphological conversion of cell cultures by avian myeloblastosis virus. Virology *15:* 185.

BALUDA, M. A. and D. P. NAYAK. 1970. DNA complementary to viral RNA in leukemic cells induced by avian myeloblastosis. Proc. Nat. Acad. Sci. *66:* 329.

BASSIN, R. H., P. J. SIMONS, F. C. CHESTERMAN and J. J. HARVEY. 1968. Murine sarcoma virus (Harvey): Characteristics of focus formation in mouse embryo cell cultures, and virus production by hamster tumor cells. Int. J. Cancer *3:* 265.

BASSIN, R. H., N. TUTTLE and P. J. FISCHINGER. 1970. Isolation of murine sarcoma virus-transformed mouse cells which are negative for leukaemia virus from agar suspension cultures. Int. J. Cancer *6:* 95.

———. 1971a. Rapid cell culture assay technique for murine leukaemia viruses. Nature *229:* 564.

BASSIN, R. H., L. A. PHILLIPS, M. J. KRAMER, D. K. HAAPALA, P. T. PEEBLES, S. NOMURA and P. J. FISCHINGER. 1971b. Transformation of mouse 3T3 cells by murine sarcoma virus: Release of virus-like particles in the absence of replicating murine leukemia helper virus. Proc. Nat. Acad. Sci. *68:* 1520.

BATHER, R., A. LEONARD and J. YANG. 1968. Characteristics of the *in vitro* assay of murine sarcoma virus (Moloney) and virus infected cells. J. Nat. Cancer Inst. *40:* 551.

BAUER, H. 1966. Untersuchungen uber das myeloblastose-virus des huhnes (BAI Stamm

A): II. Isolierung und charakterisierung der im virus enthaltenen nukleinsaure. Z. Naturforsch. *21b:* 453.

BAUER, H. and H. G. JANDA. 1967. Group-specific antigen of avian leukosis virus. Virus specificity and relation to an antigen contained in Rous mammalian cells. Virology *33:* 483.

BEARD, J. W. 1963. Avian virus growths and their etiologic agents. Advance. Cancer Res. *7:* 1.

BENTVELZEN, P. 1972. Hereditary infections with mammary tumor viruses in mice. In *RNA viruses and host genome in oncogenesis*, ed. P. Emmelot and P. Bentvelzen, p. 309. North-Holland, Amsterdam.

BENTVELZEN, P., A. TIMMERMANS, J. H. DAAMS and A. VAN DER GUGTEN. 1968. Genetic transmission of mammary tumor inciting viruses in mice: Possible implications for murine leukaemia. Bibliotheca Haematol. *31:* 101.

BERNHARD, W. 1958. Electron microscopy of tumor cells and tumor viruses. A review. Cancer Res. *18:* 491.

———. 1960. The detection and study of tumor viruses with the electron microscope. Cancer Res. *20:* 712.

BERNHARD, W. and A. BAUER. 1955. Mise en évidence de corpuscules d'aspect virusal dans des tumours mammaires de la étude au microscope electronique. Compt. Rend. Acad. Sci. (Paris) *240:* 1380.

BERNHARD, W., R. A. BONAR, D. BEARD and J. W. BEARD. 1958. Ultrastructure of viruses of myeloblastosis and erythroblastosis isolated from plasma of leukaemic chickens. Proc. Soc. Exp. Biol. Med. *97:* 48.

BISHOP, J. M., W. E. LEVINSON, N. QUINTRELL, D. SULLIVAN, L. FANSHIER and J. JACKSON. 1970a. The low molecular weight RNAs of Rous sarcoma virus. I. The 4S RNA. Virology *42:* 182.

———. 1970b. The low molecular weight RNAs of Rous sarcoma virus. II. The 7S RNA. Virology *42:* 927.

BISWAL, N., B. MCCAIN and M. BENYESH-MELNICK. 1971. The DNA of murine sarcoma-leukemia virus. Virology *45:* 706.

BITTNER, J. J. 1936. Some possible effects of nursing on the mammary gland tumor incidence in mice. Science *84:* 162.

———. 1942. Hormones, susceptibility and milk influence in cancer. Cancer Res. *2:* 710.

BLAIR, P. B. 1960. Serologic comparison of mammary tumor viruses from 3 strains of mice. Proc. Soc. Exp. Biol. Med. *103:* 188.

———. 1963. Neutralization of the mouse mammary tumor virus by rabbit antisera against C3Hf tissue. Cancer Res. *23:* 381.

BOLOGNESI, D. P. and H. BAUER. 1970. Polypeptides of avian RNA tumor viruses. I. Isolation and physical and chemical analysis. Virology *42:* 1097.

BONAR, R. A. and J. W. BEARD. 1959. Virus of avian myeloblastosis: XII. Chemical constitution. J. Nat. Cancer Inst. *23:* 183.

BONAR, R. A., L. SVTRAK, D. P. BOLOGNESI, A. J. LANGLOIS, D. BEARD and J. W. BEARD. 1967. Ribonucleic acid components of BAI strain (myeloblastosis) avian tumor virus. Cancer Res. *27:* 1138.

BOOT, L. M. and O. MÜHLBOCK. 1956. The mammary tumour incidence in the C3H mouse-strain with and without the agent (C3H; C3Hf; C3He). Acta Unio Intern. Contra Cancrum *12:* 569.

BOOT, L. M., G. RÖPCKE and H. G. KWA. 1972. Hormonal factors in the origin of mammary tumors. In *RNA viruses and host genome in oncogenesis*, ed. P. Emmelot and P. Bentvelzen, p. 275. North-Holland, Amsterdam.

BOYSE, E. A., L. J. OLD and E. STOCKERT. 1972. The relation of linkage group IX to leukemogenesis in the mouse. In *RNA viruses and host genome in oncogenesis*, ed. P. Emmelot and P. Bentvelzen, p. 171. North-Holland, Amsterdam.

BRODEY, R. S., S. K. MCDONOUGH, F. L. FRYE and W. D. HARDY. 1970. Epidemiology of feline leukemia (lymphosarcoma). In *Comparative leukemia research 1969*, ed. R. M. Dutcher, p. 333. Karger, Basel.

BRYAN, W. R., H. KAHLER, M. B. SHIMKIN and H. B. ANDERVONT. 1942. Extraction and ultracentrifugation of mammary tumor inciter of mice. J. Nat. Cancer Inst. *2:* 451.

CALAFAT, J. and P. HAGEMAN. 1968. Some remarks on the morphology of virus particles of the B-type and their isolation from mammary tumors. Virology *36:* 308.

CHANG, R. S., H. D. GOLDEN and B. HARROLD. 1970. Propagation in human cells of a filterable agent from the ST feline sarcoma. J. Virol. *6:* 599.

CHEUNG, K-S., R. E. SMITH, M. P. STONE and W. K. JOKLIK. 1972. Comparison of immature (rapid harvest) and mature Rous sarcoma virus particles. Virology *50:* 851.

CHOPRA, H. C. and M. M. MASON. 1970. A new virus in a spontaneous mammary tumor of a rhesus monkey. Cancer Res. *30:* 2081.

CHOPRA, H. C., N. J. WOODSIDE and A. E. BOGDEN. 1970. Virus particles in rat leukemias. Cancer Res. *30:* 1544.

CHUBB, R. C. and P. M. BIGGS. 1968. The neutralization of Rous sarcoma virus. J. Gen. Virol. *3:* 87.

COLE, R. K. and J. FURTH. 1941. Experimental studies on the genetics of spontaneous leukemia in mice. Cancer Res. *1:* 937.

COOK, K. M. 1969. Cultivation of a filterable agent associated with Marek's disease. J. Nat. Cancer Inst. *43:* 203.

CRITTENDEN, L. B. 1968. Observations on the nature of genetic cellular resistance to avian tumor viruses. J. Nat. Cancer Inst. *41:* 145.

CRITTENDEN, L. B. and J. V. MOTTA. 1969. A survey of genetic resistance to leukosis-sarcoma viruses in commercial stocks of chickens. Poultry Sci. *48:* 1751.

CRITTENDEN, L. B., W. OKAZAKI and R. H. REAMER. 1964. Genetic control of responses to Rous sarcoma and strain RPL12 viruses in the cells, embryos, and chickens of two inbred lines. Nat. Cancer Inst. Monogr. *17:* 161.

CRITTENDEN, L. B., H. A. STONE, R. H. REAMER and W. OKAZAKI. 1967. Two loci controlling genetic cellular resistance to avian leukosis-sarcoma viruses. J. Virol. *1:* 898.

CRITTENDEN, L. B., W. F. BRILES and H. A. STONE. 1970. Susceptibility to an avian leukosis-sarcoma virus: Close association with an erythrocyte isoantigen. Science *169:* 1324.

CRITTENDEN, L. B., E. J. WENDEL and J. V. MOTTA. 1973. Interaction of genes controlling resistance to RSV(RAV-O). Virology. In press.

DAVIS, N. L. and R. R. RUECKERT. 1972. Properties of ribonucleoprotein particle isolated from nonidet P-40-treated Rous sarcoma virus. J. Virol. *10:* 1010.

DAWSON, P. J., R. B. TACKE and A. H. FIELDSTEEL. 1968. Relationship between Friend virus and an associated lymphatic leukaemia virus. Brit. J. Cancer *22:* 569.

DERINGER, M. K. 1965. Occurrence of mammary tumors, reticulative neoplasms, and pulmonary tumors in strain BALB/cAmDe breeding female mice. J. Nat. Cancer Inst. *35:* 1047.

DE-THÉ, G. 1964. Localization and origin of the adenosine-triphophatase activity of avian myeloblastosis virus. Nat. Cancer Inst. Monogr. *17:* 651.

———. 1966. Ultrastructural cytochemistry of enzymes associated with murine leukaemia viruses: I. Adenosine triphosphatase. Int. J. Cancer *1:* 119.

DE-THÉ, G. and T. E. O'CONNOR. 1966. Structure of a murine leukemia virus after disruption with Tween ether and comparison with two myxoviruses. Virology *28:* 713.

DE-THÉ, G., C. BECKER and J. W. BEARD. 1964. Virus of avian myeloblastosis (BAI strain A): XXV. Ultracytochemical study of virus and myeloblast phosphatase activity. J. Nat. Cancer Inst. *32:* 201.

DMOCHOWSKI, L. 1954. Progress in mammalian genetics and cancer. J. Nat. Cancer Inst. *15:* 785.

———. 1966. Present status of viruses as causative agents in cancer. Southern Med. Bull. *54:* 65.

DMOCHOWSKI, L. and C. E. GREY. 1957. Electron microscopy of tumours of known and suspected viral etiology. Texas Rep. Biol. Med. *15:* 256.

DMOCHOWSKI, L., T. YUMOTO and C. E. GREY. 1967. Electron microscopic studies of human leukemia and lymphoma. Cancer *20:* 760.

DOUGHERTY, R. M. and H. S. DI STEFANO. 1965. Virus particles associated with "non-producer" Rous sarcoma cells. Virology *27:* 351.

DOUGHERTY, R. M. and R. RASMUSSEN. 1964. Properties of a strain of Rous sarcoma virus that infects mammals. Nat. Cancer Inst. Monogr. *17:* 337.

DUC-NGUYEN, H. 1968. Enhancing effect of diethylaminoethyl dextran on the focus forming titers of a murine sarcoma virus (Harvey strain). J. Virol. *2:* 643.

DUESBERG, P. H. 1968. Physical properties of Rous sarcoma virus RNA. Proc. Nat. Acad. Sci. *60:* 1511.

———. 1970. On the structure of RNA tumor viruses. Current Topics Microbiol. Immunol. *51:* 79.

DUESBERG, P. H. and W. S. ROBINSON. 1966. Nucleic acid and proteins isolated from the Rauscher mouse leukemia virus (MLV). Proc. Nat. Acad. Sci. *55:* 219.

DUESBERG, P. H. and P. K. VOGT. 1970. Differences between the ribonucleic acids of transforming and nontransforming avian tumor viruses. Proc. Nat. Acad. Sci. *67:* 1673.

DUESBERG, P. H., G. S. MARTIN and P. K. VOGT. 1970. Glycoprotein components of avian and murine RNA tumor viruses. Virology *41:* 631.

DUESBERG, P. H., P. K. VOGT and E. CANAANI. 1971. Structure and replication of avian tumor virus RNA. Lepetit Colloq. Biol. Med. *2:* 154. North-Holland, Amsterdam.

DUESBERG, P. H., E. CANAANI, K. VAN DER HELM, M. M. C. LAI and P. K. VOGT. 1973. News and views on avian tumor virus RNA. In *Possible episomes in eukaryotes*, ed. L. G. Silvestri. North-Holland, Amsterdam.

DUFF, R. G. and P. K. VOGT. 1969. Characteristics of two new avian tumor virus subgroups. Virology *39:* 18.

EAST, J., P. R. PROSSER, E. J. HOLBORROW and H. JAQUET. 1967. Autoimmune reactions and virus-like particles in germ-free NZB mice. Lancet *1:* 755.

ECKERT, E. A., R. ROTT and W. SCHAFER. 1963. Myxovirus-like structure of avian myeloblastosis virus. Z. Naturforsch. *18b:* 339.

EL-FIKY, S. M., N. H. SARKAR and D. H. MOORE. 1970. Ultrastructural cytochemistry of enzymes associated with mouse mammary tumor virus (MTV). I. Adenosine triphosphatase (ATPase). Microscopie Electronique *3:* 923.

ELLERMAN, V. and O. BANG. 1908. Experimentelle leukamia bei huhnern. Centralbl. Baketriol. Abt. 1 (Orig.) *46:* 595.

———. 1909. Experimentelle leukemia bei huhnern. Z. Hyg. Infekt. *63:* 231.

ERIKSON, E. and R. L. ERIKSON. 1970. Isolation of amino acid acceptor RNA from purified avian myeloblastosis virus. J. Mol. Biol. *52:* 387.

———. 1971. Association of 4S ribonucleic acid with oncornavirus ribonucleic acids. J. Virol. *8:* 24.

ERIKSON, E., R. L. ERIKSON, B. HENRY and N. R. PACE. 1973. Comparison of oligonucleotides generated by RNase T_1 from 7S RNA from oncornaviruses and uninfected cells. Virology. In press.

ERIKSON, R. L. 1969. Studies on the RNA from avian myeloblastosis virus. Virology *37:* 124.

FERRER, J. F., N. D. STOCK and P-S. LIN. 1971. Detection of replicating C-type viruses in continuous cell cultures established from cows with leukemia: Effect of the culture medium. J. Nat. Cancer Inst. *47:* 613.

FIELDSTEEL, A. H., C. KURAHARA and P. J. DAWSON. 1969. Moloney leukemia virus as a helper in retrieving Friend virus from a non-infectious reticulum cell sarcoma. Nature *223:* 1274.

FINKEL, M. P., B. O. BISKIS and P. B. JINKINS. 1966. Virus induction of osteosarcomas in mice. Science *151:* 698.

FISCHINGER, P. J. and T. E. O'CONNOR. 1968. Tissue culture assay of helper activity of murine leukemia virus for mouse sarcoma virus. J. Nat. Cancer Inst. *40:* 1199.

———. 1969. Viral infection across species barriers: Reversible alteration of murine sarcoma virus growth in cat cells. Science *165:* 714.

FLEISSNER, E. 1971. Chromatographic separation and antigenic analysis of proteins of the oncornaviruses. I. Avian leukemia-sarcoma viruses. J. Virol. *8:* 778.

FOULDS, L. 1956. The histologic analysis of mammary tumors of mice. I. Scope of investigations and general principles of analysis. J. Nat. Cancer Inst. *17:* 701.

FRANKS, W. R., S. MCGREGOR, M. M. SHAW and J. SKUBLICS. 1959. Development of leukosis by cell-free filtrates of solid tumors or in mice surviving immune to these tumors. Proc. Amer. Assoc. Cancer Res. *3:* 19.

FRIEND, C. 1957. Cell-free transmission in adult Swiss mice of a disease having the character of a leukemia. J. Exp. Med. *105:* 307.

FRIEND, C. and G. B. ROSSI. 1968. The phenomenon of differentiation in murine virus-induced leukemic cells. Can. Cancer Conf. *8:* 171.

FRIEND, C., W. SCHER, J. G. HOLLAND and J. SATO. 1971. Hemoglobin synthesis in murine virus-induced leukemic cells *in vitro:* Stimulation of erythroid differentiation by dimethyl sulfoxide. Proc. Nat. Acad. Sci. *68:* 378.

FRIIS, R. R., K. TOYOSHIMA and P. K. VOGT. 1971. Conditional lethal mutants of avian sarcoma viruses. I. Physiology of *ts* 75 and *ts* 149. Virology *43:* 375.

FUJINAMI, A. and K. INAMOTO. 1914. Ueber geschwülste bei Japanischen haushühnern insbesondere über einen transplatablen tumor. Z. Krebsforsch. *14:* 94.

GALLO, R. C., P. S. SAVIN, P. T. ALLEN, W. A. NEWTON, E. S. PRIORI, J. M. BOWER and L. DMOCHOWSKI. 1971. Reverse transcriptase in type C virus particles of human origin. Nature New Biol. *232:* 140.

GARDNER, M. B., R. W. RONGEY, P. ARNSTEIN, J. D. ESTES, R. J. HUEBNER and L. G. RICKARD. 1970. Experimental transmission of feline fibrosarcoma to cats and dogs. Nature *226:* 807.

GAY, F. W., J. K. CLARKE and E. DERMOTT. 1970. Morphogenesis of Bittner virus. J. Virol. *5:* 801.

GEERING, G., T. AOKI and L. J. OLD. 1970. Shared viral antigen of mammalian leukaemia viruses. Nature *226:* 265.

GILDEN, R. V., Y. K. LEE, S. OROSZLAN and J. L. WALKER. 1970. Reptilian C-type virus: Biophysical, biological and immunological properties. Virology *41:* 187.

GILDEN, R. V., W. P. PARKS, R. J. HUEBNER and G. J. TODARO. 1971. Murine leukaemia virus group specific antigen in the C-type virus-containing human cell line, ESP-1. Nature *233:* 102.

GILLESPIE, D., S. MARSHALL and R. C. GALLO. 1972. RNA of RNA tumor viruses contains poly A. Nature New Biol. *236:* 227.

GILLESPIE, D., S. GILLESPIE, R. C. GALLO, J. L. EAST and L. DMOCHOWSKI. 1973. Genetic origin of RD114 and other RNA tumor viruses. Submitted for publication.

GRAF, T. 1972. A plaque assay for avian RNA tumor viruses. Virology *50:* 567.

GRAFFI, A. 1957. Chloroleukemia of mice. Ann. N.Y. Acad. Sci. *68:* 540.

GRAFFI, A., H. RIELKA, F. FEY, F. SCHARSACH and R. WEISS. 1955. Gahänftes Anftreten von leukämien nach injektion von sarkom-filtraten. Wiener Med. Wochenschr. *105*: 61.

GRAFFI, A., T. SCHRAMM, E. BENDER, J. GRAFFI, K-H. HORN and D. BIERWOLF. 1968. Cell-free transmissible leukoses in Syrian hamsters, probably of viral aetiology. Brit. J. Cancer *22:* 577.

GRANBOULAN, N., J. HUPPERT and F. LACOUR. 1966. Examen au microscope electronique du RNA du virus de la myeloblastose aviate. J. Mol. Biol. *16:* 571.

GROSS, L. 1951. "Spontaneous" leukemia developing in C3H mice following inoculation in infancy, with A-K leukemic extracts, or AK embryos. Proc. Soc. Exp. Biol. Med. *76:* 27.

———. 1960. Development of myeloid (chloro-) leukemia in thymectomized C3H mice following inoculation of lymphatic leukemia virus. Proc. Soc. Exp. Biol. Med. *103:* 509.

———. 1964. Attempt at classification of mouse leukemia viruses. Acta Haematol. *32:* 81.

———. 1970. *Oncogenic viruses*, 2nd ed. Pergamon, Oxford.

GROUPÉ, V., V. C. DUNKEL and R. A. MANAKER. 1957. Improved pock counting method for the titration of Rous sarcoma virus in embryonated eggs. J. Bacteriol. *74:* 409.

GULATI, S. C., R. AXEL and S. SPIEGELMAN. 1972. Detection of RNA-instructed DNA polymerase and high molecular weight RNA in malignant tissue. Proc. Nat. Acad. Sci. *69:* 2020.

HACKETT, A. J. and S. S. SYLVESTER. 1972. Cell line derived from BALB/3T3 that is transformed by murine leukaemia virus: A focus assay for leukaemia virus. Nature New Biol. *239:* 164.

HAGEMAN, P. C., J. LINKS and P. BENTVELZEN. 1968. Biological properties of B particles from C3H and C3Hf mouse milk. J. Nat. Cancer Inst. *40:* 1319.

HAGEMAN, P. C., J. CALAFAT and J. H. DAAMS. 1972. The mouse mammary tumor viruses. In *RNA viruses and host genome in oncogenesis*, ed. P. Emmelot and P. Bentvelzen, p. 283. North-Holland, Amsterdam.

HANAFUSA, H. 1964. Nature of the defectiveness of Rous sarcoma virus. Nat. Cancer Inst. Monogr. *17:* 543.

———. 1965. Analysis of the defectiveness of Rous sarcoma virus. III. Determining influence of a new helper virus on the host range and susceptibility to interference of RSV. Virology *25:* 248.

HANAFUSA, H. and T. HANAFUSA. 1966. Determining factor in the capacity of Rous sarcoma virus to induce tumors in mammals. Proc. Nat. Acad. Sci. *55:* 532.

———. 1967. Interaction among avian tumor viruses giving enhanced infectivity. Proc. Nat. Acad. Sci. *58:* 818.

HANAFUSA, H., T. HANAFUSA and H. RUBIN. 1963. The defectiveness of Rous sarcoma virus. Proc. Nat. Acad. Sci. *49:* 572.

———. 1964. Analysis of the defectiveness of Rous sarcoma virus. II. Specification of RSV antigenicity by helper virus. Proc. Nat. Acad. Sci. *51:* 41.

HANAFUSA, T., T. MIYAMOTO and H. HANAFUSA. 1970a. A type of chick embryo cell that fails to support formation of infectious RSV. Virology *40:* 55.

HANAFUSA, T., H. HANAFUSA and T. MIYAMOTO. 1970b. Recovery of a new virus from apparently normal chick cells by infection with avian tumor viruses. Proc. Nat. Acad. Sci. *67:* 1797.

HARDY, W. D. 1970. Immunodiffusion studies of feline leukemia and sarcoma. J. Amer. Vet. Med. Ass. *158:* 1060.

HARTLEY, J. W. and W. P. ROWE. 1966. Production of altered cell foci in tissue culture by defective Moloney sarcoma virus particles. Proc. Nat. Acad. Sci. *55:* 780.

HARTLEY, J. W., W. P. ROWE, W. I. CAPPS and R. J. HEUBNER. 1965. Complement fixation and tissue culture assays for mouse leukemia viruses. Proc. Nat. Acad. Sci. *53:* 931.

———. 1969. Isolation of naturally occurring viruses of the murine leukemia virus group in tissue culture. J. Virol. *3:* 126.

HARTLEY, J. W., W. P. ROWE and R. J. HUEBNER. 1970. Host range restrictions of murine leukemia viruses in mouse embryo cell cultures. J. Virol. *5:* 221.

HARVEY, J. J. 1964. An unidentified virus which causes the rapid production of tumors in mice. Nature *204:* 1104.

HARVEY, J. J. and J. EAST. 1971. The murine sarcoma virus (MSV). Int. Rev. Exp. Path. *10*: 265.

HATANAKA, M., R. J. HUEBNER and R. V. GILDEN. 1970. DNA polymerase activity associated with RNA tumor viruses. Proc. Nat. Acad. Sci. *68:* 1844.

HEHLMANN, R., D. KUFE and S. SPIEGELMAN. 1972a. RNA in human leukemic cells related to the RNA of a mouse leukemia virus. Proc. Nat. Acad. Sci. *69:* 435.

———. 1972b. Viral related RNA in Hodgkin's disease and other human lymphomas. Proc. Nat. Acad. Sci. *69:* 1727.

HESTON, W. E. 1958. Mammary tumors in agent-free mice. Ann. N. Y. Acad. Sci. *71:* 931.

HILGERS, J., J. H. DAAMS and P. BENTVELZEN. 1971. The induction of precipitating antibodies to mammary tumor virus in several inbred mouse strains. Israel J. Med. Sci. *7:* 154.

HORST, J., J. KEITH and H. FRAENKEL-CONRAT. 1972. Characteristic two-dimensional patterns of enzymatic digests of oncorna and other viral RNAs. Nature New Biol. *240:* 105.

HUEBNER, R. J., J. W. HARTLEY, W. P. ROWE, W. T. LANE and W. I. CAPPS. 1966. Rescue of the defective genome of Moloney sarcoma virus from a noninfectious hamster tumor and the production of pseudotype sarcoma viruses with various leukemia viruses. Proc. Nat. Acad. Sci. *56:* 1164.

HUNG, P. P., H. L. ROBINSON and W. S. ROBINSON. 1971. Isolation and characterization of proteins from Rous sarcoma virus. Virology *43:* 251.

IMAI, T., H. OKANO, A. MATSUMOTO and A. HORIE. 1966. The mode of virus elaboration in C3H mouse mammary carcinoma as observed by electron microscopy in serial thin sections. Cancer Res. *part 1:* 443.

ISHIZAKI, R. and T. SHIMIZU. 1970a. Observations on the envelope properties of RSV(O). Virology *40:* 415.

———. 1970b. Heterogeneity of strain R avian (erythroblastosis) virus. Cancer Res. *30:* 2827.

ISHIZAKI, R. and P. K. VOGT. 1966. Immunological relationships among envelope antigens of avian tumor viruses. Virology *30:* 375.

JAINCHILL, J. L., S. A. AARONSON and G. J. TODARO. 1969. Murine sarcoma and leukemia viruses: Assay using clonal lines of contact-inhibited mouse cells. J. Virol. *4:* 549.

JARRETT, O., H. M. LAIRD, D. HAY and G. W. CRIGHTON. 1968. Replication of cat leukaemia virus in cell cultures. Nature *219:* 521.

JARRETT, O., H. M. LAIRD and D. HAY. 1969. Growth of feline leukaemia virus in human cells. Nature *224:* 1208.

JARRETT, O., J. D. PITTS, J. M. WHALLEY, A. E. CLASON and J. HAY. 1971. Isolation of the nucleic acid of feline leukaemia virus. Virology *43:* 317.

JARRETT, O., H. M. LAIRD and D. HAY. 1972. Restricted host range of a feline leukaemia virus. Nature *238:* 220.

———. 1973. Determinants of the host range of feline leukaemia viruses. J. Gen. Virol. In press.

JARRETT, W. F. H. 1971. Feline leukaemia. Int. Rev. Exp. Path. *10:* 243.

JARRETT, W. F. H., W. B. MARTIN, G. W. CRIGHTON, R. G. DALTON and M. F. STEWART. 1964. Leukaemia in the cat: Transmission experiments with leukaemia (lymphosarcoma). Nature *202:* 566.

JARRETT, W. F. H., M. ESSEX and L. J. MACKEY. 1973a. Feline leukemia virus-associated antibodies in normal and leukemic cats. J. Nat. Cancer Inst. In press.

JARRETT, W. F. H., O. JARRETT, L. J. MACKEY, H. LAIRD, W. D. HARDY and M. ESSEX. 1973b. Horizontal transmission of leukemia virus and leukemia in the cat. J. Nat. Cancer Inst. In press.

JUNGEBLUT, C. W. and H. KODZA. 1962. Studies of leukemia L_2C in guinea-pigs. Arch. Ges. Virusforsch. *12:* 537.

KAKEFUDA, T. and J. P. BADER. 1969. Electron microscopic observations on the ribonucleic acid of murine leukaemia virus. J. Virol. *4:* 460.

KAWAI, S. and H. HANAFUSA. 1972a. Plaque assay for some strains of avian leukosis. Virology *48:* 126.

———. 1972b. Genetic recombination with avian tumor virus. Virology *49:* 37.

KAWAKAMI, T., S. D. HUFF, P. M. BUCKLEY, D. L. DUNGWORTH, S. P. SNYDER and R. V. GILDEN. 1972. C-type virus associated with gibbon lymphosarcoma. Nature New Biol. *235:* 170.

KELLOFF, G., R. J. HUEBNER, N. H. CHANG, Y. K. LEE and R. V. GILDEN. 1970a. Envelope antigen relationships among three hamster-specific sarcoma viruses and a hamster-specific helper virus. J. Gen. Virol. *9:* 19.

KELLOFF, G., R. J. HUEBNER, S. OROSZLAN, R. TONI and R. V. GILDEN. 1970b. Immunological identity of the group-specific antigen of hamster specific C-type viruses and an indigenous hamster virus. J. Gen. Virol. *9:* 27.

KEOGH, E. V. 1938. Ectodermal lesions produced by the virus of Rous sarcoma. Brit. J. Exp. Path. *19:* 1.

KIRSTEN, W. H. and L. A. MAYER. 1967. Morphological responses to a murine erythroblastosis virus. J. Nat. Cancer Inst. *39:* 311.

KLEIN, E. and G. KLEIN. 1964. Mouse antibody production test for the assay of the Moloney virus. Nature *204:* 339.

KLEMENT, V. and P. VESELY. 1965. Tumour induction with Rous sarcoma virus in hamsters and production of infectious virus in a heterologous host. Neoplasma *12:* 147.

KLEMENT, V., J. W. HARTLEY, W. P. ROWE and R. J. HUEBNER. 1969a. Recovery of a hamster-specific, focus-forming, and sarcomagenic virus from a "non-infectious" hamster tumor induced by the Kirsten mouse sarcoma virus. J. Nat. Cancer Inst. *43:* 925.

KLEMENT, V., W. P. ROWE, J. W. HARTLEY and W. E. PUGH. 1969b. Mixed culture cytopathogenicity: A new test for growth of murine leukemia viruses in tissue culture. Proc. Nat. Acad. Sci. *63:* 753.

KUFE, D., R. HEHLMANN and S. SPIEGELMAN. 1972. Human sarcomas contain RNA related to the RNA of a mouse leukaemia virus. Science *175:* 182.

LAI, M. M. C. and P. H. DUESBERG. 1972. Adenylic acid-rich sequence in RNAs of Rous sarcoma virus and Rauscher mouse leukaemia virus. Nature *235:* 383.

LEVINSON, W., J. M. BISHOP, N. QUINTRELL and J. JACKSON. 1970. Presence of DNA in Rous sarcoma virus. Nature *227:* 1023.

LEVY, J. A. and T. PINCUS. 1970. Demonstration of biological activity of a murine leukemia virus of New Zealand black mice. Science *170:* 326.

LILLY, F. 1967. Susceptibility to two strains of Friend leukemia virus in mice. Science *155:* 461.

———. 1968. The effect of histocompatibility-2 type on resistance to the Friend leukemia virus in mice. J. Exp. Med. *127:* 465.

———. 1970. Fv-2: Identification and location of a second gene governing the spleen focus response to Friend leukemia in mice. J. Nat. Cancer Inst. *45:* 163.

LILLY F. and T. PINCUS. 1972. Genetic control of murine viral leukemogenesis. Advanc. Cancer Res. In press.

LIVINGSTON, D. M. and G. J. TODARO. 1973. Endogenous type C virus from a cat cell clone with properties distinct from previously described feline type C viruses. Virology. In press.

LUFTIG, R. B. and S. S. KILHAM. 1971. An electron microscope study of Rauscher leukemia virus. Virology *46:* 277.

MCALLISTER, R. M., J. MELNYK, J. Z. FINKLESTEIN, G. C. ADAMS and M. B. GARDNER. 1969. Cultivation *in vitro* of cells derived from a human rhabdomyosarcoma. Cancer *24:* 520.

MCALLISTER, R. M., W. A. NELSON-REES, W. A. JOHNSON, E. Y. RONGEY and M. B. GARDNER. 1971. Disseminated rhabdomyosarcomas formed in kittens by cultured human rhabdomyosarcoma cells. J. Nat. Cancer Inst. *47:* 603.

MCALLISTER, R. M., M. NICOLSON, M. B. GARDNER, R. W. RONGEY, S. RASHEED, P. S. SARMA, R. J. HUEBNER, M. HATANAKA, S. OROSZLAN, R. V. GILDEN, A. KABIGTING and L. VERNON. 1972. C-type virus released from cultured human rhabdomyosarcoma cells. Nature New Biol. *235:* 3.

MCDONALD, R., L. G. WOLFE and F. DEINHARDT. 1972. Feline fibrosarcoma virus: Quantitative focus assay, focus morphology and evidence for a "helper" virus. Int. J. Cancer *9:* 57.

MACDOWELL, E. C. and M. N. RICHTER. 1935. Mouse leukemia. IX. The role of heredity in spontaneous cases. Arch. Pathol. *20:* 709.

MACHALA, O., L. DONNER and J. SVOBODA. 1970. A full expression of the genome of Rous sarcoma virus in heterokaryons formed after fusion of virogenic mammalian cells and chicken fibroblasts. J. Gen. Virol. *8:* 219.

MACPHERSON, I. A. 1965. Reversion in hamster cells transformed by Rous sarcoma virus. Science *148:* 1731.

MACPHERSON, I. A. and L. MONTAGNIER. 1964. Agar suspension culture for the selective assay of cells transformed by polyoma virus. Virology *23:* 291.

MANAKER, R. A. and V. GROUPÉ. 1956. Discrete foci of altered chicken embryo cells associated with Rous sarcoma virus in tissue culture. Virology *2:* 838.

MANNING, J. S., F. L. SCHAFFER and M. E. SOERGEL. 1972. Correlation between murine sarcoma virus buoyant density, infectivity and viral RNA electrophoretic mobility. Virology *49:* 804.

MARTIN, G. S. 1970. Rous sarcoma virus: A function required for maintenance of the transformed state. Nature *227:* 1021.

MARTIN, G. S. and P. H. DUESBERG. 1972. The *a* subunit in the RNA of transforming avian tumour viruses. Virology *47:* 494.

MELLORS, R. C. and C. Y. HUANG. 1966. Immunopathology of NZB/BL mice. V. Viruslike (filtrable) agent separable from lymphoma cells and identifiable by electron microscopy. J. Exp. Med. *124:* 1031.

METCALF, D., J. FURTH and R. F. BUFFETT. 1959. Pathogenesis of mouse leukemia caused by Friend virus. Cancer Res. *19:* 52.

MIRAND, E. A. 1966. Erythropoietic response of animals infected with various strains of Friend virus. Nat. Cancer Inst. Monogr. *22:* 483.

MIRAND, E. A. and J. T. GRACE. 1962. Induction of leukemia in rats with Friend virus. Virology *17:* 364.

MIZUTANI, S. and H. M. TEMIN. 1971. Enzymes and nucleotides in virions of Rous sarcoma virus. J. Virol. *8:* 409.

MOLONEY, J. B. 1960. Biological studies on a lymphoid leukemia virus extracted from Sarcoma S37. I. Origin and introductory investigations. J. Nat. Cancer Inst. *24:* 933.

———. 1966. A virus-induced rhabdomyosarcoma of mice. Nat. Cancer Inst. Monogr. *22:* 139.

MONTAGNIER, L., A. GOLDÉ and P. VIGIER. 1969. A possible subunit structure of Rous sarcoma virus RNA. J. Gen. Virol. *4:* 449.

MOORE, D. H., J. CHARNEY, B. KRAMARSKY, E. Y. LASFARGUES, N. H. SARKAR, M. J. BRENNAN, J. H. BURROWS, S. M. SIRSAT, J. C. PAYMASTER and A. B. VAIDYA. 1971. Search for human breast cancer virus. Nature *229:* 611.

MORGAN, H. R. and W. TRAUB. 1964. Origin of Rous sarcoma strains. Nat. Cancer Inst. Monogr. *17:* 392.

MORONI, C. 1972. Structural proteins of Rauscher leukemia virus and Harvey sarcoma virus. Virology *47:* 1.

MORTON, D. L., W. T. HALL and R. A. MALMGREN. 1969. Human liposarcomas: Tissue cultures containing foci of transformed cells with viral particles. Science *165:* 813.

MOSCOVICI, C. and M. ZANETTI. 1970. Studies on single foci of hematopoietic cells transformed by avian myeloblastosis virus. Virology *42:* 61.

MÜHLBOCK, O. 1965. Note on a new inbred mouse-strain GR/A. European J. Cancer *1:* 123.

———. 1972. The value of experimental cancer research for the understanding of the human disease. In *RNA viruses and host genome in oncogenesis*, ed. P. Emmelot and P. Bentvelzen, p. 339. North-Holland, Amsterdam.

MUNROE, J. S. and W. E. WINDLE. 1963. Tumors induced in primates by chicken sarcoma virus. Science *140:* 1415.

NANDI, S. and K. B. DEOME. 1965. An interference phenomenon associated with resistance to infection with mouse mammary tumor virus. J. Nat. Cancer Inst. *35:* 299.

NANDI, S., M. HANDIN and L. YOUNG. 1966. Strain-specific mammary tumor virus activity in blood of C3H and BALB/cf.C3H mice. J. Nat. Cancer Inst. *36:* 803.

NERMUT, M. V., F. HERMANN and W. SCHAFER. 1972. Properties of mouse leukemia viruses. III. Electron microscopic appearance as revealed after conventional preparation techniques as well as freeze-drying and freeze-etching. Virology *49:* 345.

NOWINSKI, R. C., L. J. OLD, N. H. SARKAR and D. H. MOORE. 1970. Common properties of the oncogenic RNA viruses (oncornaviruses). Virology *42:* 1152.

NOWINSKI, R. C., E. FLEISSNER, N. H. SARKAR and T. AOKI. 1972a. Chromatographic separation and antigenic analysis of proteins of the oncornaviruses. II. Mammalian leukemia-sarcoma viruses. J. Virol. *9:* 359.

NOWINSKI, R. C., E. FLEISSNER and N. H. SARKAR. 1972b. Structural and serological aspects of the oncornaviruses. Perspectives in Virology. In press.

O'CONNOR, T. E. and P. J. FISCHINGER. 1968. Titration patterns of a murine sarcoma-leukemia virus complex: Evidence for existence of competent sarcoma virions. Science *159:* 325.

———. 1969. Physical properties of competent and defective states of a murine sarcoma (Moloney) virus. J. Nat. Cancer Inst. *43:* 487.

ODAKA, T. 1969. Inheritance of susceptibility to Friend mouse leukaemia virus. J. Virol. *3:* 543.

ODAKA, T. and T. YAMAMOTO. 1962. Inheritance of susceptibility to Friend mouse leukemia virus. Japan. J. Exp. Med. *32:* 405.

OLD, L. J., E. A. BOYSE and E. STOCKERT. 1964. Typing of mouse leukemias by serological methods. Nature *201:* 777.

OROSZLAN, S., C. FOREMAN, G. KELLOFF and R. V. GILDEN. 1971a. The group-specific antigen and other structural proteins of hamster and mouse C-type viruses. Virology *43:* 665.

OROSZLAN, S., M. HATANAKA, R. V. GILDEN and R. J. HUEBNER. 1971b. Specific inhibition of mammalian ribonucleic acid C-type virus deoxyribonucleic acid polymerases by rat antisera. J. Virol. *8:* 816.

OROSZLAN, S., M. M. H. WHITE, R. V. GILDEN and H. P. CHARMAN. 1972a. A rapid direct radioimmunoassay for type-C virus group specific antigen and antibody. Virology *50:* 294.

OROSZLAN, S., T. COPELAND, M. SUMMERS and R. V. GILDEN. 1972b. Amino terminal sequences of mammalian type C RNA tumor virus group-specific antigens. Biochem. Biophys. Res. Comm. *48:* 1549.

PADGETT, F. and A. S. LEVINE. 1966. Fine structure of the Rauscher leukemia virus as revealed by incubation in snake venom. Virology *30:* 623.

PARKMAN, R., J. A. LEVY and R. C. TING. 1970. Murine sarcoma virus: The question of defectiveness. Science *168:* 387.

PASTERNAK, G. 1967. Differentiation between viral and new cellular antigens in Graffi leukaemia of mice. Nature *214:* 1364.

———. 1969. Antigens induced by the mouse leukemia viruses. Advanc. Cancer Res. *12:* 1.

PAYNE, L. N. and P. M. BIGGS. 1964. Differences between highly inbred lines of chickens in the response to Rous sarcoma virus of the chorioallantoic membrane and of embryonic cells in tissue culture. Virology *24:* 610.

———. 1966. Genetic basis of cellular susceptibility to the Schmidt-Ruppin and Harris strains of Rous sarcoma virus. Virology *29:* 190.

———. 1970. Genetic resistance of fowl to MH2 reticuloendothelioma virus. J. Gen. Virol. *7:* 177.

PAYNE, L. N. and P. K. PANI. 1971. Evidence for linkage between genetic loci controlling response of fowl to subgroup A and subgroup C sarcoma viruses. J. Gen. Virol. *13:* 253.

PAYNE, L. N., P. K. PANI and R. A. WEISS. 1971. A dominant epistatic gene which influences cellular susceptibility to RSV (RAV-O). J. Gen. Virol. *13:* 455.

PAYNE, L. N., L. B. CRITTENDEN and R. A. WEISS. 1973. A brief definition of host genes which influence infection by avian RNA tumour viruses. In *Possible episomes in eukaryotes*, ed. L. G. Silvestri. North-Holland, Amsterdam.

PERK, K., M. V. VIOLA, K. L. SMITH, N. A. WIVEL and J. B. MOLONEY. 1969. Biologic studies on hamster tumors induced by the murine sarcoma virus (Moloney). Cancer Res. *29:* 1089.

PINCUS, T., J. W. HARTLEY and W. P. ROWE. 1971a. A major genetic locus affecting resistance to infection with murine leukemia viruses. I. Tissue culture studies of naturally occurring viruses. J. Exp. Med. *133:* 1219.

PINCUS, T., W. P. ROWE and F. LILLY. 1971b. A major genetic locus affecting resistance to infection with murine leukemia viruses. II. Apparent identity to a major locus described for resistance to Friend murine leukemia virus. J. Exp. Med. *133:* 1234.

PINKEL, D., K. YOSHIDA and K. SMITH. 1966. Studies of Moloney and Rauscher leukemia viruses in cell cultures by immunofluorescence. Nat. Cancer Inst. Monogr. *22:* 671.

PIRAINO, F. 1967. The mechanism of genetic resistance of chick embryo cells to infection by Rous sarcoma virus-Bryan strain. Virology *32:* 700.

PLUZNICK, D. H. and L. SACHS. 1964. Quantitation of a murine leukemia virus with a spleen colony assay. J. Nat. Cancer Inst. *33:* 535.

PONTÉN, J. and I. MACPHERSON. 1966. Interference with Rous sarcoma virus focus formation by a mycoplasmalike factor present in human cell cultures. Ann. Med. Exp. Fenn. *44:* 260.

POPE, J. H. 1962. The isolation of a mouse leukemia virus resembling Friend virus. Australian J. Exp. Biol. Med. *40:* 263.

PRICE, P. M., K. HIRSCHHORN, N. G. ABELMAN and S. WAXMAN. 1973. *In situ* hybridization of RD-114 virus RNA with human metaphase chromosomes. Proc. Nat. Acad. Sci. *70:* 11.

PRIORI, E. S., L. DMOCHOWSKI, B. MYERS and J. R. WILBUR. 1971. Constant production of type C virus particles in a continuous tissue culture derived from pleural effusion cells of a lymphoma patient. Nature New Biol. *232:* 61.

QUIGLY, J. P., D. B. RIFKIN and E. REICH. 1971. Phospholipid composition of Rous sarcoma virus, host cell membranes and outer enveloped RNA viruses. Virology *46:* 106.

RAO, P. R., R. A. BONAR and J. W. BEARD. 1966. Lipids of the BAI strain A avian tumor virus and of the myeloblast host cell. Exp. Mol. Pathol. *5:* 374.

RAUSCHER, F. J. 1962. A virus-induced disease of mice characterized by erythrocytopoiesis and lymphoid leukemia. J. Nat. Cancer Inst. *29:* 515.

RICKARD, C. G., J. E. POST, F. NORONHA and L. M. BARR. 1969. A transmissible virus-induced lymphocytic leukemia of the cat. J. Nat. Cancer Inst. *42:* 937.

RIFKIN, D. B. and R. W. COMPANS. 1971. Identification of the spike proteins of Rous sarcoma virus. Virology *46:* 485.

RIMAN, J. and G. S. BEAUDREAU. 1970. Viral DNA-dependent DNA polymerase and the properties of thymidine labelled material in virions of an oncogenic RNA virus. Nature *228:* 427.

ROBINSON, W. S. and M. A. BALUDA. 1965. The nucleic acid from avian myeloblastosis virus compared with the RNA from the Bryan strain of Rous sarcoma virus. Proc. Nat. Acad. Sci. *54:* 1686.

ROBINSON, W. S., A. P. PITKANEN and H. RUBIN. 1965. The nucleic acid of the Bryan strain of Rous sarcoma virus: Purification of the virus and isolation of the nucleic acid. Proc. Nat. Acad. Sci. *54:* 137.

ROBINSON, W. S., H. L. ROBINSON and P. H. DUESBERG. 1967. Tumour virus RNAs. Proc. Nat. Acad. Sci. *58:* 825.

ROKUTANDA, M., H. ROKUTANDA, M. GREEN, K. FUJINAGA, R. K. RAY and C. GURGO. 1970. Formation of viral RNA-DNA hybrid molecules by the DNA polymerase of sarcoma-leukaemia viruses. Nature *227:* 1026.

ROSS, J., Y. IKAWA and P. LEDER. 1972. Globin messenger-RNA induction during erythroid differentiation of cultured leukemia cells. Proc. Nat. Acad. Sci. *69:* 3620.

ROUS, P. 1911. A sarcoma of the fowl transmissible by an agent from the tumor cells. J. Exp. Med. *13:* 397.

ROWE, W. P. and I. BRODSKY. 1959. A graded-response assay for the Friend leukemia virus. J. Nat. Cancer Inst. *23:* 1239.

ROWE, W. P., W. E. PUGH and J. W. HARTLEY. 1970. Plaque assay techniques for murine leukemia viruses. Virology *42:* 1136.

ROWSON, K. E. and I. B. PARR. 1970. A new virus of minimal pathogenicity associated with Friend virus. I. Isolation by endpoint dilution. Int. J. Cancer *5:* 96.

RUBIN, H. 1955. Quantitative relations between causative virus and cell in the Rous No. 1 chicken sarcoma. Virology *1:* 445.

———. 1957. The production of virus by Rous sarcoma cells. Ann. N.Y. Acad. Sci. *68:* 459.

———. 1960a. An analysis of the assay of Rous sarcoma cells in vitro by the infective center technique. Virology *10:* 29.

———. 1960b. A virus in chick embryos which induces resistance *in vitro* to infection with Rous sarcoma virus. Proc. Nat. Acad. Sci. *46:* 1105.

———. 1961. The nature of virus-induced cellular resistance to Rous sarcoma virus. Virology *13:* 170.

———. 1964. Virus defectiveness and cell transformation in the Rous sarcoma. J. Cell. Comp. Physiol. *64* (suppl. 1): 173.

———. 1965. Genetic control of cellular susceptibility to pseudotypes of Rous sarcoma virus. Virology *26:* 270.

———. 1966. The inhibition of chick embryo cell growth by medium obtained from cultures of Rous sarcoma cells. Exp. Cell Res. *41:* 149.

RUBIN, H. and P. K. VOGT. 1962. An avian leukosis virus associated with stocks of Rous sarcoma virus. Virology *17:* 184.

SARKAR, N. H. and D. H. MOORE. 1968. Internal structure of mouse mammary tumour virus as revealed after Tween ether treatment. J. de Microscopie *7:* 539.

———. 1972. On the possibility of a human breast cancer virus. Nature *236:* 103.

SARKAR, N. H., D. H. MOORE and R. C. NOWINSKI. 1972. Symmetry of the nucleocapsid of the oncornaviruses. In *RNA viruses and host genome in oncogenesis*, ed. P. Emmelot and P. Bentvelzen, p. 71. North-Holland, Amsterdam.

SARMA, P. S. and T. LOG. 1971. Viral interference in feline leukemia-sarcoma complex. Virology *44:* 352.

SARMA, P. S., H. C. TURNER and R. J. HUEBNER. 1964. An avian leukosis group-specific complement fixation reaction. Application for the detection and assay of non-cyto-pathogenic leukosis viruses. Virology *23:* 313.

SARMA, P. S., M. CHEONG, J. W. HARTLEY and R. J. HUEBNER. 1967. A viral interference test for mouse leukemia viruses. Virology *33:* 180.

SARMA, P. S., T. LOG and R. J. HUEBNER. 1970a. Trans-specific rescue of defective genomes of murine sarcoma virus from hamster tumor cells with helper feline leukemia virus. Proc. Nat. Acad. Sci. *65:* 81.

SARMA, P. S., R. J. HUEBNER, J. F. BASKER, L. VERNON and R. V. GILDEN. 1970b. Feline leukemia and sarcoma viruses: Susceptibility of human cells to infection. Science *168:* 1098.

SARMA, P. S., R. V. GILDEN and R. J. HUEBNER. 1971a. A complement-fixation test for feline leukemia and sarcoma viruses (the COCAL test). Virology *44:* 137.

SARMA, P. S., J. F. BASKER, R. V. GILDEN, M. B. GARDNER and R. J. HUEBNER. 1971b. *In vitro* isolation and characterization of the GA strain of feline sarcoma virus. Proc. Soc. Exp. Biol. Med. *137:* 1333.

SARMA, P. S., T. LOG and G. H. THEILEN. 1971c. ST feline sarcoma virus. Biological characteristics and *in vitro* propagation. Proc. Soc. Exp. Biol. Med. *137:* 1444.

SARMA, P. S., A. L. SHARAR and S. MCDONOUGH. 1972. The SM strain of feline sarcoma virus. Biological and antigenic characterization of the virus. Proc. Soc. Exp. Biol. Med. *140:* 1365.

SCHAFER, W., P. J. FISCHINGER, J. LANGE and L. PISTER. 1972a. Properties of mouse leukemia viruses: Characterization of various antisera and serological identification of viral components. Virology *47:* 197.

SCHAFER, W., J. LANGE, P. J. FISCHINGER, H. FRANK, D. P. BOLOGNESI and L. PISTER. 1972b. Properties of mouse leukemia viruses. II. Isolation of viral components. Virology *47:* 210.

SCHEELE, C. M. and H. HANAFUSA. 1971. Proteins of helper-dependent RSV. Virology *45:* 401.

SCHLOM, J. and S. A. SPIEGELMAN. 1971. DNA polymerase activities and nucleic acid components of virions isolated from a spontaneous mammary carcinoma from a rhesus monkey. Proc. Nat. Acad. Sci. *68:* 1613.

SCHLOM, J., S. SPIEGELMAN and D. MOORE. 1971. RNA-dependent DNA polymerase activity in virus-like particles isolated from human milk. Nature *231:* 97.

———. 1972. Detection of high-molecular weight RNA in particles from human milk. Science *175:* 542.

SCHMIDT, F. 1955. Über filtraversuche mit mäusetransplantations-tumoren. Krebsforsch. *60:* 445.

SCHMIDT-RUPPIN, K. N. 1959. Versuche zur heterologen transplantation mit frischen und gefriergetrockneterm material des Rous sarcoma. Krebsforsch. Krebsebekamf. *3:* 26.

SCHOOLMAN, H. M., W. SPURRIER, S. O. SCHWARTZ and P. B. SZANTS. 1957. Studies in leukemia. VII. The induction of leukemia in Swiss mice by means of cell-free filtrates of leukemic mouse brain. Blood *12:* 694.

SCOLNICK, E. M., W. P. PARKS, G. J. TODARO and S. A. AARONSON. 1972a. Immunological characterization of primate C-type virus reverse transcriptases. Nature New Biol. *235:* 35.

———. 1972b. Reverse transcriptase of primate viruses as immunological markers. Science *177:* 1119.

SCOLNICK, E. M., W. P. PARKS and D. M. LIVINGSTON. 1972c. Radioimmunoassay of mammalian type C viral proteins. I. Species specific reactions of murine and feline viruses. J. Immunol. *109:* 570.

SHANMUGAM, G., G. VECCHIO, D. ATTARDI and M. GREEN. 1972. Immunological studies on viral polypeptide synthesis in cells replicating murine sarcoma-leukemia virus. J. Virol. *10:* 447.

SHARP, D. G., E. A. ECKERT, D. BEARD and J. W. BEARD. 1952. Morphology of the virus of avian erythromyeloblastic leucosis and a comparison with the agent of Newcastle disease. J. Bacteriol. *63:* 151.

SHEVLYAGHIN, V. Y., T. I. BIRYULINA, Z. N. TIKHONOVA and N. V. KARAZAS. 1969. Activation of Rous virus in the transplanted golden hamster tumor with the aid of artificial heterokaryon formation. Int. J. Cancer *4:* 42.

SHIGEMATSU, T., E. S. PRIORI, L. DMOCHOWSKI and J. R. WILBUR. 1971. Immunoelectron microscopic studies of type C virus particles in ESP-1 and HEK-1-HRLV cell lines. Nature *234:* 412.

SIEGLER, R. and M. A. RICH. 1966. Pathogenesis of murine leukemia. Nat. Cancer Inst. Monogr. *22:* 525.

ŠIMKOVIČ, D., N. VALENTOVA and V. THURZO. 1962. An *in vitro* system for the detection of the Rous sarcoma virus in the cells of the rat tumour XC. Neoplasma *8:* 104.

ŠIMKOVIČ, D., M. POPOVIC, J. SVEC, M. GROFOVA and N. VALENTOVA. 1969. Continuous production of avian sarcoma virus B77 by rat tumor cells in tissue culture. Int. J. Cancer *4:* 80.

SMIDA, J. and V. SMIDOVA. 1971. Host range differences among subgroup C members of avian tumour viruses. Neoplasma *18:* 555.

SMITH, R. and C. MOSCOVICI. 1969. The oncogenic effects of non-transforming viruses from avian myeloblastosis virus. Cancer Res. *29:* 1356.

SNYDER, S. P. and G. H. THEILEN. 1969. Transmissible feline lymphosarcoma. Nature *221:* 1074.

SOMERS, K. D. and W. H. KIRSTEN. 1968. Focus formation by murine sarcoma virus: Enhancement by DEAE-dextran. Virology *36:* 155.

SPIEGELMAN, S., A. BURNY, M. R. DAS, J. KEYDAR, J. SCHLOM, M. TRAVNICEK and K. WATSON. 1970. DNA-directed DNA polymerase activity in oncogenic RNA viruses. Nature *227:* 1029.

SPIRIN, A. S. 1963. Some problems concerning macromolecular structure of RNA. In *Progress in nucleic acid research*, ed. J. N. Davidson and W. E. Cohn, vol. 1, p. 301. Academic Press, New York.

STECK, F. T. and H. RUBIN. 1966a. The mechanism of interference between an avian leukosis virus and Rous sarcoma virus. I. Establishment of interference. Virology *29:* 628.

———. 1966b. The mechanism of interference between an avian leukosis virus and Rous sarcoma virus. II. Early steps of infection by RSV of cells under conditions of interference. Virology *29:* 642.

STEEVES, R. A. and R. J. ECKNER. 1970. Host-induced changes in infectivity of Friend spleen focus-forming virus. J. Nat. Cancer Inst. *44:* 587.

STEEVES, R. A., R. J. ECKNER, E. A. MIRAND and R. L. PRIORI. 1971. Rapid assay of murine leukemia virus helper activity for Friend spleen focus-forming virus. J. Nat. Cancer Inst. *46:* 1219.

STEPHENSON, J. R. and S. A. AARONSON. 1971. Murine sarcoma and leukemia viruses: Genetic differences determined by RNA-DNA hybridization. Virology *46:* 480.

STEWART, S. E., J. LANDON and E. LOVELACE. 1964. Viruses in cultures of human leukemic cells. In *Att. Del Simposio Sul. Tena. I. Virus nelle leucemie dei mammiferi*. Acad. Nat. dei Lincei. Rome *364:* 271.

STEWART, S. E., G. KASNIC, C. DRAYCOTT, W. FELLER, A. GOLDEN, E. MITCHELL and T. BEN. 1972a. Activation *in vitro* by 5-iododeoxyuridine of a latent virus resembling C-type virus in a human sarcoma cell line. J. Nat. Cancer Inst. *48:* 273.

STEWART, S. E., G. KASNIC, C. DRAYCOTT and T. BEN. 1972b. Activation of viruses in human tumors by 5-iododeoxyuridine and dimethyl sulfoxide. Science *175:* 198.

STÜCK, B., L. J. OLD and E. A. BOYSE. 1964a. Occurrence of soluble antigen in the plasma of mice with virus-induced leukemia. Proc. Nat. Acad. Sci. *52:* 950.

———. 1964b. Antigenic conversion of established leukemias by an unrelated leukemogenic virus. Nature *202:* 1016.

STÜCK, B., E. A. BOYSE, L. J. OLD and E. A. CARSWELL. 1964c. ML—A new antigen found in leukemias and mammary tumours of the mouse. Nature *203:* 1033.

SVET-MOLDAVSKY, G. J. 1958. Sarcoma in albino rats treated during the embryonic stage with Rous virus. Nature *182:* 1452.

SVOBODA, J. 1960. Presence of chicken tumour virus in the sarcoma of the adult rat inoculated after birth with Rous sarcoma virus. Nature *186:* 980.

SVOBODA, J. and I. HLOŽANEK. 1970. Role of cell association in virus infection and virus rescue. Advanc. Cancer Res. *13:* 217.

SVOBODA, J., P. P. CHYLE, D. ŠIMKOVIČ and I. HILGERT. 1963. Demonstration of the absence of infectious Rous virus in rat tumor XC whose structurally intact cells produce Rous sarcoma when transferred to chicks. Folia Biol. (Prague) *9:* 77.

SVOBODA, J., O. MACHALA and I. HLOŽANEK. 1967. Influence of Sendai virus in RSV formation in mixed culture of virogenic mammalian cells and chicken fibroblasts. Folia Biol. (Prague) *13:* 155.

SVOBODA, J., O. MACHALA, L. DONNER, and V. SOVOVA. 1971. Comparative study of RSV rescue from RSV-transformed mammalian cells. Int. J. Cancer *8:* 391.

TANAKA, H. and D. H. MOORE. 1967. Electron microscopic localization of viral antigens in mouse mammary tumors by ferritin-labeled antibody. I. The homologous systems. Virology *33:* 197.

TEMIN, H. M. 1962. Separation of morphological conversion and virus production in Rous sarcoma virus infection. Cold Spring Harbor Symp. Quant. Biol. *27:* 407.

———. 1963. Further evidence for a converted non-virus producing state of Rous sarcoma virus infected cells. Virology *20:* 235.

———. 1971. Mechanism of cell transformation by RNA tumor viruses. Ann. Rev. Microbiol. *25:* 609.

TEMIN, H. M. and D. BALTIMORE. 1972. RNA-directed DNA synthesis and RNA tumor viruses. Advanc. Virus. Res. *17:* 129.

TEMIN, H. M. and H. RUBIN. 1958. Characteristics of an assay for Rous sarcoma virus and Rous sarcoma cells in tissue culture. Virology *6:* 669.

TENNANT, J. R. 1962. Derivation of a murine lymphoid leukemia virus. J. Nat. Cancer Inst. *28:* 1291.

TENNANT, J. R. and G. D. SNELL. 1968. The H-2 locus and viral leukemogenesis as studied in congenic strains of mice. J. Nat. Cancer Inst. *41:* 597.

THEILEN, G. H., R. F. ZEIGEL and M. J. TWIEHAUS. 1966. Biological studies with RE virus (strain T) that induces reticulo-endotheliosis in turkeys, chickens and Japanese quail. J. Nat. Cancer Inst. *37:* 731.

THEILEN, G. H., D. GOULD, M. FOWLER and D. L. DUNGWORTH. 1971. C-type virus in tumor tissue of a woolly monkey (lagothrix) with fibrosarcoma. J. Nat. Cancer Inst. *47:* 881.

THOMAS, J. A., E. HOLLANDE, E. HENRY and M. DUCROS. 1969. Organisation et differentiations de la particule mature du virus de la tumeur mammaire de la Souris. Compt. Rend. Acad. Sci. (Paris) *269:* 2471.

TIMMERMANS, A., P. BENTVELZEN, P. C. HAGEMAN and J. CALAFAT. 1969. Activation of a mammary tumour virus in 020 strain mice by irradiation and urethane. J. Gen. Virol. *4:* 169.

TING, R. C. 1968. Biological and serological properties of viral particles from a non-producer rat neoplasm induced by murine sarcoma virus (Moloney). J. Virol. *2:* 865.

TOYOSHIMA, K. and P. K. VOGT. 1969. Enhancement and inhibition of avian sarcoma viruses by polycations and polyanions. Virology *38:* 414.

TOYOSHIMA, K., R. R. FRIIS and P. K. VOGT. 1970. The reproductive and cell-transforming capacities of avian sarcoma virus B77: Inactivation with UV light. Virology *42:* 163.

TRAVNIČEK, M. 1968. RNA with amino acid-acceptor activity isolated from an oncogenic virus. Biochim. Biophys. Acta *166:* 757.

VARMUS, H. E., W. E. LEVINSON and J. M. BISHOP. 1971. Extent of transcription by the RNA-dependent DNA polymerase of Rous sarcoma virus. Nature New Biol. *233:* 19.

VIGIER, P. 1970. RNA oncogenic viruses: Structure, replication and oncogenicity. Progr. Med. Virol. *12:* 240.

VISSCHER, M. B., R. G. GREEN and J. J. BITTNER. 1942. Characterisation of milk influence in spontaneous mammary carcinoma. Proc. Soc. Exp. Biol. Med. *49:* 94.

VOGT, P. K. 1964. Fluorescence microscopic observations on the defectiveness of Rous sarcoma virus. Nat. Cancer Inst. Monogr. *17:* 527.

VOGT, P. K. 1965a. Avian tumor viruses. Advanc. Virus Res. *11:* 293.
———. 1965b. A heterogeneity of Rous sarcoma virus revealed by selectively resistant chick embryo cells. Virology *25:* 237.
———. 1967a. Phenotypic mixing in the avian tumor virus group. Virology *32:* 708.
———. 1967b. DEAE-dextran: Enhancement of cellular transformation induced by avian sarcoma viruses. Virology *33:* 175.
———. 1967c. A virus released by "non-producing" Rous sarcoma cells. Proc. Nat. Acad. Sci. *58:* 801.
———. 1969. Focus assay of Rous sarcoma virus. In *Fundamental techniques of virology*, ed. K. Habel and N. P. Salzman, p. 198. Academic Press, New York.
———. 1970. Envelope classification of avian RNA tumor viruses. In *Comparative leukemia research 1969*, ed. R. M. Dutcher, p. 153. Karger, Basel.
———. 1971a. Spontaneous segregation of nontransforming viruses from cloned sarcoma viruses. Virology *46:* 939.
———. 1971b. Genetically stable reassortment of markers during mixed infection with avian tumor viruses. Virology *46:* 947.
VOGT, P. K. and R. R. FRIIS. 1971. An avian leukosis virus related to RSV(O): Properties and evidence for helper activity. Virology *43:* 223.
VOGT, P. K. and R. ISHIZAKI. 1965. Reciprocal patterns of genetic resistance to avian tumor viruses in two lines of chickens. Virology *26:* 664.
———. 1966. Patterns of viral interference in the avian leukosis and sarcoma complex. Virology *30:* 368.
WARE, L. M. and A. A. AXELRAD. 1972. Inherited resistance to N- and B-tropic murine leukemia viruses *in vitro:* Evidence that congenic mouse strains SIM and SIM.R differ at the Fv-1 locus. Virology *50:* 339.
WATSON, J. D. 1971. The structure and assembly of murine leukemia virus: Intracellular viral RNA. Virology *45:* 586.
WEISS, R. A. 1967. Spontaneous virus production from "non-virus producing" Rous sarcoma cells. Virology *32:* 719.
———. 1969a. The host range of Bryan strain Rous sarcoma virus synthesized in the absence of helper virus. J. Gen. Virol. *5:* 511.
———. 1969b. Interference and neutralization studies with Bryan strain Rous sarcoma virus synthesized in the absence of helper virus. J. Gen. Virol. *5:* 529.
———. 1970. Studies on the loss of growth inhibition in cells infected with Rous sarcoma virus. Int. J. Cancer *6:* 333.
———. 1973. Transmission of cellular genetic elements by RNA tumor viruses. In *Possible episomes in eukaryotes*, ed. L. G. Silvestri. North-Holland, Amsterdam.
WEISS, R. A. and P. M. BIGGS. 1972. Leukosis and Marek's disease viruses of feral red jungle fowl and domestic fowl in Malaya. J. Nat. Cancer Inst. *49:* 1713.
WEISS, R. A., R. R. FRIIS, E. KATZ and P. K. VOGT. 1971. Induction of avian tumor viruses in normal cells by physical and chemical carcinogens. Virology *46:* 920.
WOLFE, L. G., F. DEINHARDT, G. H. THEILEN, H. RUBIN, T. KAWAKAMI and L. K. BUSTAD. 1971. Induction of tumors in marmoset monkeys by simian sarcoma virus type 1 (lagothrix): A preliminary report. J. Nat. Cancer Inst. *47:* 1115.
WYKE, J. A. and M. LINIAL. 1973. Temperature-sensitive avian sarcoma viruses: A physiological comparison of twenty mutants. Virology. In press.
YOSHIKURA, H. 1973. Host range conversion of murine sarcoma-leukemia complex. J. Gen. Virol. In press.
ZAVADA, J. and I. A. MACPHERSON. 1970. Transformation of hamster cell lines *in vitro* by a hamster sarcoma virus. Nature *225:* 24.

ZEIGEL, R. F. and H. F. CLARK. 1969. Electron microscopic observations on a "C"-type virus in cell cultures derived from a tumor-bearing viper. J. Nat. Cancer Inst. *43:* 1097.

ZILBER, L. A. 1946. On the filtrability of tumors induced by 1, 2, 5, 6-dibenzanthracene. Amer. Rev. Soviet Med. *4:* 100.

———. 1964. Some data on the function of Rous sarcoma virus with mammalian cells. Nat. Cancer Inst. Monogr. *17:* 261.

11

Replication of Nondefective RNA Tumour Viruses

Unlike the papova viruses and adenoviruses, the RNA tumour viruses do not usually kill the cells in which they replicate, and replication and transformation are often concomitant events.

In 1958 Temin and Rubin reported a focus assay for the transformation by Rous sarcoma virus of chick embryo fibroblasts growing in culture. This was the first assay of transformation in vitro to be devised for any tumour virus, and since 1958 most quantitative studies of the mechanism of replication of the RNA tumour viruses have involved the use of RSV and cultivated chick fibroblasts. The adsorption and penetration of Rous sarcoma viruses (and avian leukosis viruses) are governed by the availability of suitable cell surface receptors and the absence of interfering viruses (see Chapter 10). Once within a susceptible fibroblast, which survives the infection, the nondefective sarcoma virus is uncoated and its genome causes the transformation of the cell as it replicates infectious progeny virus. Many of the newly acquired properties of the RSV-transformed chick fibroblast resemble those acquired by mammalian fibroblasts transformed by papova or adenoviruses (compare Tables 6.1 and 11.1).

Table 11.1. PROPERTIES OF CELLS TRANSFORMED BY RNA TUMOUR VIRUSES

Morphological and Behavioural Changes
Cells round up and/or lose orientation
Foci arise by cell multiplication and/or recruitment by reinfection
Form colonies in suspension culture
Grow on contact inhibited cell layer
Grow to high or indefinite saturation densities
Motility altered
Enhanced tumourigenicity in susceptible animals
Biochemical Changes
Altered, usually decreased, serum requirement
Increased anaerobic glycolysis
Increased hyaluronate synthesis
Increased sugar transport, decreased concentration of cAMP
Increased susceptibility to agglutination by lectin
Acquire fibrinolysin T activity
Increased susceptibility to dibucaine (RSV only)
Changes in glycoproteins and glycolipids
Evidence of Virus
Infectious virus produced
Noninfectious virus produced
Virus rescued by heterokaryon formation or superinfection
GS antigens present
Virus-induced transplantation immunity
Virus-specific RNA
Virus interference
Viral control of focus morphology

Cells transformed by RNA tumour viruses show many, if not all, of these properties which are not shared by untransformed parental cells.

By contrast, avian leukosis viruses do not transform susceptible fibroblasts but will replicate and produce progeny virions, usually without killing the host cell. Under certain conditions, however, leukosis viruses of subgroups B and D do cause cell lysis and they can therefore be assayed by plaque formation (Graf, 1972; Kawai and Hanafusa, 1972). Leukosis viruses, such as avian myeloblastosis virus, will transform certain differentiated cell types growing in cultures; these cells are presumably the precursors of normal granulocytic cells, and foci of myeloblasts which produce virus have been observed (Baluda, 1962).

In short, leukosis viruses may produce one of three effects on sensitive cells, depending on the conditions: (1) infection and replication without serious harmful effects on the cells, (2) cell lysis concomitant with virus replication, or (3) cell transformation and virus replication.

ESTABLISHMENT OF INFECTION: DEPENDENCE ON STATE OF CELLS

Unlike DNA tumour viruses, RNA tumour viruses are unable to induce a round of DNA replication in host cells that have ceased to divide either because they are in a confluent layer (Yoshikura, 1970; Weiss, 1971) or because they are starved of serum (Temin, 1967). This fact no doubt explains why freshly replated cultures of cells are more efficiently transformed by and support more extensive replication of RSV than do aged cultures of cells (Rubin, 1960). Progeny viruses are produced only when the infected cells proceed, not only through an S phase, but also through mitosis (Humphries and Temin, 1972); if mitosis is blocked, the production of progeny viruses is inhibited (Temin, 1967). These data establish that the replication of RSV is dependent upon certain events in the normal cell cycle. Currently we do not know the precise nature of these crucial events. However, one round of mitosis is both necessary and sufficient to establish transformation (Rubin and Colby, 1968; Nakata and Bader, 1968; Weiss, 1970).

Inhibitors of DNA Synthesis Block Infection

Examination in the electron microscope of thin sections of susceptible chick cells at various times after exposure to RSV has revealed that virus particles in vacuoles are rapidly transported across the cytoplasm to the vicinity of the nucleus (Dales and Hanafusa, 1972). At this stage (1 to 2 hours post infection) intact parental virus RNA and apparently intact virions are found associated with the cell nucleus, and events essential for the establishment of infection and replication of the virus can be blocked with inhibitors of DNA synthesis.

The following inhibitors of DNA synthesis—amethopterin,

5-bromodeoxyuridine, 5-iododeoxyuridine, 5-fluorodeoxyuridine, 5-fluorodeoxythymidine, excess thymidine and cytosine arabinoside—block infection of susceptible cells by Rous sarcoma virus if they are present at the time of infection (Temin, 1964a, 1968; Bader, 1965, 1966). For example, when 12-hour pulses of cytosine arabinoside are added to a culture of stationary, susceptible cells starved of serum at 12 hours before infection with Rous sarcoma virus, at infection or 12 hours after infection, and then serum is added, only the 12-hour pulse concomitant with exposure to the virus blocks infection (Temin, 1968; Murray and Temin, 1970). The results of such experiments suggest that during the first 12 hours of infection some step involving DNA synthesis is required, and if this step is inhibited, the infection aborts.

The interpretation of this type of experiment is, however, complicated by the requirement of cell DNA synthesis and mitosis for the replication of the virus. Clearly inhibitors of DNA synthesis might block infection indirectly by preventing cell division rather than directly by blocking the metabolism of viral nucleic acid. However, these inhibitor experiments, together with the observation that cells deprived of serum and therefore not synthesizing DNA do not support infection by RSV (Humphries and Temin, 1972), prove that some DNA synthesis is an early requirement of infection by RNA tumour viruses.

In 1970 Boettiger and Temin and Balduzzi and Morgan further investigated the effect of 5-bromodeoxyuridine on infections by RNA tumour viruses. If cells that are starved of serum are exposed to 5-bromodeoxyuridine and simultaneously infected with Rous sarcoma virus, subsequent exposure to light, after the addition of serum, aborts the infection. Examination of single-cell clones rules out the possibility that the virus is stimulating synthesis of cellular DNA and that the stimulated cells, rather than the sarcoma virus genomes, are then being inactivated by the 5-bromodeoxyuridine and light.

When Boettiger repeated this experiment using two multiplicities of infection, 0.1 and 3.0 focus-forming units per cell, he found infections at the higher multiplicity were significantly more resistant to abortion by 5-bromodeoxyuridine and light than the infections at the lower multiplicity. This suggests that the number of copies of the DNA molecule essential for infection depends on the number of infecting virus genomes. The DNA made early in infection is therefore likely to be specified by the viral RNA genome.

Previous to these experiments Duesberg and Vogt (1969) had found that when a cell infected by one avian tumour virus is superinfected by a second virus, successful superinfection depends upon an early pulse of DNA synthesis. This observation suggests, of course, that each infecting virus must direct its own DNA synthesis and therefore that the DNA being made is virus specific.

Actinomycin D Blocks Synthesis of Viral Genomes

Actinomycin D does not inhibit the replication of most non-oncogenic RNA viruses. However at both early and late times during the replication cycle of RSV, the production of progeny virus particles is inhibited by actinomycin D (Bather, 1963; Temin, 1963; Bader, 1964; Vigier and Goldé, 1964).

Labeling with precursors of RNA, protein and lipid before and after adding actinomycin D to leukemic myeloblasts from chicks infected with avian myeloblastosis virus (Baluda and Nayak, 1969) or cells transformed by Rauscher murine leukemia virus or Rous sarcoma virus (Bader, 1970) indicates that actinomycin D probably directly inhibits viral RNA synthesis rather than some other step in the maturation of progeny virus. Since actinomycin D inhibits DNA-dependent synthesis of RNA, one interpretation of these experiments is that DNA is the template for synthesis of viral RNA.

THE PROVIRUS HYPOTHESIS

To account for these unexpected data, Temin (1964b) proposed his provirus hypothesis. He suggested that a DNA copy, the provirus, of the infecting single-stranded RNA genome was the replicative intermediate of the RNA tumour viruses. By postulating that an early event in the infection of cells by this class of viruses was the synthesis of a double-stranded DNA provirus containing all the genetic information present in the single-stranded viral RNA genome, he could account for the inhibition of virus replication caused by both inhibitors of DNA synthesis and actinomycin D. Moreover, by postulating that the DNA provirus becomes integrated into host cell chromosomes, in a manner analogous to the integration of papova virus and adenovirus DNAs into the chromosomes of cells they transform, Temin could account for the stable inheritance of the transformed state.

RNA-dependent DNA Polymerase: Reverse Transcriptase

The provirus hypothesis demands the existence of an enzyme capable of using the viral RNA as a template for the synthesis of a double-stranded DNA provirus. In 1964, when Temin put forward his hypothesis, no such enzyme had been detected, and the possibility that RNA might act as a template for DNA was not envisaged by those who summarized the central dogma in the equation ↻DNA→ RNA→ protein. However as several nononcogenic viruses were characterized, conventional nucleic acid polymerases were detected in their virions. For example, in 1967 Kates and McAuslan detected DNA-dependent RNA polymerase in pox virus particles, and numerous other examples have since been reported (Baltimore, 1971). In short, the precedent for virion nucleic acid polymerases was established. Moreover, it was shown that virus-specific DNA synthesis occurs in the absence of protein synthesis when chick cells are infected with avian tumour viruses (Mizutani, cited by Temin, 1970). This result meant the putative RNA-dependent DNA polymerase responsible for provirus synthesis was either an integral component of the RNA tumour virus particle or existed in the uninfected cell.

In 1970 Baltimore and Temin and Mizutani reported that after freezing and thawing or exposure to nonionic detergent, the virions of certain avian and murine tumour viruses incorporate deoxynucleoside triphosphates into DNA. This reaction is sensitive to ribonuclease, which indicates that the template for DNA synthesis is RNA. Because the nascent DNA molecules can be recovered in association with 70S RNA and because the DNA can be hybridized to the 70S RNA, the template for DNA synthesis in the virions must be the 70S RNA (Garapin et al., 1970; Rokutanda et al., 1970; Spiegelman et al., 1970a; Manly et al., 1971). This reaction, in which the DNA polymerase copies 70S RNA in disrupted virions, is called the endogenous reaction. It is generally assumed that when RNA tumour viruses infect susceptible cells, the virion RNA-dependent DNA polymerase, called here reverse transcriptase, synthesizes a DNA provirus by this reaction.

Since 1970 numerous groups have searched for reverse transcriptase activity in all available RNA tumour viruses and in many nononcogenic viruses. At the same time the structure and properties of the enzyme from avian and murine RNA tumour viruses have been most intensively investigated. These data have been reviewed recently (Temin and Baltimore, 1972) and are more briefly summarized here.

Distribution of Viral Reverse Transcriptase

Reverse transcriptase occurs in the virions of all members of a group of viruses which have the following characteristics: 60S–70S single-stranded RNA genome, relative resistance to ultraviolet light inactivation compared to other RNA viruses (for example, Newcastle disease virus), need for DNA synthesis early in the growth cycle and sensitivity to actinomycin D at all times in the growth cycle (see Temin and Baltimore, 1972). The viruses in this group include all of the transforming (sarcoma) C-type viruses, the nontransforming (leukemia) C-type viruses, and the mammary tumour (B-type) virus. Also included are visna, maedi, and progressive penumonia viruses which cause slow, degenerative diseases, and the foamy (syncytium-forming) viruses which cause cell fusion and lysis in vitro but which have no known effects in vivo (Parks et al., 1971; Stone et al., 1971; Parks and Todaro, 1972). Even though visna and progressive pneumonia viruses can cause transformation of mouse cells in vitro (Takemoto and Stone, 1971), the occurrence of reverse transcriptase in fundamentally virulent viruses suggests that the enzyme is not limited to oncogenic viruses. Rather, it appears that all oncogenic RNA viruses are members of a group of viruses which have reverse transcriptase and therefore the enzyme may be necessary but not sufficient for an RNA virus to be oncogenic.

Purification of Reverse Transcriptase

Because of the limited supply of 70S viral RNA, the purification of the reverse transcriptase has exploited the enzyme's ability to transcribe templates other than 70S RNA. Numerous reactions have been used to assay reverse transcriptase during its purification. Among these are the copying of nuclease-treated DNA, poly(dA-dT), poly(C) and poly(A). The amount of DNA synthesized in the presence of any of these polymers by detergent-treated virions is 10- to 1000-fold greater than the amount synthesized in their absence when the endogenous 60–70S viral RNA is template (Spiegelman et al., 1970b, c; Mizutani et al., 1970; Riman and Beaudreau, 1970; Baltimore and Smoler, 1971a, b). This reverse transcriptase activity is mostly soluble and only a small portion of the enzyme sediments with the detergent-resistant residue of virions. The soluble enzyme can be purified by chromatography, gradient centrifugation and other techniques (Kacian et al., 1971; Hurwitz and Leis, 1972; Faras et al., 1972; Livingston et al., 1972; Gerwin and Milstein, 1972; Verma and Baltimore, 1973). The polymerase activity is quite stable and yields are quite high if a sulfhydryl reagent, glycerol and, most important, a nonionic detergent are present throughout the purification. Reverse transcriptase comprises less than one percent of the virion protein and, from their molecular weight data, Kacian et al. calculate that each virion contains ten enzyme molecules.

Avian myeloblastosis virus particles have been the chief source of this enzyme, but large amounts can also be obtained from cultures infected by RSV chicken cells. Murine enzymes have been purified but the starting material, murine RNA tumour virus particles, is in short supply.

Structure of the Enzyme

The enzymes purified from different viruses have different molecular weights, and until more data are available it is difficult to attach significance to these variations. The best characterized enzyme, that from avian myeloblastosis virus, has two polypeptide chains, one of 110,000 daltons and one of 69,000 daltons (Kacian et al., 1971). The enzyme purified from Rauscher murine leukemia virus is smaller and has only a single polypeptide of 70,000 daltons (Ross et al., 1971; Hurwitz and Leis, 1972; Tronick et al., 1972). The polypeptides of the purified enzyme from AMV do not correspond to any of the major virion proteins (Kacian et al., 1971; Mölling et al., 1971a; Hatanaka et al., 1970; Aaronson et al., 1971; Oroszlan et al., 1971).

Initiation of DNA Synthesis

DNA polymerases are primer-dependent enzymes which cannot initiate new chains of DNA de novo but which can only extend preformed chains (Bollum, 1967; Goulian et al., 1968; Harwood and Wells, 1970). Like other DNA polymerases, the reverse transcriptase is primer-dependent (Baltimore and Smoler, 1971a, b; Smoler et al., 1971; Goodman and Spiegelman, 1971; Hurwitz and Leis, 1972; Leis and Hurwitz, 1972a; Wells et al., 1972).

The role of the primer during the synthesis of poly(dT) by AMV reverse transcriptase using poly(A) as a template has been investigated in detail.* Poly(A), by itself, is virtually inactive as a template for the AMV reverse transcriptase, but the addition of a small amount of an oligomer of dT stimulates extensive poly(dT) synthesis (Baltimore and Smoler, 1971a, b). The oligo(dT) primer is physically incorporated at the 5′ end of the product (Smoler et al., 1971). All of the other homopolymer templates that have been studied also require a primer in order to support DNA synthesis (Baltimore and Smoler, 1971a, b; Hurwitz and Leis, 1972; Wells et al., 1972). Either oligomers or polymers can be used as primers, but oligodeoxyribonucleotides are generally much better primers than oligoribonucleotides (Goodman and Spiegelman, 1971; Hurwitz and Leis, 1972; Leis and Hurwitz, 1972a; Wells et al., 1972). This specificity is surprising because a polyribonucleotide is the natural primer in the 60–70S RNA (see below).

The direction of poly(dT) synthesis on a poly(A) template is 5′ to 3′. Using the inhibitor dideoxythymidine triphosphate (ddTTP), which can only be incorporated if DNA synthesis is 5′ to 3′ (Atkinson et al., 1969), it has been shown that the copying of poly(dA-dT) and poly(A), as well as the copying of 70S RNA in the endogenous reaction, all occur 5′ to 3′ and that the enzyme therefore probably always synthesizes in this direction (Smoler et al., 1971). Reverse transcriptase thus resembles all other known polymerases in its direction of synthesis.

Template and Primer Specificity

The extensive investigation of various homopolymers, DNAs and RNAs as template for the avian and murine reverse transcriptases and

* "Poly(A)" designates a polymer of riboadenylic acid. Deoxyribopolymers have prefixes of "d" as in poly(dA). Polymers are generally molecules consisting of more than 100 units, while oligomers are less than 20 units in length.

for *E. coli* and mammalian cell DNA polymerase has been reviewed by Temin and Baltimore (1972). They can be summarized as follows: the purified DNA polymerase from virions of RNA tumour viruses is a primer-dependent enzyme capable of copying both DNA and RNA. Ribohomopolymer templates combined with complementary deoxyribooligomer primers support high rates of DNA synthesis by the virion DNA polymerase, whereas other known DNA polymerases are relatively inactive with most of the polymer-oligomer combinations. The combination poly(C)-oligo(dG) is the most specific template-primer known for identifying reverse transcriptase.

Ribonuclease H

Hausen and Stein (1970) first demonstrated that cells contain an enzyme that degrades the RNA strand of a DNA-RNA hybrid but does not degrade either free RNA or double-stranded RNA. They called the enzyme ribonuclease H. Mölling et al. (1971b) found that virions of AMV contain ribonuclease H and that, unlike other nucleases in the virion, ribonuclease H activity does not separate from the AMV DNA polymerase during purification of the polymerase.

The most sensitive assay for detecting the ribonuclease H of AMV appears to be degradation of the [^{3}H]poly(A) portion of a [^{3}H]poly(A)-poly(dT) hybrid. Using this assay, Baltimore and Smoler (1972) confirmed the occurrence of ribonuclease H in purified AMV DNA polymerase. Keller and Crouch (1972), Leis et al. (1973) and Flugel et al. (1973) have also found ribonuclease H activity associated with the virion DNA polymerase, and none of these groups has been able to separate nuclease from polymerase, although Leis et al. believe that the activities may be separable. The nuclease can be inhibited by 7.5 mM sodium fluoride without affecting polymerase activity (Flugel et al., 1973), indicating that different active sites are involved for the two reactions.

The AMV ribonuclease H degrades [^{3}H]poly(A) mainly to oligonucleotides, most of which are at least six bases in length, and the termini at the points of scission are 3′-hydroxyl and 5′-phosphoryl (Baltimore and Smoler, 1972; Keller and Crouch, 1972; Leis et al., 1973). Although the nuclease would appear from these results to be an endonuclease, it requires a free end in the RNA substrate in order to initiate degradation (Keller and Crouch, 1972; Leis et al., 1973). It is therefore topologically an exonuclease, and Leis et al. find that once the enzyme begins degradation of a molecule of RNA, it finishes with that molecule before beginning on another; therefore it is a processive exonuclease. The possible role of ribonuclease H in reverse transcription is discussed below.

Stimulatory Protein

Leis and Hurwitz (1972b) observed that during the purification of the AMV DNA polymerase, the ability of the enzyme to copy a 70S viral RNA template was progressively lost relative to its ability to copy poly(dA-dT). They were able to separate from the enzyme a protein that stimulated DNA synthesis directed by 70S RNA. The factor also stimulates ribonuclease H activity in the AMV DNA polymerase (Leis et al., 1973). The exact function of the stimulatory protein is not yet known.

Reverse Transcription of 60–70S RNA

So far, we have only considered the reactions of reverse transcriptase with artificial templates. The natural template for the enzyme, however, is the 60–70S genomic RNA. Denaturation of the 60–70S RNA produces 35S RNA, plus a small amount of 4S RNA (Erikson and Erikson, 1971) and possibly other small molecules. The 60-70S RNA is therefore complex, and elucidation of its structure will probably be necessary before its behaviour as a template for the DNA polymerase can be fully understood. There is, however, some information about its template properties.

Reverse transcription of 60–70S RNA can be studied either in unfractionated virions treated with detergent (the endogenous reaction) or by adding 60–70S RNA to purified DNA polymerase (the reconstructed system). Where tests have been made, these two systems appear to be very similar (Taylor et al., 1972). Actually almost all our data come from study of the avian myeloblastosis virus or Rous sarcoma virus enzymes, because the enzymes purified to date from murine viruses copy homopolymer templates or DNA but not 60–70S RNA.

The virion polymerase does not seem to have any specificity for its homologous RNA (Duesberg et al., 1971). For example, the AMV polymerase transcribes more of the murine leukemia virus RNA or hamster leukemia virus RNA than of the AMV RNA (Spiegelman et al., 1971; Verma et al., 1972a).

Of the RNAs tested in the absence of an added primer, the 60–70S RNAs have been found to be the best templates for the virion DNA polymerase, giving the highest rate and extent of reaction (Duesberg et al., 1971; Goodman and Spiegelman, 1971; Spiegelman et al., 1971; Verma et al. 1972a; Leis and Hurwitz, 1972a). The amount of DNA synthesized when 60–70S RNA is the template is generally reported to be a few percent of the amount of input RNA (Verma et al., 1972a). In one case a yield of 40 percent was reported (Leis and Hurwitz, 1972a). The ability of the 60–70S RNA to act as a good template without any added primer suggests that it contains a primer.

Primer in 60–70S RNA

Evidence for the existence of a primer in the 60–70S RNA complex comes from studies of the initial product that is formed when this RNA is used as a template. By centrifugation of the denatured product to equilibrium in Cs_2SO_4, Verma et al. (1971) obtained evidence that virtually all of the newly formed DNA was covalently bonded to RNA, suggesting that an RNA acts as a primer for the initiation of DNA synthesis on the 60–70S RNA. Leis and Hurwitz (1972a) found similar results even after the product DNA was heated in formaldehyde. Only after treatment with alkali or ribonuclease does the product band in Cs_2SO_4 at the density of pure DNA. From the change in buoyant density of the denatured product as the reaction proceeds, Verma et al. suggested that the primer was a short molecule of RNA, about the size of transfer RNA, and this idea has received general support (Leis and Hurwitz, 1972a; Taylor et al., 1972; Canaani and Duesberg, 1972; Bishop et al., 1973b).

One prediction of this model is that a ribonucleotide-deoxyribonucleotide linkage should exist between the 3′ end of the RNA primer and the 5′ end of the DNA product. Using the principal of nearest-neighbour analysis (Josse et al., 1961), Flugel and Wells (1972),

Verma et al. (1972b) and Bishop et al. (1973b) have identified DNA-RNA linkages in the reverse transcriptase product.

Bishop et al. (1973b) have determined the size of the primer; it is a 4S RNA molecule. This is consistent with the suggestion that it might be transfer RNA (Verma et al., 1971).

Canaani and Duesberg (1972) have observed that heating 60–70S RNA removes a low molecular weight component needed for initiation of DNA synthesis. Removal of the primer from 70S RNA is a separate event from the conversion of 70S RNA to 35S RNA, because the transition from 60–70S to 35S occurs at a temperature lower than that required to remove the initiator. Furthermore, much 4S RNA is released from 70S RNA before the initiator is removed, suggesting that there is much more 4S RNA in the 60–70S RNA than is used as primer.

When 60–70S RNA is heated to a high enough temperature for removal of the natural primer, addition of a synthetic primer oligonucleotide will stimulate extensive DNA synthesis. Thus oligo(dT) or oligo(dG) are able to stimulate DNA synthesis on heated viral RNA (Duesberg et al., 1971; Verma et al., 1972a; Bishop et al., 1973b). Oligo(dT) probably initiates by virtue of its ability to bind to the poly(A) component present in the viral RNA subunits (Harel et al., 1966, 1967; Lai and Duesberg, 1972; Gillespie et al., 1972; Green and Cartas, 1972).

Complexity of the Product

When 60–70S RNA is used as template, the product is a mixture of double-stranded and single-stranded DNA (Taylor et al., 1972). By purifying the double-stranded product, melting it and studying its reannealing kinetics, it has been possible to determine its genetic complexity (the number of nucleotides in classes of unique sequences of nucleotides in the product; Britten and Kohne, 1968). Two classes of product have been identified: a large proportion of the product re-anneals rapidly and corresponds to only a small fraction of the genome (5%), while a small proportion (15%) corresponds to as much as 30% of the genome (Varmus et al., 1971; Gelb et al., 1971; Bishop et al., 1973b). Thus the product is mainly a transcript of a very limited portion of the viral RNA.

Although most of the product is copied from a limited part of the RNA, all portions of the RNA are represented by at least some RNA transcripts. Duesberg and Canaani (1970) found that over 90 percent of ^{32}P-labeled 60–70S Rous sarcoma virus RNA could be protected from ribonuclease digestion by hybridizing it with the DNA product made by virions of Rous sarcoma virus. A vast excess of DNA over labeled RNA was needed, however, in order to get complete protection of the RNA. These data also indicate that most of the DNA is synthesized from only a small part of the viral RNA and is of limited complexity. Recently Bishop et al. (1973b) found that if actinomycin D is present during the reaction, virions of RSV will synthesize a much more representative product. Actinomycin D restricts synthesis to a single-stranded DNA copy of the RNA template (McDonnell et al., 1970; Manly et al., 1971) but there is no obvious explanation of the increase in the average complexity of the product.

Size of the Product

The size of the product formed, either in the endogenous or the reconstructed system, is less than 150 nucleotides as measured by sedimentation rate in sucrose gradients (Taylor et al., 1972). By determining the ratio of internal to initiating nucleotides, values of 100 or less were found (Verma et al., 1972b). The DNA polymerase, however, is able to form larger polymers using other templates; with poly(A) as template the product can be 10,000 nucleotides in length (Baltimore and Smoler, 1971a), and with globin mRNA or α-crystallin mRNA as template it can be 500 nucleotides long (Verma et al., 1972a; Kacian et al., 1972; Ross et al., 1972). The limited size of the DNA made from a 60–70S viral RNA template therefore results from the structure of the template, not from an inherent property of the enzyme. The fact that the enzyme is able to extensively copy messenger RNAs provided biochemists with an invaluable tool for synthesising DNA complementary to these messengers. This DNA can be used as a probe in hybridization experiments designed to locate and quantitate genes and gene transcripts.

Overall Mechanism of Copying 60–70S RNA

The evidence available at the moment does not allow a complete description of the endogenous or the reconstructed reactions. However, from the evidence presented above and with the proviso that no relation of these reactions to the in vivo reaction can be assumed, a number of conclusions can be made:

1. All DNA synthesis by the reverse transcriptase proceeds from 5′ to 3′.
2. The 60–70S RNA complex contains short primer RNAs.
3. All DNA molecules made during at least the first part of the reaction are covalently linked to the RNA primer.
4. The product, especially when synthesized in the presence of actinomycin, contains copies of all regions of the 60–70S RNA template.
5. The product consists of very short DNA molecules (about 100 nucleotides).
6. The amount of DNA synthesized is, at best, a few percent of the input RNA.

It appears that primers must exist at a large number of sites on the 60–70S RNA for the reverse transcriptase to produce a product which is a complete copy of the 60–70S RNA but consists of short, covalently bonded RNA-DNA molecules. Whether any one RNA molecule has sufficient primers to allow its complete transcription into DNA is not known; because the product DNA represents only a few percent of the input template RNA, it could arise by partial transcription of a large number of RNA molecules. If the product of reverse transcription of viral RNA is molecules which have a 4S RNA primer (about 80 nucleotides) and a short DNA chain (less than 200 nucleotides), and if the primer is totally hydrogen-bonded to the template, about one-third of the template would be covered by primer and two-thirds by DNA. Excision of the primer and linkage of the DNA would then be necessary to generate a whole DNA transcript of the genomes (or of a 35S genome subunit). No evidence exists for such a process of excision and linkage, except for the report of a ligase activity in the virion (Mizutani et al., 1971; Mizutani and Temin, 1971). Ribonuclease H would not appear to play a part in such a process because it does not degrade double-stranded RNA.

This discussion has dealt only with the synthesis of a single-stranded DNA copy of the viral RNA, but much of the endogenous reverse transcription product is actually double-stranded DNA. These duplexes are about the same length as the single-stranded DNA product, and their synthesis has not been investigated in much detail. They presumably arise by displacement of RNA from an initial DNA-RNA hybrid. One could imagine that ribonuclease H would play a role here, degrading RNA after it has been copied into DNA. Ribonuclease H is an exonuclease which yields 3′ hydroxy end groups of RNA molecules hydrogen-bonded to DNA. It could therefore digest the RNA template strand of an RNA-DNA hybrid molecule produced by reverse transcription to yield a small RNA primer for the initiation of synthesis of a complementary DNA chain and hence a double-stranded DNA molecule.

All of these considerations apply only to the in vitro system. Because no one has yet identified and studied DNA made following infection by an RNA tumour virus, it is impossible to compare the in vitro results with the in vivo process they are supposed to elucidate. This lack of data on the in vivo situation makes further speculations futile.

Role of Reverse Transcriptase in Replication and Transformation

Reverse transcriptase was originally sought because indirect evidence suggested that viral RNA was copied into a specific DNA as an early step in viral growth and transformation, and there is evidence that reverse transcriptase is a necessary component of the virion, a fact which strengthens the belief that a DNA provirus is formed by the enzyme.

One line of evidence for an essential role for the reverse transcriptase is the lack of detectable DNA polymerase in a noninfectious mutant of Rous sarcoma virus called RSV α (Hanafusa and Hanafusa, 1968). This mutant cannot infect cells even in the presence of helper virus or inactivated Sendai virus, but the virions contain 70S RNA (Hanafusa and Hanafusa, 1971). RSV α particles lack reverse transcriptase activity (Hanafusa and Hanafusa, 1971; Hanafusa et al., 1972; Robinson and Robinson, 1971). Also, no immunologically cross-reacting material could be found in RSV α using an antiserum against AMV reverse transcriptase (Hanafusa et al., 1972). The lack of infectivity of RSV α therefore could result from the absence of reverse transcriptase, implying that the polymerase has an indispensible role in the infectious cycle of the virus.

Probably a better argument for an essential role for reverse transcriptase is the recent isolation of Rous sarcoma virus mutants which are temperature sensitive in a function required early in viral infection (Wyke, 1973; Linial and Mason, 1973). These mutants show normal adsorption to and penetration of cells at the nonpermissive temperature, but they are unable to replicate themselves or to transform the cells at this temperature. However, if cultures are maintained at the permissive temperature for as little as 18 hours after infection, a shift to nonpermissive conditions does not affect subsequent replication or transformation. Conversely, cultures infected at the nonpermissive temperature rapidly lose the ability to commence virus replication and cell transformation on a shift to permissive temperature. Thus these viruses are temperature sensitive in a function which is essential for the early establishment of infection, but is not required for the subsequent maintenance of replicative and transforming functions. The mutant virions are heat labile, as is their reverse transcriptase activity in both endogenous and exogenous reactions. The reverse transcriptase has recently been isolated from one of these mutants, and the purified enzyme itself appears to be temperature sensitive (Mason, personal communication). Revertants of these mutants are being sought, and if they possess normal reverse transcriptase, this will confirm that the enzyme is essential to the growth of Rous sarcoma virus.

Origin of Virion Reverse Transcriptase

The isolation of temperature-sensitive mutants of RSV that contain a temperature-sensitive reverse transcriptase implies that at least part of this enzyme is specified by the viral genome. Consonant with this idea are the findings that a reverse transcriptase activity, with all the characteristics of the virion enzyme, is present in extracts of cells infected by RNA tumour viruses but is absent from extracts of uninfected cells and cells infected by RSVα (Ross et al., 1971; Weissbach et al., 1972; McCaffrey and Baltimore, unpublished data). These data do not exclude the possibility that part of the reverse transcriptase molecule is specified by host cells; but in the light of Keller and Crouch's

(1972) report that chick cell RNase H is distinct from both the polypeptides which comprise the reverse transcriptase/RNase H complex of avian myeloblastosis virus, it seems probable that the complete enzyme is specified by the virus.

Numerous attempts have been made to identify reverse transcriptase-like enzymes in uninfected cells. Some groups have isolated particulate (membrane-bound?) preparations which show endogenous, ribonuclease-sensitive DNA synthesis (Coffin and Temin, 1971; Kang and Temin, 1972; Sarngadharan et al., 1972; Temin et al., 1973). From this particulate material a DNA polymerase can be extracted, but it does not have the characteristics of virion reverse transcriptase, although it is immunologically distinct from the major cellular DNA polymerase activities (Kang and Temin, 1972). A particulate preparation from human leukemic cells has, however, been claimed to have many of the hallmarks of a reverse transcriptase (Sarngadharan et al., 1972).

The second type of "reverse transcriptase" found in uninfected cells is an enzyme which copies poly(A)-oligo(dT) to give a poly(dT) product (Fridlender et al., 1972; Maia et al., 1971). This enzyme occurs in many different cell types from numerous species, including man, and it is apparently a normal constituent of cells (McCaffrey et al., 1973). It responds poorly to synthetic polymer templates other than poly(A)-oligo(dT) and this it transcribes only in the presence of manganese ions and the absence of magnesium ions. This enzyme seems to be unrelated to virion reverse transcriptase.

DIRECT EVIDENCE OF DNA PROVIRUSES

Hybridization Experiments

Reverse transcriptase has the properties expected of an enzyme that is to synthesize a DNA provirus using the 60–70S viral RNA as a template. But neither the existence of this enzyme in RNA tumour virus particles nor the properties of mutants lacking reverse transcriptase activity or containing a temperature-sensitive enzyme proves that the replication of these viruses involves the synthesis of a provirus. Proof of Temin's hypothesis demands the direct demonstration of the existence of a provirus in infected cells.

Temin (1964c) initially attempted to demonstrate the presence of new viral DNA in infected cells by measuring the extent of hybridization of labeled viral RNA to the DNA of infected and uninfected cells. The results of these experiments were ambiguous and considerable amounts of RNA hybridized to uninfected as well as infected cell DNA. Harel et al. (1966) and Wilson and Bauer (1967) also reported equally inconclusive data. The reason for the ambiguity of these experiments is now obvious: the preparations of viral RNA used as a probe contained cellular tRNAs (see Chapter 10) and the uninfected chick cells that were controls contained proviruses of endogenous avian leukosis virus (see Chapter 12). Nevertheless Rosenthal et al. (1971) and Baluda (1972) persevered with the sort of experiments that Temin had done, using partially purified 60–70S viral RNA as a probe for proviral DNA in infected cells, and they obtained suggestive results that can be summarized as follows: First, the amount of RSV RNA and RAV RNA which hybridizes to the DNA of both uninfected and infected chick cells is 3- to 10-fold more than the amounts which hybridize to *E. coli*, hamster and calf thymus DNAs; second, infection of chick fibroblasts by RSV increases the amount of viral RNA that hybridizes to cell DNA. Baluda, for example, claims that uninfected chick cells contain between two and five genome equivalents of viral DNA per cell, whereas chick fibroblasts infected with RSV and chick leukemic cells contain four to 13 genome equivalents of viral DNA. All that can be said of these data is that they are at least consistent with the provirus hypothesis and with other evidence that proves that chick cells inherit an endogenous provirus of an avian leukosis virus (see Chapter 12).

Recently evidence of the existence of DNA proviruses has been sought using a more precise hybridization technique and exploiting reverse transcriptase. Double-stranded RSV DNA was obtained from the endogenous reaction of reverse transcriptase in RSV virions. DNA and RNA from uninfected chick and mammalian cells and from chick and mammalian cells infected by RSV were then assayed for their ability to accelerate the reassociation of the labeled RSV DNA obtained by reverse transcription (Varmus et al., 1972, 1973). From changes in the rate of reassociation of RSV DNA in these experiments, it can be calculated that uninfected chick fibroblasts, regardless of whether or not they are expressing chick helper factor (see Chapter 12) and contain the group-specific antigens of the avian RNA tumour viruses, have in their genome about 10–15 genome equivalents of RSV DNA (Varmus et al., 1972). These results independently confirm that all chick cells inherit the provirus of an avian leukosis virus, and they indicate that the RNAs and DNAs of avian leukosis and sarcoma viruses have extensive homology. It should be emphasized, however, that the RSV DNA used in these experiments does not contain all the sequences found in the 60–70S viral RNA because reverse transcriptase preferentially transcribes parts of the viral RNA template. In addition it is possible that the 60–70S RNA extracted from populations of virions may contain contaminating cellular RNA sequences.

Varmus and his colleagues failed to detect an increase in the concentration per chick cell of RSV DNA following infection by RSV, but similar experiments with the DNAs of rat, mouse and hamster cells transformed by RSV do indicate that upon transformation these cells acquire 1–2 genome equivalents of RSV DNA (Varmus et al., 1973). Before infection such mammalian cells do not contain detectable RSV DNA. This result provides more convincing direct evidence in support of the provirus hypothesis; it also suggests why the group failed to detect any increase in the amount of RSV DNA after the infection of chick cells; the background concentration of endogenous leukosis virus DNA in chick cells probably masks any increase resulting from infection by RSV. The discrepancy between these data and those of Baluda remain to be explained.

Finally, since the cellular DNA which contains the RSV DNA sequences has a molecular weight considerably in excess of 24×10^6 daltons—which is the maximum estimated molecular weight of an RSV provirus, assuming all the 60–70S RNA is transcribed into a single DNA duplex—the provirus is probably covalently integrated into cellular DNA (Varmus et al., 1973).

Recovery of Infectious DNA

A quite independent and more satisfying proof that cells infected with RSV acquire a DNA provirus containing the genetic information necessary to specify infectious progeny RSV particles has been obtained by Hill and Hillova (1972). They extracted DNA from rat cells that had been transformed by the Prague strain of RSV and from hamster cells that had been transformed by a temperature-sensitive mutant strain of RSV. As a control they extracted DNA from untransformed rat and hamster cells, which do not contain endogenous proviruses of avian tumour viruses. They then exposed cultures of appropriate susceptible chick cells to these DNAs in the presence of DEAE dextran to facilitate uptake of the DNA. The cultures of chick cells were then incubated and subcultured. Within 8 to 20 days Hill and Hillova detected foci of transformed chick fibroblasts in the cultures that had been exposed to DNA from transformed cells but not in the cultures exposed to DNA from uninfected cells. The progeny RSV particles liberated by the transformed chick cells had the expected subgroup antigenicities and host ranges, and those stemming from the proviral DNA of the temperature-sensitive mutant were temperature sensitive.

These experiments, which have been confirmed (Svoboda et al., 1972), prove that DNA proviruses are made when mammalian cells are infected by RSV. Unfortunately, however, the efficiency of this infectious DNA assay is very low and it cannot readily be used to investigate the structure of the provirus and the life cycle of the virus.

SITE OF SYNTHESIS OF THE PROVIRUS

The virions of RNA tumour viruses contain the enzyme reverse transcriptase, which is able to synthesize provirus, and they contain the template, 60–70S viral RNA. Synthesis of a provirus does not, however, occur within extracellular virions. Two attempts have been made using autoradiographic techniques to detect the site of synthesis of the provirus in infected cells. Hatanaka et al. (1971) have reported that soon after mouse cells are infected with murine sarcoma or leukemia viruses, DNA synthesis above background levels can be detected by [^{3}H]thymidine autoradiography in the infected cell cytoplasm. By contrast Dales and Hanafusa (1972) failed to detect by this technique cytoplasmic DNA synthesis above background in chick fibroblasts infected with RSV or an avian leukosis virus. They did, however, observe evidence of the rapid transport of virions within vacuoles from the cell surface to the cell nucleus, and they believe that in this avian system proviruses are synthesized within the nucleus.

But in any event the hybridization experiments of Varmus et al. (1972, 1973) and the experiments of Hill and Hillova (1972) indicate that the provirus resides within the nucleus and probably is covalently integrated into host chromosomes.

Transcription of the Provirus

The establishment of infection by RNA tumour viruses depends upon the synthesis of a DNA provirus soon after infection, but, as experiments with actinomycin D indicate, the production of progeny virions depends upon the transcription of the proviral DNA. This transcription, which as far as we know is totally asymmetric, is controlled at least in part by cellular factors. For example, when stationary chick cells are infected with RSV, a provirus is synthesized, but the synthesis of viral RNA, viral proteins and progeny virions cannot be detected until the cells divide after infection (Humphries and Temin, 1972). And when mammalian cells are transformed by RSV, a provirus is synthesized and probably integrated into the cellular genome, the cells continue to divide and viral RNA may even be transcribed, but progeny virions are not produced. Mouse 3T3 cells transformed by B77 sarcoma virus and rat cells transformed by Schmidt-Ruppin RSV, for example, contain respectively about 10 and 30–50 copies of the viral RNA (Bishop et al., 1973a), but these cells do not liberate progeny virus. No viral RNA can be detected in 3T3 cells transformed by SR.-RSV and these cells, unlike those which contain viral RNA, do not yield progeny virus even after fusion with chick cells. Either RNA is not transcribed or it is transcribed but promptly degraded.

Unfortunately these various data are too sparse to indicate anything more than that transcription of proviruses is a controlled process, the control mechanism probably depending on an interplay between cell and virus-specific factors (see Chapter 12).

The 60–70S RNA in virions contains 3′ terminal tracts of poly(A), and it is claimed that AMV RNA acts as a messenger in an *E. col* cell-free system, programming the synthesis of group-specific antigens

(Siegert et al., 1972); the RNAs of Rauscher murine leukemia virus and mouse mammary tumour virus also apparently stimulate amino acid incorporation in *E. coli* cell-free systems (Gielkens et al., 1972). These data indicate, as expected, that the genomic RNA of RNA tumour viruses also serves as viral messenger RNA. Nothing, however, is known about the synthesis of viral proteins in vivo, although 35S RNAs have been found associated with the polysomes of infected cells. One can speculate that the viral RNA specifies perhaps six structural proteins, reverse transcriptase and perhaps as many as five functions required for transformation but not replication (see Chapter 13).

Maturation of Progeny Virions

From electron micrographs of thin sections of cells supporting the replication of C-type RNA tumour viruses and mouse mammary tumour virus, models of the morphological aspects of virion maturation have been devised (see Chapter 10). Apparently the amount of progeny virions produced by synchronized cultures of chick fibroblasts transformed by RSV varies as the cells pass through the various stages of the cell cycle (Leong et al., 1972), but the molecular biology of the maturation of progeny virions is quite obscure.

SUMMARY OF THE LIFE CYCLE OF RNA TUMOUR VIRUSES

Figure 11.1 is a schematic diagram of the life cycle of a typical RNA tumour virus. It is based on the information discussed in this Chapter, most of which comes from experiments with nondefective strains of RSV. We assume, however, that this information is of general relevance to all RNA tumour viruses and that, for example, during the replication of defective sarcoma viruses, helper leukemia viruses merely donate particular gene products during the maturation process or at earlier stages in the replication cycle.

(1) Infecting virions bind to specific cell receptors, penetrate the cell and (2) are transposed to the nucleus, being uncoated while in transit or upon arrival. (3) Virion reverse transcriptase uses 60–70S viral RNA and endogenous primers to synthesize a complementary

DNA strand and (4) a DNA strand with the same base sequence as the viral RNA is synthesized. (5) Duplex viral DNA is integrated by an unknown mechanism into host cell DNA. We do not know if this takes place before or after cell division. (6) After cell division transcription of the provirus is initiated. Whether initiation of transcription is directly dependent upon cell division or indirectly dependent is unknown. For example, it may be that integration is dependent upon cell division

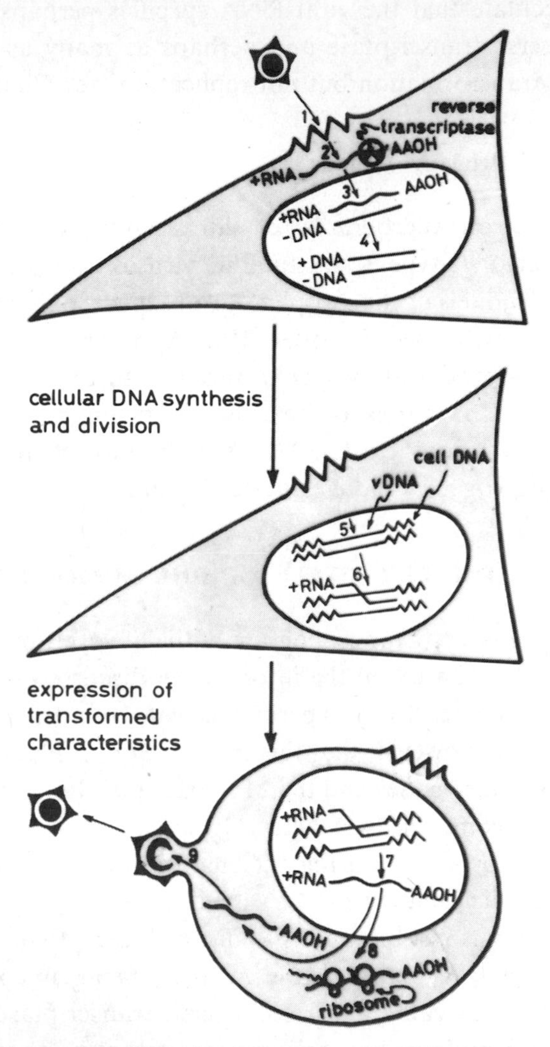

Figure 11.1. Diagram of the life cycle of an RNA tumour virus. See text for description.

and that initiation of transcription is dependent in turn upon integration. (7) Transcripts of the provirus migrate to the cytoplasm where they must first (8) act as messengers for specifying virion proteins. As the concentration of virion proteins increases, some viral RNA molecules are sequestered into maturing progeny particles, either directly (9) or after they have functioned as mRNA (9a). (10) Maturing virions move to the cell surface, from which they bud, acquiring an outer envelope. Further maturation steps probably occur in the extracellular particles. As depicted in Figure 11.1 the infected cell assumes a transformed phenotype after it has divided and transcription of the provirus has begun. As experiments with mutants prove, transformation is not essential to the replication of a sarcoma virus (see Chapter 13).

Readers aware of the sophisticated controls that regulate the replication of viruses as simple as the RNA bacteriophages will realize that this scheme at best presents only the vaguest of outlines of the life cycle of these viruses, about which we know as yet very little.

Literature Cited

AARONSON, S. A., W. P. PARKS, E. M. SCOLNICK and G. J. TODARO. 1971. Antibody to the RNA-dependent DNA polymerase of mammalian C-type RNA tumor viruses. Proc. Nat. Acad. Sci. *68:* 920.

ATKINSON, M. R., M. P. DEUTSCHER, A. KORNBERG, A. F. RUSSELL and J. G. MOFFATT. 1969. Enzymatic synthesis of DNA: Termination of chain growth by a 2′,3′-dideoxyribonucleotide. Biochemistry *8:* 4897.

BADER, J. P. 1964. The role of DNA in the synthesis of Rous sarcoma virus. Virology *22:* 462.

———. 1965. The requirement for DNA synthesis in the growth of Rous sarcoma and Rous-associated virions. Virology *26:* 253.

———. 1966. Metabolic requirements for infection by Rous sarcoma virus. I. Transient requirement for DNA synthesis. Virology *29:* 444.

———. 1970. Synthesis of the RNA-containing tumor viruses. I. The interval between synthesis and envelopment. Virology *40:* 494.

BALDUZZI, P. and H. R. MORGAN. 1970. Mechanisms of oncogenic transformation by Rous sarcoma virus. II. Intracellular inactivation of cell-transforming ability of Rous sarcoma virus by 5-bromodeoxyuridine and light. J. Virol. *5:* 470.

BALTIMORE, D. 1970. RNA-dependent DNA polymerase in virions of RNA tumor viruses. Nature *226:* 1209.

———. 1971. Expression of animal virus genomes. Bacteriol. Rev. *35:* 235.

BALTIMORE, D. and D. SMOLER. 1971a. Primer requirement and template specificity of the RNA tumor virus DNA polymerase. Proc. Nat. Acad. Sci. *68:* 1507.

———. 1971b. Template and primer requirements for the avian myeloblastosis DNA polymerase. In *Nucleic acid-protein interaction and nucleic acid synthesis in virus infection* (ed. D. W. Ribbons, J. F. Woessner and J. Schultz). Miami Winter Symposia, vol. 2, p. 328–332. North-Holland, Amsterdam.

———. 1972. Association of an endoribonuclease with the avian myeloblastosis virus DNA polymerase. J. Biol. Chem. *247:* 7282.

BALUDA, M. A. 1962. Properties of cells infected with avian myeloblastosis virus. Cold Spring Harbor Symp. Quant. Biol. *27:* 415.

———. 1972. Widespread presence in chickens of DNA complementary to the RNA genome of avian leukosis viruses. Proc. Nat. Acad. Sci. *69:* 576.

BALUDA, M. A. and D. P. NAYAK. 1969. Incorporation of precursors into ribonucleic acid, protein, glycoprotein and lipoprotein of avian myeloblastosis virus. J. Virol. *4:* 554.

BATHER, R. 1963. Influence of 5-fluorodeoxyuridine and actinomycin D on the production of Rous sarcoma virus in vitro. Proc. Amer. Assoc. Cancer Res. *4:* 4

BISHOP, J. M., N. JACKSON, N. QUINTRELL and H. E. VARMUS. 1973a. Transcription of RNA tumour virus genes in normal and infected cells. Lepetit Colloq. Biol. Med. *4.* North-Holland, Amsterdam. In press.

BISHOP, J. M., A. J. FARAS, A. C. GARAPIN, H. M. GOODMAN, W. E. LEVINSON, J. STAVNEZER, J. M. TAYLOR and H. E. VARMUS. 1973b. Characteristics of the transcription of RNA by the DNA polymerase of Rous sarcoma virus. In *DNA synthesis in vitro*, the Steenbock Symposium (ed. R. Inman and R. D. Wells). In press.

BOETTIGER, D. and H. M. TEMIN. 1970. Light inactivation of focus formation by chicken embryo fibroblasts infected with avian sarcoma virus in the presence of 5-bromodeoxyuridine. Nature *228:* 621.

BOLLUM, F. J. 1967. Enzymatic replication of polydeoxynucleotides. In *Genetic elements, properties and function* (ed. D. Shuger) p. 3–15. Academic Press, N.Y.

BRITTEN, R. J. and D. E. KOHNE. 1968. Repeated sequences in DNA. Science *161*: 520.

CANAANI, E. and P. DUESBERG. 1972. Role of subunits of 60 to 70S avian tumor virus RNA in its template activity for the viral DNA polymerase. J. Virol. *10:* 23.

COFFIN, J. M. and H. M. TEMIN. 1971. Ribonuclease-sensitive DNA polymerase activity in uninfected rat cells and rat cells infected with Rous sarcoma virus. J. Virol. *8:* 630.

DALES, S. and H. HANAFUSA. 1972. Penetration and intracellular release of the genomes of avian RNA tumor viruses. Virology *50:* 440.

DUESBERG, P. H. and E. CANAANI. 1970. Complementarity between Rous sarcoma virus (RSV) RNA and the *in vitro*-synthesized DNA of the virus-associated DNA polymerase. Virology *42:* 783.

DUESBERG, P. H. and P. K. VOGT. 1969. On the role of DNA synthesis in avian tumor virus infection. Proc. Nat. Acad. Sci. *64:* 939.

DUESBERG, P., K. V. D. HELM and E. CANAANI. 1971. Comparative properties of RNA and DNA templates for the DNA polymerase of Rous sarcoma virus. Proc. Nat. Acad. Sci. *68:* 2505.

ERIKSON, E. and R. L. ERIKSON. 1971. The association of 4S RNA with oncornavirus RNA's. J. Virol. *8:* 254.

FARAS, A. J., J. M. TAYLOR, J. P. MCDONNELL, W. E. LEVINSON and J. M. BISHOP. 1972. Purification and characterization of the DNA polymerase associated with Rous sarcoma virus. Biochemistry *11:* 2334.

FLUGEL, R. M. and R. D. WELLS. 1972. Nucleotides at the RNA-DNA covalent bonds formed in the endogenous reaction by the avian myeloblastosis virus DNA polymerase. Virology *48:* 394.

FLUGEL, R. M., J. E. LARSON, P. F. SCHENDEL, R. W. SWEET, T. R. TAMBLYN and R. D. WELLS. 1973. RNA-DNA bonds formed by DNA polymerases from bacteria and RNA tumor viruses. In *DNA synthesis in vitro*, the Steenbock Symposium (ed. R. Inman and R. D. WELLS). In press.

FRIDLENDER, B., M. FRY, A. BOLDEN and A. WEISSBACH. 1972. A new synthetic RNA-dependent DNA polymerase from human tissue culture cells. Proc. Nat. Acad. Sci. *69:* 452.

GARAPIN, A. C., J. P. MCDONNELL, W. E. LEVINSON, N. QUINTRELL, L. FANSHIER and J. M. BISHOP. 1970. DNA polymerase associated with Rous sarcoma virus and avian myeloblastosis virus: Prcperties of the enzyme and its product. J. Virol. *6:* 589.

GELB, L., S. A. AARONSON and M. MARTIN. 1971. Heterogeneity of murine leukemia virus in vitro DNA: Detection of viral DNA in mammalian cells. Science *172:* 1353.

GERWIN, B. J. and J. B. MILSTEIN. 1972. An oligonucleotide affinity column for RNA-dependent DNA polymerase from RNA tumor viruses. Proc. Nat. Acad. Sci. *69:* 2599.

GIELKENS, A. L. J., M. H. L. SALDEN, H. BLOEMENDAL and R. N. H. KONINGS. 1972. Translation of oncogenic viral RNA and eukaryotic messenger RNA in the *E. coli* cell-free system. FEBS Letters *28:* 348.

GILLESPIE, D., S. MARSHALL and R. C. GALLO. 1972. RNA of RNA tumor viruses contains poly (A). Nature New Biol. *236:* 227.

GOODMAN, N. C. and S. SPIEGELMAN. 1971. Distinguishing reverse transcriptase of an RNA tumor virus from other known DNA polymerases. Proc. Nat. Acad. Sci. *68:* 2203.

GOULIAN, M., Z. J. LUCAS and A. KORNBERG. 1968. Enzymatic synthesis of DNA. XXV. Purification and properties of DNA polymerase induced by infection with phage $T4^{+}$. J. Biol. Chem. *243:* 627.

GRAF, T. 1972. A plaque assay for avian RNA tumor viruses. Virology *50:* 567.

GREEN, M. and M. CARTAS. 1972. The genome of RNA tumor viruses contains polyadenylic acid sequences. Proc. Nat. Acad. Sci. *69:* 791.

HANAFUSA, H. and T. HANAFUSA. 1968. Further studies on RSV production from transformed cells. Virology *34:* 630.

———. 1971. Noninfectious RSV deficient in DNA polymerase. Virology *43:* 313.

HANAFUSA, H., D. BALTIMORE, D. SMOLER, K. F. WATSON, A. YANIV and S. SPIEGELMAN. 1972. Absence of polymerase protein in virions of alpha-type Rous sarcoma virus. Science *177:* 1188.

HAREL, L., J. HAREL, F. LACOUR and J. HUPPERT. 1966. Homologie entre genome du virus de la myeloblastose avaire (AMV) et genome cellulaire. Compt. Rend. Acad. Sci. (Paris) *263:* 616.

HAREL, L., J. HAREL and J. HUPPERT. 1967. Partial homology between RNA from Rauscher mouse leukaemia virus and cellular DNA. Biochem. Biophys. Res. Comm. *28:* 44.

HARWOOD, S. J. and R. D. WELLS. 1970. *Micrococcus luteus* DNA polymerase. Studies on the initiation of DNA synthesis in vitro. J. Biol. Chem. *245:* 5625.

HATANAKA, M., R. J. HUEBNER and R. V. GILDEN. 1970. DNA polymerase activity associated with RNA tumor viruses. Proc. Nat. Acad. Sci. *67:* 143.

HATANAKA, M., T. KAKEFUDA, R. V. GILDEN and E. A. O. CALLEN. 1971. Cytoplasmic DNA synthesis induced by RNA tumor viruses. Proc. Nat. Acad. Sci. *68:* 1844.

HAUSEN, P. and H. STEIN. 1970. Ribonuclease H: An enzyme degrading the RNA moiety of DNA-RNA hybrids. Eur. J. Biochem. *14:* 278.

HILL, M. and J. HILLOVA. 1972. Recovery of the temperature-sensitive mutant of Rous sarcoma virus from chicken cells exposed to DNA extracted from hamster cells transformed by the mutant. Virology *49:* 309.

HUMPHRIES, E. H. and H. TEMIN. 1972. Cell cycle-dependent activation of Rous sarcoma virus-infected stationary checken cells: Avian leukosis virus group-specific antigens and ribonucleic acid. J. Virol. *10:* 82.

HURWITZ, J. and J. P. LEIS. 1972. RNA-dependent DNA polymerase activity of RNA tumor viruses. I. Directing influence of DNA in the reaction. J. Virol. *9:* 116.

JOSSE, J., A. D. KAISER and A. KORNBERG. 1961. Enzymatic synthesis of DNA. VII. Frequencies of nearest neighbor base sequences in DNA. J. Biol. Chem. *236:* 864.

KACIAN, D. L., K. F. WATSON, A. BURNY and S. SPIEGELMAN. 1971. Purification of the DNA polymerase of avian myeloblastosis virus. Biochem. Biophys. Acta *246:* 365.

KACIAN, D. L., S. SPIEGELMAN, A. BANK, M. TERADA, S. METAFORA, L. DOW and P. A. MARKS. 1972. In vitro synthesis of DNA components of human genes for globins. Nature New Biol. *235:* 167.

KANG, C. and H. M. TEMIN. 1972. Endogenous RNA-directed DNA polymerase activity in uninfected chicken embryos. Proc. Nat. Acad. Sci. *69:* 1550.

KATES, J. R. and B. R. MCAUSLAN. 1967. Messenger RNA synthesis by a "coated" viral genome. Proc. Nat. Acad. Sci. *57:* 314.

KAWAI, S. and H. HANAFUSA. 1972. Plaque assay for some strains of avian leukosis virus. Virology *48:* 126.

KELLER, W. and R. CROUCH. 1972. Degradation of DNA·RNA hybrids by ribonuclease H and DNA polymerases of cellular and viral origin. Proc. Nat. Acad. Sci. *69:* 3360.

LAI, M. M. C. and P. H. DUESBERG. 1972. Adenylic acid-rich sequence in RNAs of Rous sarcoma virus and Rauscher mouse leukaemia virus. Nature *235:* 383.

LEIS, J. P. and J. HURWITZ. 1972a. RNA-dependent DNA polymerase activity of RNA tumor viruses. II. Directing influence of RNA in the reaction. J. Virol. *9:* 130.

———. 1972b. Isolation and characterization of a protein that stimulates DNA synthesis from avian myeloblastosis virus. Proc. Nat. Acad. Sci. *69:* 2331.

LEIS, J. P., I. BERKOWER and J. HURWITZ. 1973. Mechanism of action of ribonuclease H isolated from avian myeloblastosis virus and *E. coli.* Proc. Nat. Acad. Sci. *70:* 466.

LEONG, J. A., W. LEVINSON and J. M. BISHOP. 1972. Synchronization of Rous sarcoma virus production in chick embryo cells. Virology *47:* 133.

LINIAL, M. and W. S. MASON. 1973. Characteristics of two conditional early mutants of Rous sarcoma virus. Virology, submitted for publication.

LIVINGSTON, D. M., E. M. SCOLNICK, W. P. PARKS and G. J. TODARO. 1972. Affinity chromatography of RNA-dependent DNA polymerase from RNA tumor viruses on a solid phase immunoadsorbent. Proc. Nat. Acad. Sci. *69:* 393.

MAIA, J. C. C., F. ROUGEON and F. CHAPEVILLE. 1971. Chick embryo poly(rA:dT)-dependent DNA polymerase. FEBS Letters *18:* 130.

MANLY, K., D. F. SMOLER, E. BROMFELD and D. BALTIMORE. 1971. Forms of DNA produced by virions of the RNA tumor viruses. J. Virol. *7:* 106.

MCCAFFREY, R., D. SMOLER and D. BALTIMORE. 1973. Terminal deoxynucleotidyl transferase in a case of childhood acute lymphoblastic leukemia. Proc. Nat. Acad. Sci. *70:* 521.

MCDONNELL, J. P., A. GARAPIN, W. E. LEVINSON, N. QUINTRELL, L. FANSHIER and J. M. BISHOP. 1970. DNA polymerases of Rous sarcoma virus: Delineation of two reactions with actinomycin. Nature *228:* 433.

MIZUTANI, S. and H. M. TEMIN. 1971. Enzymes and nucleotides in virions of Rous sarcoma virus. J. Virol. *8:* 409.

MIZUTANI, S., D. BOETTIGER and H. M. TEMIN. 1970. A DNA-dependent DNA polymerase and a DNA endonuclease in virions of Rous sarcoma virus. Nature *228:* 424.

MIZUTANI, S., H. TEMIN, M. KODAMA and R. T. WELLS. 1971. DNA ligase and exonuclease activities in virions of Rous sarcoma virus. Nature *230:* 232.

MÖLLING, K., D. P. BOLOGNESI and H. BAUER. 1971a. Polypeptides of avian RNA tumor viruses. III. Purification and identification of a DNA synthesizing enzyme. Virology *45:* 298.

MÖLLING, K., D. P. BOLOGNESI, H. BAUER, N. BUSEN, H. W. PLASSMANN and P. HAUSEN. 1971b. Association of the viral reverse transcriptase with an enzyme degrading the RNA moiety of RNA-DNA hybrids. Nature New Biol. *234:* 240.

MURRAY, R. K. and H. M. TEMIN. 1970. Carcinogenesis by RNA sarcoma viruses. XIV. Infection of stationary cultures with murine sarcoma virus (Harvey). Int. J. Cancer *5:* 320.

NAKATA, Y. and J. P. BADER. 1968. Studies on fixation and development of cellular transformation by Rous sarcoma virus. Virology *36:* 401.

OROSZLAN, S., M. HATANAKA, R. V. GILDEN and R. J. HUEBNER. 1971. Specific inhibition of mammalian RNA C-type virus DNA polymerases by rat antisera. J. Virol. *8:* 816.

PARKS, W. P. and G. J. TODARO. 1972. Biological properties of syncytium-forming ("foamy") viruses. Virology *47:* 673.

PARKS, W. P., E. M. SCOLNICK, G. J. TODARO and S. A. AARONSON. 1971. RNA-dependent DNA polymerase in primate syncytium-forming ("foamy") viruses. Nature *229:* 258.

RIMAN, J. and G. S. BEAUDREAU. 1970. Viral DNA-dependent DNA polymerase and the properties of thymidine-labeled material in virions of an oncogenic RNA virus. Nature *228:* 427.

ROBINSON, W. A. and H. L. ROBINSON. 1971. DNA polymerase in defective Rous sarcoma virus. Virology *44:* 457.

ROKUTANDA, M., H. ROKUTANDA, M. GREEN, K. FUJINAGA, R. K. RAY and C. GURGO. 1970. Formation of viral RNA-DNA hybrid molecules by the DNA polymerase of sarcoma-leukaemia viruses. Nature *227:* 1026.

ROSENTHAL, P. N., H. L. ROBINSON, W. S. ROBINSON, T. HANAFUSA and H. HANAFUSA. 1971. DNA in uninfected and virus-infected cells complementary to avian tumor virus RNA. Proc. Nat. Acad. Sci. *68:* 2336.

ROSS, J., E. M. SCOLNICK, G. J. TODARO and S. A. AARONSON. 1971. Separation of murine cellular and murine leukaemia virus DNA polymerases. Nature New Biol. *231:* 163.

ROSS, J., H. AVIV, E. SCOLNICK and P. LEDER. 1972. In vitro synthesis of DNA complementary to purified rabbit globin mRNA. Proc. Nat. Acad. Sci. *69:* 264.

RUBIN, H. 1960. Analysis of the assay of Rous sarcoma cells in vitro by the infective center technique. Virology *10:* 29.

RUBIN, H. and C. COLBY. 1968. Early release of growth inhibition in cells infected with Rous sarcoma virus. Proc. Nat. Acad. Sci. *60:* 482.

SARNGADHARAN, M. G., P. S. SARIN, M. S. REITZ and R. C. GALLO. 1972. Reverse transcriptase activity of human acute leukaemic cells: Purification of the enzyme, response to AMV 70S RNA, and characterization of the DNA product. Nature New Biol. *240:* 67.

SIEGERT, W., R. N. H. KONINGS, H. BAUER and P. H. HOFSCHNEIDER. 1972. Translation of avian myeloblastosis virus RNA in a cell-free lysate of *E. Coli*. Proc. Nat. Acad. Sci. *69:* 888.

SMOLER, D., I. MOLINEUX and D. BALTIMORE. 1971. Direction of polymerization of the avian myeloblastosis virus DNA polymerase. J. Biol. Chem. *246:* 7697.

SPIEGELMAN, S., A. BURNY, M. R. DAS, J. KEYDAR, J. SCHLOM, M. TRAVNICEK and K. WATSON. 1970a. Characterization of the products of RNA-directed DNA polymerases in oncogenic RNA viruses. Nature *227:* 563.

———. 1970b. DNA-directed DNA polymerase activity in oncogenic RNA viruses. Nature *227:* 1029.

———. 1970c. Synthetic DNA-RNA hybrids and RNA-RNA duplexes as templates for the polymerases of the oncogenic RNA viruses. Nature *228:* 430.

SPIEGELMAN, S., K. WATSON and D. L. KACIAN. 1971. Synthesis of DNA complements of natural RNA's: A general approach. Proc. Nat. Acad. Sci. *68:* 2843.

STONE, L. B., K. K. TAKEMOTO and M. A. MARTIN. 1971. Physical and biochemical properties of progressive pneumonia virus. J. Virol. *8:* 573.

SVOBODA, J., I. HLOZANEK and O. MACH. 1972. Detection of chicken sarcoma virus after transfection of chicken fibroblasts with DNA isolated from mammalian cells transformed with Rous virus. Folia Biologica (Praha) *18:* 149.

TAKEMOTO, K. K. and L. B. STONE. 1971. Transformation of murine cells by two "slow viruses," visna and progressive pneumonia virus. J. Virol. *7:* 770.

TAYLOR, T. M., A. J. FARAS, H. E. VARMUS, W. E. LEVINSON and J. M. BISHOP. 1972. RNA-directed DNA synthesis by the purified DNA polymerase of Rous sarcoma virus: Characterization of the enzymatic product. Biochemistry *11:* 2343.

TEMIN, H. 1963. The effects of actinomycin D on growth of Rous sarcoma virus in vitro. Virology *20:* 577.

———. 1964a. The participation of DNA in Rous sarcoma virus production. Virology *23:* 486.

———. 1964b. Nature of the provirus of Rous sarcoma. Nat. Cancer Inst. Monogr. *17:* 557.

———. 1964c. Homology between RNA from Rous sarcoma virus and DNA from Rous sarcoma virus-infected cells. Proc. Nat. Acad. Sci. *52:* 323.

———. 1967. Studies on carcinogenesis by avian sarcoma viruses. J. Cell. Physiol. *69:* 53.

———. 1968. Carcinogenesis by avian sarcoma viruses. Cancer Res. *28:* 1835.

———. 1970. Malignant transformation of cells by viruses. Perspect. Biol. Med. *14:* 11.

TEMIN, H. and D. BALTIMORE. 1972. RNA-directed DNA synthesis and RNA tumor viruses. Advance Virus Res., vol. 17. Academic Press.

TEMIN, H. and S. MIZUTANI. 1970. RNA-dependent DNA polymerase in virions of Rous sarcoma virus. Nature *226:* 1211.

TEMIN, H. and H. RUBIN. 1958. Characteristics of an assay for Rous sarcoma virus and Rous sarcoma cells in tissue culture. Virology *6:* 669.

TEMIN, H., C. T. KANG and S. MIZUTANI. 1973. RNA-directed DNA synthesis in viruses and cells. Lepetit Colloq. Biol. Med. *4.* North-Holland, Amsterdam. In press.

TRONICK, S. R., E. M. SCOLNICK and W. P. PARKS. 1972. Reversible inactivation of the DNA polymerase of Rauscher leukemia virus. J. Virol. *10:* 885.

VARMUS, H. E., W. E. LEVINSON and J. M. BISHOP. 1971. Extent of transcription by the RNA-dependent DNA polymerase of Rous sarcoma virus. Nature New Biol. *233:* 19.

VARMUS, H. E., R. A. WEISS, R. R. FRIIS, W. E. LEVINSON and J. M. BISHOP. 1972. Detection of avian tumor virus-specific nucleotide sequences in avian cell DNA's. Proc. Nat. Acad. Sci. *69:* 20.

VARMUS, H. E., C. B. HANSEN, E. MEDEIROS, C. T. DENG and J. M. BISHOP. 1973. Detection and characterization of RNA tumor virus-specific nucleotide sequences in cell DNA. Lepetit Colloq. Biol. Med. *4.* North-Holland, Amsterdam. In press.

VERMA, I. M. and D. BALTIMORE. 1973. Purification of the RNA-directed DNA polymerase from avian myeloblastosis virus and its assay with polynucleotide templates. Methods in Enzymology. In press.

VERMA, I. M., N. L. MEUTH, E. BROMFELD, K. F. MANLY and D. BALTIMORE. 1971. A covalently linked RNA-DNA molecule as the initial product of the RNA tumor virus DNA polymerase. Nature New Biol. *233:* 131.

VERMA, I. M., G. F. TEMPLE, H. FAN and D. BALTIMORE. 1972a. In vitro synthesis of DNA complementary to rabbit reticulocyte 10S RNA. Nature New Biol. *235:* 163.

VERMA, I. M., N. L. MEUTH and D. BALTIMORE. 1972b. The covalent linkage between RNA primer and DNA product of the avian myeloblastosis virus DNA polymerase. J. Virol. *10:* 622.

VIGIER, P. and A. GOLDÉ. 1964. Effects of actinomycin D and of mitomycin C on the development of Rous sarcoma virus. Virology *23:* 511.

WEISS, R. A. 1970. Studies on the loss of growth inhibition in cells infected with Rous sarcoma virus. Int. J. Cancer *6:* 333.

———. 1971. Cell transformation induced by Rous sarcoma virus: Analysis of density dependence. Virology *46:* 209.

WEISSBACH, A., A. BOLDER, R. MULLER, H. HANAFUSA and T. HANAFUSA. 1972. DNA polymerase activities in normal and leukovirus-infected chicken embryo cells. J. Virol. *10:* 321.

WELLS, R. D., R. M. FLUGEL, J. E. LARSON, P. F. SCHENDEL and R. W. SWEET. 1972. Comparison of some reactions catalyzed by DNA polymerase from avian myeloblastosis virus, *E. coli* and *M. luteus.* Biochemistry *11:* 621.

WILSON, D. E. and H. BAUER. 1967. Hybridization of avian myeloblastosis virus RNA to DNA from chick embryo cells. Virology *33:* 754.

WYKE, J. 1973. The selective isolation of temperature-sensitive mutants of Rous sarcoma virus. Virology. In press.

YOSHIKURA, H. 1970. Dependence of murine sarcoma virus infection on the cell cycle. J. Gen. Virol. *6:* 183.

12

Genetic Transmission of RNA Tumour Viruses

The idea that cancer induces viruses is not new. Andrewes (1939) speculated on the possible activation of latent virus infections in cancerous tissues and it was postulated by Darlington (1948) that such viruses could arise from cellular genetic elements, which he named "proviruses." Gross (1958) and Lieberman and Kaplan (1959) observed that lymphoid tumours induced in mice by X rays contained murine leukemia virus (called RadLV), which induced similar tumours when inoculated into unirradiated mice. More recently the induction of RNA tumour viruses following treatment of animals (usually mice) with physical or chemical carcinogens has been reported many times (see Gross, 1970). Apparently mice and other animals are commonly infected with, or at least contain genetic elements capable of developing into, RNA tumour viruses, but these viruses normally remain latent in the host.

There are two chief ways in which carcinogenic agents probably cause these latent viruses to become active. First, it is well known that the immunological mechanisms of the host are particularly sensitive to carcinogens, and under immunosuppressed conditions any latent virus infections would have an increased probability of becoming virulent. Second, it was suggested (Lwoff, 1960; Latarjet and Duplan, 1962; Bentvelzen et al., 1968) that the viruses may exist in a proviral state that can be activated by carcinogens or X rays in the same way that temperate bacteriophages are activated in lysogenic bacteria. Recently chemical evidence for the presence of proviruses has been obtained, and murine and avian C-type RNA viruses have been induced from cultures of fibroblasts. These and other studies have led to the oncogene hypothesis (Bentvelzen et al., 1968; Huebner and Todaro, 1969; Todaro and Huebner, 1972) and the protovirus hypothesis (Temin, 1971, 1972), new speculations about the role in cancer of latent RNA tumour virus genomes. These hypotheses will be discussed in Chapter 13; in this chapter discussion is restricted to the evidence for the existence of inherited RNA tumour viruses and the factors that control their expression.

TRANSMISSION OF RNA TUMOUR VIRUSES

In Chapters 10 and 11 we discussed the replication of RNA tumour viruses as infectious agents that are released by budding from one cell, enter other cells through specific receptor sites and form DNA proviruses, replicative intermediates which probably integrate into the host genome. The infectious virus particles may be transmitted from one host animal to another by contact or proximity, but a frequent mode of transmission is from parent to offspring. Gross (1944, 1970) distinguished these two routes of transmission from one organism to another as horizontal and vertical transmission, respectively. We now recognize two modes of vertical transmission, congenital infection and genetic transmission (Weiss, 1973b), which are quite different at the molecular level and are consequently dependent on different biological controls.

Congenital infection occurs when infectious virus particles released by the mother infect the offspring. The virus may infect the ovum directly or be transmitted via the placenta or milk, but in each case the virus is transmitted as an infectious particle bearing an RNA genome. It must be released from a maternal cell and then enter a cell of the progeny, where a DNA provirus is presumably transcribed and integrated into the cell genome. Thus the same host-range restrictions will apply to congenital infection as to horizontal transmission.

During genetic transmission, on the other hand, the viral genome is vertically transmitted from one generation to the next as a DNA provirus, integrated as part of the genetic complement of the gametes. Bentvelzen (1972) has called this viral genome a "germinal provirus" to distinguish it from "somatic proviruses" acquired by infection. While congenital infection usually occurs through the mother, genetic transmission may be either maternal or paternal, and, because the viral genome is not transmitted as a virus particle, genetic transmission bypasses host-range restrictions.

INHERITANCE OF AVIAN TUMOUR VIRUSES

Avian RNA tumour viruses are commonly transmitted from one host to another by all three routes, horizontal infection, congenital infection and genetic transmission. Figure 12.1 depicts these modes of transmission and their consequences for the host. Horizontal infection with leukosis viruses appears to be common in commerical flocks in all countries where it has been studied, and the viruses involved almost always belong to subgroups A and B, subgroup A viruses being most prevalent (see Weiss, 1973b). When chicks or adult birds become infected with a virus, they usually develop neutralizing antibodies specific to the viral envelope antigens. As a result of this immune response, viremia is only transient and there is little chance of the bird developing leukemia (leukosis) (Rubin et al., 1962). If, on the other hand, the virus is transmitted congenitally (Burmester et al., 1955), the chick typically becomes viremic during embryonic development and remains immunologically tolerant to the viral antigens throughout life (Rubin et al., 1961, 1962). The bird grows normally but frequently develops leukemia when adult. Such birds are a major source of horizontal infection, as well as further congenital infection, because they are continually shedding virus (Zeigel et al., 1964).

In a morphological study, Di Stefano and Dougherty (1966) observed very large concentrations of virus particles in the female reproductive organs, including the ovarian follicles and the oviducts, so the gametes and zygote must be exposed to high concentrations of virus. Budding particles were not, however, observed on the germ cells

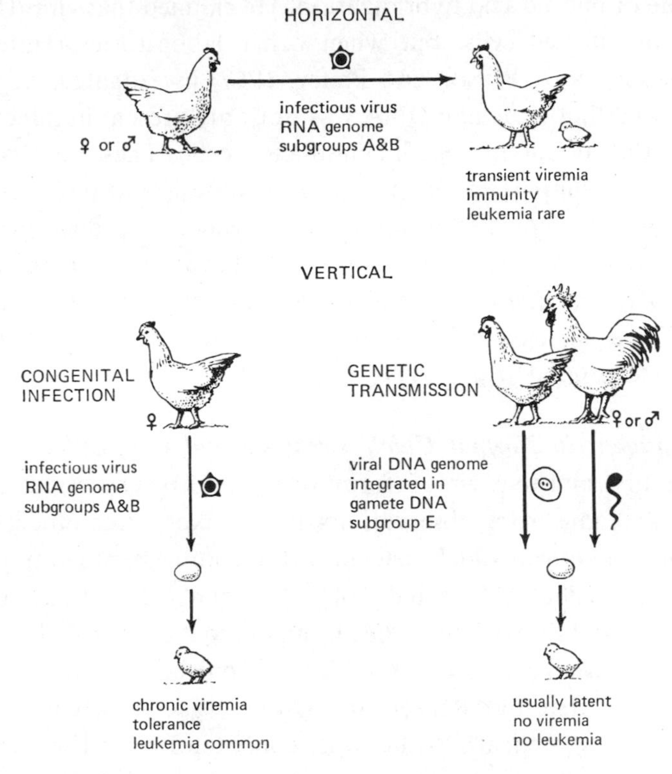

Figure 12.1. Modes of transmission of avian leukosis virus.

themselves. If the germ cells are already infected, the virus could be carried in the form of a provirus from one generation to the next, but this is unlikely because the congenital transmission of subgroup A and B viruses via the male has not been recorded. A subgroup E viral genome is, however, transmitted genetically (Weiss, 1972) and appears to be present in all chickens, including their feral counterpart, the red jungle fowl (Weiss and Biggs, 1972).

Evidence for Genetic Transmission of ALV

Virus-like DNA in Normal Cells

In an attempt to find evidence to support the provirus hypothesis, Temin (1964) looked for viral DNA in infected cells using the technique of nucleic acid hybridization. He claimed that viral DNA was present in infected cells, but when other laboratories (Harel et al., 1966; Bader, 1967; Wilson and Bauer, 1967) investigated the problem they showed that virus-like DNA was not only present in infected cells, but also that it was present in uninfected cells. These results did not seem to lend support to the provirus hypothesis and their real significance was not immediately appreciated. Since introduction of more sophisticated hybridization techniques we have, however, realized that several DNA genomes or partial genomes of avian leukosis virus exist in normal uninfected cells, apparently integrated into the host cell genome (Rosenthal et al., 1971; Baluda, 1972; Varmus et al., 1972a).

Viral Antigens in Normal Chick Embryos and Cell Cultures

The first hint that viral information might be expressed in normal avian cells came from observations that leukosis-free chick embryos contained an antigen which reacted in the complement fixation test for the group-specific (gs) antigen of the avian RNA tumour viruses (Dougherty and Di Stefano, 1966; Dougherty et al., 1967). In studies of gs antigens in leukosis-free, inbred lines of chickens, Payne and Chubb (1968) found that the Reaseheath C-line was consistently negative, whereas the Reaseheath I-line was consistently positive for the gs antigen although no mature virus could be detected. In cross-breeding experiments they found that the F_1 hybrids were all gs positive and backcrosses to the C-line were 50 percent gs positive. Moreover the gs positive trait was not sex-linked. These results showed that gs antigen was determined by a single, autosomal, Mendelian locus with a dominant allele for gs antigen expression. The cellular gs antigen was indistinguishable chemically and immunologically from the gs antigen of virus particles. It could be argued that the gs antigens of virus particles are coded by cellular genes, and that it is no more surprising to find gs antigens in uninfected cells than other cellular proteins. However, it seems likely that the viral gs antigens are coded by viral genes because they appear in gs negative cells infected with ALV and also in the progeny virus from such cells; furthermore mammalian cells transformed by RSV synthesize avian gs antigens (Sarma et al., 1964; Bauer and Janda, 1967). Thus the chick cell possesses a gene which determines a viral antigen, and Payne and Chubb (1968) suggested that this autosomal locus might represent the site of an integrated ALV genome, possibly a defective genome because complete virus was not released.

Chick Helper Factor

A second marker for viral functions became evident when it was found (Weiss, 1969a; T. Hanafusa et al., 1970a) that cells from certain normal chick embryos could complement the defectiveness of Bryan strain Rous sarcoma virus (BH.RSV) (see Chapters 10 and 13). Following the discovery of BH.RSV(0) (Vogt, 1967; Weiss, 1967), which was later assigned to subgroup E, Weiss (1969a) realized that this apparently nondefective virus was released from cells derived from gs^+ chick embryos. When the virus was recloned in cells of the gs^- C-line, infectious progeny particles were not released. T. Hanafusa et al. (1970a) also observed that cells from some chick embryos, but not others, permitted the release of infectious particles, and they later correlated the release of infectious progeny with the presence of gs antigen (H. Hanafusa et al., 1970). Both Weiss (1969a) and T. Hanafusa et al. (1970a) found that noninfectious particles were released from gs^- cells and that these

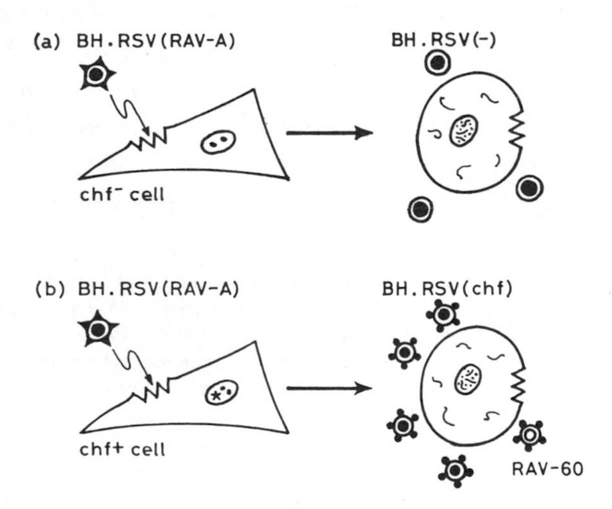

Figure 12.2. Complementation of defective BH.RSV by chick helper factor (chf) and the rescue of RAV-60.

particles were competent for infection and transformation of cells if introduced into cells with inactivated Sendai virus. Thus there appeared to be an agent in gs^+ cells which acted like a helper virus for BH.RSV by providing envelope antigens (Figure 12.2). The agent was named chick helper factor (*chf*) (H. Hanafusa et al., 1970), and because infectious BH.RSV(0) is, in fact, a pseudotype, it is now designated BH.RSV(*chf*) or BH.RSV(*f*), whereas the noninfectious form is designated BH.RSV(–).*

In studies of the Reaseheath C- and I-lines and their crosses, Weiss and Payne (1971) found that *chf* activity always segregated with the *gs* locus. However, recent studies with other chicken lines (T. Hanafusa et al., 1972; Weiss, 1973a; Vogt et al., 1973) revealed that this is not invariably the case; gs^- chf^+ embryos frequently occur, and more seldom, gs^+ chf^- embryos can be found. T. Hanafusa et al. (1972) also noted quantitative differences in the titre of *chf* in different embryos. The defect in BH.RSV appears to be an inability to synthesize envelope glycoproteins (Scheele and Hanafusa, 1971) rather than gs antigen. Therefore the presence of gs antigen and of *chf* represents two distinct markers for virus-like genes in normal chick embryo cells.

Rescue of Endogenous Viral Genes by Exogenous Viral Infection

BH.RSV(*chf*) readily infects cells which are susceptible to subgroup E viruses whether or not they express *chf*. When susceptible chf^- cells, for example Japanese quail cells, were infected with high titres of BH.RSV(*chf*), the *chf* determinant replicated, leading to the release of progeny BH.RSV(*chf*) pseudotypes (Weiss, 1969a; H.

* Some confusion has arisen in the literature over the nomenclature of BH.RSV(0). Hanafusa and Hanafusa (1968) first described infectious and noninfectious forms of BH.RSV(0), which they named β and α variants, respectively. Weiss (1969a, b) adopted these terms as phenotypic designations, but T. Hanafusa et al. (1970a, b) restricted the use of α to a genetic variant which lacks reverse transcriptase activity (Hanafusa and Hanafusa, 1971) and which is not complemented by *chf*. The Hanafusas introduced a new term, β', for the noninfectious variant of BH.RSV released from chf^- cells. Thus Weiss's α appears to be equivalent to the Hanafusas's β'. It has since been agreed to use α for the reverse transcriptase mutant only; β is now called BH.RSV(*chf*) or BH.RSV(*f*) and β' is called BH.RSV(-). Following the identification of RAV-0 and RAV-60, BH.RSV(*chf*) has also been called BH.RSV(RAV-0) and BH.RSV(RAV-60). These latter two terms should, perhaps, be restricted to pseudotypes produced by deliberate superinfection of chf^- cells releasing BH.RSV(-) with RAV-0 and RAV-60.

Hanafusa et al., 1970; Vogt and Friis, 1971; Weiss and Payne, 1971). Thus the infectious pseudotype could be propagated indefinitely on *chf*⁻ cells, provided a relatively high multiplicity of infection was maintained. Whenever the BH.RSV was propagated at low titre or cloned in *chf*⁻ cells, the foci reverted to releasing only noninfectious, BH.RSV(–) particles. The dose-response relation for propagation of BH.RSV (*chf*) suggested that a *chf*⁻ cell must be doubly infected with BH.RSV (*chf*) and an indcpendent particle bearing the *chf* determinant in order to obtain infectious progeny (Weiss, 1969a; Vogt and Friis, 1971). This indicated that propagation of BH.RSV in *chf*⁺ cells led not only to the release of *chf* pseudotypes, but also activated the release of a helper virus bearing the *chf* specificity, although the helper virus was not present in excess of the RSV (Weiss, 1969b). A helper virus of this kind has not, in fact, been isolated and characterized from BH.RSV(*chf*) stocks, but T. Hanafusa et al. (1970b) isolated a virus, which they named RAV-60, from stocks of RAV-2 harvested from *chf*⁺ cells and propagated in quail cells that are resistant to infection with RAV-2. RAV-60 has similar envelope properties to BH.RSV (*chf*) and interferes with BH.RSV(*chf*) infection (T. Hanafusa et al., 1970b). Since RAV-60 and BH.RSV(*chf*) differ from the known subgroups of exogenous viruses (Vogt, 1967; Weiss, 1969a, b; T. Hanafusa et al., 1970b), RAV-60 and *chf* have been allocated to a new subgroup, E.

Because complete helper virus of subgroup E was not released from chick cells unless the cells were infected with another avian tumour virus, T. Hanafusa et al. (1970b) suggested that the endogenous viral genome might itself be defective and that RAV-60 may be a genetic recombinant between RAV-2 and the *chf* genes. We now know that complete virus (RAV-0), which belongs to subgroup E, can be released from certain normal chick cells (Vogt and Friis, 1971), but this virus replicates less efficiently than RAV-60, which may indeed be a recombinant. With the discovery of genetic reassortment between nondefective strains of RSV and RAV (Vogt, 1971; Kawai and Hanafusa, 1971; see Chapter 13), recombinants between the *chf* markers of the inherited viral genome and SR.RSV and PR.RSV were sought and found (Weiss et al., 1973). Therefore it would not be surprising if genetic recombination also took place between RAV-2 and *chf*.

Spontaneous Release of RAV-O

Most chickens are resistant to infection with subgroup E viruses and as a consequence are unlikely to become viremic with genetically transmitted subgroup E virus, even if the inherited genome becomes activated to specify complete virus particles. However, Vogt and Friis (1971) made a special study of embryos of Line 7 chickens from the Regional Poultry Center, East Lansing, Michigan, because these embryos were unusual in being both gs positive and susceptible to infection. They found that about 20 percent of the embryos after 11 days' incubation were viremic with a leukosis virus possessing similar envelope properties to BH.RSV(*chf*). This virus was named RAV-0 because it resembled the helper agent (*chf*) of BH.RSV(0). The RAV-0 released by these embryos did not appear to be a result of congenital infection from a viremic mother because embryos producing RAV-0 occurred sporadically in the Line 7 flock without obvious maternal influence. Vogt and Friis (1971) concluded that RAV-0 represented a spontaneously activated form of the genetically transmitted genome that provides *chf* for BH.RSV. Thus normal cells derived from Line 7 chickens contained the information to produce complete, infectious virus, and it seemed likely that gs^+ cells from other strains of chicken also contained the same information.

RAV-0 alone did not replicate efficiently in Japanese quail fibroblasts, although in the presence of coinfecting BH.RSV it did replicate (Vogt and Friis, 1971). Recently Friis (1972) has shown that some other nondefective strains of avian tumour virus, such as the sarcoma virus B77, which replicate efficiently in chick cells, only abortively infect quail cells, and this may be the case with RAV-0. It is noteworthy, however, that RAV-60, which was selected by propagation in quail cells, replicates to high titres in these cells (T. Hanafusa et al., 1970b). The observation that RAV-0 more readily elicited neutralizing antibodies in ring-necked pheasants than in quails (Vogt and Friis, 1971) led Weiss et al. (1971) to test the replication of RAV-0 in pheasant embryo cells, which proved to be permissive.

The chicken Line 100 is closely related by backcross mating to Line 7; cultivated fibroblasts of all the embryos of Line 100 spontaneously release RAV-0, and the embryos, which are susceptible to spreading infection, become viremic and remain so throughout adult life (Crittenden et al. 1973a, b). Most other lines of chickens that are susceptible to subgroup E virus, however, do not become viremic and it appears that Line 7 and Line 100 chickens carry a dominant *V* gene which predisposes the endogenous viral genome to be fully expressed (see below). Whether RAV-0 has any leukemogenic potential remains to be seen but it is pertinent that Line 100 chickens do not have a high incidence of leukosis although 50 percent of the birds are viremic with RAV-0.

Induction of ALV Release by Mutagens and Carcinogens

The heritable nature of the viral genome specifying gs antigen and *chf* led Weiss and Payne (1971) to attempt to activate virus production by X-irradiation, which activates bacteriophages in lysogenic bacteria. Their results, based on a helper virus assay, were inconclusive, but with the discovery that RAV-0 replicates efficiently in pheasant cells, Weiss et al. (1971) used these cells to amplify the titre of virus that might be released from gs^+ nonproducer cells following X-ray treatment. They found that chemical mutagens and carcinogens, as well as ionizing radiations, could induce release of avian leukosis virus belonging to subgroup E: this virus was named induced leukosis virus (ILV).

In summary, these six lines of evidence prove that some chickens carry a complete viral genome, in DNA form, which is not usually expressed to the extent that mature virus particles form, but which may specify gs antigens in the host cell and complement the envelope defect of BH.RSV. The endogenous viral genes are rescued by infection with RSV or RAV and may be activated to release infectious virus by chemical and physical inducing agents.

The presence of tumour virus genes in species of birds other than the chicken has not been proven. Japanese quails do not synthesize gs antigen that reacts with antisera prepared against the chicken virus gs antigen, but embryos of ring-necked pheasants are frequently gs positive. Weiss (1972) was unable to demonstrate an active helper factor for BH.RSV in gs^+ pheasant cells, but he observed that after long-term culture of cells transformed by BH.RSV many clones initially releasing noninfectious virus began to release infectious RSV. These findings have recently been confirmed by Hanafusa and Hanafusa (1973), who found that the infectious RSV contains a new helper virus, RAV-61, which they have designated to subgroup F because it possesses a different host range from viruses of the other subgroups (A-E).

Presence of Viral Genes in* gs$^-$ *Chickens

The simple genetic control of gs antigen and *chf* (Payne and Chubb, 1968; Weiss and Payne, 1971) suggested that *gs*$^+$ chickens might inherit a viral genome not inherited by *gs*$^-$ chickens. Alternatively, the viral genome might be present in all chickens, but only be expressed in *gs*$^+$ chickens. One approach to this question was to repeat the nucleic acid hybridization studies using DNA extracted from embryos which were carefully characterized for presence of gs antigen and *chf*. Rosenthal et al. (1971) found no significant difference in the extent of hydridization of viral RNA to DNA prepared from *gs*$^+$ and *gs*$^-$ embryos. Likewise, Varmus et al. (1972a) studied the kinetics of annealing of viral DNA (prepared in vitro from viral RNA) in the presence of excess DNA extracted from *gs*$^+$ and *gs*$^-$ embryos and estimated that cells of both kinds of embryo contained 10–12 genome equivalents of viral DNA per diploid cell genome. These results, and those of Baluda (1972), indicated that there were equivalent numbers of genetically transmitted viral genomes in all chickens, irrespective of the alleles at the *gs* locus. This view was substantiated by the successful induction of ILV in *gs*$^-$ *chf*$^-$ cells (Weiss et al., 1971) and by the rescue of *chf* genes by prolonged cultivation of RSV or RAV in *chf*$^-$ cells (T. Hanafusa et al., 1972).

Regulation of Expression of Avian Endogenous Virus

gs *and* chf *Expression*

In the Reaseheath lines of chicken, *gs* and *chf* expression are controlled by a single dominant gene at the *gs* locus (Payne and Chubb, 1968; Weiss and Payne, 1971). It is not clear whether this is the case with other lines of chicks. The differential expression of gs antigen (Weiss, 1972; Vogt et al., 1973), of high and low titres of *chf* (T. Hanafusa et al., 1972) and of the stage in development when gs antigen appears (Payne, personal communication) suggests that other factors may influence the expression of these viral markers or at least modify the penetrance of the gs^+ allele. There are four or more separate components of the gs antigen of the ALV virion (see Table 10.2) and with the relatively crude complement fixation tests used so far to detect gs antigens, it has not been possible to distinguish which gs antigen components are expressed in normal chicken tissues.

Since complete viral genomes are present in both gs^+ and gs^- cells (Weiss et al., 1971; T. Hanafusa et al., 1972), the *gs* locus appears to be a regulatory gene rather than a structural gene for gs antigen. Gs antigen and *chf* expression are found in a majority of commerical flocks of chickens and also in red jungle fowl (Weiss and Biggs, 1972), so partial expression of a genetically transmitted virus does not seem to be of selective disadvantage to the host. Weiss et al. (1971) found that while induction of ILV in chick cells by carcinogens was a rare event, gs antigen and *chf* were induced in a large proportion of cells in gs^- chf^- cultures. It has been repeatedly suggested (Latarjet and Duplan, 1962; Bentvelzen et al., 1968; Todaro and Huebner, 1972) that endogenous viral genomes remain dormant, like temperature bacteriophages, because of the action of a specific repressor product, and that inducing agents interfere with the repressor. However, if a repressor normally inhibits viral gene expression, it is significant that the *gs* locus (and the *V* locus, see below) is dominant for viral gene expression in heterozygotes, suggesting that the putative repressor might act only on cis genes. Furthermore, the *gs* locus does not appear to influence the replication of nondefective exogenous viruses, although it has a profound effect on the expression of endogenous virus. It is probable, therefore, that the regulatory genes are structurally linked to the endogenous genome that they govern, and they may be viral genes rather than host genes (Vogt et al., 1973). In this case the numerous viral genomes estimated by nucleic acid hybridization to be present in "uninfected" chick cells could represent different genomes that may not all belong to subgroup E. The ILVs from cultures of C-line cells (Weiss et al., 1971) were not restricted to the host range of subgroup E and may therefore include other endogenous viruses of chick cells, although the wider host range of the induced viruses may have been acquired during their passage in pheasant cells.

DNA obtained by reverse transcription in vitro of the 70S RNA of RSV has been used as a probe for viral RNA in chick cells. Viral RNA sequences cannot be detected by such hybridization experiments in *gs*$^-$ *chf*$^-$ cells, but *gs*$^+$ *chf*$^+$ cells contain about 30–50 genome equivalents of viral RNA, whereas cells infected by avian RNA tumour viruses and supporting virus replication contain 3000–5000 genome equivalents of viral RNA (Leong et al., 1972; Hayward and Hanafusa, 1973; Bishop et al., 1973). Furthermore, studies of the genetic complexity of viral RNA in *gs*$^+$ *chf*$^+$ cells indicate that only about 50–80 percent of the viral genome is transcribed; this result is in accord with other evidence indicating that in *gs*$^+$ *chf*$^+$ cells only part of the endogenous viral genome is expressed. These various data suggest either that in *gs*$^-$ *chf*$^-$ cells viral RNA is synthesized but rapidly degraded, or that the expression of the endogenous viral genome is regulated at the transcriptional level.

One product that is not found in *gs*$^+$ *chf*$^+$ cells is viral reverse transcriptase (Weissbach et al., 1972; Kang and Temin, 1972). This finding is consistent with studies of complementation of mutants defective for reverse transcription. The α variant of BH.RSV (Hanafusa and Hanafusa, 1968) which lacks reverse transcriptase (Hanafusa and Hanafusa, 1971; H. Hanafusa et al., 1972) is not complemented by or phenotypically mixed in *chf*$^+$ cells, although coinfecting wild-type ALV will complement the enzyme defect of BH.RSVα by phenotypic mixing (see Chapter 13). Neither does BH.RSVα rescue RAV-60; this led T. Hanafusa et al. (1970b) to suggest that the endogenous genome might be unable to specify reverse transcriptase. Weiss (1973a) also found that propagation in *chf*$^+$ cells of a mutant of PR.RSV, which is temperature sensitive for reverse transcription (Linial and Mason, 1973), did not produce progeny phenotypically mixed for reverse transcriptase. The same type of *chf*$^+$ cell could, however, be induced to release ILV possessing functional reverse transcriptase (Weiss et al., 1971), indicating that the gene for the enzyme is present but not expressed in *chf*$^+$ cells. Weiss (1972) also showed that some other mutants of RSV that are complemented by exogenous helper viruses are not complemented in *chf*$^+$ cells.

As more mutants of viral replication become better defined, they will prove useful as probes for the differential expression of endogenous viral genes. However, it should be pointed out that anyone investigating the genetics of avian RNA tumour viruses would be well advised to screen the chick embryo cells they use for the presence of *chf*; fairly simple and rapid *chf* assays have been described by T. Hanafusa et al. (1972) and Weiss et al. (1973).

RAV-0 Expression

Vogt and Friis (1971) found spontaneous production of RAV-0 in embryo cells of Line 7 chickens. Crittenden et al. (1973a, b) have recently shown that there is high incidence of RAV-0 viremia in Line 7 and Line 100 chickens because they carry a dominant gene(s) predisposing the cells to spontaneous activation of the endogenous viral genomes. The Line 7 subline and Line 100 chickens studied by Crittenden et al. do not, in fact, carry the gs^+ allele, and the embryonic tissues and cultures lack gs antigen until the time that fully infectious virus appears. Viremia in Line 100 chickens occurs in all embryos and birds that are susceptible to infection by subgroup E viruses, but not in birds which are resistant (see below). However, substantial titres of RAV-0 develop when resistant cells from Line 100 and susceptible cells from another line (15I) are cocultivated, indicating that viremia is controlled in resistant birds by preventing spread of virus from cell to cell, rather than preventing activation. Even in populations of cells containing readily activable viral genomes it appears that only a small proportion of cells in fact release virus. This situation is analogous to that of endogenous viremia in mice (Rowe and Hartley, 1972; see below). Crosses between Line 100 and Line 6 (which is gs^+ and resistant to subgroup E viruses) again revealed RAV-0 only after cocultivation with susceptible cells (Crittenden et al., 1973a). This result indicates that the virus activation gene, *V*, allows full expression of virus, even in the presence of the gs^+ allele which allows only partial expression.

The location and number of *V* genes is not yet known. So far they have been found only in lines related to Line 7, although Weiss et al. (unpublished data) have observed spontaneous RAV-0 production in certain crosses between the I- and C-lines. It is possible that the "host" genes controlling RAV-0 production are endogenous viral genes (Vogt et al., 1973), and the *V* and *gs* loci may be allelic.

Susceptibility to Spreading Infection

Once a genetically transmitted virus has become activated, its propagation by horizontal transmission to further cells as an exogenous virus will depend on host-range controls, which were irrelevant to its genetic transmission. Susceptibility to subgroup E viruses is complex. Studies on the Reaseheath C- and I-lines and crosses between them (Payne et al., 1971) show that both parent lines and their F_1 hybrids are resistant to infection, but about 19 percent of the F_2 hybrids and 25 percent of the first generation backcross to the C-line are susceptible. Payne et al. (1971) proposed that two independently segregating loci control susceptibility. One locus, *tv-e*, determines the receptor for subgroup E viruses and has a dominant allele, *tv-*e^s (or e^s), for susceptibility, analogous to cellular genes controlling susceptibility to viruses of the other subgroups. The other locus has a dominant allele, I^e, and a recessive allele, i^e. The I^e allele has an epistatic inhibitory effect on the e^s allele; that is, in the presence of I^e even cells bearing an e^s allele are resistant to infection. It was proposed that the C-line has the genotype $e^re^ri^ei^e$ and is resistant because it lacks receptors, while the I-line has the genotype $e^se^sI^eI^e$ and is resistant because the presence or availability of the receptors coded by e^s is somehow prevented by the dominant inhibitor gene I^e.

The F_1 hybrids with genotype $e^se^rI^ei^e$ are also resistant for the same reason, but a proportion of the F_2 hybrids and progeny of the backcrosses to the C-line have the genotypes $e^se^si^ei^e$ or $e^se^ri^ei^e$ and are therefore susceptible to infection. In other words, susceptibility to subgroup E viruses depends on both the presence of a receptor and the absence of an inhibitor (see Table 12.1).

The nature of the inhibitor of susceptibility is not yet firmly established, but Payne et al. (1971), noting the I-line carried both gs^+ and I^e alleles, suggested that the resistance to exogenous infection by subgroup E viruses might be connected with the partial expression of the endogenous viral genome. This idea is supported by Crittenden et al., 1973b and Weiss et al. (unpublished data), who observed in a limited number of crosses that the gs^+ gene and I^e gene are genetically linked and that in several strains of chicken segregating chf^+ and chf^- embryos, only

Table 12.1. Chick Cell Genotypes Influencing Control of Susceptibility to RAV-0

Genotype	Phenotype		
$e^re^ri^ei^e$	resistant	chf^-	
$e^re^rI^e$–	resistant	chf^+	
e^s–i^ei^e	susceptible	chf^-	
e^s–I^e	resistant	chf^+	
$b^rb^ri^ei^e$	resistant	chf^-	R_1ag^-
$b^rb^rI^e$–	resistant	chf^+	R_1ag^-
b^{s1}–i^ei^e	susceptible	chf^-	R_1ag^+
b^{s1}–I^e–	susceptible	chf^+	R_1ag^+
$b^{s2}b^ri^ei^e$	susceptible	chf^-	R_1ag^-
$b^{s2}b^rI^ei^e$	resistant	chf^+	R_1ag^-

cultures of *chf*⁻ embryo cells are susceptible to infection with RSV (RAV-0). Weiss et al. have proposed that the inhibitory effect of I^e may be caused by blocking the receptors with subgroup E envelope antigens synthesized by the endogenous viral genome, analogous to viral interference between two exogenous viruses of the same subgroup (see Figure 12.3). This seems plausible because subgroup E envelope antigens were shown, by their capacity to adsorb neutralizing antiserum specific to RAV-0, to be present on the surface of gs^+ chf^+ cells but not on gs^- chf^- cells. Thus the spread of activated RAV-0 may be prevented by the inhibitory influence of the partially expressed genome (Figure 12.3).

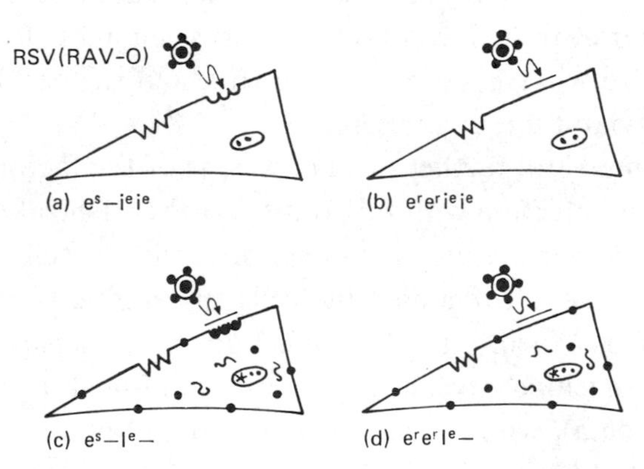

Figure 12.3. Possible mechanisms controlling susceptibility and resistance of chick cells to RAV-0 (subgroup E virus).

Crittenden et al. (1973a, b) and Weiss et al. (unpublished data) have shown that all chickens (but not quails or pheasants) that are resistant to infection with subgroup B viruses are also resistant to RSV (RAV-0). However, the converse is not always true, for there are many strains, for example, the C-line, which are susceptible to subgroup B viruses but resistant to subgroup E viruses. Also, (see Chapter 10) preinfection with ALV-B causes interference to infection by RSV-(RAV-0) (Vogt, 1967; Weiss, 1969b), whereas infection with ALV-E does not cause interference to RSV-B infection (T. Hanafusa et al., 1970b; Weiss et al., 1971). It is possible that susceptibility alleles at two loci are required to form surface receptors for subgroup E viruses and render the cells susceptible. Alternatively, the *tv-e* locus may be identical with the *tv-b* locus, in which case both the e^s and e^r alleles specify receptors for subgroup B viruses. In addition, Crittenden et al. (1973a, b) have distinguished two susceptibility alleles for both subgroup B and E viruses at the *tv-b* locus, where b^{s1} is associated with the R_1 erythrocyte antigen. The I^e gene inhibits susceptibility to subgroup E viruses of cells carrying the b^{s2} allele without influencing susceptibility to subgroup B viruses. However, the influence of I^e on the susceptibility of the b^{s1} genotype to subgroup E virus is negligible, so that even in the presence of the I^e allele, the cells are susceptible to infection with RSV (RAV-0) (Crittenden et al., 1973a, b). In other words $I^e b^{s1}$ cells are gs^+ but susceptible to subgroup E. Further studies are required to clarify the functions and interactions of the various genes that affect the expression of genetically transmitted avian tumour virus genomes. Tables 12.1 and 12.2 summarize what is currently known about the genetic control of endogenous avian tumour viruses.

To confound still further our knowledge of the factors controlling susceptibility to infection with subgroup E viruses, Ishizaki and Shimizu (1970) observed that preinfection of certain resistant cells with leukosis viruses of subgroup A rendered the cells susceptible to infection with RSV (*chf*). This "facilitation" of infection to virus particles carrying subgroup E envelope glycoproteins was confirmed by Weiss et al. (unpublished data), who found that both subgroup A and subgroup C viruses had this effect (viruses of subgroups B and D interfere with viruses of subgroup E), but that facilitation of infection was restricted

Table 12.2. GENES AFFECTING ENDOGENOUS AVIAN LEUKOSIS VIRUS EXPRESSION

Location	Designation	Allele	Expression	Phenotype	Reference
viral?	*gs*	gs^+	dominant	gs antigen expression, chick helper factor	Payne and Chubb, 1968; Weiss and Payne, 1971
		gs^-	recessive	no *gs* or *chf* expression	
viral?	*V*	*V*	dominant	RAV-0 release	
		v	recessive	no RAV-0 release, no *gs* or *chf* expression	Crittenden et al., 1973a
host	*tv–e*	e^s	dominant	susceptible to E	Payne et al., 1971
		e^r	recessive	resistant to E	
host	*tv–b*	b^{s1}	dominant	R_1 erythrocyte antigen susceptible to B and E	
		$b^{s2}(e^s?)$	dominant to b^r recessive to b^{s1}	susceptible to B and E no R_1 antigen	Crittenden et al., 1973b
		$b^{s3}?(e^r?)$	dominant to b^r	susceptible to B but resistant to E	
		b^r	recessive	resistant to B and E	
viral?	I^e	$I^e(gs^+?)$	dominant	epistatic inhibitor of e^s and b^{s2}	Payne et al., 1971; Crittenden et al., 1973b
		$i^e(gs^-?)$	recessive	no epistatic inhibition	

to cells which were resistant as a result of the activity of the I^e gene. Cells which lacked the appropriate receptor gene were not facilitated by heterologous leukosis viruses, and the susceptibility of *tv-b*^{s1} genotypes was not further enhanced. The mechanism of facilitation is not known, but if the effect of I^e is one of interference resulting from blocked receptor sites, we may surmise that the heterologous leukosis viruses somehow clear the receptor sites, perhaps by reducing the pool of excess endogenous subgroup E envelope antigens by phenotypic mixing with the progeny ALV particles. At the same time, the exogenous ALV will activate the endogenous genomes to produce RAV-60, and phenotypically mixed particles will result and probably genetic recombinants as well. Thus infection of *chf*$^+$ cells with ALV may have manifold consequences for the activation and spread of the genetically transmitted genome.

INHERITANCE OF MURINE LEUKEMIA VIRUSES

When Gross (1951) found that the leukemia of the high leukemia strain AKR mice could be transmitted by cell-free filtrates to C3H mice, he showed not only that murine leukemia was a viral disease, but that the causative virus is transmitted vertically in its natural strain—hence the existence of high and low incidence strains of mice. Experimental studies show that MLV can be transmitted horizontally (see Gross, 1970; Mirand and Mirand, 1969; McKissick et al., 1970), and vertical transmission can occur by congenital infection, usually through the milk, and through the semen or uterus (Buffet et al., 1969). The usual mode of transmission, however, is genetic, and in AKR mice the virus becomes evident around birth or soon after (Rowe and Pincus, 1972), whereas in low leukemia strains, virus may not appear at all, or only late in life. Congenital infection or early activation of genetically transmitted virus usually renders chickens immunologically tolerant; by contrast the later appearance of MLV in mice typically induces a host immune response. However, the production of viral antigens is excessive and the host antibodies are usually unable to check the proliferation of virus in the host tissues. In some strains of mice, for example, the New Zealand black and white strains, the formation of large amounts of antigen-antibody complexes (Oldstone and Dixon, 1971) leads to glomerulonephritis, which can be fatal (Mellors et al., 1971; Dixon et al., 1971).

Evidence for the existence of endogenous MLV in normal cells and tissues of the mouse has been gathered from experiments similar, though less extensive, to those done with chicken cells. The presence of virus-like sequences in the DNA of uninfected murine cells was reported by Harel et al. (1967). Viral gs antigens were found in embryonic and adult tissues (Huebner et al., 1970; Taylor et al., 1971) and another MLV associated antigen (Gross$_{IX}$) has been detected in certain normal mouse strains (Stockert et al., 1971). A "mouse helper factor" that allows nonproducing cells to complement MSV has not, however, been described. Neither are there any precise data on the rescue of endogenous virus by exogenous infection. The spontaneous appearance in vivo of MLV in strains with a high leukemia incidence has been described, but the first convincing evidence that the MLV was released as a result of the activation of inherited viral genomes came from experiments which showed the spontaneous or induced appearance of MLV in previously nonproducing cultures of fibroblasts (Aaronson et al., 1969; Rowe et al., 1971). Lowy et al. (1971) found that the halogenated analogues of thymidine, 5-bromodeoxyuridine (BrdU) and 5-iododeoxyuridine (IdU), were the most efficient inducing agents in AKR cells. Some strains of mice inherit N-tropic virus (for example AKR) and others B-tropic (for example B10, BR), and some inherit both kinds of virus (for example Balb/c) (Pincus et al., 1971a; Rowe and Hartley, 1972).

Genetic Regulation of Murine Endogenous Virus Expression

Recent research on the spontaneous appearance of MLV in mice and in murine cell cultures is beginning to clarify what had appeared to be an impenetrably complex genetic control of murine leukemia (see reviews by Lilly and Pincus, 1972; Boyse et al., 1972).

Genes on Linkage Group IX

Three loci, *H-2*, *Tla* and G_{IX}, in linkage group IX all determine cell surface antigens and influence endogenous viral leukemogenesis. The

H-2 locus affects the dose of MLV needed to induce leukemia and the latent period of leukemogenesis in experimentally inoculated hosts (Boyse et al., 1972; Lilly and Pincus, 1972). The K pole of this complex locus is the determining factor, and $H\text{-}2^k$ mice are more susceptible than $H\text{-}2^b$ mice. The locus also influences leukemogenesis by genetically transmitted virus because leukemia is considerably delayed in the AKR/$H\text{-}2^b$ strain of mice, which is congenic to the pure-bred AKR strain, which has the $H\text{-}2^k$ allele. How the *H-2* locus influences viral leukemia is not known; it may affect the development of lymphoid neoplasms rather than act on the virus directly, just as the *Fv-2*, *Sl* and *W* loci affect the development of erythroid leukemia (see Lilly and Pincus, 1972; and Chapter 10).

The TL antigens, determined by the *Tla* locus (Boyse and Old, 1969), appear only on leukemic cells and on normal thymocytes of certain mouse strains. The structural genes for TL antigens 1 and 2 appear to be present in both TL^- and TL^+ mice because they always appear on leukemic cells. The locus on linkage group IX appears to be an "expression" (regulatory) gene, although the structural genes for the antigens may be closely linked. It has been suggested (Boyse and Old, 1969) that the *Tla* locus may represent a viral genome because of its association with viral leukemia, but this suggestion is admittedly tenuous.

The G_{IX} (Gross IX) locus (Stockert et al., 1971) determines a cell surface antigen which is confined to thymocytes, lymphocytes and lymphoid leukemia cells. Like TL antigens, G_{IX} antigen may be present on leukemic cells of G_{IX}^- mice, but murine leukemic cells are not invariably G_{IX}^+. Segregation studies between G_{IX}^+ (strain 129) and G_{IX}^- (strain C57BL/6) mice (Boyse et al., 1972) indicate that two unlinked genes are required for G_{IX} antigen expression on normal thymocytes. One is fully dominant and is unmapped; the other is semidominant and is in linkage group IX. Inoculation of mouse cells producing Gross-type MLV into W/Fu rats converts the rat lymphoid cells from G_{IX}^- to G_{IX}^+ and the rats produce cytotoxic antibody to G_{IX} antigen. This phenomenon occurs even when MLV is serially passaged in W/Fu rats, suggesting that the MLV itself can code for or induce G_{IX} antigen. It is unlikely, however, that G_{IX} antigen is a structural component of the virion because antiserum to G_{IX} antigen does not cross react with virion antigens; neither do antisera against gs antigens cross react with normal thymocytes of strain 129 mice. Nevertheless, it was suggested (Stockert et al., 1971; Boyse et al., 1972) that G_{IX} may represent a viral gene.

Genetic Control of gs Antigen Expression

The control of gs antigen in the absence of infectious virus is not so clear in mice as in chickens, perhaps because gs^+ cells and virus-producing cells have not always been carefully distinguished. Taylor et al. (1971) described two independent, dominant loci for gs antigen expression in the spleens of AKR mice. Limited tests showed that complete virus was also present in the spleens, and it seems possible that the loci are the same as the V_1 and V_2 loci defined by Rowe (1972) for complete virus production (see below). However, in multiple inbred lines derived from F_2 crosses of AKR and C57BL/6, Taylor et al. (1971) found only one dominant locus for gs antigen expression and postulated another for production of complete, infectious virus. Recently Taylor et al. (1973) have identified a new locus, *Mlv-1*, in the 58N mouse strain that is recessive for gs antigen expression, but they have not attempted to relate this locus to the dominant genes they detected in earlier studies. The new locus is not closely linked with the *Fv-1* locus and infectious virus was not present in the gs^+ spleens they examined. Finally, Hilgers (1972) found that gs antigen in the mouse spleen was controlled by a single semidominant gene.

Genetic Control of MLV Activation

The knowledge that Gross-type MLV in AKR strain mice was transmitted not by congenital infection but as a latent genome (Rowe et al., 1971), which does not appear as active virus until birth or later (Rowe and Pincus, 1972), led to a study of genetic factors which control the activation of MLV in mouse strains with high and low incidences of leukemia. The activable genomes of AKR mice are N-tropic and Rowe (1972) made a detailed examination of the appearance of virus in the tissues of mice resulting from crosses with low incidence strains. He studied crosses which yielded mice that were homozygous for the $Fv\text{-}1^n$ allele, so that in all cases the mice would be permissive to spreading infection of activated N-tropic virus. His observations on the F_1,

F_2, and first and second backcross generation hybrids indicated that AKR mice carry two unlinked, autosomal loci, named V_1 and V_2, either of which were sufficient to activate MLV and both of which were dominant for MLV expression. V_1 was assigned to linkage group I, and V_2 remains to be assigned but it is not associated with the *Fv-1* locus (on linkage group VIII, Rowe et al., 1973) or to linkage group IX. In a further study (Rowe and Hartley, 1972) AKR mice were crossed with strains carrying $Fv\text{-}1^b$ alleles and having a low intrinsic incidence of MLV early in life or producing B-tropic virus. The occurrence of both N- and B-tropic genomes in some of these mice allowed Rowe and Hartley to determine whether the dominant V_1 and V_2 alleles contributed by the AKR parent activated genetically transmitted viruses of both host-range types. They found that only N-tropic virus was activated and therefore they suggest that the V_1 and V_2 loci may be located in the viral genome itself.

The dominant *V* loci of AKR mice which determine spontaneous activation of MLV in mice or mouse embryo cell cultures also predispose cells to virus induction by halogenated pyrimidines (Rowe, 1972). Studies with mice of the high leukemia strain C58 and with hybrid mice obtained from crosses between C58 mice and mice of low leukemia strains (Stephenson and Aaronson, 1972a, b) indicate that dominant genes similar to V_1 and V_2 determine virus inducibility. Balb/c mice appear to carry a single gene for virus induction, which Stephenson and Aaronson named the *ind* locus.

Susceptibility to Spreading Infection by Endogenous MLV

The horizontal spread of activated endogenous MLV depends on the host-range genes that control exogenous viruses. The chief host-range control is exerted by the *Fv-1* locus (Pincus et al., 1971a, b; see Chapter 10), which determines whether cells are permissive to the replication of N-tropic or B-tropic viruses. After activation the virus will spread through the tissues of the host or the cell culture only if the cells are susceptible to infection with the activated virus (Rowe and Hartley, 1972; Stephenson and Aaronson, 1972b). Mice that develop early viremia and leukemia must carry at least one activation gene and must be homozygous for the appropriate allele of the *Fv-1* locus. For example, the AKR strain is $Fv\text{-}1^{nn}$ and carries the V_1 and V_2 alleles for high incidence activation of endogenous, N-tropic viruses. Viremia will not occur or it will be significantly delayed if either (1) the mice are homozygous for the recessive alleles at the activation loci or (2) the spread of activated virus is restricted by the *Fv-1* locus. In outbred populations that are heterozygous at the *Fv-1* locus, spreading infection of either N- or B-tropic MLV will be restricted, because $Fv\text{-}1^{nb}$ cells are relatively resistant to both types of virus.

Stephenson and Aaronson (1972a) described a highly inducible virus that is genetically transmitted in C58 mice. This virus, once activated, replicates readily in $Fv\text{-}1^{nb}$ cells and therefore has some attributes of NB-tropic viruses.

The control of MLV viremia in mice by *V* or *ind* genes and by host-range controls resembles the control of RAV-0 viremia in chickens (see above), although the mechanism of host-range control itself is different. However, nothing equivalent to the I^e gene of chickens has been described for MLV. Table 12.3 summarizes the genetic factors which influence endogenous MLV expression.

Analysis of MLV Induction

Rowe et al. (1971) showed that cultivated fibroblasts derived from AKR embryos could be induced to release N-tropic MLV by a variety of agents, including ionizing radiations. The virus was detected by an indirect XC plaque assay in which NIH-Swiss mouse cells were added to the AKR cells to amplify the induced MLV, and XC cells were then added to the culture to reveal the foci of MLV proliferation. Rowe et al. established two cloned "permanent" cell lines from the AKR embryonic cells that did not spontaneously release MLV but could be induced to release virus even more readily than the embryonic fibroblasts in culture. Using these cell lines, Lowy et al. (1971) found that BrdU and IdU were much more efficient inducing agents than any other drugs or treatments. Whereas X rays induced about one in 10^4–10^5 cells to release MLV, BrdU induced one in 10^2 cells. Klement

Table 12.3. GENES AFFECTING ENDOGENOUS MURINE LEUKEMIA VIRUS EXPRESSION

Linkage group	Location	Designation	Allele	Expression	Phenotype	Mouse strain	Reference
I	viral?	V_1	V_1	dominant	N-tropic MLV release	AKR	Rowe, 1972
			v_1	recessive	no MLV release		
unknown not I, VIII, IX	viral?	V_2	V_2	dominant	N-tropic MLV release	AKR	Rowe and Hartley, 1972
			v_2	recessive	no MLV release		
unknown, not VIII	viral?	*Ind*	Ind^+	dominant	N-tropic MLV release	Balb/c	Stephenson and Aaronson, 1972b
			Ind^-	recessive	no MLV release	NIH-Swiss	
unknown	?	none	+	dominant	gs antigen	AKR	Taylor et al., 1971
			−	recessive	no gs antigen	C57L	
unknown	?	none	+	dominant	MLV release if gs^+ gene is present	AKR	Taylor et al., 1971
			−	recessive	no MLV release		
unknown	?	*Mlv*–1	$Mlv\text{–}1^a$	recessive	gs antigen	B10, D2	Taylor et al., 1973
			$Mlv\text{–}1^b$	dominant	no gs antigen		

VIII	host	*Fv–1*	$Fv\text{–}1^n$	recessive	susceptibility to N-tropic MLV	AKR, C58	Rowe and Hartley, 1972; Stephenson and Aaronson, 1972b
			$Fv\text{–}1^b$	recessive	susceptibility to B-tropic MLV	C57BL/6 Balb/c	
IX	host	*H–2*	$H\text{–}2^k$	semidominant	early leukemia	AKR	Boyse et al., 1972*
			$H\text{–}2^b$	semidominant	late leukemia	AKR/$H\text{–}2^b$	
IX	viral?	*Tla*	Tla^+	dominant	epistatic for TL antigen expression	A, C58	Boyse and Old, 1969*
			Tla^-	recessive	no TL antigen	AKR, C57BL/6	
IX	viral?	G_{IX}	G_{IX}^+	semidominant	G_{IX} antigen	129	Boyse et al., 1972*
			G_{IX}^-	semidominant	no G_{IX} antigen		
unknown, not IX	viral?	none	+	dominant	G_{IX} antigen	129	Boyse et al., 1972*
			–	recessive	no G_{IX} antigen		

* It is not known whether the *H–2*, *Tla* and G_{IX} loci affect MLV replication directly or whether they affect leukemogenesis induced by MLV. Both G_{IX} loci are unlikely to represent structural genes for G_{IX} antigens; either or both loci may have epistatic effects on the structural gene.

et al. (1971) also noticed that BrdU was effective in inducing the release of infectious MSV pseudotypes from nonproducer transformed cells. The release of infectious MSV from such cells proved to be a result of the activation of endogenous MLV, which then acted as a helper virus for the MSV (Klement et al., 1971; Aaronson, 1971b). The discovery of high efficiency induction by BrdU and IdU led other investigators (Aaronson et al., 1971b; Silagi et al., 1972; Stephenson and Aaronson, 1972a, b; Todaro, 1972) to use these agents on many kinds of normal and neoplastic cells in culture. Aaronson et al. (1971b) quickly found that BrdU efficiently induced C-type virus particles (assayed by reverse transcriptase activity) from Balb/3T3 cells, but Stephenson and Aaronson (1972a) were unable to induce and recover virus from NIH-Swiss mouse cells.

It is interesting to note that established cell lines are more readily induced to release endogenous MLV than are new cultures and that cell lines that have become transformed spontaneously or by SV40 are more inducible than nontransformed cell lines (Rowe et al., 1971; Todaro, 1972). Indeed, neoplastic transformation predisposes cells to spontaneous release of endogenous virus and this could be of selective advantage to the host because it presumably makes the neoplastic cells more immunogenic.

Teich et al. (1973) have suggested that the incorporation of BrdU or IdU into the host cell DNA is necessary for virus induction, which is then significantly enhanced by treating the cells with ultraviolet or blue light. Incorporation of IdU into one strand of host DNA is sufficient to induce virus. MLV appears to persist in induced cultures only if the cells are susceptible to spreading infection and amplification (Aaronson et al., 1971b; Rowe and Hartley, 1972; Stephenson and Aaronson, 1972a, b). If N-tropic MLV is induced by BrdU in B-type cells (e.g. Balb/c cells), virus release reaches a peak 2–3 days after treatment and then disappears if the BrdU is removed. More virus can be induced by treating the culture with BrdU again. It is not clear whether virus production is repressed once again after removal of the inducing agent, or whether the induced cells are killed. All these data suggest, but do not prove, that induction of C-type RNA viruses involves the excision of the integrated viral DNA genomes. If this is the case, the induction of these viruses resembles the induction of other integrated viral DNA genomes, for example bacteriophage lambda genomes and SV40 genomes, which depends on an excision step. It is not obvious, however, why the induction of C-type RNA viruses should depend on excision of the DNA provirus because in theory induction of transcription of the integrated provirus ought to be sufficient to lead to the production of infectious virus particles.

ENDOGENOUS C-TYPE RNA TUMOUR VIRUSES IN OTHER SPECIES

Only the avian and murine RNA tumour viruses have been studied in sufficient detail to characterize genetically transmitted viruses and the mechanisms which control their activation. There is, however, evidence that cells of other species also carry endogenous "leukemia" virus genomes. The induction of infectious pseudotypes of MSV and companion helper viruses in nonproducer, transformed Syrian hamster cells (Klement et al., 1969) and rat cells (Aaronson, 1971a; Klement et al., 1972) is reminiscent of the chick helper factor and activation of RAV-60. Following the discovery of MLV induction by BrdU and IdU, many attempts were made to recover C-type viruses from human tumour cells or cell lines, but there is only one report (Stewart et al., 1972) of suggestive evidence that virus particles are induced and these particles have not been characterized (see Chapter 10). Recently Livingston and Todaro (1973) reported the spontaneous appearance and induction of C-type virus from an established feline cell line, CCC. This virus is most interesting because its reverse transcriptase and gs-1 antigen do not cross react immunologically with those of previously described feline leukemia and sarcoma viruses, which are transmitted horizontally (see Chapter 10). The virus does, however, show a close relationship to RD-114 virus, isolated from human rhabdomyosarcoma (RD) cells following passage through a fetal cat (McAllister et al., 1972; see Chapter 10). The virus isolated from CCC cells does not replicate efficiently in feline cells but grows to high titres in primate cells, including human cells. The endogenous virus of chicken cells also grows in cells of other avian species (which have suitable receptors and no

I^e inhibitor gene). Therefore, one may postulate that genetically transmitted viruses of one species can horizontally infect cells of other species, and it may be profitable to search for zoonoses of this kind in nature (Weiss, 1973b). But not all endogenous viruses are restricted to horizontal infection of other species or strains. For instance, endogenous viruses of the Syrian hamster and the rat propagate well in Syrian hamster and rat cells (Kelloff et al., 1970; Aaronson, 1971a); neither is the susceptibility of the mouse to N- or B-tropic MLV governed by the tropism of its endogenous virus, although the evolution of resistant populations may be favoured by natural selection (see Weiss, 1973b).

The adaptation of viruses to new host strains and species has sometimes been interpreted as resulting from the selection of recombinants between the infecting virus and endogenous viral genes (Altaner and Temin, 1970; Aaronson, 1971b). Recombinants between exogenous and endogenous avian viruses have been reported recently (Weiss et al., 1973), but in this case it is interesting to note that passage of RSV in foreign species such as quail or pheasant selects for recombinants of RSV and the chick endogenous virus, which already exist as a minor component of the RSV stock. Interaction between the RSV genome and viral genes, endogenous in pheasant cells, may however also occur (Hanafusa and Hanafusa, 1973).

Todaro et al. (1973) have described the production of a C-type virus (AT-124) in human RD cells after inoculation of the cells into immunosuppressed NIH-Swiss mice, an experiment which parallels the appearance of RD-114 virus in the same cell line passaged in a fetal cat's brain (McAllister et al., 1972). The AT-124 virus possesses murine gs antigen and reverse transcriptase specificity, but has the host range of RD-114 virus, which does not grow in murine cells. Since NIH-Swiss mice have not previously produced a virus, even after IdU induction (Stephenson and Aaronson, 1972a), Todaro et al. (1973) suggest that AT-124 represents either a virus from NIH-Swiss mice that replicates in human cells or a recombinant virus of murine and human origin. The development of NB-tropic virus from N- or B-tropic MLV (Lilly and Pincus, 1972) is another example of host-range adaptation, which is not understood but may be caused by mutation or by recombination with endogenous viral genes.

MURINE MAMMARY TUMOUR VIRUSES

Inheritance of MMTV

The "milk factor" for mammary carcinogenesis (Bittner, 1936), which was shown to be a virus (Bittner, 1942; Andervont and Bryan, 1944), is the classical example of an RNA tumour virus that is transmitted vertically by congenital infection (see Gross, 1970 for a review of early studies on MMTV transmission). Bittner virus (MMTV-S) is normally transmitted in the milk but not, apparently, in the egg or sperm. Other strains of MMTV (see Table 10.11) are transmitted in the gametes as well as the milk, or are transmitted exclusively in the gametes. In GR mice, for example, MMTV-P is transmitted in milk but also via eggs and sperm, as was shown by genetic crosses (Bentvelzen, 1968) and by transplantation of fertilized eggs (Zeilmaker, 1969). If, however, MMTV-P is introduced into mice of other strains, it is transmitted via the milk only (Mühlbock and Bentvelzen, 1969). MMTV-L is transmitted by the germ cells in C3Hf mice (Pitelka et al., 1964) but when it is inoculated into females of the highly susceptible Balb/c strain, it is transmitted exclusively in the milk. Yet the Balb/c strain itself carries MMTV-O (Bentvelzen et al., 1970), which is not transmitted by milk but presumably by gametes. There is evidence that MMTV, which is transmitted in the gametes, is present in the form of an integrated viral genome, a "germinal provirus" (Bentvelzen, 1972), rather than a mature virus particle.

Because no one has yet developed cell culture methods for the infection, propagation and rescue of MMTV, the biology of these viruses remains largely unknown. However, experiments similar to those that are easily performed with leukemia viruses in vitro have been attempted with MMTV in vivo. Timmermans et al. (1969) found that cells of mammary tumours induced in 020 strain mice (which appeared to be free of MMTV) by combined exposure to X-irradiation and urethane produced MMTV-X. By contrast, tumours induced in 020 strain mice by extreme hormonal stimulation did not release MMTV. Even in C57BL strain mice, which have a low incidence of mammary tumours and from which no MMTV has been recovered, MMTV-Y antigens are detectable in mammary tumours induced by X-irradiation followed by hormonal stimulation (Bentvelzen et al., 1968). Thus it appears that all mouse strains, whether of high or low incidence for mammary tumours, probably carry genetically transmitted MMTV genomes as well as MLV genomes. Nucleic acid hybridization studies confirm this view; Varmus et al. (1972b) have demonstrated that mice of strains showing both high and low incidences of mammary tumours contain the same amounts of MMTV DNA in their genomes.

Regulation of MMTV Expression

The expression of MMTV and the development of mammary tumours is hormone dependent (Gross, 1970; Boot et al., 1972; Mühlbock, 1972) and estrogens play a prominent role. MMTV appears in mammary tumours and in lactating mammary tissue, and force-bred females are most susceptible to oncogenesis. But genetic factors also play a large part in MMTV expression (Dux, 1972; Bentvelzen, 1972; Bentvelzen et al., 1972) and the genetic factors do not necessarily act on hormone activity. In GR strain mice the expression of MMTV is controlled by a single Mendelian locus, *Ms*, with a dominant allele Ms^E for MMTV-P expression (Bentvelzen, 1968; Bentvelzen and Daams, 1969). Bentvelzen and Daams propose that other strains of mice carry different alleles at the *Ms* locus. Thus *Ms-1* in Balb/c mice and *Ms-2* in C3Hf mice permit only a limited expression of the MMTVs in these strains, and the 020 and C57BL strains carry the Ms^+ allele, which confers resistance to MMTV expression, resistance believed to be mediated by a repressor. Bentvelzen and his colleagues have suggested that the *Ms* locus is actually part of or closely linked to the germinal provirus, the endogenous MMTV genome.

Daams et al. (1968) and Bentvelzen and Daams (1969) further suggested that mice carrying the Ms^+ allele are resistant to superinfection with MMTV-S and MMTV-P because the same repressor that inhibits the release of endogenous MMTV-O or MMTV-Y also interferes with the replication of the superinfecting virus. It is noteworthy that the model for resistance to superinfection by endogenous MMTV is different from the epistatic inhibitor model for avian RAV-0 (discussed above), which operates at the level of virus penetration.

The model for genetic transmission and control of MMTV proposed by Bentvelzen and his colleagues (Bentvelzen, 1968; Bentvelzen et al., 1968; Daams et al., 1968; Mühlbock and Bentvelzen, 1968) was based on that of lysogeny by temperate bacteriophages, and Bentvelzen et al. (1968) suggested that it might also apply to C-type leukemia viruses. When Huebner and Todaro (1969) adopted this model, they named it the oncogene hypothesis.

The control of endogenous MMTV expression has recently been examined using immunological techniques (Bentvelzen et al., 1972; Hilgers et al., 1972) and molecular hybridization (Bishop et al., 1973). The presence of gs antigens of MMTV, detected by immunofluorescence, parallels the presence of virus in different mouse strains and in different tissues of positive strains (Hilgers et al., 1972). However molecular hybridization of DNA, reverse-transcribed in vitro in MMTV particles, to RNA extracted from murine tissues yielded unexpected results (Bishop et al., 1973). Significant numbers of genome equivalents of MMTV-like RNA were found in lactating mammary gland tissue and other tissues of mice of strains with a low incidence of mammary tumours (for example C57BL/6 and 129) even when no MMTV or viral antigens could be detected. However, the amount of MMTV RNA in mice of different strains correlated with the incidence of mammary tumours in and the susceptibility to MMTV infection of those strains.

Progeny of backcross matings of C3H $\times$ C57BL to the C57BL parent (which should segregate 50 percent Ms^{+}/Ms^{+} homozygotes for repressed MMTV and 50 percent $Ms^{+}/Ms\text{-}2$ heterozygotes which may produce some MMTV proteins) fell into two distinct classes when the amount of viral RNA in lactating mammary gland tissue was measured by nucleic acid hybridization. Mammary gland tissue containing MMTV antigens contained more than 2000 genome equivalents of MMTV RNA, whereas tissues that lacked antigen contained only about 100 genome equivalents of MMTV RNA. The genetic complexity of the RNA molecules in both classes of animals was the same, indicating that the extent of expression of the endogenous genome was the same; clearly, the idea that in cells homozygous for Ms^+ only the MMTV repressor gene is transcribed (Bentvelzen et al., 1970, 1972) is therefore incorrect. The results of the experiments of Bishop et al. (1973) also indicate that, in addition to an apparently quantitative control of the transcription of the endogenous MMTV genome, post-transcriptional controls also operate.

Bishop et al. (1973) also studied the thermal stability (Tm) of hybrids between MMTV DNA made in vitro and MMTV RNA in cells of different strains of mice. All these DNA/RNA hybrids displayed a high level of thermal stability (Tm 60–68°C), but there were small (4–8°C) but consistent differences in the Tm of MMTV-S DNA and the MMTV RNAs present in different strains of mice. These differences indicate about 5 percent mismatching between the nucleotide sequences involved, suggesting that the endogenous MMTV genomes of different strains of mice are not entirely homologous. Such differences in RNA sequences may be the origin of the antigenic differences between MMTV from different strains of mice (Hageman et al., 1972; Hilgers et al., 1971).

ENDOGENOUS VIRUSES AND CARCINOGENESIS

It seems probable that the genomes of all species of mammals and birds will prove to contain elements capable of specifying C-type RNA viruses, and since at least some strains of mice inherit proviruses of both N-tropic and B-tropic MLV, as well as MMTV, it would not be surprising if many species genetically transmit the proviruses of more than one type of RNA "tumour" virus. Precisely how the repression and expression of these endogenous viral genomes is controlled in different species remains to be seen. But it is already obvious from what we know about RAV-0 in chickens and endogenous MLVs and MMTVs in mice that the expression and host range of endogenous viruses can be controlled in more than one way at more than one level in the replicative cycle of these viruses. This means, of course, that we should be extremely cautious when we attempt to extrapolate from one species to another; what holds for mice does not hold for chickens and may well not hold for man. Furthermore only the endogenous viruses of inbred strains of mice have been shown to have an oncogenic potential. RAV-0 has not yet been shown to cause leukosis in chickens; neither have the C-type particles induced from cells of several mammalian species been shown to be oncogenic.

Literature Cited

AARONSON, S. A. 1971a. Isolation of a rat-tropic helper virus from M-MSV-0. Virology *44:* 29.

———. 1971b. Chemical induction of focus-forming virus from nonproducer cells transformed by murine sarcoma virus. Proc. Nat. Acad. Sci. *68:* 3069.

AARONSON, S. A., J. W. HARTLEY and G. J. TODARO. 1969. Mouse leukemia virus: "Spontaneous" release by mouse embryo cells after long-term in vitro cultivation. Proc. Nat. Acad. Sci. *64:* 87.

AARONSON, S. A., G. J. TODARO and E. M. SCOLNICK. 1971. Induction of murine C-type viruses from clonal lines of virus-free BALB/3T3 cells. Science *174:* 157.

ALTANER, C. and H. M. TEMIN. 1970. Carcinogenesis by RNA sarcoma viruses. XII. A quantitative study of infection of rat cells in vitro by avian sarcoma viruses. Virology *40:* 118.

ANDERVONT, H. B. and W. R. BRYAN. 1944. Properties of the mouse mammary-tumor agent. J. Nat. Cancer Inst. *5:* 143.

ANDREWES, C. H. 1939. Latent virus infections and their possible relevance to the cancer problem. Proc. Roy. Soc. Med. *33:* 75.

BADER, J. P. 1967. In *Subviral carcinogenesis*, ed. Y. Ito, pp. 144–155, Nagoya.

BALUDA, M. A. 1972. Widespread presence, in chickens, of DNA complementary to the RNA genome of avian leukosis viruses. Proc. Nat. Acad. Sci. *69:* 576.

BAUER, H. and H. G. JANDA. 1967. Group-specific antigen of avian leukosis virus. Virus specificity and relation to an antigen contained in Rous mammalian cells. Virology *33:* 483.

BENTVELZEN, P. 1968. Genetical control of the vertical transmission of the Mühlbock mammary tumour virus in the GR mouse strain. Ph.D. thesis. Amsterdam, Hollandia.

———. 1972. Hereditary infections with mammary tumor viruses in mice. In *RNA viruses and host genome in oncogenesis*, ed. P. Emmelot and P. Bentvelzen, pp. 309–337. North-Holland, Amsterdam.

BENTVELZEN, P. and J. H. DAAMS. 1969. Hereditary infections with mammary tumor viruses in mice. J. Nat. Cancer Inst. *43:* 1025.

BENTVELZEN, P., A. TIMMERMANS, J. H. DAAMS and A. VAN DER GUGTEN. 1968. Genetic transmission of mammary tumor inciting viruses in mice: Possible implications for murine leukemia. Bibl. Haematol. *31:* 101.

BENTVELZEN, P., J. H. DAAMS, P. HAGEMAN and J. CALAFAT. 1970. Genetic transmission of viruses that incite mammary tumor in mice. Proc. Nat. Acad. Sci. *67:* 377.

BENTVELZEN, P., J. H. DAAMS, P. HAGEMAN, J. CALAFAT and A. TIMMERMANS. 1972. Interactions between viral and genetic factors in the origin of mammary tumors in mice. J. Nat. Cancer Inst. *48:* 1089.

BISHOP, J. M., N. JACKSON, N. QUINTRELL and H. E. VARMUS. 1973. Transcription of RNA tumor virus genes in normal and infected cells. In *Possible episomes in eukaryotes*, ed. L. G. Silvestri. North-Holland, Amsterdam. In press.

BITTNER, J. J. 1936. Some possible effects of nursing on the mammary gland tumor incidence in mice. Science *84:* 162.

BITTNER, J. J. 1942. Hormones, genetic susceptibility and milk influence in cancer. Cancer Res. *2:* 710.

BOOT, L. M., G. RÖPCKE and H. G. KWA. 1972. Hormonal factors in the origin of mammary tumors. In *RNA viruses and host genome in oncogenesis*, ed. P. Emmelot and P. Bentvelzen, pp. 275–282. North-Holland, Amsterdam.

BOYSE, E. A. and L. J. OLD. 1969. Some aspects of normal and abnormal cell surface genetics. Ann. Rev. Genetics *3:* 269.

BOYSE, E. A., L. J. OLD and E. STOCKERT. 1972. The relation of linkage group IX to leukemogenesis in the mouse. In *RNA viruses and host genome in oncogenesis*, ed. P. Emmelot and P. Bentvelzen, pp. 171–185. North-Holland, Amsterdam.

BUFFET, R. F., J. T. GRACE, L. A. DIBERARDINO and E. A. MIRAND. 1969. Vertical transmission of murine leukaemia virus. Cancer Res. *29:* 588.

BURMESTER, B. R., R. F. GENTRY and N. F. WATERS. 1955. The presence of the virus of visceral lymphomatosis in embryonated eggs of normal appearing hens. Science *34:* 609.

CRITTENDEN, L. B., E. J. SMITH, R. A. WEISS and P. S. SARMA. 1973a. Host gene control of endogenous avian leukosis virus production. Virology. In press.

CRITTENDEN, L. B., E. J. WENDEL and J. V. MOTTA. 1973b. Interaction of genes controlling resistance to RSV(RAV-0). Virology. In press.

DAAMS, J. H., A. TIMMERMANS, A. VAN DER GUGTEN and P. BENTVELZEN. 1968. Genetical resistance of mouse strain C57BL to mammary tumor viruses. II. Resistance by means of a repressed, related provirus. Genetica *38:* 400.

DARLINGTON, C. D. 1948. The plasmagene theory of the origin of cancer. Brit. J. Cancer *2:* 118.

DISTEFANO, H. S. and R. M. DOUGHERTY. 1966. Mechanisms for congenital transmission of avian leukosis virus. J. Nat. Cancer Inst. *37:* 869.

DIXON, F. J., M. B. A. OLDSTONE and G. TONIELTI. 1971. Pathogenesis of immune complex glomerulonephritis of New Zealand mice. J. Exp. Med. *134:* 65s.

DOUGHERTY, R. M. and H. S. DISTEFANO. 1966. Lack of relationship between infection with avian leukosis virus and the presence of COFAL antigen in chick embryos. Virology *29:* 586.

DOUGHERTY, R. M., H. S. DISTEFANO and F. ROTH. 1967. Virus particles and virus antigens in chicken tissues free of infectious avian leukosis virus. Proc. Nat. Acad. Sci. *58:* 808.

DUX, A. 1972. Genetic aspects in the genesis of mammary cancer. In *RNA viruses and host genome in oncogenesis*, ed. P. Emmelot and P. Bentvelzen, pp. 301–308. North-Holland, Amsterdam.

FRIIS, R. R. 1972. Abortive infection of Japanese quail cells with avian sarcoma viruses. Virology *50:* 701.

GROSS, L. 1944. Is cancer a communicable disease? Cancer Res. *4:* 293.

———. 1951. "Spontaneous" leukemia developing in C3H mice following inoculation in infancy, with AK leukemic extracts, or AK embryos. Proc. Soc. Exp. Biol. Med. *76:* 27.

———. 1958. Attempt to recover a filterable agent from X-ray-induced leukemia. Acta Haematol. *19:* 353.

———. 1970. *Oncogenic viruses*, 2nd ed. Pergamon Press, Oxford.

HAGEMAN, P., J. CALAFAT and J. H. DAAMS. 1972. The mouse mammary tumor viruses. In *RNA viruses and host genomes in oncogenesis*, ed. P. Emmelot and P. Bentvelzen, pp. 283–300. North-Holland, Amsterdam.

HANAFUSA, H. and T. HANAFUSA. 1968. Further studies on RSV production from transformed cells. Virology *34:* 630.

Hanafusa, H. and T. Hanafusa. 1971. Noninfectious RSV deficient in DNA polymerase. Virology *43:* 313.

Hanafusa, H., T. Miyamoto and T. Hanafusa. 1970. A cell-associated factor essential for formation of an infectious form of Rous sarcoma virus. Proc. Nat. Acad. Sci. *66:* 314.

Hanafusa, H., D. Baltimore, D. Smoler, K. F. Watson, A. Yaniv and S. Spiegelman. 1972. Absence of polymerase protein in virions of alpha-type Rous sarcoma virus. Science *177:* 1188.

Hanafusa, T. and H. Hanafusa. 1973. Isolation of leukosis-type virus from pheasant embryo cells: Possible presence of viral genes in cells. Virology *51:* 247.

Hanafusa, T., T. Miyamoto and H. Hanafusa. 1970a. A type of chick embryo cell that fails to support formation of infectious RSV. Virology *40:* 55.

Hanafusa, T., H. Hanafusa and T. Miyamoto. 1970b. Recovery of a new virus from apparently normal chick cells by infection with avian tumor viruses. Proc. Nat. Acad. Sci. *67:* 1797.

Hanafusa, T., H. Hanafusa, T. Miyamoto and E. Fleissner. 1972. Existence and expression of tumor virus genes in chick embryo cells. Virology *47:* 475.

Harel, L., J. Harel, F. Lacour and J. Huppert. 1966. Homologie entre genome du virus de la myeloblastose aviare (AMV) et genome cellulaire. Comp. Rend. Acad. Sci. *263:* 616.

Harel, L., J. Harel and J. Huppert. 1967. Partial homology between RNA from Rauscher mouse leukemia virus and cellular DNA. Biochem. Biophys. Res. Comm. *28:* 44.

Hayward, W. S. and H. Hanafusa. 1973. Detection of avian tumor virus RNA in uninfected chick embryo cells. J. Virol. *11:* 157.

Hilgers, J. 1972. Immunological detection of oncornaviruses in normal and neoplastic tissues from individual animals. Thesis, University of Amsterdam.

Hilgers, J., J. H. Daams and P. Bentvelzen. 1971. The induction of precipitating antibodies to mammary tumor virus in several inbred mouse strains. Israel J. Med. Sci. *7:* 154.

Hilgers, J., M. Beya, G. Geering, E. A. Boyse and L. J. Old. 1972. Expression of MuLV-gs antigen in mice of segregating populations: Evidence for mendelian inheritance. In *RNA viruses and host genome in oncogenesis*, ed. P. Emmelot and P. Bentvelzen, pp. 187–192. North-Holland, Amsterdam.

Huebner, R. J. and G. J. Todaro. 1969. Oncogenes of RNA tumor viruses as determinants of cancer. Proc. Nat. Acad. Sci. *64:* 1087.

Huebner, R. J., G. J. Kelloff, P. S. Sarma, W. T. Lane, H. C. Turner, R. V. Gilden, S. Oroszlan, H. Meier, D. D. Myers and R. L. Peters. 1970. Group-specific antigen expression during embryogenesis of the genome of the C-type RNA tumor virus: Implications for ontogenesis and oncogenesis. Proc. Nat. Acad. Sci. *67:* 366.

Ishizaki, R. and T. Shimizu. 1970. Observations on the envelope properties of RSV(0). Virology *40:* 415.

Kang, C-Y. and H. M. Temin. 1972. Endogenous RNA-directed DNA polymerase activity in uninfected chicken embryos. Proc. Nat. Acad. Sci. *69:* 1550.

Kawai, S. and H. Hanafusa. 1972. Genetic recombination with avian tumor virus. Virology *49:* 37.

Kelloff, G., R. J. Huebner, N. H. Chang, Y. K. Lee and R. V. Gilden. 1970. Envelope antigen relationships among three hamster-specific sarcoma viruses and a hamster-specific helper virus. J. Gen. Virol. *9:* 19.

Klement, V., J. W. Hartley, W. P. Rowe and R. J. Huebner. 1969. Recovery of a hamster-specific, focus-forming and sarcomagenic virus from a "non-infectious" hamster tumor induced by the Kirsten mouse sarcoma virus. J. Nat. Cancer Inst. *43:* 925.

KLEMENT, V., M. D. NICHOLSON and R. J. HUEBNER. 1971. Rescue of the genome of focus forming virus from rat non-productive lines by 5-bromodeoxyuridine. Nature *234:* 12.

KLEMENT, V., M. D. NICHOLSON, R. V. GILDEN, S. OROSZLAN, P. S. SARMA, R. W. RONGEY and M. B. GARDNER. 1972. Rat C-type virus induced in rat sarcoma cells by 5-bromodeoxyuridine. Nature New Biol. *238:* 234.

LATARJET, R. and J.-F. DUPLAN. 1962. Experiment and discussion on leukaemogenesis by cell-free extracts of radiation-induced leukaemia in mice. Int. J. Radiation Biol. *5:* 339.

LEONG, J. A., A. C. GARAPIN, N. JACKSON, L. FANSHIER, W. LEVINSON and J. M. BISHOP. 1972. Virus-specific ribonucleic acid in cells producing Rous sarcoma virus: Detection and characterization. J. Virol. *9:* 891.

LIEBERMAN, M. and H. S. KAPLAN. 1959. Leukemogenic activity of filtrates from radiation-induced lymphoid tumors of mice. Science *130:* 387.

LILLY, F. and T. PINCUS. 1972. Genetic control of murine viral leukemogenesis. Adv. Cancer Res. (in press).

LINIAL, M. and W. S. MASON. 1973. Characteristics of two conditional early mutants of Rous sarcoma virus. Virology. In press.

LIVINGSTON, D. M. and G. J. TODARO. 1973. Endogenous type C virus from a cat cell clone with properties distinct from previously described feline type C viruses. Virology. In press.

LOWY, D. R., W. P. ROWE, N. TEICH and J. W. HARTLEY. 1971. Murine leukemia virus: High-frequency activation in vitro by 5-iododexoyuridine and 5-bromodeoxyuridine. Science *174:* 155.

LWOFF, A. 1960. Tumor viruses and the cancer problem: A summation of the conference. Cancer Res. *20:* 820.

MCALLISTER, R. M., M. NICHOLSON, M. B. GARDNER, R. W. RONGEY, S. RASHEED, P. S. SARMA, R. J. HUEBNER, M. HATANAKA, S. OROSZLAN, R. V. GILDEN, A. KABIGTING and L. VERNON. 1972. C-type virus released from cultured human rhabdomyosarcoma cells. Nature New Biol. *235:* 3.

MCKISSICK, G. E., R. A. GRIESEMER and R. L. FARRELL. 1970. Aerosol transmission of Rauscher murine leukemia virus. J. Nat. Cancer Inst. *45:* 625.

MELLORS, R. C., T. SHIRAI, T. AOKI, R. J. HUEBNER and K. KRAWCZYNSKI. 1971. Wild-type Gross leukemia virus and the pathogenesis of the glomerulonephritis of New Zealand mice. J. Exp. Med. *129:* 123.

MIRAND, A. G. and E. A. MIRAND. 1969. Transmission of Rauscher leukemia in mice. Experientia *25:* 829.

MÜHLBOCK, O. 1972. Role of hormones in the etiology of breast cancer. J. Nat. Cancer Inst. *48:* 1213.

MÜHLBOCK, O. and P. BENTVELZEN. 1969. The transmission of mammary tumor viruses. Perspect. Virol. *6:* 75.

OLDSTONE, M. B. A. and F. J. DIXON. 1971. Immune complex disease in chronic viral infections. J. Exp. Med. *134:* 34s.

PAYNE, L. N. and R. CHUBB. 1968. Studies on the nature and genetic control of an antigen in normal chick embryos which reacts in the COFAL test. J. Gen. Virol. *3:* 379.

PAYNE, L. N., P. K. PANI and R. A. WEISS. 1971. A dominant epistatic gene which inhibits cellular susceptibility to RSV(RAV-0). J. Gen. Virol. *13:* 455.

PINCUS, T., J. W. HARTLEY and W. P. ROWE. 1971a. A major genetic locus affecting resistance to infection with murine leukemia viruses. I. Tissue culture studies of naturally occurring viruses. J. Exp. Med. *133:* 1219.

PINCUS, T., W. P. ROWE and F. LILLY. 1971b. A major genetic locus affecting resistance to infection with murine leukemia viruses. II. Apparent identity to a major locus described for resistance to Friend murine leukemia virus. J. Exp. Med. *133:* 1234.

PITELKA, D. R., H. A. BERN, S. NANDI and K. B. DEOME. 1964. On the significance of virus-like particles in mammary tissues of C3Hf mice. J. Nat. Cancer Inst. *33:* 867.

ROSENTHAL, P. N., H. L. ROBINSON, W. S. ROBINSON, T. HANAFUSA and H. HANAFUSA. 1971. RNA in uninfected and virus infected cells complementary to avian tumor virus RNA. Proc. Nat. Acad. Sci. *68:* 2336.

ROWE, W. P. 1972. Studies of genetic transmission of murine leukemia virus by AKR mice. I. Crosses with Fv-1^n strains of mice. J. Exp. Med. *136:* 1272.

ROWE, W. P. and J. W. HARTLEY. 1972. Studies of genetic transmission of murine leukemia virus by AKR mice. II. Crosses with Fv-1^b strains of mice. J. Exp. Med. *136:* 1286.

ROWE, W. P. and T. PINCUS. 1972. Quantitative studies of naturally occurring murine leukemia virus infection of AKR mice. J. Exp. Med. *135:* 429.

ROWE, W. P., J. W. HARTLEY, M. R. LANDER, W. E. PUGH and N. TEICH. 1971. Non-infectious AKR mouse embryo cell lines in which each cell has the capacity to be activated to produce infectious murine leukemia virus. Virology *46:* 866.

ROWE, W. P., J. B. HUMPHREY and F. LILLY. 1973. A major genetic locus affecting resistance to infection with murine leukemia viruses. III. Assignment of the Fv-1 locus to linkage group VIII of the mouse. J. Exp. Med. *137:* 850.

RUBIN, H., A. CORNELIUS and L. FANSHIER. 1961. The pattern of congenital transmission of an avian leukosis virus. Proc. Nat. Acad. Sci. *47:* 1058.

RUBIN, H., L. FANSHIER, A. CORNELIUS and W. F. HUGHES. 1962. Tolerance and immunity in chickens following congenital and contact infection with avian leukosis virus. Virology *17:* 143.

SARMA, P. S., H. C. TURNER and R. J. HUEBNER. 1964. An avian leucosis group-specific complement fixation reaction. Application for the detection and assay of non-cytopathogenic leucosis viruses. Virology *23:* 313.

SCHEELE, C. M. and H. HANAFUSA. 1971. Proteins of helper-dependent RSV. Virology *45:* 401.

SILAGI, S., D. BEJU, J. WRATHALL and E. DEHARVEN. 1972. Tumorigenicity, immunogenicity and virus production in mouse melanoma cells treated with 5-bromodeoxyuridine. Proc. Nat. Acad. Sci. *69:* 3443.

STEPHENSON, J. R. and S. A. AARONSON. 1972a. Genetic factors influencing C-type RNA virus induction. J. Exp. Med. *136:* 175.

———. 1972b. A genetic locus for inducibility of C-type virus in BALB/c cells: The effect of a non-linked regulatory gene on detection of virus after chemical activation. Proc. Nat. Acad. Sci. *69:* 2798.

STEWART, S., G. KASNIC, C. DRAYCOTT and T. BEN. 1972. Activation of viruses in human tumors by 5-iododeoxyuridine and dimethyl sulfoxide. Science *175:* 198.

STOCKERT, E., L. J. OLD and E. A. BOYSE. 1971. The G_{IX} system. A cell surface alloantigen associated with murine leukemia virus; implications regarding chromosomal integration of the viral genome. J. Exp. Med. *133:* 1334.

TAYLOR, B. A., H. MEIER and D. D. MEYERS. 1971. Host gene control of C-type RNA tumor virus: Inheritance of the group-specific antigen of murine leukemia virus. Proc. Nat. Acad. Sci. *68:* 3190.

TAYLOR, B. A., H. MEIER and R. J. HUEBNER. 1973. Genetic control of the group-specific antigen of murine leukaemia virus. Nature *241:* 184.

TEICH, N., D. R. LOWY, J. W. HARTLEY and W. P. ROWE. 1973. Studies of the mechanism of induction of infectious murine leukemia virus from AKR mouse embryo cell lines by 5-iododeoxyuridine and 5-bromodeoxyuridine. Virology. *51:* 163.

TEMIN, H. M. 1964. Homology between RNA from Rous sarcoma virus and DNA from Rous sarcoma virus-infected cells. Proc. Nat. Acad. Sci. *52:* 323.

———. 1971. The protovirus hypothesis: Speculations on the significance of RNA-directed DNA synthesis for normal development and for carcinogenesis. J. Nat. Cancer Inst. *46:* 3.

———. 1972. The protovirus hypothesis and cancer. In *RNA viruses and host genome in oncogenesis*, ed. P. Emmelot and P. Bentvelzen, pp. 351–363. North-Holland, Amsterdam.

TIMMERMANS, A., P. BENTVELZEN, P. C. HAGEMAN and J. CALAFAT. 1969. Activation of a mammary tumour virus in 020 strain mice by irradiation and urethane. J. Gen. Virol. *4:* 169.

TODARO, G. J. 1972. "Spontaneous" release of type C viruses from clonal lines of "spontaneously" transformed BALB/3T3 cells. Nature New Biol. *240:* 157.

TODARO, G. J. and R. J. HUEBNER. 1972. The viral oncogene hypothesis: New evidence. Proc. Nat. Acad. Sci. *69:* 1009.

TODARO, G. J., P. ARNSTEIN, W. P. PARKS, E. H. LENETTE and R. J. HUEBNER. 1973. A type C virus in human rhabdomyosarcoma cells after inoculation into antithymocyte serum-treated NIH Swiss mice. Proc. Nat. Acad. Sci. *70:* 859.

VARMUS, H. E., R. A. WEISS, R. R. FRIIS, W. LEVINSON and J. M. BISHOP. 1972a. Detection of avian tumor virus-specific nucleotide sequences in avian cell DNA. Proc. Nat. Acad. Sci. *69:* 20.

VARMUS, H. E., J. M. BISHOP, R. C. NOWINSKI and N. H. SARKAR. 1972b. Detection of mammary tumor virus-specific nucleotide sequences in the DNA of high and low incidence mouse strains. Nature New Biol. *238:* 189.

VOGT, P. K. 1967. A virus released by "non-producing" Rous sarcoma cells. Proc. Nat. Acad. Sci. *58:* 801.

———. 1971. Genetically stable reassortment of markers during mixed infection with avian tumor viruses. Virology *46:* 947.

VOGT, P. K. and R. R. FRIIS. 1971. An avian leukosis virus related to RSV(0): Properties and evidence for helper activity. Virology *43:* 223.

VOGT, P. K., R. R. FRIIS and R. A. WEISS. 1973. Cell genetics and growth of endogenous viruses. Cancer. In press.

WEISS, R. A. 1967. Spontaneous virus production from "non-virus producing" Rous sarcoma cells. Virology *32:* 719.

———. 1969a. The host range of Bryan strain Rous sarcoma virus synthesized in the absence of helper virus. J. Gen. Virol. *5:* 511.

———. 1969b. Interference and neutralization studies with Bryan strain Rous sarcoma virus synthesized in the absence of helper virus. J. Gen. Virol. *5:* 529.

———. 1972. Helper viruses and helper cells. In *RNA viruses and host genome in oncogenesis*, ed. P. Emmelot and P. Bentvelzen, pp. 117–135. North-Holland, Amsterdam.

———. 1973a. Transmission of cellular genetic elements by RNA tumor viruses. In *Possible episomes in eukaryotes*, ed. L. G. Silvestri, North-Holland, Amsterdam. In press.

———. 1973b. Ecological genetics of RNA tumour viruses and their hosts. In *Analytic and experimental epidemiology of cancer*, ed. T. Hirayama, Univ. of Tokyo Press, Tokyo. In press.

WEISS, R. A. and P. M. BIGGS. 1972. Leukosis and Marek's disease viruses of feral red jungle fowl and domestic fowl in Malaya. J. Nat. Cancer Inst. *49:* 1713.

WEISS, R. A. and L. N. PAYNE. 1971. The heritable nature of the factor in chicken cells which acts as a helper virus for Rous sarcoma virus. Virology *45:* 508.

WEISS, R. A., R. R. FRIIS, E. KATZ and P. K. VOGT. 1971. Induction of avian tumor viruses in normal cells by physical and chemical carcinogens. Virology *46:* 920.

WEISS, R. A., W. S. MASON and P. K. VOGT. 1973. Genetic recombinants and heterozygotes derived from endogenous and exogenous avian RNA tumor viruses. Virology. In press.

WEISSBACH, A., A. BOLDEN, R. MULLER, H. HANAFUSA and T. HANAFUSA. 1972. Deoxyribonucleic acid polymerase activities in normal and leukovirus-infected chicken embryo cells. J. Virol. *10:* 321.

WILSON, D. E. and H. BAUER. 1967. Hybridization of avian myeloblastosis virus RNA with DNA from chick embryo cells. Virology *33:* 754.

ZEIGEL, R. F., B. R. BURMESTER and F. J. RAUSCHER. 1964. Comparative morphologic and biologic studies of natural and experimental transmission of avian tumor viruses. Nat. Cancer Inst. Monogr. *17:* 711.

ZEILMAKER, G. H. 1969. Transmission of mammary tumor virus by female GR mice: Results of egg transplantation. Int. J. Cancer *4:* 261.

13

Genetics of RNA Tumour Viruses

The genetic analysis of RNA tumour viruses is in its infancy and questions such as how many genes these viruses have and how the functions they specify bring about virus replication and cell transformation have yet to be answered. Over the past few years, however, a steadily increasing number of reports of the isolation and characterization of mutant strains of RNA tumour viruses have appeared, and preliminary analyses of recombination between these viruses have been made. In short, the genetic analysis of RNA tumour viruses has begun.

The classical approach to the genetic analysis of the small genomes of viruses is to isolate mutants which can be characterized both physiologically and genetically. Such mutants fall into two classes: (1) conditional mutants, which are particularly suited to physiological experiments because the gene containing the mutation continues to function in permissive environments but fails to function in nonpermissive environments—this means that the gene function can be identified and the time at which it acts during infection determined; (2) nonconditional mutants, which lack this advantage. The gene bearing the nonconditional lesion is inactive under all conditions. But such mutants have one advantage—they are generally more stable than conditional mutants and are therefore suited for the fine analysis of the genome.

At present we know a great deal more about the genetics of the avian RNA tumour viruses than the mammalian RNA tumour viruses (see reviews by Vogt, 1972 and Vogt et al., 1972), not least because the relationship between the avian viruses and their host cells has been extensively characterized and nondefective transforming viruses are available (see Chapter 10). In this chapter, therefore, we discuss first the avian RNA tumour viruses and then draw comparisons between these viruses and the RNA tumour viruses of other animals.

CONDITIONAL MUTANTS OF AVIAN VIRUSES: PHYSIOLOGICAL GENETICS

A number of conditional mutants of avian sarcoma viruses have been isolated; all of them are temperature sensitive (*ts*) with a permissive temperature of 35–37°C and a nonpermissive temperature of 40–41°C. With one exception (Temin, 1971a, b) these mutants were isolated from stocks of mutagen-treated virus. Mutagenesis was performed either by treating virus preparations with nitrosoguanidine (Martin, 1970; Kawai et al., 1972) or by permitting virus replication in the presence of the RNA mutagens 5-azacytidine (Toyoshima and Vogt, 1969; Wyke, 1973a) or 5-fluorouracil (Biquard and Vigier, 1970; Kawai and Hanafusa, 1971). A surprisingly high yield of *ts* mutants has also been found in the early harvests following virus infection in the presence of the DNA mutagen 5-bromodeoxyuridine (Bader and Brown, 1971). Most mutants were isolated by a process of random cloning and testing, but Wyke (1973a) has described methods of enriching for mutants with desired properties.

Table 13.1 lists the *ts* mutants of avian sarcoma viruses reported up to January 1973, the virus strains from which they originated, and the workers responsible for their isolation. (We hope that a standard nomenclature can be agreed upon before too many more mutants are isolated.) By virtue of their known properties, these mutants have been tabulated in a scheme modified from Wyke and Linial (1973). The subdivisions of Table 13.1 are not intended to represent separate viral gene functions, and they may not even reflect meaningful functional groupings because the behaviour of several of these mutants varies with different conditions of infection, notably the multiplicity of infection. However,

Table 13.1. Temperature-Sensitive Mutants of Avian Sarcoma Viruses

Mutant Class[a]	Mutant Category	Description of *ts* Lesion[a]	Mutant Designation	Strain Designation[b]
T	T-1	*late*, continuously required	T1 to T6[c] FU-19[d] Ts 10, Ts 19, Ts 68[e] Ta[f] *ts* 20, *ts* 21[g] *ts* 22 to 29, 31 to 35[h]	SR SR SR BH B77 PR
	T-2	*early*, but required continuously	*ts* 30[h]	PR
C	C-1	*early*, required only transiently	*ts* 335[h], *ts* 337[j]	PR
	C-2	*early*, but continuously required for transformation; *late* and continuously required for replication	*ts* 338[h], *ts* 343[k]	PR
	C-3	*early*, but required continuously for focus formation; *early*, and required only transiently for agar colony formation and replication	*ts* 336[m] (previously *ts* 149)	B77
	C-4	*late*, and required continuously for focus formation and replication. No defect in agar colony formation	*ts* 334[m] (previously *ts* 75)	B77
	?	Not enough known to permit categorization	*ts* 339, *ts* 340[g]	B77
R	?	Not enough known to permit categorization	*ts* 667, *ts* 668[g] *ts* 669 to *ts* 680[g]	B77 PR

Modified from the scheme of Wyke and Linial (1973).

[a] See text.

[b] Strain designations: BH, Bryan high titre strain of Rous sarcoma virus (RSV); PR, Prague strain of RSV; SR, Schmidt-Ruppin strain of RSV; B77, avian sarcoma virus Bratislava 77.

[c] Martin, 1970, 1971. [d] Biquard and Vigier, 1970. [e] Kawai and Hanafusa, 1971; Kawai et al., 1972. [f] Bader and Brown, 1971; Bader, 1972. [g] Friis, unpublished data; Vogt et al., 1972. [h] Wyke, 1973a, b. [j] Linial and Mason, 1973. [k] Wyke and Linial, 1973. [m] Toyoshima and Vogt, 1969.

Table 13.1 gives an idea of the potential complexity of sarcoma virus genomes and it provides a rational framework for further discussion and experimentation.

In the table mutant avian sarcoma viruses are divided into three classes: (1) those which replicate at the nonpermissive temperature but are defective in transforming functions (class T); (2) those which are coordinately temperature sensitive in at least some functions involved both in transformation and replication (class C); and (3) those which appear to transform competently but cannot replicate infectious virus at the nonpermissive temperature (class R). Mutants in these classes can be subdivided into physiological categories by virtue of their behaviour when infected cultures are shifted to or from the permissive temperature 24 hours after infection.

If a temperature-sensitive property of a cell infected by a mutant is not restored after a shift from nonpermissive to permissive temperature 24 hours post infection, the mutated function of the virus is defined as early. Early functions are, therefore, those that are required during the first 24 hours after infection to achieve some form of stable association between the virus and the cell. These early functions tend to be required only transiently after infection, so that if the cells infected by early mutants are kept for the first 24 hours post infection at the permissive temperature and then switched to the nonpermissive temperature, the infection resembles a wild-type infection. There may, however, be a continuous requirement for some early viral functions.

By contrast, if a mutant virus transforms its host and replicates when infected cells are shifted to the permissive temperature after a first 24 hours at the nonpermissive temperature, the *ts* function is defined as a late function. With all late mutants isolated to date, the maintenance of the *ts* property of the infected cells seems to require the continuing action of the mutant function. When cells infected by late *ts* mutants are shifted from permissive to nonpermissive temperatures, phenotype traits controlled by the *ts* viral gene product are lost. A late function is, however, defined simply as one required after the virus has become stably associated with the cell, and it is possible to imagine late functions that are required only transiently, although none have yet been identified.

It has not been possible, on the basis of data available, to categorize in Table 13.1 all the known *ts* mutants of avian sarcoma viruses. In particular Temin (1971a, b) isolated *ts* mutants from nonmutagenized stocks of virus at the relatively high frequency of 3 percent. These mutants apparently include representatives of all the three chief physiological classes, but since little information about these viruses is available they are not discussed further.

Mutant Category T-1

For unknown reasons the majority of *ts* mutants so far isolated fall into this group. These mutants replicate in chick cells at the nonpermissive temperature but transformation is temperature sensitive, which proves that transformation is under the continuous control of viral genes (Biquard and Vigier, 1970, 1972; Martin, 1970, 1971; Bader, 1972; Kawai et al., 1972; Wyke, 1973a). The *ts* functions can be recognized, not only by their effect on cell morphology and cell growth under agar and in agar suspension, but also by the conditional expression of other parameters of transformation. Thus, when the properties discussed below are assayed, cells infected by T-1 mutants behave like uninfected or leukosis virus-infected cells at the nonpermissive temperature.

Temperature-sensitive Properties of Cells Infected by T-1 Mutants

1. The rate of sugar uptake is increased several-fold in transformed cells (Hatanaka and Hanafusa, 1970; Kawai and Hanafusa, 1971; Martin et al., 1971).

2. Susceptibility to agglutination by wheat germ agglutinin or concanavalin A is increased following transformation (Burger and Martin, 1972).

3. Hyaluronic acid synthesis is also increased in transformed cells (Bader, 1972).

4. The amount of particular glycoproteins and glycolipids in cells transformed by Martin's (1971) mutant T5 and growing at the permissive temperature resembles the amount of these molecules in cells transformed by wild-type virus, but at the nonpermissive temperature the amount of these molecules resembles that in uninfected cells (Warren et al., 1972a) and in cells infected with leukosis virus (Hakomori et al., 1971). Furthermore Warren et al. (1972b) report that the activity of a sialyl transferase in cells infected by mutant T5 is not detectable at the restrictive temperature.

5. The concentration of cyclic adenosine 3′,5′-monophosphate (cyclic AMP) is lower in transformed cells than in untransformed cells, and in cells infected by the Ta mutant isolated by Bader (1972) it is far

lower at the permissive than the nonpermissive temperature (Otten et al., 1972). Moreover, the analogue dibutyryl cyclic AMP will prevent the appearance of transformation when infected cultures are shifted from the nonpermissive to the permissive temperature.

6. Unkeless et al. (1973) have described an enzyme, fibrinolysin T, which is present in cells of solid tumours and in cells infected by Rous sarcoma virus, but not in those infected with avian leukosis virus. This enzyme activity is not detectable in cells which are infected by the T5 mutant of Martin (1971) and maintained at the nonpermissive temperature.

7. Tumours produced on the chorioallantoic membrane of embryonated eggs by T-1 mutants are temperature sensitive (Biquard and Vigier, 1972), and Kawai and Hanafusa (1971) found that in chicks (whose body temperature is nonpermissive for the mutants) T-1 mutants induced solid tumours much more slowly than wild-type virus.

Heterogeneity of T-1 Mutants

The general similarity between cells infected by leukosis viruses and cells infected by these T-1 mutants and maintained at the nonpermissive temperature does not, however, justify the assumption that all the T-1 mutants are similar and behave as leukosis viruses in a restrictive environment. Leukosis viruses are, after all, capable of transforming cells of the hemopoietic and reticuloendothelial systems, and it is not known whether their tropism is encoded in, or can be acquired by, the genome of sarcoma viruses. Thus, some of the T-1 *ts* sarcoma viruses may be completely nonpathogenic, or they may produce unusual clinical symptoms. A study of such mutants may help to reveal the means by which the expression of the virus genome is controlled in different cell types. Unconditionally defective transformation mutants, which are nonpathogenic, might also have prophylactic applications because of their ability to interfere with other members of their subgroup (see Chapter 10).

Some T-1 mutants are known to produce effects in tissue culture which suggest that they still possess functions that are not found in leukosis viruses. Kawai and Hanafusa (1972a) and Wyke and Linial (1973) have noted that confluent cultures infected at high multiplicities under nonpermissive conditions with certain T-1 mutants of subgroups A and C support plaque production by superinfecting leukosis viruses of subgroups B or D. This phenomenon may result from the activity of residual transforming functions in the T-1 mutant genomes, since Wyke and Linial (1973) observed that their plaque-supporting mutants stimulated increased cell growth at the nonpermissive temperature even though there was no obvious change in cell morphology. Plaque formation may alternatively be a trivial effect produced by some undefined stress in doubly infected cells. This second possibility is favoured by Graf's (1972) observation that leukosis viruses of subgroups B and D are able to produce plaques in solitary infection under appropriate culture conditions. Moreover, Bader (1972) noticed that "foci" produced by his mutant of Bryan high titre strain Rous sarcoma virus at nonpermissive temperature were themselves plaque-like. Biquard and Vigier (1972) also found that transformation by their mutant was not entirely defective, for the addition of dextran sulfate to the medium produced a sixfold enhancement of focus formation under nonpermissive conditions.

Other differences between mutants in the T-1 category have been detected by determining the time scale and metabolic requirements for the transformation or reversion of transformation which occur upon shifts in temperature of incubation. The usual parameters for quantifying cell transformation have been changes in cell morphology and rate of sugar uptake. On a shift to permissive temperature some mutants produce 50 percent transformation in 5–6 hours, and inhibitor studies indicate that this transformation requires protein synthesis but not synthesis of DNA or RNA (Kawai and Hanafusa, 1971; Biquard and Vigier, 1972). Thus transformation will occur in the presence of actinomycin D even though this drug greatly reduces virus production per cell. The equally rapid disappearance of transformation at nonpermissive conditions seems to require no macromolecular synthesis at all (Biquard and Vigier, 1972). The mutant studied by Bader (1972) differs in the great rapidity (less than one hour) with which morphological transformation appears after a shift to the permissive temperature even in the presence of inhibitors of protein synthesis. The concomitant decrease in the concentration of cyclic AMP, another parameter of transformation, is also very rapid (Otten et al., 1972). However, the change in the rate of hexose uptake in cells infected by this mutant is slower than the morphological change and is dependent on protein and, apparently, RNA synthesis. The T5 mutant of Martin (1971) may be different yet again. The appearance of the transformed cell fibrinolysin T on a shift to permissive conditions seems to require both messenger RNA and protein synthesis (Unkeless et al., 1973), but we do not know whether other parameters of transformation have the same requirements.

The variable results from these inhibitor studies, comprising as they do work on individual mutants in several laboratories, are not the best evidence for heterogeneity among the T-1 mutant category. More convincing evidence comes from tests for complementation of cell transforming ability at nonpermissive temperatures. Kawai et al. (1972) have divided three T-1 mutants into two complementation groups, and Wyke (1973b), in a study of 14 mutants, has detected at least four complementation groups. Unfortunately it is not yet possible to correlate these two sets of results, but it seems likely that there are at least four complementation groups, the functions of which are required only for cell transformation. Although it has not yet been proven that these four complementation groups correspond to four distinct genes, it seems likely that the mutants in the T-1 category are not all defective in one and the same function. Clearly further investigations of these mutants should provide valuable information about the mechanism of transformation.

Mutant Category T-2

The sole mutant in this category resembles the T-1 mutants but possesses, in addition, a *ts* lesion in an early function that is required for the establishment of transformation (Wyke and Linial, 1973). Cells infected with this mutant and held for 24 hours at the nonpermissive temperature are not transformed when returned to the permissive temperature, but cells infected at the permissive temperature are transformed, although they lose the transformed cell phenotype when shifted to 41°C. The mutant virus particles are heat labile, and when held at 41°C they lose their ability subsequently to transform cells at 35°C. When the subgroup specificity of this virus is altered by genetic recombination with an appropriate leukosis virus, the virions are no longer heat labile and the establishment of transformation is not temperature sensitive. In short, the mutant behaves as a member of the T-1 category (Wyke, 1973b). This suggests that the original isolate was a double mutant, and it is possible that other mutants with a lesion(s) in a function(s) that is anomalously required both early in infection and continuously during infection may also prove to be double mutants. It also seems that some factor present in, or closely associated with, the T-2 mutant virus particle is needed for successful cell transformation but not for virus replication. Clearly further study of this mutant could prove rewarding.

Mutant Category C-1

The two mutants in this category are defective in early functions that are transiently required soon after infection for both virus replication and establishment of transformation. The virions of these two mutants are heat labile and the reverse transcriptase activity of the virions is temperature sensitive (see Chapter 11 and Linial and Mason, 1973).

Mutant Categories C-2 and C-3

Mutants in these two categories are coordinately *ts* viruses: they neither replicate nor transform at the nonpermissive temperature, the virions are not heat labile, and the physiology of the mutants is a complex combination indicating lesions in early and late functions (Friis et al., 1971; Wyke and Linial, 1973 and see Table 13.1). These properties suggest that more than one gene of these mutants is mutated but there is as yet no direct evidence that this is the case. Obviously as these mutants are more fully characterized, they should throw light on both early and late stages of transformation and replication.

Mutant Category C-4

The sole mutant in this category, *ts* 334, has been studied and compared with the early mutant *ts* 336 (category C-3) by Friis et al.

(1971). Both viruses resemble one another in their coordinate temperature sensitivity and their inability to produce tumours in chickens. The virions are no more heat labile than wild-type virus particles. The C-4 mutant, however, behaves as a late mutant in temperature-shift experiments, and this correlates with the presence of virus-specific RNA and increased amounts of the group-specific antigen complex in infected cells at the nonpermissive temperature. This mutant can also induce colony formation in agar at 41°C. However, cells infected by *ts* 334 at 41°C are not morphologically transformed (see Figure 13.1) and foci cannot be detected. This suggests that focus formation and growth in agar may be controlled by different viral genes.

Envelope antigens and noninfectious C-type particles are not detected in cultures under nonpermissive conditions, but on a shift to permissive temperature, synthesis of infectious virus is rapidly resumed. This maturation requires protein synthesis, and data of Katz and Vogt (1971) suggest that one of the viral envelope glycoproteins, and possibly also one of the group-specific antigens, is not synthesized at the nonpermissive temperature. Cells infected by all T-class mutants probably produce these virion structural proteins, so, although they may be necessary, they are not sufficient to induce transformation; and these proteins may even be irrelevant to transformation because Toyoshima (personal communication) finds that the mutant *ts* 334 possesses more than one mutation.

Mutant Class R

Mutants of this class produce foci at the nonpermissive temperature but no infectious virus (Friis, unpublished data; Vogt et al., 1972). Virus-specific RNA is detected at the restrictive temperature and this, presumably, is at least partially translated since transforming functions are expressed (Friis, personal communication). Available physiological data do not permit further subdivision of mutants in this class, but Friis (personal communication) suggests that these mutants may be heterogeneous because some of them are complemented by endogenous chick helper factor (Chapter 12) whereas others are not.

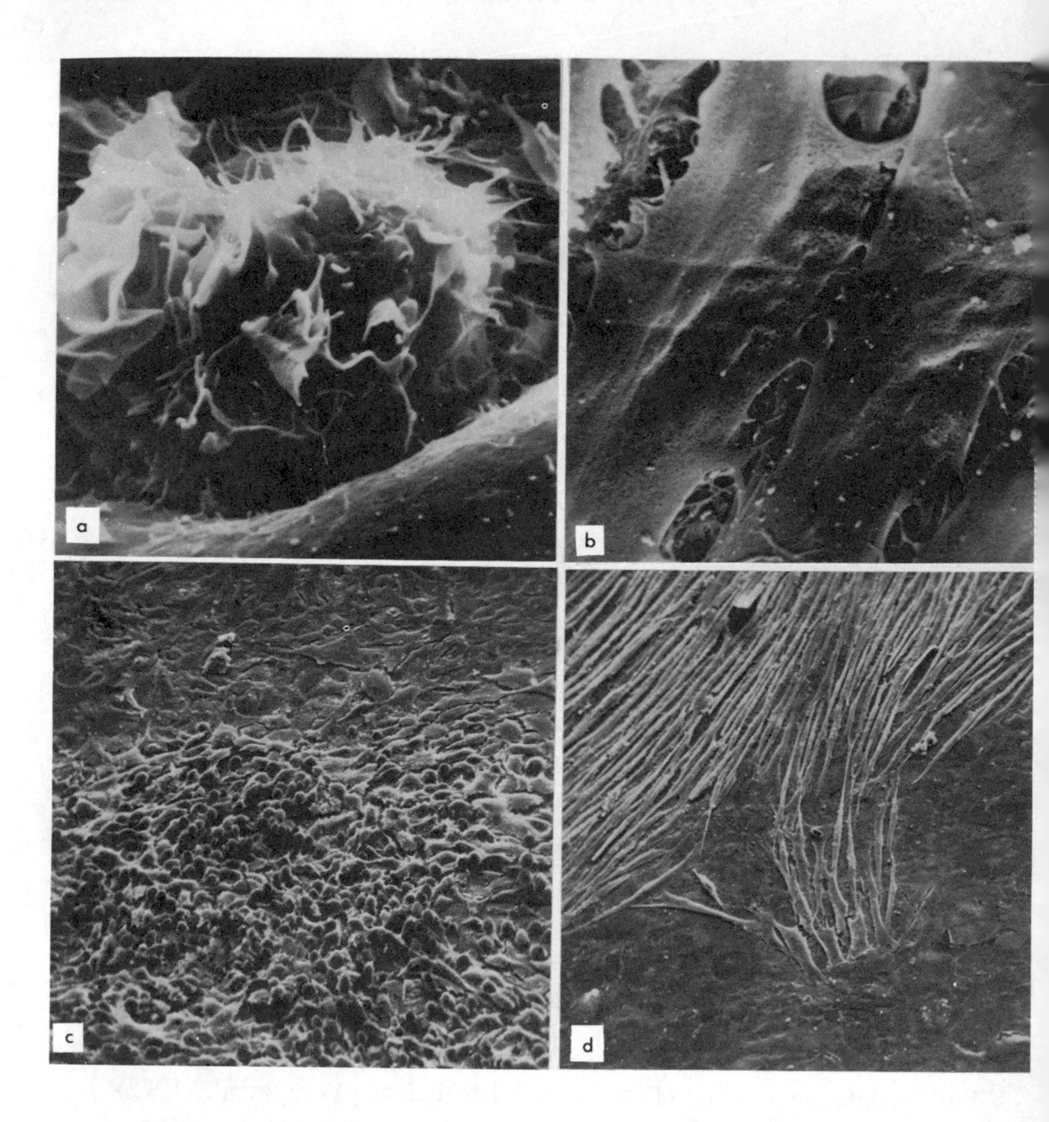

Figure 13.1. a, b: Chick cells infected with a temperature-sensitive mutant of avian sarcoma virus B77 (*ts* 334, originally *ts* 75). ×5000, scanning electron micrographs by A. Boyde and R. Weiss. **(a)** At permissive temperature (35°C), cell transformed; **(b)** at nonpermissive temperature (41°C), cells not transformed
c, d: Quail cells infected with morphological variants of Bryan high titre strain RSV. ×200, scanning electron micrographs by A. Boyde and R. Weiss. **(c)** Round cell (morph^{r}) variant of BH.RSV; **(d)** fusiform cell (morph^{f}) variant of BH.RSV.

Complementation between *ts* Mutants and Wild-type Viruses

Attempts have been made to identify the lesions in various *ts* mutants by testing for complementation with either wild-type leukosis or wild-type sarcoma viruses. As expected, leukosis viruses do not complement lesions in *ts* sarcoma virus mutants that replicate in, but do not transform, chick fibroblasts at 41°C (Wyke and Linial, 1973), but leukosis viruses can complement R-class sarcoma mutants and allow them to replicate at 41°C (Friis, unpublished data; Vogt et al., 1972). In the light of this finding it is perhaps surprising that wild-type leukosis viruses fail to complement any defect of C-class mutants at the non-permissive temperature (Friis et al., 1971; Linial and Mason, 1973; Wyke and Linial, 1973). One might have expected complementation at least for virus replication. Although complementation tests with C-2 mutants have not yet been done, it may well prove to be the case that gene products of leukosis viruses cannot rescue sarcoma viruses with lesions in early functions. By contrast, Friis et al. (1971) claim that wild-type sarcoma viruses can complement the replication of both *ts* 334 and *ts* 336, mutants of categories C-4 and C-3, upon coinfection at 41°C. Complementation of transforming functions obviously cannot be tested by coinfection with a transforming virus. Since *ts* 336 possesses a lesion in an early function transiently required for virus synthesis, this leads to the intriguing possibility that sarcoma and leukosis viruses differ in an early step that is essential for virus replication. More extensive complementation tests by both coinfection and superinfection, applied to a wider range of mutants and linked with recombination studies (see below), are needed to confirm or refute such speculations.

Complexity of Avian Sarcoma Virus Genome

To summarize, the properties of these various *ts* mutants indicate several of the functions specified by avian sarcoma virus genomes. The behaviour of C-1 mutants, which have a temperature-sensitive, reverse transcriptase activity, indicate that at least part of this enzyme is probably coded by the virus and that its activity is essential for the establishment of infection. The C-3 mutant, virions of which have an active reverse transcriptase (Katz, personal communication) and are not particularly thermolabile, is apparently defective in a distinct and as yet unidentified early function also required for the establishment of infection. The mutants in categories C-2 and T-2 are defective in early functions essential for the subsequent establishment of transformation, but C-2 mutant virions are not temperature sensitive whereas T-2 mutant virions are; so although the physiological sequelae are similar, the genetic lesions may differ. A function with a *ts* lesion in mutants of the C-2 category is also required continuously to maintain both transformation and replication. This may be a function (physiologically similar to the function mutated in C-4 mutants) whose late nature is obscured by a second *ts* lesion in an early function. It would be informative to see if C-2 and C-4 mutants have any other similarities, for instance, in the transcription of virus-specific RNA under nonpermissive conditions. Not surprisingly, the other late mutants described are defective in either replication or transformation. Again, differences between mutants can be detected within these categories, notably among the T-class mutants. Physiological differences among the T-1 mutants, based on inhibitor studies, differential ability to support leukosis virus plaques and complementation tests, suggest that at least four functions may be associated with transformation and yet be seemingly irrelevant to virus replication.

Taken at face value the different physiological properties of all these *ts* mutants indicate that the viral genome specifies about a dozen functions, but differences in such physiological properties may be trivial or even artifactual. So far only the C-1 lesion in a gene for reverse transcriptase, and possibly the C-4 lesion in a gene for an envelope glycoprotein, has been correlated with specific biochemical defects. There are not only many "orphan" mutants but also many structural molecules of the virion which are not known to be specifically altered in mutants isolated to date. Table 13.1 is simply a summary of current data; it is certain to be rendered obsolete as new classes of mutants are isolated and as currently available mutants are more thoroughly characterized.

NONCONDITIONAL MUTANTS OF AVIAN VIRUSES

The nonconditional mutants of avian RNA tumour viruses include those with absolute defects in some virus function and those showing variation in a virus-specific marker.

These mutants, many of which have been isolated from stocks of virus after chemical mutagenesis or irradiation, tend to be more stable than temperature-sensitive mutants, and many defective viruses probably arise by genetic deletion (Duesberg and Vogt, 1970; Martin and Duesberg, 1972). Their reversion frequency is low and they do not suffer from the leakiness, the tendency for a mutant function to behave as wild type under nonpermissive conditions, shown by all *ts* mutants to a greater or lesser extent.

Virus Variants

The two chief virus markers for which variants have been detected are those determining the morphology of the transformed cells and the envelope properties of the virus particles (see Chapter 10). To these may be added the differences between virus strains, often manifest as variations in focus architecture (compact or diffuse, monolayered or multilayered) (Vogt, 1967b).

The most common morphological variants are those inducing fusiform transformed cells (Temin, 1960, 1961; Yoshii and Vogt, 1970; Martin, 1971). Temin named these mutants "morph^{f}," distinguishing them from "morph^{r}" viruses which induce transformed cells of the more usual round cell type (see Figure 13.1). Martin (1971) found that cells infected by the *ts* mutant T5 and held at temperatures intermediate between the permissive and the nonpermissive show a fusiform transformation. He suggested that the fusiform cell morphology represents an intermediate degree of transformation, and it is interesting that Friis et al. (1971) noted that the fusiform mutant they worked with was also temperature sensitive.

Variations in virus envelope properties, reflected in differences in host range (see Chapter 10), occur in several strains of nondefective (helper independent) avian sarcoma viruses (Duff and Vogt, 1969). In addition, Altaner and Temin (1970), in a study of the infection of mammalian cells by avian sarcoma virus B77, found that virus recovered from rat cells has stably increased plating efficiency on these cells. This change is accompanied by an alteration in envelope antigenicity but no concomitant reduction in the virus's ability to infect avian cells. However, they also presented evidence arguing that the broadened host range is not solely the result of these changes in envelope properties.

Altaner and Temin proposed that B77 acquired this ability to efficiently infect rat cells after genetic recombination, by an unknown mechanism, between the portion of the avian virus genome specifying virion envelope properties and some component of the rat cell genome. It has now been shown that the determinants of the envelope properties of avian sarcoma viruses can recombine, not only with genes in other independently replicating viruses (Vogt, 1971b; Kawai and Hanafusa, 1972b), but also with the endogenous viral genomes of host cells, which have been described in Chapter 12 (Weiss et al., 1973). This phenomenon will be discussed later.

Classes of Nonconditional Mutants

The nonconditional mutants of RNA tumour viruses, like the conditional mutations, can be divided into three chief classes: (1) those defective for transformation but not replication. Such mutants are variously designated in the literature as nonconverting (NC) or nontransforming (NT) viruses. (2) Those able to transform cells but unable to replicate. Cells infected by these viruses have been described as nonproducer (NP) cells. (3) Those with a coordinate defect in both cell transformation and virus replication. We refer to these viruses as transformation defective (*td*), replication defective (*rd*) and coordinately defective (*cd*), respectively.

Transformation-defective Viruses

Nonconditional mutants of this type arise spontaneously in nonmutagenized stocks of helper-independent avian sarcoma viruses (Dougherty and Rasmussen, 1964; Hanafusa and Hanafusa, 1966; Duff and Vogt, 1969; Vogt, 1971a; Martin and Duesberg, 1972). Vogt

(1971a) has estimated the incidence of these transformation-defective viruses, which are called Rous associated viruses, to be about 4–17% in various cloned stocks of sarcoma viruses. It is possible, but not proven, that this incidence is increased by treatment with physical or chemical mutagens (Goldé, 1970; Toyoshima et al., 1970; Graf et al., 1971).

The *td* viruses possess the same envelope properties as the sarcoma viruses they accompany, and they arise with the same incidence in sarcoma virus stocks grown in cells possessing endogenous chick helper factor (Chapter 12) as in stocks propagated in cells lacking this factor (Vogt, 1971a). It seems, therefore, that they are derived from sarcoma viruses and their origin does not normally involve recombination between the genes of sarcoma viruses and those of endogenous viruses.

The behaviour of these defective viruses in cultures of fibroblasts resembles that of avian leukosis viruses, and, where examined, their RNAs possess the electrophoretic mobilities of those of leukosis viruses (Duesberg and Vogt, 1970; Martin and Duesberg, 1972). In other words, *td* sarcoma virus particles lack the larger *a* RNA subunit typical of transforming sarcoma viruses (Chapter 10) and are probably deletion mutants. The formal possibility that the *a* unit is missing because of some constraint applied by replication in untransformed cells was made less likely by the findings of Martin and Duesberg (1972) that a sarcoma virus temperature sensitive for transformation but not for replication retains the *a* subunit even when propagated at the nonpermissive temperature.

It is dangerous to suppose, however, that all transformation defective variants are analogous to leukosis viruses, even though the resemblances between some *td* sarcoma viruses and leukosis viruses may be of evolutionary significance. Graf et al. (1971) claim that one of their *td* mutants can rescue transforming functions of Rous sarcoma virus rendered defective by irradiation. Such rescue is not usually effective when attempted with leukosis viruses (Levinson and Rubin, 1966; Toyoshima et al., 1970), and this suggested to Graf et al. that their variant retains focus-forming activity which is not present in true leukosis agents. Such mutants should exist, for it would be surprising if all *td* sarcoma viruses arose by deletions to the complete exclusion of point mutations. On the other hand, it has already been pointed out in the section on conditional mutants that leukosis viruses themselves transform appropriate cells, and we do not know how many of the *td* sarcoma viruses retain this capability. Biggs et al. (1973) find that some *td* sarcoma viruses do cause leukosis in chicks, but it is equally likely that many do not.

Replication-defective Viruses

An example of this class, the Bryan high titre strain of Rous sarcoma virus (BH.RSV) (Hanafusa et al., 1963), has long been known. This virus is able to transform cells, but in the absence of a helper virus it can only produce noninfectious particles. A surface glycoprotein of these noninfectious particles appears to be defective (Scheele and Hanafusa, 1971) and this defect can be compensated by phenotypic mixing with the envelope proteins of other viruses (see below). When the endogenous viral genomes of chick cells are sufficiently active to complement the coat defect of Bryan virus, the cells are said to be expressing their chick helper factor activity and to be "chick helper factor positive" (see Chapter 12). Replication-defective viruses that may have lesions similar to BH.RSV have been detected in stocks of the leukosis viruses avian myeloblastosis virus (Moscovici and Zanetti, 1970) and MC29 (Ishizaki et al., 1971; Langlois et al., 1971). However, it is in the mammalian RNA tumour viruses that defectiveness of this kind is most widespread, and this will be discussed later.

In recent years studies of both mutagenized and nonmutagenized stocks of sarcoma virus have revealed many more replication-defective variants (Goldé, 1970; Kawai and Yamamoto, 1970; Toyoshima et al., 1970; Weiss, 1972). Graf et al. (1971), however, could find very few such viruses, and cells transformed by those they did detect rapidly lost the transformed phenotype. By contrast to other workers, Graf et al. suggested that the whole of the sarcoma virus genome is essential, directly or indirectly, for transformation and that defects in replication incur defects in transformation. Defects in replicating and transforming properties may indeed be linked, particularly if there is some polarity in the expression of the virus genome. However, the well-documented existence of both nonconditional and *ts* replication-defective mutants that can transform fibroblasts (Vogt et al., 1972) makes it very unlikely that all viral genes are involved in transformation. Possibly Graf et al. failed to detect such mutants because of helper activity in their cells which complemented all or most of the functions that were required for replication alone.

Kawai and Yamamoto (1970) and Weiss (1972) were able to classify a number of *rd* mutants on the basis of their behaviour in superinfection and rescue experiments. The categories of Kawai and Yamamoto are:

Type I. The infected cells are still susceptible to infection by virus of the same subgroup as the variant; that is, there is no interference (see Chapter 10). The transformation markers of the variant can be rescued by superinfection with leukosis virus.

Type II. The infected cells exert viral interference, but even leukosis virus that is not interfered with fails to rescue the transformation marker.

Type III. There is no interference, but neither is marker rescue achieved.

Weiss's categories cannot be directly compared with these, for he used different distinguishing criteria, as follows:

1. Mutants that are not complemented by leukosis virus or chick helper factor (Chapter 12). The mutants isolated by Goldé (1970) and Toyoshima et al. (1970) seem to fall into this group.
2. Mutants that are complemented by leukosis virus but not by chick helper factor.
3. Mutants that are complemented by both leukosis virus and chick helper factor. The behaviour of these mutants resembles that of the Bryan strain RSV, though it is not known whether they produce noninfectious particles.

Coordinately Defective Viruses

Viruses of this type may occur frequently, but they are seldom recognized because of the extreme nature of their defect. However, coordinate lesions often affect early events in the virus life cycle (see section on conditional mutants) and if such early functions can be complemented, the virus might subsequently transform or replicate and hence be detected. This is the case with the α mutant of Bryan high titre strain of Rous sarcoma virus (Hanafusa and Hanafusa, 1968). This mutant not only bears the envelope defect of the Bryan strain, but it also lacks a functional reverse transcriptase (Hanafusa and Hanafusa, 1971; Robinson and Robinson, 1971; Hanafusa et al., 1972 and see Chapter 11). Studies of virus with a *ts* lesion in the activity of this enzyme (Linial and Mason, 1973) confirm that it is necessary for both replication and transformation. Thus the RSVα mutant is coordinately defective and can only spread from cell to cell by extensive phenotypic mixing with competent viruses, as described in the next section.

In short, a large number of defective sarcoma viruses have been isolated, but they have not yet been used to any great extent in experiments. Physiological differences have been detected between different defective mutants, but these mutants should prove more useful in genetic studies than investigations of virus physiology. One such application is to use the presumed deletion mutants to map the functions of *ts* mutants (Vogt et al., 1972). For example, mutants defective in a particular property and with a presumed deletion could be tested for their ability to complement various mutants which are temperature sensitive in that property.

Whatever the experimental uses of defective viruses, the concept of defectiveness is fundamental to an understanding of the biology of RNA tumour viruses. The reasons why defective viruses might accumulate in a population are discussed later in this chapter, but one aspect of defectiveness will be considered now. It is not clear whether the ability to transform a particular cell type confers a selective advantage on an RNA tumour virus. Thus one cannot say whether a transformation-defective virus suffers any disadvantage compared with its competent brethren. However, a defect in replication is obviously disadvantageous and one would expect such variants to be rapidly eliminated from the replicating population. That this does not occur is due to the complementation of defective functions which result from phenotypic mixing.

INTERACTIONS BETWEEN VIRUSES

Phenotypic Mixing

When two viruses infect the same cell, progeny may be formed that possess the genome of one parent but the structural proteins, and hence the phenotype, of either or both parents. Such phenotypic mixing was first demonstrated among avian tumour viruses in studies on the defective Bryan high titre strain of Rous sarcoma virus (BH.RSV).

BH.RSV contains an excess of a leukosis virus called Rous associated virus (RAV) (Rubin and Vogt, 1962). Solitary infection of cells with BH.RSV results in transformation but no production of infectious virus (Temin, 1962; Hanafusa et al., 1963). Transforming virus can be rescued from these nonproducer cells by superinfection with an RAV, and this rescued sarcoma virus bears the envelope properties of the rescuing RAV (Hanafusa et al., 1964; Hanafusa, 1965; Vogt, 1965; see also Chapter 10). Solitary infections by the rescued sarcoma virus again result in transformed nonproducer cells, so the BH.RSV is still defective and presumably has not acquired any envelope genes from the leukosis virus (Hanafusa et al., 1963). BH.RSV can, in fact, only be transferred from cell to cell as a pseudotype (Rubin, 1965), its own envelope defect being complemented by the nondefective envelope components of the helper virus. The nature of a BH.RSV pseudotype is indicated in abbreviations by including the helper virus type in parentheses, for example, BH.RSV (RAV-2).

Subsequent work showed that phenotypic mixing of envelope properties could also occur among nondefective sarcoma and leukosis viruses (Hanafusa and Hanafusa, 1966; Vogt, 1967a). The behaviour of the phenotypically mixed progeny virions suggested that their envelopes are a mosaic of materials acquired from both parents (Vogt, 1967a). It is also probable that phenotypic mixing of other virion structures occurs; there is, for example, strong evidence that this is the case with the virion reverse transcriptase. The mutant BH.RSVα, which lacks reverse transcriptase, can be rescued in infectious form from transformed cells by superinfection with RAV (Hanafusa and Hanafusa, 1968). The rescued virus must therefore be a pseudotype for the enzyme as well as for envelope glycoproteins. A similar reverse transcriptase pseudotype can be obtained of a mutant whose enzyme activity is temperature sensitive (Linial, personal communication).

Phenotypic mixing can be maintained only under conditions where cells are infected by at least two viruses and where the viruses involved do not significantly interfere with one another. If a phenotypically mixed virus population is grown under conditions of solitary infection, the parental phenotypes should rapidly reappear. If this does not happen, then the existence of some form of genotypic mixing, either recombination or heterozygosis, must be suspected.

Genotypic Mixing

The phenomenon of phenotypic mixing and its biological importance delayed the demonstration of any form of genotypic mixing among avian RNA tumour viruses. Eventually careful cloning and analysis of the progeny of mixed infections enabled Vogt (1971b) and Kawai and Hanafusa (1972b) to demonstrate the occurrence of genetic recombination between markers among these viruses. The markers used in both laboratories were transformation and host range (envelope properties). Vogt found that sarcoma viruses acquire the host range of coinfecting leukosis (nontransforming) viruses at a high frequency, suggesting that transformation and host-range markers are unlinked and reassort independently. Studies on *ts* mutants, which are temperature sensitive for transformation (class T), suggest that the *ts* markers also segregate with host-range markers in an unlinked fashion (Kawai et al., 1972; Wyke, 1973b). In a study of 13 class-T mutants, Wyke (1973b) found that recombinant *ts* sarcoma viruses could be obtained readily in every case, indicating the *ts* lesions are not closely linked to envelope properties. Preliminary recombination studies with other *ts* markers give less clearcut results. Five host-range recombinants of the T-2 mutant, *ts* 30, all behave like mutants in group T-1, suggesting that the early lesion in this mutant might be linked to host-range properties

(Wyke, 1973b). The putative reverse transcriptase mutants of RSV, *ts* 335 and *ts* 337, will recombine with leukosis virus to yield wild-type sarcoma virus which retains its original host range (Linial and Mason, 1973). A few host-range recombinants are obtained, and these also seem to be predominantly wild type (Linial, unpublished data). These results are difficult to interpret, but further experiments should reveal whether or not the gene(s) for reverse transcriptase is linked to those controlling host range or those for transformation, or reassorts independently of both. In fact, there are probably now enough mutants in sufficiently diverse functions to attempt a preliminary genetic mapping of these viruses by recombination studies. If linkage groups exist, it should be possible to detect them by such experiments. It will also be instructive to compare any data so obtained with those derived from the deletion mapping experiments outlined in the section on nonconditional mutants.

Interesting light on the possible mechanism of recombination between avian C-type viruses has been shed by the experiments of Weiss et al. (1973). They showed that the host-range marker of the endogenous virus of chick cells, a leukosis-type virus of subgroup E (see Chapter 12), will recombine with infecting sarcoma viruses. Such recombination occurs only in cells in which the endogenous virus envelope markers are expressed and are available for phenotypic mixing with defective BH.RSV, in other words, cells positive for chick helper factor (*chf*). No recombinants are obtained from cells which are negative for helper factor and possess endogenous viral DNA but little or no virus-specific RNA (Leong et al., 1972). Cloned recombinants possess dual host range for more than one generation and on repeated cloning will segregate viruses of parental or recombinant genotypes. Thus, these viruses are genotypically mixed and are most probably unstable "heterozygotes." These results have the following implications.

1. An endogenous viral genome is only available for recombination if it is transcribed into RNA. This transcription is usually incomplete in *chf* positive cells (Bishop et al., 1973; Hayward and Hanafusa, 1973) so only certain endogenous viral genes will recombine. The genes specifying endogenous virus envelope proteins are expressed in *chf* positive cells since both phenotypic mixing and recombination of these markers are observed, and such cells specifically absorb antiserum

prepared against subgroup E envelope proteins (Weiss, Friis, Vogt, unpublished).

By contrast, the genes specifying the reverse transcriptase of the endogenous virus are not expressed in such helper factor positive cells: viral reverse transcriptase is not found in *chf* positive cells (Weissbach et al., 1972) and mutants lacking the enzyme are not complemented in these cells (Hanafusa and Hanafusa, 1971; Weiss, 1973); neither do they recombine with the endogenous virus to yield wild-type viruses with functional reverse transcriptase (Weiss, 1973). However, leukosis virus synthesized when the endogenous provirus in *chf* positive and *chf* negative cells is induced to replicate contains functional reverse transcriptase (Weiss et al., 1971), so the endogenous provirus contains a gene for this enzyme, even though it is not expressed in most *chf* positive cells.

Weiss (1973) has also demonstrated that *chf* positive cells will complement one of his replication-defective mutants (*rd* L12) but not another (*rd* L5). Thus, by performing complementation and recombination studies with a variety of other mutants, it should be possible to further identify the viral functions which are active during the partial expression of endogenous viruses. Finally, if recombination between endogenous and exogenous viruses occurs at the RNA level, the same is presumably also true of recombination between pairs of exogenous viruses.

2. The formation of unstable "heterozygotes" may be necessary for recombination because particles with dual host range are common in early recombinant clones. Possible ways in which recombinants can arise from heterozygotes are discussed below. Necessary or not, such heterozygote particles will prove a stumbling block to any genetic analysis of RNA tumour viruses, for it will be unwise to assume that any virus is genotypically stable until it has been cloned several times.

3. Recombination with endogenous virus provides a rationale for instances in which host-range modification of a virus occurs after passage through apparently uninfected cells (e.g., see Altaner and Temin, 1970). The failure of some viruses to recombine is possibly of great importance in maintaining from generation to generation certain defective viral traits, and this is discussed in the section on the genome of RNA tumour viruses.

OTHER ONCOGENIC RNA VIRUSES

Although we currently know more about the genetics and biology of the avian sarcoma and leukemia viruses than we know about the corresponding viruses of mammals such as mice and cats, the mammalian systems offer certain advantages to the experimenter. Chief among these is the availability of stable lines of host mammalian cells that can be cloned (stable lines of chick cells have not yet been selected and experiments perforce are done with primary and secondary cell cultures; see Chapter 2). Because stable lines of mammalian cells exist, large-scale biochemical experiments with cloned viruses and cloned cells are feasible, as are investigations of the genetic basis of the modifications of host cells consequent upon virus infection. Furthermore, information obtained from investigations of mammalian RNA tumour viruses and their host cells is probably of greater relevance to oncogenesis in man than is equivalent information about avian systems.

The mammalian viruses that have been most studied by molecular geneticists are those of the mouse leukemia/sarcoma complex.

Defectiveness of Murine Sarcoma Virus (MSV)

Nondefective, helper-independent murine sarcoma viruses, resembling the helper-independent strains of RSV, have never been isolated.

Early work showed that the formation of foci of transformed cells by MSV follows a two-hit dose response, and it was suggested that MSV can form foci of transformants only when the cells are infected with both MSV and murine leukemia virus (MLV) (Hartley and Rowe, 1966; Huebner et al., 1966; Parkman et al., 1970). Subsequently, however, it was shown that the transformation by MSV of cells of the highly contact-inhibited Balb/3T3 mouse line and NRK rat cells growing in monolayers followed a one-hit dose response (Aaronson and Rowe, 1970; Parkman et al., 1970). Since MSV also induces colony formation by mouse 3T3 cells suspended in agar with one-hit kinetics (Bassin et al., 1970), it seems likely that a solitary MSV particle can transform a mouse or rat fibroblast. Parkman et al. suggested that the apparent two-hit kinetics of transformation of some mouse cells growing in monolayer reflects the inability of a single transformed cell of some lines to multiply fast enough to form a detectable clone focus. They further suggested that the formation of foci by such slowly dividing cells depends upon the infection and transformation of cells adjacent to the original transformant by progeny MSV released by the original transformant. Since all strains of MSV require a helper activity of a coinfecting MLV if they are to replicate infectious progeny, the formation of recruitment foci would be a two-hit event.

The fact that Balb/3T3 cells and NRK cells can be transformed by MSV in the absence of a helper MLV facilitated investigations of lines of MSV-transformed cells, and it rapidly became apparent that the extent of the defectiveness of the MSV genome in various transformed cell lines is variable.

Some mouse and also hamster and rat cells transformed by MSV produce no C-type particles and no RNA virus group-specific antigen and are termed nonproducer (NP) cells (Table 13.2). However, the sarcoma genome persists in these nonproducer cells because it can be rescued by superinfection with MLV (Huebner et al., 1966; Klement

Table 13.2. PROPERTIES OF CELLS TRANSFORMED BY SOLITARY INFECTION WITH MURINE SARCOMA VIRUS

Property	Nonproducer (NP)	Sarcoma-positive Leukemia-negative (S+ L−)
Transformed cell morphology	+	+
Release of infectious MSV	−	−
Release of infectious MLV	−	−
Release of visible virus-like particles	−	+
Production of material incorporating uridine at density 1.16 g/ml	−	+
Presence of MLV gs antigens	−	+
Focus/plaque formation with MLV	−	+
Infectious MSV genome rescuable by added MLV	+	+

See text for sources of data.

et al., 1969b; Aaronson and Rowe, 1970; Klement et al., 1971). Production of focus-forming virus can also be induced in these cells by treatment with chemicals such as 5-bromodeoxyuridine (Aaronson, 1971; Klement et al., 1971). Such chemical induction is known to activate endogenous C-type viruses in mammalian cells (Lowy et al., 1971; Aaronson et al., 1971) (see Chapter 12), and the properties of the activated sarcoma genome suggest that it is helped by these endogenous viruses.

Other cloned, MSV-transformed cells produce C-type particles, but these viruses are not infectious (Bassin et al., 1970, 1971b, 1973; Gazdar et al., 1971). The cells are said to be sarcoma-positive, leukemia-negative or "S+ L−" (Table 13.2). The viruses produced by these cells thus bear a superficial resemblance to the Bryan strain of RSV which lacks an envelope glycoprotein, but closer investigation has not confirmed this similarity. The noninfectious virus produced in small quantities by S+ L− cells contains some group-specific antigens and, on the basis of indirect antibody adsorption tests, appears to possess an altered envelope specificity (Fischinger et al., 1972a). However, unlike Bryan strain RSV these murine viruses cannot infect cells treated with inactivated Sendai virus and they lack the 60–70S RNA typical of RNA tumour viruses. In its place are found three RNA species which sediment at 28S, 18S and 4S, respectively (Bassin et al., 1973). Molecular hybridization experiments indicate that most of this RNA is cellular, but a small proportion is virus-specific (Bassin, personal communication). Moreover, the particles produced by S+ L− cells are deficient in reverse transcriptase, although a low level of activity can be detected using synthetic templates (Bassin et al., 1973; Bassin, personal communication). Another noninfectious virus produced in larger amounts by hamster tumour cells induced by MSV shows a low infectivity after cosedimentation with MLV and it contains 60S–70S viral RNA. This virus, however, also appears to be relatively deficient in reverse transcriptase activity (Gazdar et al., 1971) and it may thus be analogous to RSVα (Hanafusa and Hanafusa, 1971).

Rescue of an infectious murine sarcoma virus from all these MSV-transformed rodent cells requires replication of the rescuing MLV (Peebles et al., 1971), and defective MLV is unable to complement the replication defect(s) of MSV (Nomura et al., 1972a). Aaronson et al. (1972) have studied the properties of MSV rescued from both NP and S+ L− cells; they found the viruses to be genetically stable through at least two cycles of rescue and transmission to fresh cells. They detected no defective helper viruses in the S+ L− cells and concluded that the virus rescued from S+ L− cells contains information required for viral replication that was not present in the genome rescued from NP cells.

It seems, therefore, that MSV shows a spectrum of defectiveness. The same spectrum of defectiveness has not yet been found in the avian RNA tumour viruses, where helper-independent strains have received the most attention, but it seems likely that viruses with analogous defects could arise in the avian group. It is not known whether the defects in various strains of MSV are stable, or whether virus populations show a polymorphism for these lesions and are continuously generating new defective particles. Preliminary data suggest that MSV does not recombine with rescuing MLV (Peebles et al., 1971), but further studies with virus from cloned transformed cells will be needed to show whether this bar to recombination applies to all or only some of the virus's defects. It is noteworthy that leukemia viruses themselves can show defects which might be complemented by a transforming virus. For instance, Sinkovics et al. (1966) have reported that a strain of Rauscher leukemia virus loses its pathogenicity on prolonged passage through cells in culture. This virus becomes leukemogenic again if fresh cells are added to the culture or if it undergoes an in vivo passage. In this case, at least, it seems that viruses of variable defectiveness coexist and can possibly arise de novo in a population subject to fluctuating environmental selections.

Defective sarcoma viruses are propagated by coinfection and phenotypic mixing with helper leukemia virus. It is worth mentioning here an unusual example of phenotypic mixing, in which murine leukemia viruses or avian myeloblastosis virus can donate envelope material to vesicular stomatitis virus (VSV) (Závada, 1972a,b). It seems that analogous pseudotypes can also be obtained from some human mammary carcinoma cells (Závada et al., 1972), and it is suggested that this is evidence for the presence of viral antigens in the tumour cells. Virus has not been detected by other means in these human mammary tumour cells so, if confirmed, phenotypic mixing with VSV could prove a useful means of investigating otherwise latent viruses.

Temperature-sensitive Mutants of the Murine Leukemia/Sarcoma Complex

Murine leukemia viruses can readily be assayed by their ability to form syncytia on the XC line of RSV-transformed rat cells (Klement et al., 1969a; Rowe et al., 1970). This has facilitated a search for nitrosoguanidine- and 5-bromodeoxyuridine-induced *ts* mutants of these viruses, and a number have been found which show markedly reduced replication at the nonpermissive temperature (Stephenson et al., 1972). It is interesting that the yield of *ts* mutants following 5-bromodeoxyuridine mutagenesis was far lower than that reported for transformation-defective RSV mutants (Bader and Brown, 1971). The mutants can so far be divided into four groups on the basis of their physiology and complementation (Aaronson, 1973).

Wong et al. (1973) have isolated several *ts* mutants of MLV by using a selection procedure on nonmutagenized virus stocks. S+ L− cells round up and become only loosely attached to the substrate after superinfection with MLV (Bassin et al., 1971a). Such foci of rounded cells may be readily removed by agitation, and if this procedure is applied at nonpermissive temperature, the remaining attached cells yield an enriched proportion of *ts* MLV after a switch to permissive conditions. Three mutants isolated in this way have been found to fall into two complementation groups (Wong and McCarter, 1973), although the complementation between mutants is only slight. Cells which are coinfected by mutants in different complementation groups yield progeny in which about 8 percent of the virus is phenotypically wild type. Though some of these "wild-type" viruses may be heterozygotes, similar to those described by Weiss et al. (1973) in studies of avian RNA tumour viruses, it seems that many are true recombinants. The significance of the relatively high recombination frequency is hard to assess at present. Wong and McCarter find that recombination is negligible between viruses belonging to the same complementation group, and the recombination frequencies they detect between viruses of different complementation groups are almost as high as those described by Vogt (1971b) and Kawai and Hanafusa (1972b) for recombination between the unlinked or very loosely linked markers of transformation and viral envelope of avian sarcoma viruses.

Clearly a promising start has been made on the genetics of MLV utilizing *ts* mutants. It may now be possible to detect similar mutants of avian leukosis viruses using the plaque assay of Graf (1972), and we look forward to an expansion of studies on the physiology and genetics of the replicative functions of oncogenic viruses.

Temperature-sensitive mutants of MSV have also been isolated recently from stocks of virus treated with 5-azacytidine or 5-bromodeoxyuridine (Scolnick et al., 1972). Clones of cells infected with these mutants are transformed only under permissive conditions, but they show no evidence of release of virus or of group-specific antigens at either temperature. Two of the mutants can be rescued by MLV, and their *ts* behaviour persists during superinfection and in subsequent cell infections. By contrast, the *ts* lesion of a third mutant is apparently complemented at the nonpermissive temperature by superinfecting MLV, although the reason for this effect is unknown. (It is worth recalling here that some S+ L− cells can also produce foci or plaques upon superinfection with MLV [Bassin et al., 1970, 1971a; Nomura et al., 1972a; Wong et al., 1973] and some sublines of 3T3 cells are susceptible to transformation by MLV alone [Hackett and Sylvester, 1972].) Virus-specific RNA is found in cells infected by all three mutants under nonpermissive conditions, so they would probably be classified as late mutants in the scheme proposed for *ts* mutants of avian sarcoma viruses. Moreover, since a switch to nonpermissive conditions results in reversion of transformed characters, the *ts* lesions are in functions which are continuously required for transformation to be expressed. However, the MSV mutant complemented by MLV seems to have no counterpart among any of the known avian sarcoma viruses with mutations in late continuously required functions.

Host-cell Influences on the Expression of Viral Functions

The effect of cellular alterations on virus infection can be conveniently studied on readily cloned mammalian cell lines. Cellular factors are known to influence the expression of transformation by DNA tumour viruses (see Chapter 6), and in at least some cases revertants of cells transformed by DNA tumour viruses resume the transformed cell phenotype upon superinfection with an RNA tumour virus (Wyke, 1971; Renger, 1972). It now appears that RNA tumour viruses can themselves become suppressed by cellular factors. Fischinger et al. (1972b) find that flat revertants of MSV-transformed mouse cells lose the ability to yield infectious sarcoma virus upon superinfection with MLV and also lose reverse transcriptase activity. However, the revertants retain murine group-specific antigens and show an enhanced susceptibility to MSV and MLV infection. These revertants will also retransform spontaneously, and in such cells the murine group-specific antigens are also lost (Nomura et al., 1972b). The mechanisms operative here are quite obscure and deserve further study, because it seems possible that expression of the group-specific antigen might in some way be related to the control of expression of the transformed phenotype.

THE GENOME OF RNA TUMOUR VIRUSES—ITS NATURE AND ORIGIN

Two subjects have recently been examined in the light of our present knowledge of the genetics of these viruses: the nature of the virus genome (Vogt, 1973) and the origin and significance of viral defectiveness (Weiss, 1973). A consideration of either topic is of necessity speculative but nonetheless valuable, for the present hypotheses involve not only a broad synthesis of available information but also make predictions that should be experimentally testable.

The Virus Genome

The genome of RNA tumour viruses is probably the single-stranded 60–70S viral RNA, which weighs about 10^7 daltons and comprises three or four 35S subunits, each of about 3×10^6 daltons (see Chapter 10). These subunits, as judged from their electrophoretic mobilities, fall into two size classes, *a* subunits being larger than *b* subunits. Uncloned avian leukosis viruses and *td* mutants of avian

a b c d e f g h i j k l m n o (1) Sarcoma virus

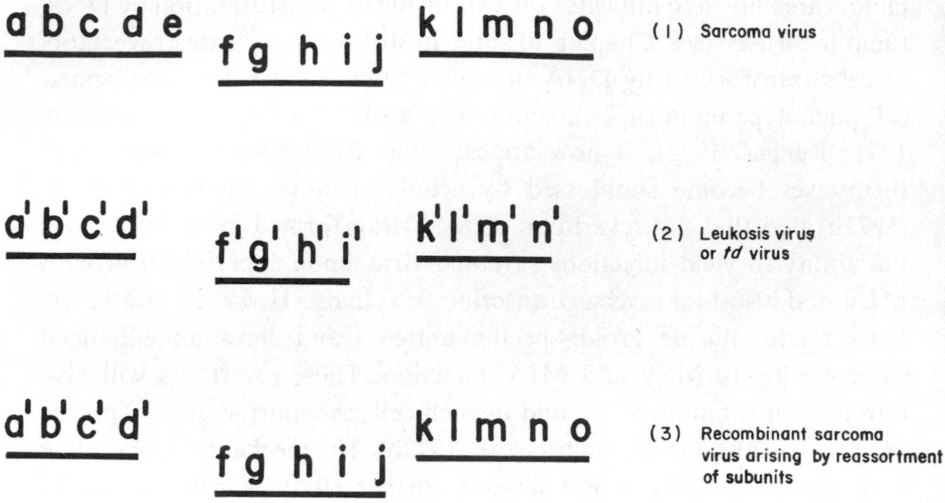

a' b' c' d' f' g' h' i' k' l' m' n' (2) Leukosis virus or *td* virus

a' b' c' d' f g h i j k l m n o (3) Recombinant sarcoma virus arising by reassortment of subunits

Figure 13.2. Haploidy model of RNA tumour virus genome.

sarcoma viruses contain only the smaller *b* subunits, whereas uncloned sarcoma viruses contain variable proportions of both *a* and *b* subunits. However, cloned sarcoma viruses derived from a single infectious event do not contain detectable *b* subunits (Duesberg et al., 1973). Since sarcoma and leukosis viruses share a number of functions in common, this suggests that at least some of the genetic information which resides in *a* subunits in cloned sarcoma viruses is found in *b* subunits in leukosis viruses and *td* mutants. Oligonucleotide fingerprinting and hybridization studies on viral RNAs likewise suggest that *a* and *b* subunits contain large amounts of similar RNA sequences (Lai et al., 1973, quoted in Duesberg et al., 1973).

In the light of these data Vogt (1973) has argued that there are only two viable models of the structure of the viral genome.

The Haploid Genome Model

Each 35S RNA subunit is genetically distinct, and *td* variants must arise by a reduction in size of each of the *a* subunits. There are several conceivable mechanisms for this size reduction (for example, premature termination of transcription or cleavage of mature molecules at the formation of either provirus DNA or new viral RNA) and they may affect all subunits simultaneously or different subunits in different infectious generations. Simultaneous diminution of all subunits would allow transforming functions to be associated with only one unit, if only with a portion of this unit, but the genetic material lost from the other subunits must be presumed to be irrelevant to any virus functions observed in vitro. A sequential reduction in subunit size would predict that transforming functions are present on each *a* subunit, and whereas gene functions distributed over all subunits are needed to induce growth in agar, possibly only one *a* subunit is necessary to transform cells in monolayer and therefore *td* variants only arise when all three are diminished. Available data do not favour this hypothesis, for whereas growth in agar is often an inefficient event compared with focus formation, a *ts* mutant (*ts* 334) will induce agar colonies but not foci under nonpermissive conditions. Recombinant sarcoma viruses (Vogt, 1971b; Kawai and Hanafusa, 1972b) could arise by reassortment of subunits (see Figure 13.2). This model predicts that such recombinants would possess both *a* and *b* subunits and thus may show some defects in transformation.

The Polyploid Genome Model

This model (see Figure 13.3) proposes that the 35S RNA subunits are allelic; viruses derived from double infections could thus be heterozygous as described by Weiss et al. (1973). Heterozygotes would not tend to persist if at some stage during replication the viral genome is normally reduced to one 35S equivalent. However, recombinants could arise from heterozygotes by a copy-choice reverse transcription or by breakage and reunion among DNA transcripts prior to the continuing replication of the single genome equivalent. Moreover, if a deletion occurred at some stage during formation, integration or transcription of this provirus, the progeny virions would contain uniformly smaller RNA subunits, and in this way *td* variants could arise.

The avian RNA tumour viruses are remarkably radiation resistant (Rubin and Temin, 1959; Levinson and Rubin, 1966). If the true genome of the virus is smaller than the total virion RNA, this resistance would be partially explained, and the possibility of recombination

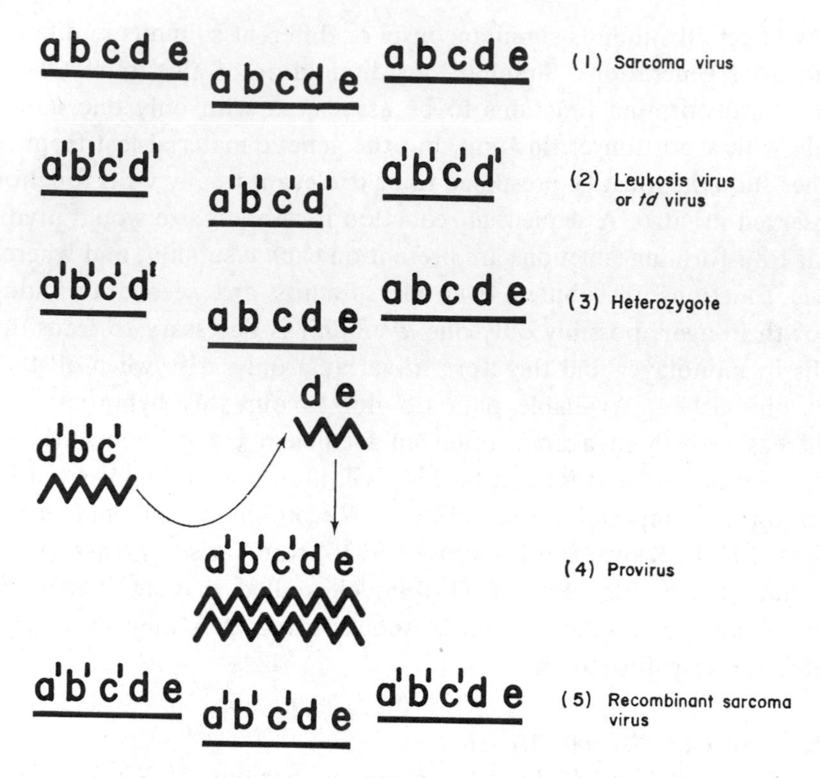

Figure 13.3. Polyploidy model of RNA tumour virus genome.

events prior to provirus formation could also increase radiation resistance. Inactivation of the virus by chemical and physical agents follows single-hit kinetics in most investigations (Rubin and Temin, 1959; Rubin, 1960; Friesen and Rubin, 1961; Graf and Bauer, 1970; Toyoshima et al., 1970; Friis, 1971). If the virus is polyploid, then these results suggest that the integrity of all or a part of one of the genomes is essential to the functioning of all the genomes in a virion.

This polyploid model thus accounts for the available experimental observations, but it has one major drawback. It greatly reduces the genetic complexity of these viruses, for it envisages that a single 35S RNA subunit carries all the genetic information for transformation and replication. It is clearly important to decide which of these genome models is correct. This could be done by analyzing the genetic complexity of the viruses by molecular hybridization. However, data on the physical nature of the viral RNA is always suspect, for very little of the RNA extracted from virus preparations is actually derived from infectious particles. It is therefore preferable to bolster these biochemical data with genetic experiments on the nature of heterozygotes and the possible existence of linkage groups.

Defectiveness of Sarcoma Viruses

Defective viruses are intrinsically difficult to work with, but an understanding of their origin and the nature of their defectiveness might well shed light on the genesis of oncogenic viruses and their role in "spontaneous" neoplasia.

Murine and feline sarcoma viruses and some strains of chick sarcoma viruses, for example BH.RSV, can transform fibroblasts but cannot replicate infectious progeny in the absence of a helper leukemia virus. These replication-defective sarcoma viruses are stable and they do not acquire by recombination with their helper viruses the ability to replicate independently. This suggests first, that some restriction of recombination with a helper virus may be the basis of stable defectiveness and second, that the helper-independent chick sarcoma viruses are exceptional. The defect in recombination of at least some sarcoma viruses does not seem to apply to the whole viral genome, for there is indirect evidence that BH.RSV can recombine with respect to the reverse transcriptase gene or genes (Hanafusa and Hanafusa, 1968, 1971). The portion of the genome that fails to recombine with helper is therefore probably strongly linked in some way to transforming functions; Figure 13.4 illustrates one explanation for such a recombination defect. The numbers in Figure 13.4 symbolize genes involved in virus replication; the letters represent genes involved in fibroblast transformation. Each segment represents a gene linkage group which, without prejudice to either of the genome models discussed above, could be defined by the physical nature of the nucleic acid, some characteristic of reverse transcription or some other mechanism. If 8 and 9 represent the loci controlling synthesis of RSV envelope glycoprotein, then BH.RSV, which is defective in this property, would have a deletion or mutation at these sites. Moreover, in BH.RSV, unlike the

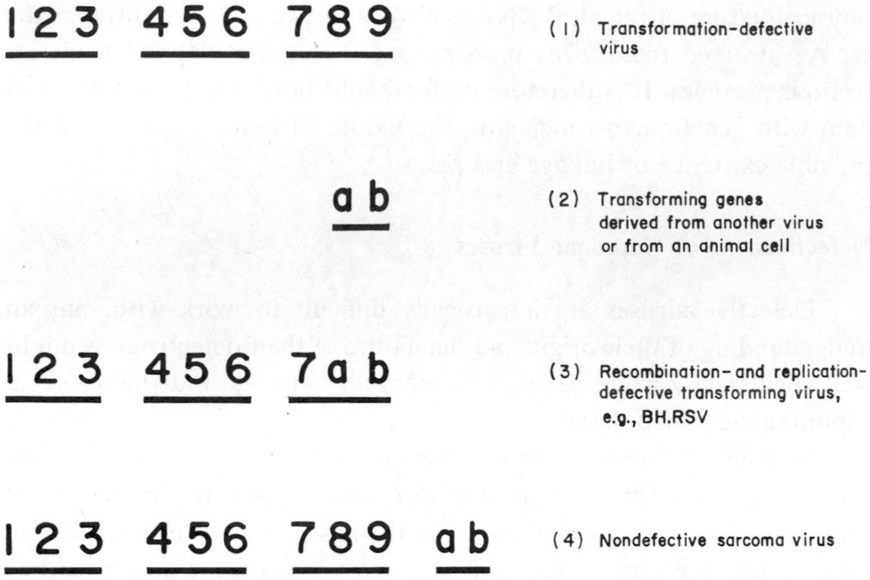

123 456 789 (1) Transformation-defective virus

ab (2) Transforming genes derived from another virus or from an animal cell

123 456 7ab (3) Recombination- and replication-defective transforming virus, e.g., BH.RSV

123 456 789 ab (4) Nondefective sarcoma virus

Figure 13.4. A possible model for a defective sarcoma virus genome.

helper-independent sarcoma viruses, these loci would be linked to transforming functions and could not recombine with the equivalent linkage group in the helper virus without loss of transforming ability. The reverse transcriptase locus, on the other hand, is envisaged as being somewhere on sites 1 to 6 where it is available for recombination. A similar model could be applied to the defective murine sarcoma viruses and in these cases the extreme deficiency of virus replicative functions that is sometimes observed would be correlated with a smaller genome size (Gazdar et al., 1971; Bassin et al., 1973; Bassin, personal communication).

A linkage between replication defects and transforming genes can thus account for the stability of defective viruses, but how could such a linkage arise? Replication-defective variants do occur in stocks of helper-independent sarcoma viruses, but when tested these prove to be capable of recombination with helper virus (Weiss, 1973). Further experiments on such variants are necessary to determine if recombination is the rule, but the infrequent occurrence of helper-independent transforming viruses in nature suggests that they are not the usual progenitors of defective sarcoma viruses. Another possibility is that replicating but nontransforming viruses might incorporate genetic material of host origin into their virions (Vogt, 1972; Weiss, 1973). This material, possibly messenger RNA, could be transcribed into DNA by the virion reverse transcriptase and may thus be transduced to other cells. The "transduction" of oncogenic information by the viruses may be a special case of a generalized phenomenon that is not recognized in many cases because expression of the foreign information is efficiently restricted by the new host, or because the expressed information results in no noticeable difference in host cell behaviour. This transduction would probably be a haphazard affair unless the new information becomes stably linked in some way with the viral genetic material. Such linkage, which is probably a very rare event, may occur during the copy-choice reverse transcription or recombination between DNA proviruses that is postulated in the polyploid genome model. The linkage would thus normally involve elimination of some viral genes, resulting in a replication-defective virus (Figure 13.4), whose acquired transforming functions can only be propagated by the helper action of competent viruses. An additive rather than a substitutive linkage, which gives rise to helper-independent transforming viruses, is probably an even rarer event that would result in a larger than usual viral genome.

The concept that transforming functions are acquired from host cells has several attractions (Vogt, 1972; Weiss, 1973). It explains why helper-independent RSV appear to have a larger genome than their transformation-defective derivatives and why this apparent excess of genetic material results in a cell transformation which is superfluous to the requirements of virus replication. It also provides a rationale for the fact that these viruses show a tropism for the cells they transform, even though they replicate in a far wider range of cell types. The transforming functions acquired by different virus strains may differ or be modified on passage, and this could account for the failure to observe complementation for transformation between *ts* mutants in different strains (Wyke, 1973b). Though one might expect the viruses to show differences in their transforming capacities, their replicative functions should be fairly uniform. Future investigations of transforming functions should reveal whether these do in fact differ from strain to strain. It will also be useful to study a variety of defective transforming viruses to determine how their defects differ. This may reveal how these viruses have arisen.

Oncogenes and Protoviruses

Current ideas on the origin of RNA tumour viruses have been formulated in two concepts: the oncogene hypothesis of Huebner and Todaro (1969) and the protovirus hypothesis of Temin (1971a,b). These two hypotheses are not mutually exclusive and workers such as Bentvelzen (1972) have independently proposed mechanisms for viral carcinogenesis that combine features of both. However, these two hypotheses emphasize different aspects of the relationship between RNA tumour viruses and their host cells. Since the hypotheses should prove heuristically valuable, it is worth considering briefly their postulates.

The oncogene theory proposes that all cells contain in their DNA the information necessary to specify the complete genome of an RNA tumour virus (the virogene). This virus-specific DNA is vertically transmitted from generation to generation, and certain virogene functions may be normally expressed, for example, in development. The virogene includes oncogenic information (the oncogene) which is responsible for most tumours. This information is normally repressed, but a breakdown of the cellular regulation of these inherited viral genes may result in expression of the virogene and oncogene and therefore in neoplasia and the release of RNA tumour viruses by the neoplastic cells. Obviously this hypothesis can be elaborated to account for neoplastic cells that do not release viruses (activation of only the oncogenic part of the virogene) or for cells that are not transformed but do release nononcogenic C-type particles (activation of that part of the virogene which is not oncogenic).

The protovirus hypothesis stresses the importance of information flow from DNA to RNA and back to DNA as a process of gene expression, gene duplication and gene modification. A protovirus would arise if genetic information at some stage in this cycle acquired a stable phase distinct from its integrated DNA progenitor. If these protoviruses contained genes relating to cell multiplication, movement or differentiation, then mutation or recombination events in the protovirus genome could result in the appearance of genes capable of causing neoplastic transformation. The protovirus hypothesis thus differs from the oncogene hypothesis in that it postulates that information transfers by transcription and reverse transcription, coupled with genetic accidents, continually generate oncogenes and thereby neoplastic events. The oncogene hypothesis, on the other hand, suggests that stable oncogenes that evolved in the distant past are inherited through the germ line. Transfer of information from RNA to DNA was presumably important in the establishment of viral genes in the ancestral germ line but plays no part in the activation of oncogenes leading to "spontaneous" tumours. Reverse transcription may be necessary for the induction of tumours by infectious RNA tumour viruses, but such events are, according to supporters of the oncogene hypothesis, relatively rare in the natural population.

The available data are insufficient to indicate which, if either, of the hypotheses is correct. On the one hand, complete viral genomes can be activated from normal cells as predicted by the oncogene hypothesis (Chapter 12), but these endogenous viruses have not yet been shown to be oncogenic. On the other hand, experiments on defective viruses suggest that genetic recombination may be important in acquiring oncogenic information, a possible corollary of the protovirus hypothesis. Further experiments on these lines should elucidate the nature of the relationship between viral and host cell genomes and this, one hopes, should lead to an understanding of virus-induced oncogenesis.

Literature Cited

Aaronson, S. A. 1971. Chemical induction of focus-forming virus from nonproducer cells transformed by murine sarcoma virus. Proc. Nat. Acad. Sci. *68:* 3069.

———. 1973. The biology of mammalian C-type RNA tumor viruses. In *Possible episomes in eukaryotes*, ed. L. Silvestri. North-Holland, Amsterdam. In press.

Aaronson, S. A. and W. P. Rowe. 1970. Nonproducer clones of murine sarcoma virus-transformed BALB/3T3 cells. Virology *42:* 9.

AARONSON, S. A., G. J. TODARO and E. M. SCOLNICK. 1971. Induction of murine C-type viruses from clonal lines of virus-free BALB/3T3 cells. Science *174:* 157.

AARONSON, S. A., R. H. BASSIN and C. WEAVER. 1972. Comparison of murine sarcoma viruses in nonproducer and S+ L−-transformed cells. J. Virol. *9:* 701.

ALTANER, C. and H. M. TEMIN. 1970. Carcinogenesis by RNA sarcoma viruses. XII. A quantitative study of infection of rat cells *in vitro* by avian sarcoma viruses. Virology *40:* 118.

BADER, J. P. 1972. Temperature-dependent transformation of cells infected with a mutant of Bryan Rous sarcoma virus. J. Virol. *10:* 267.

BADER, J. P. and N. R. BROWN. 1971. Induction of mutations in an RNA tumour virus by an analogue of a DNA precursor. Nature New Biol. *234:* 11.

BASSIN, R. H., N. TUTTLE and P. J. FISCHINGER. 1970. Isolation of murine sarcoma virus-transformed mouse cells which are negative for leukemia virus from agar suspension cultures. Int. J. Cancer *6:* 95.

———. 1971a. Rapid cell culture assay technique for murine leukaemia viruses. Nature *229:* 564.

BASSIN, R. H., L. A. PHILLIPS, M. J. KRAMER, D. K. HAAPALA, P. T. PEEBLES, S. NOMURA and P. J. FISCHINGER. 1971b. Transformation of mouse 3T3 cells by murine sarcoma virus: Release of virus-like particles in the absence of replicating murine leukemia helper virus. Proc. Nat. Acad. Sci. *68:* 1520.

———. 1973. Properties of 3T3 cells transformed by murine sarcoma virus in the absence of replicating murine leukemia helper virus. In *Unifying concepts of leukemia* (ed. R. M. Dutcher and P. Chieco-Bianchi) Bibl. Haemat. No. 39. Karger, Basel. In press.

BENTVELZEN, P. 1972. Hereditary infections with mammary tumor viruses in mice. In *RNA viruses and the host genome in oncogenesis* (ed. P. Emmelot and P. Bentvelzen) p. 309. North-Holland, Amsterdam.

BIGGS, P. M., B. S. MILNE, T. GRAF and H. BAUER. 1973. Oncogenicity of non-transforming mutants of avian sarcoma viruses. J. Gen. Virol. *18:* 399.

BIQUARD, J.-M. and P. VIGIER. 1970. Isolement et étude d'un mutant conditionnel du virus de Rous a capacité transformante thermosensible. Compt. Rend. Acad. Sci. Ser. D. *271:* 2430.

———. 1972. Characteristics of a conditional mutant of Rous sarcoma virus defective in ability to transform cells at high temperature. Virology *47:* 444.

BISHOP, J. M., J. JACKSON, N. QUINTRELL and H. E. VARMUS. 1973. Transcription of RNA tumor virus genes in normal and infected cells. In *Possible episomes in eukaryotes*, ed. L. Silvestri. North-Holland, Amsterdam. In press.

BURGER, M. M. and G. S. MARTIN. 1972. Agglutination of cells transformed by Rous sarcoma virus by wheat germ agglutinin and concanavalin A. Nature New Biol. *237:* 9.

DOUGHERTY, R. M. and R. RASMUSSEN. 1964. Properties of a strain of Rous sarcoma virus that infects mammals. Nat. Cancer Inst. Monogr. *17:* 337.

DUESBERG, P. H. and P. K. VOGT. 1970. Differences between the ribonucleic acids of transforming and nontransforming avian tumour viruses. Proc. Nat. Acad. Sci. *67:* 1673.

DUESBERG, P. H., E. CANAANI, K. VON DER HELM, M. M. C. LAI and P. K. VOGT. 1973. News and views on avian tumor virus RNA. In *Possible episomes in eukaryotes*, ed. L. Silvestri. North-Holland, Amsterdam. In press.

DUFF, R. G. and P. K. VOGT. 1969. Characteristics of two new avian tumor virus subgroups. Virology *39:* 18.

FISCHINGER, P. J., W. SCHAFER and E. SEIFERT. 1972a. Detection of some murine leukemia virus antigens in virus particles derived from 3T3 cells transformed only by murine sarcoma virus. Virology *47:* 229.

FISCHINGER, P. J., S. NOMURA, P. T. PEEBLES, D. K. HAAPALA and R. H. BASSIN. 1972b. Reversion of murine sarcoma virus transformed mouse cells: Variants without a rescuable sarcoma virus. Science *176:* 1033.

FRIESEN, B. and H. RUBIN. 1961. Some physiochemical and immunological properties of an avian leukosis virus (RIF). Virology *15:* 387.

FRIIS, R. R. 1971. Inactivation of avian sarcoma viruses with UV light: A difference between helper dependent and helper independent strains. Virology *43:* 521.

FRIIS, R. R., K. TOYOSHIMA and P. K. VOGT. 1971. Conditional lethal mutants of avian sarcoma viruses. I. Physiology of *ts* 75 and *ts* 149. Virology *43:* 375.

GAZDAR, A. F., L. A. PHILLIPS, P. S. SARMA, P. T. PEEBLES and H. C. CHOPRA. 1971. Presence of sarcoma genome in a "non-infectious" mammalian virus. Nature New Biol. *234:* 69.

GOLDÉ, A. 1970. Radio-induced mutants of the Schmidt-Ruppin strain of Rous sarcoma virus. Virology *40:* 1022.

GRAF, T. 1972. A plaque assay for avian RNA tumor viruses. Virology *50:* 567.

GRAF, T. and H. BAUER. 1970. Studies on the relative target size of various functions in the genome of RNA tumor viruses. In *Defectiveness, rescue and stimulation of oncogenic viruses,* ed. M. Boiron. Edition C.N.R.S., Paris, 87.

GRAF, T., H. BAUER, H. GELDERBLOM and D. P. BOLOGNESI. 1971. Studies on the reproductive and cell-converting abilities of avian sarcoma viruses. Virology *43:* 427.

HACKETT, A. J. and S. S. SYLVESTER. 1972. Cell line derived from BALB/3T3 that is transformed by murine leukaemia virus: A focus assay for leukaemia virus. Nature New Biol. *239:* 164.

HAKOMORI, S.-I., T. SAITO and P. K. VOGT. 1971. Transformation by Rous sarcoma virus: Effects on cellular glycolipids. Virology *44:* 609.

HANAFUSA, H. 1965. Analysis of the defectiveness of Rous sarcoma virus. III. Determining influence of a new helper virus on the host range and susceptibility to interference of RSV. Virology *25:* 248.

———. 1970. Virus production by Rous sarcoma cells. Current Topics Microbiol. Immunol. *51:* 114.

HANAFUSA, H. and T. HANAFUSA. 1966. Determining factor in the capacity of Rous sarcoma virus to induce tumors in mammals. Proc. Nat. Acad. Sci. *55:* 532.

———. 1968. Further studies on RSV production from transformed cells. Virology *34:* 630.

———. 1971. Noninfectious RSV deficient in DNA polymerase. Virology *43:* 313.

HANAFUSA, H., T. HANAFUSA and H. RUBIN. 1963. The defectiveness of Rous sarcoma virus. Proc. Nat. Acad. Sci. *49:* 572.

———. 1964. Analysis of the defectiveness of Rous sarcoma virus. II. Specification of RSV antigenicity by helper virus. Proc. Nat. Acad. Sci. *51:* 41.

HANAFUSA, H., D. BALTIMORE, D. SMOLER, K. F. WATSON, A. YANIV and S. SPIEGELMAN. 1972. Absence of polymerase protein in virions of alpha-type Rous sarcoma virus. Science *177:* 1188.

HARTLEY, J. W. and W. P. ROWE. 1966. Production of altered cell foci in tissue culture by defective Moloney sarcoma virus particles. Proc. Nat. Acad. Sci. *55:* 780.

HATANAKA, M. and H. HANAFUSA. 1970. Analysis of a functional change in membrane in the process of cell transformation by Rous sarcoma virus: Alteration in the characteristics of sugar transport. Virology *41:* 647.

HAYWARD, W. S. and H. HANAFUSA. 1973. Detection of avian tumor virus RNA in uninfected chick embryo cells. J. Virol. *11:* 157.

HUEBNER, R. J. and G. J. TODARO. 1969. Oncogenes of RNA tumor viruses as determinants of cancer. Proc. Nat. Acad. Sci. *64:* 1087.

HUEBNER, R. J., J. W. HARTLEY, W. P. ROWE, W. T. LANE and W. I. CAPPS. 1966. Rescue of the defective genome of Moloney sarcoma virus from a noninfectious hamster tumor and the production of pseudotype sarcoma viruses with various murine leukemia viruses. Proc. Nat. Acad. Sci. *56:* 1164.

ISHIZAKI, R., A. J. LANGLOIS, J. CHABOT and J. W. BEARD. 1971. Component of strain MC29 avian leukosis virus with the property of defectiveness. J. Virol. *8:* 821.

KATZ, E. and P. K. VOGT. 1971. Conditional lethal mutants of avian sarcoma viruses. II. Analysis of the temperature-sensitive lesion in *ts* 75. Virology *46:* 745.

KAWAI, S. and H. HANAFUSA. 1971. The effects of reciprocal changes in temperature on the transformed state of cells infected with a Rous sarcoma virus mutant. Virology *46:* 470.

———. 1972a. Plaque assay for some strains of avian leukosis virus. Virology *48:* 126.

———. 1972b. Genetic recombination with avian tumor virus. Virology *49:* 37.

KAWAI, S. and T. YAMAMOTO. 1970. Isolation of different kinds of non-virus producing chick cells transformed by Schmidt-Ruppin strain (subgroup A) of Rous sarcoma virus. Japan. J. Exp. Med. *40:* 243.

KAWAI, S., C. E. METROKA and H. HANAFUSA. 1972. Complementation of functions required for cell transformation by double infection with RSV mutants. Virology *49:* 302.

KLEMENT, V., W. P. ROWE, J. W. HARTLEY and W. E. PUGH. 1969a. Mixed culture cytopathogenicity: A new test for growth of murine leukemia viruses in tissue culture. Proc. Nat. Acad. Sci. *63:* 753.

KLEMENT, V., J. W. HARTLEY, W. P. ROWE and R. J. HUEBNER. 1969b. Recovery of a hamster specific focus forming and sarcomagenic virus from a "non-infectious" hamster tumor induced by the Kirsten mouse sarcoma virus. J. Nat. Cancer Inst. *43:* 925.

KLEMENT, V., M. D. NICHOLSON and R. J. HUEBNER. 1971. Rescue of the genome of focus forming virus from rat non-productive lines by 5′-bromodeoxyuridine. Nature New Biol. *234:* 12.

LANGLOIS, A. J., L. VEPREK, D. BEARD, R. B. FRITZ and J. W. BEARD. 1971. Isolation of a non-focus forming virus from stock MC29 avian leukosis strain by end point dilution. Cancer Res. *31:* 1010.

LEONG, J. A., A. C. GARAPIN, N. JACKSON, L. FANSHIER, W. LEVINSON and J. M. BISHOP. 1972. Virus-specific ribonucleic acid in cells producing Rous sarcoma virus: Detection and characterization. J. Virol. *9:* 891.

LEVINSON, W. and H. RUBIN. 1966. Radiation studies of avian tumor viruses and of Newcastle disease virus. Virology *28:* 533.

LINIAL, M. and W. S. MASON. 1973. Characteristics of two conditional early mutants of Rous sarcoma virus. Virology, submitted for publication.

LOWY, D. R., W. P. ROWE, N. TEICH and J. W. HARTLEY. 1971. Murine leukemia virus: High-frequency activation *in vitro* by 5-iododeoxyuridine and 5-bromodeoxyuridine. Science *174:* 155.

MARTIN, G. S. 1970. Rous sarcoma virus: A function required for maintenance of the transformed state. Nature *227:* 1021.

———. 1971. Mutants of the Schmidt-Ruppin strain of Rous sarcoma virus. In *The biology of oncogenic viruses*, ed. L. Silvestri, p. 320. North-Holland, Amsterdam.

MARTIN, G. S. and P. H. DUESBERG. 1972. The *a* subunit in the RNA of transforming avian tumor viruses. I. Occurrence in different virus strains. II. Spontaneous loss resulting in non-transforming variants. Virology *47:* 494.

MARTIN, G. S., S. VENUTA, M. WEBER and H. RUBIN. 1971. Temperature-dependent alterations in sugar transport in cells infected by a temperature-sensitive mutant of Rous sarcoma virus. Proc. Nat. Acad. Sci. *68:* 2739.

MOSCOVICI, C. and M. ZANETTI. 1970. Studies on single foci of hematopoietic cells transformed by avian myeloblastosis virus. Virology *42:* 61.

NOMURA, S., R. H. BASSIN, W. TURNER, D. K. HAAPALA and P. J. FISCHINGER. 1972a. Ultraviolet inactivation of Moloney leukaemia virus: Relative target size required for virus replication and rescue of "defective" murine sarcoma virus. J. Gen. Virol. *14:* 213.

NOMURA, S., P. J. FISCHINGER, C. F. T. MATTERN, P. T. PEEBLES, R. H. BASSIN and G. P. FRIEDMAN. 1972b. Revertants of mouse cells transformed by murine sarcoma virus. 1. Characterization of flat and transformed sublines without a rescuable murine sarcoma virus. Virology *50:* 51.

OTTEN, J., J. P. BADER, G. J. JOHNSON and I. PASTAN. 1972. A mutation in a Rous sarcoma virus gene that controls adenosine-3′5′-monophosphate levels and transformation. J. Biol. Chem. *247:* 1632.

PARKMAN, R., J. A. LEVY and R. C. TING. 1970. Murine sarcoma virus: The question of defectiveness. Science *168:* 387.

PEEBLES, P. T., R. H. BASSIN, D. K. HAAPALA, L. A. PHILLIPS, S. NOMURA and P. J. FISCHINGER. 1971. Rescue of murine sarcoma virus from a sarcoma-positive leukemia-negative cell line: Requirement for replicating leukemia virus. J. Virol. *8:* 690.

RENGER, H. C. 1972. Retransformation of *ts* SV40 transformants by murine sarcoma virus at non-permissive temperature. Nature New Biol. *240:* 19.

ROBINSON, W. S. and H. L. ROBINSON. 1971. DNA polymerase in defective Rous sarcoma virus. Virology *44:* 457.

ROBINSON, W. S., A. PITKANEN and H. RUBIN. 1965. The nucleic acid of the Bryan strain of Rous sarcoma virus: Purification of the virus and isolation of the nucleic acid. Proc. Nat. Acad. Sci. *54:* 137.

ROWE, W. P., W. E. PUGH and J. W. HARTLEY. 1970. Plaque assay techniques for murine leukemia viruses. Virology *42:* 1136.

RUBIN, H. 1960. Growth of Rous sarcoma virus in chick embryo cells following irradiation of host cells or free virus. Virology *11:* 28.

———. 1965. Genetic control of cellular susceptibility to pseudotypes of Rous sarcoma virus. Virology *26:* 270.

RUBIN, H. and H. M. TEMIN. 1959. A radiological study of cell-virus interaction in the Rous sarcoma. Virology *7:* 75.

RUBIN, H. and P. K. VOGT. 1962. An avian leukosis virus associated with stocks of Rous sarcoma virus. Virology *17:* 184.

SCHEELE, C. M. and H. HANAFUSA. 1971. Proteins of helper-dependent RSV. Virology *45:* 401.

SCOLNICK, E. M., J. R. STEPHENSON and S. A. AARONSON. 1972. Isolation of temperature sensitive mutants of murine sarcoma virus. J. Virol. *10:* 653.

SINKOVICS, J. G., B. A. BERTIN and C. D. HOWE. 1966. Occurrence of low leukemogenic but immunising mouse leukemia virus in tissue culture. Nat. Cancer Inst. Monogr. *22:* 349.

STEPHENSON, J. R., R. K. REYNOLDS and S. A. AARONSON. 1972. Isolation of temperature-sensitive mutants of murine leukemia virus. Virology *48:* 749.

TEMIN, H. M. 1960. The control of cellular morphology in embryonic cells infected with Rous sarcoma virus *in vitro*. Virology *10:* 182.

———. 1961. Mixed infection with two types of Rous sarcoma virus. Virology *13:* 158.

———. 1962. Separation of morphological conversion and virus production in Rous sarcoma virus infection. Cold Spring Harbor Symp. Quant. Biol. *27:* 407.

———. 1971a. The protovirus hypothesis: Speculations on the significance of RNA-directed DNA synthesis for normal development and for carcinogenesis. J. Nat. Cancer Inst. *46:* 3.

———. 1971b. The role of DNA provirus in carcinogenesis by RNA tumor viruses. In *The biology of oncogenic viruses*, ed. L. Silvestri, p. 176. North-Holland, Amsterdam.

TOYOSHIMA, K. and P. K. VOGT. 1969. Temperature sensitive mutants of an avian sarcoma virus. Virology *39:* 930.

TOYOSHIMA, K., R. R. FRIIS and P. K. VOGT. 1970. The reproductive and cell-transforming capacities of avian sarcoma virus B77: Inactivation with UV light. Virology *42:* 163.

UNKELESS, J. C., A. TOBIA, L. OSSOWSKI, J. P. QUIGLEY, D. B. RIFKIN and E. REICH. 1973. An enzymatic function associated with transformation of fibroblasts by oncogenic viruses. J. Exp. Med. *137:* 85.

VOGT, P. K. 1965. A heterogeneity of Rous sarcoma virus revealed by selectively resistant chick embryo cells. Virology *25:* 237.

———. 1967a. Phenotypic mixing in the avian tumor virus group. Virology *32:* 708.

———. 1967b. Virus-directed host responses in the avian leukosis and sarcoma complex. In *Perspectives in virology* (ed. M. Pollard) vol. 5, p. 199. Harper and Row, New York.

———. 1971a. Spontaneous segregation of nontransforming viruses from cloned sarcoma viruses. Virology *46:* 939.

———. 1971b. Genetically stable reassortment of markers during mixed infection with avian tumor viruses. Virology *46:* 947.

———. 1972. The emerging genetics of RNA tumor viruses. J. Nat. Cancer Inst. *48:* 3.

———. 1973. The genome of avian RNA tumor viruses: A discussion of four models. In *Possible episomes in eukaryotes*, ed. L. Silvestri. North-Holland, Amsterdam. In press.

VOGT, P. K., J. A. WYKE, R. A. WEISS, R. R. FRIIS, E. KATZ and M. LINIAL. 1972. Avian tumor viruses: Mutants, markers and genotypic mixing. M. D. Anderson Symposium on Fundamental Cancer Research. Houston, Texas.

WARREN, L., D. CRITCHLEY and I. MACPHERSON. 1972a. Surface glycoproteins and glycolipids of chicken embryo cells transformed by a temperature-sensitive mutant of Rous sarcoma virus. Nature *235:* 275.

WARREN, L., J. P. FUHRER and C. A. BUCK. 1972b. Surface glycoproteins of normal and transformed cells: A difference determined by sialic acid and a growth-dependent sialyl transferase. Proc. Nat. Acad. Sci. *69:* 1838.

WEISS, R. A. 1972. Helper viruses and helper cells. In *RNA viruses and the host genome in oncogenesis* (ed. P. Emmelot and P. Bentvelzen.) p. 117. North-Holland, Amsterdam.

———. 1973. Transmission of cellular genetic elements by RNA tumour viruses. In *Possible episomes in eukaryotes*, ed. L. Silvestri. North-Holland, Amsterdam. In press.

WEISS, R. A., R. R. FRIIS, E. KATZ and P. K. VOGT. 1971. Induction of avian tumor viruses in normal cells by physical and chemical carcinogens. Virology *46:* 920.

WEISS, R. A., W. S. MASON and P. K. VOGT. 1973. Genetic recombination between endogenous and exogenous avian RNA tumor viruses. Virology. In press.

WEISSBACH, A., A. BOLDEN, R. MULLER, H. HANAFUSA and T. HANAFUSA. 1972. Deoxyribonucleic acid polymerase activities in normal and leukovirus-infected chicken embryo cells. J. Virol. *10:* 321.

WONG, P. K. Y. and J. A. MCCARTER. 1973. Genetic studies of temperature sensitive mutants of Moloney murine leukemia virus. Submitted for publication.

WONG, P. K. Y., L. J. RUSS and J. A. MCCARTER. 1973. Rapid, selective procedure for isolation of spontaneous temperature-sensitive mutants of Moloney leukemia virus. Virology. *51:* 424.

WYKE, J. 1971. Phenotypic variation and its control in polyoma-transformed BHK21 cells. Exp. Cell Res. *66:* 209.

———. 1973a. The selective isolation of temperature sensitive mutants of Rous sarcoma virus. Virology. In press.

———. 1973b. Complementation of transforming functions by temperature sensitive mutants of avian sarcoma virus. Virology. Submitted for publication.

WYKE, J. A. and M. LINIAL. 1973. Temperature sensitive avian sarcoma viruses: A physiological comparison of twenty mutants. Virology. In press.

YOSHII, S. and P. K. VOGT. 1970. A mutant of Rous sarcoma virus (type O) causing fusiform cell transformation. Proc. Soc. Exp. Biol. Med. *135:* 297.

ZĂVADA, J. 1972a. Pseudotypes of vesicular stomatitis virus with the coat of murine leukaemia and of avian myeloblastosis virus. J. Gen. Virol. *15:* 183.

———. 1972b. VSV pseudotype particles with the coat of avian myeloblastosis virus. Nature New Biol. *240:* 122.

ZĂVADA, J., Z. ZĂVADOVÁ, A. MALIR and A. KOCENT. 1972. VSV pseudotype produced in cell line derived from human mammary carcinoma. Nature New Biol. *240:* 124.

Reference Index

Numbers in italics refer to pages on which the complete references are listed.

D

G

H

I

L

M

N

O

P

T

U

V

W

X

Y

Z

Subject Index

B

C

D

E

F

G

I

K

L

M

R

S

T

U

V

W

X

Y